Reactivity and Structure
Concepts in Organic Chemistry

Volume 11

W0225748

Editors:

Klaus Hafner   Jean-Marie Lehn
Charles W. Rees   P. von Ragué Schleyer
Barry M. Trost   Rudolf Zahradník

New Syntheses with Carbon Monoxide

Edited by J. Falbe

With Contributions by H. Bahrmann
B. Cornils, C. D. Frohning, A. Mullen

With 118 Figures and 127 Tables

Springer-Verlag
Berlin Heidelberg New York 1980

Professor Dr. Jürgen Falbe
Executive Vice-President of Ruhrchemie AG
Postfach 13 01 35
D-4200 Oberhausen 13

List of Editors

Professor Dr. Klaus Hafner
Institut für Organische Chemie der TH Darmstadt
Petersenstr. 15, D-6100 Darmstadt

Professor Dr. Jean-Marie Lehn
Institut de Chimie, Université de Strasbourg
1, rue Blaise Pascal, B.P. 296/R8, F-67008 Strasbourg-Cedex

Professor Dr. Charles W. Rees, F.R.S. Hofmann
Professor of Organic Chemistry, Department of Chemistry
Imperial College of Science and Technology
South Kensington, London SW7 2AY, England

Professor Dr. Paul v. Ragué Schleyer
Lehrstuhl für Organische Chemie der Universität Erlangen-Nürnberg
Henkestr. 42, D-8520 Erlangen

Professor Barry M. Trost
Department of Chemistry, The University of Wisconsin
1101 University Avenue, Madison, Wisconsin 53706, U.S.A

Professor Dr. Rudolf Zahradník
Tschechoslowakische Akademie der Wissenschaften
J.-Heyrovský-Institut für Physik. Chemie und Elektrochemie
Máchova 7, 121 38 Praha 2, C.S.S.R.

ISBN-13: 978-3-642-67454-9   e-ISBN-13: 978-3-642-67452-5
DOI: 10.1007/978-3-642-67452-5

Library of Congress Cataloging in Publication Data. Main entry under title: New syntheses with carbon monoxide. (Reactivity and structure; v. 11). Edition of 1970 by J. Falbe published under title: Carbon monoxide in organic synthesis. Bibliography: p. Includes index. 1. Chemistry, Organic-Synthesis. 2. Carbon monoxide. I. Falbe, Jürgen. II. Falbe, Jürgen. Carbon monoxide in organic synthesis. III. Series. QD262.F313. 1980. 547'.2. 79-23156.

© by Springer-Verlag Berlin Heidelberg 1980

Softcover reprint of the hardcover 1st edition 1980

2152/3140-543210

Preface

More than a decade has passed since the appearance of the last edition of the book "Carbon Monoxide in Organic Synthesis". During the intervening period, the significance of carbon monoxide chemistry has become even more evident. The oil crisis and the ever-present awareness of the constant depletion of the oil reserves have both contributed to a surge in activity. The fact that coal will once again replace oil in the near future as a feedstock source has been an impetus for new efforts in the field of carbon monoxide chemistry.

Moreover, older almost neglected processes such as the Fischer-Tropsch synthesis have become a focal point for new activities.

Therefore, the time was ripe to fundamentally revise "Carbon Monoxide in Organic Synthesis" and to complete its coverage by introducing the chapters "Homologation" and "Carbon Monoxide Hydrogenation".

However, other important sectors of carbon monoxide chemistry such as the conversion of synthesis gas to methanol or higher alcohols were excluded. These are treated in the book "Chemierohstoffe aus Kohle", which devotes considerable attention to these syntheses.

I would like to thank Drs. Bahrmann, Cornils, Frohning, Mullen and Tummes, who had the task of compiling and critically evaluating the large volume of literature which has appeared during the last decade.

The editor is also indebted to Prof. H. Behrens (Universität Erlangen-Nürnberg), Dr. K. Bott (BASF), Prof. P. Chini (Università di Milano), Prof. J. Ellermann (Universität Erlangen-Nürnberg), Prof. B. Fell (TH Aachen), Prof. M. Herberhold (TU München), Dr. W. Himmele (BASF), Dr. N. Imjanitow (Allunious-Institut Leningrad), K.H. Keim (UK-Wesseling), Dr. R. Kummer (BASF), Prof. R. Lai (Université de Droit), Prof. L. Markó (Hungarian Academy of Sciences), Dr. D.E. Morris (Monsanto), Dr. F.J. Müller (BASF), Prof. M. Orchin (University of Cincinnati), Dr. G. Pályi (Hungarian Academy of Sciences), Prof. F. Piacenti (Università di Firenze), Prof. P. Pino (TH Zürich), Dr. R.L. Pruett (UCC), Prof. H. Schulz (TH Karlsruhe), Prof. B.L. Shaw (The University of Leeds), Dr. H. Siegel (BASF), Prof. W. Strohmeier (Universität Würzburg), Dr. R. Whyman (ICI), and Sir Geoffrey Wilkinson (Imperial College London) for providing him with information about newly published data and the latest developments in the various fields of carbon monoxide chemistry as well as for their valuable discussions and assistance in critically reviewing various sections of the manuscript.

The publisher must also be thanked for his cooperation and assistance in the revision of the manuscripts and ensuring the best possible presentation.

Oberhausen, March 1980 J. Falbe

Table of Contents

1. Hydroformylation
Oxo Synthesis, Roelen Reaction

B. Cornils

1.1 Introduction

In 1978, the Oxo synthesis, which was discovered by Otto Roelen in the Ruhrchemie laboratories at Oberhausen-Holten, W. Germany, had its fortieth jubilee [1]. Based on comments made at the end of its third decade [2, 3], the capacity had increased threefold [4] — or even sixfold according to other sources [5] — to 5 million t/a during the last ten years. However, no new comprehensive treatise has been published covering all aspects of carbon monoxide chemistry [6].

This particularly applies to the hydroformylation reaction (Oxo synthesis, Roelen reaction) where a considerable number of publications deal with individual aspects of the hydroformylation, e.g., transition metals as Oxo catalysts [7–25], mechanism [7, 9–12], comparison of various catalysts [26–28], Oxo synthesis as an important outlet for lower olefins [29, 30], hydroformylation reaction as a preparative method [31], economic significance of the products from the Oxo process and their derivatives [32–35], or future developments in the Oxo synthesis [36–42, 1919].

The new developments made during the last decade covering the whole spectrum of carbon monoxide chemistry made the compilation of a comprehensive review appear worthwhile.

These important advances in carbon monoxide chemistry included the hydroformylation reaction. This work is concerned with modification of the Oxo catalysts on varying the central atom or the ligands, or via change of application phase. Progress has also been made in reaction control and associated technology. Some of this work received attention in 1973, when Geoffrey Wilkinson and Ernst Otto Fischer were awarded the Nobel prize for chemistry as a result of their work in advancing organometallic chemistry [49].

Moreover, the Oxo products are of even greater significance today than ten years ago. In addition to the "classical" applications, e.g., of solvents [33, 43], plasticizers [33, 44], paint raw materials and siccatives [33, 45], lubricants [33, 46, 269], detergents [50], and intermediates [16, 47, 48, 1969] which have been treated in comprehensive reports, other applications are being given serious consideration. These include structurally different and — for the first time — quantitatively important fields of application such as the substitution of native raw materials with synthetic Oxo products (e.g., use of synthetic fatty acids in lubricant manufacture or synthetic Oxo alcohols for detergents). Figure 1.1 gives an impression of many products and synthetic routes opened by the Oxo synthesis.

1

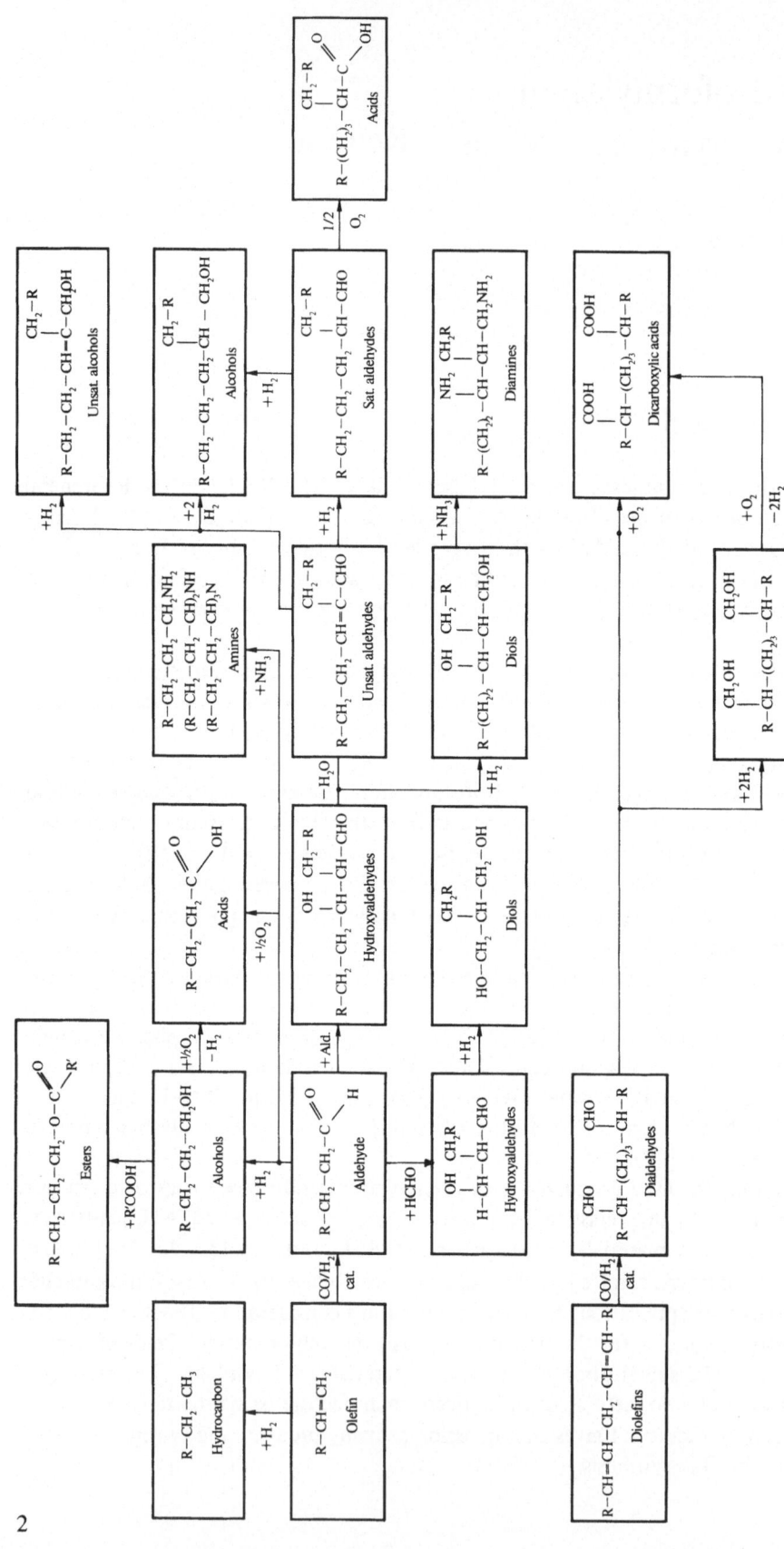

Fig. 1.1. Product spectrum based on the Oxo synthesis

Furthermore, recent activities in the field of coal conversion will lead to modern economic coal gasification processes making syn gas available from coal instead of petroleum. As well as stimulating activity in syn gas chemistry as a whole, this will mean for the economically significant hydroformylation reaction [Eq. (1)] that in molar terms at least 2/3 of the feedstocks can be obtained from coal instead of oil. This aspect has been comprehensively treated elsewhere [51].

$$R-CH=CH_2 + CO + H_2 \rightarrow R-CH_2-CH_2-CHO \tag{1}$$
$$\text{Olefin} \qquad \text{Syn gas} \qquad \text{Oxo aldehyde}$$

Finally, a whole series of "Oxo analogue" reactions has been described which employ transition metal carbonyls as catalysts, and carbon monoxide and/or syn gas as reactants thereby extending the spectrum of products even more. The more important reactions are:

● Homologation (cf Chap. 2 as well as review [52])

$$R-CH_2OH + CO + 2\,H_2 \xrightarrow[\text{catalyst}]{} R-CH_2-CH_2OH + H_2O \tag{2}$$

● Variants of the hydrosilation [53, 54]

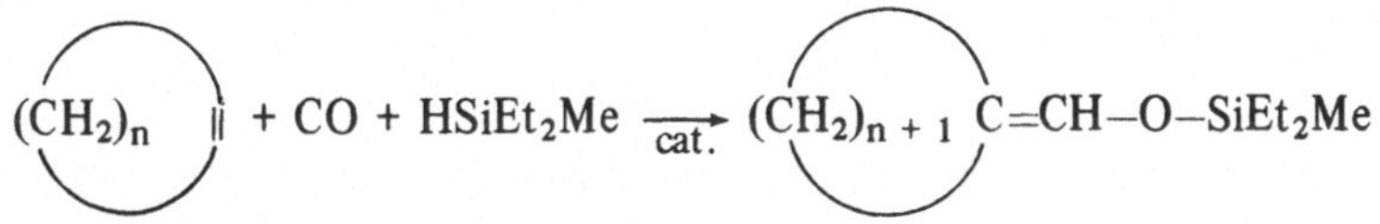

$$(CH_2)_n \parallel + CO + HSiEt_2Me \xrightarrow[\text{cat.}]{} (CH_2)_{n+1} \, C=CH-O-SiEt_2Me \tag{3}$$

● Variants of the hydroboration [55–59]

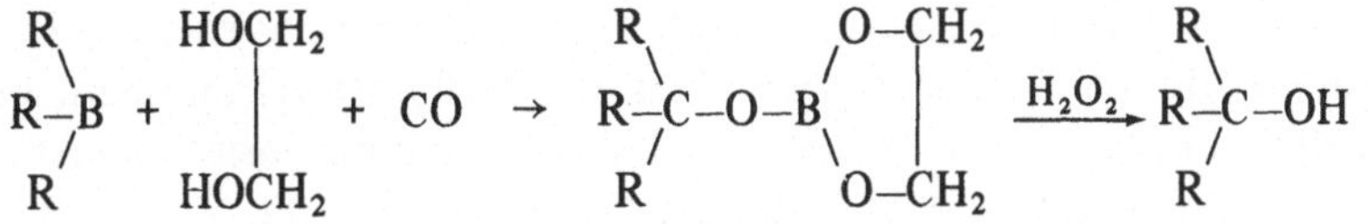

$$R-B + \underset{HOCH_2}{\overset{HOCH_2}{|}} + CO \rightarrow R-C-O-B\underset{O-CH_2}{\overset{O-CH_2}{|}} \xrightarrow{H_2O_2} R-C-OH \tag{4}$$

$$R-B + CO \rightarrow R-\overset{-}{B}-\overset{+}{C}\equiv O \rightleftharpoons \overset{O}{\underset{||}{B-C-R}} + MH \rightarrow \overset{OM}{\underset{|}{B-CH-R}} \tag{5}$$
$$2\,ROH + RCHO$$

● Carbonylation of organomercuric intermediates [68, 69]

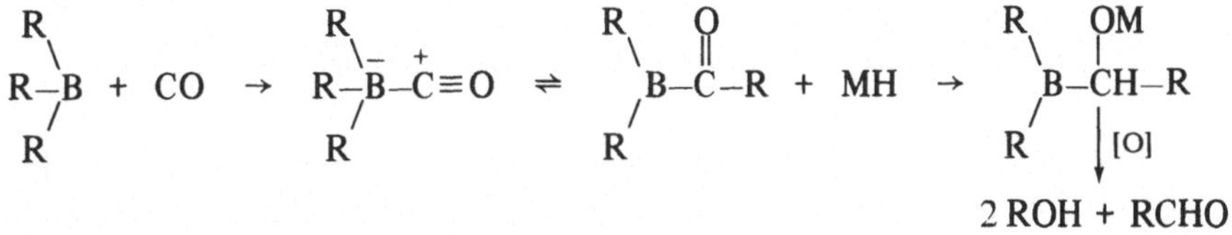

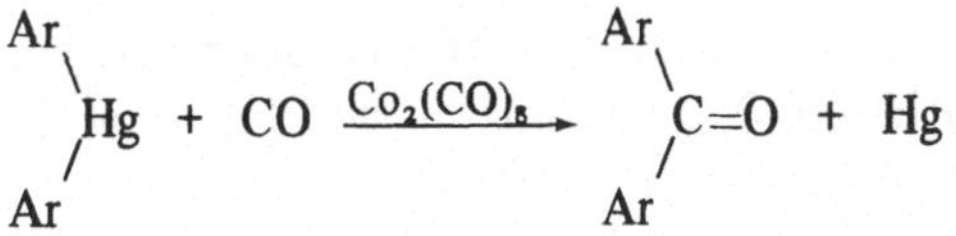

$$\underset{Ar}{\overset{Ar}{}}Hg + CO \xrightarrow{Co_2(CO)_8} \underset{Ar}{\overset{Ar}{}}C=O + Hg \tag{6}$$

● as well as Oxo analogue reactions of alkyl orthoformates which allow conclusions to be drawn about the hydroformylation mechanism [70–72]

3

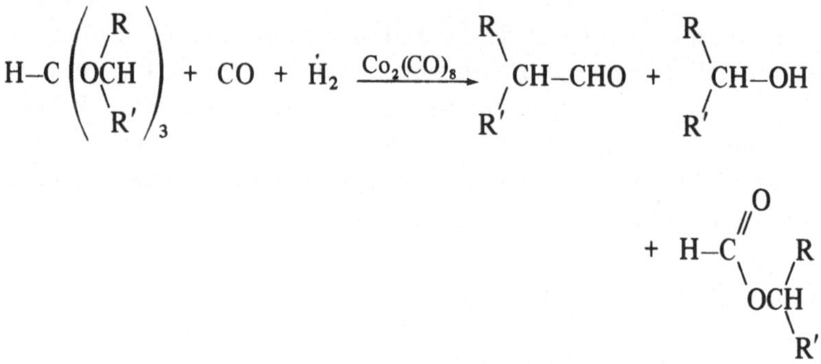

$$H-C\left(\underset{R'}{\overset{R}{\underset{|}{O\overset{|}{C}H}}}\right)_3 + CO + \overset{.}{H}_2 \xrightarrow{Co_2(CO)_8} \underset{R'}{\overset{R}{\diagdown}}CH-CHO + \underset{R'}{\overset{R}{\diagdown}}CH-OH$$

$$(7)$$

$$+ H-C\overset{\diagup\!\!\diagup O}{\underset{O\overset{|}{C}H}{\diagdown}}\underset{R'}{\overset{R}{\diagdown}}$$

In addition, transition metal carbonyls catalyze hydroqinone or anthraquinone formation via cyclizations [60–62], ketone formation via hydrozirconation [63], or – via reaction with aluminium alkyls and CO – the aldehydes/alcohols formation [64]. Other syntheses are also dealt with in the literature [65–67].

In this treatise, it will be attempted to present the new developments made since the publication of the German edition of "Carbon Monoxide in Organic Syntheses" [6] in 1967.

1.2 Hydroformylation Mechanism

Although a better understanding of the hydroformylation mechanism led to only minor progress in the reaction control of the Oxo process or to advances in hydroformylation chemistry, the study of mechanistic aspects has always enjoyed priority among chemists working on the Roelen reaction. Nevertheless millions of tons of Oxo products were manufactured. The production was performed in empirically developed reactors along with partially false ideas about the reaction mechanismens. This background helps to explain the considerable number of comprehensive reports are dealing with the mechanism of the Oxo reaction [6–12, 15, 42, 73, 75–79].

1.2.1 The Mechanism According to Heck and Breslow

After a series of theoretical approaches which are summarized elsewhere [6, 7], Heck and Breslow in 1960 presented their currently generally accepted mechanism [80], which is based on the postulated formation of $HCo(CO)_4$ from $Co_2(CO)_8$ and hydrogen and its subsequent dissociation into $HCo(CO)_3$ and CO. The hydrocarbonyl and the olefin form a π-olefin complex which rearranges to an alkyl cobalt carbonyl. Thereafter, an acyl complex results via insertion of CO between the C atom of the alkyl group and the cobalt atom. After possible formation of a tetracarbonyl complex, a reaction ensues with hydrogen or $HCo(CO)_4$, resulting in the formation of aldehydes – the primary products of the Oxo synthesis.

$$Co_2(CO)_8 + H_2 \rightleftharpoons 2\,HCo(CO)_4 \tag{8}$$

$$HCo(CO)_4 \rightleftharpoons HCo(CO)_3 + CO \tag{9}$$

$$R-CH=CH_2 + HCo(CO)_3 \rightleftharpoons R-CH=CH_2 \rightleftharpoons R-CH_2-CH_2-Co(CO)_3$$
$$\downarrow$$
$$HCo(CO)_3 \tag{10}$$

$$R-CH_2-CH_2-Co(CO)_3 + CO \rightleftharpoons R-CH_2-CH_2-Co(CO)_4 \tag{11}$$

$$R-CH_2-CH_2-Co(CO)_4 \rightleftharpoons R-CH_2-CH_2-CO-Co(CO)_3 \tag{12}$$

$$R-CH_2-CH_2-CO-Co(CO)_3 + CO \rightleftharpoons R-CH_2-CH_2-CO-Co(CO)_4 \tag{13}$$

$$R-CH_2-CH_2-CO-Co(CO)_4 \begin{cases} \xrightarrow{HCo(CO)_4} R-CH_2-CH_2-CHO + Co_2(CO)_8 \ (14) \\ \qquad\qquad\quad \text{Aldehyde} \\ \xrightarrow{+H_2} R-CH_2-CH_2-CHO + HCo(CO)_4 \ (15) \end{cases}$$

$$R-CH_2-CH_2-CO-Co(CO)_3 + HCo(CO)_4 \rightarrow R-CH_2-CH_2-CHO + Co_2(CO)_7 \tag{16}$$

$$Co_2(CO)_7 + CO \rightleftharpoons Co_2(CO)_8 \tag{17}$$

The reaction sequence of the Heck-Breslow mechanism is shown in Figs. 1.2 and 1.3 as a cyclic homogeneously catalyzed process.

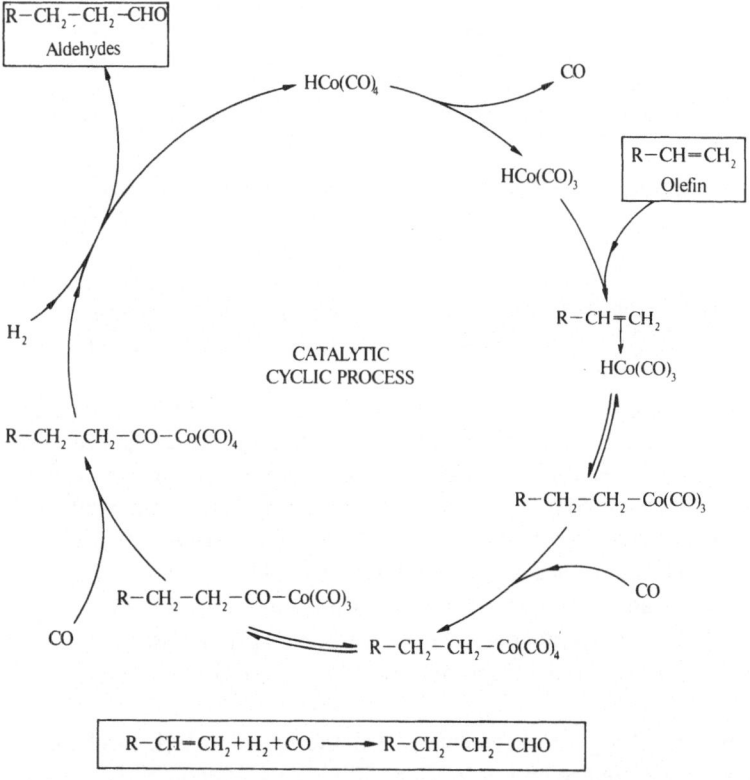

Fig. 1.2. Oxo mechanism according to Heck-Breslow – homogeneously catalyzed cyclic process (simplified version)

5

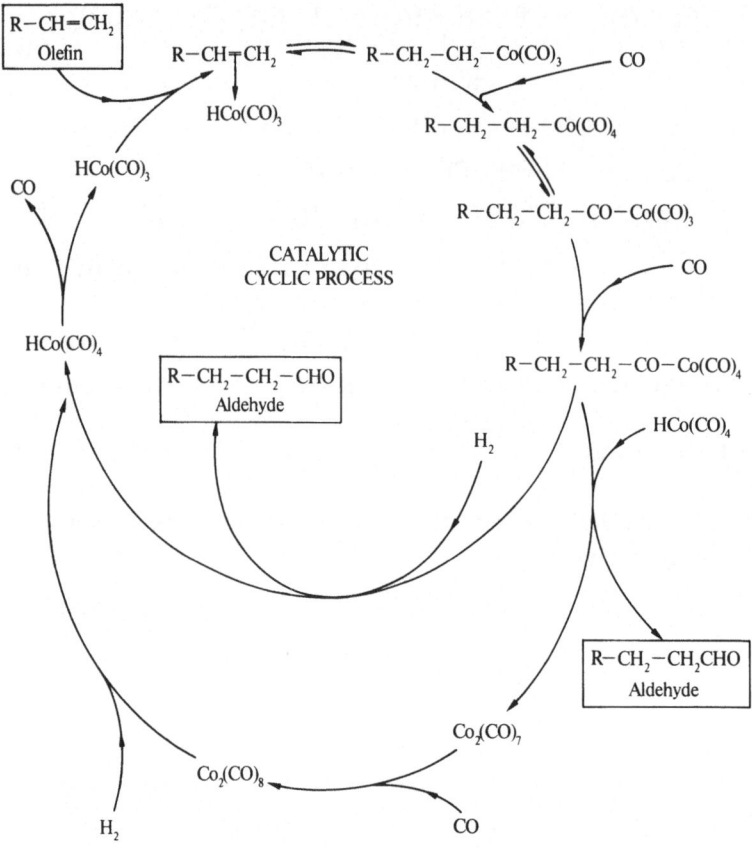

Fig. 1.3. Oxo mechanism according to Heck-Breslow – homogeneously catalyzed cyclic process with alternative routes

It is assumed that this mechanism also applies to the hydroformylation with other metal carbonyls [32, 75, 88, 106], e.g., on exchanging the central atom (Co) with Rh.

The most important steps of this mechanism have been shown to be probable [81–94, 201, 202, 204, 205, 462, 474] from a series of specially designed experiments, simulations and analogies together with IR studies at the pressure and temperature conditions of the industrial Oxo process. Particular attention was paid to Eq. (8) (behavior of octacarbonyldicobalt/formation of the hydrocarbonyl [83, 93, 95–99]), Eq. (9) (dissociation of tetracarbonylhydridocobalt [83, 86, 100, 1921]), Eq. (10) (π-olefin complex and formation of the alkyl-Co compounds [72, 73, 100–106, 1922]), Eqs. (12)/(13) (formation of acyl-metal carbonyls [80, 81, 86, 106–112, 462, 1925]), Eqs. (14)–(16) (hydrogenolysis of the acyl metal carbonyls [86, 93, 113, 210]) and Eq. (17) [114].

Altough there is currently general agreement that the Breslow-Heck mechanism best explains the reaction sequence and is confirmed by many experimental findings, Pino [115] maintains that several aspects have not been sufficiently considered: e.g., a) the formation of isomeric aldehydes (n/iso problem) during the hydroformylation of an olefin with more than two carbon atoms, and b) the effect of carbon monoxide partial pressure on reaction velocity and isomer distribution.

More recent interpretations and theories are dealt with in the next section.

1.2.2 Recent Interpretations

The Heck-Breslow mechanism [Eqs. (8)–(17)] is presented in Fig. 1.4 taking into account the considerations of Pino and Piacenti [116]. This simplified version helps to portray all possible and discussed isomerization steps from the olefin feedstock to the acyl complex.

Generally the hydroformylation of olefins, which are not asymmetric and/or do not form isomers via double bond migration (e.g., ethylene, cyclopentene etc.), leads to the formation of isomeric aldehyde mixtures.

$$2 \text{ R–CH=CH}_2 + 2 \text{ CO} + 2 \text{ H}_2 \xrightarrow{\text{cat.}} \underset{\text{``n-aldehyde''}}{\text{R–CH}_2\text{–CH}_2\text{–CHO}} + \underset{\text{``iso-aldehyde''}}{\overset{\overset{\text{CH}_3}{|}}{\text{R–CH–CHO}}} \quad (18)$$

However, the hydroformylation of isomeric olefins sometimes yields identical reaction products [73, 841]:

$$\begin{array}{l} \text{CH}_3\text{–CH}_2\text{–CH}_2\text{–CH=CH}_2 \\ \text{or} \\ \text{CH}_3\text{–CH}_2\text{–CH=CH–CH}_3 \end{array} \xrightarrow[\text{cat.}]{\text{+CO/H}_2} \begin{array}{l} \text{CH}_3\text{–CH}_2\text{–CH}_2\text{–CH}_2\text{–CH}_2\text{–CHO} \\ \overset{\overset{\text{CH}_3}{|}}{+ \text{ CH}_3\text{–CH}_2\text{–CH}_2\text{–CH–CHO}} \\ + \text{ CH}_3\text{–CH}_2\text{–}\underset{\underset{\text{C}_2\text{H}_5}{|}}{\text{CH}}\text{–CHO} \end{array} \quad (19)$$

In some cases, products are obtained in which the formyl group is not attached to one of the carbon atoms of the original double bond. The following steps have been discussed in order to assist in the interpretation of the paths leading to the various reaction products.

The *isomerization of the olefinic starting material* (equilibrium I in Fig. 1.4), can explain the formation of the isomeric aldehydes. But this is only the case if the reaction sequence following the first addition step procedes without any further isomerization. The following reactions provide experimental support:

a) The hydroformylation of deuterated propylene [117] (at high and low CO partial pressures) gives a deuterium distribution in the Oxo products corresponding largely to the statistical distribution. Apparently no $HCo(CO)_4$[or $DCo(CO)_4$] is released (cf Refs. [83, 111, 119, 133] regarding $HCo(CO)_4$ or $DCo(CO)_4$ induced olefin isomerization; cf Refs. [120, 121] regarding stoichiometrical isomerization).

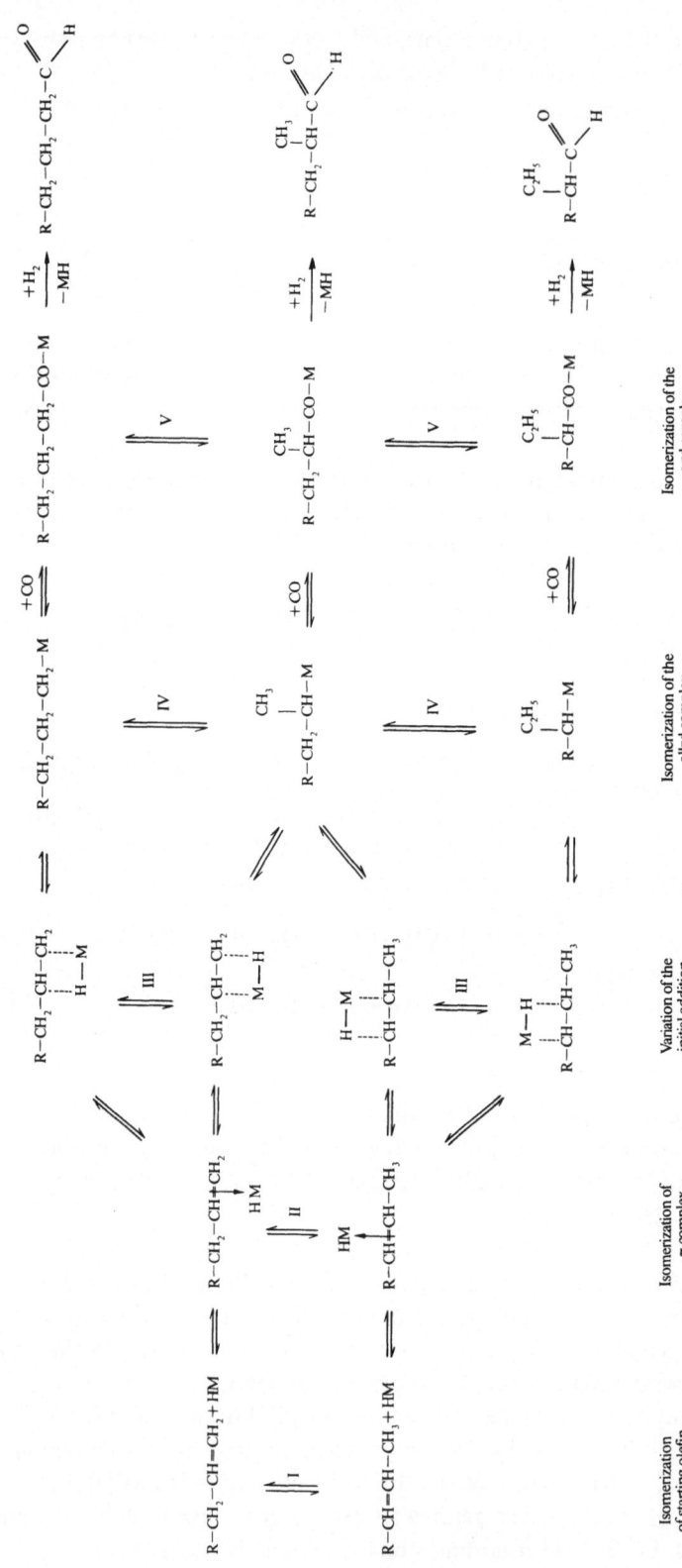

Fig. 1.4. Mechanism of the hydroformylation with possible isomerization steps (HM = HMe(CO)$_m$)

b) The hydroformylation of 1-^{14}C propylene [122] gives mixtures of 2-^{14}C-n-butanal, 4-^{14}C-n-butanal, and 2-methyl-3-^{14}C-propanal.

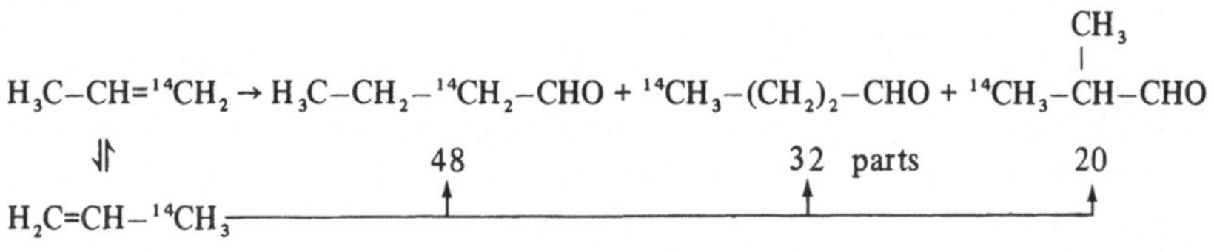

$$H_3C-CH={}^{14}CH_2 \rightarrow H_3C-CH_2-{}^{14}CH_2-CHO + {}^{14}CH_3-(CH_2)_2-CHO + {}^{14}CH_3-\underset{\underset{CH_3}{|}}{C}H-CHO$$

$$\Updownarrow \qquad\qquad 48 \qquad\qquad\qquad 32 \text{ parts} \qquad\qquad 20 \tag{20}$$

$$H_2C=CH-{}^{14}CH_3$$

This suggests a previous olefin isomerization (see below for explanation via variation of direction of addition).

c) The hydroformylation of 1-pentene or 2-pentene which produces very similar products [cf Eq. (19)] [73, 124] under standardized reaction conditions (e.g. 110 °C).

d) The hydroformylation of hexenes and other olefins [123, 125, 126, 1230].

e) The reaction involving C_{14}–C_{18} olefins after isomerization [46, 1895] which is of commercial interest.

f) In the reaction of methylpentenes [135], e.g. (+)(S)-3-methyl-1-pentene, (−) (S)-4-methyl-1-hexene or (+) (S)-5-methyl-1-heptene [104, 127, 128, 138]

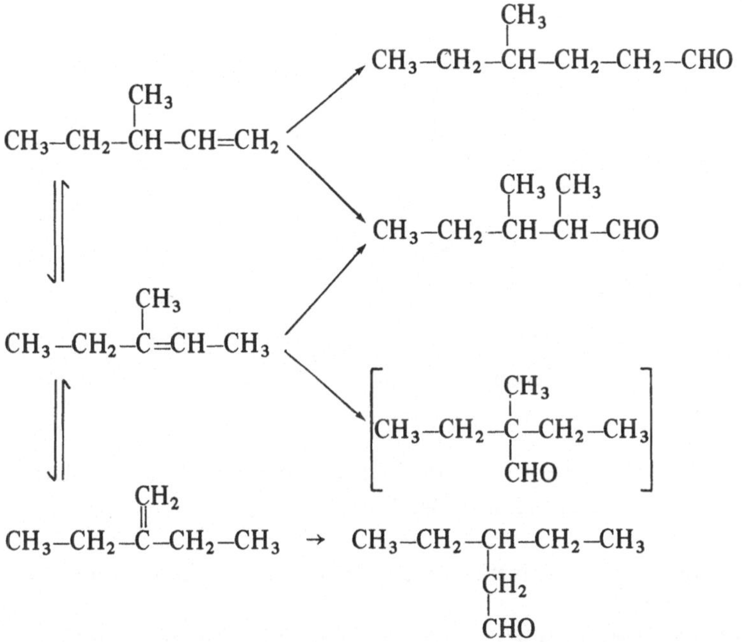

$$CH_3-CH_2-\underset{\underset{CH_3}{|}}{C}H-CH=CH_2 \nearrow CH_3-CH_2-\underset{\underset{CH_3}{|}}{C}H-CH_2-CH_2-CHO \tag{21}$$

$$\searrow CH_3-CH_2-\underset{\underset{CH_3}{|}}{C}H-\underset{\underset{CH_3}{|}}{C}H-CHO$$

$$\Updownarrow$$

$$CH_3-CH_2-\underset{\underset{CH_3}{|}}{C}=CH-CH_3 \nearrow \tag{22}$$

$$\searrow \left[CH_3-CH_2-\underset{\underset{CHO}{|}}{\overset{\overset{CH_3}{|}}{C}}-CH_2-CH_3\right]$$

$$\Updownarrow$$

$$CH_3-CH_2-\overset{\overset{CH_2}{||}}{C}-CH_2-CH_3 \rightarrow CH_3-CH_2-\underset{\underset{\underset{CHO}{|}}{CH_2}}{C}H-CH_2-CH_3 \tag{23}$$

in which 3-ethylpentanal appears from 3-methyl-1-pentene and suggests the formal hydroformylation of the original 3-methyl group [129].

g) The reactions of allyl alcohols [130], allylbenzene [131], methyl oleate [132], dodecene [134], or of intermediate olefins resulting from ortho-formates [136, 137].

The various olefin isomerization mechanisms–1,3-hydride migration, 1,3-allyl exchange or 1,2 addition/elimination–have been dealt with by Orchin and Rupilius [73].

The effect of high CO partial pressure on suppressing the olefin isomerization has also been studied [7, 73, 104]. The observation, that high CO partial pressures (at moderate temperatures) facilitate "isomerization free" hydroformylations is of commercial interest [203] (cf Sect. 1.3).

In addition, the *equilibria between the π-complexes* postulated as the initial intermediates (equilibria **II** in Fig. 1.4) was also considered, especially in such cases where the olefin isomerization can be excluded from thermodynamic or kinetic considerations [124, 139, 140, 213].

Newer concepts about the *type and direction of the initial addition* of metal hydrocarbonyls to the olefinic double bond have been developed. Mainly they have been based on the results of the hydroformylation of deuterated [113, 141], ^{14}C-labelled-olefins [122], or of olefins with an asymmetric carbon atom [104, 141–144]. These concepts are much more superior to the simple Markownikoff/anti-Markownikoff rule. Orchin and Rupilius suggested an σ-π-interconversion mechanism to explain the formation of 3% (R)-3-(1-d-ethyl)-hexanal from (+)-S-3-methyl-1-hexen-3-d (cf equilibria **III** in Fig. 1.4):

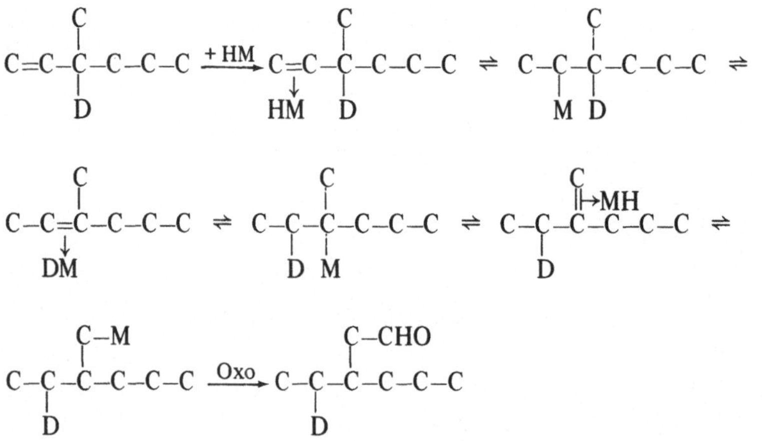

$$
\text{(24)}
$$

3-(1d-ethyl)-hexanal

After studying the stereochemistry of the hydroformylation, Pino et al. developed a concept based on a 1,2-hydrogen shift [144, 145].
The assumptions were:

the isomerization of the π-catalyst complex does not involve formation of true σ-cobalt-carbon bonds, and

the equilibrium between the olefins and π-complex formation, if it is attained, is largely displaced towards the olefin-catalyst complex.

The stereochemistry of the attack of CO/H$_2$ at the double bond (from the cis position with cobalt as well as with rhodium catalysts) has been discussed by various authors [81, 144, 165]. The same is true for MO interpretations which postulate a preceding trans state [160] (cf also Pino [144] and the equilibria **II** in Fig. 1.4):

$$
\begin{array}{c}
\overset{|}{C}H-[M] \\
\overset{|}{C}H_2 \\
\overset{*}{\underline{C}}D \\
\overset{|}{C}H_2 \\
|
\end{array}
\qquad
\begin{array}{c}
\overset{|}{C}H_2 \\
\overset{|}{C}H-M \\
\overset{*}{C} \\
\overset{|}{C}H_2 \\
|
\end{array}
$$

$$\uparrow \qquad\qquad\qquad\qquad \uparrow \qquad\qquad\qquad\qquad\qquad\qquad\qquad \uparrow$$

$$
\begin{array}{c}
\overset{|}{C}H \\
\Vert\!\!\rightarrow MH \\
\overset{|}{C}H \\
-\overset{*}{\underline{C}}D \\
\overset{|}{C}H_2
\end{array}
\;\rightleftharpoons\;
\begin{array}{c}
\overset{|}{C}H\text{------}H \\
\overset{|}{C}H \quad M \\
-\overset{|}{C}\text{-------}D \\
\overset{|}{C}H_2
\end{array}
\;\rightleftharpoons\;
\begin{array}{c}
\overset{|}{C}H_2 \\
\overset{|}{C}H\!\!\rightarrow MD \\
-\overset{|}{C} \\
\overset{|}{C}H_2 \\
\downarrow
\end{array}
\;\rightleftharpoons\;
\begin{array}{c}
\overset{|}{C}H_2 \\
\overset{|}{C}H\text{------}D \\
-\overset{|}{C}\quad M \\
\overset{|}{C}H\text{------}H
\end{array}
\;\rightleftharpoons\;
\begin{array}{c}
\overset{|}{C}H_2 \\
\overset{|}{C}HD \\
-\overset{|}{C}\!\!\rightarrow MH \\
\overset{|}{C}H \\
\downarrow
\end{array}
\qquad (25)
$$

$$
\begin{array}{c}
\overset{|}{C}H_2 \\
\overset{|}{C}H-M \\
\overset{*}{\underline{C}}D \\
\overset{|}{C}H_2
\end{array}
\qquad\qquad
\begin{array}{c}
\overset{|}{C}H_2 \\
\overset{|}{C}HD \\
\overset{*}{\underline{C}}-M \\
\overset{|}{C}H_2
\end{array}
\qquad\qquad\qquad\qquad
\begin{array}{c}
\overset{|}{C}H_2 \\
\overset{*}{C}HD \\
-\overset{|}{C}H \\
\overset{|}{C}H-M
\end{array}
$$

Should the direction of addition of the reactants (CO and H₂) and thus the formation of various isomers not be determined by either isomerization of the olefin skeleton, by the π-complex as the initial intermediate, or by steric factors, then the *isomerizations of the alkylcobalt carbonyls* must be considered (equilibrium **IV** in Fig. 1.4). However, all control experiments have shown that alkylcobalt carbonyls do not isomerize under the conditions of the industrial Oxo synthesis (Falbe et al. [146], Piacenti et al. [147], Pino [72]).

Another possible interpretation of the formation of isomeric aldehydes *is the isomerization of the acylcobalt complexes* of the final intermediates before their hydrogenolysis to the aldehydes [Eqs. (14) and (15) as well as equilibria **V** in Fig. 1.4]. There is abundant evidence for a direct transfer of the "unbranched" acyl complex to the "branched" isomer at relatively low partial CO pressure and low temperatures. The latter deviate considerably from the industrial conditions of the Oxo synthesis [148–152]; however, there is no support for this direct transfer at higher partial pressures and temperatures [97, 110, 153, 1925].

Discussion of the steps in this intermediate stage of the Oxo mechanism is impeded by the disproportionation of the acylcobalt complex under certain conditions (low CO partial pressure/absence of CO) to aldehydes and olefins with one carbon less than the original acyl-complex. The newly-formed olefin can then undergo hydroformylation once again thereby ensuring the impression that isomerization has ensued (Rupilius, Orchin [152, 154]).

$$2\ R-CH_2-CH_2-CO-Co(CO)_4 \;\rightarrow\; R-CH_2-CH_2-CHO + R-CH{=}CH_2 + Co_2(CO)_8 + CO$$

$$(26)$$

The discussions about this irreversible reaction of the acylcobalt complex and its implications for indirect conclusions about the Oxo mechanism have not been completed.

The last reaction once again testifies the great significance of the CO partial pressure – besides the effect it has on the stability of the actual Oxo catalyst $HCo(CO)_m$ (cf Sect. 1.3.2 as well as detailed treatment in the Ref. [6]. Wender and Pino have summarized the effects of the CO partial pressure as shown in Table 1.1 [155]:

Table 1.1. Effect of high or low CO partial pressure on the hydroformylation and its side reactions

Effect on	High P_{CO} (100–300 bar)	Low P_{CO} (2–10 bar)
relative reaction velocity	minor	greater
relative rate of olefin isomerization	minor	greater
isomer distribution	more unbranched products	more branched products

It is thought that the marked effect of CO partial pressure on both the rate of hydroformylation and isomer distribution is due to the existence of two active catalytic species. Two compounds – in agreement with Eqs. (8)–(17) – come in question: tetra- and tricarbonylhydridocobalt.

$$HCo(CO)_4 \;\rightleftharpoons\; HCo(CO)_3 + CO \tag{9}$$

According to the above, $HCo(CO)_4$ should be more stable at high CO partial pressures as well as being responsible for other effects at higher CO partial pressures (i.e., the favored formation of unbranched compounds). This can be readily interpretated as compared to the tricarbonyl, the more sterically specific $HCo(CO)_4$ adds preferentially to the α-position in olefinic compounds yielding unbranched aldehydes via the steps π-complex → alkylcobalt carbonyl → acylcobalt carbonyl. On the other hand, tricarbonyl-hydridocobalt is thought to be the more reactive Oxo catalyst, especially in the absence of free carbon monoxide – particularly in the stoichiometric hydroformylation with $HCo(CO)_m$ instead of CO/H_2 mixtures – at lower CO partial pressures. This also explains the observation that in the stoichiometric hydroformylation of propylene (in the absence of CO), the primary addition of the cobalt catalyst apparently ensues in accordance with the Markownikoff rule. This reaction yields 70% isobutyraldehyde and only 30% n-butyraldehyde [117, 119, 156]. The catalytic reaction (as well as the stoichiometric hydroformylation in the presence of CO) yields 70% n-butyraldehyde and 30% isobutyraldehyde [122] – an anti-Markownikoff addition. This can be explained by means of the reaction sequences in Figs. 1.5 and 1.6 with the additional assumption that $k_1 \approx k_2$ and $k_4 > k_3$ [157].

It also becomes clearer as to why the unmodified *rhodium* hydridocarbonyls produce more branched products than their cobalt counterparts – $HRh(CO)_4$ has a greater

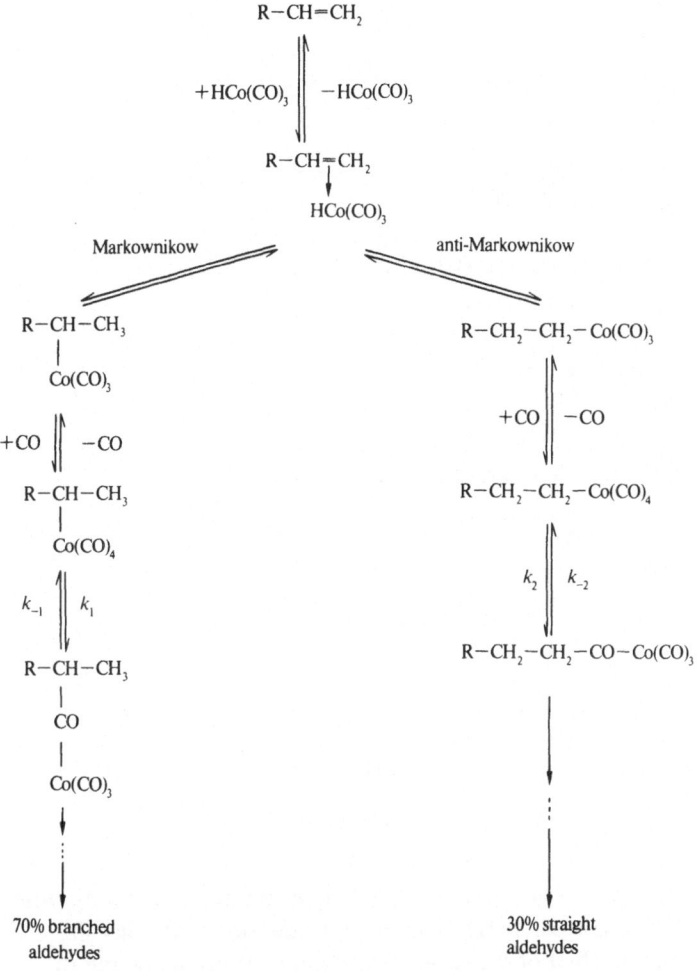

Fig. 1.5. Reaction sequence of the stoichiometric hydroformylation in the absence of CO, $k_1 \approx k_2$

tendency than $HCo(CO)_4$ to release a CO ligand. This results in the equilibrium

$$HRh(CO)_4 \rightleftharpoons HRh(CO)_3 + CO \qquad (27)$$

being less markedly displaced to the left than with the cobalt complex [75].

Presented in Fig. 1.6 are the findings that with the Oxo catalyst $HCo(CO)_m$, one or more CO ligands can be replaced by phosphines, phosphites, arsines etc. (cf Sect. 1.3); resulting in the new complex retaining the Oxo activity and in many cases preferentially yielding unbranched compounds: The bulkier, more sterically demanding, new ligand effects a preferential attack at the α-position of the olefin, producing even more linear aldehydes – assuming $k_4 \gg k_3$.

Besides this "steric" interpretation of the effect of ligand-modified Oxo catalysts, electronic theories and explanations involving both effects have been proposed [6, 78,

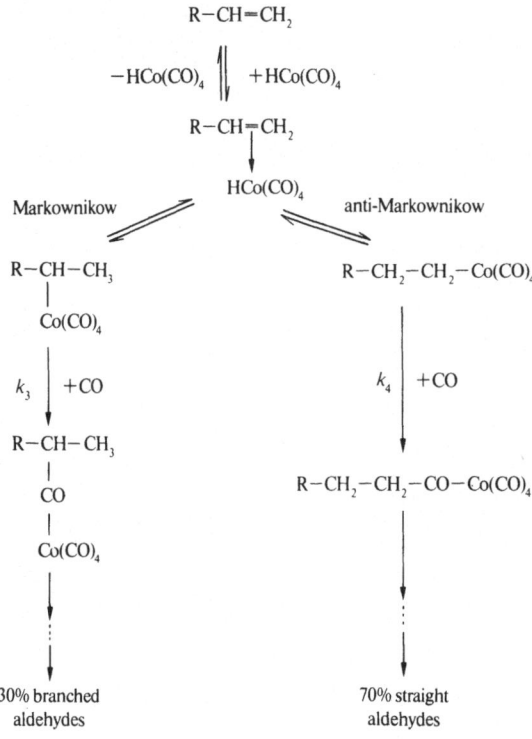

Fig. 1.6. Reaction sequence of the stoichiometric hydroformylation in the presence of CO, $k_4 > k_3$

80, 82, 108, 158−161]. The relative inactivity of triphenylphosphine compared to other phosphines with the same or even lesser steric requirements does not contradict the steric interpretation. The minor effect of the triphenylphosphine is probably due to its lower nucleophilicity compared to other phosphines [19, 162, 163] causing the equilibrium (28) of the modified catalyst to be displaced to the left and the effect of the unmodified $HCo(CO)_4$ to be more pronounced than with other phosphines [164, 620].

$$HCo(CO)_4 + PR_3 \rightleftharpoons HCo(CO)_3PR_3 + CO \tag{28}$$

This is in agreement with the fact that with triphenylphosphine greater amounts of unbranched compounds are obtained if it is used in excess. A typical example is the Oxo process operated by the group Union Carbide/Davy Powergas/Johnson Matthey [26] (cf Sect. 1.6).

That which has been written about the cobalt catalysts also basically applies to the rhodium-catalyzed hydroformylation [78, 81, 82, 86, 88, 106]. The individual steps of the mechanism − olefin isomerization [103, 106, 130, 138, 166−168], π-complexes [103], or alkyl/acyl complexes [36, 88, 92, 106, 144] have already been described in a manner similar to Eqs. (8)−(17). $HRh(CO)_mL_n$ (n + m = 4; L = CO) is assumed to be the active catalyst under industrial hydroformylation conditions.

Analogous to Eq. (28), on substituting one or more CO ligands in $HRh(CO)_4$ with phosphines, phosphites etc. (cf Sect. 1.3) complexes of the type $HRh(CO)_mL_n$ ($n + m = 4$; $L = PPh_3$, PBu_3, etc.) can be obtained. The first complex of this type – $HRh(CO)$ $[P(C_6H_5)_3]_3$ – was introduced as a hydroformylation catalyst by Wilkinson in 1966 [169, 170]. This complex facilitated the Oxo synthesis to be conducted at low temperatures and pressures. Most mechanistic studies [19, 92, 171–174] of the hydroformylation reaction with ligand modified complexes have been carried out with this catalyst. According to these results, $HRh(CO)_2L_2$ can be readily formed from $HRh(CO)L_3$ and CO, $HRh(CO)_2L_2$ being the actual active species capable of reacting with the olefinic starting material. $[Rh(CO)_2L_2]_2$, which results simultaneously, reacts with H_2 to form "monomeric" $HRh(CO)_2L_2$ [169, 839].

$$[Rh(CO)_2L_2]_2 + H_2 \rightarrow 2 HRh(CO)_2L_2 \tag{29}$$

The presumed mechanism of the hydroformylation with modified rhodium hydridocarbonyls is presented in Fig. 1.7 as a cyclic catalytic process.

Wilkinson denotes the partial mechanism I as the *associative pathway* i.e., addition of the olefin to the complex $HRh(CO)_2L_2$ leading to $R–CH_2–CH_2–Rh(CO)_2L_2$.

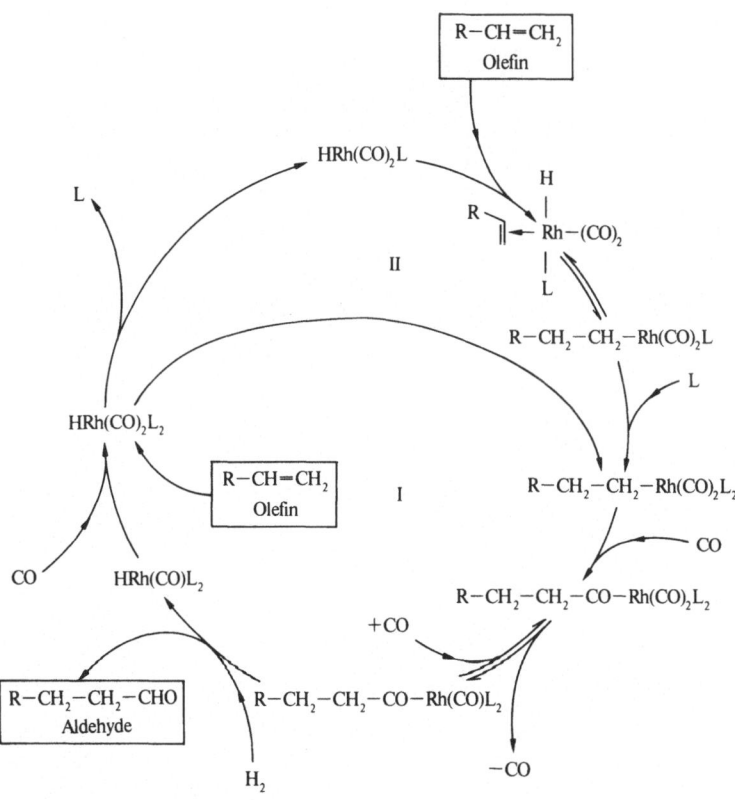

Fig. 1.7. Homogeneously catalyzed cyclic process with modified rhodium hydridocarbonyls

15

At catalyst concentrations greater than $6 \cdot 10^{-3}$ mol/l, the tetra-coordinated acyl complex is probably responsible for the oxidative addition of H_2 [75]. This is possibly also due to the greater activity of rhodium compared to cobald at lower Rh concentrations. It is thought that route II dominates. The latter is called the *dissociative pathway* after the dissociation of one of the ligands (L). The stereochemical representation shown in Fig. 1.8 is based on Pino's work [175].

Kinetic studies of the above pathways are discussed in Sect. 1.3.3.1.2 and elsewhere [207].

1.2.3 Kinetics

Natta et al.'s relationship established in 1955 and shown in Eq. (30) best represents all effects observed in the applied pressure and temperature range [176, 179] (cf also Martin [177] and Iwanaga [178]).

$$\frac{d\,(aldehyde)}{dt} = k[\text{olefin}]\,[Co]\,\frac{P_{H_2}}{P_{CO}} \tag{30}$$

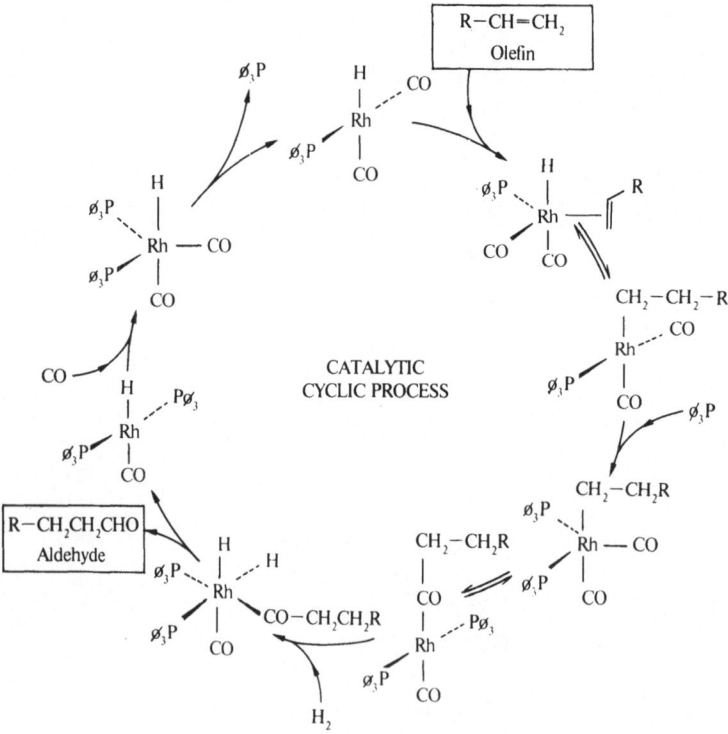

Fig. 1.8. Homogeneously catalyzed cyclic process with $HRh(CO)_2L$ [92, 175] (associative pathway)

According to Natta, the reaction velocity is independent of the total pressure when the $CO:H_2$ ratio is $1:1$ — this is due to the opposing effects of CO and H_2 partial pressures. While high H_2 partial pressures increase the reaction velocity, they are confined to limited values owing to the instability of cobalt hydridocarbonyls [161] at constant total pressure (cf Fig. 1.9 [161]).

According to the above, raising the CO partial pressure increases the stability of the cobalt hydridocarbonyls thereby causing the equilibrium in Eq. (9) to be displaced to the left hand side. The increased effect of tetracarbonylhydridocobalt supports the tendency for unbranched compounds to form (cf Sect. 1.2.2) while concurrently — according to Natta — lowering the reaction velocity.

On choosing the optimal reaction parameters, for an industrial hydroformylation process — the opposing effects of high or low H_2 and CO partial pressures must be considered.

Marko et al. have established a relationship, based on Natta's equation, which can be applied when using rhodium catalysts [88, 181, 209]

$$\frac{d\,(\text{aldehyde})}{dt} = k[\text{olefin}]^x \cdot [Rh]^y \cdot \frac{P_{H_2}}{P_{CO}} \tag{31}$$

$x = 0.1$ (depending on olefin), $y = 1.0-0.7$ [197] (for other interpretations cf Ref. [75, 182, 186]).

The orders of reaction for the hydroformylation with Co or Rh catalysts are shown in Table 1.2.

Further details and discussions about kinetic studies of the Oxo reaction can be found in the Ref. [7, 10, 36, 73, 75, 87, 106, 161, 162, 187–190, 192, 193].

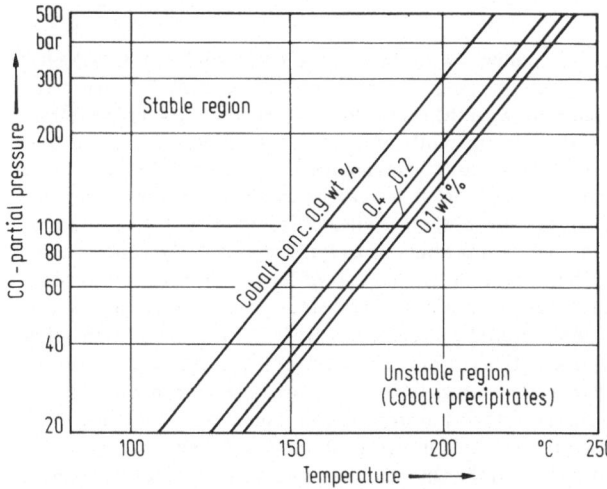

Fig. 1.9. Stability of the hydroformylation catalysts $HCo(CO)_4 + Co_2(CO)_8$ in relation to temperature and CO partial pressure [161, 281] (for rate of formation of $HCo(CO)_4$ cf Ref. [93, 99, 114, 180]).

Table 1.2. Orders of reaction for the hydroformylation [31, 84, 88, 166, 181, 183, 185–187, 196–200]

Order of reaction relative to:	Hydroformylation catalyst	
	$Co_2(CO)_8{}^a$	$Rh_4(CO)_{12}{}^b$
olefin concentration	1	0 or 1
H_2 partial pressure	1	1
CO partial pressure	1^c to -1^d	1 to -1
concentration of metal	1^e or 0.5^f	1 to 0.16

[a] or $HCo(CO)_4/HCo(CO)_3$ [b] or $HRh(CO)_4$ [c] at $P_{CO} < 10$ bar   [d] at $P_{CO} > 10$ bar
[e] at $P_{CO} \approx 100$ bar   [f] according to Natta–Ercoli

1.2.4 Potential Industrial Significance

Seelig suggested that the oscillation states of chemical systems or of substrate inhibition could also be applied to the Oxo synthesis [191]. According to this theory, substrate S reacts with catalyst E forming a complex of type SE or SE_2 via an enzyme-analogue conversion. Thereafter, the complex decomposes releasing both catalyst and desired reaction product:

$$2\,S + E \;\leftrightharpoons\; ES_2 \;\leftrightharpoons\; ES + S \;\to\; product + E \qquad (32)$$

According to Eq. (32) the substrate S combines with the catalyst E to form ES_2 from which S is either released or reacts further as the intermediate ES. This is analogous to the Oxo synthesis in that the tricarbonylhydridocobalt can react with the substrate CO to yield the less active tetracarbonylhydridocobalt (cf Sect. 1.2.2) which is apparently inhibited by the substrate i.e., it is transformed into a less active form. This is another way of portraying Natta's statements in Eq. (30). The result of this inhibition is that, on forming $ES_2[HCo(CO)_4]$, an active catalyst is removed from the process causing the conversion rate to fall despite constant feeding of reactants (in the continuous process). A new, lower steady conversion rate results effecting a rapid increase in the concentration (or partial pressures) of the reactants. This in turn, due to the Law of Mass Action, effects a rise in the conversion rate and the conversion as well as an increase in substrate inhibition. If the reaction rate (v) is represented as a function of the substrate concentration [S] (obtained via addition – Fig. 1.10, 3rd curve – of corresponding representations of the function v = f [S] for the usual chemical conversion according to the Law of Mass Action – Fig. 1.10.1, 1st curve – and assuming substrate inhibition (10.2, 2) and after rotating the axes) then a hysteresis loop results as shown in Fig. 1.10.

If the reaction velocity (v) is constantly increased in Fig. 1.10 (curve *a*, for example via increase in CO partial pressure at constant feed rate) then the Oxo reaction system becomes unstable on exceeding the critical point *1* and rapidly reverts to the upper "dead" hysteresis branch *(b)* i.e., from S_1 to S_2. On the other hand, with decreasing values for v (curve *c*), the S concentration falls from S_3 to S_4 *(d)*. As separate values of S for hydroformylations with various conversions must be correlated, different conversion rates result, and due to the exothermic reaction, various temperatures.

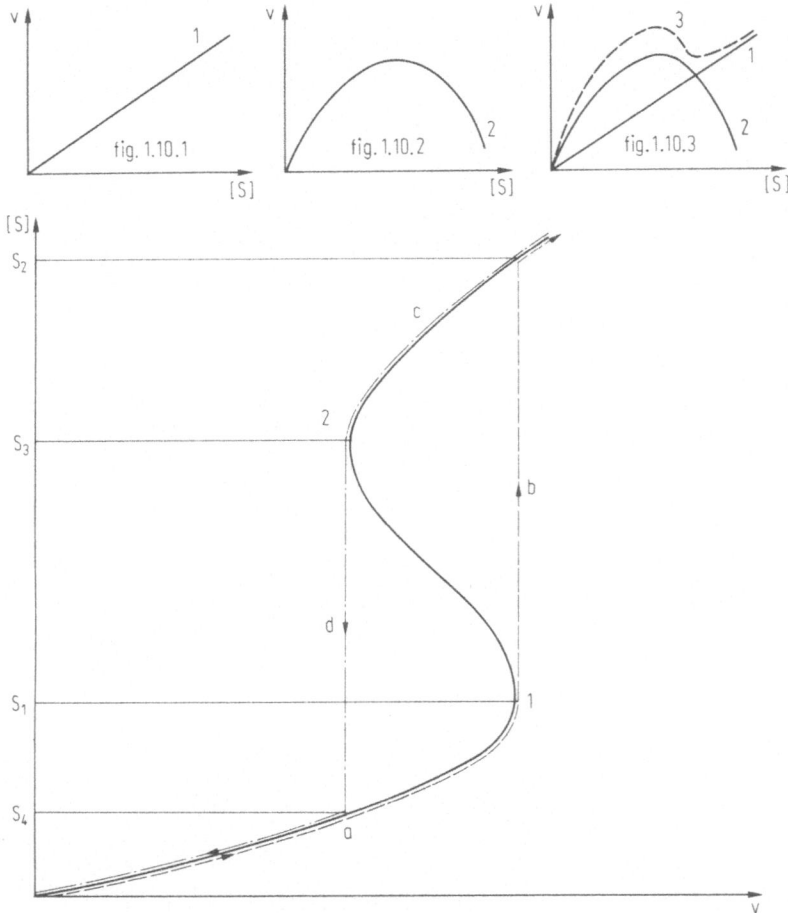

Fig. 1.10. Hysteresis loop of the inverted function v = f [S]

These must cause periodic temperature effects with different conversions and must be correlated to stable states. The accuracy of these explanations — which must be clearly differentiated from temperature effects from dynamic and control processes [194]— has not been established so far. Periodic temperature oscillations during the Oxo reaction are known, e.g., Fig. 1.11. Their cause is being investigated as it could lead to important information about possibly dangerous operating conditions thereby facilitating optimal operation [195].

The basic equilibria used to represent a catalytic cyclic process are shown in Fig. 1.12 [191]. A series of differential equations can be derived from these equations or rate constants which finally yield the qualitatively correct relationship [Eq. (33)] for the net production (A) of aldehydes.

$$A = j \cdot K \cdot \frac{[CO]}{[catalyst]} \tag{33}$$

j = total available CO (pool of CO)   K = quotient of rate constants k_0 and k_1

19

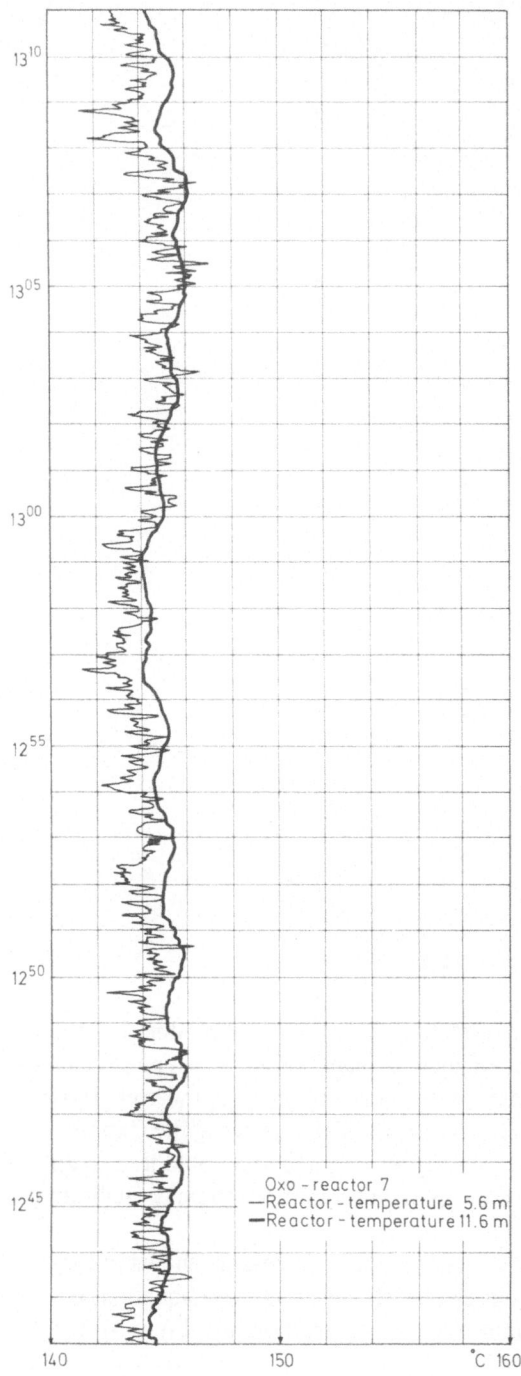

Fig. 1.11. Temperature oscillations in industrial Oxo reactors [195]

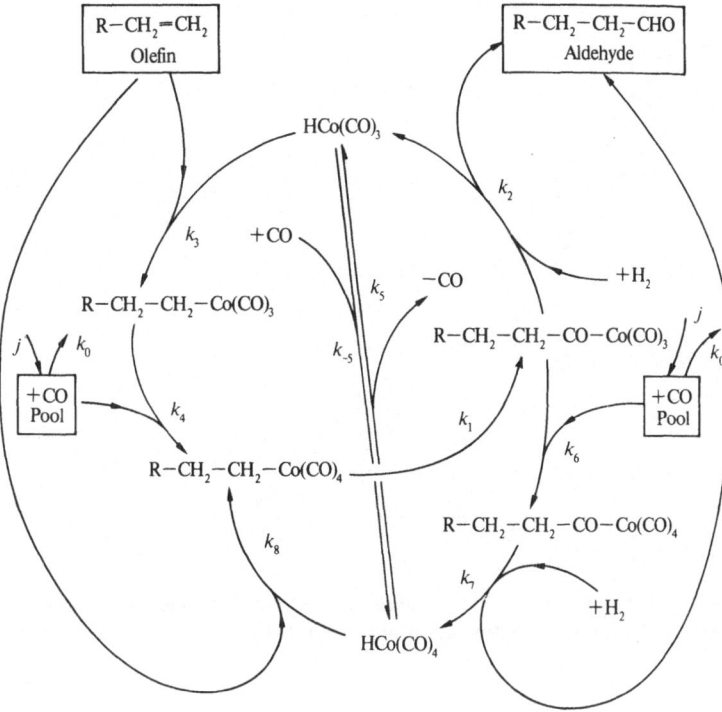

Fig. 1.12. Seelig's model [191] for the hydroformylation reaction

The optimal control of the Oxo process is discussed in Sect. 1.3.7 and elsewhere [206].

1.3 Effect of Reaction Conditions on Conversion, Selectivity and Operation of the Oxo Synthesis

1.3.1 Temperature

1.3.1.1 Unmodified Catalysts

While the *stoichiometric hydroformylation* takes place even at room temperature, the *catalytic* reaction requires minimum CO partial pressures (cf Sect. 1.3.2) as well as minimum temperatures which are between 25 and 30 °C depending on the olefin [6, 211, 212].

On utilizing olefins with more than two carbon atoms, there is a discord between the generally necessary high reaction velocity, which increases with rising temperature (cf Table 1.3 and Fig. 1.13), and the decreasing tendency towards the formation of the favored unbranched aldehydes with rising temperature [77, 213, 216–220].

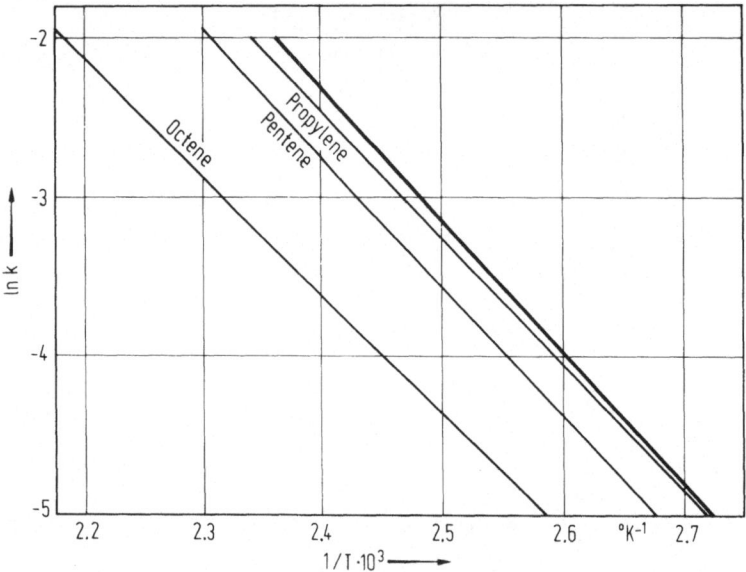

Fig. 1.13. Effect of reaction temperature on the rate constants of the hydroformylation of various olefins [77] – Rate constant of the forward reaction according to Eq. (8).

Table 1.3. Effect of temperature on reaction velocity [213]

Reaction temperature (°C)	Relative reaction velocity
90	0.01
100	0.04
120	0.20
140	1.00

The dependence of the n : iso ratio of the reaction products of the propylene hydroformylation on reaction temperature is shown in Fig. 1.14 [213, 216–220, 228, 229].

The diverse relationships found by various authors are due to the differing degrees of mixing in the test reactors. A small or large quantity CO influences the position and rate at which equilibrium is attained in Eq. (9) and thus the formation of the unbranched aldehydes via a small or large $[HCo(CO)_4]$ concentration. It therefore appears that at high reaction temperatures there is a CO deficiency which can only be partially balanced –and not eliminated–by extremely good mixing (agitated autoclaves, ⊙ values in Fig. 1.14) or by special construction features in the Oxo reactors (○ values in Fig. 1.14 represent continously operated nonstirred pilot plant reactors as measured by Ruhrchemie AG). This information has influenced not only reaction control techniques but also the construction of the industrial Oxo reactors [221, 223–227] (cf Sect. 1.3.7). On the other hand, the rate of formation of the actual Oxo catalyst $HCo(CO)_m$ from the precursor octacarbonyldicobalt probably has no effect. According to Fig. 1.13, its rate of formation is always higher than that of the rate of olefin hydroformylation [77, 180]. The effect of promoters on the formation of $HCo(CO)_4$ from cobalt compounds

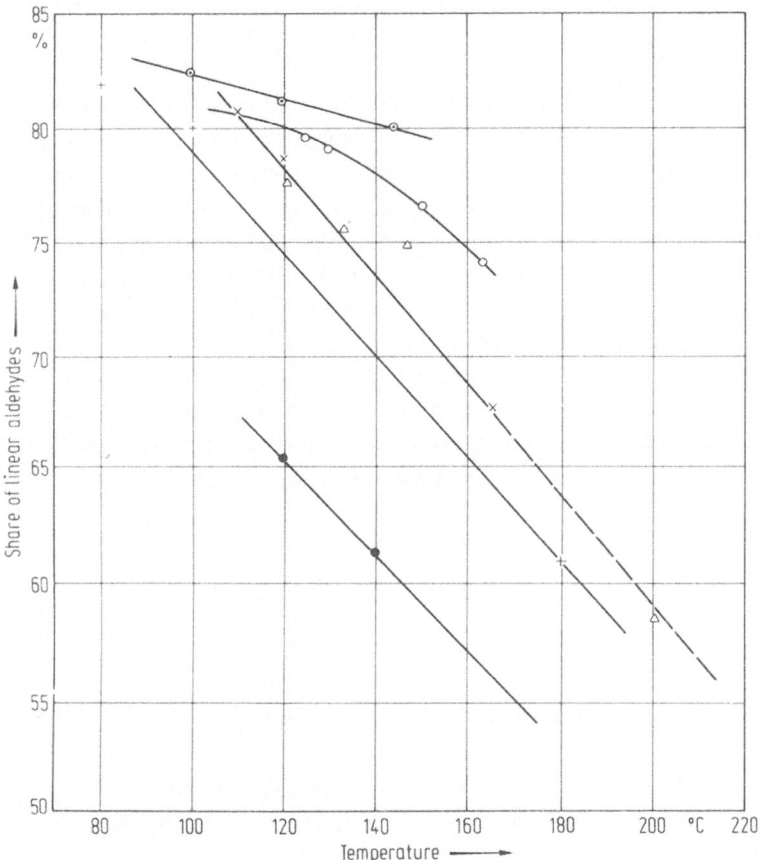

Fig. 1.14. Dependence of n : iso ratio of the reaction products of the propylene hydroformylation on reaction temperature [213, 216–221]

○ pilot plant results by Ruhrchemie AG [217]   + Hughes, Kirshenbaum [213]
× ICI Ltd. [218]   ● Macho [216]
△ Alekseeva et al. [219]   ⊚ Pino, Piacenti [220]

other than $Co_2(CO)_8$ has been studied by various workers [222, 223, 274] (cf Sect. 1.3.3.1.4).

The pattern of decreasing n : iso ratios of the resulting isomeric aldehydes with rising temperature applies more or less to all olefins. In the temperature range of 80–180 °C, with α-olefins the n : iso ratio drops from around 3–5 to 0.8–2 [213, 228–232, 282]. Strongly deviating ratios of unbranched : branched aldehydes are found only when steric factors play an important role [233–235] (cf Sect. 1.4). The most well-known example is the hydroformylation of isobutene which produces an n : iso ratio of around 40 : 1 with regard to the favored formation of 3-methylbutanal compared to 2,2-dimethylpropanal [236–238].

$$\underset{CH_3-\overset{\overset{\displaystyle CH_3}{|}}{C}=CH_2}{} + CO/H_2 \xrightarrow{\text{cat.}} H_3C-\overset{\overset{\displaystyle CH_3}{|}}{C}H-CH_2-CHO + H_3C-\overset{\overset{\displaystyle CH_3}{|}}{\underset{\underset{\displaystyle CH_3}{|}}{C}}-CHO \qquad (34)$$

23

"Keulemans' Law" — the apparent incapacity of quaternary C atom structures to result during the hydroformylation — has been dealt with by various workers [239, 266] (cf Sect. 1.4).

The formation of the various isomeric aldehydes on employing olefins with internally located double bonds (internal olefins) is also determined by their rate of isomerization with cobalt catalysts. The rate of isomerization of the olefins at temperatures below around 150 °C is usually lower than the rate of hydroformylation. Consequently, α-olefins and 'internal olefins' preferentially yield unbranched and branched aldehydes, respectively, (e.g., heptene [213], industrial application [40]). On the other hand, rhodium carbonyls are not only highly active hydroformylation catalysts, but also intensify the rate of isomerization of the olefins (cf Sect. 1.3.3.1.1).

The tendency for branched aldehydes to form with increasing temperature can reverse in cases where the olefinic substrate has a particular electronic configuration. The most well-known examples are the unsaturated esters which preferentially yield unbranched products at high Oxo temperatures (and low pressures) [36, 240−243, 1077].

$$
H_2C=\overset{\overset{\displaystyle CH_3}{|}}{C}-C\overset{\displaystyle O}{\underset{\displaystyle OR}{\diagup}} \quad \xrightarrow[\text{cat.}]{CO/H_2} \quad OHC-CH_2-\overset{\overset{\displaystyle CH_3}{|}}{CH}-C\overset{\displaystyle O}{\underset{\displaystyle OR}{\diagup}} \quad + \quad H_3C-\overset{\overset{\displaystyle CH_3}{|}}{\underset{\displaystyle CHO}{C}}-C\overset{\displaystyle O}{\underset{\displaystyle OR}{\diagup}} \tag{35}
$$

β-isomer $\qquad\qquad\qquad$ α-isomer

This phenomena, also dealt with in Fig. 1.15, has been treated in Ref. [6].

A superposition of electronic and structural effects can occur with longer unsaturated alkyl groups and/or with longer ester groups [244].

On choosing the most suitable hydroformylation temperature, the following points must be considered:

1) The rate of reaction — which in batch processes determines the residence time and in continuous processes determines the reactor volume.

2) Effect of the product range — required n : iso ratio of products.

3) Factors determing aldehyde selectivity.

As will be fully discussed in Sect. 1.5, the extent of parallel and consecutive reactions increases with rising hydroformylation temperature : hydrocarbons result via olefin hydrogenation, alcohols via hydrogenation of the aldehydic product, and alkyl formates and higher condensation products via further consecutive reactions (condensation, aldolization, acetal formation etc.), cf Fig. 1.16 [245−247]. With particular substrates, desired deviations from the expected product range can ensue, e.g., diethyl ketone from ethylene at high reaction temperatures [248, 249, 268] or lactones from appropriately substituted unsaturated esters [241, 250, 251]. Even when large amounts of branched products are acceptable or desired high reaction temperatures are only permissible when alcohols are the required final products and suitable units are available to process by-products and heavy ends (cf Sect. 1.5.4).

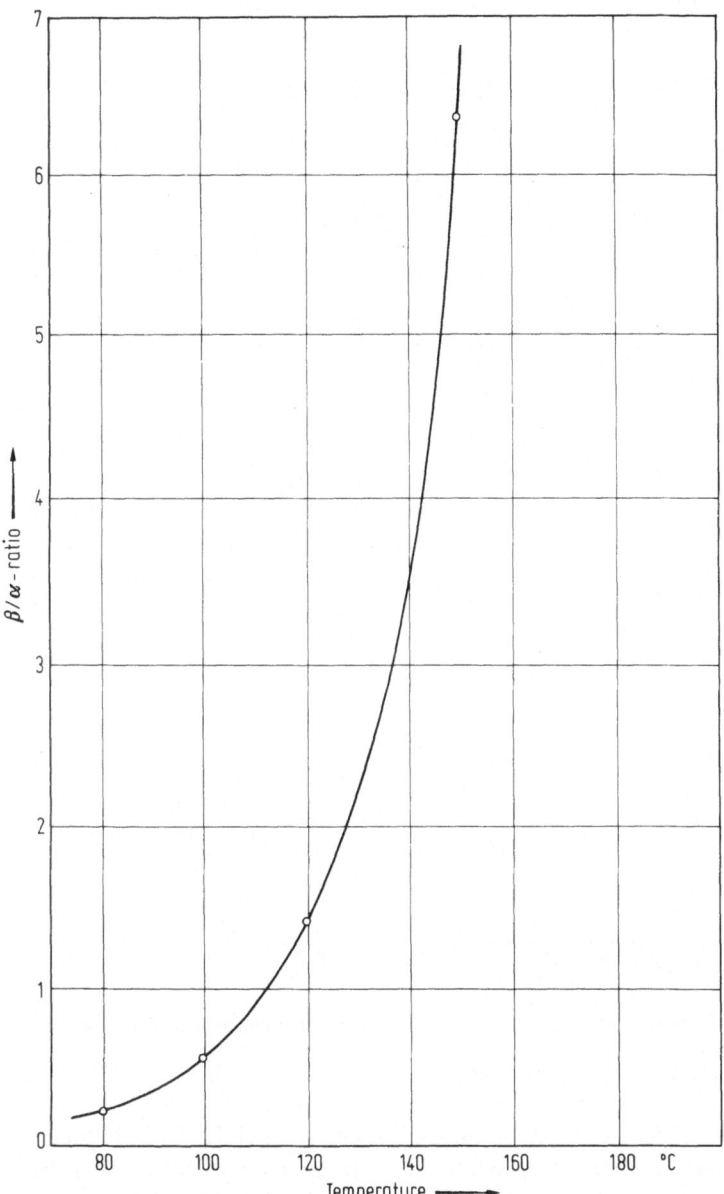

Fig. 1.15. Effect of reaction temperature on the n: iso ratio (= $\beta:\alpha$ quotient) of the reaction products from the hydroformylation of methyl methacrylate [240]

This could well be the case in the production of higher plasticizer or detergent alcohols where the total olefin feedstock is to be converted into alcohols and the alde-hyde stage is by-passed. The frequently proposed "adiabatic operation" with final temperatures around 300 °C will only be of interest for these types of reactions [252, 253, 267, 275, 944] if safe operation at such high hydroformylation tempera-tures can be ensured, e.g., only at insufficient CO partial pressures, cf Fig. 1.9 [254].

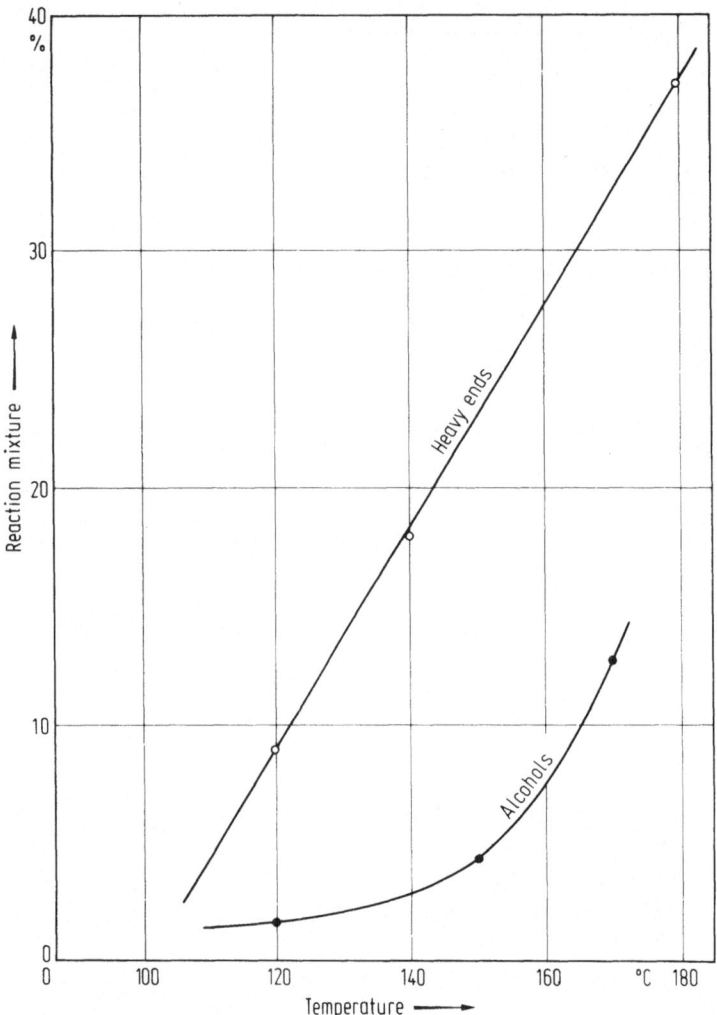

Fig. 1.16. Temperature dependence of by-product formation in the propylene hydroformylation [216]

1.3.1.2 Modified Catalysts

After preparatory work [255–257], Slaugh and Mullineaux of Shell [258–261] prepared the way for a modification of the classical cobalt catalyst via substitution of the
CO ligands by phosphines. The results of this modification (cf Sect. 1.3.3.1.2) are,
besides an increase in the selectivity of the hydroformylation, an increase in the
thermal stability of the catalyst-$HCo(CO)_m L_n$-and a rise in the hydrogenation activity
as well as a decrease in Oxo activity [26, 40, 47]. The drop in Oxo activity means that
reaction temperatures had to be maintained at around 180 °C and, even then, five- to
sixfold reactor volumes were necessary.

The effect of reaction temperature on the ratio of unbranched to branched compounds is largely determined by the type of ligand in the catalyst. As a consequence of
the strong hydrogenation activity, the reaction products with the phosphine substituted
cobalt carbonyls are usually not aldehydes but alcohols. Besides phosphines, which
hardly affect the n:iso ratio [15, 262] (cf Fig. 1.17), with P-heterocycles aldehydes are

obtained, as the former cause a lowering of hydrogenation activity of the Oxo catalysts of type $HCo(CO)_m L_n$. Clear relationships between the n : iso ratio and the reaction temperatures were also confirmed [263] (cf Fig. 1.18).

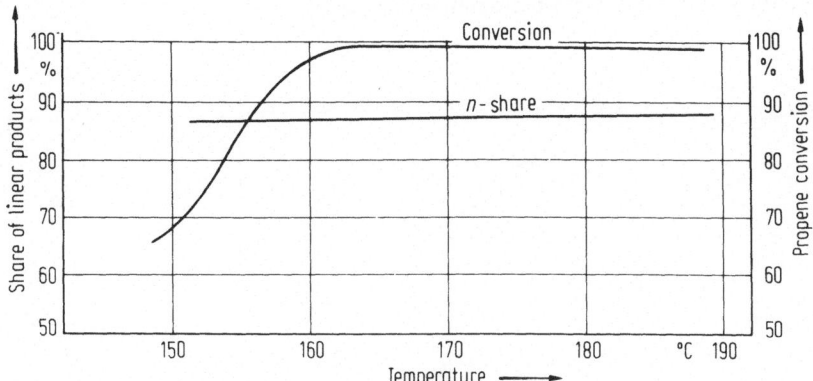

Fig. 1.17. Effect of reaction temperature on conversion and n : iso ratio in the propylene hydroformylation using P-modified Co catalysts [262]

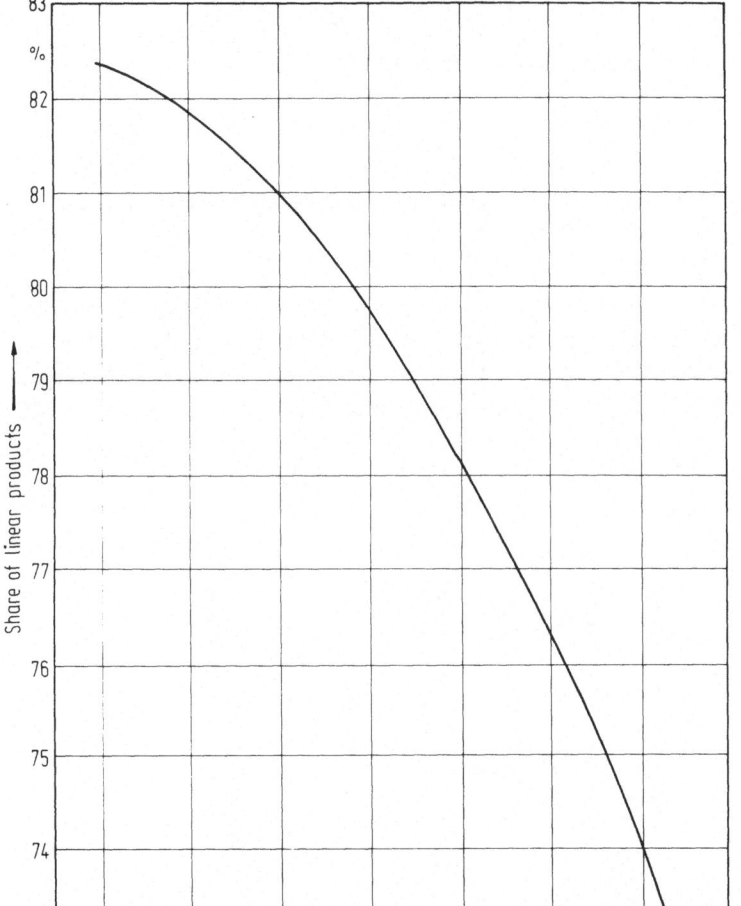

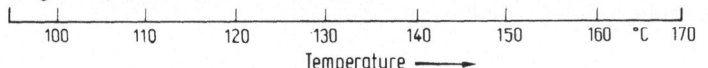

Fig. 1.18. Dependence of the formation of unbranched compounds on the reaction temperature during the hydroformylation of propylene with P-modified Co catalysts [263]

27

There exist similiar types of relationships for the formation of alcohols and other by-products, as to unmodified Co catalysts [264]. Figure 1.19 shows typical conditions on modifying with 1,2-diphosphacylopent-5-en-4-one [263].

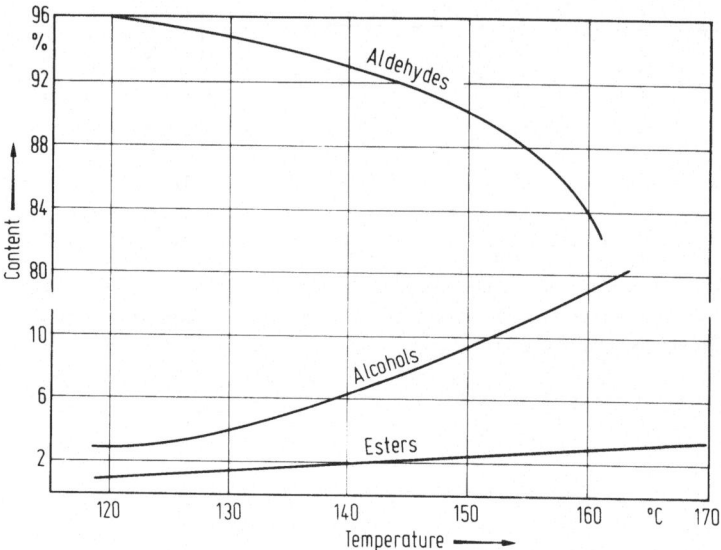

Fig. 1.19. Composition of product from the propylene hydroformylation with P-modified Co catalysts in relation to reaction temperature [263]

Modified Rh catalysts also undergo a similar loss in activity [26]. In addition, the degree of unbranched aldehyde formation is so strongly dependent on the reaction temperature that industrially important variants (LPO process – UCC/Davy Powergas/ Johnson-Matthey cf Sect. 1.6.2.2.2) are obliged to operate at temperatures below 120 °C [26, 265].

More details can be found in Sects. 1.3.3.1.1 and 1.3.3.1.2.

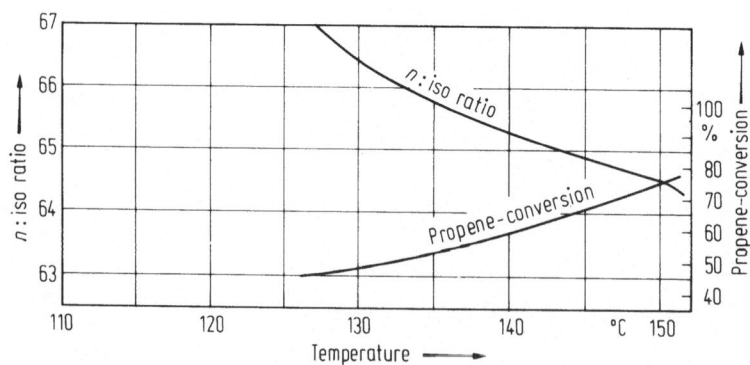

Fig. 1.20. Dependence of the n:iso ratio and the conversion of the propylene hydroformylation on the reaction temperature (catalyst – Rh(PAr$_3$)$_2$(CO)Cl [265])

1.3.2 Total Pressure; CO and H_2 Partial Pressures

1.3.2.1 Unmodified Catalysts

In the hydroformylation reaction, minimum temperatures as well as minimum pressures are necessary. As long as a CO : H_2 ratio of 1 : 1 (in syn gas) and a minimum CO partial pressure is maintained to ensure carbonyl stability [123, 161, 277–281, 1933, 1934] (cf Fig. 1.9) then the reaction velocity is independent of the total pressure in accordance with Natta's equation [Eq. (30)]. In industrial processes employing unmodified hydridocarbonyls, the usual parameters are > 100 °C and 80–300 bar total pressure [252, 277, 282–293] (cf Sect. 1.6.2). Patent disclosures about total pressures of 2000 bar [267, 278, 294–299] have not found any actual technical application and are probably only significant for patent litigation. However, several firms have described their process operation at high total pressures with apparative details.

In accordance with Eq (30), the partial pressures of CO and H_2 have different effects on the reaction velocity. Low CO partial pressures initially increase the rate of reaction before it once again drops off. This occurs with both the cobalt catalyzed reaction and that catalyzed by Rh carbonyls. Apparently, the actual catalyst is initially unsaturated with regard to CO and first forms under a certain CO partial pressure (with Co around 15 bar, with Rh around 40 bar) [88, 99, 177, 181, 225, 300, 302, 311]. An obvious supposition is to make Eq. (9) responsible for the above and to correlate the fivefold coordinated species $HCo(CO)_4$ [or $HRh(CO)_4$] with the decreasing reaction rate and to regard $HCo(CO)_3/HRh(CO)_3$ as the more active forms of the catalysts [113, 117, 123] (for MO-SCF calculations see Ref. [160]).

Increasing the H_2 partial pressure raises the reaction velocity [177, 282, 300, 303], this fact is exploited in a number of process variants by using syn gas with a CO : H_2 ratio > 1. It is claimed that induction periods [304] for example are eliminated by this means. However, it must be kept in mind that many deviations from the effects described here may be due to inadequate experimental control (poor mixing, mass transfer problems).

The above interpretation concerning the concentration of the catalytic species $HMe(CO)_4$ increasing with CO partial pressure is in agreement with the observation – on employing olefins with > 2 carbon atoms, e.g., propylene–that the share of the unbranched isomers increases in the same manner (Fig. 1.21) [113, 123, 147, 217, 219, 270–272, 305].

The group of curves in Fig. 1.21 shows, despite a few deviations in position and gradient (e.g., Pino et al. [270], Brewis [271]), that the share of the unbranched compounds – on maintaining an adequate reaction velocity – can only be raised on increasing the total pressure. The rate of increase flattens out above around 300 bar. Total pressures of 300–350 bar are usual pressure levels in industrial Oxo processes. In this range the yield of "n-products" is estimated to increase by around 0.5 kg/100 kg propylene per 10 bar (CO:H_2=1:1) [306].

According to Brewis [271, 299] (cf Table 1.4), n-butyraldehyde formation actually decreases again when the total pressure exceeds this optimal value. One possible explanation is the neglected amount of n-butanol which ought to be taken into account to-

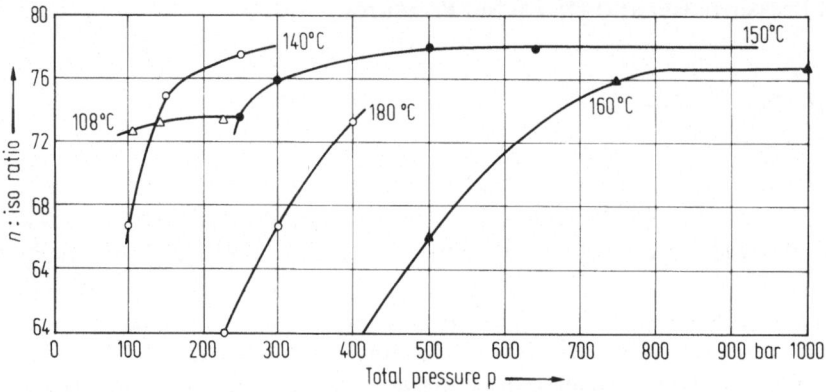

Fig. 1.21. Effect of total pressure (CO:H$_2$=1:1) on isomer ratio in the propylene hydroformylation

- Ruhrchemie AG (CO/H$_2$ = 1:1) [217]   △ Pino et al. (p$_{H_2}$ = 80 bar) [270]
- Alekseeva (CO/H$_2$ = 1:1) [219]   ▲ Brewis (CO/H$_2$ = 1:1) [271]

gether with the n-butyraldehyde share. It also appears feasible that polynuclear Co complexes result which have a different effect on the n:iso ratio compared to Co$_2$(CO)$_8$/ HCo(CO)$_4$ [334, 335] (for Rh carbonyls cf Ref. [336]).

Table 1.4: Effect of total pressure (CO:H$_2$ =1:1) and reaction temperature on n-butyraldehyde formation [271]

Temp. (°C)	Catalyst conc. (wt.%)	% n-Butyraldehyde at following pressures (bar)							
		250	500	750	1000	1250	1500	2000	2500
100	5.0	81.4	83.1	–	84.0	–	80.7	79.6	79.0
130	1.0	74.0	80.1	82.1	80.9	79.9	–	71.3	70.1
160	0.1	56.5	66.1	76.0	76.7	76.7	73.0	69.7	67.2
200	0.01	55.2	57.1	61.4	70.5	72.4	70.9	69.9	67.5
250	0.001	–	55.5	–	61.9	–	64.9	67.5	68.1

Besides the partial pressure of CO, that of H$_2$ also has a similar effect on the isomer ratio even if it is slightly less noticeable (cf Fig 1.22) [144, 234, 307].

The effect of the CO or H$_2$ partial pressure on the extent of the side reactions (cf Sect. 1.5) is also worth examining. Lower CO partial pressures generally increase the degree of the hydrogenation reactions taking place in situ thereby encouraging alcohol and hydrocarbon formation (via reduction of the olefins [218, 282, 308–310, 555]). In commercial reactors for the propylene hydroformylation, the share of the unbranched product diminishes by around 0.3 kg/100 kg propylene on reducing the p$_{CO}$ by 15 bar at constant total pressure in the range of p$_{CO}$ = 125–150 bar. A marked increase in the butanol yield can also be expected [306]. In some cases, such extensive hydrogenation to the butanols is the actual aim of the process variant [308–310].

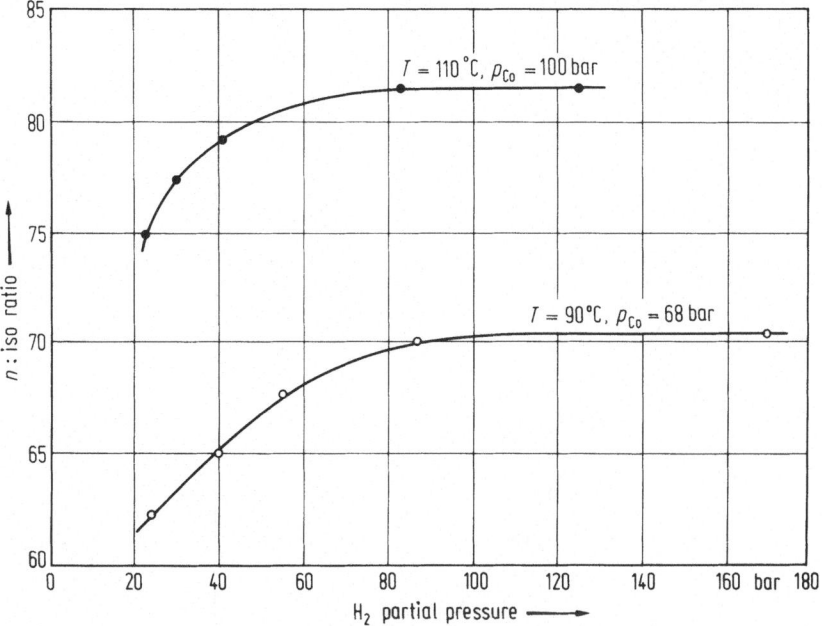

Fig. 1.22: Effect of H_2 partial pressure on isomer distribution in the products of the propylene hydroformylation [234]

The effect of total pressure or varying $CO:H_2$ ratios on the isomer distribution with other olefins has been dealt with by several workers e.g., higher α-olefins [282, 319], olefins with internally located double bonds [144], cycloolefins [320–322], and terpenes [323].

When examining the effect of the total and partial pressures on the isomer distribution, quite different results to those obtained with the olefins are found with unsaturated substrates exhibiting special electronic effects. For example, on hydroformylating methyl methacrylate (MMA) in the presence of rhodium hydridocarbonyl (cf Fig. 1.23 [240]) the β:α ratio (comparable to n:iso ratio with olefins) drops on raising the total pressure. Other reactions of this type can also be found in Ref. [36, 178, 242, 312–318].

The highest possible reaction velocity results via high H_2 partial pressures and the greatest possible yield of unbranched isomers via high CO partial pressures. Coupled with the desire for the best possible quantitative exploitation of the syn gas feed — according to Eq. (36) — this means that a molar $CO:H_2$ ratio of $1:1$ has to be employed to achieve optimum processing conditions for aldehyde production. As alcohol formation is encouraged by a $CO:H_2$ ratio of $1:2$ [Eq. (37)], this has led to a number of variants with gas recycles [324–327, 434, 435].

$$R-CH=CH_2 + CO + H_2 \xrightleftharpoons{cat.} R-CH_2-CH_2-CHO \qquad (36)$$

$$R-CH=CH_2 + CO + 2\,H_2 \xrightleftharpoons{cat.} R-CH_2-CH_2-CH_2OH \qquad (37)$$

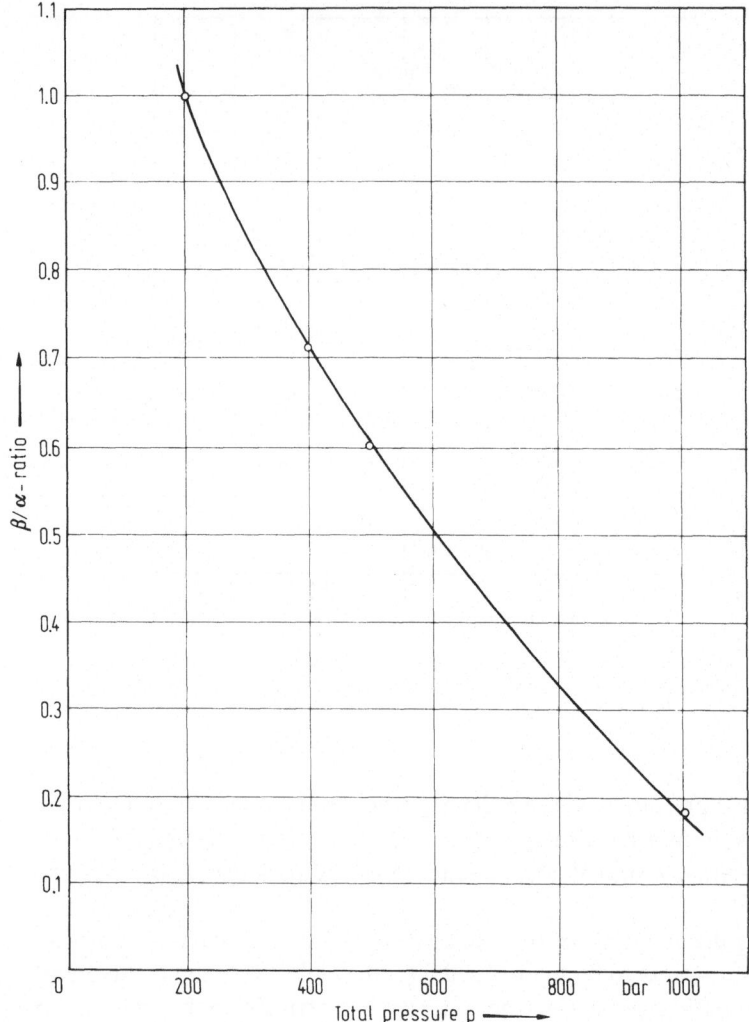

Fig. 1.23. $\beta:\alpha$ quotient in the hydroformylation of MMA as a function of the total pressure [240]

1.3.2.2 Modified Catalysts

The position of the equilibrium [Eq. (28)] can be influenced by the nucleophilicity of the ligands as well as by the CO partial pressure. Accordingly, a reduction in the CO partial pressure should displace the equilibrium to the right hand side leading to greater formation of the unbranched isomers compared to $HCo(CO)_4$. This is also actually observed (Table 1.5 and Fig. 1.24) [73, 209, 262, 273].

Piacenti et al. recently showed with the system $Co_2(CO)_6[P(C_4H_9)_3]_2$ (or $HCo(CO)_3$ $[P(Bu)_3]$) that CO partial pressures in the region 3.5 to around 10 bar lead to the expected and above described drop in formation of unbranched compounds. However, on further increasing CO partial pressure, the effect of $HCo(CO)_4$ dominates and the n:iso

Table 1.5. Relation between n:iso ratio and total pressure with HRh(CO)[P(OAr)$_3$]$_3$ (CO/H$_2$= 1:1, 1-octene) [276]

Total pressure		n:iso ratio
(psig)	(bar)	
100	7	87
850	17.5	80
500	35	75
1000	70	72
1500	105	71
2000	140	70
2500	175	69

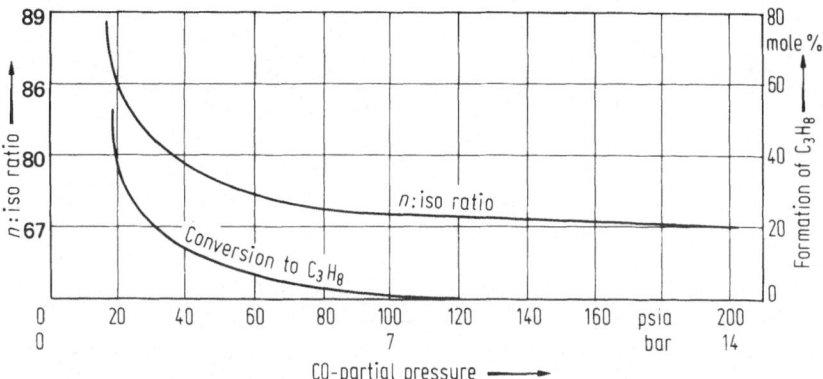

Fig. 1.24. n:iso ratio and propane formation versus CO-partial pressure at constant total pressure using HRh(CO) (PAr$_3$)$_3$ and propylene [273]

ratio (roughly corresponding to Fig. 1.21) increases once again [272]. Some of Piacenti et al.'s measurements are shown in Table 1.6.

Table 1.6. Hydroformylation of propylene with Co$_2$(CO)$_6$ [P(C$_4$H$_9$)$_3$]$_2$ at various CO partial pressures, P$_{H_2}$ = 40 bar [272]

P$_{CO}$ (bar)	Propane formation (mol %)[a]	Alcohol formation (mol %)[a]	n:iso ratio in	
			Aldehyde	Alcohol
3.5	16.6	36.4	85.0	85.0
8	5.4	15.1	68.0	68.0
10	3.9	11.3	60.0	60.0
15	2.3	8.0	63.7	64.0
20	1.6	6.0	67.0	67.0
30	1.0	3.7	68.2	68.0
64	1	1.2	68.5	68.5
90	1	1	69.0	69.0
133	1	1	70.0	70.0

[a] Relative to converted propylene

33

Cavalieri d'Oro, Andreetta et al. [207] proposed using kinetic equations based on Happel's work [208] to explain the unsatisfactory n:iso ratio of approx. 1:1 with unmodified Rh carbonyls —compared to ratios of > 10:1 on employing phosphine–modified Rh catalysts (cf Refs. [308, 26] and Sect. 1.3.3). Equations (38) and (39) result from the equilibria in Fig. 1.25.

$$V_n = (V_+^{(1,2,3)} + V_+^{(12,13)} + V_+^{(14,15,16)}) \cdot \frac{V_+^{(4,5,6)}}{V_+^{(4,5,6)} + V_-^{(1,2,3)} + V_-^{(12,13)} + V_-^{(14,15,16)}}$$

$$(38)$$

$$V_{iso} = (V_+^{(\overline{1},\overline{2},\overline{3})} + V_+^{(\overline{12},\overline{13})} + V_+^{(\overline{14},\overline{15},\overline{16})}) \cdot \frac{V_+^{(\overline{4},\overline{5},\overline{6})}}{V_+^{(\overline{4},\overline{5},\overline{6})} + V_-^{(\overline{1},\overline{2},\overline{3})} + V_-^{(\overline{12},\overline{13})} + V_-^{(\overline{14},\overline{15},\overline{16})}}$$

$$(39)$$

V_n and V_{iso} represent the rate of formation of the n- and iso-aldehydes, respectively, V_+ and V_- represent the forward and retro reactions, of the conversions in Fig. 1.25.

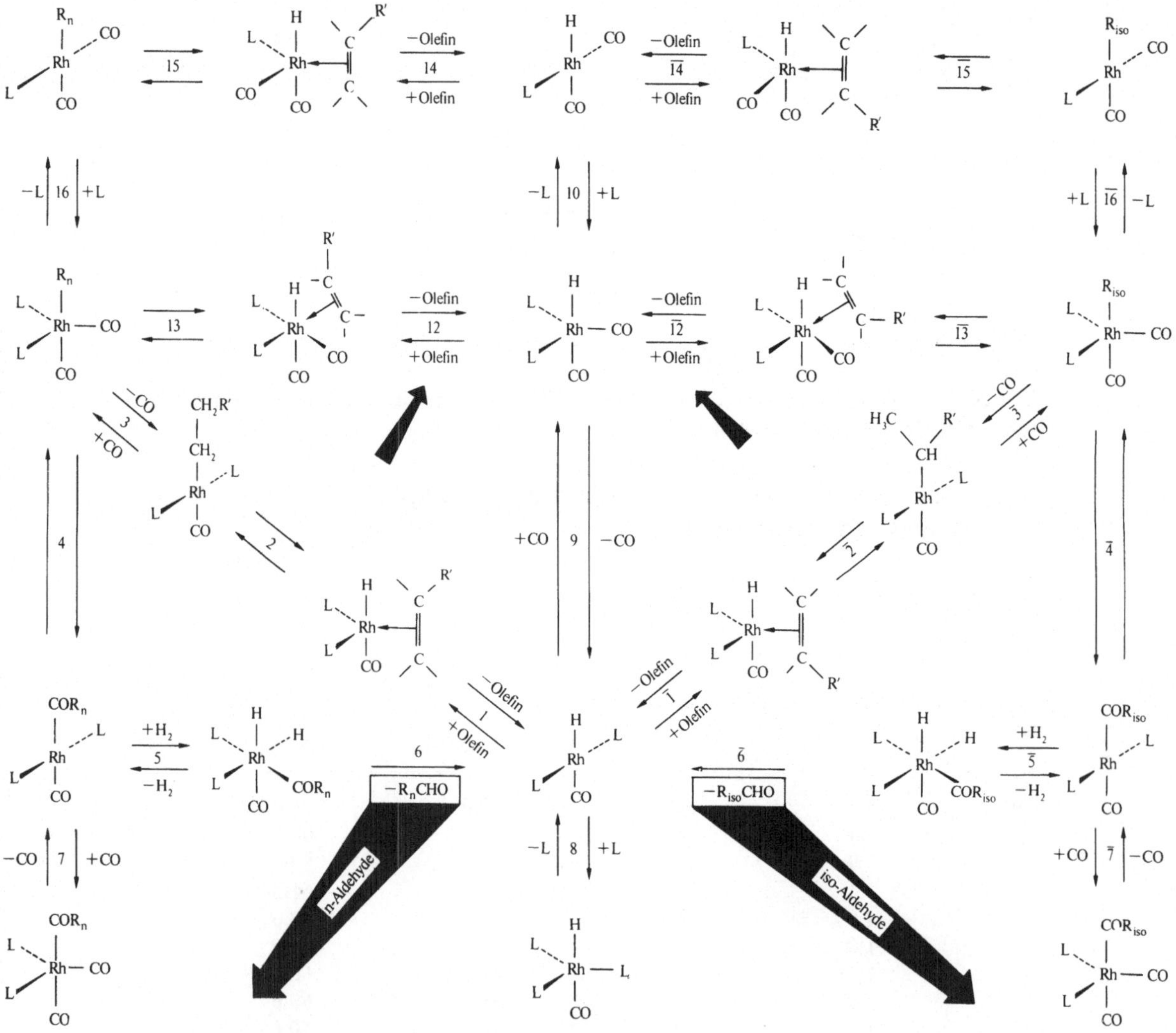

Fig. 1.25. Dissociative and associative pathway for the formation of n- and iso-aldehydes resp.

The expressions in parenthesis in Eqs. (38) and (39) relate to the formation of the n- or iso-alkyl derivates $R_nRh(CO)_2L_2$ or $R_{iso}Rh(CO)_2L_2$, respectively. The fractions represent the probability of formation of the corresponding aldehydes from the alkyl derivatives.

An expression for R (the n:iso ratio) as a function of the relation [CO]:[L] (ratio of p_{CO} or CO concentration to ligand concentration) is given by Eq. (40) which is derived from the above equations:

$$R = \frac{V_n}{V_{iso}} = R_1 \cdot \frac{1 + \dfrac{R_2}{R_1}\dfrac{1+R_1}{1+R_2}\dfrac{K_2}{K_1}K_9^e[CO] + \dfrac{R_3}{R_1}\dfrac{1+R_1}{1+R_3}\dfrac{K_3}{K_1}K_9^e K_{10}^e \cdot \dfrac{[CO]}{[L]}}{1 + \dfrac{1+R_1}{1+R_2}\dfrac{K_2}{K_1}K_9^e[CO] + \dfrac{1+R_1}{1+R_3}\dfrac{K_3}{K_1}K_9^e K_{10}^e \cdot \dfrac{[CO]}{[L]}} \tag{40}$$

where

$$K_1 = K_+^{1,2} + K_+^{\overline{1,2}}; \qquad K_2 = K_+^{12,13} + K_+^{\overline{12,13}}; \qquad K_3 = K_+^{14,15} + K_+^{\overline{14,15}}$$

$$R_1 = K_+^{1,2}/K_+^{\overline{1,2}}; \qquad R_2 = K_+^{12,13}/K_+^{\overline{12,13}}; \qquad R_3 = K_+^{14,15}/K_+^{\overline{14,15}}$$

and $K_+^{i,j} = K_{+i}K_{+j}/K(_{-i} + K_{+j})$; where K_{+i}, K_{-i} and K_{+j} represent the kinetic constants of the basic steps and K_9^e and K_{10}^e are the equilibrium constants for the steps 9 and 10 fo the scheme in Fig. 1.25.

According to Eq. (40), R tends to R_1 when [CO] (corresponding to the CO partial pressure) and the ratio [CO]:[L] becomes smaller. This is the n:iso ratio which would result on using the dissociative pathway 1, 2, 3, 4, 5, 6 and $\overline{1}, \overline{2}, \overline{3}, \overline{4}, \overline{5}, \overline{6}$. At higher [CO] values and smaller values for [CO] : [L], R tends to R_2 which corresponds to using the associative pathway 9, 12, 13, 4, 5, 6 (and $\overline{9}, \overline{12}, \overline{13}, \overline{4}, \overline{5}, \overline{6}$). The same applies to the combination consisting of low [CO] and a high [CO]:[L] ratio.

P. Cavalieri d'Oro et al. [207] produced Fig. 1.26 which stemps from a comparison of the calculated n:iso ratios [from Eq. (40)] with the experimentally determined values.

The agreement between measured and calculated values [based on Eq. (40)] for R presents another ready explanation as to why the n:iso ratio approaches 1:1 on passing from modified Rh carbonyls to unmodified catalysts (cf Sects. 1.3.3.1.1 and 1.3.3.1.2).

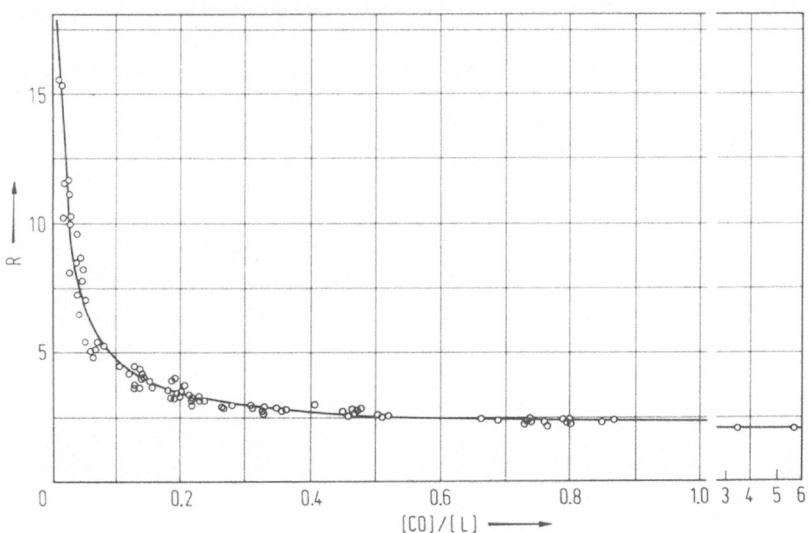

Fig. 1.26. Relationship between formation of n- to iso-butyraldehyde ($\equiv$R) and the [CO] : [Ligand] ratio, experimental values ($\circ$), – calculated from simplified equation (40)

In analogy to the unmodified Oxo catalysts, increasing H_2 partial pressure (lowering of CO : H_2 ratio at constant total pressure) raises not only the activity of the modified Co or Rh carbonyls [15, 169, 209, 262, 307, 328, 329, 332, 620], but also their hydrogenation tendency [164, 209, 262, 265, 330, 333]. Therefore, if it is attempted to improve the activity of the initially inactive phosphine-modified Co or Rh complexes via hydrogen-rich synthesis gas, then increased aldehyde hydrogenation as well as growing amounts of saturated hydrocarbons (via olefin hydrogenation) are to be expected (cf Table 1.6 and Figs. 1.27 and 1.28).

The same basically applies to other quite different catalyst systems e.g., $PtCl_2(PAr_3)_2$-$SnCl_2$ [331].

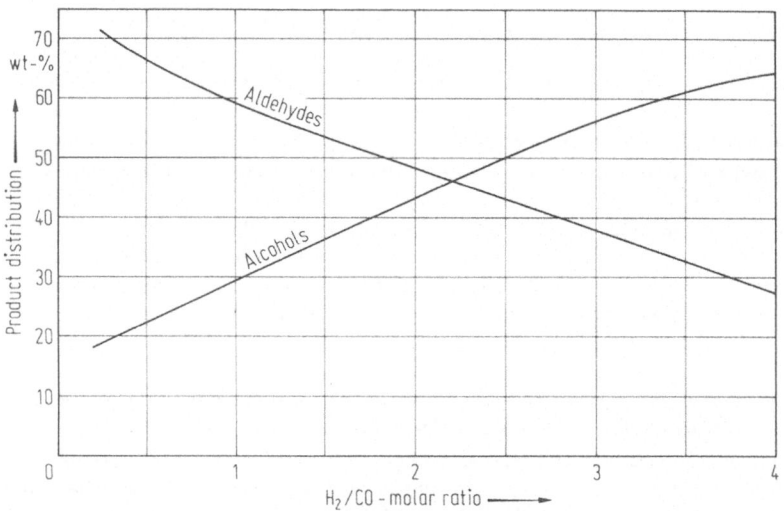

Fig. 1.27. Effect of CO and H_2 partial pressures on aldehyde/alcohol distribution in the propylene hydroformylation [15, 262]

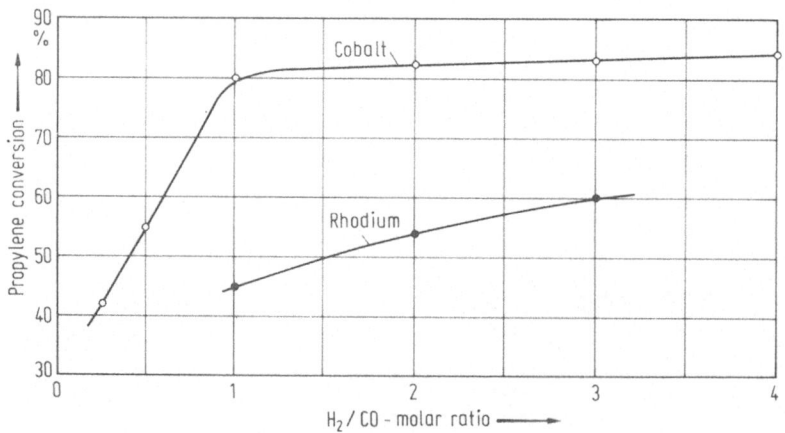

Fig. 1.28. Relation between olefin conversion and H_2 : CO ratio on using $Rh[P(Ar)_3]_2(CO)Cl$ [265] (●) or $Co_2(CO)_6[P(Bu)_3]_2$ [620] (○)

1.3.3 Catalysts

1.3.3.1 Hydroformylation Catalysts and their Variants

Despite the highly developed Oxo technology and the overriding position of the catalytic species $HCo(CO)_4$, there has been no lack of attempts to improve the total yield or the selectivity of the reaction. In addition, efforts were also directed at increasing the activity or stability of the catalysts via variation or modification. The variants outlined in Fig. 1.29 have undergone basic studies.

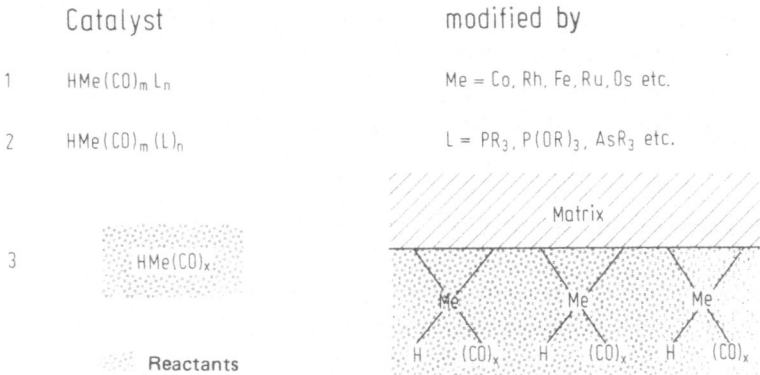

Fig. 1.29. Potential modifications of the hydroformylation catalysts

It is possible to replace the central atom (Me) of the Oxo catalyst $HMe(CO)_m L_n$ with other transition metals as shown in Fig. 1.29. The most well-known examples are the new Oxo catalysts with Rh as central atom (cf. Sect. 1.3.3.1.1). The carbonyl-forming transition metals listed in Fig. 1.30 are theoretically possible and some are actual catalysts.

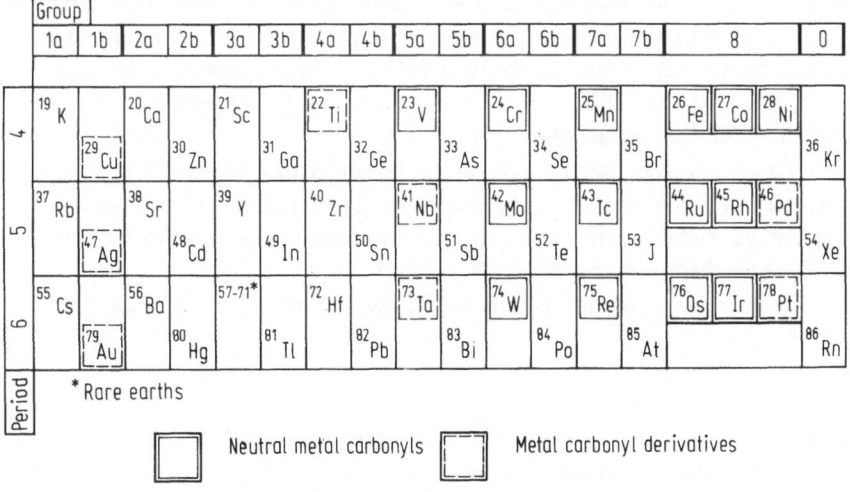

Fig. 1.30. Carbonyl-forming transition metals [108]

Furthermore, one or more of the CO ligands of the standard catalyst ($HCo(CO)_4$) can be replaced by organic electron donors such as amines (NR_3), phosphines (PR_3), phosphites ($P[OR]_3$), arsanes (AsR_3) etc. (cf Sect. 1.3.3.1.2). "Ligand modification" has led to industrially relevant developments in the catalyst recycle as well as in the Oxo process itself (Sect. 1.3.3.4, 1.3.7, and 1.6.2).

A third possibility is to vary the application phase of the active Oxo catalyst (normally used in liquid phase) via heterogenization. The latter involves either converting the actual homogeneous species into an immobilized heterogenized form or operating a gas phase hydroformylation (cf Sect. 1.3.3.1.3)

It is of course possible to combine the effects mentioned above for example by substituting the central atom of $HCo(CO)_4$ with rhodium and then to substitute one or more of the CO ligands of the new Oxo catalyst $HRh(CO)_4$ with phosphines etc. Catalysts of the type $HRh(CO)_m L_n$ result. Oxo catalysts of the general formula $HCo(CO)_4$, $HRh(CO)_4$ as well as $HMe(CO)_m L_n$ can all be heterogenized.

1.3.3.1.1 Via Variation of Central Atom

Tetracarbonylhydridocobalt is the standard catalyst of the industrial Oxo process. In 1978 alone, it was employed in processes which manufactured approx. 4.5 million t of Oxo products. The various process engineering solutions to cobalt recycle problems in large scale Oxo plants are summarized in Sect. 1.3.3.4 and 1.6.2. A recent report covering all aspects of the cobalt catalyzed hydroformylation can be found in the literature [6]. Newer studies on the Oxo synthesis with Co catalysts involve not only mechanistic aspects (cf Sect. 1.2) but, above all, new information about improvements in $HCo(CO)_4$ manufacture in situ [520−523, 582], sometimes in the presence of promoters [524] (cf Sect. 1.3.3.1.4). Processes for the manufacture of $HCo(CO)_4$ and its intermediates are also available industrially i.e., not merely as a section of complete Oxo plants [525].

For some time, many other carbonyl-forming metals have been claimed to be active in the Oxo reaction e.g., Mn, Re, Cr, Cu, Mo, and even Na and Ca. After cobalt, rhodium is the most active and well-known central atom in complex hydroformylation catalysts (for state of the art up to 1970 see Ref. [6]). Comparative studies involving the Oxo mechanism and the activity of Oxo catalysts include the central atoms Ru, Os, and Ir [7, 26, 36, 75, 144, 183, 467, 477, 500−506, 526, 527, 549]. The results of comparisons conducted by Marko [36] are shown in Table 1.7.

For comparison purposes, several hydroformylations with ligand-modified Oxo catalysts are listed in the latter part of the Table. The typical effects of this type of modification can be readily appreciated − longer reaction times − and favored the formation of unbranched isomers (cf Sect. 1.3.3.1.2).

If it is assumed in the equilibrium state of the Oxo catalysts [Eq. (9)] that the

$$HMe(CO)_m \rightleftharpoons HMe(CO)_{m-1} + CO \tag{9}$$

activity is determined by the coordinatively unsaturated species ($HCo(CO)_3$ in the case of cobalt) then the higher coordinated form − on account of its greater steric requirement−encourages the formation of the unbranched isomers. When this is correlated

Table 1.7. Hydroformylation of propylene using Co-, Rh-, Ir-, Ru- or Os-carbonyls as catalyst [36, 75, 576, 577, 1927]

Catalyst precursor	Catalyst conc. (mg-Atom·1^{-1})	Temp. (°C)	Partial Press. (bars) H₂	CO	Reaction time (hours)	Aldehydes Yield(%)	n/i-ratio
$Co_2(CO)_8$	23.4	110	75	75	1	93.7	4.0
$Rh_4(CO)_{12}$	5.3	110	110	110	1	80.0	1.1
$Ru_3(CO)_{12}$	23.4	110	75	75	7	40.2	2.8
$Os_3(CO)_{12}$	18.5	180	80	80	2	74.0	1.0
$Ir_4(CO)_{12}$		125					1.8
$Co_2(CO)_6[P(Bu)_3]_2$	23.4	140	30	15	3	80.0	6.1
$HRh(CO)[P(Ph)_3]_3$		120	< 2	< 1	5		> 10
$Ir(CO)[P(Ph)_3]_2Cl$	9.3	130	15	15	6	67.6	1.8

with the atomic radii of the central atoms listed in Table 1.7, then the central atoms of greater atomic radii should tend to "dissociate" in accordance with Eq. (9) and therefore be more active. Thus, they will be more inclined to catalyze the formation of branched aldehydes (cf Fig. 1.31).

Therefore, in qualitative agreement with Table 1.7, the application of Co carbonyls should result in the formation of particularly high shares of the unbranched isomers and with $Os_3(CO)_{12}$ in a great amount of branched aldehydes. This theory cannot be applied without limitation as is shown by the low activity of the Os carbonyl (cf review [75]). The particular activity of the Pt carbonyls (cf Fig. 1.31) will be discussed on page 45.

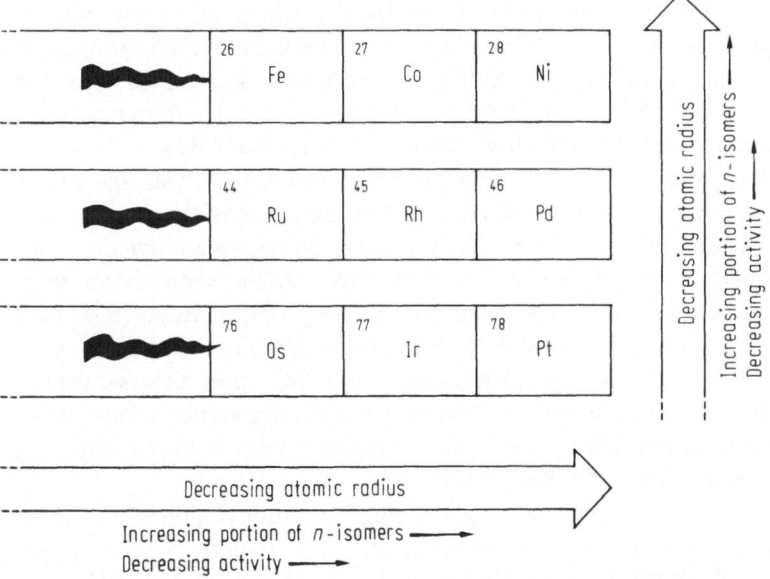

Fig. 1.31. Correlation between atomic radius and hydroformylation activity/selectivity

Russian workers, in particular, support the theory that the Oxo activity of the metal hydridocarbonyls – $HMe(CO)_m$ – can be correlated to the acidity of the hydrogen atom [578, 579] (cf Table 1.8).

Table 1.8. Acidic properties of compounds with hydrogen-transition metal bonds

Compound	Qual. characteristic	Diss. constant
$HCo(CO)_4$	strong acid	1
$H_2Fe(CO)_4$	weak acid	$4 \cdot 10^{-5}$
$HCo(CO)_3[P(Bu)_3]$	very weak acid	$1.08 \cdot 10^{-7}$
$HMn(CO)_5$	very weak acid	$0.8 \cdot 10^{-7}$
$HRe(CO)_5$	extremely weak acid	
$HRe(C_6H_5)_2$	weak base	
K_2ReH_9	hydride	

According to Imyanitov [578, 579], an increase in acidity should accelerate the initial reaction as well as improving the reactivity of the acyl metal carbonyl intermediate. The acidic properties can be altered by introducing electron-releasing or electron-attracting ligands. In metal hydridocarbonyls, replacing CO molecules by triphenylphosphine leads to a lowering of the acidity, a drop in activity and a decrease in the share of the Markownikoff reaction products.

The rhodium hydridocarbonyls, which were first recognized as being Oxo active in 1952 [528], are of particular interest. Their activity is 10^2 to 10^4 times greater than that of the cobalt hydridocarbonyls [36, 273, 319, 532–535]. This rhodium-based catalyst was first reported by Schiller in 1952 [528], then intensive research was conducted by Esso from 1959 (cf patents [536]), the first scientific publications being written by Imyanitov and Rudkovski [537]. Unmodified rhodium hydridocarbonyls are thus highly active but – in agreement with the greater atomic radius and displacement of resultant equilibrium in Eq. (9) to the right hand side – lead to the formation of more branched reaction products than the cobalt catalysts (cf Fig. 1.32).

According to the above, at high CO partial pressures with Co catalysts, the 1-pentene feedstock yields a n:iso product ratio of over 4:1, only 18% of the reaction products being branched. In the case of a Rh catalyst, the n:iso ratio is around 1.6:1, the share of the branched products being 38% [529]. As expected, the isomerization tendency falls with decreasing CO partial pressure. Wakamatsu [530] and Evans [538] have published data about the isomeric ratio in the hydroformylation of terminally and internally unsaturated olefins using Co or Rh catalysts. Another report discusses the effect of CO partial pressure and temperature on the product composition of the Rh catalyzed hydroformylation. The former is more complicated due to the simultaneously occurring isomerization steps (cf Ref. [531]).

Some examples of recent publications concerning hydroformylation with unmodified Rh compounds are as follows: synthesis of defined Rh carbonyls [539, 540], general reviews covering hydroformylation chemistry [14, 26, 36, 75, 78, 171, 531, 541–548], stereochemistry of the reaction [144, 551], hydroformylation of ethylene [552,

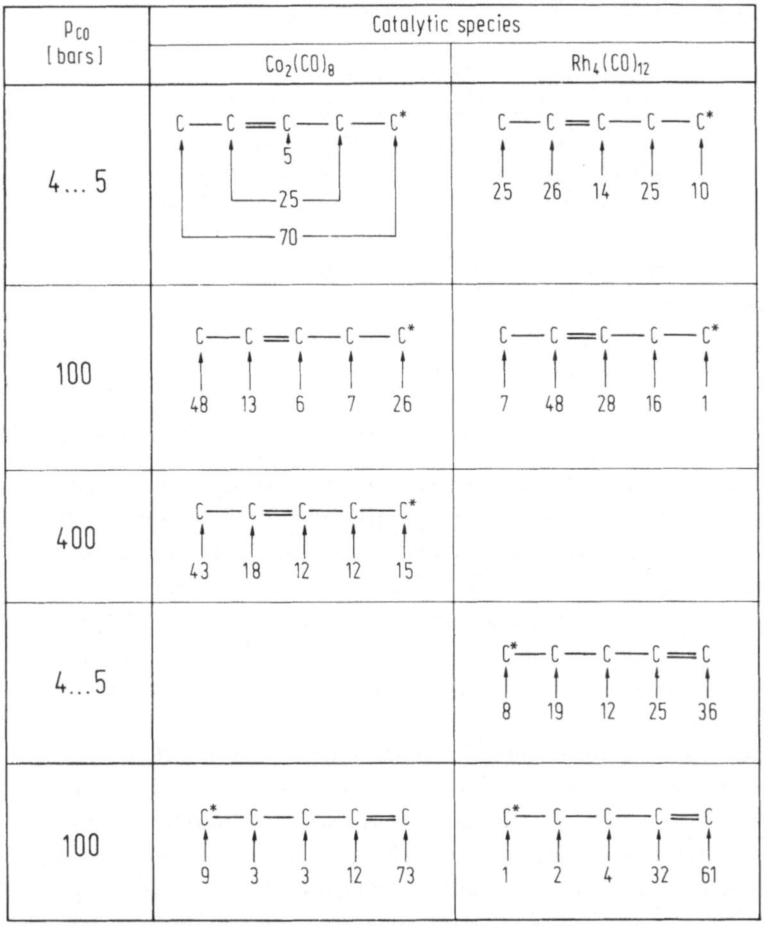

p_{CO} [bars]	Catalytic species	
	$Co_2(CO)_8$	$Rh_4(CO)_{12}$
4...5	C—C=C—C—C* 5, 25, 70	C—C=C—C—C* 25 26 14 25 10
100	C—C=C—C—C* 48 13 6 7 26	C—C=C—C—C* 7 48 28 16 1
400	C—C=C—C—C* 43 18 12 12 15	
4...5		C*—C—C—C=C 8 19 12 25 36
100	C*—C—C—C=C 9 3 3 12 73	C*—C—C—C=C 1 2 4 32 61

Fig. 1.32. Hydroformylation of labelled linear pentenes with $Co_2(CO)_8$ or $Rh_4(CO)_{12}$. Extent of formylation at each carbon atom [183, 529]

553], propylene [172, 554], hexene [532, 555–557], octene [558, 559], cyclohexene [196, 535, 560], styrene or substituted styrenes [556, 561, 562], dicyclopentadiene, norbornene and other olefins [563–565], unsaturated functional derivatives [242, 312, 314, 566–572] or of acetylenes, dienes or allenes [573–575].

Due to the high activity of rhodium hydridocarbonyls only small quantities of this catalyst are required. Consequently, side reactions such as aldehyde hydrogenation or formation of high boiling substances or the parallel reaction involving hydrogenation of the olefin feedstock are catalyzed to a much lesser extent. This enables substrates to be hydroformylated with minor amounts of catalyst as would be the case with the more inactive Co (cf Table 1.9) which would lead to hydrogenation taking place.

The advantages of operating a process with minute amounts of rhodium hydrido-carbonyls are however coupled with certain disadvantages. On the one hand, Rh carbonyls tend to catalyze double bond isomerization – this feature is industrially exploited [46, 550] – and the price. While rhodium is around 10^3 times more active than cobalt it is also a thousand times more expensive (mid 1978: Co 40 DM/kg, Rh 40,000 DM/kg).

Table 1.9. Selectivity of Rh or Co in hydroformylation [36, 103]

| Substrate | Yield (%) of products formed by | | | |
| | Hydroformylation | | Hydrogenation | |
	Rh	Co	Rh	Co
$H_3C–CH=CH_2$	99	98	1	2
 ⬡–CH=CH–COOR	73	8	26	91

$H_2C=C–COOR$
    $\|$
    CH_3
	82	51	10	42

Thus, losses of 1 mg catalyst metal per kg Oxo product cost 0.004 German pfennigs in the case of cobalt in contrast to 4 German pfennigs with rhodium. Thus, with rhodium catalyzed hydroformylations catalyst losses cannot be merely accepted but necessitate the implementation of a rhodium recycle to keep losses lower than the ppm level. This is only feasible when sophisticated techniques are utilized (cf Sect. 1.3.3.4). Other work relating to the disadvantages of the rhodium catalyzed hydroformylation involved maintaining the Rh charge at a relatively low level and suppressing the isomerization tendency via ligand modification (cf Sect. 1.3.3.1.2).

Besides the industrially significant central atom-modified catalysts [$HCo(CO)_4$ and $HRh(CO)_4$] a number of other metal carbonyls have been reported to be Oxo active, e.g., ruthenium has been recommended for the Oxo sythesis since 1960 [6, 433]. It is also suspected that under certain circumstances the Ru cluster $H_4Ru_4(CO)_{12}$ is an active Oxo catalyst [465]. Schulz and Bellstedt published detailed studies of the Ru catalyzed propylene hydroformylation in 1973 [466]. According to their work, the most favorable results (approx. 60% conversion, n : iso ratio 2–2.5 : 1, up to 80% selectivity to C_4-aldehydes) were obtained at 120–130 °C/100–300 bar and a catalyst concentration of $4.2 \cdot 10^{-3}$ mol Ru/mol propene. The dependence of conversion, n : iso ratio as well as aldehyde selectivity on pressure or residence time qualitatively correspond to the known relationships with Co (cf Fig. 1.33).

Other technical Ru catalyzed hydroformylations including reactions with other olefins, have also been investigated [467–473, 843], with ligand modified Ru-catalysts, cf [1750]. Detailed studies have been conducted by Pino et al. [469] in particular, as well as by Marko and Chini [468].

In addition to Ref. [6], several transition metal carbonyls, which have been described as being Oxo active, are summarized in Table 1.10.

Table 1.10. Oxo active transition metal carbonyls

Metal (carbonyl)	Hydroformylation of	Ref.
Mn	tetradecene	[474]
	octene	[475, 476]
	various	[477, 478]

Table 1.10 (continued)

Metal (carbonyl)	Hydroformylation of	Ref.
Fe	ethylene	[479, 480, 481, 484]
	propylene	[479, 480, 482, 483]
	butene	[480]
	cyclohexene	[479, 480]
Cu (Ag, Au)	propylene	[384, 485–488]
	methylpentene	[487]
	hexadecene	[487]
Pd	ethylene	[489, 490]
	propylene	[489]
	butenes	[491]
	various	[492, 494]
Re	heptene	
	decene	[476, 493]
Os	various	[507, 844, 845]
Ir	ethylene	[495]
	hexene, heptene	[496]
	dodecene	[497]
	various	[498–506]
Pt	various	[501, 507, 508, 1928, 1929]

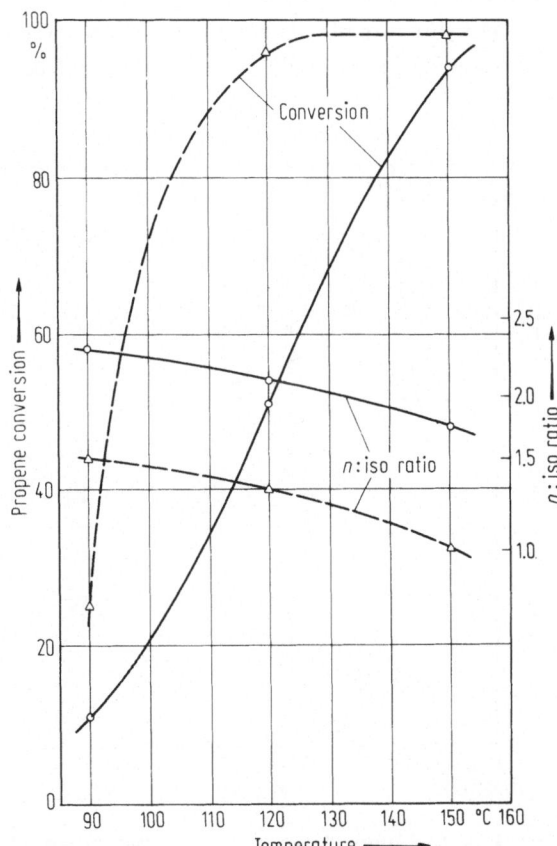

Fig. 1.33. Olefin conversion and n:iso ratio as a function of reaction temperature in presence (—) and absence (– –) of γ-irradation [466]

The previously mentioned misgivings about the reproductiveness of the results (deposits of highly active noble metals in autoclaves? contamination of apparatus?) should perhaps be repeated here with regard to Table 1.10.

Many patents [500–506, 535] disclose the application of complexes with more than one central atom including the mixed catalysts e.g., Co-Rh or Co-Pt [103, 183, 509, 510, 518, 535, 826, 1945], Co-Fe [479], Co-Se [511], Co–Ni [510] Rh-Fe [1930], Rh-Mo [1931] etc. Co-Al or Co-Cu systems [384] form the transition to heterogeneously applicable skeleton catalysts (cf Sect. 1.3.3.1.3). Mixed catalysts are said to exhibit particular effects; besides an increase in activity [479] in the main selectivity improvements [509] have been reported. However, in every case it must be examined whether both metals of the mixed catalyst would be compatible in a single catalyst recycle or whether the different properties of the metals would require two seperate and thus costly recycles.

This does not apply when the various metals of a mixed catalyst from a polynuclear cluster which does not exhibit the combined properties of both central atoms but leads to special effects due to the "synergic effect". Pino [183] assumes the existence of this type of cluster in the hydroformylation of diketene with Co/Rh mixed catalysts. While no reaction ensues with Co catalysts, 14% methylsuccinic anhydride are obtained with Rh catalysts, and with a Co:Rh ratio approx. 12:1 in the catalyst yields up to 89% of the aforementioned anhydride result [512] (cf Fig. 1.34). Similar effects are also found with Fe-Co catalyst systems [479].

The structure of a new complex, which was described by Orchin and Hsu [513], ICI researchers [514, 519], Schwager and Knifton [331, 515, 824] and others [501, 742,

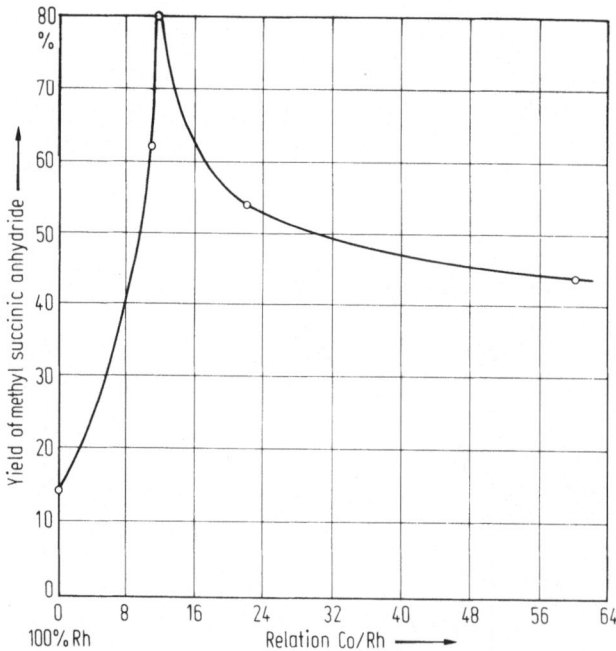

Fig. 1.34. Hydroformylation of diketene using mixed catalysts [183, 512]

822, 823, 942, 1134, 1226, 1914, 1962] is still uncertain. It consists of Pt-Sn central atoms and metal halides with central atoms from groups IVB/VB (cf Ref. [517] for corresponding Co complexes). The structures are discussed in Fig. 1.35.

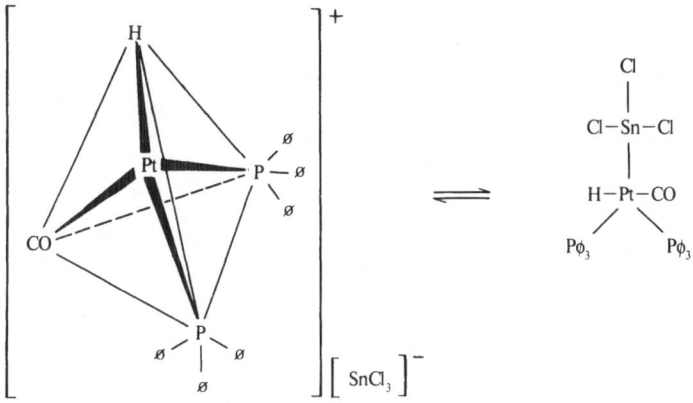

Fig. 1.35. Proposed structures of ligand-modified Pt/Sn catalysts [513]

According to Orchin, these catalysts do not undergo the loss in activity experienced by rhodium hydridocarbonyls when ligand modification ensues (for contradictory results see Schwager and Knifton [331]). Furthermore, high and greater selectivities to unbranched compounds are attainable. $Rh(CO)Cl(PPh_3)_2$ was however used for comparison purposes and not the more selective phosphine-modified rhodium hydridocarbonyls of the type $HRh(CO)_m(PPh_3)_n$ [331].

Conversion, activity and selectivity are affected by changes in the catalyst and reaction parameters in a similar manner to the classical Oxo catalysts (cf Table 1.11).

According to the above, the effectiveness of the Co-catalysts recedes in the following sequences $SnCl_2 > GeCl_2 > PbCl_2 \approx SiCl_4$ (positions 1–5, cf Sect. 1.3.3.1.4). However, the selectivity (positions 8,1,6 and 7) can be increased via ligand modification with phosphines, arsanes or stibanes, phosphines increasing the selectivity to unbranched aldehydes to a greater extent than arsanes or stibanes (positions 1,6 and 7). An increase in the $H_2:CO$ ratio effects an increase in conversion and a growth in the hydrogenation tendency of the system (positions 9 and 10). Raising the reaction temperature causes an increase in conversion, isomerization tendency and the extent of the hydrogenation of the olefin feedstock and effects a drop in selectivity to n-aldehydes (positions 12–14). The effect of the total pressure has not been clearly established. The heterogenization of this type of catalyst system has been discussed in Sect. 1.3.3.1.3.1 [1229, 1949].

1.3.3.1.2 Via Variation of Ligands

Introduction

It was already shown in Section 1.2 that one or more CO ligands of the complex hydroformylation catalyst $HMe(CO)_n$ can be substituted by electron-donating ligands. Further reaction steps can ensue after the initial substitution [Eq. (41)] takes place [86, 92, 174, 187, 498, 610, 619, 620, 839].

Table 1.11. Hydroformylation of 1-heptene using ligand-stabilized Pt/Sn complexes [331]

Pos.	Catalyst	Temp. (°)	Total pressure (bar)	H_2/CO ratio (syngas)	Heptene conversion (%)	Total yield[a] (mol-%)	Selectivity[b] (mol-%)	Isomerization[c] (mole-%)	Hydrogenation[d] (mole-%)
1	$PtCl_2[P(Ph)_3]_2 \cdot SnCl_2$	66	105	1:1	100	85	90	3.6	2.7
2	$PtCl_2[P(Ph)_3]_2 \cdot GeCl_2$	66	105	1:1	14	14	98	–	–
3	$PtCl_2[P(Ph)_3]_2 \cdot SnCl_4$	66	105	1:1	100	50	84	6.5	8.5
4	$PtCl_2[P(Ph)_3]_2 \cdot PbCl_2$	66	105	1:1	0	–	–	–	–
5	$PtCl_2[P(Ph)_3]_2 \cdot SiCl_4$	66	105	1:1	0	–	–	–	–
6	$PtCl_2[As(Ph)_3]_2 \cdot SnCl_2$	66	105	1:1	100	46	75	10	9
7	$PtCl_2[Sb(Ph)_3]_2 \cdot SnCl_2$	66	105	1:1	94	64	75	12	8
8	$K_2PtCl_4 \cdot SnCl_2$	66	105	1:1	99	60	73	13	10
9	$PtCl_2[P(Ph)_3]_2 \cdot SnCl_2$	78	88	1:1	21	18	99	3	18
10	$PtCl_2[P(Ph)_3]_2 \cdot SnCl_2$	78	88	30:1	100	56	93	26	
12	$PtCl_2[P(Ph)_3]_2 \cdot SnCl_2$	24	105	1:1	58	57	91	1.1	0.3
13	$PtCl_2[P(Ph)_3]_2 \cdot SnCl_2$	66	105	1:1	100	85	90	3.6	2.7
14	$PtCl_2[P(Ph)_3]_2 \cdot SnCl_2$	93	105	1:1	100	66	77	5.6	3.2
15	$PtCl_2[P(Ph)_3]_2 \cdot SnCl_2$	78	7	1:1	100	25	95	70	5
16	$PtCl_2[P(Ph)_3]_2 \cdot SnCl_2$	66	70	1:1	100	90	91	7.3	2.7
17	$PtCl_2[P(Ph)_3]_2 \cdot SnCl_2$	78	88	1:1	98	85	91	9.5	3.5
18	$PtCl_2[P(Ph)_3]_2 \cdot SnCl_2$	66	105	1:1	100	85	90	3.6	2.7
19	$PtCl_2[P(Ph)_3]_2 \cdot SnCl_2$	66	210	1:1	100	88	89	5.0	3.8

[a] to C_8-aldehydes   [b] to n-C_8-aldehydes   [c] to 2,3-heptenes   [d] to n-heptane

$$HMe(CO)_m + L \rightleftharpoons HMe(CO)_{m-1}L + CO \qquad (41)$$

$$HMe(CO)_{m-1}L + L \rightleftharpoons HMe(CO)_{m-2}L_2 + CO \qquad (42)$$

$$HMe(CO)_{m-2}L_2 + L \rightleftharpoons HMe(CO)_{m-3}L_3 + CO \qquad (43)$$

This "ligand modification", which when $HMe(CO)_m = HCo(CO)_4$, leads to the species $HCo(CO)_3L$, $HCo(CO)_2L_2$, and $HCo(CO)L_3$, was used by Reppe et al. for the first time in 1941 for stabilizing catalysts [583]. After the basic research conducted by Slaugh, Mullineaux, and Wilkinson [618] around 1962, this ligand modification was employed to directly influence catalyst properties. Ligand-modified cobalt (Shell process of Sect. 1.6.2.1) and rhodium catalyst (UCC and Mitsubishi processes, cf. Sect. 1.6.2.2) have both found industrial application.

The ligand modification has a drastic effect on Oxo catalysts where the central atom is Co or Rh. The operation and technology of such processes can be decisively affected (for detailed summaries see Ref. [6, 7, 15, 24, 26, 30, 78, 82, 171]. The following aspects will receive particular attention:

Stability: – The metal hydridocarbonyls modified with electron donators – such as phosphines PR_3, phosphites $P(OR)_3$, arsanes AsR_3, stibanes (SbR_3; R either alkyl, aryl, cycloalkyl or component of a homo- or heterocyclic ring system) – are more stable than the starting material. This is due to the superior σ donor and weaker π acceptor properties of the aforementioned modifiers compared to CO. The result is an increase in electron density at the central atom which improves the metal-CO bond and thus the stability of the complex $HMe(CO)_mL_n$ via stronger electron back-donation [108, 585–588]. This leads to a greater resistance of such modified Oxo catalysts with regard to thermal strain at lower CO partial pressures. Thus, in the industrial operation of an Oxo process the hydroformylation can ensue at lower CO partial pressures. Consequently, pressurized reactors and compression energy for syn gas can be dispensed with. In addition, the hydroformylation products can be distillatively separated from the modified catalyst, the homogeneous non-decomposed catalyst being simply recycled (cf Sect. 1.3.3.4 [589–597].

Activity: The catalyst activity drops when the stability increases, as in accordance with Figs. 1.2, 1.3 or 1.7 – catalytic cyclic process – the necessary dissociation of a CO ligand will be impeded. Similar mechanisms to those applying to unmodified catalysts are proposed for the ligand-modified Oxo catalysts, cf Sect. 1.2, and Figs. 1.7. and 1.8 [15, 160, 198, 598–600, 602]. The lower activity (compared to the unmodified catalyst) makes a greater reactor volume necessary if the same productivity is to be achieved [26, 209, 601–604]. This effect becomes, as expected, more noticeable [606–609] with increasing ratio of ligand to central atom. In Sect. 1.3.4, the effect of increasing ligand concentration is discussed, the ligand serves initially as solvent for the Oxo catalyst [73, 437, 605].

Hydrogenation activity: Another possible consequence of the higher electron density at the central atom – which can result in the H atom of the hydridocomplex possessing a stronger hydride character – is that the ligand modified Oxo catalysts are more active in hydrogenations than their unmodified counterparts. Therefore, lower yields are to be expected (via hydrogenation of the olefin feedstock) [15] along with a drop in selectivity (via hydrogenation of the aldehyde product). When tertiary phosphines or amines function as ligand with cobalt as central atom then they exhibit particular activity [54, 329, 584, 585, 611–617, 620]. Other interpretations are based on the strong hydrogenation activity of metal clusters which have been proved to form in situ during hydroformylation [86].

Selectivity: The actual motive for studying ligand-modified Oxo catalysts was a consequence of their property of repressing isomerizations [166, 167, 203, 603, 620, 641, 646] and encouraging the formation of the–usually desired–unbranched isomers. Various interpretations have been presented involving steric and/or electronic effects of the modified complexes [6]. Tucci [330] correlated the p_{ka} value of modified phosphines with the activity of the resulting catalysts and the n : iso ratio of the Oxo products (cf Table 1.12 and Fig. 1.36).

Table 1.12. Hydroformylation of 1-hexene with $Co(CO)_3PR_3$, effect of phosphine basicity [330]

No.	PR_3=	p_{ka}-value	Linear product % of total product
1	$P(i\text{-}C_3H_7)_3$	9.4	85.0
2	$P(C_2H_5)_3$	8.7	89.6
3	$P(n\text{-}C_3H_7)_3$	8.6	89.5
4	$P(n\text{-}C_4H_9)_3$	8.4	89.6
5	$P(n\text{-}C_8H_{17})_3$	8.4	90.2
6	$P(C_2H_5)_2 (Ph)$	6.3	84.6
7	$P(C_2H_5) (Ph)_2$	4.9	71.7
8	$P(Ph)_3$	2.7	62.4

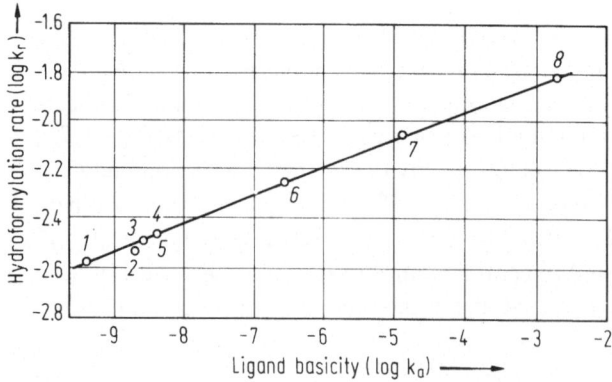

Fig. 1.36. Correlation between relative hydroformylation rate of 1-hexene and p_{ka} value of phosphine ligand (Nos. refer to Table 1.12)

It is obvious, while the reaction velocity is dependent on the basicity of the modified ligand, this is not the case with the n : iso ratio of the reaction products. More detailed treatment of correlations can be found in the literature [621, 622]. The suggested combination (Bahrmann, Fell et al.) of Tolman's [19] x_i value with a "cone angle" [621, 624] would appear to be more applicable than the "Δ-HNP value" [623, 832] as introduced by UCC. Discussions of dipole moments, spectroscopic data and steric structures etc. can be found in the literature [621, 625, 626, 642].

Susceptibility to poisons: As mentioned in Sect. 1.3.3.1.1, the activity of the rhodium hydridocarbonyls exceeds that of the cobalt hydridocarbonyls by a factor between 10^2 and 10^4. This comparison also applies to the ligand-modified Co or Rh complexes and, in addition, means that Rh catalyzed industrial Oxo processes should employ the lowest possible amount of Rh. Thus, due to the relatively small quantity of Rh present which cannot be continuously regenerated as in the Co process (cf Sect. 1.3.3.4) – catalyst poisons present in the feedstocks [627–635] must be counteracted. This necessitates employing special analysis and purification steps – involving increased costs [627, 629, 632–634, 726] (cf Sect. 1.3.3.2).

The influence of reaction parameters on selectivity and activity has an equally large effect on the reaction control and technology of the Oxo process with ligand-modified catalysts. This will be discussed below using several typical examples. Equations (41)–(43) indicate special features of these catalysts compared to the unmodified species.

The effect of reaction temperature on the activity and selectivity of the hydroformylation with ligand-modified catalysts corresponds to that in the synthesis with unmodified catalysts (cf Fig. 1.20 and 1.37).

For example, with the Rh catalyst used in Fig. 1.20 it is clear that to achieve high shares of the unbranched isomers, temperatures below 120 °C must be maintained. However, at these low reaction temperatures only partial conversions are possible, this is a notable feature of all industrial variants of the Oxo process with modified Rh catalysts [187, 209, 264, 265, 640]. The corresponding effects with modified Co catalysts are outlined in Figs. 1.17 and 1.18 [262]. Figure 1.37 shows the hydrogenation tenden-

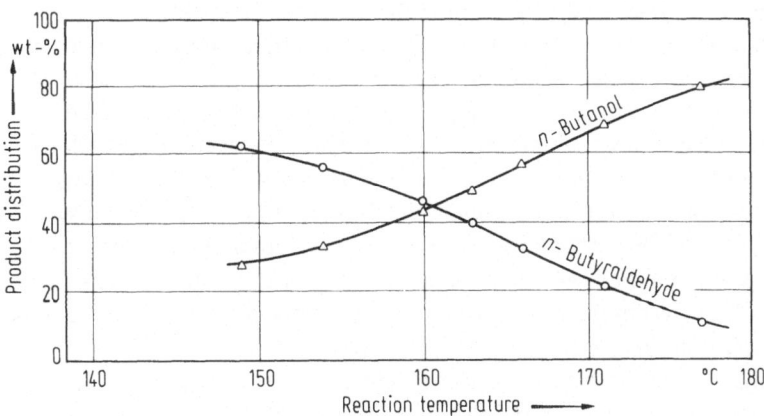

Fig. 1.37. Product distribution (aldehyde/alcohol) vs. reaction temperature (propylene hydroformylation with HCo(CO)$_3$(PBu$_3$) [264])

cy of a ligand-modified Co catalyst increases markedly with temperature, at around 160 °C the ratio n-butanal:n-butanol is already 1:1. At this temperature, the butanol share of the propylene hydroformylation products of the classical – unmodified – Oxo synthesis is max. 10% (cf Fig. 1.16 and Sect. 1.3.1). However, the hydrogenation tendency of the modified catalysts depends strongly on the type of ligands present [629] (e.g., [620], cf Fig. 1.19 [263]).

The dependence of activity and product composition on total pressure and CO partial pressure, which was already dealt with in Sect. 1.3.2, takes an unexpected course. Figure 1.38 shows the effect of total pressure on the product composition in the hydroformylation of 1-octene with $HRh(CO)[P(OAr)_3]_3$ [636]. The effect of the CO partial pressure is shown in Fig. 1.24.

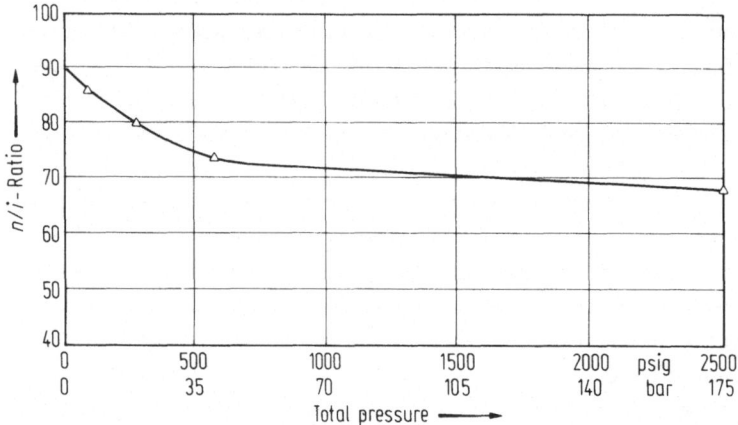

Fig. 1.38. Relationship between n∶iso ratio and CO/H_2 total pressure (1-octene hydroformylation with $HRh(CO)[P(OAr)_3]_3$ [636]

The decreasing share of the unbranched products in the reaction mixture with increasing total pressure and rising CO partial pressure is a consequence of the displacement of the equilibrium [Eq. (41)] to the left. This allows the selectivity distribution of the unmodified Oxo catalyst to become more and more prominent while repressing the selectivity improving effect of the ligand modified catalysts [15, 209].

Figure 1.24 indicates while the application of ligand-modified catalysts enables the hydroformylation to be conducted at low total or low CO partial pressures, it also introduces the disadvantage that at low CO partial pressures the parallel reaction – hydrogenation of the olefin feedstock – also increases [273]. This side reaction – which has a detrimental effect on the yield–is encouraged by the necessity of, at least slightly, improving the olefin conversion (at low temperatures) by employing H_2-rich syn gas (cf Fig. 1.28 [169, 265]). Thus, in order to achieve selective formation of unbranched compounds with ligand-modified catalysts it is necessary to operate the process by combining the effects of low reaction temperatures, low total pressures, low CO partial pressures and H_2:CO ratios in syn gas of > 1∶1 [26, 171, 187, 209, 265, 329, 330, 637].

However, it must be considered that – according to Piacenti et al. [272] – at higher CO partial pressures the counteraction $HCo(CO)_3 + CO \leftrightharpoons HCo(CO)_4$ ensues and consequently an increase in n-product share (cf Fig. 1.39). This effect is also responsible for the many diverging results.

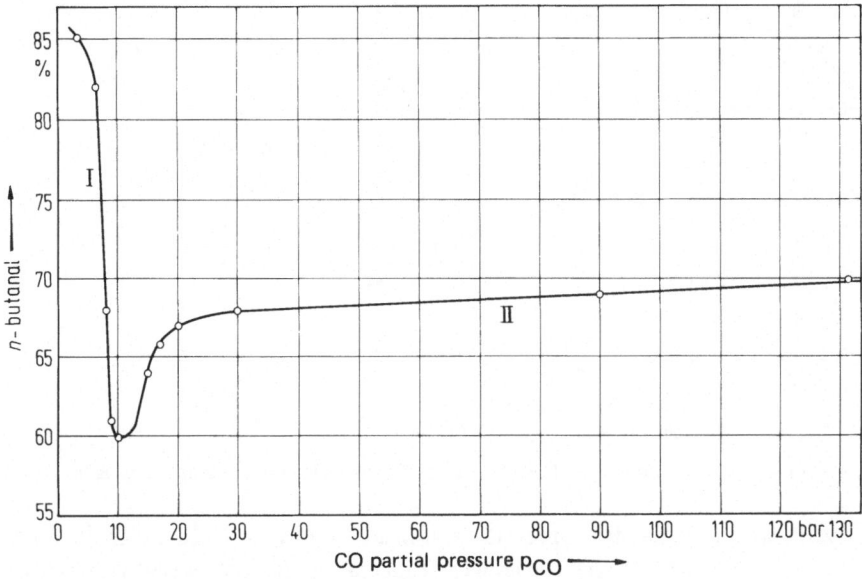

Fig. 1.39. Hydroformylation of propylene with $Co_2(CO)_6[P(Bu)_3]_2$ (150 °C, P_{H_2} = 40 bar). Influence of p_{CO} on the isomeric composition of the resulting aldehydes [272]

Thus, the course of plot I must correspond to the Eqs. (41) to (43) and plot II to Eq. (9). In addition, the ratio of the new ligand L to the central atom Me must be considered. It is to be expected from Eqs. (41) to (43) that the typical properties of the ligand-modified catalysts – in particular the high n : iso ratio – become more predominant with an increasing L : Me ratio. This is actually the case as shown in Fig. 1.40.

As dealt with above, ligand-modified Rh catalysts are preferentially employed using the ligand as solvent. It is important to note in these instances, that the large excess of the complexing agent – as indicated in Fig. 1.40 – hardly increases the n : iso ratio but causes a clear drop in activity (Fig. 1.41) [26, 169, 273, 620, 638–640]. In the region of slight excesses the activity usually increases [1811].

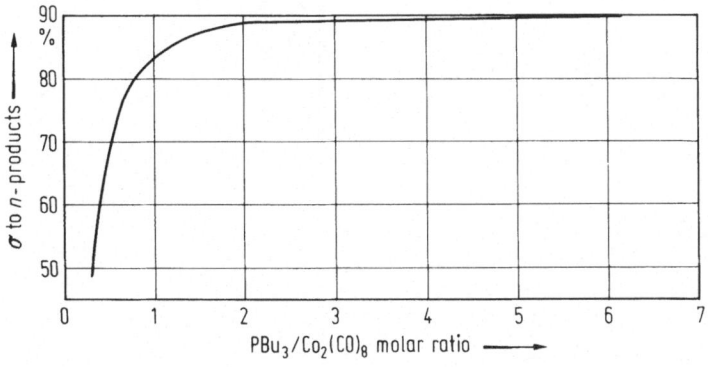

Fig. 1.40. Effect of PBu_3 concentration on selectivity σ to branched/unbranched isomers [620]

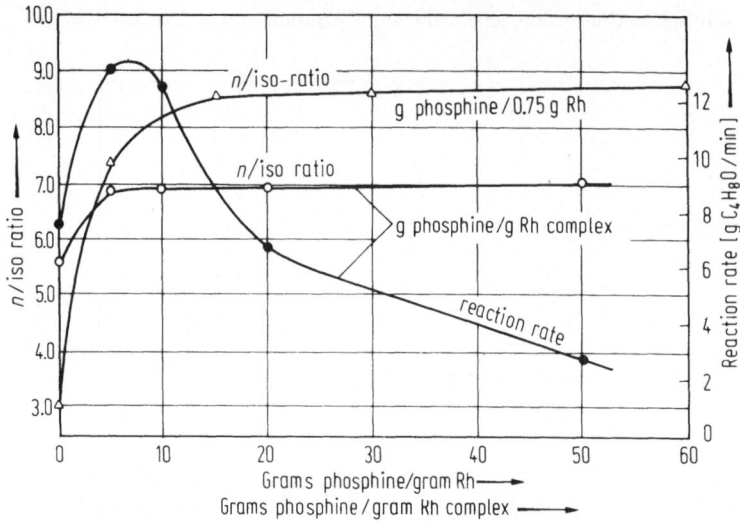

Fig. 1.41. n : iso ratio (n : iso quotient) and reaction velocity versus excess of complexing agent

As a result, with the industrial application of ligand-modified Rh catalysts the raise of the H_2 partial pressure (involving the danger of increasing the hydrogenative activity of the system) and/or the increase of the rhodium feed (danger of Rh losses due to leakages etc.) must compensate the decreases in activity via modification and the large excess of ligand (solvent).

A high ligand excess is particularly beneficial when bis-hydroformylating dienes (cf Sect. 1.4). According to Fell and Boll [574, 609] when the P : Rh ratio increases so does the degree of bis-hydroformylation and the reaction velocity [639] (cf Fig. 1.42).

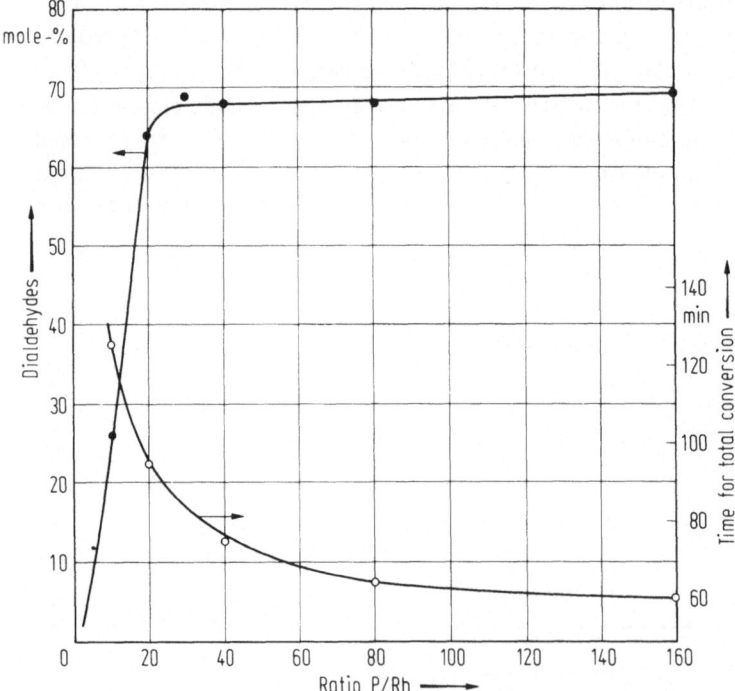

Fig. 1.42. Hydroformylation of 1,3-pentadiene. Formation of dialdehydes and period required for total diene conversion versus P : Rh ratio [609]

52

There is a similar situation when allyl alcohol [727] or alkynes [575] are hydroformylated.

This also applies to the effects of reaction temperature and total pressure or partial pressures (cf Sect. 1.3.1 and 1.3.2), when ligand-modified catalysts are used with olefinic substrates possessing distinct electronic properties then the olefins behave in an abnormal manner [242, 608]. For example, when methyl methacrylate or ethyl acrylate are hydroformylated with phosphine-modified Rh catalysts then the selectivity to the (more branched) α-isomers increases with growing P:Rh ratio [242].

Individual results, in particular regarding the effect of certain ligands and/or certain reaction parameter combinations can be found in the appropriate literature (cf next section).

Cobalt catalysts

Ligand-modified *cobalt* catalysts have already found commercial application [6, 26, 30, 40] in the Shell process. Oxo processes involving ligand-modified catalysts have the following characteristic features:

● Increased stability of Oxo catalysts, facilitating hydroformylation at low CO partial pressures and distillative separation of catalyst and reaction products.
● Reduced activity of ligand-modified catalysts necessitating markedly greater reactor volumes for the same productivity − even at higher reaction temperatures.
● The marked hydrogenation activity resulting in only alcohols and *no* aldehydes being obtained as a reaction product, furthermore around 15% of the olefin feed is hydrogenated to paraffins.
● High n:iso product ratio.
● Susceptibility towards poisons due to phosphine oxide formation from the modified catalysts.

Besides the tertiary phosphines employed in the Shell process, the ligands listed in Table 1.13 have been recommended as substituent L of the complex Oxo catalyst $HMe(CO)_m L_n$.

Table 1.13. Proposed substituents for ligand modification of complex cobalt catalysts

Ligand	Proposed by	Ref.
Tributylphosphines	Shell Dev.	[584, 643]
	Esso Res. Dev.	[644]
	Texaco Inc.	[329]
	van Boven et al.	[645]
	Paulik	[15]
	Tucci	[266, 602, 637]
	BASF AG	[646]
	Piacenti et al.	[598, 600, 647]
	Imyanitov et al.	[613, 648]
	Rupilius et al.	[603, 610]
Prim. phosphines $(R-PH_2)$	BASF AG	[834]
Sec. phosphines (R_2-PH)	BASF AG	[651]
	Shell	[652]

Table 1.13 (continued)

Ligand	Proposed by	Ref.
Trialkylphosphines (R$_2$P)	BASF AG	[649, 814, 1413]
	Esso Res. Engng.	[650]
Triphenylphosphine	Dow Chem. Corp.	[653]
	IFP	[654]
	other authors	[620, 655, 656]
Tris-(alkyl/aryl)-phosphites	Shell	[657]
	Esso	[658]
	Jefferson Chem. Co.	[659]
	IFP	[660]
	Diamond Alkali	[661]
	others	[830, 1409]
PH$_3$ and radical sources	BASF AG.	[835]
Amines/polyamines	Jefferson Chem. Co.	[662]
	Shell	[663]
	others	[517, 664, 665, 705, 1406]
Arylsulfonates	Chevron Res.	[585, 586]
Azoxy compounds	Chevron Res.	[666, 1955]
Polynitro aromatics	UOP Co.	[667]
Nitriles	Montecatini	[668]
	Azote et Prod. Chim.	[669]
Tris-(alkyl/aryl)-phosphines	Shell	[670]
	BASF (carboxy substituted)	[671, 943]
	others	[642, 672, 706]
N-alkylsalicylaldimines	–	[673, 674]
Pyridines and other heterocycles	Shell	[688]
	BASF AG	[689–692]
	Chevron	[585]
	Jap. Agency	[693]
	Lion Fat	[694, 695]
	others	[696–698, 1224, 1225, 1957]
Phosphorous as component of other heterocycles	Shell	[675, 676]
	BASF AG	[677]
	Gulf Res.	[678]
	Ruhrchemie AG	[263, 265, 687]
Pyridines as component of cyclic phosphines	Shell	[679–682, 701]
	Gulf	[611, 683, 684]
	others	[707, 836]
Tropolones etc.	Chevron Res.	[585, 586]
Phosphine oxides	Jefferson Chem. Co.	[699]
Halomethanes	ICI Ltd.	[615]
Diamino diphenylalkanes	BASF AG	[700, 829]
Chlorophosphines	Gulf Res. Co	[642]
Arsines	Shell	[702]
	Jefferson	[703]
	Esso	[704]

In addition, a series of publications or patents must also be mentioned – Shell [708–711], ICI [712, 713], BASF [714, 715], Esso (Standard Oil) [716, 717, 941], Gulf [718], Montecatini [686, 719, 720, 1402], and others [721–725, 842, 1401, 1403, 1948]. The effect of nitrogen-containing ligands is discussed in Sect. 1.3.4.

Rhodium Catalysts

Ligand-modified *Rh* catalysts can be characterized as follows [26, 328]:
• Ligand modification of rhodium hydridocarbonyl increases the stability of the catalyst to such an extent that the reaction products from the hydroformylation can be destillatively separated from the catalyst (bottom product). Morever, the low CO partial or the total pressure is sufficient to keep the modified Oxo catalyst in an active and stable state.
⊙ However, the modification also lowers the activity necessitating Rh concentrations of 0.01 to 0.1% instead of minute Rh quantities [26, 47] as it is not possible to overcome this drop in activity by increasing the reaction temperature. The higher temperatures would lead not only to a decrease in yield of unbranched compounds but also to higher hydrogenation activity (cf Figs. 1.20 and 1.24).
• A side effect of the reduced activity is a fall in the olefin conversion, necessitating unreacted olefin to be separated and recycled. The use of H_2-rich syn gas must also be regarded as an attempt to compensate for the drop in activity.
• The H_2-rich syn gas fortifies the hydrogenation tendency which must be opposed by limiting the reaction temperature.
• Large relative amounts of unbranched aldehydes can be obtained at low CO partial or total pressures with high ligand excess and at low temperatures on employing H_2-rich syn gas. $HRh(CO)[P(C_6H_5)_3$, which is obtained from $HRh(CO)_4$ via ligand substitution with a large excess of PPh_3, facilitates a virtually isomerization-free hydroformylation (Fell [749]).

These partially opposing effects demand fairly sophisticated reaction control techniques i.e., including both, reaction parameters and process engineering.

The commercial realization is being executed by a group of companies consisting of Union Carbide Corp., Davy Powergas and Johnson, Matthey & Co. Ltd. (cf Sect. 1.6.2).

Some compounds, which have been proposed for ligand modification of Rh catalysts, are listed in Table 1.14.

Table 1.14. Proposed substituents for the ligand modification of complex rhodium catalysts $HRh(A)_m(CO)_n(L)_p$ (where $m+n+p = 4$)

(A) (L)	Proposed or studied by	Ref.
(A) = (L) = triphenylphosphine	Union Oil Co.	[187, 209, 273, 605] [728, 729, 831]
m+p = 1 to 3 n = 3 to 1	Union Carbide Corp. BASF AG	[328, 636, 730, 731] [732, 733]

Table 1.14 (continued)

(A)	(L)	Proposed or studied by	Ref.
A = Cl, L = PPh$_3$, with m + n = 3		Gulf Res. Dev. Co.	[734]
		Mitsubishi	[735, 736]
		Celanese Corp.	[737]
		Monsanto Co.	[265, 640, 738]
		Shell	[739]
		BP Ltd.	[740]
		Johnson, Matthey & Co. ⎫	
		Davy Powergas ⎬	[501, 588, 741−747]
		G. Wilkinson et al. ⎭	
		Ethyl Corp.	[748]
		others	[173, 574, 609, 641] [749−765]
(A) = (L) = tributylphosphine		BP Ltd.	[766]
		Shell	[767]
		Fell	[768]
CO/Cl	diphenyl-(dimethyl-amino)phenylphosphine	Union Oil Co.	[769, 1058]
CO/Cl	methylphenylbenzyl-phosphine	Japan Agency	[770]
CO/Cl	methylcyclohexyl-anisylphosphines etc.	Monsanto Co.	[771]
		Shell	[772]
		others	[773, 774]
Cl	cyclic, bidentated phosphines	–	[775, 776]
CO	heterocyclic phosphines	Ruhrchemie AG	[777]
CO	aromatic or aliphatic phosphites	Mitsubishi	[778, 779]
		Progil S.A.	[503]
		Chisso Corp.	[596]
		Union Carbide Corp.	[780]
		others	[781−784]
CO	tertiary amines	Mitsubishi	[785]
		Ethyl Corp.	[614]
		Chevron Res.	[786]
		Mobil Oil Corp.	[616, 787]
		others	[612, 700, 788, 829,] [1407]
CO	carboxylates	BP Ltd.	[789]
		BASF AG	[587]
		Monsanto	[815]
CO	β-diketonates	BP Ltd.	[555, 790]
CO	N-alkylsalicylaldimines (and others)	BP Ltd.	[791]
		others	[792]
CO	pyridines	Ogata et al.	[793, 1132]
CO	O-, S- or N-bidentated ligands	BP Ltd.	[794, 833]
		others	[1628, 1958]

Table 1.14 (continued)

(A)	(L)	Proposed or studied by	Ref.
CO	B-containing ligands	Ethyl Corp. Monsanto Co. BP Ltd.	[795] [796, 838] [797]
Acac	trialkylphosphines	BP Ltd.	[766]
PPh₃	β-diketonates	Johnson, Matthey & Co.	[743]
PPh₃	carboxylates	BP Ltd.	[798]
Cl/CO	dienes	BP Ltd. Phillips Petr. Co. others	[601, 799] [800] [200, 801, 837, 1135, 1956]
[A] and [L] as component of cyclic phosphines		Consiglio et al. Booth et al.	[840] [812]
NO	phosphines	Mitsui	[1133]

The application of particular ligands [332, 497, 592, 802–809, 811, 813, 816, 817, 1227, 1408] as well as that of polynuclear ligand-modified Rh clusters [86, 765, 810–812, 818] is discussed elsewhere. These include the interesting, recently developed, water-soluble complexes of the type $HRh(CO)(dpm)_3$ which allow the hydroformylation to be conducted in aqueous solution and which may facilitate the catalyst recycle [805, 1130].

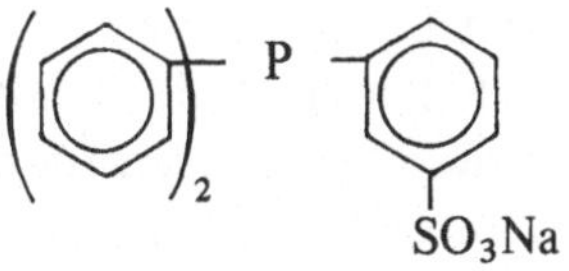

Various

The effect of ligand modification on Oxo activity and selectivity of catalysts with a central atom other than Co or Rh has not been systematically studied. The results so far can be interpreted in various ways, however preliminary summaries have been published [7, 75, 171], and more recent work is summarized in Table 1.15 (page 58).

Particularly high n:i ratios have been reported using $[(C_6H_5)_4As]^+ [H_3Ru(CO)_{12}]^-$ [1750].

1.3.3.1.3 Via Variation of Application Phase

Heterogenized (Immobilized) Catalysts

Introduction The homogeneously catalyzed hydroformylation with a metal carbonyl of the type $HMe(CO)_m L_n$ requires a step after the actual Oxo stage to separate, recover

Table 1.15. Ligand modified Oxo catalysts

Central atom	Ligand modified with	Ref.
Cu	phthalocyanines	[487]
	phosphines	[580]
	phosphites	[581]
Cr	phosphines	[943]
Mn	phosphines	[476]
Fe	N,O-heterocyclic derivatives, cyclo-	[846]
	pentadiene, subst. cyclopentadienes	[818, 847, 848]
Ni	phosphites	[849]
Ag	phthalocyanines	[487]
Ru	phosphines	[467, 470, 850–853,
	arsines	1750]
Pd	phosphines	[849, 854]
Au	phthalocyanines	[487]
Re	phosphines	[476]
Os	phosphines	[507]
Ir	phosphines	[75, 86, 507, 845, 855]
Pt	phosphines	[331, 849, 856, 857, 1404]

and reprocess the catalyst (cf Section 1.3.3.4). In order to avoid this separate step "the Diaden" or "two tower" process [337, 338, 1807] was operated on the Ruhrchemie site in 1942 in the first industrial Oxo unit. This process involved two reactors for the alternate hydroformylation with $HCo(CO)_4$ removal and Co precipitation on supports [6].

In continuation of this trend to a precipitated — "heterogenized", "anchored" or "immobilized" — supported cobalt catalyst and in connection with the studies of the Merrifield group [339] on the 'solid (or hybrid) phase synthesis', various workers attempted to replace one or more CO groups of the homogeneous complex catalyst $HMe(CO)_mL_n$ not by monomolecular ligands (cf Sect. 1.3.3.1.2) but by polymeric complexing agents. In addition, it was also attempted to precipitate transition metal complexes on to supports of large surface area (cf reviews [340–347]). The object is always to attach the metal (Me) to a carrier matrix. The underlying concept is to enable the actual homogeneously catalyzed hydroformylation to ensue at fixed active sites, thereby combining the advantages of homogeneous and heterogeneous catalysis (cf Table 1.16). However, the partially unrealistic ideas of the various researchers regarding a commercially feasible Oxo process and the parameter combinations must be kept in mind.

Various problems must be considered — the metal atoms should not be too strongly bonded to the matrix thereby causing serious mass-transfer problems and a drop in reaction velocity. On the other hand, the linkage must not be so weak that, despite

Table 1.16. Advantages and disadvantages of homogeneous and heterogeneous catalysis [24, 344, 348–351, 1919]

Process	Advantages	Disadvantages
Homogeneous catalysis	relatively resistant towards catalyst poisons	process step for catalyst separation necessary
	high activity due to very ready availability of metal	sensitive to extreme temperatures
	facile electronic and steric variability, no mass-transfer problems	
Heterogeneous catalysis	catalyst readily separable from substrate	active sites make only part of the metal content available and thus more susceptible to inactivity via poisoning
	insensitive to high temperatures	limited variability
		mass-transfer problems in pores

heterogenization, catalyst losses would occur which would necessitate reprocessing steps [24, 352].

Heterogenized Oxo catalysts can be employed in the gas phase or in the liquid phase as fixed- bed or finely divided catalysts suspended in liquids (e.g., Monsanto [353, 409], Mobil [354], Eastman Kodak [361] or BP [355, 356], cf pp. 67/68.

In heterogeneous catalysts, transitions to metal clusters (cf Section pp 68/69) are just as possible as hydroformylations in the presence of heterogeneous promoters [357, 393]. In addition, developments could involve reactions with catalysts containing metals recovered via adsorption techniques from "de-metallization" steps and, in analogy to Sect. 1.3.3.4, represent special cases of a metal recycle of a homogeneous Oxo process [358–363]. However, this must be differentiated from hydroformylations of polymers [365, 366, 827] (cf Sect. 1.4).

Production of Heterogenized Oxo Catalysts. Heterogenized Oxo catalysts with anchored metal atoms can be manufactured via:

A Polymerization or copolymerization of suitable monomers.

B Functionalization of preformed supports, e.g., polymers or silicas.

C Precipitation of metals on supports.

D Various methods.

Method A is rarely used for heterogenized Oxo catalysis. The most well-known example is the ICI procedure for the polymerization of suitably substituted bis(dialkyl-styrenephosphine) metal halides [367]:

$$n \quad \underset{Hal}{\overset{Hal}{\diagdown}}Me \to \underset{R}{\overset{R}{\diagdown}}P-\langle\rangle CH{=}CH_2 \to \underset{Hal}{\overset{Hal}{\diagdown}}Me \to \underset{R}{\overset{R}{\diagdown}}P-\langle\rangle-\overset{\displaystyle\vdash}{\underset{CH_2}{\overset{|}{CH}}}\Bigg]_n \qquad (44)$$

There is a number of standard methods for the functionalization of polymers, silica gels, zeolites or other supports according to method B. Two frequently applied techniques are shown in Eqs. (45) and (46) (**P** = polymer support e.g., polystyrene $(C_6H_5-CH-CH_2)_n$, **S** = silica [340, 343, 344, 354, 368–377, 1950]):

$$\mathbf{P}\text{-}\langle\rangle \xrightarrow{Br_2} \mathbf{P}\text{-}\langle\langle\text{/}\rangle\text{-}Br \xrightarrow{LiP(Ph)_2} \mathbf{P}\text{-}\langle\rangle\text{-}\overset{Ph}{\underset{Ph}{\overset{|}{P}}} \xrightarrow{HCo(CO)_4} \mathbf{P}\text{-}\langle\rangle\text{-}\overset{Ph\ H}{\underset{Ph}{\overset{|\ |}{P}}}\text{-}Co(CO)_4$$
$$(45)$$

$$S\text{-}OH + H_3CO\text{-}\overset{OR}{\underset{OR}{\overset{|}{Si}}}\text{-}\overset{Ph}{\underset{Ph}{\diagup\diagdown}}P \to S\text{-}O\text{-}\overset{OR}{\underset{OR}{\overset{|}{Si}}}\text{-}\overset{Ph}{\underset{Ph}{\overset{|}{P}}} \xrightarrow{HRh(CO)_4} S\text{-}O\text{-}\overset{OR}{\underset{OR}{\overset{|}{Si}}}\text{-}\overset{Ph\ H}{\underset{Ph}{\overset{|\ |}{P}}}\text{-}Rh(CO)_4 \quad (46)$$

The groups $-P(Ph_2)$ or $Si(OR)_2-P(Ph)_2-$ are termed "linking agents".

With polymeric supports, particular attention must be paid to swelling in the presence of the organic Oxo products. This effect can be eliminated via cross-linking with copolymers, on employing styrene for example with divinylbenzene (DVB [354, 356]) or 2-vinylpyridine (VP [358]) or via pre-swelling. While an increasing degree of cross-linking lowers the solubility/swellability, it also detrimentally affects the capacity to undergo functionalization. On the other hand, a decreasing degree of cross-linkage improves the ability of the copolymer to absorb the catalytically active metal and the swelling capability in the same manner [358].

The following systems are examples of a more adsorbed/chemisorbed coating of suitable supports with active metal species according to method C: Co/kieselguhr [378], $HCo(CO)_mL_n/SiO_2\text{-}Al_2O_3$ [379], Rh/zeolite [380], $HCo(CO)_4$/P-DVB [358] or $HRh(CO)_3PR_3$/active carbon or kieselguhr [353, 381, 382, 1072].

The coating of pores of porous supports with liquid complex catalysts ("supported liquid phase catalysts"–SLPC [382, 383]), the application of catalysts with Co skeleton [384], of transition metal phthalocyanines [453, 505] or the use of pre-formed Co–Al silicates [385] are examples of method D.

Table 1.17 presents a summary of the different variants of the methods A to D as well as the various combinations of support, linking agent and Oxo catalyst.

Hydroformylation with Heterogenized Oxo Catalysts. Anchoring complex Oxo catalysts to a matrix support generally lowers its activity and tendency to form linear compounds compared to the soluble complex. In addition, the isomerization tendency also increases [388, 372, 397].

Table 1.17. Hydroformylation with immobilized catalysts, DVB = Divinylbenzene

Method	Support	Linking agent	HMe(CO)$_m$L$_n$ (soluble analoge)	Ref.
A	poly-(diphenylstyrene-phosphine)	–	HCo(CO)$_4$	[367]
B	polysterene-DVB	-P(Ph$_2$)-, N(Me)-CH$_2$- etc.	HRh(CO)$_4$	[386, 390]
	polystyrene-DVB	-CH$_2$-P(Ph$_2$)-	Rh(CO)[P(Ph)$_3$]$_3$	[369]
	polystyrene-DVB	-CH$_2$-P(Ph$_2$)-	RhCl$_3$	[372]
	polystyrene-DVB	-CH$_2$-P(Ph$_2$)-, -N(Me)$_2$- etc.	Rh(CO)$_2$Cl	[354, 1950]
	polystyrene-DVB	-P(n-Bu)$_2$-	Rh(CO)$_2$Cl	[352, 387]
	polystyrene-DVB	-P-(Ph)$_2$-(CH$_2$)$_n$-P(Ph$_2$)$_2$-	HRh(CO)L	[388]
	polystyrene-DVB	-P(Et)$_2$-	HCo(CO)$_4$	[360, 389, 391]
	polystyrene-DVB	-P(Ph)$_2$-	HRh(CO)[P(Ph)$_3$]$_3$	[392, 396]
	polystyrene-DVB	-P(Ph)$_2$-	PtCl$_2$ + SnCl$_2$	[1229]
	PVC	-P(Ph)$_2$-	Rh(acac)$_2$(CO)$_2$	[356]
	PVC	-P(Ph)$_2$-	CoCl$_2$[P(Ph)$_3$]	[395]
	polybutadiene	-P(Ph)$_2$-	HCo(CO)$_4$	[376]
	polystyrene	phosphines	Mo(CO)$_5$[P(Ph)$_3$]	[394]
	silica	polystyrene-poly-DVB-P(Ph)$_2$-	RhCl$_2$(CO)$_2$	[368]
	silica	-O-Si(OMe)$_2$-P(Ph$_2$)-	RhCl$_2$(CO)$_2$	[397, 454, 1943, 1944]
	silica	-O-Si(R)$_2$-R-NR-	RhCl$_2$(CO)$_2$	[355]
	silica	$\begin{array}{c} -O \quad OSiR_3 \\ \diagdown \diagup \\ Si \quad etc. \\ \diagup \diagdown \\ -O \quad R\text{-}NR\text{-} \end{array}$	RhCl$_2$(CO)$_2$	[375, 398]
	silica	$\begin{array}{c} -O \quad Cl \\ \diagdown \diagup \\ Si \\ \diagup \diagdown \\ -O \quad R\text{-}P(Ph)_2\text{-} \end{array}$	RhCl$_2$(CO)$_2$	[400, 402]
	silica	-P(Ph)$_2$-	RhCl$_2$(CO)$_2$	[401]
	silica	silyl-phosphine	RhCl(CO)[P(Ph)$_3$]	[403, 1131]
	silica	-O-Si(OEt)$_2$-(CH$_2$)$_2$P(Ph)$_2$-	HRh(CO)$_4$	[356]
	zeolite	-O-	HCo(CO)$_4$	[399, 408]
	ion exchange resin	-SO$_2$-O-NH-NH- etc.	RhCl$_2$(CO)$_2$	[387, 404, 406]
	arom. polyamides	-P(Ph)$_2$-	RhCl[P(Ph)$_3$]$_3$	[405]
	polyphenylene	-P(Ph)$_2$-	HCo(CO)$_4$	[407]
C	polystyrene-DVB	–	HCo(CO)$_4$	[358, 363, 410, 455]
	silica	–	HRh(CO)$_4$	[411]
	silica	–	HRh(CO)$_m$L$_n$	[379, 412, 413]
	silica	–	CoO, Co$_2$O$_3$	[361, 414–416]
	silica	–	Fe(CO)$_5$	[417]
	zeolite	–	HRh(CO)$_4$	[364, 380, 418]
	zeolite	–	Co	[419]
	kieselguhr	–	Co	[378, 420, 421, 828]
	kieselguhr	–	RhCl[P(Ph)$_3$]$_3$	[353]
	Al$_2$O$_3$	–	Rh(CO)Cl[P(Ph)$_3$]$_2$	[409]
	Al$_2$O$_3$	–	Rh	[458]
	asbestos	–	HRh(CO)$_4$	[422]
	pumice	–	Co/HCo(CO)$_4$	[362, 423, 424, 1175]

Table 1.17 (continued)

Method	Support	Linking agent	HMe(CO)$_m$L$_n$ (soluble analoge)	Ref.
	charcoal	–	Rh	[425, 1946]
	charcoal	–	Co	[426]
	charcoal	–	HRh(CO)$_m$L$_n$	[381, 427, 1072]
D	cobalt framework	–	Co	[384]
	Co-Al-silicate	–	Co	[385]
	Co-aluminate	–	Co	[428]
	Al$_2$O$_3$ and other supports }	SLPC-technique	Rh(CO)Cl[P(Ph)$_3$]$_2$	[382, 429]
			Co$_2$(CO)$_6$[P(n–Bu)$_3$]$_2$	[353, 383]
			etc.	

An example of the above is the scheme presented by Lang et al. [352, 354] for the 1-hexene conversion:

$$
\begin{array}{c}
\quad + \text{CO/H}_2 \\
\hline
\end{array}
\longrightarrow \text{H}_3\text{C–CH}_2\text{–CH}_2\text{–CH}_2\text{–CH}_2\text{–CH}_2\text{–CHO} \qquad (47)
$$

1-heptanal

$$\text{H}_3\text{C–CH}_2\text{–CH}_2\text{–CH}_2\text{–CH=CH}_2 \rightleftharpoons \text{H}_3\text{C–CH}_2\text{–CH}_2\text{–CH=CH–CH}_3 \rightleftharpoons \text{H}_3\text{C–CH}_2\text{–CH=CH–CH}_2\text{–CH}_3$$

1-hexene · · · · · · · · · 2-hexene · · · · · · · · · 3-hexene (48)

+ CO   H$_2$ + CO   H$_2$ + CO   H$_2$ + CO   H$_2$

$$
\begin{array}{cc}
\text{H}_3\text{C–CH}_2\text{–CH}_2\text{–CH}_2\text{–CH–CH}_3 & \text{H}_3\text{C–CH}_2\text{–CH}_2\text{–CH–CH}_2\text{–CH}_3 \\
\qquad\qquad\qquad\qquad | & \qquad\qquad\qquad\qquad | \\
\qquad\qquad\qquad\quad \text{CHO} & \qquad\qquad\qquad\quad \text{CHO}
\end{array} \qquad (49)
$$

2-methylhexanal · · · · · · · · · 3-ethylpentanal

Figure 1.43 illustrates the superior activity of the homogeneous catalyst HRh(CO)$_x$ compared to its polymer bound analogue [354]

However, various authors also report an increase in activity on employing polymer bound homogeneous catalysts [349, 1629]. This is the case when reaction temperatures are used which approach the limit of the thermal stability of the carrier and when high metal contents can be proven to be on the support. In this instance, noticeably high space-time throughputs can be achieved (Table 1.18).

Table 1.18. Olefins converted to Oxo aldehydes in continuous flow reaction [354]

Olefin	Temp. °C	Pressure (psig/bar)	LHSV	Conversion %
propylene	118	500/ 35	1.1	96
1-octene	128	1500/105	2.5	90
1-dodecene	149	2000/140	2.5	92

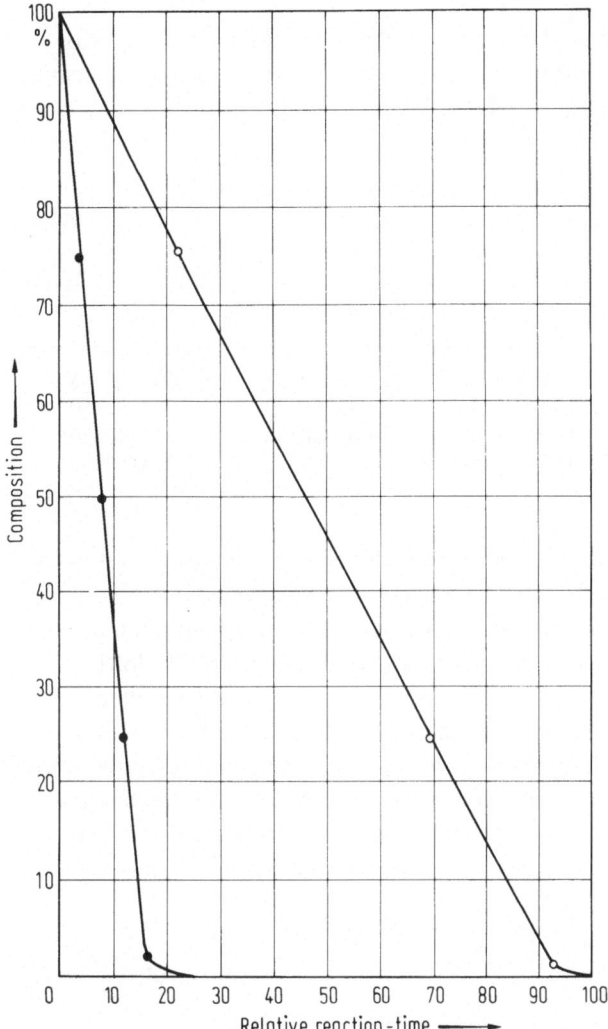

Fig. 1.43. Comparative hydroformylation of 1-hexene using ● HRh(CO)$_x$ or

$$P-P(Bu)_2-Rh\overset{\displaystyle H}{\underset{\displaystyle (CO)_x}{\diagup}} \quad (\circ) \quad [354].$$

When judging these results it must be considered that tests were conducted in small reactors (internal diameter = 1 cm) which exhibit particularly favorable heat removal features.

The higher shares of unbranched aldehydes obtained with metal complexes on phosphine-containing supports are often presented as proof that the central atom of the complex catalyst is actually bound to the support during hydroformylation and no desorption/adsorption steps ensue which would enable the complex catalyst to function homogeneously in the usual way [352].

A series of other relationships involving heterogenized Oxo catalysts are comparable with their soluble counterparts. The reaction velocity of the hydroformylation decreases in the sequence ethylene>propylene>butene [368], the conversion rate falling with shorter residence times [358]. In addition, the n:iso ratio decreases with both increasing total pressure and reaction temperature (cf Table 1.19).

Table 1.19: Hydroformylation of 1-pentene by polymer-anchored catalyst [388]

Pressure (psi/bar)	Yield/n:iso ratio at temperatures of			
	60 °C	80 °C	100 °C	120 °C
100/ 7	61%/67:33	96%/67:33	95%/69:32	100%/55:45
200/14	72%/67:33	100%/67:33	98%/68:32	100%/50:50
400/28	100%/69:31	100%/68:32	100%/68:32	99%/50:50
800/56	100%/50:50	100%/50:50	100%/68:32	99%/50:50

Figure 1.44 shows the dependence of hydroformylation and isomerization products on the conversion rate using the same heterogenized Oxo catalyst.

Not only the steric structure of anchored catalysts has been dealt with in the Ref. [430] but also the kinetics and mechanism of heterogenized catalysts [354, 368].

Industrial application of heterogenized Oxo catalysts of this type will be determined not only by their activity but also by metal discharge and their resistance to poisons. Probably as a consequence of the limited research work conducted in this sector, divergent statements have been made about the last two points. For instance, the observation that the polymer-bound HCo(CO)$_4$ decomposes Oxo poisons (possibly via hydrogenation of dienes, alkynes, etc. [358]) contrasts with the fact that Rh complexes

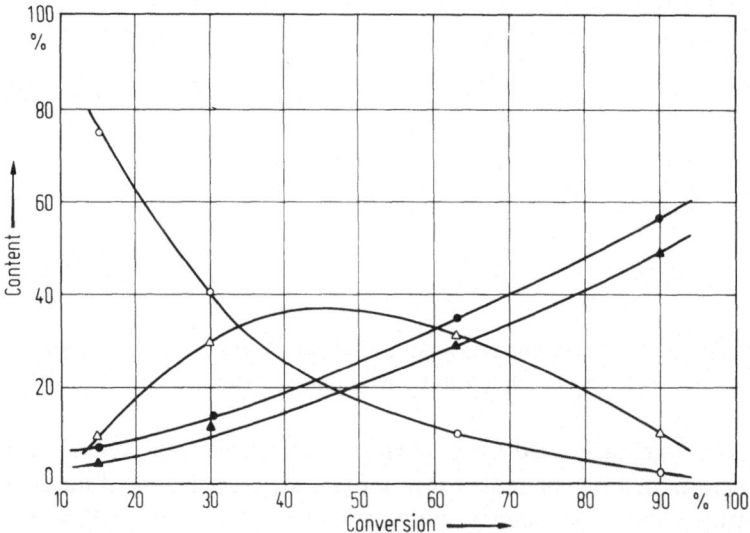

Fig. 1.44. Product distribution as a function of conversion rate on hydroformylating 1-pentene with polymer-anchored catalysts [388]

- o   1–pentene
- •   1–hexanal
- △   2–pentene
- ▲   2–methyl pentanal

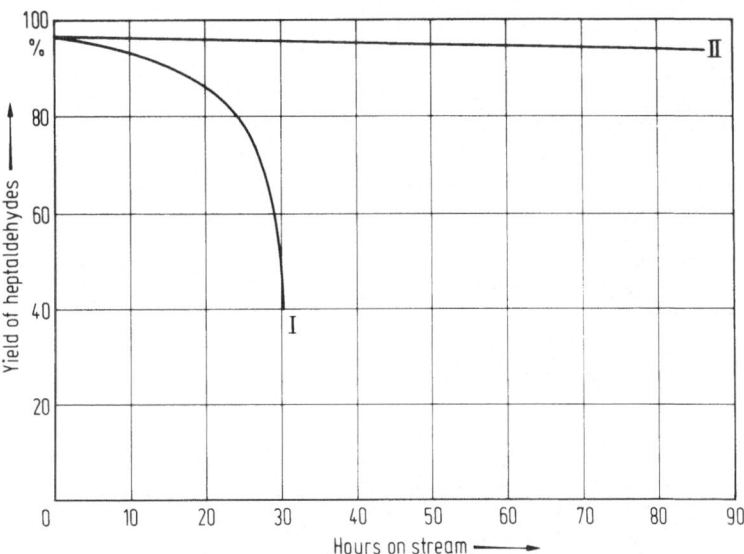

Fig. 1.45. Continuous flow liquid phase hydroformylation of 1-hexene by Rh species
$P \rightarrow CH_2-P(Ph)_2-Rh(acac)(CO)$ under normal (I) and oxygen-free (II) conditions [356]

bonded to crosslinked polystyrene are extremely sensitive to the presence of oxygen
in the reactants (cf Fig. 1.45 [356])

The drop in activity can be explained both via phosphine oxide formation (possibly
from free phosphine stemming from an intermediate stage) as well as by the cleavage of
P–C bonds (cf Sect. 1.3.3.2 [431]). The 'Turnover Number' (TON) is often used as a
measure of the relative stability of heterogenized Oxo catalysts, this value also incorpo-
rates the resistance to poisoning. With the catalyst $P \rightarrow P(Ph)_2-Rh(CO)_2Cl$, for example,
the values are 400,000 moles of olefin/g atom Rh [354]. On using propylene (mw = 42),
this corresponds to a feed/consumption of ca. 5 mg Rh/kg propylene.

Contradictory results concerning the metal discharge have also been published. Depending on the support [352] and the purity of the reactants, values <1 to 45 ppm have
been reported in liquid discharges [356, 358, 363]. Figure 1.46 gives an indication of
the transport properties of rhodium on phosphine resins in accordance with Lang et al.

Plotting the rhodium distribution in recovered, sectioned reactor beds shows that
the total rhodium content of the catalyst charge decreases and that Rh deficiency starts
at the top of the reactor arising from transport phenomena as a consequence of desorp-
tion/absorption effects. The influence of pressure, temperature and CO and H_2 partial
pressures on the Rh discharge has been discussed in the Refs. [352, 387].

As expected, the n : iso ratio is also affected by the number and type of ligands at
the heterogenized central atom. With a homogeneous $HRh(acac)(CO)_2$ catalyst, n : iso
ratios of 1.2 : 1 are obtained, with homogeneous $HRh(acac)(CO)(PPh_3)$ 2.9 : 1 and with
$P-CH_2-P(Ph)_2-Rh(acac)(CO)_2$ around 2 to 2.5 : 1 as anticipated [356, 395].

Table 1.20 lists a selection of heterogenized Oxo catalysts used with various olefin
feedstocks.

Hydroformylations with heterogenized catalysts containing a modified central
atom have been carried out by various workers [417, 433, 453, 1950]. For simultaneous
IR investigations cf [1947].

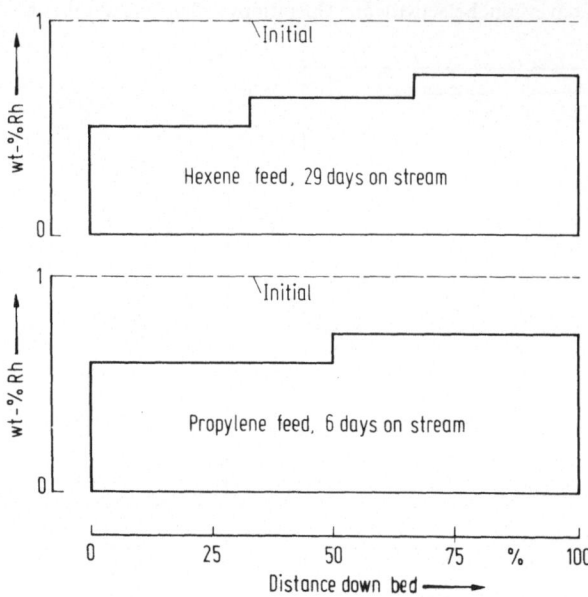

Fig. 1.46. Rhodium distribution in recovered phosphine resin catalyst beds [352]

Table 1.20. Tests conducted with various heterogenized Oxo catalysts (**P** = polymer, **Si** = silica, **T** = other support)

Olefin	Catalyst	Ref.
propylene	P–CH$_2$–P(Ph)$_2$–Rh(CO)$_2$Cl etc. P–N(CH$_3$)$_2$→Rh(CO)$_2$Cl Si–O–Si(OMe)$_2$–P(Ph)$_2$–Rh(CO)$_2$Cl$_2$	[354] [397]
propylene	Si–P–P(Ph)$_2$–Rh(CO)$_2$Cl Si–HCo(CO)$_4$/–HCo(CO)$_3$(PBu$_3$) T–Co T–Rh(CO)Cl(PPh$_3$) T–SLPC-technique	[368] [379, 423] [361, 414, 419] [409] [383]
1-butene	Si–P–P(Ph)$_2$–Rh(CO)$_2$Cl	[368]
1-pentene	Co aluminate etc.	[388, 417, 428, 432]
1-hexene	P–CH$_2$–P(Ph)$_2$–Rh(acac)(CO)$_2$ P–CH$_2$–P(Ph)$_2$–Rh(CO)$_2$Cl etc. Si–P(Ph)$_2$–Rh(CO)$_2$Cl$_2$ T–Rh(CO)$_m$L$_n$	[356] [352, 387] [401] [385, 413, 422, 427]
2-hexene	P–HCo(CO)$_4$	[363, 410]
1-heptene	P–CH$_2$–P(Ph)$_2$–RhCl$_3$ Co framework (skeleton)	[372] [384]
1-octene	polyphenylene–P(Ph)$_2$–Co(CO)$_3$	[407]
dodecene tetradecene hexadecene	T–Rh, T–Co(CO)$_4$ etc.	[408, 425]
dienes	Si–O–Si(Cl)(O)–R–P(Ph)$_2$–Rh(CO)Cl T–Co	[400] [428]

Gas Phase Hydroformylation

A further improvement in the operation of the Oxo synthesis would be — besides heterogenizing the homogeneous catalysts — to overcome the disadvantages inherent in the liquid medium (transport phenomena at the interface between phases, critical data of participating liquids, limited mixing and solubility) by conducting the reaction in the gas phase. Furthermore, cooled gas recycles would make feasible external removal of the exothermic heat of reaction [434, 435]. Consequently, even at the outset of its development, gas phase studies on the Oxo reaction were a topic of interest (cf kinetics [368, 436]).

A series of methods have been described for the operation of the hydroformylation reaction in the gas phase.
These include:

• The introduction of the gaseous reactants into a previously prepared solution of the catalyst, in some cases dissolved in excess ligand [437, 449] (cf Sects. 1.3.3.1.2 and 1.6.2). This method represents the transition to the molten salt technique. Some typical relationships are shown in Fig. 1.47.

• The SLPC (supported liquid phase catalyst) technique in which the Oxo catalysts, dissolved in high boiling solvents (in some cases the ligands themselves are used as

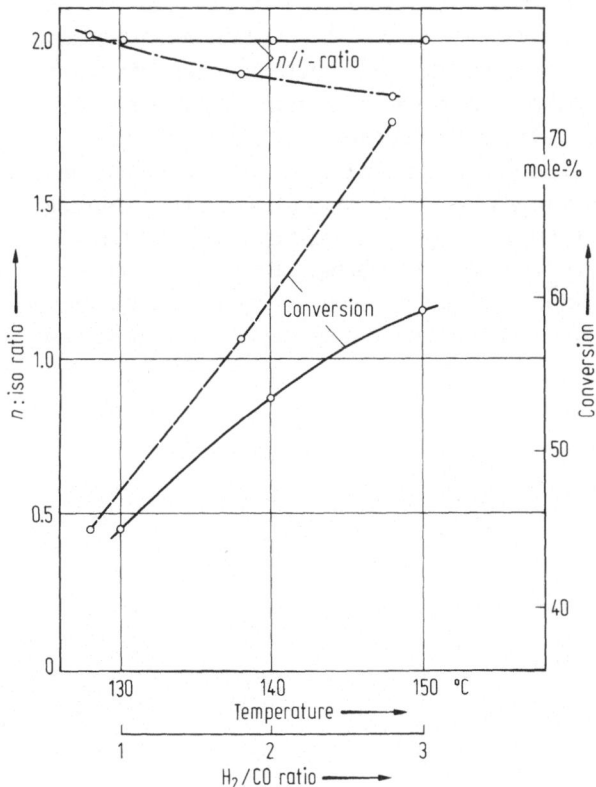

Fig. 1.47. Conversion and n : iso ratio as function of H_2 : CO ratio (—) and temperature (—.—) [449]

solvent), are deposited in the pores of suitable supports where they are made available to the gaseous reactants (alkene and CO + H$_2$) [353, 381–383, 438, 444, 757]. In 1960, Guyer describes a technical modification of the ethylene hydroformylation involving the use of the catalyst as a film on the heat transfer surface [439].

• Transition metals bound by adsorption which enable a fixed-bed reaction of the gaseous reactants [361, 381, 409, 414, 419, 440–442, 449, 825].

• The reaction of the gaseous reactants (alkene and syn gas) and the gaseous Oxo catalyst e.g., HCo(CO)$_4$ [252, 443].

As far as is known, no gas phase hydroformylation has been commercially operated.

Various Methods

The use of molten salts is one of the more recent methods being investigated for the hydroformylation of olefins. These techniques begin with the SLPC methods which represent the transition to the reaction of alkenes on ligand-modified Oxo catalysts dissolved in excess molten ligand (cf Sect. 1.3.3.1.2). In addition, phase transfer catalyzed carbonylations [446, 463] or reactions at semi-permeable membranes [447, 448] should be mentioned. The potential of these developments is currently difficult to estimate.

At the moment, Oxo active clusters — polynuclear transition metal carbonyls with Me-Me bonds — are the subject of intensive research [103, 364, 456, 457, 459–461, 819–821, 1915, 1961] as "monometallic complexes have not yet been found active for useful carbon-carbon bond cleavage, reduction of carbon monoxide etc. Since only one metal atom is present at the active site it is not surprising that metal surfaces with a variety of metal sites have reactivities heretofore unavailable to mononuclear complexes" [456]. This intensification of the effect of the transition metal complex is also exploited in hydroformylations. The results [464] are less surprising than the fact that the clusters Co$_3$(CO)$_9$ [μ_3–(C$_6$H$_5$)], I, and Co$_4$(CO)$_8$(μ_2–CO)$_2$–(μ_4–PC$_6$H$_5$)$_2$, II (cf Fig. 1.48) – can be recovered unchanged after the reaction. Thus, apparently the actual transition from homogeneous to heterogeneous reaction does not involve decomposition to the monomolecular species. Pittman et al. have

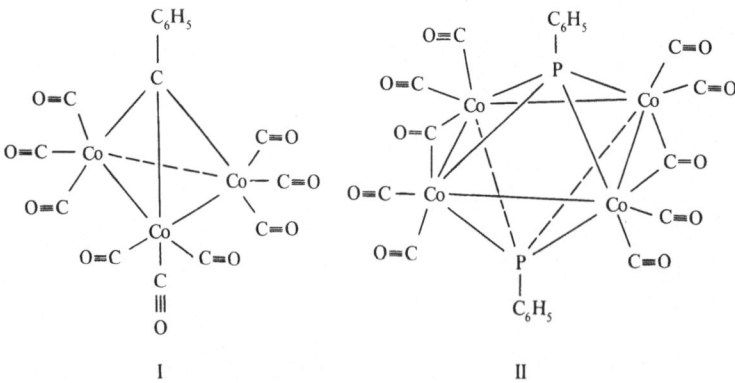

Fig. 1.48. Cobalt clusters for hydroformylation.
I: Co$_3$(CO)$_9$[μ_3–C$_6$H$_5$)], II: Co$_4$(CO)$_8$(μ_2–CO)$_2$(μ_4–PC$_6$H$_5$)$_2$

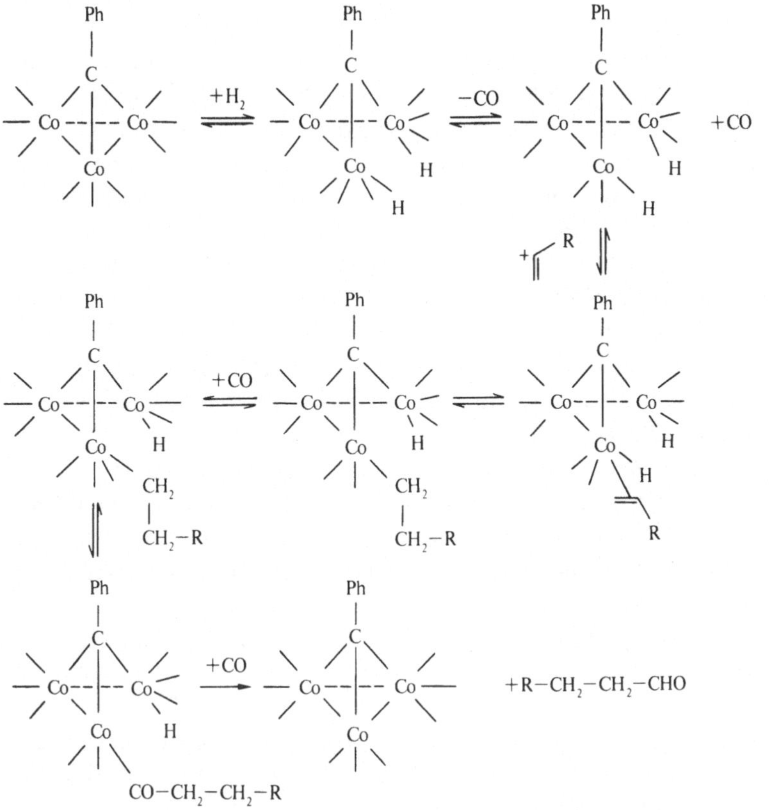

Fig. 1.49. Proposed mechanism of hydroformylation with Co-clusters [456]

postulated the mechanism shown in Fig. 1.49 which is based on the Heck-Breslow mechanism – however with semi-rigid clusters.

For hydroformylation using Rh containing clusters [161] (anchored clusters [1945]) it is difficult in this case, too, to judge the potential industrial significance of these developments.

1.3.3.1.4 Activators, Promoters and Other Catalyst Additives

Almost from the outset of the Oxo process, additives were described for improving the activity or altering the selectivity of the reaction (cf Ref. [6] for period before 1970). These additives, which usually do not form defined complexes with the actual Oxo catalyst – as in ligand modification cf Sect. 1.3.3.1.2. – can be divided into the following groups according to their effects or apparent activity:
- Additives affecting formation and stability of carbonyls
- Additives affecting activity and selectivity of catalysts
- Catalyst additives to stimulate or repress parallel or consecutive reactions.

Active carbon or zeolites are recommended for the carbonyl formation zone *to accelerate the rate of carbonyl formation* from cobalt metal, cobalt salt (solutions) or other catalyst precursors. These additives, possessing a large surface area, may be charged with Pd or Pt (or other metals) [222, 223, 245, 357, 518, 858, 859,

864, 1959] or be employed untreated [860–862]. As well as improving carbonyl formation, these types of additives should either facilitate a lowering of the hydroformylation
temperate or – combined with an extraction of formed metal hydridocarbonyls
[526] – serve to precarbonylate catalyst precursors so that carbonylation and hydroformylation ensue in separate reactors. The latter is of particular importance when
aqueous catalyst solutions are employed and the rate of formation of the carbonyls
and hydrocarbonyls represent one of the rate determining steps (cf Sect. 1.6, BASF
process). Ion-exchangers, particularly basic ones – if necessary, charged with cobalt
salts/carbonyls – are also recommended for the same purpose [862, 863, 865].

Adding alkylbenzenesulfonates is proposed for the *stabilization of previously
formed carbonyls* to prevent destructive dissociation which leads to loss of activity
and deposition of metallic cobalt on the reactor walls [866].

In the case of additives *affecting catalyst activity and selectivity* it is difficult
to correlate addition and effect. Overlapping is possible with actual ligand modifications, with mixed catalysts containing more than one catalytically active metal and lastly with catalyst poisons. While many additives have an activating effect when present
in small amounts larger quantities may terminate the Oxo reaction.

Nucleophilic additives represent the transition to ligand-modifying additives,
which – having ionization constants of $\leqslant 10^{-8}$ – accelerate the hydroformylation [6,
585, 867–869, 888, 904, 1092], e.g., pyridine, picoline, lutidine, aliphatic or aromatic amides, etc. N-methylpyrrolidone has found industrial application and concepts
have already been developed for a recycle operation [871]. Other examples of additives which improve activity or selectivity are summarized in Table 1.21.

Table 1.21. Oxo catalyst additives responsible for improving activity and/or selectivity

Additives	Effect improvement in			Ref.
	stability	activity	selectivity	
alkali(ne earth) salts of C$_1$-C$_4$ fatty acids		x		[870]
alkali metal hydroxides	x[a]	x		[831, 872, 873–875, 945]
alkaline earth oxides		x		[876]
oxygen		x		[877–879]
radical precursors	x[b]	x		[835, 880]
Ziegler type additives		x		[881, 882, 1910]
γ-irridation		x		[304, 883]
acids		x		[884, 885]
alcohols, thioalcohols, ethers, ketones etc.		x	x	[886, 887, 891, 939, 946]
alkylarylsulfonates		x	x	[805, 889]
zeolites			x	[408]
carbonate esters			x	[890]
water (not as solvent)		x	x	[275, 294, 485, 486, 520, 612, 891–893, 927]
carbon dioxide			x	[917]

[a] stability of ligand modifier   [b] to ensure application of PH$_3$ as ligand modifier

Now, a few remarks about water: H_2O is frequently used as solvent for the Co^{2+} salts which serve as precursors (cf Sect. 1.6). The advantage is facile feeding and simpler recycle operation (cf Sect. 1.3.3.4). However, it must be considered that the actual hydroformylation takes place homogeneously with the organic substrate and, in aqueous solution, $HCo(CO)_4$ is formed from Co^{2+} salts. The former must then enter the organic phase via mass transfer processes. In contrast to Co metal or organic soluble Co salts, the hydroformylation rate may be determined by the rate of carbonyl formation and the transport velocity. Moreover, the water in the reaction mixture hydrolyzes the esters, being present in the reaction mixtures. With unsaturated esters, the resulting acid leads to slow deactivation (cf Sect. 1.3.3.2) [893].

In continuously operated hydroformylation processes with olefins, hydrolysis of the resulting alkyl formates produces formic acid which causes corrosion. The extent of the above can perhaps best be represented by Table 1.22 using as example the propylene hydroformylation with or without addition of water but with Co soaps as catalyst precursor.

Table 1.22. Propylene hydroformylation with and without addition of H_2O (T = 150 °C, total pressure = 250 bar, CO : H_2 = 1 : 1, LHSV = 1.0 V/Vh) [894]

	Water added	(% rel. to propylene)	
	none	10	15
Oxo product data			
iodine number	1.6	1.5	1.6
neutralization number	6.3	14.6	22.9
saponification number	81	76	66
GLC of Oxo product (%)			
hydrocarbons	0.2	0.2	0.2
aldehydes	82.5	80.5	77.2
esters	4.9	1.9	0.9
alcohols	4.2	8.0	8.5
corrosion rates in Oxo reactor (in mm/a)			
steel No. 4541[a]	< 0.1	0.3	0.3
steel No. 4550[a]	< 0.1		0.2
steel No. 4571[a]	< 0.1	0.2	0.2

[a] Nos. refer to German Standards (DIN)

The increase in concentration of the formic acid — shown by a fall in the saponification number and an increase in the neutralization number — causes a noticeable growth in corrosion. Consequently, when the Oxo process is operated using aqueous solutions then — and *only* in this case — the reactors and apparatus must be made of high alloyed steel.

The "Aldox" process is an example where *catalyst additives are used to stimulate or repress consecutive or parallel reactions.* On introducing suitable additives to the Oxo catalysts, the aldolization of the (primary resulting) aldehydes ensues in situ and, after hydrogenating the total Oxo product, dimeric alcohols are obtained e.g., 2-EH from the propylene hydroformylation.

Besides adding basic compounds to the Co catalysts such as Mg ethylate/pyridine, compounds of Zn, Sn, Ti, Zr, Hf, Th, Pb, Cd, Hg, Al, or Cu have often been described [6, 25, 201, 895–898, 1493]. KOH is mainly recommended as an additive for phosphine-modified Co catalysts (Shell [710, 899, 900]).

While Esso's Aldox process and the Shell variant are industrially operated they are inferior to processes involving isolation of the intermediate aldehyde to be aldolized. The reasons are, on the one hand, the limited flexibility of the process—only alcohols are obtained, the aldehyde intermediate not being isolated. In addition, mixed aldolization ensues due to the presence of isomeric aldehydes in the crude Oxo product. For example, in the propene hydroformylation, besides 2-ethylhexanol, 2,2-dimethylhexanol, 2-ethyl-4-methylpentanol and 2,2,4-trimethylpentanol are formed. These isomeric alcohols are difficult to separate:

$$
\begin{array}{c}
\text{C–C–C–CHO} \\
+ \\
\underset{\overset{|}{\text{C}}}{\text{C–C–CHO}}
\end{array}
\longrightarrow
\left\{
\begin{array}{ll}
\underset{\overset{|}{\text{C–C–C–C–C–CH}_2\text{OH}}}{\overset{\text{C–C}}{}} & \text{n/n condensate} \\[1em]
+ & \\[1em]
\underset{\overset{|}{\text{C}}}{\overset{\text{C}}{\text{C–C–C–C–C–CH}_2\text{OH}}} & \text{n/iso condensate} \\[1em]
+ & \\[1em]
\underset{}{\overset{\text{C} \quad \text{C–C}}{\text{C–C–C–C–CH}_2\text{OH}}} & \text{iso/n condensate} \\[1em]
+ & \\[1em]
\underset{\overset{|}{\text{C}}}{\overset{\text{C} \quad \text{C}}{\text{C–C–C–C–CH}_2\text{OH}}} & \text{iso/iso condensate}
\end{array}
\right.
\quad (50)
$$

2-EH obtained via the above route is of inferior quality to 2-EH from the aldolization of n-butyraldehyde – this aspect is particularly important when a high quality product is required for certain applications. Moreover, in the industrial catalyst recycle, attention should not only be paid to the (Oxo active) cobalt but also to the whereabouts of the additive. Both metals ought to be recycled together (cf Chap. 1.5.2.1.2).

In the hydroformylation of unsaturated functional olefins, hydroquinone is added to prevent polymerization [243]—this is example of an *additive suppressing parallel reactions.*

1.3.3.2 Catalyst Poisons

Catalyst poisons of the classical Oxo synthesis with unmodified cobalt carbonyls have already been comprehensively treated by Falbe [6]. In addition, special studies have been conducted regarding process control in the presence of poisons [901–903, 905] and the effect of sulfur or sulfur compounds (as catalyst poisons) [201, 906–910] – particular attention being devoted to sulfur-containing carbonyls [911–916] –, carbon dioxide [201, 872] (usage in water gas shift conversion cf [1923, 1924]), dienes [299, 304, 817, 902, 918, 919, 1916], iron [920–924] and acids [893, 925]. The work which differentiates between the effect of sulfur on the actual Oxo syntheses and the subsequent metal removal step is of special industrial significance due to the development of Oxo syntheses variants which tolerate a syn gas feedstock containing sulfur [909].

As previously mentioned in Section 1.3.3.1.4, a series of catalyst poisons, when present at low concentration, effect an improvement in activity or selectivity e.g. oxygen [877, 878, 926], water [893, 927, 1123] or CO_2 [917]. Improved recycling techniques for cobalt compounds, which always involve a certain regneration of the catalyst (cf Sect. 1.3.3.4), enable the catalyst poison problem to become insignificant in the classical Oxo synthesis. Modern Oxo processes with unmodified Co catalysts are relatively resistant towards catalyst poisons.

However, this does not apply to Oxo processes with modified rhodium catalysts. In these processes, the Rh concentration is kept as low as possible for economic reasons. Thus, the minor Rh quantities are more susceptible to contamination by poisons which are being continuously introduced. In certain process variants, no rhodium recycling takes place (stationary rhodium cf Sect. 1.6.2.2.2) and therefore continuous regeneration is dispensed with. Besides the 'classical' catalyst poisons – dienes/acetylenes [627, 630, 928, 929, 1261, 1330], sulfur [627, 635, 929, 930, 1074], halogens [627, 878, 931, 1327] or carboxylic acids [106, 200, 526] – the following poisons are of particular importance:

> oxygen
> iron (carbonyl)
> thermal/chemical strain

Figure 1.50 (cf also Fig. 1.45) gives an impression of the rate of the inhibiting effect of oxygen, sulfur or acetylene on modified Rh catalysts of the type $Rh(Cl)(CO)$-$[P(C_6H_5)_3]_2$.

While the low Rh concentration is responsible for the poisoning with iron (carbonyls) [628], the poisoning resulting from oxygen with phosphine modified Rh catalysts is of a more complex nature. The effect of O_2 poisoning has been known for a long time and is assumed to stem from the reaction of phosphines with traces of O_2 (in the reactants) forming phosphine oxides [Eq. (51)] [356, 627, 629, 635, 872, 926, 932–934].

$$R\text{--}\overset{\displaystyle R}{\underset{\displaystyle R}{P}} + \tfrac{1}{2}\,O_2 \;\rightarrow\; R\text{--}\overset{\displaystyle R}{\underset{\displaystyle R}{P}}\!\rightarrow O \tag{51}$$

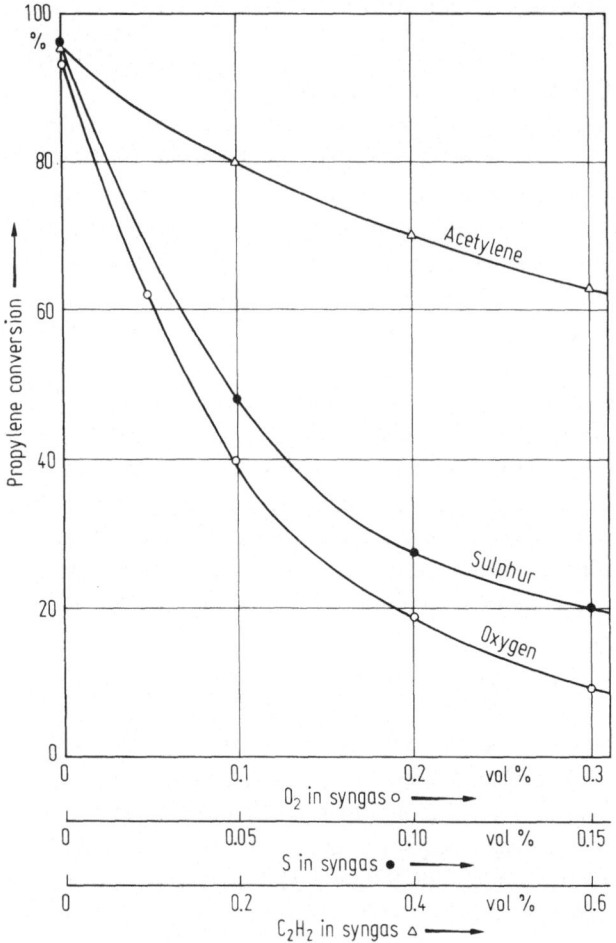

Fig. 1.50. Effect of various catalyst poisons (O_2, sulfur and acetylene) on the hydroformylation of propylene with $Rh(Cl)(CO)[P(C_6H_5)_3]_2$ [929]

This conversion reduces the activity of the modified Rh carbonyls [629, 635, 827, 932]. Gregorio et al. first showed that phosphine oxide could be formed from phosphines, aldehydes and aldols e.g., triphenylphosphine, n-butyraldehyde, and butyraldol or 2-ethylhexenal [635]. This represents the connection to the poisoning of modified Rh catalysts from thermal or chemical effects. Further information can be found in either Gregorio et al.'s work [635] or in patents filed by UCC [628, 633, 634, 878, 935].

According to Gregorio et al. [635], one reason for catalyst deactivation – as a result of consecutive reactions with triphenylphosphine – is the formation of a heterocycle **A** on heating the ligand-modified Rh complex $HRh(CO)[P(C_6H_5)_3]_3$ in the presence of H_2. The heterocycle **B**, formed by carbonylation, is converted into the intermediate **C** via cleavage of a P–C bond. Finally, the dimeric $[Rh(CO)_2P(C_6H_5)_2]_2$ and benzaldehyde result (Eq. 52 + 53):

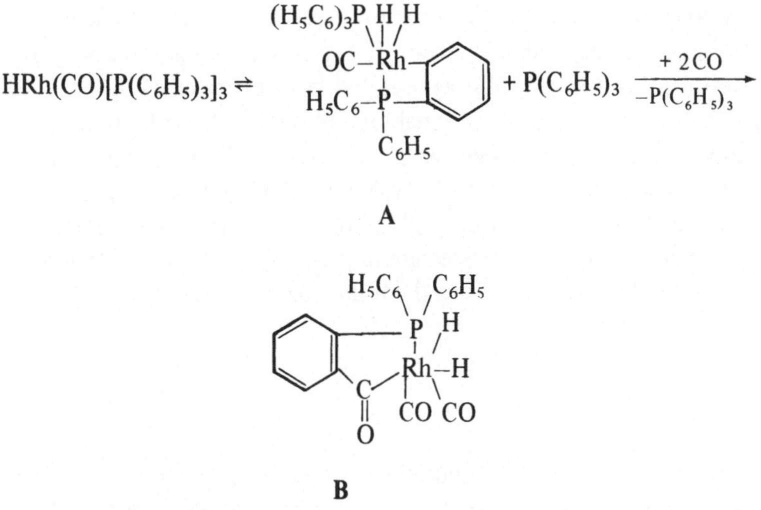

$$HRh(CO)[P(C_6H_5)_3]_3 \rightleftharpoons \quad \mathbf{A} \quad + P(C_6H_5)_3 \xrightarrow[-P(C_6H_5)_3]{+2CO}$$

A

(52)

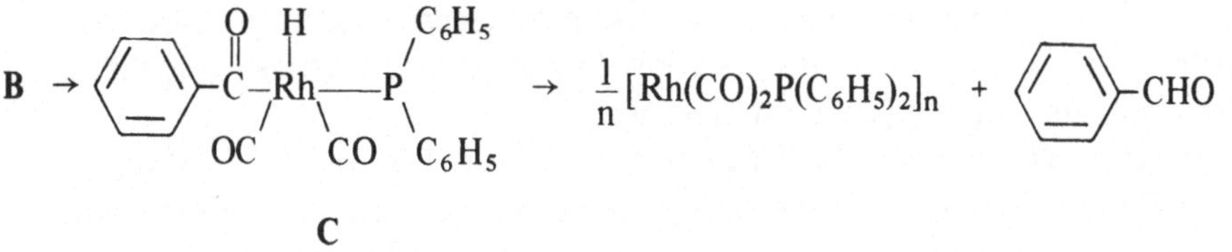

B

$$\mathbf{B} \rightarrow \quad \mathbf{C} \quad \rightarrow \frac{1}{n}[Rh(CO)_2P(C_6H_5)_2]_n + \quad \text{—CHO} \quad (53)$$

C

Benzene can also be formed instead of benzaldehyde [936, 937]. Moreover, the partial substitution of the aryl group C_6H_5 in the triphenylphosphine ligand by the less effective — and more volatile — alkyl group (propyl) can be interpreted as a consequence of the previous hydroformylation step. It can therefore be made responsible for the decrease in activity (via step **C**) [cf Eqs. (54) and (55)]

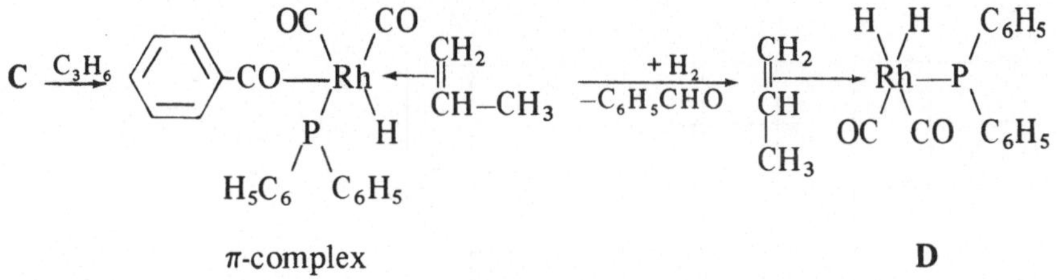

$$\mathbf{C} \xrightarrow{C_3H_6} \quad \pi\text{-complex} \quad \xrightarrow[-C_6H_5CHO]{+H_2} \quad \mathbf{D} \quad (54)$$

π-complex

D

$$\mathbf{D} \xrightarrow{P(C_6H_5)_3} \quad \quad (55)$$

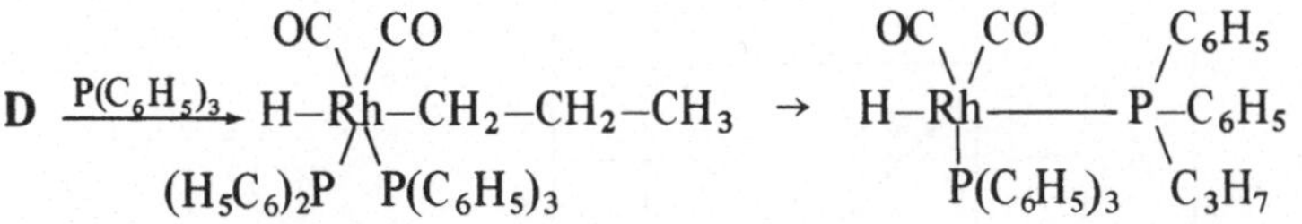

Union Carbide [628, 633, 634, 878] and others [458] postulated an internal poisoning of modified Rh catalyst systems via the combined effect of temperature, CO partial pressure and the P:Rh ratio. This could cause a poisoning of 3% per day of the Rh catalyst [634]stemming from hydroxyl-containing condensation products (aldols) from the aldolization of the isomeric aldehydes. The formation or accumulation of these deleterious by-products in the reaction medium is repressed via a controlled expulsion [935] or a controlled side stream [634] or via the combination of low reaction temperatures, low CO partial pressures and large excesses of ligand [633, 634]. 'Stability factors' represent the optimal arrangement of these three parameters. Their increasing numerical value signifies a reduction in the rate of activity loss (in % per day) to the economically acceptable value of 0.75% per day (Fig. 1.51).

Depending on the ageing rate ϵ (0.03 or 0.0075 corresponding to 3 or 0.75% per day) the catalyst life time t – defined as decrease of activity from Ao = 100% (at time t = o) to A = 10% – can be calculated using the following equation:

$$t = \frac{\log \dfrac{A}{Ao}}{\log (1 - \epsilon)} \tag{56}$$

According to Fig. 1.52, t is 76 days at 3% and 306 days at 0.75%.

Besides rigorous purification of the reactants [627–629, 631, 872, 902, 903, 928, 933, 934, 938] to prevent poisoning, it has also been recommended to restrict the combinations of reaction conditions. As a low loss in activity is strived for, the above measures could also include productivity losses.

In processes with unmodified Rh carbonyls (cf Sect. 1.6.2), the oxidation of the Rh acylcarbonyl complex to the carboxylate complex, which then reacts with CO + H$_2$ to carboxylic acid, can also be regarded as an O$_2$based poisoning reaction [106].

$$4\ \underset{\substack{\| \\ O}}{R\text{–}C}\text{–}Rh(CO)_n \xrightarrow{O_2} 2\ [(\underset{\substack{\| \\ O}}{R\text{–}C}\text{–}O)\,Rh\,(CO)_n]_2 \xrightarrow{CO/H_2} 4\ RCOOH + Rh_4(CO)_{12} \tag{57}$$

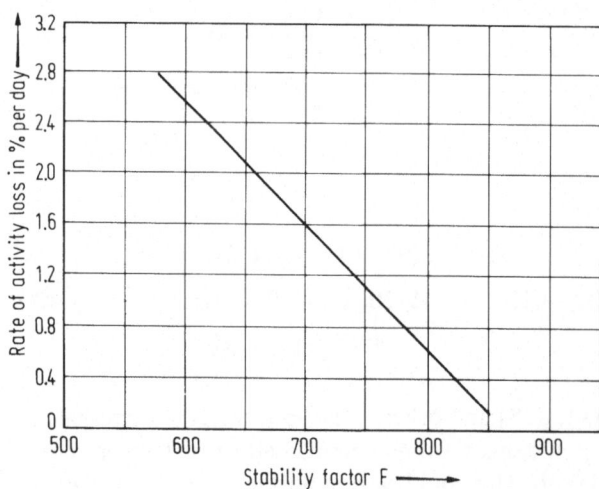

Fig. 1.51. Dependence of activity loss of modified Rh carbonyls on the stability factor F [633, 634]

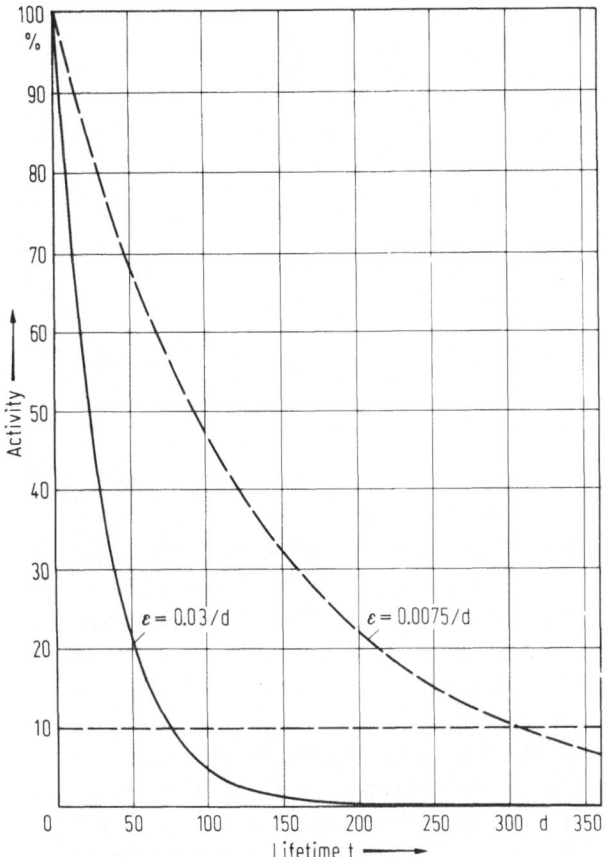

Fig. 1.52. Lifetimes of modified Rh catalysts as function of ageing rate ϵ [940]

1.3.3.3 Effect of Catalyst Concentration

The heated discussion about the effect of the cobalt concentration on the n : iso ratio of the aldehydes belong to the past [6, 215] (Hughes and Kirshenbaum [213]: high concentrations suppress formation of unbranched products; Goddard and Mansfield [948]: contradictory report; Falbe, Tummes and Weber [228] – experimental proof that, under industrial conditions, the n : iso ratio is virtually unaffected by catalyst feed). The deviating observations probably stem from a neglected or wrong interpretation as to the effect of the residence time, the role of the solvent being possibly also ignored (cf Sect. 1.3.4 and 1.3.6).

The catalyst concentration is more important for controlling the conversion and degree of by-product formation. If the hydroformylation is to be operated in two stages (cf Sect. 1.3.7) then the cobalt feed is one of the methods available (besides reaction temperature and residence time) for adjusting the conversion of the first reaction step. The sensitivity of the propylene conversion to the cobalt feed can be readily appreciated from Table 1.23.

Table 1.23. Propylene conversion as function of cobalt feed (in wt. % Co rel. to propylene, throughput 0.32 V/Vh, T = 130 °C [949])

Cobalt feed	Propylene conversion (%)
0.39	96.4
0.21	92.8
0.17	90.9
0.12	83.7
0.08	82.6
0.073	80.3
0.057	72.4
0.049	71.5
0.035	57.3

Basically similar relationships are to be found on employing special cobalt [229, 420] and other catalysts e.g., ruthenium [466].

The type and extent of the consecutive reactions of the resulting aldehydes as well as the parallel reaction (olefin hydrogenation) are influenced by the catalyst concentration. This property can be exploited if alcohols are the desired final products — as larger amounts of catalyst can increase the conversion of aldehydes to alcohols without being accompanied by any side reactions which would cause a decrease in selectivity (cf Fig. 1.53 [466, 555, 1223, 1690]).

Figure 1.53 shows however that with increasing catalyst concentration, butyl esters are formed from the aldehydes, i.e., the butanol selectivity falls. While the extent of the parallel reaction — hydrogenation of olefin feedstock — changes only slightly over the whole test range, if the double bond is in a sterically unfavourable

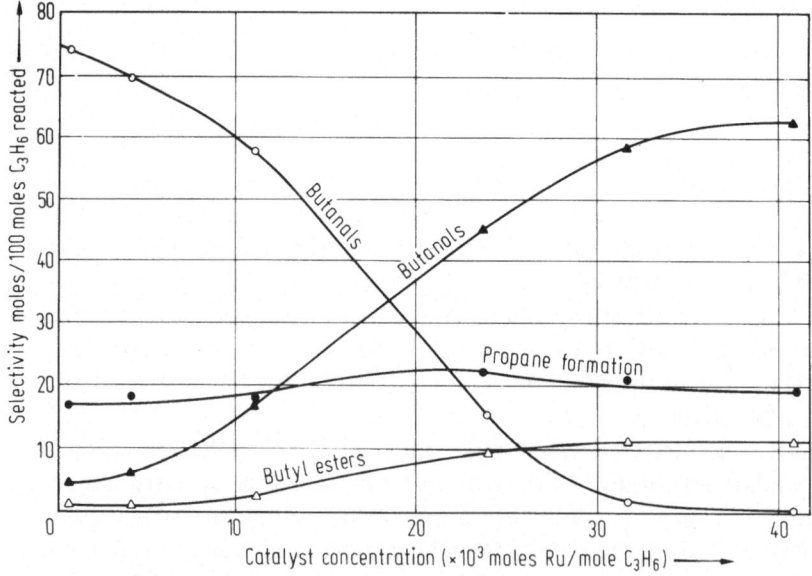

Fig. 1.53. Effect of amount of catalyst on selectivity of butanal and butanol formation (Ru, 300 bar, 120 °C [466])

position e.g., diisobutylene (2,4,4-trimethylpentene) then there will be greater paraffin formation on increasing the catalyst feed (Table 1.24 [950]).

Table 1.24. Effect of amount of catalyst on the hydroformylation of diisobutylene (T = 170 °C, Co as catalyst, total pressure 250 bar)

Catalyst conc.[a]	Olefin workup[b] to			Reaction period (min) required for 96% conversion
	paraffins	esters	heavy ends	
0.06	15.0	1.3	2.5	330
0.13	17.5	1.5	3.0	125
0.25	19.0	2.4	3.9	90
0.50	20.9	5.5	8.9	60

[a] in wt-% rel. to olefin   [b] in % of olefin feed

Contradictory results have been published for the hydroformylation with ligand modified cobalt or rhodium catalysts. According to Rupilius et al. [164], on hydroformylating 1-pentene with $HCo(CO)_3[P(Bu)_3]$, the share of unbranched products rised from 74.5% with 0.1 g catalyst/0.1 mol olefin to ca. 85% at 2 g/0.1 mol. Similar results were reported with modified ruthenium catalysts [1137]. Contrary to his expectations, Tucci [620] could not confirm any relationship between the n : iso ratio and the catalyst concentration in the propylene hydroformylation. However, the propylene conversion and butanal hydrogenation increase in the expected manner with growing catalyst concentration (Table 1.25).

Table 1.25. Effect of $HCo(CO)_3[P(Bu)_3]$ concentration on propene hydroformylation [620]

Amount of catalyst total m moles	Branched isomers (Wt.-%)	Propene conversion (%)	Formation of butanols (Wt.-%)
6.3	11.3	45.5	21.3
12.6	10.7	63.6	26.0
25.1	11.0	81.3	33.8
50.3	11.5	96.9	47.0
75.4	11.7	98.4	52.7
100.5	10.8	98.6	60.4

On hydroformylating 1-hexene with $HRh(CO)[P(C_6H_5)_3]_3$, Brown and Wilkinson [169] established a clear relationship between the n : iso ratio and the catalyst feed in the region 5—50 m moles. This partly contrasts with the proposals of the group UCC/Davy Powergas/Johnson Matthey about the process for the hydroformylation with ligand-modified catalysts which is based on Wilkinson's work. In a patent application [935] they state:

" In general, the optimal catalyst concentration depends on the concentration of the α-olefin e.g., propylene. As a rule, higher propylene concentrations necessitate lower catalyst concentrations in order to achieve good conversion rates to the aldehyde for a given reactor size. As the partial pressures and concentrations are dependent on one another, higher propylene partial pressures lead to a greater share of propylene in the off-gas. In order to remove some of the propane (if present) it might be necessary to flush part of the gas stream from the product recovery section before it is recycled to the liquid. The amount of propylene lost in the propane flush gas stream is directly proportional to the propylene content of the off-gas. It is therefore necessary to compare the economic level of the propylene losses (in the propane flush gas stream) with the capital cost saving connected with the lower catalyst concentration."

In general, Rh concentrations in the range 250 mg/l are recommended [634]. This is partly based on the connection between increasing Rh concentration and increasing conversion which would make an energy intensive recycle of the unreacted feedstocks unnecessary and the larger financial outlay for the initial charge of an Oxo plant with Rh (cf Sect. 1.6.3).

1.3.3.4 Catalyst Recycle in Industrial Oxo Process

1.3.3.4.1 Unmodified Catalysts

The reaction mixture of the homogeneously catalyzed hydroformylation contains soluble metal hydridocarbonyls as long as a CO/H_2 partial pressure is maintained over the reaction product. Before further processing the Oxo products, the metalhydrido-carbonyls should be removed (with cobalt "decobalting") from the reaction mixture for the following reasons:
- The transition metal carbonyls would be deposited in the distillation columns etc., during work up, leading to processing and quality control problems [523, 906].
- Traces of metals can initiate considerable side reaction during the distillative processing of the aldehydes (cf Sect. 1.5).
- Only an immediate separation and recovery of the spent catalyst metal (after the Oxo reactor) would facilitate a brief and economic catalyst recycle [26, 47, 57, 523].
- In several Oxo variants, catalyst excess serves to filter out the poisons introduced with the reactants. The make-up or regeneration step—which is essential before recycling the catalyst to the Oxo reactor—should ensue for economic reasons immediately after the Oxo reactor [545, 904, 920, 921, 923, 947, 952–955, 1071].
- When several catalyst components are present e.g., in the Aldox process or with mixed catalysts [896, 956, 957], then a simultaneous removal of all components can be more readily guaranteed the sooner decobalting occurs after the Oxo reactor.

Only in a few — now obsolete — special cases is the metal recovery stage situated after the down-stream steps or even at the final product stage. When doing this, the most convenient point is after oxidizing the aldehyde product from the hydroformylation [558]. If the metal-containing Oxo crude product passes the hydrogenation stage then there is danger of metal deposition on the hydrogenation catalysts. There are no additional processing and separation problems in those cases in which the Oxo and heterogeneous hydrogenation catalysts have the same metal base or if the hydrogenation is conducted in homogeneous phase with the Oxo catalyst [223, 858, 958–960]. On hydroformylating unsatured polymers, catalyst residues are removed from the final product [1070].

Various methods are used to separate the metal hydridocarbonyls from the Oxo products. With $HCo(CO)_4$, these are:

1. Recycling the formaly monovalent negative cobalt in $HCo(CO)_4$ without valency change.

2. The thermal or thermo-chemical decomposition which mainly leads to cobalt metal i.e., valency change $Co^{-1} \rightarrow Co^{\pm 0}$.

3. The valency change $Co^{1-} \rightarrow Co^{2+}$ and the conversion of cobalt into −usually water soluble−Co^{2+} salts.

Processes with "stationary" catalyst metal − either in the heterogenized form, the SLPC technique [331, 360, 364, 438, 961] (cf Sect. 1.3.3.1.3) or with fixed metal concentration from which the reaction product is removed in situ via distillation [623] − are outwith the above considerations as, in these cases, a catalyst recycle only takes place periodically over a long period of time and not within the actual Oxo plant.

The Kuhlmann process [523, 951, 962–966] is a classic example involving recycling the Co^{1-} species. The Oxo catalyst $HCo(CO)_4$ is initially converted into Na tetracarbonylcobaltate on adding an aqueous Na_2CO_3 solution:

$$2 \, HCo(CO)_4 \; + \; Na_2CO_3 \quad \rightarrow \quad 2 \, Na[Co(CO)_4] \; + \; CO_2 \; + \; H_2O \qquad (58)$$

After separating both phases − a virtually cobalt-free organic phase of Oxo products and an aqueous phase containing a sodium salt − the aqueous phase is treated with H_2SO_4. The cobalt hydridocarbonyl is released, extracted with olefin feedstock and fed to the Oxo reactor.

$$2 \, Na[Co(CO)_4] \; + \; H_2SO_4 \quad \rightarrow \quad 2 \, HCo(CO)_4 \; + \; Na_2SO_4 \qquad (59)$$

Other patents [967] have similar aims, in some cases combined with the concept that on converting the strongly acidic $HCo(CO)_4$ to the Na salt, by-product formation from the aldehyde product will be suppressed [968]. In this instance, the cycle involving $HCo(CO)_4$ extraction must not necessarily be completed.

The fundamental idea of recycling the Co^{1-} is also behind proposals that the Oxo products should be distilled under such high CO or H_2 partial pressures that during this separation step the stability of the $HCo(CO)_4$ is ensured (cf Fig. 9) [969–971]. Although − as a consequence of the higher stability of the modified metal hydridocarbonyls (cf pp. 45/47) − the ligand-modified cobalt and rhodium carbonyls can be separated from the Oxo products on heating, the unmodified carbonyls decompose on heating. Consequently, finely divided Co metal is present in the bottom products of the pressurized distillation, the atmospheric distillation [297, 593, 694, 973–977, 1051, 1076] or the vacuum distillation [594, 977, 978] i.e., a valency change from $Co^{1-} \rightarrow Co^{\pm 0}$ has occured. This distillative enrichment of the catalyst has given rise to proposals (often used in practise) that the heavy ends from the Oxo synthesis (cf Sect. 1.5) should be recycled to the synthesis to serve as catalyst supports or slurry medium [246, 287, 589, 702, 979, 980, 1034, 1085, 1086, 1273, 1625]. A distillative separation is also proposed for the crude Oxo product and catalyst from the Aldox variant [956].

Due to the somewhat extensive thermal decomposition of $HCo(CO)_4$, distillative concentration processes represent border-line cases of actual catalyst deposition by means of thermal or chemical treatment of Oxo raw product containing cobalt carbonyl. In these processes, the oxidation $Co^{1-} \rightarrow Co^{\pm 0}$ (cobalt metal) is directly carried out (for details cf [906], for process variants of the decobalting [279, 981–986, 991], chemical variants in the presence of H_2 [282, 987, 988, 1117] or oxidising agents [362]. This also includes the deposition of cobalt on supports or polymers (cf Sect. 1.3.3.1.3 and Refs. [988, 989, 992]) which can then serve as heterogenized catalysts and, under certain conditions, resemble the old "Diaden method" i.e., process operation involving alternate Co deposition and Co carbonyl generation [990, 1946].

Process variants in which sulfur is employed in the oxidation step $Co^{1-} \rightarrow Co^{\pm 0} \rightarrow Co^{2+}$ to enable the oxidation to be arrested at the cobalt metal stage [906, 907, 909, 993] result in metallic cobalt being deposited. These Oxo variants can tolerate sulfur contents > 5 ppm in syn gas – this represents a major advantage for various locations (cf Sect. 1.6).

The transfer to cobalt deposition processes involving the oxidation step $Co^{1-} \rightarrow Co^{2+}$ is represented by variants in which the Co^{2+} compounds are yielded in solid heterogeneous form e.g., cobalt formate is obtained on treating the crude Oxo product – containing $HCo(CO)_4$ – with oxygen and formic acid [906, 994–996].

Other processes are characterized by the Co^{1-} ion of cobalt hydridotetracarbonyl being converted into the – usually water soluble – Co^{2+} salts [906]. These variants are operated by BASF, whereby the oxidation is conducted using O_2 or other oxidizing agents in the presence of cation-forming organic acids, in particular, formic [860] or acetic acid [997–1000, 1056, 1057] sometimes in the presence of noble metal promoters [999], by Gulf [1001, 1009], Leuna [1002], Daicel [1003], Standard Oil Co. [1004], Hoechst-Ruhrchemie [907], Union Carbide [1005] etc. [1006–1008]. Addition of higher carboxylic acids during the oxidation stage facilitates the simultaneous formation of cobalt soaps which are suitable for re-use (Montecatini [1010], Monsanto [1011], Standard Oil [1012] and other processes [976, 1013–1016, 1626]).

The basic oxidation step Co^{1-} (or $Co^{\pm 0}$ in $Co_2(CO)_8$) $\rightarrow Co^{2+}$ can ensue via of oxidation with the water protons [Eq. (60)], via valency disproportionation [Eq. (61)] or via catalyzed reactions [Eq. (62)] [906]:

$$Co^{\pm 0} + 2 H^+ \rightarrow Co^{2+} + H_2 \tag{60}$$

$$3 Co^{\pm 0} \rightarrow 2 Co^{1-} + Co^{2+} \tag{61}$$

$$Co_2(CO)_8 + 2 Co(OH)_2 + CO \rightarrow Co[Co(CO)_4]_2 + CoCO_3 + 2 H_2O \tag{62}$$

The following potential process modifications have not been used industrially – decobalting via treatment with ion-exchangers, [896, 1017–1021] or zeolites [1022, 1023], or via precipitation with HCN [830, 1024, 1025] or maleic acid [1026]. Postdecobalting of various substrates can be achieved on using hydrazine [938]. NaOH treatment [872, 968] is less advisable as the aldolization reaction with the aldehydic product is encouraged as soon as stronger bases other than $NaHCO_3$ or Na_2CO_3 are employed. While oxidative decobalting with chlorine [1027] is effective, it tends to cause stress corrosion (from chloride ions) in the material used for the pressure reactors. Cobalt carbonyls can also be reacted with iron or iron compounds, the cobalt being removed as Fe cobaltate [1028]. In this case, two difficult-to-separate metals — Fe and Co — are brought into contact with one another. This could necessitate considerable costs in the make-up stage and even be a strong poison in processes using ligand-modified Rh catalysts (cf Sect. 1.3.3.2).

The cobalt cycle of the Oxo synthesis is completed by reintroducing the recovered cobalt. This can ensue either as Co^{1-} in the form of unchanged cobalt hydridocarbonyl, as cobalt metal, or as cobalt salt suspended in a suitable liquid. In addition, the Co^{2+} salt can be fed in aqueous acidic solution possibly after the energy-intensive volume reduction [999]. Another possibility is an involved cobalt cycle after a special conversion of deposited cobalt compounds and after passing separate make-up stages [955, 1029, 1073, 1862]. While it is assumed that there is an in situ conversion of cobalt compounds of various oxidation states to the Co^{1-} of cobalt hydridocarbonyl, a whole series of proposed processes serve to precarbonylate the catalyst in order to lead to the pre-formation of $HCo(CO)_4$:

$$Co \ + \ \frac{1}{2} H_2 \ + \ 4 \, CO \ \rightarrow \ HCo(CO)_4 \tag{63}$$

$$Co(Anion)_2 \ + \ \frac{3}{2} H_2 \ + \ 4 \, CO \ \rightarrow \ HCo(CO)_4 \ + \ 2 \, \text{H-Anion} \tag{64}$$

In the case of $Co^{\pm 0}$ the precarbonylation offers the possibility of separating the carbonylation and hydroformylation steps enabling the Oxo stage to ensue at lower temperatures. This encourages the formation of unbranched compounds. Using Co^{2+} solution, precarbonylation has the advantage of excluding water from the Oxo stage [999] — where it acts rate retarding — and to eliminate the anion from the catalyst.

The readiness with which $Co^{\pm 0}$, but not Co^{2+} salts, can be converted into $HCo(CO)_4$ is demonstrated by the virtual absence of publications about the precarbonylation of $Co^{\pm 0}$ [1030] while there are many dealings with the reaction of Co^{2+} compounds sometimes together with promoters (cf Sect. 1.3.3.1.4 and Refs. [524, 860, 999, 1031—1037]).

The unmodified Rh carbonyls are particularly difficult to recover when employed in small amounts as they must then be unequivocally brought into solution without formation of a metallic Rh mirror or deposited at one site. It is recommended that a defined deposition of the metal should ensue by means of shaking with mercury or via precipitation with other metals [1038, 1042]. In addition, thermal decomposition with steam [557, 1039—1041] is possible or reduction in the presence of hydrogen [1043]. Heat treatment during distillation [1044, 1045] and the separation of Rh-containing organic residues [1046] have also been proposed. Lastly, treatment with HCOOH [1048], halogens [1049], HCN [1047] or feeding the Rh-containing solution over ion-exchanger resins [358, 1017, 1050] have also been proposed.

Although the Rh recovery rate with these and similar processes has been claimed to be as high as 99,9% the problem of a low loss recycle of minute Rh quantities still cannot be regarded as being solved.

1.3.3.4.2 Modified Catalysts

The deposition, regeneration and recycling of modified catalysts are impeded by the presence of the readily decomposable ligand which must be recycled along with the metal component. Therefore, the preferred method is to concentrate the modified catalyst via distillation – possible due to increased thermal and chemical stability (cf pp. 47/53) – if necessary, in the presence of organic acids [592, 594, 1052] or after converting the complex $Co_2(CO)_6L_2$ into a more sparingly soluble compound [830. 1053]. A reduction in the quantity of phosphine oxides to be discharged can be achieved via treating the Oxo products with aqueous solutions of alkali metal hydroxides [872] or non oxidizing mineral acids [686, 1054] as well as via extraction [1055].

With the ligand-modified Rh catalysts a recovery stage is necessary when the "stationary rhodium" technique is not employed [1060] (for periodic processing of Rh/ligand content of such systems cf [635, 1059]). Often, very involved complex recycles are necessary and sometimes even overseas processing plants form part of catalyst reprocessing cycle [1149]. The proposed methods for treating ligand-modified Rh complexes serve to separate phosphine/phosphine oxide [1061, 1062]. Procedures employed include dialysis or reverse osmosis [1063, 1064] or treating the Rh-containing distillation residue with acids or peroxides [593, 953, 1065–1068] or other variants [589, 1069] as well as separating the modifying material and the organic substrate [1078].

When the recycled catalysts are completely poisoned – or with the stationary rhodium process when the reactor contents are contaminated – then the catalyst metal, organic substrate and modifying substances can be separated by combustion [1079].

1.3.4 Solvent Effects

Solvents are often employed in hydroformylation reactions, mainly in batch processing with autoclaves. The advantages of using solvents are as follows:

- Improved reaction control on varying the concentration of the reactants, the intermediates and final products in the reaction mixture ('dilution') and thus influencing the selectivity of such reactions which occur with a change in mol number.

- In addition, CO and H_2 are particularly soluble in several solvents thereby facilitating the mass transfer gas → liquid.

- Process engineering advantages, such as the solvent functioning as additional coolant, as catalyst support or as selective solvent for the feedstock or product.

- Several solvents react selectively with intermediates or products yielding desired substances ('reactive solvents')

- When isobutyraldehyde in particular, is used as solvent then the postulated equilibrium of the primary reaction products of the Oxo synthesis is apparently displaced to the right.
  isobutyraldehyde ⇋ n-butyraldehyde

- Finally, as several solvents exert electronic effects they can function as activators or modifiers (cf Sect. 1.3.3.1.4).

However, these advantages are outweighed by serious disadvantages which have hitherto prevented application of solvents in the industrial Oxo process — except for the temporary use of butyl ethers as catalyst transport medium in an older variant of the Kuhlmann process [1080]. It is perhaps worthwhile mentioning the additional outlay for the solvent recycle (usually necessary) — possibly combined with a regeneration stage with modified Oxo catalysts [597] — as well as the reactor volume taken up by the solvent (which is not available for synthesis). This also applies to the "in situ" solvents e.g., product recyles [1081–1083] (not however, to reactors with back-mixing, leading to special effects [1084]) or the reuse of heavy ends from the synthesis as catalyst carriers [246, 287, 702, 1034, 1085, 1086].

In favor of the application of solvents — *from reaction control viewpoint* — is the possibility of diluting the reactants and thereby varying their concentration in the reaction mixture. In this way, equilibria — which occur with a change in mol number (applies to all reactions in Oxo reactor) — can be displaced so that, on lowering the total concentration of the reactants, the total number of moles will increase (and the reverse).

Altering the total reactant concentration is thus a basic method for influencing the conversion: If the undesired side reactions (cf Sect. 1.5) have a different order of reaction to the main reaction, then the selectivity to the desired main product can be improved. Side reactions of higher order than the main reaction are at a disadvantage at lower concentrations (e.g., via dilution) compared to reactions of lower order (Ziegler's dilution principle). On considering, that the order of reaction is only equal to the molecularity in simple conversions ("mono-, bi- or tri-molecular reactions") then this relationship can be quantitatively interpreted as follows: in the presence of molecules of the diluent, the probability of three reactants colliding will be reduced to a greater extent than that for a two molecule collision. However, every consideration must take into account that with dilution in the continuous process the residence times of the olefinic substrate and products are also reduced. Often phenomena stemming from the latter are said to be due to solvent effects (and converse, cf Sect. 1.3.6).

Many studies have dealt with the relationship between increasing the rate of the Oxo reaction by solvents and its effect on the conversion. Table 1.26 stems from Pino, Piacenti and Bianchi's work [1087].

As can be appreciated from Table 1.26, the differences between the various feedstocks are greater than the maximum effect of the solvents. In general, it appears that the more polar solvents accelerate the Oxo reaction to a greater extent [1085, 1086,

Table 1.26. Effect of Solvent on the rate of hydroformylation with various feedstocks

Solvent	Specific reaction rate v (10^3 k/min) Hydroformylation of				
	1-Hexene	2-Hexene	Cyclohexene	Methyl acrylate	Acrylonitrile
benzene	32	9.2	6.7	41.8	12
acetone	34	9.1	6.1	59.5	23
methanol	54	9.2	8.9	157	80
ethanol			8.7	186	128
methyl ethyl ketone			5.7	39.1	

1088, 1089–1091]. The same applies to the application of ligand-modified Oxo catalysts [319, 554, 598, 665, 831, 1092] – in the absence of ligand excesses. When excess modifier is present, its influence normally dominates [640]. According to Craddock et al., the rate of reaction in the polar solvent butyraldehyde is fast and decreases as a function of time, whereas the reaction rate in the nonpolar benzene solvent system is slow initially and builds up a maximum (as aldehyde product is formed and polarity of the solvent system increases), then subsequently diminishes as a function of time [640] (Fig. 1.54).

This typical result for discontinuous tests must not be generalized and applied to a continuous operation. Moreover, with polar reactive solvents, reaction with the catalyst precursor $Co_2(CO)_8$ is discussed in accordance with Eq. (65)

$$3\ Co_2(CO)_8\ +\ 12\ ROH\ \rightarrow\ 2\ Co(ROH)_6[Co(CO)_4]_2\ +\ 8\ CO \tag{65}$$

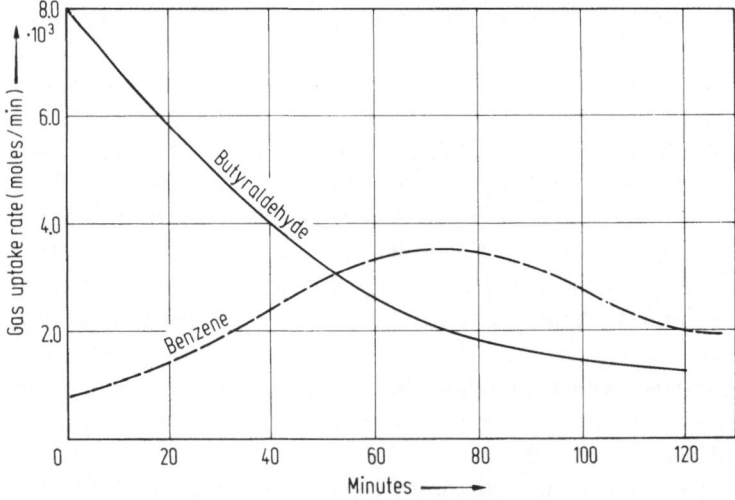

Fig. 1.54. Conversion rate of propylene as a function of time using butyraldehyde or benzene as solvent [640]

This should be readily seen from the inhibition periods as well as from the various effects with different alcohols [1093].

Influencing the selectivity of the hydroformylation by diluting the reactants with solvent has frequently been reported with propylene [268, 945, 1083, 1094, 1095] as well as with pentenes [1094, 1096], hexenes [472, 555], 1,3,5-cycloheptatriene [322] and dicyclopentadiene [200, 1097]. Polar solvents appear to be particularly effective with regard to increasing the selectivity – possibly due to their inhibiting the isomerization of the alkylcobalt carbonyls [154].

The following values summarized in Table 1.27 were given in an ICI patent relating to the hydroformylation of propylene [1083].

Table 1.27. Yields (%) obtained in the hydroformylation of propylene with $HCo(CO)_4$ using different solvents (130 °C, 220 bar CO : H_2 = 1 : 1) [1083].

| | Relation between propylene and solvent | | | | |
| | Isooctane | | Heavy ends of Oxo synthesis | | |
	9 : 1	1 : 2	8 : 1	1 : 1	1 : 2
n-butyraldehyde	65	75	66	65	74
isobutyraldehyd	18	16,5	18	16	19
heavy ends	16	7,5	14	16	4
propane	1	1	1	1	1
% propene conversion			78	75	75

Table 1.27 proves from the constant sum of the % values of n-butyraldehyde and heavy ends, that the increase in selectivity to n-butyraldehyde formation is probably due more to the suppression of side reactions – which consume the n-aldehyde – than an actual increase in formation of the desired product. Similar conclusion can be drawn from measurements made with continuously operated Oxo reactors (Table 1.28). The special effect of nonpolar solvents in a continuously operated reaction could also be confirmed.

Table 1.28. Continuous hydroformylation of propylene (T = 150 °C, p = 250 bar, CO : H_2 = 1 : 1 LHSV = 1.0) [1098]

| Composition of Oxo crude (%) | Solvent | | | |
	None	Isobutyraldehyde	Isobutanol	Benzene
hydrocarbons	0.1	0.1	0.1	0.1
isobutyraldehyde	19.3	18.8	20.5	19.4
n-butyraldehyde	65.7	62.8	65.5	71.0
esters	4.2	4.0	1.9	2.3
alcohols	4.4	6.5	6.6	2.5
heavy ends	6.3	7.8	5.4	4.7
Σ aldehydes	85.1	81.6	86.0	90.4
Σ esters + alcohols	8.7	10.5	8.5	4.0

There are various *process engineering advantages* inherent in using solvents – they can serve as heat regulating agents in many types of reaction, particularly in discontinuous autoclave experiments [597, 932, 1083, 1099, 1100, 1116, 1138] as well as in adiabatically operated syntheses [252]. This is particularly elegant when the solvent is already present in the system — stemming from the production, work-up or regeneration stage [1101] or is introduced via the catalyst make-up [524, 1031, 1034, 1066, 1080, 1102] or serves as catalyst carrier [331, 439, 630, 1103–1106, 1951] or is fed as hydrocarbon-rich olefin mixture (raffinate) [1109]. This includes processes with stationary Rh where the noble metal is dissolved in excess ligand as well as product cycles with solvents from the system or the catalyst slurry with the heavy ends from the same process [218, 246, 287, 702, 713, 832, 836, 878, 1034, 1066, 1081–1085, 1095]. If the Oxo catalyst is introduced in aqueous solution — water can serve as diluent [805, 874, 1103, 1107, 1108] — then other solvents must function as homogenizing agents [1031, 1103, 1106] as well as possibly helping to combat the corrosion encountered with aqueous catalyst systems [1106].

Solvents can also exhibit *specific solubilities for the feedstocks or products* [243, 1110, 1111] or for the catalyst complex thereby facilitating its recovery [1104]. Alcohols [472, 524, 555, 597, 1100, 1112, 1113], aldehydes [1031], dialcohols [945, 1103], aliphatic or aromatic hydrocarbons [164, 209, 273, 282, 489, 611, 665, 932, 1094,, 1098, 1114–1120, 1122], acids [1113], ethers [154, 170, 268, 319, 1121] and other solvents have been recommended for these applications — based on process technology considerations. The ratio olefin:solvent is often limited on account of the investment necessary for solvent recovery and recycle [1083, 1136].

Benzil, which is a typical *reactive solvent,* was introduced by Bott [572]. This solvent reacts with intermediates from the olefin hydroformylation yielding defined final products. For example, α-benzoylbenzyl alkyl carboxylate is obtained via a Tischtschenko–Claisen analogous reaction between the Oxo aldehyde and benzil [572]:

$$\underset{/}{\overset{\backslash}{C}}=\underset{\backslash}{\overset{/}{C}} + CO + H_2 \ \rightarrow \ \left[\underset{/}{\overset{\backslash}{C}}H-\overset{|}{C}-CHO\right] + \overset{O}{\overset{\|}{C}}-\overset{O}{\overset{\|}{C}} \rightarrow H-\overset{|}{\overset{|}{C}}-\overset{|}{\overset{|}{C}}-\overset{O}{\overset{\|}{C}}-O-\overset{|}{C}H-\overset{O}{\overset{/\!\!/}{C}} \qquad (66)$$

Water can also function as a reactive solvent when it hydrolyzes an appropriate substrate or product, the resulting acid can either act as a catalyst poison (cf Sect. 1.3.3.2) [893, 927, 1123, 1124] or facilitate the subsequent demetalization stage. However possible corrosion must not be ignored (cf Table 1.22 [894]). Other examples cf [1953].

It is maintained, on using isobutyraldehyde as solvent that the *equilibrium* n ⇌ iso-aldehyde is displaced [1125]. This theory has been refuted by various research groups cf pairs of values in Table 1.28 [1098]. Actually, the recycling of isobutyraldehyde (cf Sect. 1.5) increases the danger of side reactions which consume even n-butyraldehyde (cf Sect. 1.5) such as mixed-aldol or -acetal formation. There have been several publications [154, 219, 1094] presenting a detailed interpretation of the isomerization of the acyl-cobalt carbonyls formed before the aldehyde synthesis (cf Sect. 1.2.2).

Anisol/thioanisol [869], water [294, 486, 1126], N-methylpyrrolidone [871], nitriles or amides [559, 719, 720, 888, 1127–1129], pyridines [1075] and other compounds [640, 734] are solvents which exert *electronic effects* on the catalyst and should therefore be regarded more as activators or promoters (cf Sect. 1.3.3.1.4). Modified catalysts can frequently be identified by their characteristic IR absorption bands [720]. As shown in Table 1.29, nitrogen-containing solvents effect an overall increase in selectivity with regard to the total aldehyde yield and influence the n : iso ratio to a much lesser extent. In addition they reduce the catalytic activity.

Table 1.29. Hydroformylation of propylene with $Co_2(CO)_6[P(Bu)_3]_2$ in various solvents (154.3 mmol propylene in 60 ml solvent, t = 172 °C, p = 45 bar) [720]

Solvent	Reaction period (min.)	Aldehyde selectivity (%)[a]	Ratio aldehyde: alcohol	n : iso
dimethylformamide	57	76.5	88 : 12	86.5 : 13.5
methyl isobutyl ketone	18	66.7	76.1 : 23.9	86.7 : 13.3
toluene	24	65.9	73.2 : 26.8	86.4 : 13.6
n-heptane	27	58.3	63.6 : 36.4	85.9 : 14.1

[a] moles C_4 aldehyde from 100 moles propylene converted

1.3.5 Concentration of Reactants

Sections 1.3.2 and 1.3.3.3 dealt with the effect of changes in concentration of the catalyst and of the gaseous reactants — CO and H_2 (individually and together) — on the conversion and selectivity of the hydroformylation. The olefin concentration is the third variant available for influencing the conversion as well as the selectivity in accordance with the general schemes in Fig. 1.55.1 and 1.55.2.

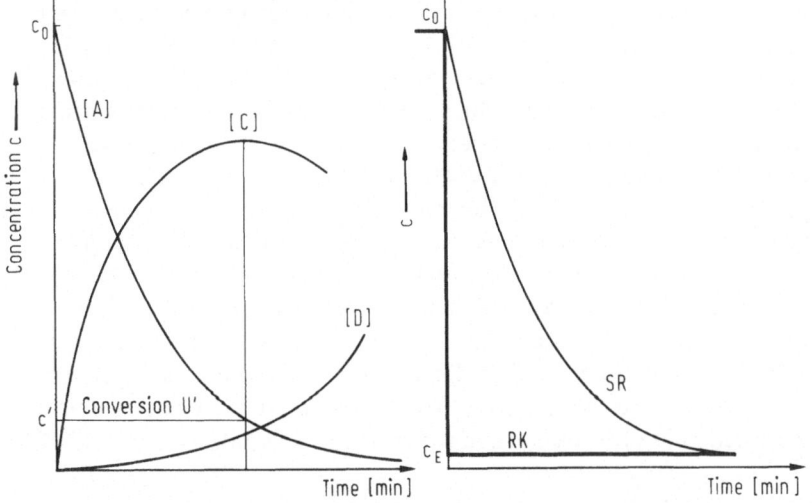

Fig. 1.55. Concentration/time profiles for the reaction A + B → C and C → D in periodically (Fig. 1.55.1) and continuously operated backmixed vessels (RK) or in continuous flow reactors (SR) (Fig. 1.55.2)

In accordance with the laws governing chemical reaction technology, in the following reaction:

$$A \xrightarrow[+B]{k_1} C \xrightarrow{k_2} D \qquad (67)$$

the concentration A falls in a periodically operated stirring vessel according to Eq. (68)

$$-\frac{dc}{dt} = k \cdot c^n \qquad (68)$$

The decrease in reactant A (olefin) shown in Fig. 1.55.1 ensues concurrently with the increase in formation of the desired reaction product C and the undesired secondary product D. Above the conversion rate U' — corresponding to a residual olefin concentration of c' — the concentration of C in the reaction mixture drops noticeably in favor of D. More information about the effect of the order of reaction and the ratio of the rate constants on the decrease in [A] and thus the position of the curves [C] and [D] can be found in standard treatise covering reaction technology. In the continuously operated stirring vesses (Fig. 1.55.2, curve RK), the concentration of A immediately drops from c_0 to c_E (concentration at the outlet). Thus, in the continuously operated stirring vessel the concentration of A (dependent on conversion) and the conversion rate (dependent on the residual concentration) are relatively low. On the other hand, in discontinuous hydroformylations the whole range c_0 to c' (or c_E) is traversed, consequently the average concentration (and thus the conversion rate) is relatively high. Therefore, the olefin concentration should not be too low if a high conversion rate (greater reactor efficiency) and a high selectivity to the desired product C are to be achieved. However, as the residual olefin present in the reaction products is often lost, its concentration should be kept as low as possible. Process technology will have to optimize these opposing demands if a solution is not found via two stage operation. In this case, partial conversions with high selectivity and conversion rate take place in the initial reactor, the residual conversion ensuing in a connected reactor. The conversion in this second reactor may occur with low selectivity (cf Refs. [275, 1084, 1139, 1177, 1271, 1808] and Sect. 1.3.7).

While, in accordance with the Law of Mass Action, one of the reactants is generally employed in excess in order to achieve a high conversion of the other components (ususally the cheaper, readily separable and recyclible syn gas [295, 1140]), several patents claim to achieve the same effect via controlling the olefin concentration in the reaction mixture [311, 331, 1141, 1142]. However, it must be kept in mind that with continuously operated stirring vessels (industrial Oxo reactors are usually reactors with the characteristics of stirred tank reactors or reactors in series with a limited number of stages [1143, 1145]) although the reactant (olefin) is continuously introduced, the olefin residual concentration, which determines the conversion rate, drops via a step function $c_0 = 100$ to $c' = c_E$ (cf Fig. 1.55.2). Table 1.30 [1144] gives an indication of the relationship between conversion and selectivity. These results were obtained using continuously operated pilot plant reactors, however the throughputs were low (LHSV = 0.75).

Other reports describe catalyst activation via limitation of the olefin concentration in the reaction mixture [1142] or improving the decobalting via alteration in the catalyst solubility in a mixture of unconverted olefin and solvent [1115].

Table 1.30. Relationship between product composition and conversion (150 °C, LHSV = 0.75 [1144])

Propylene conversion (%)	Composition of Oxo raw product (%)				
	Hydrocarbons	Aldehydes	Esters	Alcohols	Heavy ends
99.1	0.2	76.3	5.4	6.4	11.7
97.5	0.3	80.4	4.2	4.8	10.3
96.3	0.2	81.3	4.4	4.0	10.1
78.3	0.3	87.8	1.9	2.5	7.5

1.3.6 Residence Time of Reactants

Figure 1.55.1 also shows the main effects of the reaction period (residence time of the reactants) – longer reaction periods increase the conversion by lowering the residual olefin content. However, the selectivity also drops due to an increase in the number of secondary reactions. This can be clearly seen in Fig. 1.56, on increasing the reaction period from t_0 to t_1, [A] drops from c_0 to c_{E1}. The composition of the reactor content corresponds to $[A_1] + [C_1] + [D_1]$, component B (Eq. 67) being present in excess. To increase the conversion from c_{E1} to c_{E2} — a minor rise — would necessitate lengthening the residence time from t_1 to t_2 i.e., almost a third longer. At t_2, the product composition becomes $[A_2] + [C_2] + [D_2]$ and is thus much less favorable compared to t_1. The extent of change in the product content depends on the ratio of the rate constants k_1 and k_2 in Eq. 67.

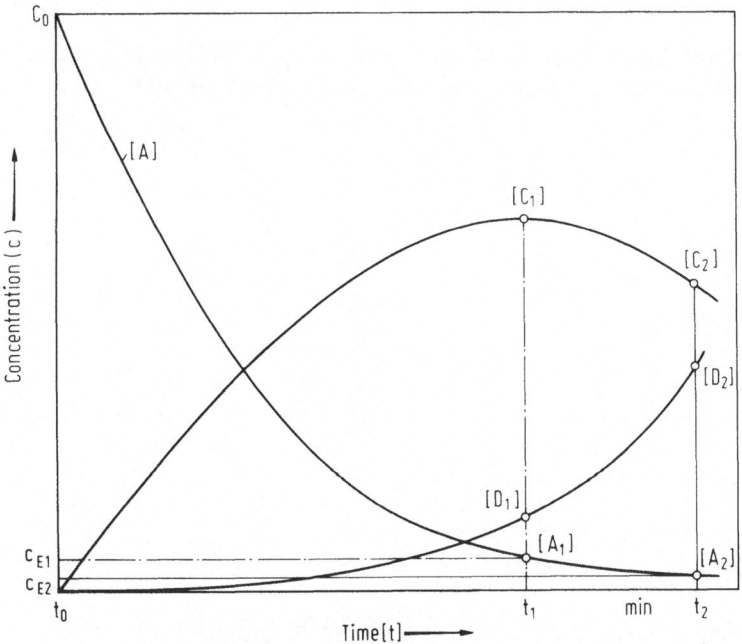

Fig. 1.56. Conversion and selectivity as a function of the residence time of the reactants [1144]

When the olefin concentration drops from c_{E1} to c_{E2}, the conversion rate falls in proportion to the residual concentration (cf Sect. 1.3.5) with simultaneous deterioration in selectivity. For this reason it is prudent *not* to conduct the conversion of the residual olefin (to avoid olefin losses) in the reactor where the main conversion takes place but in a reactor with higher conversion rates with an initial concentration c_{E1} e.g., in flow tube (cf Fig. 1.55.2). This leads to an increase in selectivity during the residual conversion for reasons which cannot be discussed here.

Two examples of the decrease in olefin concentration [A] with reaction period were already shown in Figs. 1.43 and 1.44. Table 1.30 presents data concerning decreasing aldehyde selectivity with increasing conversion. Thus, together with Fig. 1.56 it becomes clear that with batch-wise autoclave tests, the residence time is determined as much by the desired conversion as by the selectivities of the main and side reactions. The theme 'conversion' has been dealt with in many publications [83, 265, 449, 466, 1119, 1146, 1230] which have considered for example heterogenized catalysts [449] or gas-sparged reactors [265] (cf Fig. 1.57).

The effect of residence time on selectivity has also been treated in the Refs. [263, 466, 836]. With continuously operated processes, the liquid hourly space velocity (LHSV) — part by volume olefin feed per reactor volume per hour — is more suitable as a measure of the residence time τ of the reactants as a direct connection can then be made to the productivity of a reactor. LHSV and τ are connected by the following equation:

$$\tau \, [h] = \frac{1}{\text{LHSV}} \tag{69}$$

The influence of the LHSV on the continuous hydroformylation of ethylene is demonstrated in Fig. 1.58.

Similar considerations apply to other catalyst system (e.g., ruthenium [466]) and to the synthetic steps after the hydroformylation e.g., hydrogenation [1147].

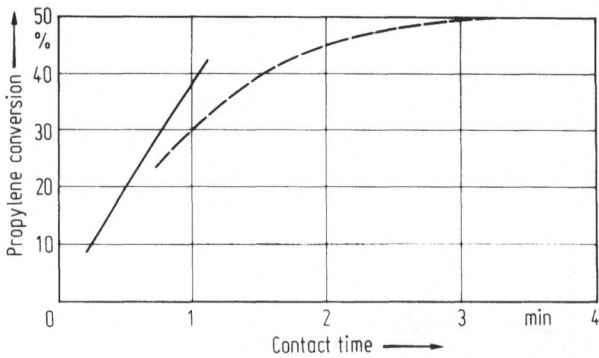

Fig. 1.57. Propylene conversion versus contact time. — Supported Rh catalyst (at 148 °C, 49 bar [449]); — — gas-sparged reactor; modified Co-catalyst (128 °C, 35 bar [265])

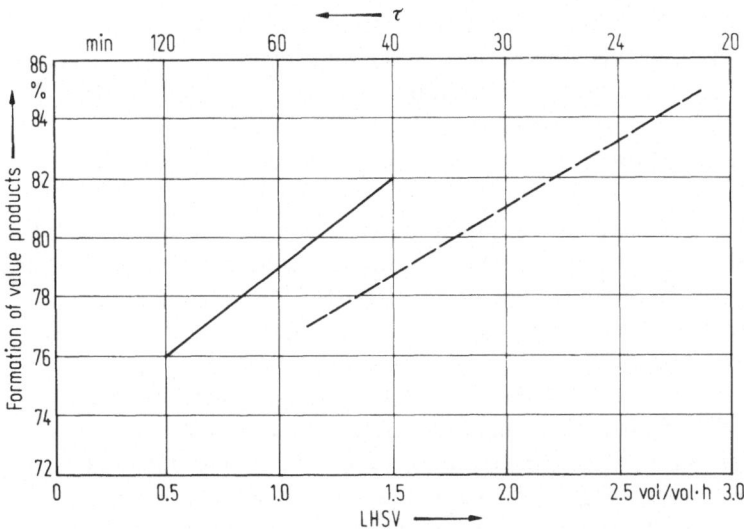

Fig. 1.58. Relationship between formation of value products (propanal and propanol) on through-put [263].
——— modified Co catalyst
– – – – unmodified Co catalyst

1.3.7 Influence of Reaction Conditions on Construction and Operation of Industrial Oxo Reactors

Commercial-scale Oxo plants represent a combination of various units: in the syn gas section CO/H_2 is made available (feedstocks are heavy fuel oil fractions or methane, in the future coal) and then compressed utilizing electrical energy. Thereafter, the syn gas is reacted with olefins in the homogeneously catalyzed Oxo stage. The heat of reaction of the exothermic hydroformylation should, if possible, be recovered as steam. The latter can then be employed as energy source in the realization of the catalyst recycle, in a possible reactant purification and in the distillative work-up of the products. After isolation of the reaction products other stages ensue e.g., aldolization, hydrogenation, oxidation etc. which must also partly take place in pressurized reactors just as in the Oxo synthesis itself. The resulting products are usually worked-up via fractional distillation. As can be appreciated from Sects. 1.3.1 to 1.3.6 compromises frequently have to be made between these demands and the equally significant effects of reaction parameters on the hydroformylation.

A high *temperature* results in an increase in reaction rate (Fig. 1.13) thereby facilitating a reduction in catalyst concentration or catalyst feed. In this way, a high olefin conversion is possible and the heat of reaction can be recovered at a high level – as steam at $\leqslant 20$ bar. Lower reaction temperatures favor the formation of unbranched products (Fig. 1.14) as well as suppressing side reactions (Fig. 1.16). Furthermore, at higher temperatures (Fig. 1.9) the transition metal carbonyls are less stable and the corrosion caused by the reaction products increases. Aside from special cases (e.g., olefinic substrates with particular electronic effects of the substituents), the general conditions with unmodified Co catalysts are 130–160 °C for propylene and 150–180 °C for higher olefins. There is a similar situation with unmodified Rh catalysts (110–150 °C) and – due to their lower activity – also with ligand-modified Co catalysts (Fig. 1.17–1.19).

With modified Rh catalysts (Fig. 1.20) there are other optimum temperatures (50–140 °C, preferably 80–110 °C) which represent a compromise between the desired conversion, the required n:iso ratio, the necessary Rh content of the reactor and thermal ageing. Currently, the heat of reaction can not be recovered as steam but must be removed via evaporation.

High *total pressures* and – with $CO:H_2$ ratios of 1:1 – high *CO partial pressures* ensure the stability of the Oxo catalysts and a high n:iso ratio (Fig. 1.21).

High H_2 *partial pressures* increase the reaction rate and thus improve the space-time yield. These advantages at high pressures are so considerable that in industrial Oxo processes with $HCo(CO)_4$ as catalyst, the disadvantages inherent in constructing the Oxo reactor with a high pressure stage and compressing the syn gas to synthesis pressure are accepted.

However, with ligand-modified Co or Rh catalysts, the drop in activity resulting from the modification is usually so considerable that even on increasing the total pressure, the space-time yields of the classical process cannot be reached. Consequently, low pressures are employed thereby exploiting – besides the advantage of energy saving via omission of a syn gas compression stage – the higher yield of unbranched products with decreasing CO partial pressures (Fig. 1.24).

In theses cases, higher H_2 partial pressures help to maintain the yield (Fig. 1.28). Hydrogen is either manufactured via conversion or is obtained from an external source as conventional heavy oil gasification plants supply *$CO:H_2$ ratios* of 1:1 and future coal gasification processes will produce syn gas of higher CO content.

Besides pressure and temperature (material quality requirements and cooling apparatus/steam recovery equipment) the *catalyst system* and *application phase* also determine the design of the Oxo reactor. The advantages and disadvantages of homogeneous and heterogeneous catalysis were outlined in Table 1.16, corresponding data relating to the gas phase reaction can be found in pp. 67/68. When choosing a suitable homogeneous catalyst system, the following factors must be considered for process optimization particularly in regard to the Oxo reactor and the catalyst recycle:

Activity and selectivity influence the reactor volume and arrangement of cooling surfaces.

Resistance to poisons – may be achieved indirectly via purification steps.

Precipitation and regeneration stages which determine the costs for decobalting and make-up steps.

Corrosiveness, which must be considered on selecting reactor material.

Loss tendency and stability which can affect the pressure stage and construction costs.

Moreover, the choice of Co or Rh catalysts is also significant as this decision will largely determine whether an internally or externally cooled tank reactor is to be employed or a backmixing reactor (cf Sect. 1.6).

Activators and promoters play a minor role in the industrial Oxo synthesis. When they are employed, design features ensure that all catalyst components pass through the reactor unchanged and are then deposited at a specific stage and recycled.

The *catalyst concentration* also determines constructive features on account of its effect on the space-time yield and the catalyst recycle, particularly when the reaction is to be conducted in two stages (to improve the selectivity). There is a similar situation with regard to the *reactant concentration*. The *application of solvents* can directly affect the design of the Oxo reactor due to the reasons mentioned in Sect. 1.3.4 or via their effect on the catalyst recycle.

The *residence time* indirectly determines the design and volume of the reactor. The type of catalyst, its activity and the choice of reaction parameters stipulate, at least quantitatively, the size of the reactor and the shape of the cooling units which determine the throughput. In addition, the type of reactor — with or without stirring, reactors in series or flow tubes — also obviously influences the constructive feature.

Table 1.31 compares the advantages and disadvantages of the above parameters and their effect on the design.

Table 1.31. Optimal reaction parameters in the Oxo synthesis [47]

	Advantages	Disadvantages
low temperature	high n : iso ratio	no steam recovery, low LHSV, low conversion
low total pressure	low investment costs for pressurized reactors and gas compression	large reactor volume necessary
low CO partial pressure[a]	high n : iso ratio	marked olefin hydrogenation
high ligand excess	high n : iso ratio	costs for expansion, low LHSV, low conversion
high Rh concentration	lowers activity loss via modifying	costs for initial investment and expansion high
syn gas H_2-rich	lowers activity loss via modifying	external H_2 source essential

[a] using ligand modified catalysts

1.4 Hydroformylation of Particular Structures

1.4.1 Monoolefins

Basically, while all olefins can be hydroformylated, they differ in reactivity. Wender et al., Heil, and Marko [106, 1673] made a systematic study of the relationship between the reaction rate and olefin structure. The observed differences in reaction rate were around the factor 50 (cf Table 1.32).

Table 1.32. Hydroformylation of olefins at 110 °C [106, 1673]

	Specific reaction rate $(10^3 k \cdot min^{-1})$ catalyst : Co	Over-all constant $(10^3 k \cdot min^{-1})$ catalyst : Rh
A) Unbranched, terminal olefins		
1-pentene	68.3	
1-hexene	66.2	55.8
1-heptene	66.8	54.2
1-octene	65.6	50.1
1-decene	64.4	40.5
1-tetradecene	63.0	
B) Unbranched olefins with internally located double bonds		
2-pentene	21.3	
2-hexene	18.1	34.4
2-heptene	19.3	40.2
3-heptene	20.0	41.9
2-octene	18.8	
C) Branched, terminal olefins		
4-methyl-1-pentene	64.3	
2-methyl-1-pentene	7.32	25.7
2,4,4-trimethyl-1-pentene	4.79	
2,3,3-trimethyl-1-butene	4.26	
camphene	2.2	
D) Branched olefins with internally located double bonds		
4-methyl-2-pentene	16.2	29.7
2-methyl-2-pentene	4.87	
2,4,4,-trimethyl-2-pentene	2.29	3.0
2,3-dimethyl-2-butene	1.35	0.7
2,6-dimethyl-3-heptene	6.23	
E) Cyclic olefins		
cyclopentene	22.4	16.2
cyclohexene	5.82	6.1
cycloheptene	25.7	
cyclooctene	10.8	
4-methyl-1-cyclohexene	4.87	
F) Styrene	–	124

The unbranched terminal olefins react most rapidly with cobalt catalysts. Their rate decreases slowly with increasing molecular weight. Nevertheless, it is possible to hydroformylate double bonds in high polymers (cf Sect. 1.4.5) [365, 1671, 1674, 1675].

The reaction rate of the unbranched olefins with internally located double bonds is usually only 1/3 of that of their terminal isomers. The position of the double bond in the chain (of the internal olefin) is not very important e.g., 2- and 3-heptene exhibit roughly the same reaction rate.

There is always a decrease in reaction rate when the olefins are branched. The most noticeable drop occurs when the olefin possesses alkyl substitution at one of the carbon atoms of the double bond. Branching at more distant carbon atom has a lesser but still noticeable effect. Internal olefins with alkyl substitution at the unsaturated linkage give rise to the lowest reaction rates.

Cyclic olefins exhibit varying behavior. While cyclohexene reacts very slowly, cyclopentene and cycloheptene react more rapidly than internal unbranched olefins. The increased reaction rate results from the tendency of these cyclic compounds to react in a manner which eliminates the strain in the ring caused by the double bond.

There have been a number of reports about the reactivity of various olefins with rhodium catalysts. Wakematsu [530] claimed, in the presence of rhodium catalysts, internal olefins react more rapidly than their terminal isomers. In contrast, Heil and Marko [106] presented the following sequence of reaction rates of various olefinic structures with rhodium hydridocarbonyls:

styrene $\gg$ linear α-olefins > linear internal olefins > mono-branched olefins > multi-branched olefins (cf Table 1.32).

Fell and Geurts' results [612] concur with the above. They found with rhodium carbonyls in the presence of N-methylpyrrolidine that the conversion rate of n-1-octene was 3–5 times greater than with trans-n-2-octene. It was also established with the two cis-/trans-n-2-octene isomers that the cis isomer was more reactive in the hydroformylation. As double bond isomerization can be rapidly catalyzed by the hydroformylation catalyst, misleading conclusions may be drawn about the actual reactivity during the hydroformylation of olefins – particularly with those olefins with double bonds located at the center of the chain [841].

The distribution of reaction products from the hydroformylation depends to a considerable extent on the structure of the olefin feedstock as well as on the applied reaction conditions. Olefins which cannot be isomerized, e.g., ethylene, or olefins which do not change their structure after double bond isomerization, e.g., unsubstituted cyclic mono-olefins, always produce uniform reaction products. Thus, only propionaldehyde is obtained from ethylene, cyclopentanaldehyde and cyclopentyl-carbinol from cyclopentene and cyclooctylcarbinol from cyclooctene [451, 1676, 1677].

In all other cases, isomeric aldehyde mixtures result. Section 1.3 deals with the relationship between the content of the individual isomeric mixtures and reaction conditions in general and in several special cases. Table 1.33 summarizes the reaction products from various olefins on using certain catalysts.

According to the above, the hydroformylation of unbranched terminal olefins – under standard conditions of the Oxo process (90–150 °C, 100–300 bar, cobalt catalysts) – leads to the favored formation of unbranched aldehydes (55–80%). When the chain length of the n-olefins is increased and the other conditions are kept constant there is a drop in the share of the linear aldehydes. It is still unclear whether this is due to statistics – a consequence of the greater probability for the formation of branched isomeric products – or to steric and/or electronic effects.

At high temperatures, isomeric unbranched internal olefins give rise to almost the same product distribution as the α-olefins. At lower temperatures however, markedly

Table 1.33. Hydroformylation of asymmetric olefins or olefins with double bonds capable of migration (analysis made in part after hydrogenation)

Feedstock	Product	Yield %	Ref.
propene	butyraldehyde and isobutyraldehyde	80	[1, 211, 248, 267, 278, 294, 451, 1672, 1683–1687]
1-butene	n-pentanal and 2-methylbutanal	96.8	[451]
2-butene	n-valeraldehyde, 2-methylbutanal	96	[222]
isobutene	3-methyl-1-butanol and neopentyl alcohol	82	[212, 237–239]
1-pentene	hexanol + 2-methylpentanol + 2-ethylbutanol (5:4:1)		[239]
	C_6-aldehydes, C_6-alcohols	92	[260]
2-pentene	C_6-aldehydes	75	[1483, 1679]
2-methyl-1-butene	4-methylpentanol, 3-methylpentanol and 2,3-dimethylbutanol (11:9:1)		[239]
2-methyl-2-butene	4-methylpentanol, 3-methylpentanol and 2,3-dimethylbutanol		[239]
3-methyl-1-butene	4-methylpentanol, 3-methylpentanol and 2,3-dimethylbutanol		[239]
2,3-dimethyl-1-butene and 2,3-dimethyl-2-butene	3,4-dimethylpentanol		[239]
3,3-dimethyl-1-butene	4,4-dimethylpentanol		[239]
2-ethyl-1-butene	3-ethylvaleraldehyde	55	[1483]
1-hexene	n-heptylaldehyde and 2-methylhexylaldehyde	90	[1120]
2-methyl-3-pentene	5-methylhexanol, 3-methylhexanol and 2,4-dimethylpentanol (4:3:3)		[239]
(+) (S)-3-methyl-1-pentene	4-methylhexanal	92.1	[233]
isoheptene	isooctanols	74.6	[1678]
1-octene	isononanols + n-nonanol (75.6%)	85.3	[1688]
cis-n-4-octene	isononanols + n-nonanol (53.9%)	78.4	[1688]
trans-n-4-octene	isononanols + n-nonanol (57%)	85.5	[1688]
di-n-butene	3-ethylheptanol	39	[1689]
2-ethyl-1-hexene	C_9-aldehydes	23	[1684]
diisobutene	3,5,5-trimethylhexanol	90	[239, 1690–1692]
styrene	phenylpropanals	> 50	[1483, 1679, 1681, 1686, 1870, 1871, 1885, 1964, 1966]
α-methylstyrene	3-phenyl-3-methyl-propanals and isopropyl benzene	40	[247, 1680]
α-pinene	2- or 3-formyl-2,6,6-trimethyl-bicyclo [3.1.1] heptane		[1681, 1825]
biphenyl ethylene	aryl propionaldehydes		[1965]
1-vinylnaphthalene	methyl (1-naphthyl) acet-aldehyde	29	[1483]
octadecene	C_{19}-aldehydes	54	[1483, 1693]
allyl benzene	phenylbutanals		[1869, 1966]

higher shares of unbranched aldehydes are obtained from the α-olefins compared to the internal olefins [6].

There are certain peculiarities with the branched olefins. As recognized at an early stage by Nienburg [236] and A.L.M. Keulemans [239], during the olefin hydroformylation − at least under the previously mentioned standard conditions − hardly any quaternary carbon atoms are formed i.e., the formyl group is not found at the carbon atom where the branching takes place. For example, isobutylene forms 3-methylbutanal (I) [237] almost exclusively − only ca. 5% of the isolated product being pivalaldehyde (II). Even under extreme conditions (temperatures of 220 °C and pressures of 420 bar) which normally favor the formation of branched structures, the pivalaldehyde yield amounts to only 8%.

$$
\begin{array}{lll}
 & \underset{H_3C}{\overset{H_3C}{>}}CH-CH_2-CHO & \text{(I)} \\[2em]
\underset{H_3C}{\overset{H_3C}{>}}C=CH_2 + CO/H_2 \rightarrow & + & \text{(70)} \\[2em]
 & H_3C-\underset{CH_3}{\overset{CH_3}{\underset{|}{\overset{|}{C}}}}-CHO & \text{(II)}
\end{array}
$$

Using modern techniques, Bianchi et al. [266] studied another rule postulated by Keulemans i.e., with 3,3-dimethyl-1-alkene (e.g., 3,3-dimethyl-1-butene), the formyl group becomes exclusively bonded to the carbon atom one (C_1). The result (cf Table 1.34) confirms Keulemans' rule to a great extent. Both effects can be interpretated by the +I effect of the two methyl groups (Baker−Nathan effect, hyperconjugation) which causes the α-carbon atom of the double bond to have a higher negative charge thereby facilitating the electrophilic attack of the $[Co(CO)_3]$ group.
Tetramethylethylene forms after rearrangement 3,4-dimethylpentanal almost exclusively. Even when branching is near one of the carbon atoms of the double bond, this hinders the formation of the formyl group at these carbon atoms [239].

Generally, during the Oxo process the formation of several isomeric aldehydes is undesired. Thus, there have been numerous studies aiming at producing a uniform reaction product on altering reaction conditions The effect of various reaction parameters such as temperature, pressure, partial pressure etc. are discussed in Sect. 1.3.

A distinct alteration in Oxo product distribution can be achieved when rhodium hydridocarbonyls or ligand-modified cobalt or rhodium catalysts replace the cobalt hydridocarbonyl catalyst. Usually, higher yields of branched aldehydes are obtained on employing Rh carbonyl or Co/Rh carbonyl mixtures instead of Co carbonyl (cf Sect. 1.3.3 and Refs. [450, 451, 530, 1120, 1680, 1694, 1695].

The reactivity of the pure cobalt and rhodium carbonyls can be strongly affected on being modified.

These modified catalysts contain trialkylphosphines or other trivalent organophosphorus, -arsenic or -antimony compounds which function as electron donors. For example, cobalt carbonyl complexes with tert. phosphines catalyze the hydroformyla-

Table 1.34. Hydroformylation of tetra-alkyl substituted olefins with $Co_2(CO)_8$ as catalyst and benzene as solvent, 110 °C, 160 bar [266]

Olefin	Distribution of isomers in hydroformylation product. Addition of formyl group to carbon atom	
	C_1	C_2
CH₃ | CH₃–C–CH=CH₂ | CH₃	99.2	0.8
H₃C | CH₃–C–C=CH₂ | | H₃C CH₃	99.2	0.8
CH₃ | CH₃–CH₂–CH₂–C–CH=CH₂ | CH₃	100	traces
CH₃ | CH₃–CH₂–C–CH₂–CH=CH₂ | CH₃	96.6	3.4

tion of olefins (at 180 °C) below 30 bar and produce particularly high yields of un-branched aldehydes and alcohols with an α-olefin feedstock [259–261, 592, 702, 1696]. Catalysts of this type find application in the Shell Oxo processes (Sect. 1.6.2.1.4) and are especially suitable for alcohol manufacture as they immediately hydrogenate the primary resulting aldehydes to alcohols at the high reaction temperatures (180–200 °C) generally present. The powerful hydrogenation tendency of these cobalt carbonyl complexes with tert. phosphines also has a less attractive side – the olefin feedstock undergoes hydrogenation to the corresponding paraffin in noticeable amounts. In contrast, the hydroformylation with cobalt carbonyl-containing tert. phosphites as ligands stops at the aldehydes stage and, even at higher temperatures, no alcohols result [1697]. Rhodium/tert. phosphine catalysts exhibit an even greater tendency than cobalt hydrido-carbonyls (modified with tert. phosphines) to produce unbranched products on hydro-formylating unbranched α-olefins. As these Rh complex catalysts are catalytically active even in small amounts and, in addition, a lower degree of hydrogenation of the aldehyde and olefin feedstock occurs, they are industrially significant hydroformylation catalysts (cf Sect. 1.3.3).

1.4.2 Di- and Triolefins

1.4.2.1 Di- and Triolefins with Isolated Double Bonds

Besides obtaining monoaldehydes, dialdehydes can also result on hydroformylating olefins with nonconjugated double bonds. The greater the distance between the double bonds, the higher are the dialdehyde yields (Table 1.35). This particularly applies to diolefins with terminal double bonds.

Table 1.35. Hydroformylation of nonconjugated acyclic dienes

Olefinic starting material	Pressure (bar)	Temp. (°C)	Catalytic metal	Product	Yield (%)	Ref.
1,4-pentadiene	150	60–100	Rh	isomeric heptane-dials	19	[1698, 1699]
1,5-hexadiene	150	100	Rh	isomeric octane-dials		[1698, 1699]
2,5-dimethyl-1,5-hexadiene				nonanol and 3,6-dimethyl-1,8-octane-diol (65:35)		[1678]
1,7-octadiene	280	120	Co	nonanol, 1,10-decane-diol and 2-methyl-1,9-nonanediol (70:30) }	70	[1700]
1,9-decadiene	40	80	mod. Rh	undec-10-enal 2-methyldec-9-enal (60:40) }	80	[772]
2,6-dimethyl-1,5-heptadiene or -2,5-heptadiene	180	150	Co	3,7-dimethyl-1-octanal	73	[1701]
2,6-dimethyl-1,5-heptadiene	700	100	Rh	3,7-dimethyloct-6-en-1-al	74	[573]
2,6-dimethyl-1,5-octadiene	700	100	Rh	3,7-dimethylnon-6-en-1-al	63	[573]
3-methyl-2,6-octadiene	85	100	Rh	2,6-dimethyloct-6-enal and 2-ethyl-5-methyl-hept-5-enal (70:30)	78	[1702]

If one double bond is located at the α-carbon atom and the other is situated internally, then it is facile to solely hydroformylate the terminal double bond to yield compounds with interesting olfactory properties [573].

Cyclic olefins with isolated double bonds also produce mixtures of mono- and dialdehydes (cf Table 1.36). Composition of the hydroformylation product can be extensively varied on modifying the reaction conditions especially in relation to the catalyst used. When 1,5-cyclooctadiene is hydroformylated with $Co_2(CO)_8$ as catalyst in the usual manner (110 °C and 200 bar), a product results which, after hydrogenation, consists of 63% hydroxymethylcyclooctane and 26% bis-(hydroxymethyl)-cyclooctane [1706].

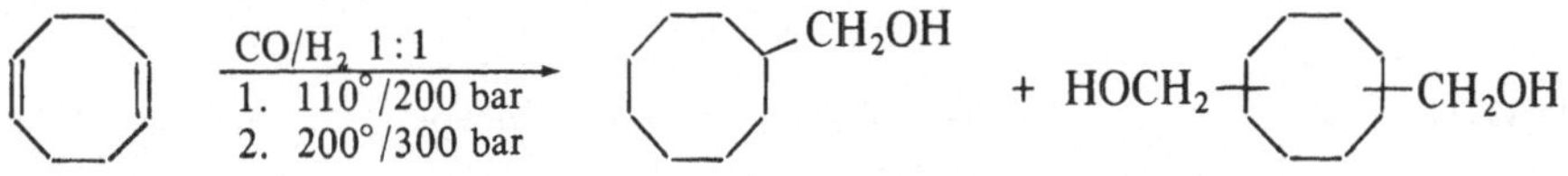

$$\text{(71)}$$

Higher diol yields can be obtained on employing Rh as catalyst under special conditions. It is of decisive importance that the isomerization of the double bond in the conjugated system can be repressed on lowering the temperature. On raising the pressure, the hydroformylation rate can be markedly improved. Thus, with Rh_2O_3 as catalyst precursor, at 90 °C/100 bar a product consisting of 13% hydroxymethylcyclooctane and 81% bis-(hydroxymethyl)-cyclooctane was obtained [1706].

On the other hand, on employing Rh catalysts modified with tert. phosphines the double hydroformylation of 1,5-cyclooctadiene was more strongly impeded, causing the monoformylated products to dominate [722, 815, 1703].

In contrast, the hydroformylation of 1,4-cyclohexadiene yields mainly the di-aldehydes. In this case the effect of the ring size is clearly recognizable [815].

The hydroformylation of 1,5,9-cyclododecatriene − a cyclic olefin with three nonconjugated double bonds − was also studied. The aim was to produce monoformyl-cyclododecane which can be readily converted into cyclododecanecarboxylic acid − a nylon 12 precursor [722]. Syn gas with a high $H_2:CO$ ratio (4−2:1) is preferably employed with a cobalt catalyst in order to reach extensive hydrogenation of the remaining double bonds. With usual cobalt concentrations, a reaction product was obtained in which the mono- and diformyl compounds were present in the ratio 6−8:1 [322]. Lowering the cobalt concentration facilitates the further suppression of diformylcyclododecane formation. However, considerable amounts of cyclododeca-triene are converted into cyclododecene [1704]. The selectivity to monoformyl-cyclododecane can be raised even further on employing ligand-modified cobalt hydrido-carbonyls [1705].

Table 1.36. Hydroformylation of nonconjugated alicyclic dienes

Olefinic starting material	Catalyst (precursor)	Temp. (°C)	Pressure (bar)	Product	Yield (%)	Lit. Ref.
1,5-cyclooctadiene	$Co_2(CO)_8$	1. 100	200	a) hydroxymethyl-cyclooctane	72	[1706]
		2. 200	200−300	b) bis(hydroxymethyl)-cyclooctane a:b = 70:30	19	
1,5-cyclooctadiene	Rh_2O_3	1. 100	200	a:b = 70:30	a) 62	[1706]
		2. 210	300		b) 16	[1905]
1,5-cyclooctadiene	Rh_2O_3	1. 90	1000	a:b = 14:86	a) 13 b) 81	[1706]
cyclododecatriene	Co	130[a]	175	hydroxymethyl-cyclododecane	84	[322]
				bis(hydroxymethyl)-cyclododecane	12	
cyclododecatriene	Co/[P(C₄H₉)₃]	199	211	hydroxymethyl-cyclododecane	82	[1705]
dipentene (limonene)	Rh/[P(C₄H₉)₃]			3-(4-methyl-3-cyclohexene-1-yl)-butyraldehyde	45−52	[1707]
vinylcyclohexene	Co	120−134	475−720	mixture of mono- and dialdehydes	65	[294]

[a] CO/H_2 ratio 1:2,5

With polycyclic dienes in which the double bonds belong to various ring systems and where no double bond isomerization is possible, dialdehydes or diols can be prepared

Table 1.37. Hydroformylation of polycyclic dienes

Olefinic starting material	Catalysts (precursor)	Temp. (°C)	Pressure (bar)	Products	Yield (%)	Ref.
dicyclopentadiene	Co	110	260	tricyclodecanedialdehyde, tricyclodecanemono-aldehyde	41 8	[1708]
dicyclopentadiene	Rh₂O₃	155	180–195	tricyclodecane-dialdehyde	67	[1509, 1904]
2-vinyl-5-norbornene	Rh	120–125	450	bis(formylethyl)-norbornane	74.4	[563]
1,5-bicyclohepta-diene	Rh/[P(Ph)₃]	80	171	2,5(6)-bis(formyl)-bicyclo[2.2.1]-heptane	88	[780]

via hydroformylation with rhodium or cobalt catalysts (Table 1.37). The hydroformylation of dicyclopentadiene has been particularly intensively studied [565, 1509, 1708–1710, 1909].

The formation of saturated monoaldehydes or monoalcohols with the cobalt catalyzed reaction [1708, 1709] strongly recedes when rhodium catalysts are employed [565, 1509, 1709]. Consequently, virtually only dialdehydes – or diols at higher temperatures – result.

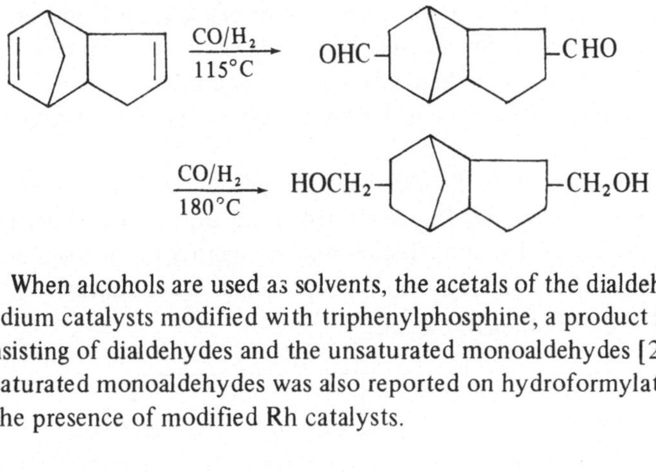

$$(72)$$

When alcohols are used as solvents, the acetals of the dialdehydes are produced. With rhodium catalysts modified with triphenylphosphine, a product mixture is obtained consisting of dialdehydes and the unsaturated monoaldehydes [200]. The formation of unsaturated monoaldehydes was also reported on hydroformylating limonene [1707] in the presence of modified Rh catalysts.

1.4.2.2 Conjugated Systems and Allenes

The hydroformylation of conjugated dienes with unmodified cobalt and rhodium catalysts takes place slowly yielding exclusively saturated monoaldehydes or mono-alcohols. The presence of dialdehydes could not be ascertained [1698, 1711–1716, 1908] (see Table 1.38).

According to studies made by Fell et al., the hydroformylation of butadiene with cobalt, modified cobalt carbonyls or with rhodium alone as catalyst yields only the monoaldehydes or the monoalcohols. This reaction ensues via a semi-hydrogenation of

Table 1.38. Hydroformylation of conjugated dienes with cobalt catalysts

Starting material	Pressure (bar)	Temp. (°C)	Product	Yield (%)	Ref.
butadiene	212–282	145–175	n-pentanal and 2-methylbutanal (1:1), dibutyl ketone	29	[1711, 1717]
2-methylbutadiene	212–282	150	hexanal-isomeric mixture	16	[1711]
cyclopentadiene	260	145–155	cyclopentanaldehyde	37	[1711]
2,5-dimethyl hexadiene			nonyl alcohol	high	[1713]
cycloheptatriene	270	140	cycloheptanemethylal	70	[1714, 1715]

the diene to the monoolefin [574]. Heck and Breslow [1676] reported that a 1,4-addition of cobalt hydridocarbonyl to butadiene takes place at 0 °C. Later studies with cobalt deuterocarbonyl and butadiene confirmed the presence of deuterium solely at the terminal carbon atom [1719].

On the other hand, Fell and Rupilius [1718] showed that dialdehydes could be prepared on hydroformylating conjugated diolefins with ligand-modified rhodium hydridocarbonyls. As it is known that ligand-modified Rh catalysts effect a hydroformylation virtually free of isomerization, they were employed in the hydroformylation of butadiene and isoprene. In this way, the formation of dialdehydes from conjugated dienes [1719] was established for the first time. Work on the optimization of the conversion of butadiene and isoprene to dialdehydes showed that a very high excess of triphenylphosphine (with the catalyst $HRh(CO)_m[P(Ph)_3]_n$) was particularly useful.

Under suitable reaction conditions, the following Oxo product yields were obtained: with butadiene 80% (ratio mono:dialdehyde = 52:48) and with 1,3-pentadiene 76% (mono:dialdehyde = 32:68). The formation as well as quantity of the reaction products stemming from butadiene are shown in the following reaction scheme:

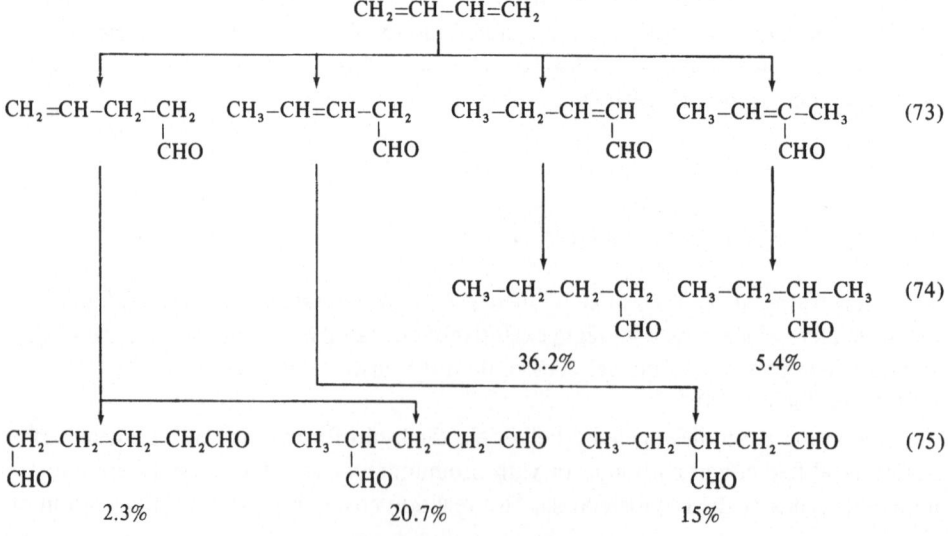

Dialdehydes with formyl groups at the 1,4-position readily undergo an intramolecular aldol addition with subsequent aldol condensation in accordance with Eqs. (76) and (77). Thus, with longer reaction periods, raised temperatures and increasing phosphine concentration, the share of the 1,4-diformyl compounds decreases while the content of the cyclic products increases [1720, 1906].

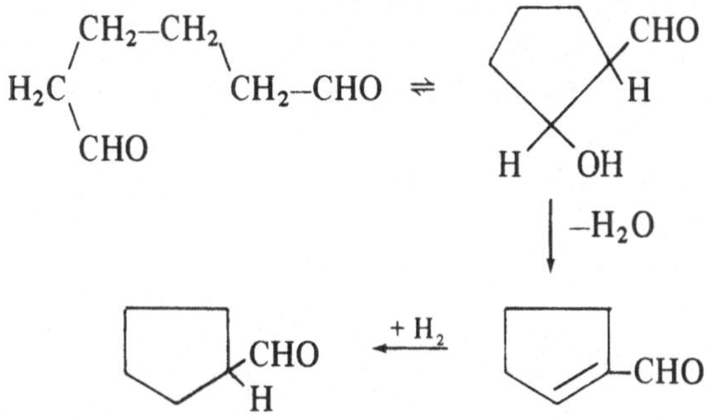

$$
\begin{array}{cc}
\underset{\substack{\nearrow \text{CH}_2-\text{CH}_2 \\ \text{H}_2\text{C} \quad\quad \searrow \\ \searrow \\ \text{CHO}}}{\text{CH}_2-\text{CHO}} \quad \rightleftharpoons \quad &
\end{array} \tag{76}
$$

$$-\text{H}_2\text{O}$$

$$
\overset{+\,\text{H}_2}{\longleftarrow} \tag{77}
$$

Branched 1,3-dienes such as e.g., isoprene, 2,3-dimethylbutadiene and 1,2-dimethylene-cyclohexane can also be bis-hydroformylated with ligand-modified Rh catalysts. However, with isoprene and 2,3-dimethylbutadiene the aldehyde yields are not so high as with the unbranched 1,3-dienes. Moreover, with isoprene the 1,3-diformyl compound is largely converted to the corresponding cyclic aldol addition products on account of the long reaction period [1720].

The effect of the structure of the ligand-modifying phosphines on the reaction rate and the product distribution from the hydroformylation of conjugated dienes has been discussed in the Refs. [621, 622].

1,2-Butadiene is the only allene which has been subject to hydroformylation studies [751]. Rhodium complex catalysts were employed in analogy to the successful reaction of alkynes and conjugated dienes. The conversion produces only moderate yields of mono- and bis-hydroformylation products. As proven by analytical studies, the reaction occurs via an unsaturated aldehyde intermediate. The reaction of polycyclic dienes is discussed in the lit. [1907].

1.4.3 Acetylenes

When Roelen initially conducted the hydroformylation of acetylene [1503] with cobalt catalysts, small amounts of acrolein resulted as reaction product. Later, studies were carried out by Greenfield et al. [1721] on n-1-pentyne, the isomeric mixture n-hexanol/2-methylpentanol being obtained as product. Fell and Beutler showed that on hydroformylating 1-butyne in the presence of Rh_2O_3, even under relatively drastic Oxo conditions, 1-butene formation only took place very slowly. Rapid hydroformylation of 1-butene initially occurs after 1-butyne has been totally converted [751]. In contrast, with phosphine-modified catalysts, the hydroformylation of 1-alkenes (to saturated aldehydes) takes place at relatively low temperatures.

$$R-C \equiv CH \xrightarrow{CO/H_2} R-CH_2-CH_2-CHO \ + \ R-\underset{\underset{CH_3}{|}}{CH}-CHO \tag{78}$$

If the conversion is conducted without ligand excess, then with 1-hexyne, for example, the corresponding aldehydes – 2-heptanal and 2-methylhexanal – are obtained in only 15% yield [1721]. However, with a large excess of phosphine, the reaction takes place quantitatively [170].

This alkyne conversion to saturated aldehydes can be interpreted using two possible reaction paths. For instance, hydrogen could be added to the triple bond followed by hydroformylation of the olefin intermediate:

$$R-C \equiv C-R' \xrightarrow{H_2} R-CH=CH-R' \xrightarrow{CO/H_2} R-\underset{\underset{CHO}{|}}{CH}-CH_2R' + R-CH_2-\underset{\underset{CHO}{|}}{CH}-R' \tag{79}$$

In addition, it is also feasible that the acetylenic compound is hydroformylated forming α/β unsaturated aldehydes which are then hydrogenated to the saturated aldehydes.

$$R-C \equiv C-R' \xrightarrow{CO/H_2}
\begin{array}{l}
\longrightarrow R-\underset{\underset{CHO}{|}}{C}=CH-R' \rightarrow R-\underset{\underset{CHO}{|}}{CH}-CH_2-R' \qquad (80) \\[2em]
\longrightarrow R-CH=\underset{\underset{CHO}{|}}{C}-R' \rightarrow R-CH_2-\underset{\underset{CHO}{|}}{CH}-R' \qquad (81)
\end{array}$$

Based on recent work [751, 1722], it has now been proven that the 2-butyne reaction must ensue via the second route as the unsaturated aldehydes were unequivocally established to be intermediates. However, when 1-butyne was hydroformylated, the intermediate was found to be 1-butene and not the unsaturated aldehyde. Thus with 1-alkynes the first route is the most probable. Other examples are summarized in Table 1.39.

2,4-Hexadiyne – a typical compound possessing two triple bonds – was also hydroformylated. Besides obtaining the expected dialdehydes, other isomeric dials were also produced. The presence of the latter can only be explained via positional isomerism [751] (cf Table 1.39) – a phenomenon which has hitherto not been observed with alkynes.

1.4.4 Functionally Substituted Olefins

1.4.4.1 Unsaturated Alcohols

Unsaturated alcohols, particularly those similar to allyl alcohol, do not exhibit normal behavior when being hydroformylated in the presence of cobalt catalysts [1483, 1679, 1681, 1724].

Table 1.39. Hydroformylation of alkynes and allenes

Olefinic starting material	Catalyst	Temp. (°C)	Pressure (bar)	Product(s)	Yield (%)	Ref.
acetylene	Rh/P(Ph)$_3$ in P(Ph)$_3$	140	210	propanal	98	[751]
1-butyne	Rh/P(Ph)$_3$ in P(Ph)$_3$	130	200	n-pentanal, 2-methyl-butanal, n:iso ratio = 3:1	99	[751]
1-butyne	Rh$_2$O$_3$	180	170	17% n-pentanal 12% 2-methylbutanal 7% n-pentanol 10% 2-methylbutanol		[751]
2-butyne	Rh/P(Ph)$_3$ in P(Ph)$_3$	130	200	2-methylbutanal	75	[751]
1-pentyne	HCo(CO)$_4$	185	200	n-hexanol and 2-methylpentanol	11.5	[1721]
1-hexyne	RhCl[P(Ph)$_3$]$_3$	110	120	n-heptanal and 2-methylhexanal	15	[1723]
1-octyne	HRh(CO)[P(Ph)$_3$]$_3$ in (−)-DIOP	95	85	n-nonanal and (S)-2-methyloctanal (2, 8:1)		[1722]
phenylacetylene	HRh(CO))[P(Ph)$_3$]$_3$ in (−)-DIOP	95	85	3-phenylpropanal and R-2-phenylpropanal (1,6:1)		[1722]
2,4-hexadiyne	HRh(CO))[P(Ph)$_3$]$_3$ in P(Ph)$_3$	150	230	dialdehyde yield 23% isomeric distribution in dial fraction 2-propyl-3-methylbutane-dial 45% 2,3-diethylbutanedial 24% 2-ethyl-4-methylpentan-edial 14% 2-butylbutanedial 17% 2,5-dimethylhexanedial < 1%	23	[751]
1,2-butadiene	HRh(CO)[P(Ph)$_3$]$_3$ in P(Ph)$_3$ Rh:P = 1:50	130	220	n-pentanal 2-methylbutanal } 16% 2,3-dimethylpentane-dial 2-methylpentanedial } 11% 2-ethylbutanedial	27	[751]

For example, allyl alcohol yields less than 30% of expected isomeric hydroxyalde-hydes − 4-hydroxy-1-butanal and 3-hydroxy-2-methyl-1-propanal:

$$HO-CH_2-CH=CH_2 + CO/H_2 \begin{array}{c} \nearrow HO-CH_2-CH_2-CH_2-CHO \\ \\ \searrow HO-CH_2-\underset{\underset{CH_3}{|}}{CH}-CHO \end{array} \qquad (82)$$

Besides forming numerous by-products, allyl alcohol is mainly converted to propion-aldehyde. This is due to the tendency of the cobalt hydridocarbonyls to isomerize the double bond during hydroformylation. The formation of propionaldehyde ensues in accordance with the following Eqs. [1725, 1726]:

$$HO-CH_2-CH=CH_2 \rightarrow HO-CH=CH-CH_3 \rightarrow OHC-CH_2-CH_3 \qquad (83)$$

When the homologue pent-2-en-1-ol [727] is hydroformylated, n-valeraldehyde is almost exclusively obtained along with 2% hydroxyaldehyde. The share of hydroxy-aldehydes increases with distance between double bond and hydroxyl group (Table 1.40):

Table 1.40. Cobalt catalyzed hydroformylations of the n-pentenols [727]

Pentenol	Yield of hydroxyaldehyde fraction[a] (wt.%)	Yield of expected hydroxyaldehydes[a] (wt.%)
2-penten-1-ol	11	2
3-penten-1-ol	15	7
4-penten-1-ol	18	12

[a] calculation based on pentenol feed (yield 100%)

If besides there being a greater distance between the double bond and the hydroxyl group there is also branching present in the connecting carbon chain, then the isomerization is even further suppressed during hydroformylation. Thus, even with cobalt or rhodium catalysts large amounts of formyl alcohols or their hemiacetals result which after hydrogenation yield dialcohols. For example, with tetrahydrobenzyl alcohol and $Co_2(CO)_8$ as catalyst, a 70% yield of the four isomeric bis(hydroxymethyl)-cyclohexanes is obtained [1694]:

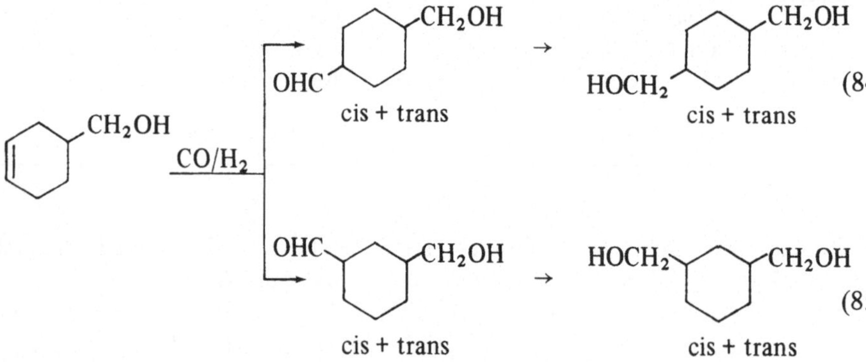

In the same way, the branched unsaturated aliphatic alcohols — 2-methyl-3-buten-2-ol and 3-methyl-3-buten-1-ol — can be converted in good yield to formyl alcohols or their cyclic hemiacetals [1727]:

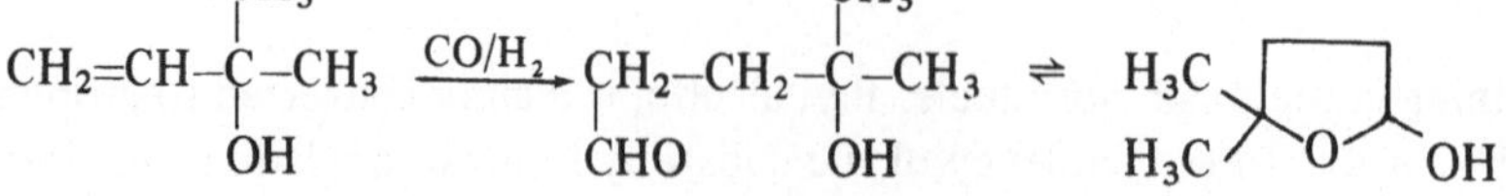

$$\underset{\substack{| \\ \text{CH}_2=\text{C}-\text{CH}_2-\text{CH}_2\text{OH}}}{\overset{\text{CH}_3}{}} \xrightarrow{\text{CO/H}_2} \underset{\substack{| \\ \text{CH}_2-\text{CH}-\text{CH}_2-\text{CH}_2\text{OH} \\ \text{CHO}}}{\overset{\text{CH}_3}{}} \rightleftharpoons \quad (87)$$

Under hydroformylation conditions, rhodium carbonyls usually have a more powerful isomerization effect on the double bonds compared to the cobalt carbonyls. However, this isomerization is not quite so marked when hydroformylating unsaturated alcohols. It was confirmed with rhodium catalysts that when the distance between the OH group and the double bond is increased, the degree of isomerization falls — causing an increase in yield of Oxo products [727]. If the rhodium catalyst is modified with triphenyl-phosphine, then the double bond isomerization can be virtually suppressed on increasing the phosphine concentration. This can be readily appreciated from the increase in yield of Oxo products from allyl alcohol [727] (cf Fig. 1.59).

Isomeric hydroxyaldehydes from the hydroformylation normally react either inter- or intra-molecularly producing cyclic acetals. The γ-hydroxyaldehydes yield 2-hydroxy-tetrahydrofuran derivatives and the β-hydroxyaldehydes give rise to m-dioxan derivatives [727, 1864]. With modified Rh catalysts, crotyl alcohol yields 2-hydroxy-3-methyltetra-hydrofuran as well as a highly viscous m-dioxan derivative.

$$\text{CH}_3-\text{CH}=\text{CH}-\text{CH}_2\text{OH} \xrightarrow[\text{HRhCO}[\text{P}(\text{Ph})_3]_3]{\text{CO/H}_2} \left[\begin{array}{l} \rightarrow \underset{\substack{| \\ \text{CH}_3-\text{CH}-\text{CH}_2-\text{CH}_2\text{OH} \\ \text{CHO}}}{} \rightleftharpoons \\ 85\% \text{ Yield} \\ \rightarrow \underset{\substack{| \\ \text{CH}_3-\text{CH}_2-\text{CH}-\text{CH}_2\text{OH} \\ \text{CHO}}}{} \end{array} \right. \quad (88)$$

$$\underset{\substack{| \\ \text{CH}_3-\text{CH}_2-\text{CH} \\ \text{CH}_2\text{OH}}}{} \overset{O}{\underset{O}{\big<\quad\big>}} -\text{CH}_2\text{CH}_3 \quad (89)$$
$$\text{OH}$$

As the hydroxyaldehydes and their hemiacetals react further, the hydroformylation of unsaturated alcohols usually leads to a large number of products making purification of individual compounds difficult. Table 1.41 gives the yields of the reactions producing hydroxyaldehydes along with the hemiacetals — the typical reaction products (cf p. 111).

1.4.4.2 Unsaturated Aldehydes and Ketones

Conjugated unsaturated aldehydes are quantitavely hydrogenated to saturated aldehydes during the cobalt catalyzed hydroformylation.

A series of unsaturated aldehydes were reacted under standard conditions using stoichiometrical amounts of cobalt hydridocarbonyl. The results of this work conducted by Orchin et al. [1731] are shown in Table 1.42.

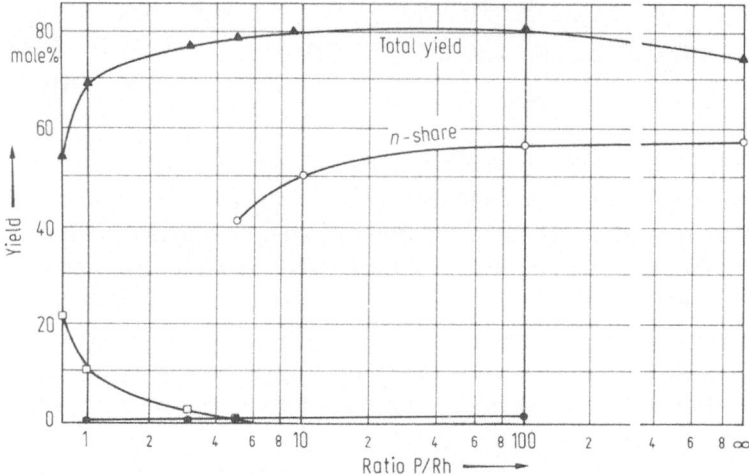

Fig. 1.59. Hydroformylation of allyl alcohol — relationship between yield and P:Rh ratio (110 °C, 200 bar, $CO:H_2 = 1:1$)

▲ total yield   □ propionaldehyde content of product
○ n-share   ● share of C_4 aldehyde in product

Table 1.42. Reaction of conjugated unsaturated aldehydes with cobalt catalysts [1731, 1732]

Starting material	Pressure (bar)	Temp. (°C)	Products	Yield (%)
acrolein	1	25	propionaldehyde	93
crotonaldehyde	1	25	butyraldehyde	80
α-methylcrotonaldehyde	1	25	α-methylbutyraldehyde	15
cinnamic aldehyde	1	25	3-phenylpropionaldehyde	97
2-hexenal			hexanal	

In these cases only hydrogenation took place. According to M. Orchin et al. this unsuccessful hydroformylation attempt was due to the formation an intermediate π-allyl analogue complex which then reacted further.

In analogy to conjugated unsaturated aldehydes, the unsaturated ketones produce saturated ketones [80] (Table 1.43).
The conjugated unsaturated aldehydes and ketones can be indirectly hydroformylated in good yield via their acetals (cf Sect. 1.4.4.4).

If the double bond is not conjugated to a carbonyl oxygen atom then the dialdehydes can be obtained in good yield when isomerization to the α/β-unsaturated aldehyde is hindered. With readily isomerizable unsaturated aldehydes possessing isolated double bonds, besides dialdehyde formation, there is also increased hydrogenation to the monoaldehyde. Even on using modified Rh catalysts with β/α unsaturated pent-3-en-1-al only 30 to 36% is hydroformylated to the dialdehyde while with γ/δ unsaturated pent-4-en-1-al the figure is 78% [621]. With δ/ε unsaturated pent-5-en-1-al, rhodium catalysts lead to the exclusive formation of the dialdehydes. On the other hand, with

Table 1.41. Hydroformylation of unsaturated alcohols

Unsaturated alcohol	Catalyst	Press. (bar)	Temp. (°C)	Reaction products	Hydroxy-aldehyde yield (%)	Ref.
allyl alcohol	Co	150	100	$CH_2\text{–}CH_2\text{–}CH_2OH$ (–CHO) $\rightleftharpoons$ (ring, OH)		[727, 1860, 1963]
				$+$ $2\ CH_3\text{–}CH\text{–}CH_2OH$ (–CHO) $\rightleftharpoons$ $CH_3\text{–}CH$ (ring, CH_3, OH, CH_2OH)	17	[1917]
allyl alcohol	Rh, Rh/P(Ph)$_3$	200	110	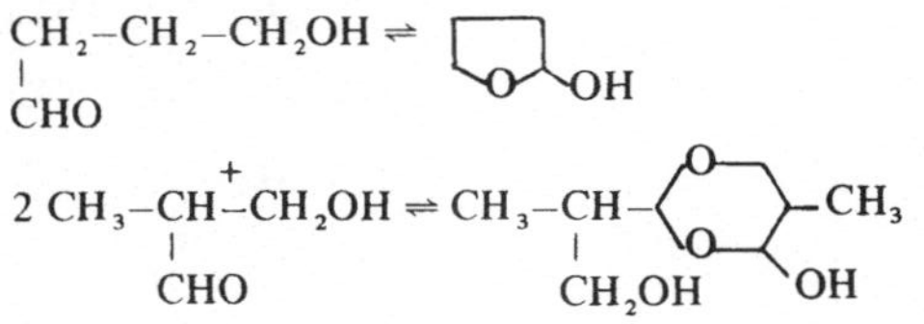 $CH_2\text{–}CH_2\text{–}CH_2OH$ (–CHO) $\rightleftharpoons$ (ring, OH) $+$ $2\ CH_3\text{–}CH\text{–}CH_2OH$ (–CHO) $\rightleftharpoons$ $CH_3\text{–}CH$ (ring, CH_3, OH, CH_2OH)		[837, 1727, 1728, 1859, 1861, 1879]
crotyl alcohol	Rh	200	110	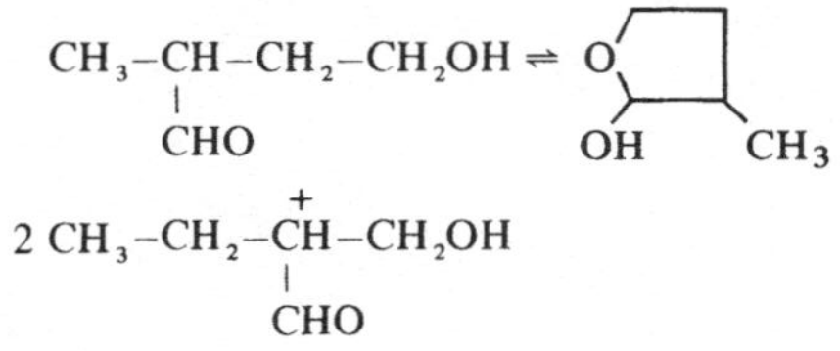 $CH_3\text{–}CH\text{–}CH_2\text{–}CH_2OH$ (–CHO) $\rightleftharpoons$ (ring, OH, CH_3) $+$ $2\ CH_3\text{–}CH_2\text{–}CH\text{–}CH_2OH$ (–CHO)	60	[727]
	Rh/P(Ph)$_3$	200	110	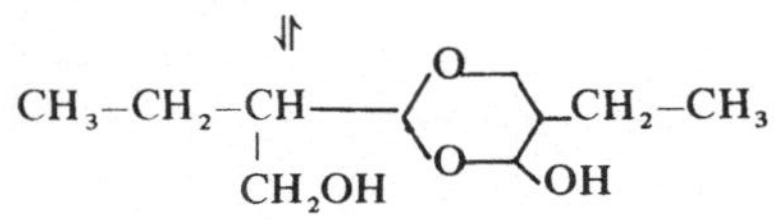 $CH_3\text{–}CH_2\text{–}CH$ (ring, $CH_2\text{–}CH_3$, OH, CH_2OH) $\rightleftharpoons$	85	[727]

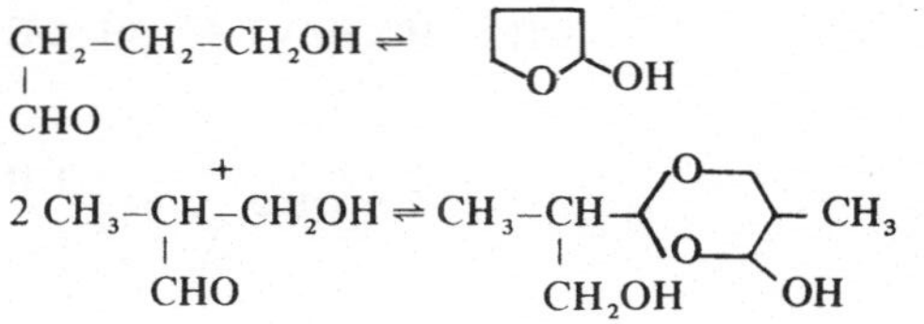

111

Table 1.41 (continued)

Unsaturated alcohol	Catalyst	Press. (bar)	Temp. (°C)	Reaction products	Hydroxy-aldehyde yield (%)	Ref.
pent-2-en-1-ol	Rh, Rh/P(Ph)$_3$	200	110	corresponding hydroxyaldehydes	80	[727]
pent-3-en-1-ol	Rh, Rh/P(Ph)$_3$	200	110		83	[727]
pent-4-en-1-ol	Rh, Rh/P(Ph)$_3$	200	110		85	[727]
3-methylbut-2-en-1-ol	Rh	200	110	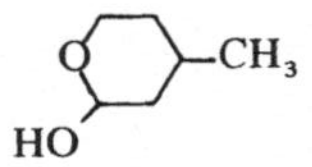 (tetrahydropyran ring with CH$_3$; HO)	30	[727]
3-methylbut-2-en-1-ol	Rh/P(Ph)$_3$	200	110	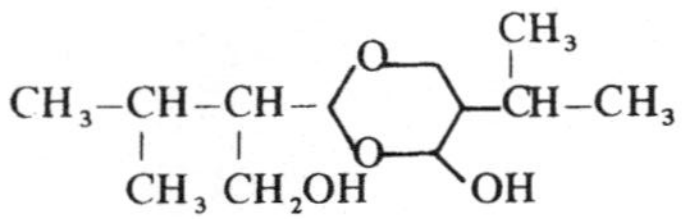 CH_3-CH-CH- ; CH_3 CH_2OH ; (dioxane ring: CH_3, CH-CH_3, OH)	10	[727]
2-methylbut-3-en-2-ol	Co, Rh	200	110	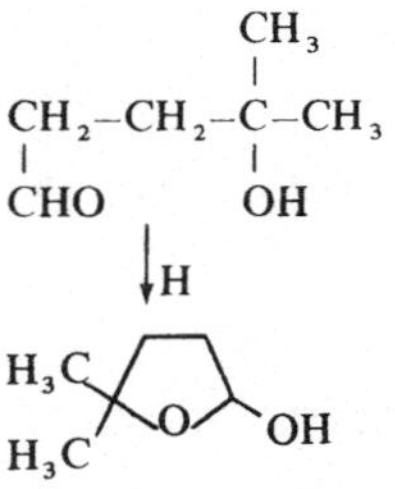 CH_2-CH_2-C-CH_3 ; CHO ; CH_3 $-$ OH $\longrightarrow$ (furan ring with OH, H_3C, H_3C)	> 80	[727]
3-methylbut-3-en-1-ol	Rh	200	110	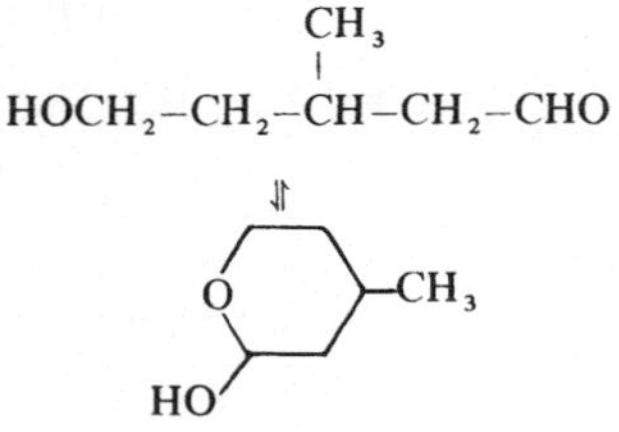 $HOCH_2$-CH_2-CH-CH_2-CHO ; CH_3 $\rightleftharpoons$ (tetrahydropyran ring with CH$_3$; HO)	70	[727]
3-methylbut-3-en-1-ol	Rh/P(Ph)$_3$	200	110		82	[837]
	Rh/P(Ph)$_3$	250	80		88	

112

Reactant	Catalyst			Product	Yield	Ref.
alkylhept-1-en-6-ols alkyl = 2,6-dimethyl- = 2,3,6-trimethyl-	Rh Rh-diolefin complex	700	100	3,7-dimethyloctan-1-al-7-ol 3,4,7-trimethyloctan-1-al-7-ol	90 70	[499] [499]
= 2-methyl- alkyloct-1-en-6-ols = 2,6-dimethyl- = 2,6,7-trimethyl-				3-methyloctan-1-al-7-ol 3,7-dimethylnonan-1-al-7-ol 3,7,8-trimethylnonan-1-al-7-ol	73 85 55	[499] [499] [499]
2-ethylhex-2-en-1-ol	Rh/P(Ph)$_3$	50	70	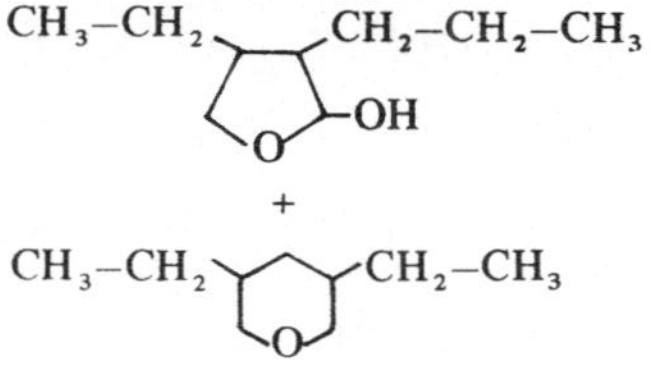 CH$_3$–CH$_2$ ⟨ring⟩ CH$_2$–CH$_2$–CH$_3$ / OH + CH$_3$–CH$_2$ ⟨ring⟩ CH$_2$–CH$_3$	80	[733]
CH$_2$=C–CH$_2$OH, R	RhCl(CO)P(Ph)$_3$	80	80	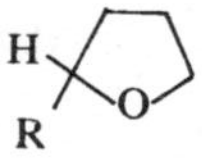 H ⟨ring⟩ R (after treatment)		[1728]
R = CH$_3$ = iso–C$_3$H$_7$ = C$_4$H$_9$ = sec–C$_4$H$_9$					85 80 89 87	
Isophytol CH$_2$=CH–C(CH$_3$)(OH)–iso–C$_{16}$H$_{33}$	Rh	200	110	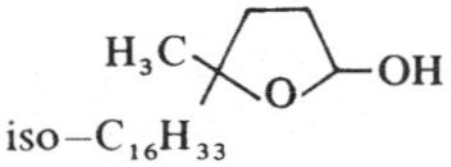 H$_3$C ⟨ring⟩ OH, iso–C$_{16}$H$_{33}$	> 80	[727]

Table 1.41 (continued)

Unsaturated alcohol	Catalyst	Press. (bar)	Temp. (°C)	Reaction products	Hydroxy-aldehyde yield (%)	Ref.
$HOCH_2-CH=CH-CH_2OH$	Rh/P(Ph)$_3$			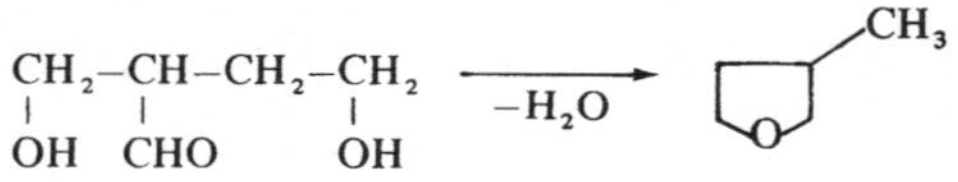 $\underset{OH}{CH_2}-\underset{CHO}{CH}-CH_2-\underset{OH}{CH_2} \xrightarrow{-H_2O}$ (methyl-furanyl)		[1729]
$HOCH_2-CH=\underset{CH_3}{C}-CH_2-CH_2OH$ +	Rh/amine			$\begin{array}{l} CH_2-CH_2OH \\ CH-CH_2-CH_2OH \\ CH_2-CH_2OH \end{array}$ + $\begin{array}{l} CH_2-CH_2OH \\ CH_3-CH \\ CH_2-CH_2OH \end{array}$		[786]
$HOCH_2-CH_2-\underset{\overset{\|}{CH_2}}{C}-CH_2-CH_2OH$						
3,7-dimethyloct-1-en-7-ol	Rh/P(Ph)$_3$	85	100	4,8-dimethylnonan-1-al-8-ol	72	[739]
(2-methyltetrahydropyran-CH$_2$OH)	Co			$HOCH_2$—(tetrahydropyranyl)—CH_2OH		[1730]
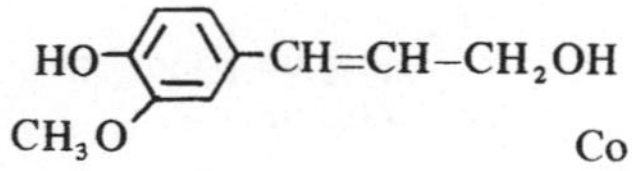 HO—(CH$_3$O-phenyl)—CH=CH—CH$_2$OH	Co	100	170	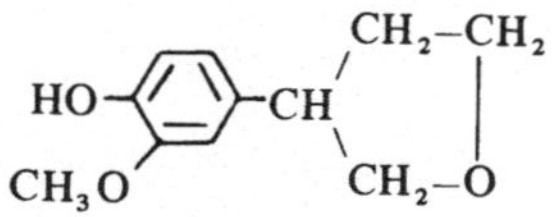 (dioxane-substituted aromatic, after treatment with H$_2$)	24	[1864]

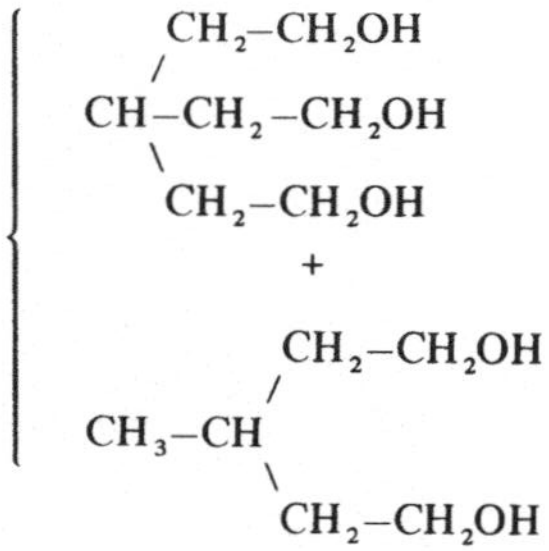

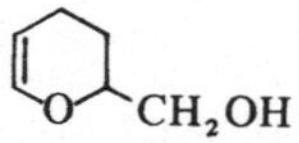

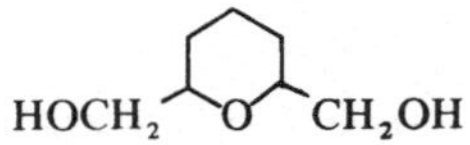

114

Table 1.43. Hydroformylation of conjugated unsaturated ketones

Starting material	Ketone: HCo(CO)$_4$	Products	Yield (%)	Ref.
CH$_2$=CH–CO–CH$_3$	1:1	CH$_3$–CH$_2$–CO–CH$_3$		[1483, 1528]
(CH$_3$)$_2$C=CH–CO–CH$_3$		(CH$_3$)$_2$CH–CH$_2$–CO–CH$_3$		[1483]
CH$_2$=CH–CO–CH$_3$	1:1	CH$_3$–CH$_2$–CO–CH$_3$	70	[1731]
C$_6$H$_5$CH=CH–CO–CH$_3$	1:1	C$_6$H$_5$–CH$_2$–CH$_2$–CO–CH$_3$	56	[1731]
(CH$_3$)$_2$C=CH–CO–CH$_3$	1:1	(CH$_3$)$_2$CH–CH$_2$–CO–CH$_3$	24	[1731]

C$_2$H$_5$–CH=C–CO–CH$_3$
        |
       CH$_3$
		C$_2$H$_5$–CH$_2$–CH–CO–CH$_3$
		          |
		         CH$_3$
				[1528]

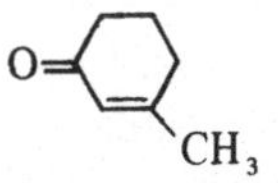		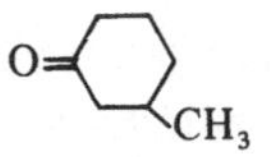		[1528]

cobalt hydrodocarbonyl, which encourages isomerization, only 38% of the reaction products [1733] were based on the dialdehydes.

The formation of the dialdehydes is favored when the carbon chain of the aldehydic starting material exhibits branching between the carbonyl group and the double bond. Tetrahydrobenzaldehyde can thus be converted with a rhodium catalyst to the dialdehydes in high yields [1694]. At higher temperatures the diols are obtained.

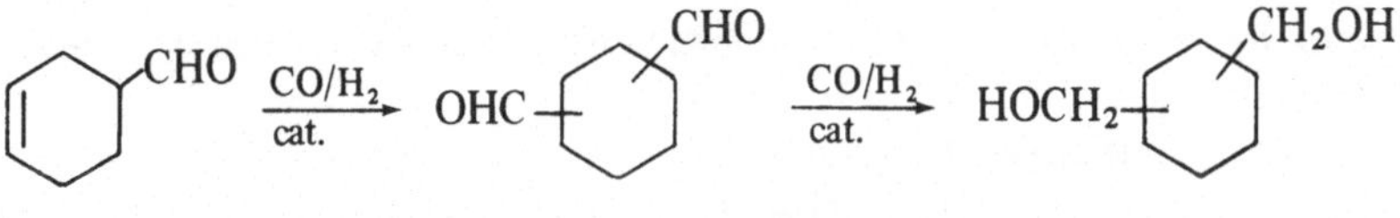

(90)

The hydroformylation of the mixture consisting of endo-endo-bicycloheptenaldehyde was studied in detail [1734]. This cyclic compound can be prepared via the Diels-Alder reaction between acrolein and cyclopentadiene. The primary resulting dialdehydes can be converted into the more stable dialcohols. In the rhodium catalyzed reaction, the four possible stereoisomeric dialcohols are produced in 71% yield [1509, 1896]. According to Stockhausen's work [1735], the possible formation of endo-endo-diformyl compounds can be excluded.

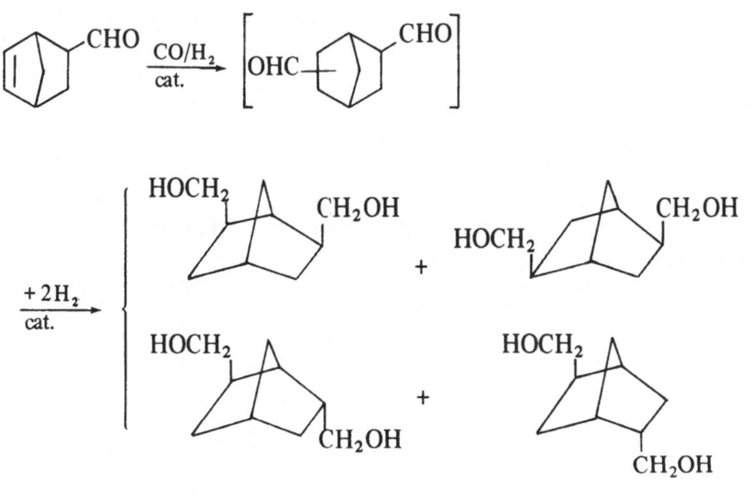

(91)

115

Even unsaturated aldehydes with the pyran structure – such as 2-formyl-3,4-dihydro-2-H-pyran from the dimerization of acrolein – can be readily converted into di-formyltetrahydropyrans using a rhodium catalyst [1736].

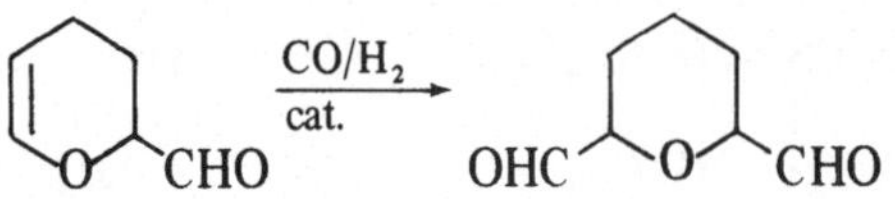

$$\text{(92)}$$

The tendency of the double bond to isomerize to α/β unsaturated aldehydes mainly determines the composition of the hydroformylation products. Table 1.44 shows that unsaturated aldehydes with double bonds which readily isomerize to α/β unsaturated aldehydes, can only be successfully hydroformylated to dialdehydes with modified Rh catalysts. Unsaturated aldehydes with double bonds not or only slightly capable of isomerization, form dialdehydes even with unmodified cobalt or rhodium catalysts.

1.4.4.3 Unsaturated esters

It must be differentiated between two types of unsaturated esters – those based on unsaturated acids and those based on unsaturated alcohols. The first type produces mainly formylcarboxylic acids on being hydroformylated. After hydrogenating the aldehyde group, hydroxycarboxylates result which usually react to form lactones.

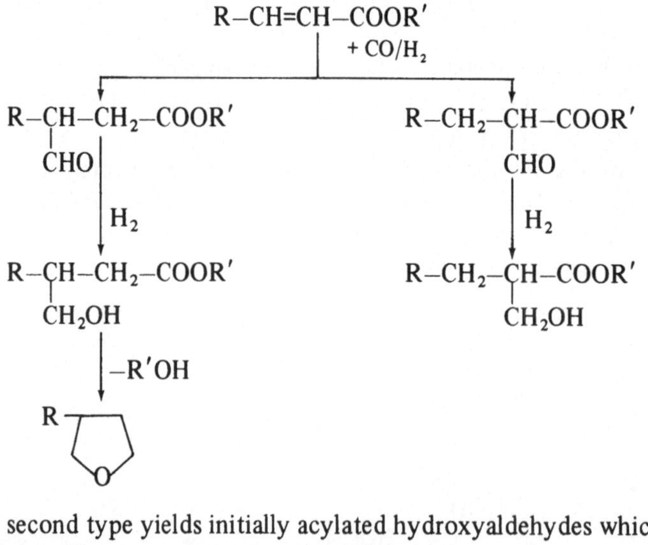

$$R-CH=CH-COOR' \quad \text{(93)}$$

$$R-\underset{|}{\overset{}{CH}}-CH_2-COOR' \qquad R-CH_2-\underset{|}{\overset{}{CH}}-COOR'$$
$$\quad CHO \qquad\qquad\qquad CHO \qquad\text{(94)}$$

$$R-\underset{|}{\overset{}{CH}}-CH_2-COOR' \qquad R-CH_2-\underset{|}{\overset{}{CH}}-COOR'$$
$$\quad CH_2OH \qquad\qquad\qquad CH_2OH$$

$$-R'OH \qquad\qquad\qquad\qquad\qquad\qquad \text{(95)}$$

The second type yields initially acylated hydroxyaldehydes which, after hydrogenation and saponification, can be converted into diols.

Table 1.44. Hydroformylation of unsaturated aldehydes with an isolated double bond (analysis of reaction products after subsequent hydrogenation)

Starting material	Catalyst	Press. (bar)	Temp. (°C)	Products	Yield (%)	Ref.
pent-3-en-1-al	Rh_2O_3 + $P(Ph)_3$	220	130	*C$_5$-Monooxo content*	45	[621]
	(Rh_2O_3 + $PO(Et)_3$)	220	130	8% 2-methylbutanol 92% 1-pentanol		
				C$_6$-Dioxo content 27% 2-ethyl-1,4-butanediol 71% 2-methyl-1,5-pentanediol 2% hexanediol	30	
pent-4-en-1-al	Rh_2O_3 + $P(Ph)_3$	200	135	*C$_5$-Monooxo content* *C$_6$-Dioxo content* 64% hexanediol 33% 2-ethyl-1,4-butanediol 3% undetermined products	– 78	[621]
pent-5-en-1-al	Rh_2O_3 + $P(Ph)_3$	200	135	*Dialdehydes* 54% hexanedial 32% 2-ethyl-pentanedial 2% 2-formylcyclopentanol[a] 3% undetermined products	100	[1733]
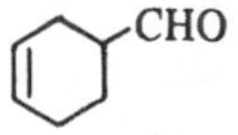 CHO	Co on kiesel-guhr	200	180			[1737]
	$Co_2(CO)_8$	130–240	200–300	hexahydrobenzyl alcohol bis(hydroxymethyl)-cyclohexane	18 52	[1694]
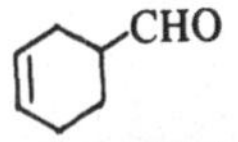 CHO	Rh_2O_3	120–200	200–300	hexahydrobenzaldehyde bis(hydroxymethyl)-cyclohexane	3 80	[1694]
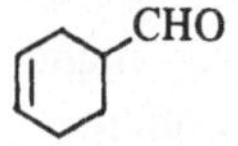 CHO	Rh_2O_3	250	100	*Composition of Oxo products* 2% monoaldehyde 71% dialdehyde 14% hydroxyaldehyde 13% heavy ends		[535]
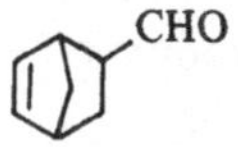 CHO	Rh_2O_3	200–300	130–240	bis(hydroxymethyl)-bicycloheptane	71	[1509]
CHO						[1738]
CHO						[1738]
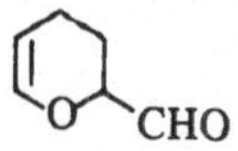 CHO	Rh_2O_3	240–450	130–150	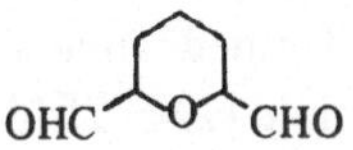	ca. 70	[1736]

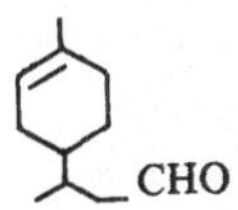

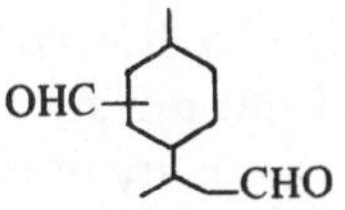

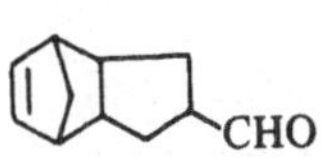

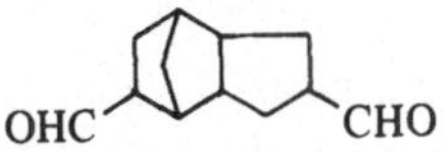

[a] result from 2-ethylpentanedial

Other hydroformylations are reported elsewhere [1897].

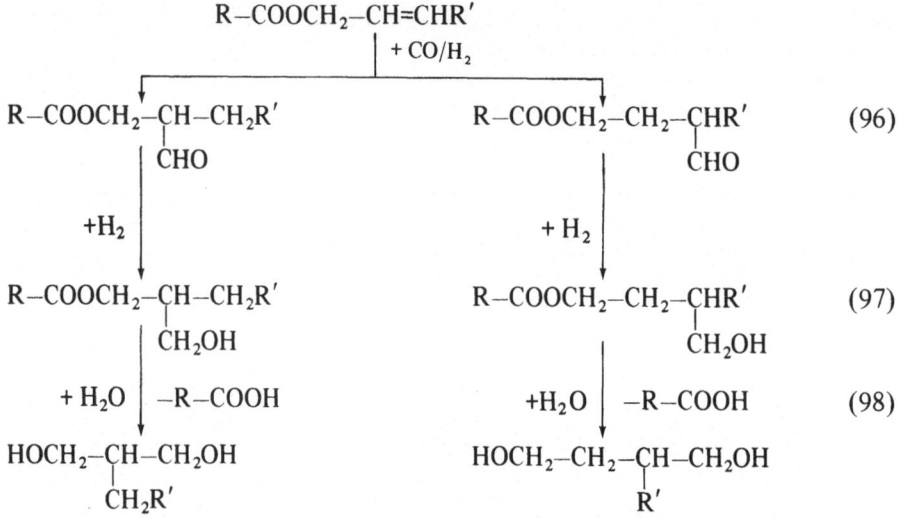

$$R-COOCH_2-CH=CHR'$$

with + CO/H₂ branching to:

$$R-COOCH_2-CH-CH_2R' \quad | \quad CHO \qquad\qquad R-COOCH_2-CH_2-CHR' \quad | \quad CHO \qquad (96)$$

$$+H_2 \qquad\qquad\qquad +H_2$$

$$R-COOCH_2-CH-CH_2R' \quad | \quad CH_2OH \qquad\qquad R-COOCH_2-CH_2-CHR' \quad | \quad CH_2OH \qquad (97)$$

$$+H_2O \quad | \quad -R-COOH \qquad\qquad +H_2O \quad | \quad -R-COOH \qquad (98)$$

$$HOCH_2-CH-CH_2OH \quad | \quad CH_2R' \qquad\qquad HOCH_2-CH_2-CH-CH_2OH \quad | \quad R'$$

The conjugation between the double bond and the carbonyl group is weaker in α,β-unsaturated esters compared to the corresponding unsaturated aldehydes (resonance energy of crotonaldehyde 2.4 kcal/mol higher than ethyl crotonate) [1739]. Consequently, the esters of α,β-unsaturated alcohols (in contrast to the α,β-unsaturated aldehydes) give good yields on being hydroformylated in the presence of cobalt catalysts. However, hydroformylation is more difficult with esters in which the double bond of the unsaturated acid is also in conjugation with another unsaturated system e.g., esters of cinnamic acid, maleic acid or fumaric acids. In these cases, the hydrogenation of the double bond dominates, causing a decrease in yield of the hydroformylation products (cf Table 1.45). For example, only 8% of cinnamic acid ester was hydroformylated, while 91% was hydrogenated [241].

If rhodium is used as catalyst instead of cobalt, then the hydrogenation ensues to a lesser extent due to the lower quantities of metal employed leading to generally higher yields of Oxo products [1694]. This effect can be fortified by modifying the rhodium catalyst with tert. phosphines (cf Table 1.46).

Besides suppressing side reactions, modified rhodium catalysts also lead to a more selective formation of a particular isomer of the possible hydroformylation products.

Selective formation can be achieved when hydroformylating esters of α,β-unsaturated acids. As there is an more uneven distribution of electron density at the carbon atoms of the double bond of this acid (the α-C atom possessing the greater negative charge) compared to the olefins, the possibility of introducing the formyl group at a particular C atom is presented.

A particular isomer can be synthesized via an appropriate choice of pressure and temperature as well as via choice of catalyst. At low temperatures and high pressures, α-formyl derivates are preferentially formed while at high temperatures and low pressures β-formyl compounds dominate [241, 251] (cf Sects. 1.3.1 and 1.3.2).

The choice of catalyst presents the possibility of selectively preparing a certain isomer. This is, for example, the case when hydroformylating methacrylic acid esters (MMA) with Rh. While the cobalt catalyzed hydroformylation of MMA always yields mixtures consisting mainly of β-formyl products, modified Rh catalysts enable the

Table 1.45. Cobalt catalyzed hydroformylation of esters of unsaturated acids

Starting material	Syn gas mixture CO:H$_2$	Pressure (bar)	Temp. (°C)	Products	Yield (%)	Ref.
methyl acrylate	1:1	200	120	methyl β-formyl-propionate	85	[1740, 1889]
ethyl acrylate	1:1	200–300	120–125	ethyl α- and β-formylpropionate	75	[312, 1483, 1741]
ethyl crotonate	1:1	200–300	120–125	ethyl α-, β- and γ-formylbutyrate	71	[1483, 1741]
butyl crotonate	1:1	145–210	140	butyl β-formyl-butyrate	78	[1742]
methyl undecylenate	1:1	200–300	120–125	methyl formyl-undecanoate	71	[1483]
diethyl fumarate	1:1	200–300	120–125	diethyl formyl-succinate	51	[313, 1743]
diethyl maleate	1:1	183–218	140–155	diethyl formyl-succinate	65	[313, 1742]
diethyl itaconate	1:1	140–210	140	diethyl formyl-methylsuccinate	56	[1742]
methyl oleate	1:2	600–750	140–145	mixture of aldehyde-esters	72	[294, 1682, 1888, 1890]
esters of 3-pentenoic acid	1:1	250	200	β-, γ- and δ-formyl-pentanoic acid methyl ester (β:γ:δ = 1:2, 4:5)	80	[1744] 1892]
esters of 4-pentenoic acid	1:1	100–800	150–350	δ-caprolactone		[1745]

α-formyl product to be obtained in 94% yield. The β-isomer can be prepared in 80–90% yield with rhodium alone or on adding pyridine [1694, 1880].

Ligand-modified Rh catalysts facilitate a bis-hydroformylation of double unsaturated esters to form a diformyl ester. This bis-hydroformylation has been conducted with linoleic acid [608]. So far there have only been unsuccessful attempts at similarly reacting the two conjugated double bonds of an ester of sorbic acid to produce the diformyl esters. The reaction led — as with unmodified Co catalysts — to the formation of monoformyl esters [1751].

The (primary resulting) formylcarboxylic acid esters from the hydroformylation of unsaturated esters can be hydrogenated to the hydroxycarboxylic acid esters [1694]. The β- and γ-hydroxy esters can be readily transformed into the γ- and δ-lactones resp. It is, however, not necessary when recovering these lactones to hydrogenate the formyl-carboxylic acid esters in a special step as lactone formation can ensue in a single stage under hydroformylating conditions. It is merely essential to raise the temperature to 200–240 °C at the end of the hydroformylation reaction, causing the hydroformyla-tion catalyst (Co or Rh carbonyls) to act as hydrogenation catalyst thereby converting the aldehyde group into the hydroxy group via homogeneous catalysis. Thus, in the presence of acidic cobalt hydridocarbonyl, the γ- for δ-lactones are obtained in high yield from the hydroxy esters (cf Falbe [63] and e.g., [1880, 1883]).

Table 1.46. Rh catalyzed hydroformylation of esters of unsaturated acids

Starting material	Catalyst	Pressure (bar)	Temp. (°C)	Products	Ratio of isomers	Yield (%)	Ref.
methyl acrylate	Rh	280	110	a) methyl α-formylpropionate b) methyl β-formylpropionate	a:b = 5,7:1	77	[159]
methyl acrylate	Rh/P(Ph)₃ (mol ratio 1:10)	280	110	a) methyl α-formylpropionate b) methyl β-formylpropionate	a:b = 50:1	93	[159]
methyl methacrylate	Rh	250	70	a) methyl α-formylisobutyrate b) methyl β-formylisobutyrate	a:b = 4,4:1	90.6	[884, 1694, 1882]
methyl methacrylate	Rh	250	130	a) methyl α-formylisobutyrate b) methyl β-formylisobutyrate	1:4,1	85	[884]
methyl methacrylate	Rh/pyridine	270	165	a) methyl α-formylisobutyrate b) methyl β-formylisobutyrate	1:10	95	[685]
methyl methacrylate	Rh/P(Ph)₃	100	110	methyl α-formylisobutyrate		94	[1694]
ethyl crotonate	Rh	200	135	a) α-formyl b) β-formyl } isomers c) γ-formyl	a:b:c = 1:22:3	72	[1694]
ethyl cinnamate	Rh	200	120	β-formyl product (β-phenyl-γ-butyrolacton)		73	[1694]
methyl oleate	Rh/P(Ph)₃	180	110	methyl 9(10)-formylstearate	1:1	90−99	[608, 757, 887, 1891]
methyl linolate	Rh/P(Ph)₃	250	190	17.7% methyl monoformylstearate 71.4% methyl bisformylstearate		99	[608]

Esters with unsaturated alcohol components are — except for vinyl esters — usually smoothly hydroformylated. The resulting acylated hydroxyaldehydes [cf Eqs. (96–98)] present a route to the industrial production of the commercially interesting 1,3- and 1,4-diols. The manufacture of 1,4-butanediol from allyl acetate is particularly important. Allyl acetate itself can be readily produced from propylene, acetic acid, and oxygen [1754, 1755, 1884].

References/Equ.

$$CH_3-COOH + CH_3-CH = CH_2$$
$$\downarrow O_2$$
$$CH_3-COO-CH_2-CH = CH_2$$
$$\downarrow CO/H_2 \qquad [1756, 1757,\ 1887]$$
$$CH_3-COO-CH_2-CH_2-CH_2-CHO$$
4-acetoxybutan-1-al
$$+$$
$$CH_3-COO-CH_2-\underset{\underset{CH_3}{|}}{CH}-CHO$$
3-acetoxy-2-methylpropan-1-al
$$+$$
$$CH_3-COO-\underset{\underset{CH_2-CH_3}{|}}{CH}-CHO$$
2-acetoxy-2-ethylacetaldehyde

$$\downarrow H_2 \qquad [1746] \qquad (99)$$

$$CH_3-COO-CH_2-CH_2-CH_2-CH_2OH$$
$$+$$
$$CH_3-COO-CH_2-\underset{\underset{CH_3}{|}}{CH}-CH_2OH$$
$$+$$
$$CH_3-COO-\underset{\underset{CH_2-CH_3}{|}}{CH}-CH_2OH$$

$$(-CH_3COOH) \quad \Big| +H_2O \qquad [1747, 1748,\ 1887]$$

$$HOCH_2-CH_2-CH_2-CH_2OH$$
$$+$$
$$HOCH_2-\underset{\underset{CH_3}{|}}{CH}-CH_2OH$$
$$+$$
$$HOCH_2-CHOH-CH_2-CH_3$$

The hydroformylation of allyl acetate has been intensively studied in order to increase the share of the unbranched 4-acetoxybutan-1-al and thus to improve the yield of the desired 1,4-butanediol.

The undesired by-products — 3-acetoxy-2-methylpropionaldehyde and 2-acetoxy-butyraldehyde — can be cleaved — dehydroformylated — leading to the formation of, in the main, allyl acetate and 1-propenyl acetate [1749].

$$CH_3-COO-\underset{\underset{CHO}{|}}{CH}-CH_2-CH_3 \xrightarrow{-CO/H_2} CH_3-COO-CH=CH-CH_3 \qquad (100)$$

$$CH_3-COO-CH_2-\underset{\underset{CH_3}{|}}{CH}-CHO \xrightarrow{-CO/H_2} \begin{array}{l} CH_3-COO-CH=CH-CH_3 \\ CH_3-COO-CH_2-CH=CH_2 \\ CH_2=\underset{\underset{CH_3}{|}}{C}-CHO + CH_3COOH \end{array} \qquad (101)$$

As in the Co catalyzed hydroformylation, 1-propenyl acetate and allyl acetate give rise to virtually only one Oxo product (with the same composition), 1-propenyl acetate and allyl acetate (from the cleavage) can be reintroduced to the hydroformylation. Consequently, only 4-acetoxybutyraldehyde – the 1,4-butanediol precursor – is obtained as product [1749].

Cracking of an undesired by-product from the Oxo synthesis via dehydroformylation and reintroduction of recovered olefin was conducted for the first time by Falbe et al. using isobutyraldehyde from the propylene hydroformylation (cf Sect. 1.5.2.3.2). The hydroformylation of bis acyloxy propene is discussed in [1881, 1886].

With the allyl esters, a study was made of the effect of acid components of varying alkyl chain length on the addition of the formyl group to the double bond. It was found, with increasing chain length the yield of 4-acyloxybutanals decreased while that of the 2-acyloxybutanals increased [244].

The hydroformylation of the vinyl esters – using vinyl acetate as an example – leads to the formation of the two isomeric α- and β-acyloxypropionaldehydes (31) [1752, 1820].

$$CH_3-COOCH=CH_2 \xrightarrow{CO/H_2} CH_3-COO\underset{\underset{CH_3}{|}}{CH}-CHO + CH_3-COOCH_2-CH_2-CHO \quad (102)$$

The α-isomer dominates particularly when using Rh catalysts [1752, 1753] (cf Table 1.47). The high reactivity of the acyloxypropionaldehydes – cleavage to acetic acid and acrolein [Eq. (101)] – markedly affects the formation and isolation of the pure acyloxypropionaldehydes.

The best yield of acyloxypropionaldehydes (83%, α:β-ratio 95:5) was achieved with a rhodium carbonyl/N-methylpyrrolidine catalyst system at 60 °C and 200 bar [1753].

Table 1.47. Hydroformylation of vinyl acetate with various catalysts [1752, 1753]

Catalyst (precursor)	Temp. (°C)	Products a) = α-acetoxypropionaldehyde b) = β-acetoxypropionaldehyde	Yield (%)
Co$_2$(CO)$_8$	120	a)	26
		b)	23
Rh$_3$(CO)$_{12}$	80	a)	72
		b)	< 2
HRh(CO)$_4$/N-pyrrolidine/H$_2$O	60	a)	79
		b)	4–5

Literature on the hydroformylation of unsaturated acids can be found in Ref. [1893, 1894]. In this case resulting saturated carboxylic acids (byproducts) can act as catalyst poisons (cf Sect. 1.3.3.2).

1.4.4.4 Unsaturated Ethers and Acetals

When unsaturated ethers are hydroformylated under standard conditions, the formyl group preferentially bonds to the olefinic carbon atom nearest the ether bridge. Therefore, the main products from the hydroformylation of butyl vinyl ether and ethyl allyl ether are α-n-butoxypropionaldehyde and β-ethoxybutyraldehyde, respectively [1483].

$$CH_2=CH-O-C_4H_9 \quad \longrightarrow CH_3-\underset{\underset{CHO}{|}}{CH}-O-C_4H_9 \tag{103}$$

$$CH_2=CH-CH_2-O-C_2H_5 \longrightarrow CH_3-\underset{\underset{CHO}{|}}{CH}-CH_2-O-C_2H_5 \tag{104}$$

With dihydropyrans, the addition of the formyl group to the C atom 2 is also strongly favored [1730, 1736, 1759–1762, 1900, 1901].

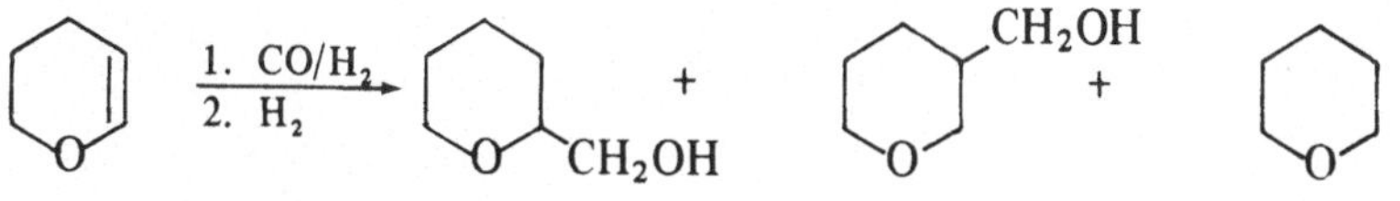

(105)

    78%   8%   3%

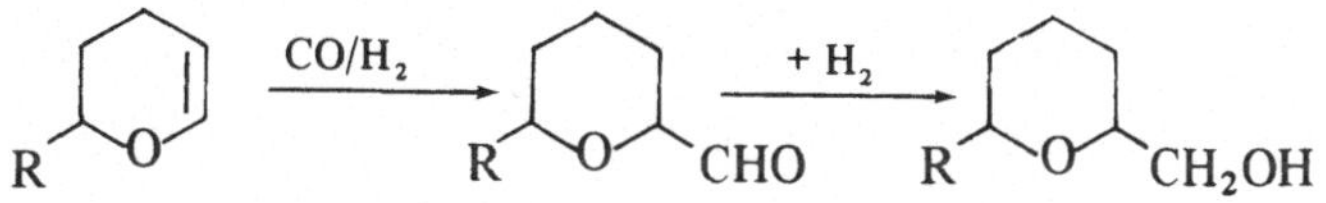

(106)

$$R= -\underset{\underset{H}{\backslash}}{\overset{\overset{O}{\parallel}}{C}} \quad [1736] \text{ or } -CH_2OH \text{ [1730, 1763]}$$

Hydroformylation ensues only at the C atom 3 when substitution is present at position 2 [1759].

    Phenyl allyl ethers and 2,4-dichlorophenyl allyl ethers depart from the above rule, yielding mainly isomers of γ-phenoxybutyraldehyde [1764] i.e., the formyl group is introduced at the more distant C atom. The same effect would appear to result from higher temperatures, as on hydroformylating methyl vinyl ether at 160–175 °C only β-methoxypropionaldehyde was obtained [278]. Results of the hydroformylation of other unsaturated ethers are shown in Table 1.48.

Table 1.48. Hydroformylation of unsaturated ethers and acetals

Starting materials	Catalyst	Press. (bar)	Temp. (°C)	Product(s)	Yield (%)	Ref.
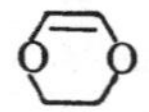 (dihydropyran ring)	Co soap			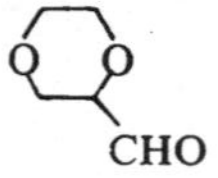 (tetrahydropyran–CHO)	86	[1767]
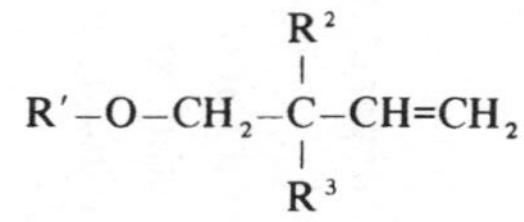 $R'-O-CH_2-\overset{R^2}{\underset{R^3}{C}}-CH=CH_2$	Co			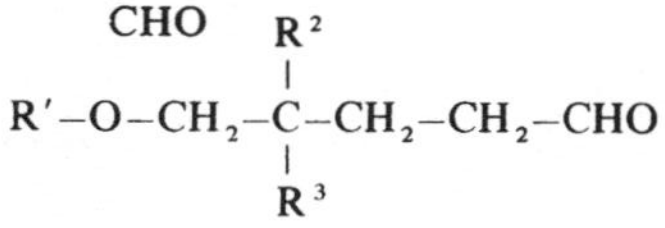 $R'-O-CH_2-\overset{R^2}{\underset{R^3}{C}}-CH_2-CH_2-CHO$		[1771]
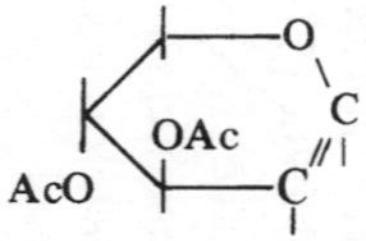 (acetal ring with OAc, AcO, C≡C)	$Co_2(CO)_8$			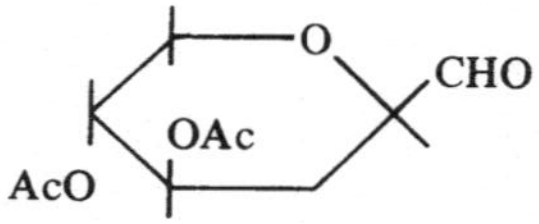 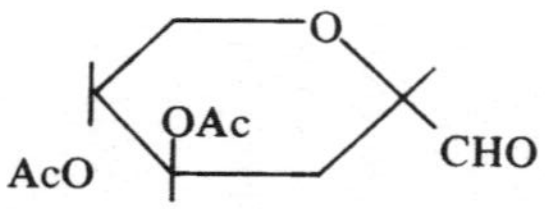 (acetal ring with OAc, AcO, CHO) + (acetal ring with OAc, AcO, CHO)		[1762, 1772]
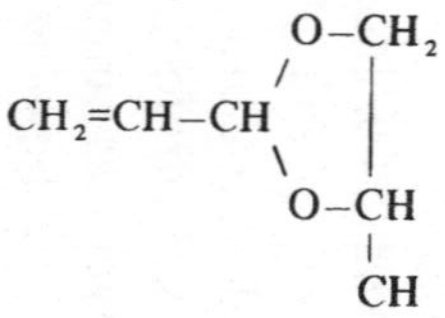 $CH_2=CH-CH\big(O-CH_2,\ O-CH\big)$ (dioxolane, CH)	$RhCl_3$-complex	600	100	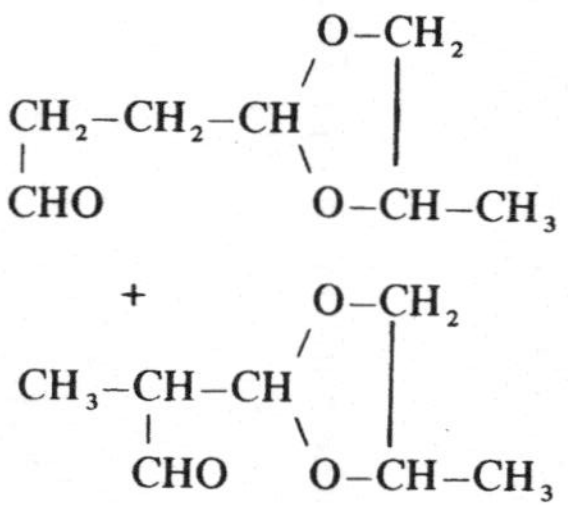 $\underset{CHO}{CH_2-CH_2-\overset{O-CH_2}{\underset{O-CH-CH}{CH}}}$ + $\underset{CHO}{CH_3-CH-CH}\overset{O-CH_2}{\underset{O-CH-CH_3}{}}$	57	[1766]
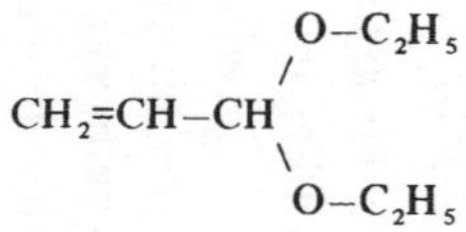 $CH_2=CH-CH\big(O-C_2H_5,\ O-C_2H_5\big)$	Rh			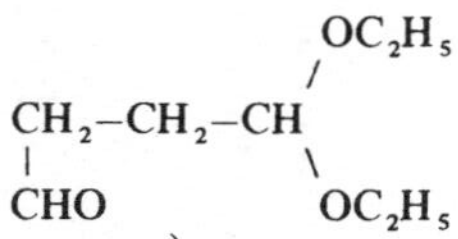 $\underset{CHO}{CH_2-CH_2-CH}\overset{OC_2H_5}{\underset{OC_2H_5}{}}$ a) 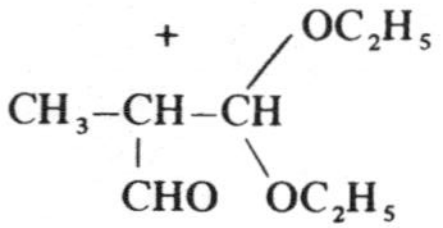 + $\underset{CHO}{CH_3-CH-CH}\overset{OC_2H_5}{\underset{OC_2H_5}{}}$ a:b = 1:2		[1773]

Substrate	Catalyst		Temp.	Product		Yield	Ref.
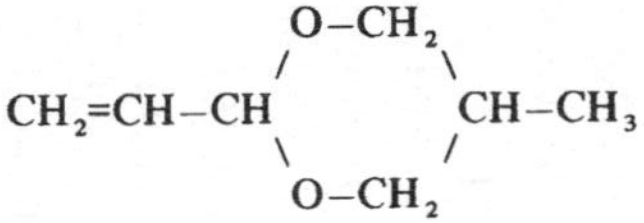 CH₂=CH-CH< with O-CH₂ / O-CH₂ ring, CH-CH₃	Co/PR₃	70–80	150	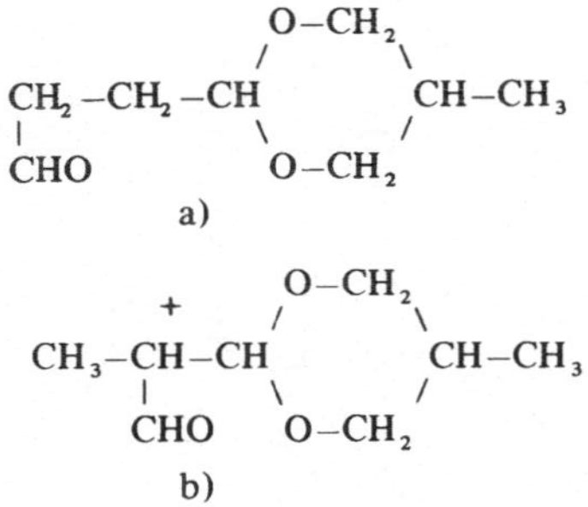 a) $CH_2-CH_2-CH<$ ... CHO ... $O-CH_2$ / $O-CH_2$, $CH-CH_3$ b) $CH_3-CH-CH<$ ... CHO ... $O-CH_2$ / $O-CH_2$, $CH-CH_3$	a : b = 80 : 20	89	[649]
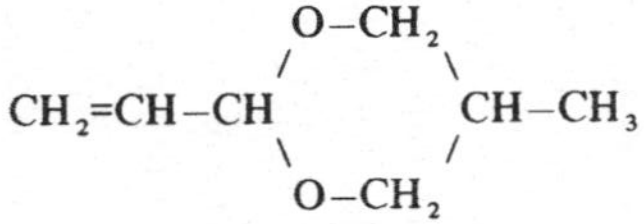 CH₂=CH-CH< with O-CH₂ / O-CH₂ ring, CH-CH₃	Rh/P(OCH₃)₃		150	87% linear product		98	[1769, 1770]
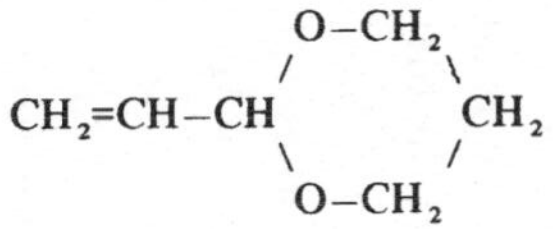 CH₂=CH-CH< with O-CH₂ / O-CH₂ ring, CH₂	Rh/P(Ph)₃	20–40	140	72% linear product		89	[649]
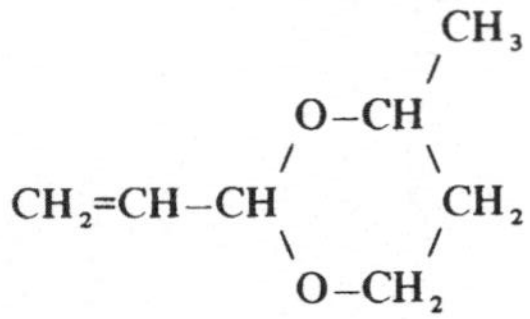 CH₂=CH-CH< with O-CH / O-CH₂ ring, CH₃ / CH₂	Co/P(Ph)₃	70–80	150	81% linear product			[649]
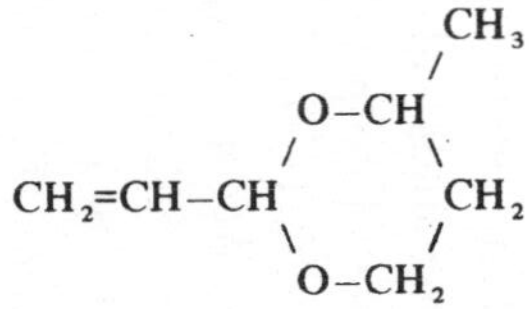 CH₂=CH-CH< with O-CH / O-CH₂ ring, CH₃ / CH₂	Rh/P(OPh)₃	6	110	85% linear product		89	[1768]

125

Unsaturated acetals can als be hydroformylated just as easily as the unsaturated ethers [1737, 1765] thereby presenting the possibility of directly hydroformylating otherwise difficult-to-react α/β unsaturated aldehydes and ketones to formyl products (cf Sect. 1.4.4.2).

For example, acrolein – in its acetal form – can be converted into Oxo products which after reductive cleavage yield butanediols [649, 1766, 1898, 1967, 1968].

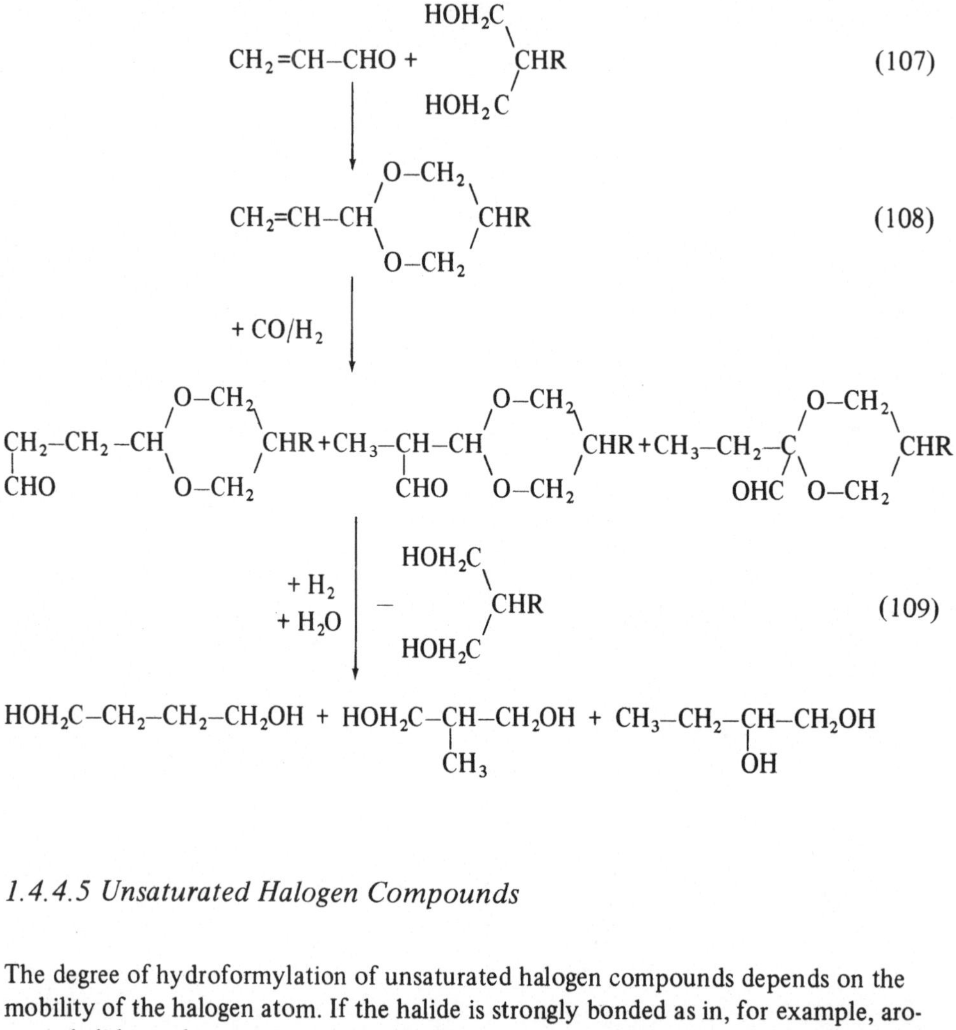

$$CH_2=CH-CHO + \begin{array}{c} HOH_2C \\ \diagdown \\ \diagup \\ HOH_2C \end{array} CHR \tag{107}$$

$$\downarrow$$

$$CH_2=CH-CH \begin{array}{c} O-CH_2 \\ \diagup \qquad \diagdown \\ \diagdown \qquad \diagup \\ O-CH_2 \end{array} CHR \tag{108}$$

$$+\ CO/H_2 \downarrow$$

$$\underset{CHO}{CH_2-CH_2-CH}\begin{array}{c}O-CH_2\\\diagup\quad\diagdown\\\diagdown\quad\diagup\\O-CH_2\end{array}CHR + \underset{CHO}{CH_3-CH-CH}\begin{array}{c}O-CH_2\\\diagup\quad\diagdown\\\diagdown\quad\diagup\\O-CH_2\end{array}CHR + \underset{OHC}{CH_3-CH_2-C}\begin{array}{c}O-CH_2\\\diagup\quad\diagdown\\\diagdown\quad\diagup\\O-CH_2\end{array}CHR$$

$$\begin{array}{c} +\ H_2 \\ +\ H_2O \end{array} \Bigg| \quad - \quad \begin{array}{c} HOH_2C \\ \diagdown \\ \diagup \\ HOH_2C \end{array} CHR \tag{109}$$

$$\downarrow$$

$$HOH_2C-CH_2-CH_2-CH_2OH\ +\ \underset{CH_3}{HOH_2C-CH-CH_2OH}\ +\ \underset{OH}{CH_3-CH_2-CH-CH_2OH}$$

1.4.4.5 Unsaturated Halogen Compounds

The degree of hydroformylation of unsaturated halogen compounds depends on the mobility of the halogen atom. If the halide is strongly bonded as in, for example, aromatic halides and in per- or partially fluorinated olefins [1774, 1764,1775], then the double bond can be hydroformylated in the usual manner (cf Table 1.49).

Compounds with halogens located at the olefinic carbon atoms, e.g., vinyl chloride or 1-bromo-1-propene, gave good yields of 2-haloalkanals at high $Co_2(CO)_8$ concentrations [1776, 1777]. Formation of the theoretically possible 3-haloalkanals was not observed. According to studies conducted by Macho and Stresinka [1764], the usual propylene hydroformylation is largely suppressed by the presence of vinyl chloride. Allyl chloride produces the same effect and thus this compound does not undergo hydroformylation. The reason lies with the formation of inactive complexes consisting of Co or Rh carbonyls with allyl chloride which lead to the total poisoning of the hydroformylation catalyst (cf Sect. 1.3.3.2) [1764, 1857].

Table 1.49. Hydroformylation of unsaturated halo-compounds

Starting material	Products	Yield (%)	Ref.
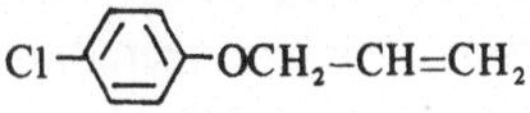 Cl—⟨⟩—OCH$_2$–CH=CH$_2$	aldehydes	11	[1764, 1774, 1775]
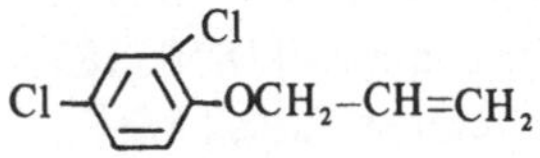 Cl—⟨⟩—OCH$_2$–CH=CH$_2$ (Cl)	aldehydes	69,7	
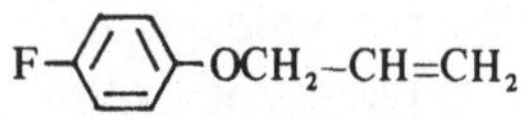 F—⟨⟩—OCH$_2$–CH=CH$_2$	aldehydes	55	
hexafluoropropylene	hexafluoropropane alcohols, aldehydes	50/40 5–8	[1778]
F$_3$C(CF$_2$)$_7$CH=CH$_2$	polyfluoroundecanal or polyfluoroundecanol	71,8 87	[1779]
vinyl chloride	α-monochloropropionaldehyde acrolein propionaldehyde ethyl chloride	85–90 10	[1776, 1777]
1-bromo-1-propene	2-bromobutyraldehyde		[1776]
1,1-dichloroethylene	dichloroether α-monochloropropionaldehyde α-chloroacrolein		[1777]
allyl chloride	no reaction		[1764, 1857]

1.4.4.6 Unsaturated Nitrogen Compounds

Conjugated and nonconjugated unsaturated nitriles can also be hydroformylated. The hydroformylation of acrylonitrile has undergone close study as this reaction would open up a commercial pathway to glutamic acid [1112, 1780, 1813, 1817, 1818, 1899], i.e., by subjecting the initially formed β-cyanopropionaldehyde to the Strecker reaction.

$$NC–CH=CH_2 \xrightarrow{CO/H_2} NC–CH_2–CH_2–CHO + \text{by-products} \qquad (110)$$

$$NC–CH_2–CH_2–CHO \xrightarrow{NH_3 + HCN} NC–CH_2–CH_2–\underset{NH_3}{CH}–CN \qquad (111)$$

$$NC–CH_2–CH_2–\underset{NH_2}{CH}–CN \xrightarrow[-NH_3]{NaOH + H_2O} NaOOC–CH_2–CH_2–\underset{NH_2}{CH}–COONa \qquad (112)$$

127

Free *dl*-glutamic acid can be prepared from the crude disodium salt. After selective crystallization of the monosodium salt, the desired monosodium salt of L-glutamic acid can be obtained in high yield [1025, 1112, 1817, 1819].

The hydroformylation of acrylonitrile does not proceed smoothly due to the conjugated double bond system and the very high reactivity of the initially resulting cyanopropionaldehydes [1781]. While it leads to the preferential formation of β-cyanopropionaldehyde, the α-isomer is also produced (Ono et al. [1781]). However, as it is unstable, only indirect identification is possible i.e., via the derivative − α-hydroxy-methylpropionitrile [1782]. One of the yield reducing side reactions − the hydrogenation of acrylonitrile to propionitrile − is favored by the conjugated unsaturated system. This reduction in yield can however be suppressed on employing other catalysts [1810, 1812, 1815, 1819] or maintaining particular reaction conditions − low temperatures [1787], solvents or diluents [243, 315, 1783, 1784, 1788, 1789, 1816]. Other by-products are propionaldehyde, propylamine and NH_3 [67, 1785, 1790, 1791]. On raising the hydroformylation temperature, the unstable β-cyanopropionaldehyde is converted into the more stable γ-hydroxybutyronitrile [1786].

Various additives have been proposed to improve the acrylonitrile hydroformylation reaction e.g., small amounts of oxygen [877] or tert. amines [1792, 1793] as well as modified cobalt carbonyls (as catalyst) [1794, 1795] or Rh_2O_3 as catalyst precursor [1796].

The isolation of the unstable intermediate − β-cyanopropionaldehyde − from the acrylonitrile hydroformylation would lead to considerable losses. Therefore, in the industrial manufacture of glutamic acid, the hydroformylation product is subjected to the Strecker reaction immediately after decomposing the cobalt carbonyls [1797, 1798]. This conversion involves reacting the β-cyano derivative with HCN in the presence of basic substances [1025, 1802]. The resulting products can then be further processed to glutamic acid in good yield [1112].

Allyl cyanide, a nonconjugated nitrile, reacts to form γ-cyanobutyraldehyde in 36% yield [294, 1682], while 5-cyano-2-methyl-1-pentene gives a 45% yield of 6-cyano-3-methylhexanal [1799].

1,4-Dicyanobutene produces a 55% yield of the dimethyl acetal of 1,4-dicyano-2-formylbutane using $Co_2(CO)_8$ as catalyst in the presence of methanol [1800].

The hydroformylation of N-alkenyl- and N-cycloalkenylcarboxylic acid amides with no hydrogen at the nitrogen atom produces good yields of the acylated α- or β-aminoaldehydes. Starting from N-vinylphthalimide, a 78% yield of the 3-phthalimido-propionaldehyde (I) and 2-phthalimidopropionaldehyde (II) in ratio 2.5 : 1 is obtained [1801]:

$$(113)$$

With N-allylphthalimide, where the double bond is one carbon atom further away from the nitrogen atom, an 87.2% yield is obtained. The product consists mainly of 4-phthalimidobutanal [1801].

Other N-alkenylcarboxylic acid imides (cf Table 1.50) give rise to mainly N-acylated β-aminoaldehydes. On the other hand, the hydroformylation of monoacylated alkenylamines such as N-allylacetoamine, does not lead to the aldehydic product [1801].

Rh catalysts cause the formyl group to preferentially add to the C atom nearest to the nitrogen. This tendency becomes more marked when modified rhodium carbonyls are employed at low temperatures (80–100 °C) and high pressures (300–800 bar) [1821].

Unsaturated compounds containing a nitro group undergo the hydroformylation reaction with retention of the nitro function. For example, with a Rh catalyst, p-nitrostyrene can be converted into a mixture of 2- and 3-(p-nitrophenyl)-propanal, and 3-nitropropylene produces a mixture consisting of 3-nitroisobutanal and 4-nitrobutanal [571].

Table 1.50. Hydroformylation of N-acyclated aminoolefins

Starting material	Catalyst (precursor)	Temp. (°C)	Pressures (bar)	Products	Yield (%)	Ref.
N-vinylphthalimide	Co$_2$(CO)$_8$	120	170	3-phthalimido-propionaldehyde a) 2-phthalimido-propionaldehyde b) a:b = 2.5:1	78	[1801]
N-allylphthalimide	Co$_2$(CO)$_8$	120	170	4-phthalimido-butanal	87.2	[1801]
N-diacetylallylamine	Co$_2$(CO)$_8$	120	170	N-diacetyl-γ-amino-butanal a) N-diacetyl-β-amino-isobutanal b) a:b = 6.7:1	86	[1801]
N-vinylsuccinimide	Rh/olefin complex	90	700	α-(N-succinimido)-propionaldehyde	90	[1821]
N-vinylpyrrolidone	Rh$_2$O$_3$ + phenothiazine	95	700	2-(N-pyrrolidonyl)-1-propanal	70	[1821]
N-allylpyrrolidone	Rh/pheno-thiazine	95	700	2-(N-pyrrolidonyl)-1-butanal	52	[1821]
N-cyclopentenyl-pyrrolidone	Rh/pheno-thiazine	95	700	2-(N-pyrrolidonyl)-cyclopentyl aldehyde	49	[1821]

1.4.4.7 Hydroformylation of Special Compounds

As *aromatic hydrocarbons* such as benzene, toluene and xylene do not undergo hydroformylation, they can be employed as solvents for the hydroformylation (cf Sect. 1.3.4). However, certain condensed aromatic ring systems e.g., indene and certain heterocycles with weak aromatic character can be hydroformylated to a greater or lesser extent.

For example with cobalt carbonyl complexes, indene mainly yields indane (via hydrogenation) along with a smaller amount of 1- and 2-formylindane. Good yields of 1- and 2-formylindane can be produced using ligand-modified Rh catalysts, the 1-formyl isomer dominating [1822, 1856].

2-Tetrahydrofurfuryl alcohol or 2,5-dimethyl-3-tetrahydrofurfuryl alcohol are produced on hydroformylating furan or 2,5-dimethylfuran, respectively [1823].

On the other hand, thiophene and its derivatives, as well as pyridines, are hydrogenated to tetrahydrothiophenes or piperidines or react even further [1823]. 2-Methyl- and 2-phenylindol are not hydroformylated to any extent but are partially hydrogenated (26% and 9%, resp.) [1824].

Natural products or their degradation products possessing one or more double bonds and belonging to various product groups have often been studied with the aim of employing them as a hydroformylation feedstock. For example the reactivity of the terpenes – dipentene, α-pinene, α-terpinene and myrcene – in the hydroformylation reaction was determined along with the distribution of their reaction products [1825]. The differing behavior of α-pinene in the hydroformylation with Co or with Rh catalysts was studied by Himmele and Siegel. According to their work, while α-pinene undergoes a rearrangement in the presence of Co catalysts to the bornane structure, with Rh catalysts the pinene structure remains intact [323, 1969].

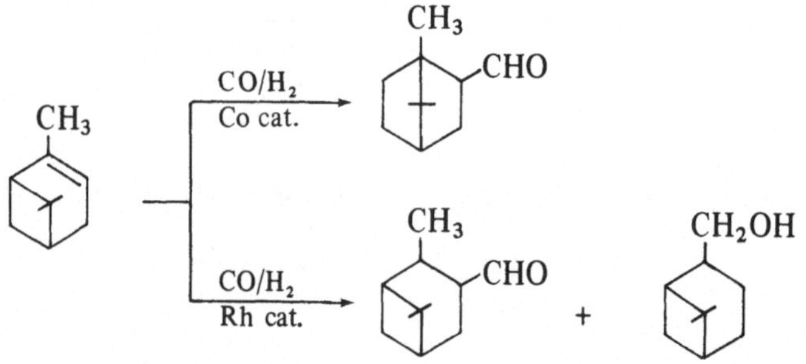

$$(114)$$

$$(115)$$

Camphene is hydroformylated in 65% yield – without rearrangement – to a uniform aldehydic product with a cobalt carbonyl catalyst [1826]:

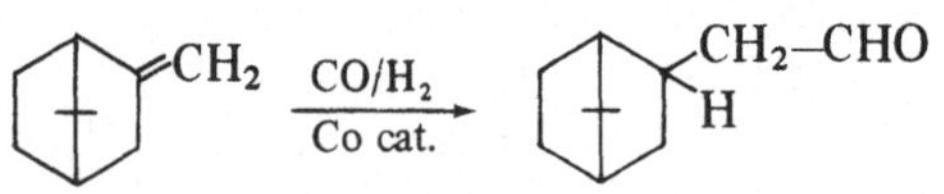

$$(116)$$

Allocime, which is manufactured via thermal cracking of terpene, possesses triple unsaturation. This substance can be hydroformylated to give a 70% yield of 4,8-dimethylnonanal [1827, 1828].

The isolated or conjugated double bonds present in complicated ring systems such as pimaric acid or abietic acid [1829] or the steroids [1830, 1831, 1833] also undergo hydroformylation:

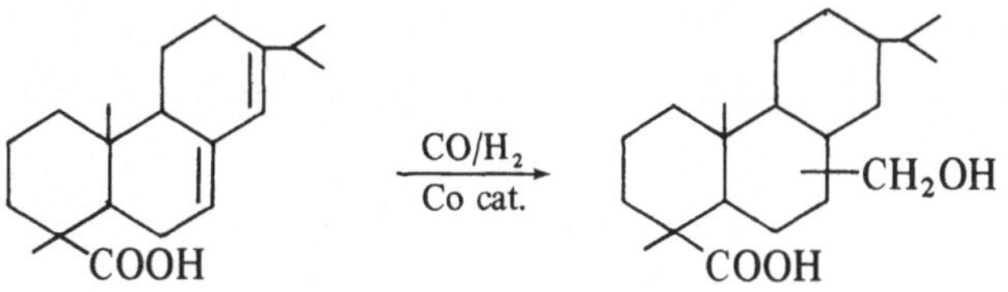

$$\text{(117)}$$

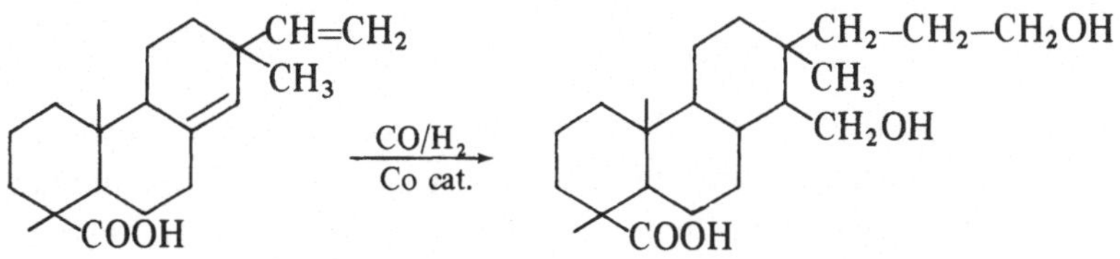

$$\text{(118)}$$

The hydroformylation of coniferyl alcohol is described in Refs. [1864, 1867], the reaction of other natural products in [74, 1245, 1867, 1868, 1877, 1878, 1892, 1969].

The reaction involving trimethylvinyl silane and trimethylallylsilane showed that organosilicon compounds possessing double bonds could be hydroformylated [1832].

Numerous attempts have been made to subject *oxiranes* (*epoxides*) to the hydroformylation reaction to produce hydroxyaldehydes [80, 1834–1837, 1903].

$$R-\overset{O}{\overset{/\backslash}{CH}}-CH_2 \xrightarrow{CO/H_2} R-\underset{OH}{\underset{|}{CH}}-CH_2-CHO + R-\underset{CHO}{\underset{|}{CH}}-CH_2OH \qquad (119)$$

On account of the greater reactivity, not only of the epoxides but also of the resulting hydroxyaldehydes, the epoxide hydroformylation generally leads to the formation of a mixture of products and thus unsatisfactory yields.

By means of stoichiometric reactions of epoxides with cobalt carbonyls, Heck showed that this conversion took place via on acylcobalt carbonyl step [1838] i.e., similar to the olefin conversion. Thereafter, the acyl compounds are probably hydrogenated to the hydroxyaldehydes, in analogy to the olefin-based Oxo mechanism [cf Eqs. (14) and (15)].

$$H_2\overset{O}{\overset{/\backslash}{C}}-CH_2 + HCo(CO)_4 \rightarrow HOCH_2-CH_2-Co(CO)_4 \xrightarrow{+CO} HOCH_2-CH_2-CO-Co(CO)_4$$
$$\text{(120)}$$

Yokokawa et al. [1837] established that there was also a strong relationship between product content and reaction temperature in the oxirane hydroformylation. For example, the hydroformylation of propylene oxide with $Co_2(CO)_8$ as catalyst only produces noticeable amounts of β-hydroxy-n-butyraldehyde when the temperature lies between 80–100 °C. When the temperature is higher than 100 °C, propylene oxide is isomerized to acetone and in addition, secondary reactions of the hydroxyaldehyde occur. The β-hydroxy-n-butyraldehyde content thus falls to a few per cent of the total yield.

The hydroxyaldehydes can be readily dehydrated to unsaturated alcohols which are then either hydrogenated to saturated aldehydes or polymerized to higher boiling substances [1837]:

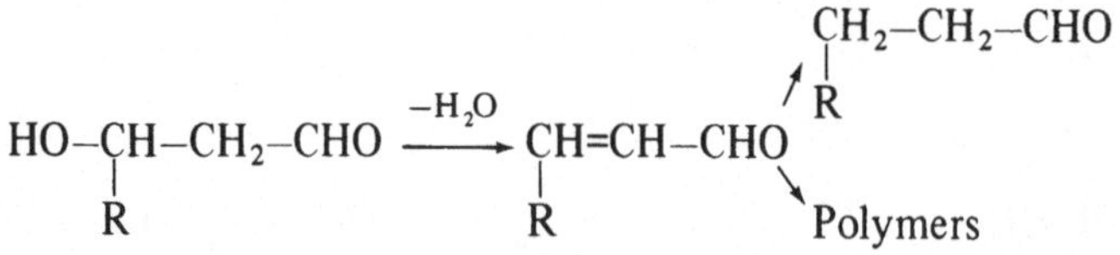

$$
\begin{array}{c}
CH_2-CH_2-CHO \\
| \\
R
\end{array} \quad (121)
$$

$$
\underset{\underset{R}{|}}{HO-CH}-CH_2-CHO \xrightarrow{-H_2O} \underset{\underset{R}{|}}{CH}=CH-CHO \begin{array}{c} \nearrow \\ \searrow \end{array} \begin{array}{c} \\ Polymers \end{array}
$$

$$(122)$$

By-product formation is particularly marked in the hydroformylation of ethylene oxide, as here both the starting material and the primary hydroformylation product are more reactive than was the case with propylene oxide [1836]. Attempts have been made to improve this conversion by adding small amounts of alcohols, ethers, ketones and esters [1839, 1843]. The ethylene oxide reaction apparently ensues more smoothly with modified Co catalysts [1842], however, unusually high catalyst concentrations were employed i.e., ca. 1 mol catalyst to 10 mol epoxide.

The hydroformylation becomes more selective with decreasing activity of the oxiranes. According to studies conducted by Niederhauser et al. [1840], the yield of hydroformylation products increases from 20% with an ethylene oxide feedstock, to over 40% with propylene oxide, to 60% with octylene oxide and finally to over 90% with methyl 9,10-epoxystearate.

The primary resulting hydroxyaldehydes readily form cyclic hemiacetals which can then be converted into 1,3-diols via reductive cleavage [1841, 1843, 1902].

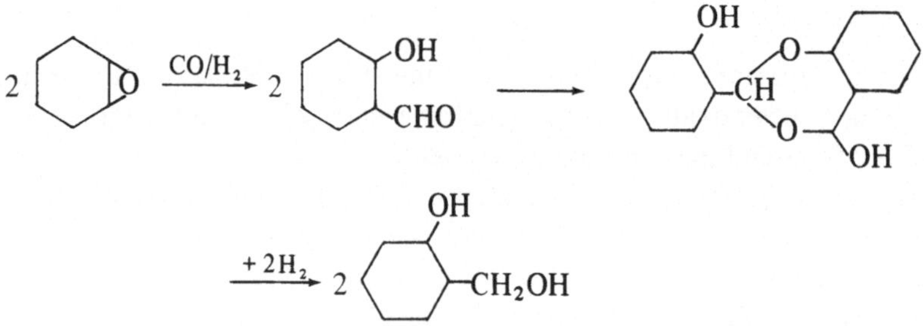

$$(123)$$

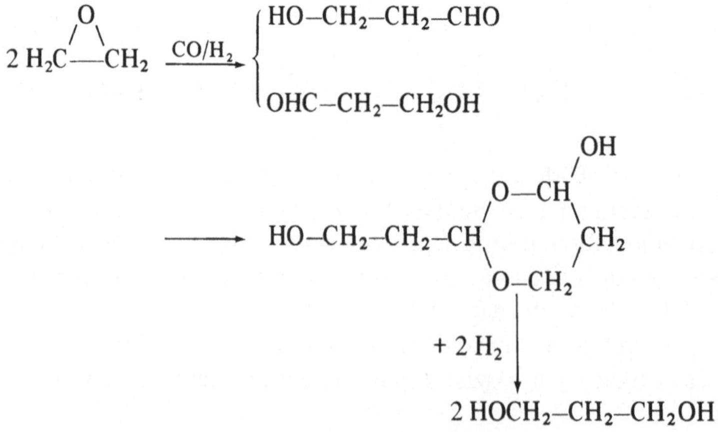

$$(124)$$

1.4.5 Hydroformylation of Polymers

Mertzweiller, Cull, and Tenney's (Esso) work is probably the most significant in the polymer hydroformylation field [365, 827, 1670]. The products, which are still largely unidentified, were prepared by reacting polyolefins or polydiolefins or unsaturated hydrocarbon resins with CO/H_2 in the presence of unmodified or ligand-modified Co or Rh catalysts. The hydroformylated product — which now contains functional groups — can be converted into the corresponding derivatives via hydrogenation, oxidation, reductive amination, condensation etc. The main application is the manufacture of particularly hard coating films for metal surfaces (cf Refs. [1671, 1672].

1.4.6 Asymmetric Hydroformylation

When, on hydroformylating a suitable olefinic (prochiralic) compound a product results possessing an asymmetric carbon atom (chiralic compound) then with conventional catalysts the formation of the two enantiomers takes place at the same rate. Thus, the chiralic product is always present as a racemate [1872, 1873].

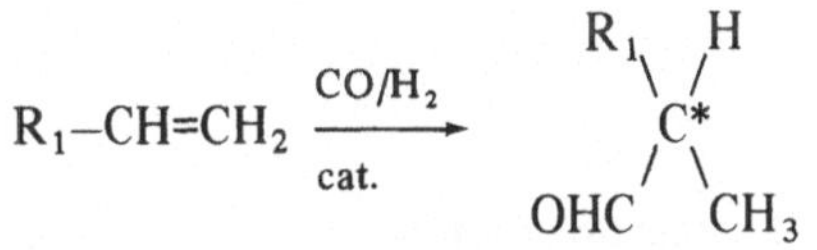

$$R_1-CH=CH_2 \xrightarrow[\text{cat.}]{CO/H_2} \begin{array}{c} R_1 \quad H \\ \backslash \ / \\ C^* \\ / \ \backslash \\ OHC \quad CH_3 \end{array} \tag{125}$$

$$\begin{array}{c} R_1 \\ \backslash \\ C=CH_2 \\ / \\ R_2 \end{array} \xrightarrow[\text{cat.}]{CO/H_2} \begin{array}{c} R_1 \quad H \\ \backslash \ / \\ C^* \\ / \ \backslash \\ R_2 \quad CH_2-CHO \end{array} \tag{126}$$

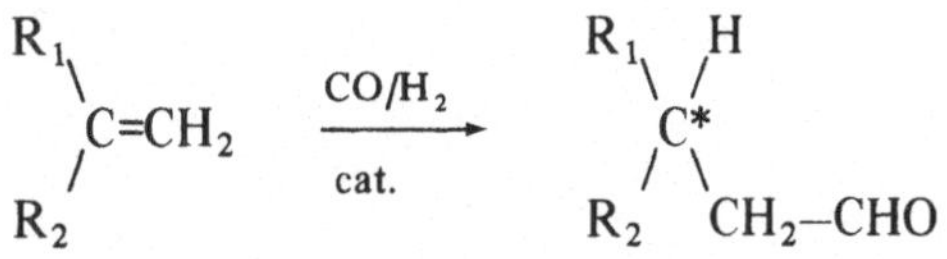

However, with modified cobalt or rhodium carbonyl catalysts containing one or more optically active ligands, the formation of one of the enantiomers may be favored. Consequently, this enantiomer will be present in the reaction product at a higher concentration possibly leading to strong optical activity [1844, 1845].

Another possibility of achieving asymmetric hydroformylation consists of hydroformylating an unsaturated racemate in which one of the enantiomers is preferentially hydroformylated.

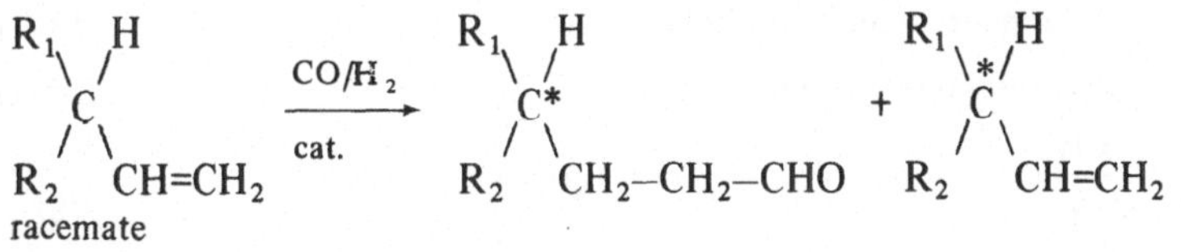

$$\begin{array}{c} R_1 \quad H \\ \backslash \ / \\ C \\ / \ \backslash \\ R_2 \quad CH=CH_2 \\ \text{racemate} \end{array} \xrightarrow[\text{cat.}]{CO/H_2} \begin{array}{c} R_1 \quad H \\ \backslash \ / \\ C^* \\ / \ \backslash \\ R_2 \quad CH_2-CH_2-CHO \end{array} + \begin{array}{c} R_1 \quad H \\ \backslash */ \\ C \\ / \ \backslash \\ R_2 \quad CH=CH_2 \end{array} \tag{127}$$

In order to achieve asymmetric hydroformylation, the usual cobalt or rhodium catalysts have been modified with various chiralic ligands (Table 1.51). Table 1.52 gives a summary of the unsaturated compounds which served as chiralic precursors.

Table 1.51. Chiralic cobalt and rhodium complex catalysts for asymmetric hydroformylation

Pos.	Catalyst	References
1	(+)-(S)-bis-(N-α-methylbenzylsalicyl-aldiminato)-cobalt(II)	[673, 1846]
1a	$Co_2(CO)_8$/(+)-(S)-N-α-methylbenzylsalicyl aldimine	[1846]
1b	Rh_2O_3 + (+)-(S)-N-α-methylbenzylsalicyl aldimine	[673, 1854]

Pos.	Catalyst	References
2	RhCl (hexadiene-1,5)$_2$ and (+)–P–CH$_3$ with C$_6$H$_5$ and CH$_2$–C$_6$H$_5$	[793, 1851]
3	[Rh(CO)$_2$Cl]$_2$ + R–(–) P with methyl, –n–propyl, phenyl	[1853]
4	[Rh(CO)$_2$Cl]$_2$ + S–(+) P with methyl, –n–propyl, phenyl	[1853]
5	RhCl(CO) P with phenyl, –phenyl, neomenthyl*	[776]
6	RhH(CO) P with phenyl, –phenyl, (2-methylbutyl)*	[1854]
7	RhH(CO)[P(Ph)$_3$]$_3$ and (–)2,3-O-isopropylidene-2,3-dihydroxy-1,4-bis-(diphenylphosphino)-butane[(–)-DIOP]	[809, 1847 –1850, 1852 1854, 1955]
8	PtCl$_2$[(–)-DIOP] SnCl$_2$	[491, 942]
9	HRh-complex in presence of 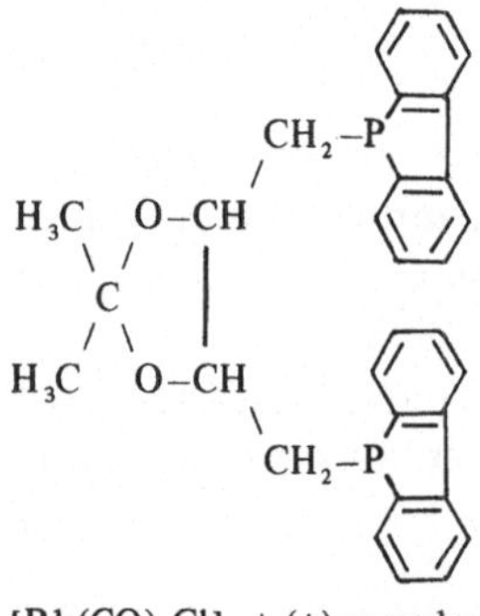	[1850]
10	[Rh(CO)$_2$Cl]$_2$ + (+)-camphor-10-sodium sulfonate	[1854]
11	Rh_2O_3 + (+)-benzenephosphorous acid dibornyl ester	[1854]
12	[Rh(CO)$_2$Cl]$_2$ + P with CH$_3$, –C$_6$H$_5$, CH$_2$–$\overset{*}{C}$H–C$_2$H$_5$ with CH$_3$	[1853]

Table 1.52. Asymmetric hydroformylation of unsaturated precursors for chiralic compounds a: published to date

Olefin	Asym. ligand (position in Table 1.51)	Chiralic product	Optical purity[a]	References
1-butene	7	(−)-(R)-2-methylbutanal	18.8	[840,
	8	(−)-(R)-2-methylbutanal		1854, 491
	9	(−)-(R)-2-methylbutanal		1850]
1-pentene	7	2-methylpentanal	19.7	[673, 840, 1854]
1-octene	7	2-methyloctanal	15.2	[673, 840, 1854]
cis-2-butene	7	(+)-(S)-2-methylbutanal	27.0	[673, 840, 1854]
trans-2-butene	7	(+)-(S)-2-methylbutanal	3.2	[673, 840, 1854]
cis-2-hexene	7	2-methylhexanal	7.6	[673, 840]
		2-ethylpentanal	5.8	
trans-2-hexene	7	2-methylhexanal	1.4	[673, 840]
		2-ethylpentanal	2.9	
3-methyl-1-butene	7	2,3-dimethylbutanal	15.2	[673, 840, 1854]
2-ethyl-1-hexene	7	2-ethylheptanal	1.1	[673]
3-methyl-1-pentene	7	4-methylhexanal	4.6	[104, 673]
2,3-dimethyl-1-pentene	8	2,3-dimethylpentanal		[491]
styrene	1	2-phenylpropanal		[673, 1846]
	1a	2-phenylpropanal		[1846]
	1b	2-phenylpropanal		[673, 1854]
	2	2-phenylpropanal		[793, 1851]
	3,4	2-phenylpropanal		[1853]
	5	2-phenylpropanal		[776]
	6	2-phenylpropanal		[1854]
	7	2-phenylpropanal		[809, 1847] [1852, 1854]
	8	2-phenylpropanal		[491]
	9	2-phenylpropanal	44.3	[1850]
	10,11	2-phenylpropanal		[1854]
	12	2-phenylpropanal		[1853]
α-methylstyrene	1	3-phenylbutanal		[1846]
	1b	3-phenylbutanal		[1854]
	7	3-phenylbutanal	1.6	[809, 1854]
	8	3-phenylbutanal		[491, 942]
trans-β-methylstyrene	7	2-phenylbutanal	14.4	[809, 1854]
α-ethylstyrene	5	3-phenylpentanal	1.0	[776]
	7	3-phenylpentanal	1.8	[809, 1854]
	8	3-phenylpentanal		[491, 942]
allylbenzene	7	3-benzylpropanal	15.5	[809, 1854]
1-vinylnaphthalene	7	2-(α-napthyl)-propanal		[1854]
4-vinylcyclohexene	7	2-(3-cyclohexenyl)-propanal		[1854]
N-vinylsuccinimide	7	2-succinimidopropanal		[1854]
norbornene	7	mixture of exo- and endo-bicyclo[2.2.1]-heptanaldehyde		[1854]

Table 1.52 (continued)

Olefin	Asym. ligand (position in Table 51)	Chiralic product	Optical purity[a]	References
indene	3	1-formylindane		[1853]
phenyl vinyl ether	5	2-phenoxypropanal	0.3	[776]
isoeugenol	4	2-(3-methoxy-4-hydroxy-phenyl)-butanal 2-methyl-3-(3-methoxy-4-hydroxyphenyl)-propanal		[1853]
cinnamic alcohol	4	2-hydroxy-3-phenyltetrahydrofuran		[1853]
cinnamic aldehyde propylene glycol acetal	4	2-phenylsuccinaldehyde monopropylene-glycol acetal		[1853]
methallyl alcohol	7	4-methyl-2,3-dihydrofuran 4-methyl-2-hydroxytetrahydrofuran		[1849] [1854]
1-buten-3-ol	7	5-methyl-2,3-dihydrofuran 5-methyl-2-hydroxytetrahydrofuran		[1849] [1854]
4-methyl-1-hexen-3-ol	7	4-sec.butyl-2,3-dihydrofuran		[1849]
3-methyl-1-buten-3-ol	7	5,5-dimethyl-2,3-dihydrofuran		[1849]
2-butyne	7	(S)-2-methylbutanal		[1848]
1-octyne	7	(S)-2-methyloctanal		[1848]
phenylacetylene	7	(R)-2-phenylpropanal		[1848]

When a cobalt carbonyl complex containing the optically active N-α-methyl benzyl-salicyl aldimine was employed in the styrene hydroformylation, the optical purity of the reaction products was only 2.9% [673, 1846]. Higher concentrations of the optically active products were obtained on using rhodium hydrocarbonyls modified with phosphines. With these components, the chiralic center can either be directly at the phosphorus itself (phosphine with three different ligands — positions 3 and 4 in Table 1.51), or in one of the ligands (optically active ligand, positions 5—12, Table 1.51). If an optically active phosphine with three different ligands is employed, then high optical purity of the products can be apparently achieved when the three ligands differ considerably in size [1853]. Suitable catalysts for the asymmetric hydroformylation can be readily prepared on using phosphines with ligands based on optically active natural products (positions 5, 7—12, Table 1.51). The individual chiralic ligands of the phosphines in the rhodium carbonyl complexes exhibit however, a widely varying selectivity towards the formation of optically active products. The chiralic diphosphine (−)-2,3-0-isopropylidene-2,3-dihydroxy-1,4-bis(diphenylphosphino)-butane, (−)-DIOP for short (position 7 in Table 1.51), has proven to be particularly selective and consequently has been employed in a large number of cases [1855].

Besides the type of chiralic center of the catalyst, the course of the asymmetric hydroformylation is quite noticeably affected by reaction conditions and the structure of the unsaturated feedstock (chiralic precursor). Moreover, as expected, these factors determine the yield of optically active products (cf Table 1.53).

Table 1.53. Effect of temperature, pressure and Rh:P ratio on the asymmetric hydroformylation of styrene.

$$\text{(catalyst: } [Rh(CO)_2Cl]_2 \cdot P \begin{array}{c} CH_3 \\ / \\ -C_6H_5 \\ \backslash \\ n-C_3H_7 \end{array} \qquad \text{reaction period: 24 hrs [1853])}$$

Pos.	Temp. (°C)	Pressure (bar)	Mol ratio Rh:P	Rotation of resulting 2- and 3-phenylpropanal mixture[a]
1	80	200	1:2	+ 37,0°
2	120	200	1:2	+ 6,8°
3	80	500	1:2	+ 27,7°
4	80	200	1:3	+ 68,3°

[a] rotation with Na_D line in 20 cm sample tube

Thus *concentration* of the optically active product clearly increases on lowering temperature and pressure. According to Pino [1854], lowering the partial pressure of the CO has a greater effect than reducing the H_2 partial pressure. Increasing the concentration of the optically active ligands of the Rh complex also effects an increase in yield of the optically active product. It was also established that with increasing conversion rate, the share of the optically active products decreased relative to the total yield [1854].

Usually, a higher conversion rate requires a longer reaction period under standard reaction conditions. Thus, a drop in yield of optically active substances with increasing conversion rate is probably connected with a partial racemization of the optically active product during the reaction. As the racemization of an optically active compound is greater at higher temperatures, the low yield of optically active substances (at higher temperatures) is possibly due to a racemization which takes place parallel to the main reaction.

The effect of reaction conditions on the *configuration* of the products could not be determined [840]. In addition, the varying yields of optically active substances with different unsaturated starting materials could possibly stem, at least in part, from the largely facile racemization.

The greatest optical purity was obtained on hydroformylating cis-2-butene with $HRh(CO)[P(Ph)_3]_3$ in (−)-DIOP under very mild reaction conditions (20 °C/0.67 bar). The trans-olefins usually give rise to a lower yield of optically active substances compared to their cis-counterparts [840]. In general, the asymmetric hydroformylation of alkynes leads to essentially lower yields of optically active substances compared to the corresponding olefins. The configuration of the similar optically active products may be different. This presents a way of studying the reaction mechanism of the alkyne hydroformylation [1848]. Other hydroformylation reactions are discussed in Refs. [1874–1876], asymmetric conversions with Pt-catalysts cf [1929].

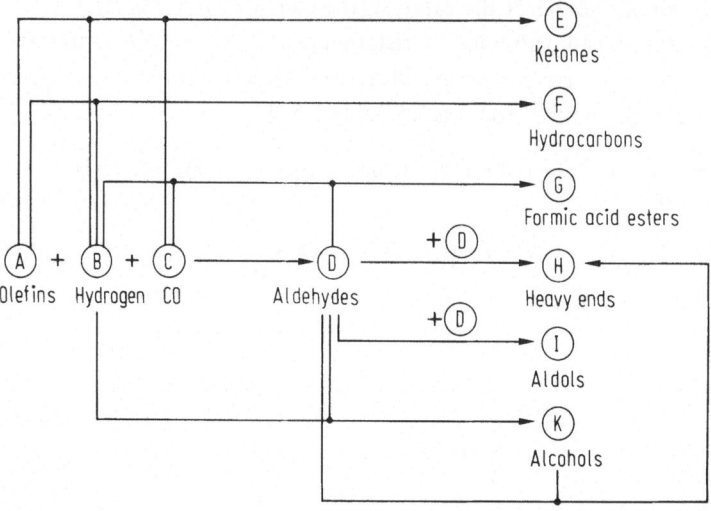

Fig. 1.60. Reactions in parallel or in series accompanying the hydroformylation reaction

1.5 Parallel and Consecutive Reactions Under Hydroformylation Conditions

Besides the desired hydroformylation reaction, a series of parallel and secondary reactions also occur under Oxo conditions. These are subdivided in Fig. 1.60 into side reactions which lower the yield or selectivity [1414—1416].

The above schematic representation shows the side reactions which lower the yield – usually parallel reactions to the hydroformylation – via conversion of the olefin feedstock to substances other than the desired Oxo products. The olefin hydrogenation to paraffins **F** and ketone formation **E** (diethyl ketone from ethylene) are two examples, the latter giving rise to the term Oxo synthesis. Alkyl formate formation **G** may also be included with reservation although they can be at least converted into the partly desired by-product alcohol via appropriate process steps. There is a similar situation with the heavy ends **H** (cf Sect. 1.5.4).

Alcohol formation **K** via hydrogenation of the aldehyde product is a selectivity reducing side reaction which still yields value products. If they react with aldehydes to acetals then the required cracking in a separated reaction step can form the transition to side reactions causing a decrease in yield.

1.5.1 Side Reactions Resulting in a Decrease in Yield

1.5.1.1 Hydrogenation of Olefins to Hydrocarbons

The hydrogenation of the olefin feed to the saturated hydrocarbon, which is homogeneously catalyzed by $Co_2(CO)_8$ or $HCo(CO)_4$ [1417—1421, 1530], is a parallel reaction (to the Oxo synthesis) which results in an irreversible decrease in yield.

$$R-CH=CH_2 \ + \ H_2 \xrightarrow[\text{cat.}]{} R-CH_2-CH_3 \tag{128}$$

According to Marko [1431], this side reaction which ensues after Eq. (8) to (17) and starts from the alkyl complex in Eq. (11), can be formulated as follows:

$$R-CH_2-CH_2-Co(CO)_4 \rightleftharpoons R-CH_2-CH_2-Co(CO)_3 \ + \ CO \tag{129}$$

$$R-CH_2-CH_2-Co(CO)_3 \ + \ H_2 \rightleftharpoons R-CH_2-CH_2-CoH_2(CO)_3 \tag{130}$$
$$\downarrow$$
$$R-CH_2-CH_3 \ + \ HCo(CO)_3 \tag{131}$$

This hydrogenation is limited to 0.2 to 1.5% of the olefin feed (cf Sect. 1.3.3.1.2) with variants of the classical Oxo synthesis employing cobalt hydridocarbonyl as catalyst. However, with ligand-modified cobalt catalysts which exhibit strong hydrogenation activity, the figure is around 15% [6, 16, 26, 30, 35, 40, 47, 1422–1426]. This marked hydrogenation tendency can be compensated with Rh carbonyls solely via the use of low catalyst charges. Consequently, even olefins which are susceptible to hydrogenation are hydroformylated and not hydrogenated [1423, 1427] e.g., compounds with double bonds conjugated to phenyl groups (styrenes), anthracenes, furans or certain unsaturated esters [1415]. The strong olefin hydrogenation can lead to difficulties during the industrial hydroformylation of propylene with ligand-modified Oxo catalysts as gas recycles of the unreacted olefinic feedstock will lead to an enrichment of the propane content. This problem can be eliminated via a propylene/propane separation step (cf Sect. 1.6.2.2).

The double bonds of the unsaturated aldehydes or ketones (possibly present in the heavy ends of the Oxo product) can also be hydrogenated along with the unsaturated bond in the olefinic feedstock [1428, 1429].

The extent to which hydrocarbon formation actually results from the parallel reaction (olefin hydrogenation) and not via a secondary reaction (decarbonylation Eq. (132) [1430]) must be determined from case to case (cf reformation of olefin from acyl complexes Eq. (26)).

$$R-CH_2-CH_2-CHO \xrightarrow[-CO]{\text{cat.}} R-CH_2-CH_3 \tag{132}$$

1.5.1.2 Formation of Formic Acid Esters

Alkyl formates result via the hydroformylation of the C=O double bond of the Oxo aldehyde product [1421, 1432–1434].

$$R-\underset{H}{\overset{}{C}}=O + HCo(CO)_4 \rightarrow R-CH_2-O-Co(CO)_4 \tag{133}$$

$$R-CH_2-O-Co(CO)_4 \rightleftharpoons R-CH_2-O-CO-Co(CO)_3 \tag{134}$$

$$R-CH_2-O-CO-Co(CO)_3 + H_2 \rightleftharpoons R-CH_2-O-COH + HCo(CO)_3 \qquad (135)$$

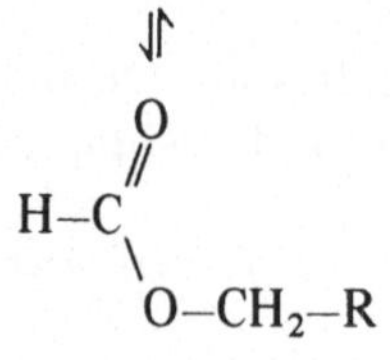

$$\updownarrow$$

$$\underset{\text{alkyl formate}}{H-C \overset{\displaystyle O}{\underset{\displaystyle O-CH_2-R}{\big\langle}}}$$

n-Butyl and isobutyl formate can be obtained from propylene in this way [1432], while 3-methylbutyl and – to a lesser extent – 2,2-dimethylpropyl formate result from isobutylene. Esters can also stem from Tishchenko analogue reactions [1436, 1437] or via other secondary reactions [695].

Alkyl formates and other esters can in subsequent process steps be at least partially converted into useful products, both the alkyl and formyl group being utilized.

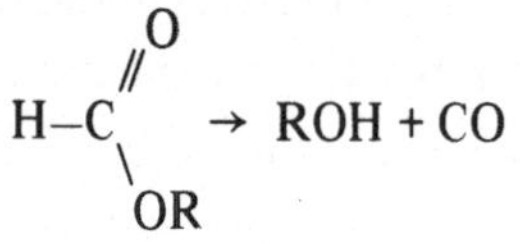

$$H-C\overset{\displaystyle O}{\underset{\displaystyle OR}{\big\langle}} \;\rightarrow\; ROH + CO \qquad (136)$$

$$H-C\overset{\displaystyle O}{\underset{\displaystyle OR}{\big\langle}} \;\xrightarrow{\;+HN\overset{R'}{\underset{R'}{\big\langle}}\;}\; ROH + H-C\overset{\displaystyle O}{\underset{\displaystyle N\overset{R'}{\underset{R'}{\big\langle}}}{\big\langle}} \qquad (137)$$

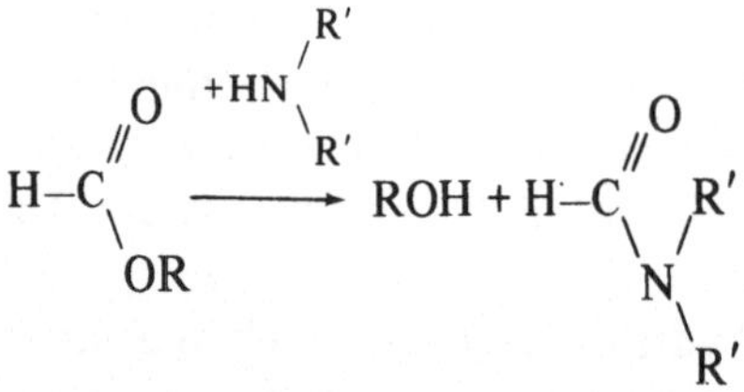

Oxo alcohols are obtained in accordance with Eq. (136) e.g., butanols from butyl formates [1439] and also via Eq. (137) on reacting alkyl formates with secondary amines, diakylformamides resulting along with the alcohols [1438]. Thus, the by-product – HCOOR – is optimally used (cf Sect. 1.5.4).

1.5.1.3 Ketone Formation

Ketone formation, e.g., diethyl ketone from ethylene, was noticed by Otto Roelen when he discovered the hydroformylation reaction [1440], giving rise to the term 'Oxo' synthesis. According to Bertrand et al. the ketone formation can be formulated as a reaction between the acyl cobalt compounds and the alkyl cobalt complex [1441–1444, 1446].

$$R-CH_2-CO-Co(CO)_4 + R-CH_2-Co(CO)_4 \;\rightarrow\; R-CH_2-\overset{\displaystyle O}{\overset{\|}{C}}-CH_2-R + Co_2(CO)_8 \quad (138)$$

This mechanism is in agreement with experimental findings that higher molecular olefins have a lesser tendency to form ketones as a result of steric hindrance. In general, the synthesis of noticeable quantities of ketones requires relatively severe conditions

[248, 249, 268, 1445, 1446]. Other ketone syntheses can be found in various reports [1447–1451, 1475]. Ketone formation from olefins higher than ethylene – which never becomes the main reaction but represents an undesired side reaction requiring special treatment in the hydrogenation stage – has also been treated in the Ref. [1452].

1.5.1.4 Formation of Heavy Ends

According to Fig. 1.60, heavy ends – the by-products of the Oxo synthesis – can result in various ways: via condensation, trimerization, aldolization, acetalization, Guerbet reaction etc. [1453–1473].

$$R-CH_2-CHO + R-CH_2-CHO \xrightarrow[-H_2O]{} R-CH_2-CH{=}\overset{\overset{\displaystyle R}{|}}{C}-CHO \tag{139}$$

$$3\ R-CH_2-CHO \longrightarrow R-CH_2-CH{\overset{O-CH{\overset{\displaystyle CH_2-R}{\diagdown}}}{\underset{O-CH{\underset{\displaystyle CH_2R}{\diagdown}}}{\bigg\langle}}}O \tag{140}$$

$$2\ R-CH_2-CHO \longrightarrow R-CH_2-\overset{\overset{\displaystyle OH}{|}}{CH}-\underset{\underset{\displaystyle R}{|}}{CH}-CHO \tag{141}$$

$$R-CH_2-CHO + 2\ R-CH_2-CH_2OH \longrightarrow R-CH_2-CH{\overset{\displaystyle O-CH_2-CH_2R}{\underset{\displaystyle O-CH_2-CH_2R}{\big\langle}}} \tag{142}$$

$$2\ R-CH_2-CH_2OH \longrightarrow R-CH_2-CH_3-\overset{\overset{\displaystyle R}{|}}{CH}-CH_2OH + H_2O \tag{143}$$

Table 1.54 presents an example of the main components of the heavy ends (boiling points higher than n-butanol) from a propylene hydroformylation.

The group of 2-ethylhexyl compounds (obtained via aldolization, sometimes followed by hydrogenation), trimeric butyraldehydes and the $C_4/C_4/C_4$ acetals are well represented [1476–1479, 1618, 1619, 1630]. Conditions are complicated and a comparison is made difficult as the type and quantity of components in the heavy ends depend not only on the hydroformylating and decobalting temperatures but also on the point at which the samples are taken (after reactor or decobalter). In general, the crude Oxo product contains more components immediately after the Oxo reactor (at all hydroformylation temperatures) than after the decobalter. Therefore, the heavy

Table 1.54. Composition of the heavy ends of the crude Oxo product from hydroformylation of propylene (values after decobalting [1474])

	% in crude Oxo product	formal n-butyraldehyde content
nonspecified components	0.5	–
2-ethylhexanal	0.1	0.1
2-ethylhexenal	0.2	0.2
2-ethylhexanol (2-EH)	0.2	0.2
2-ethylhexenol	< 0.1	< 0.1
trimeric iso/iso/iso-C_4-aldehyde	0.1	–.
trimeric iso/iso/n -C_4-aldehyde	0.2	0.07
trimeric iso/n /n -C_4-aldehyde	0.7	0.04
trimeric n /n /n -C_4-aldehyde	1.1	1.1
iso/n /n -acetal	0.1	0.06
n /iso/n -acetal	0.1	0.06
n /n /n -acetal	0.4	0.4
iso/iso/n -acetal	0.1	0.03
3-methyl-1-butanol	0.1	–
4-heptanol	0.1	–
various (higher boiling than 2-EH)[a]	3.5	1.75
Total	7.5	4.0

[a] includes higher (poly) aldols, Tishchenko esters, butyl butanoates etc.

ends are partially cracked in the decobalter. This effect becomes more noticeable with increasing temperature in the decobalter. The acetals and the trimeric aldehydes are particularly subject to this type of cracking. However, at very high decobalting temperatures there is a reversal in this trend – higher condensates are increasingly formed. These substances have higher boiling points than 2–EH.

It is assumed that these higher condensates are mainly formed via secondary reactions of the aldols with the initial reactants. According to Union Carbide [628, 633, 832, 878, 1482], the following reaction sequence is particularly important – aldolizations, Tishchenko reactions, transesterifications, and other steps:

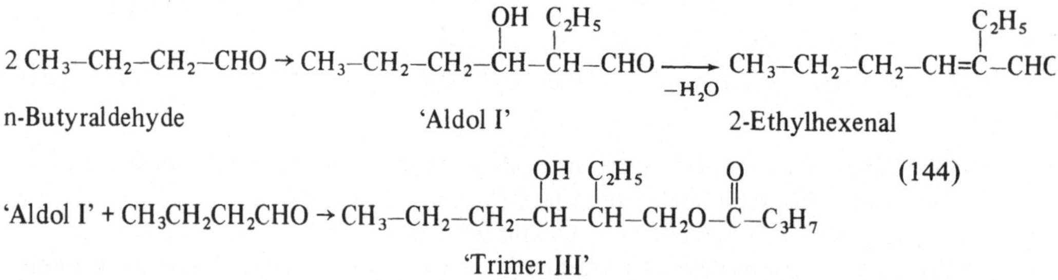

$$2 \; CH_3{-}CH_2{-}CH_2{-}CHO \rightarrow CH_3{-}CH_2{-}CH_2{-}\underset{OH}{CH}{-}\underset{C_2H_5}{CH}{-}CHO \xrightarrow[-H_2O]{} CH_3{-}CH_2{-}CH_2{-}CH{=}\underset{C_2H_5}{C}{-}CHO$$

n-Butyraldehyde 　　　　　'Aldol I' 　　　　　　　2-Ethylhexenal

$$(144)$$

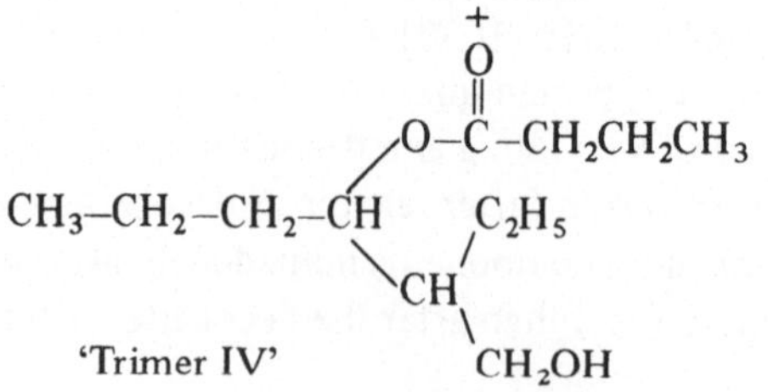

$$\text{'Aldol I'} + CH_3CH_2CH_2CHO \rightarrow CH_3{-}CH_2{-}CH_2{-}\underset{OH}{CH}{-}\underset{C_2H_5}{CH}{-}CH_2{-}O{-}\overset{O}{\underset{\|}{C}}{-}C_3H_7$$

'Trimer III'

$$+$$

$$CH_3{-}CH_2{-}CH_2{-}\underset{\diagdown}{CH} \quad \underset{\diagdown}{\overset{\diagup O{-}\overset{O}{\underset{\|}{C}}{-}CH_2CH_2CH_3}{}}$$

$$\underset{\diagdown}{CH}\underset{\diagdown}{\overset{C_2H_5}{}}$$

$$CH_2OH$$

$$(145)$$

'Trimer IV'

142

$$\underset{\text{`Trimer III'}}{CH_3CH_2CH_2-\overset{\overset{\displaystyle OH}{|}}{CH}-\overset{\overset{\displaystyle C_2H_5}{|}}{CH}-CH_2O-\overset{\overset{\displaystyle O}{\|}}{C}-C_3H_7} \quad \xrightarrow[-C_4H_8O]{\Delta} \quad CH_3CH_2CH_2-\overset{\overset{\displaystyle OH}{|}}{CH}-\overset{\overset{\displaystyle C_2H_5}{|}}{CH}-CH_2OH \quad (146)$$

$$CH_3-CH_2-CH_2-CH\underset{\displaystyle CH}{\overset{\displaystyle O-\overset{\overset{\displaystyle O}{\|}}{C}-CH_2-CH_2CH_3}{\diagdown}}\overset{\displaystyle CH_2OH}{\diagup}\underset{\underset{\displaystyle C_2H_5}{\big|}}{} \quad \xrightarrow[+C_4H_8O]{\Delta} \quad CH_3CH_2CH_2-CH\underset{\displaystyle CH}{\overset{\displaystyle O-\overset{\overset{\displaystyle O}{\|}}{C}-C_3H_7}{\diagdown}}\overset{\displaystyle C_2H_5}{\diagup}\underset{\underset{\displaystyle CH_2-O-\overset{\displaystyle C}{\underset{\displaystyle O}{\|}}-C_3H_7}{\big|}}{} \quad (147)$$

'Trimer IV'

$$2\ CH_3CH_2CH_2-\overset{\overset{\displaystyle OH}{|}}{CH}-\overset{\underset{\displaystyle C_2H_5}{|}}{CH}-CHO \xrightarrow{\Delta} CH_3-CH_2-CH_2-\overset{\overset{\displaystyle OH}{|}}{CH}-\overset{\underset{\displaystyle C_2H_5}{|}}{CH}-\overset{\overset{\displaystyle O}{\|}}{C}-O-CH_2-\overset{\overset{\displaystyle C_2H_5}{|}}{CH}-\overset{\underset{\displaystyle OH}{|}}{CH}-C_3H_7$$

'Aldol I' 'Tetramer VII' (148)

Similar reaction sequences can be formulated using isobutyraldehyde as starting material and others with the reaction between n-and isobutyraldehyde. As these types of components of the heavy ends – in particular 2-ethylhexenal – inhibit the activity of ligand-modified Rh catalysts, it is important to control their concentration in the reactor charge of "stationary" Oxo processes [623] (cf Sect. 1.6.2.2.2).

However, the activity of conventional cobalt catalysts is hardly affected by these compounds which form part of the suspension medium for the Oxo catalysts. The extent and content of the heavy ends are markedly determined by the residual cobalt present in the crude Oxo product and thus by the efficiency of the decobalting. Therefore, extensive decobalting of the crude Oxo product is not only worthwhile from loss considerations.

As Table 1.54 indicates, more than 50% of the components of the heavy ends from the propylene hydroformylation stem from n-butyraldehyde. Thus, in crude Oxo products containing approx. 80% n- and only 20% iso-compounds, the unbranched compounds are clearly under-represented in the heavy ends fraction.

The situation changes with increasing shares of heavy ends; in experiments aiming at heavy ends formation via distillation of crude Oxo products in the presence of $Co_2(CO)_8$ or Co butyrate (cf Fig. 1.61), the n:iso ratio of the residual Oxo product decreases with increasing heavy end content (Table 1.55).

Iron compounds also exert a similar negative effect. This is important for the permissible content of $Fe(CO)_5$ in syn gas (unavoidable under certain circumstances [1621, 1622]).

Only a few methods are suitable for partly recovering the unbranched products present in the heavy ends [1439]. The thermal/chemical cracking processes are still the most convenient, however, it must be ensured that the higher boiling compounds do not form undesired products. For example, the unsaturated ethers resulting from the acetals are converted into saturated ethers during the hydrogenation of the Oxo

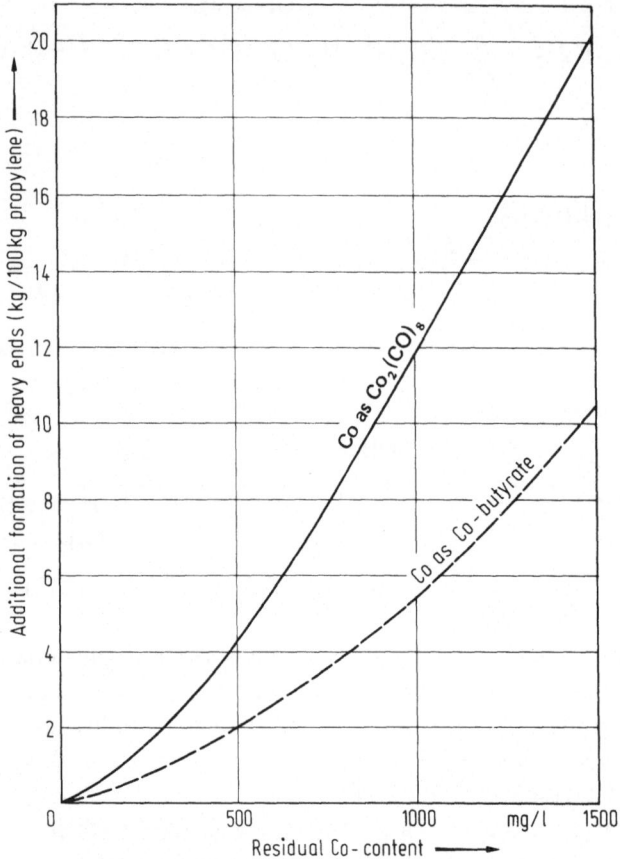

Fig. 1.61. Additional Formation of heavy ends as function of residual Co content during distillation of the C_4 Oxo raw material [1480]

Table 1.55. Dependence of the n:iso ratio of the Oxo crude product on the residual Co content [1480] (values given for Co butyrate)

Residual Co content (mg/l)	n : iso ratio in Oxo crude product
5	78.0
70	77.9
430	77.6
730	76.8
1430	76.2

product and interfere with the butanol fractionation. In addition, olefins are obtained from alcohols, cyclic products from esters, substituted acroleins from aldols via crotonization etc.

$$R-CH_2-CH \underset{O-CH_2-CH_2R}{\overset{O-CH_2-CH_2R}{<}} \xrightarrow{-RCH_2CH_2OH} R-CH=CH-O-CH_2-CH_2-R \tag{149}$$

$$R-CH_2-CH_2-\underset{\underset{R}{|}}{C}H-CH_2OH \xrightarrow{-H_2O} R-CH_2-CH_2-\underset{\underset{R}{|}}{C}=CH_2 \qquad (150)$$

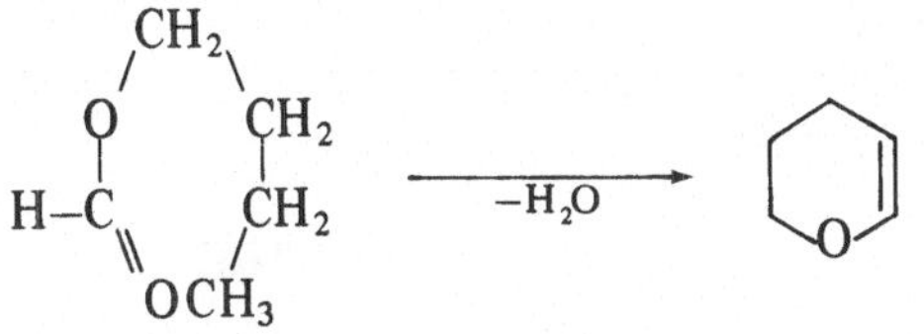

$$\xrightarrow{-H_2O} \qquad (151)$$

$$R-CH_2-\underset{\underset{R}{|}}{C}H-\overset{\overset{OH}{|}}{C}H-CHO \xrightarrow{-H_2O} R-CH_2-CH=\underset{\underset{R}{|}}{C}-CHO \qquad (152)$$

All of these compounds are not useful Oxo products and, in addition, often interfere with distillative purification of the aldehydes and alcohols.

A reduction in the content of heavy ends from the Oxo synthesis can be achieved via suitable reaction parameters (especially reaction temperature cf. Sect. 1.3.1), catalyst concentration (cf Sect. 1.3.3.3) as well as the application of solvents (cf. Sect. 1.3.4). The use of additives [1481] in hydroformylation reactions has not proven to be industrially attractive. One exception is the application of phosphines for ligand modification. Modified Oxo catalysts of this type exhibit not only a lesser tendency to cause condensations (e.g., aldolizations) but – as a result of the hydrogenative activity – lower the concentration of aldehydes available for the secondary reactions.

1.5.2 Selectivity Lowering Secondary Reactions

1.5.2.1 Aldolization of Aldehydes

1.5.2.1.1 Undesired Aldolizations

The aldolization of the Oxo aldehydes which was already mentioned in the previous section occurs under hydroformylation conditions – even without catalysts – to a fairly considerable extent [1415, 1457, 1476, 1483–1487]. As can be appreciated from Table 1.54, unbranched aldehydes are involved, too; this was the starting point for the planned aldolization of the total n-butyraldehyde yield (Aldox process, cf next chapter).

The side reactions accompanying aldol formation [cf Eq. (141)], which lead to a decrease in selectivity, are often reversible. Thus, the monomeric aldehydes can be recovered via reprocessing steps (cf Sect. 1.5.4). These side reactions can however cause a decrease in yield as soon as substituted acroleins are obtained after complete crotonization [Eq. (139)]. These unsaturated aldehydes cannot be cracked to yield useful products.

1.5.2.1.2 Controlled Aldolization (Aldox Variants)

C_{2n+2} alcohols are obtained from C_n olefins via controlled aldolization of the total aldehyde resulting from the hydroformylation (with propylene — preferably only n-butyraldehyde) followed by crotonization of the product and finally, by hydrogenation of the unsaturated aldehyde to the mono-alcohol:

$$2\ R-CH=CH_2 \xrightarrow{CO/H_2} 2\ R-CH_2-CH_2-CHO \xrightarrow[cat.]{} R-CH_2-CH_2-\overset{\overset{\displaystyle OH}{|}}{CH}-\overset{\overset{\displaystyle CH_2R}{|}}{CH}-CHO \quad (153)$$

Olefin C_n

$$\bigg\downarrow -H_2O \qquad (154)$$

$$R-CH_2-CH_2-CH_2-\underset{\underset{\displaystyle CH_2-R}{|}}{CH}-CH_2OH \xleftarrow{\ +\ 2\ H_2\ } R-CH_2-CH_2--CH=\underset{\underset{\displaystyle CH_2-R}{|}}{C}-CHO \quad (155)$$

Alcohol C_{2n+2}

Promoters must be added to the actual Oxo catalyst in order that this "dimerization" ensues (cf Sect. 1.3.3.1.4). Besides Mg ethoxide [1496], oil-soluble Zn compounds have been recommended e.g., oleate, acetate, acetylacetonate etc. [898, 1494, 1495, 1497, 1499, 1517]. Monsanto has also proposed Co–Mn systems [1488]. However, Co and Zn may only be used in a certain relationship to one another as otherwise the Oxo reaction may be suppressed [1498, 1501]. Esso quotes values between 2.8 to 10% Zn (relative to Co [898, 1495]) for this purpose. The Oxo reaction conditions lie between 120–230 °C and 100–300 bar CO + H$_2$. Esso claims that no C_{2n+2} diols are formed i.e., the partial conversion [Eq. (154)] almost goes to completion [898]. The hydrogenation reaction in [Eq. (155)] can be supported by additives such as Pd/C [1492].

Co modifiers or activators are recommended by Esso for accelerating the reaction and to increase the yield [1499, 1500]. Hg, Pb, Bi or Tl salts are all active particularly when Zn is replaced by Mg, Ba, Cd, Sr, Ca or Be salts [1500]. However, the effect is only moderate (cf Table 1.56).

Table 1.56 underlines the fact that the controlled aldolization does not ensue quantitatively. Besides the unreacted olefin (as in the hydroformylation), some of the monomeric aldehyde also remains undimerized. Both products can be recycled

Table 1.56. Influence of co-activators on the yield of C_8 and C_{16} alcohols from heptenes [1494, 1500] St = stearate, Ol = oleate, Co(Ol)$_2$:0.033 mol/l, modifier: 0.022 mol/l

Modifier	Co-activator (mol/mol modifier)	Yields (mol-%) C_8	C_{16}	Total
not present	not present	78.4	4.6	83.0
Zn(St)$_2$	not present	45.0	38.0	83.0
Zn(St)$_2$	0.25 Bi(Ol)$_2$	36.6	48.7	85.3
Zn(St)$_2$	0.55 Hg(St)$_2$	36.0	45.8	81.8
Zn(St)$_2$	0.55 Pb(Ol)$_2$	31.3	54.6	85.9
Mg(St)$_2$	not present	68.2	19.8	88.0
Sr(St)$_2$	not present	72.1	13.8	85.9
Sr(St)$_2$	5.5 Pb(St)$_2$	66.2	19.8	86.0

to the process. This is particularly advantageous on converting propylene to 2-ethylhexanol especially when the n-butyraldehyde is separated from the isobutyraldehyde and it alone is recycled [1489]. The direct aldolization of the n-aldehyde in product streams containing isomeric mixtures has been dealt with in the literature [1493]. Frequently, special catalysts and conditions must be employed when hydrogenating unsaturated aldehydes [1490].

Besides ligand-modified Co catalysts which can be made active as in the Aldox process on adding KOH (Shell process cf Sect. 1.3.3.1.4 [710, 899, 900]), modified Rh catalysts such as $[(C_6H_5)_3Rh]_2$ catalyze not only the hydroformylation but also the subsequent aldolization [1491].

The serious disadvantages of the Aldox variant of the Oxo process were already outlined in Sect. 1.3.3.1.4. These obstacles prevent a wider industrial application [896, 1488, 1492, 1493].

1.5.2.2 Hydrogenation of Aldehydes to Alcohols

1.5.2.2.1 Undesired Hydrogenation

Almost all crude Oxo products contain measurable to noticeable amounts of alcohols resulting from the homogeneously catalyzed hydrogenation of Oxo aldehydes [301, 1421, 1502–1504, 1529–1531]. Previously it was thought that the reaction was heterogeneously catalyzed by precipitated Co [1532].

$$R-CHO + H_2 \xrightarrow[cat.]{} RCH_2OH \qquad (156)$$

Assuming the stage in Eq. (133) has been reached, Marko's [301, 1521] proposed mechanism can be formulated as follows (cf [6]):

$$R-CH_2-O-Co(CO)_4 + H_2 \rightarrow R-CH_2-O-CoH_2(CO)_4 \qquad (157)$$

$$R-CH_2-O-CoH_2(CO)_4 \rightarrow R-CH_2OH + HCo(CO)_4 \qquad (158)$$
$$\text{alcohol}$$

This explains why the rate of hydrogenation initially increases with growing CO partial pressure i.e., until all the tricarbonylhydridocobalt is converted into the tetracarbonyl derivative. A corresponding equilibrium ensues before Eq. (157) and then recedes in importance (for discussion cf [6]).

The extent of this secondary reaction can be increased via raised temperatures (Sect. 1.3.1), via high hydrogen partial pressures (Sect. 1.3.2), via extremely high catalyst concentrations (Sect. 1.3.3.3) and via lengthy residence times (Sect. 1.3.6). Specially constructed Oxo reactors (multi-stage operation, reactors with increasing temperature profile along their longitudinal axis etc., cf Sect. 1.6.1) can also encourage this secondary reaction, in the extreme case leading to the complete hydrogenation of the primary Oxo aldehyde to the corresponding alcohol (cf Sect. 1.5.2.2.2). However, it must be borne in mind that the above measures could lead to not only the secondary reaction involving alcohol formation being encouraged but also to other secondary (ester and heavy ends formation) and parallel (olefin hydrogenation) reactions coming to the forefront.

This side reaction, which also takes place in the presence of Rh catalysts [1506–1509, 1803], is therefore only significant when it either occurs without loss of yield (e.g., with a higher olefin feedstock which is to produce solely Oxo alcohols) or if it leads to particular advantages (cf next Sect.).

1.5.2.2.2 Controlled Hydrogenation

Besides functioning as Oxo catalyst, tetra(tri)-carbonyl-hydridocobalt can also encourage the secondary reaction – the homogeneous hydrogenation of aldehydes to alcohols (cf pp. 5 and 6 and Ref. [1510–1513]). This fact was recognized at an early stage in the development of the Oxo process and exploited in hydroformylations with fixed-bed catalysts [294, 1514, 1515].

When hydrogenating the total aldehyde yield to Oxo alcohols the following factors must be closely controlled: –
- Reaction temperatures (should be as high as possible, up to 220 °C)
- Reaction pressure (high partial pressure of H_2, however p_{CO} must be adequate to guarantee stability of the cobalt carbonyls, cf Fig. 1.9).
- Residence times (should be as long as possible, however not too lengthy as otherwise the other secondary reactions will dominate)
- Catalyst concentration (as high as possible, but should not catalyze the olefin hydrogenation to saturated hydrocarbons).

These aspects are discussed in Sects. 1.3.1–1.3.6 as well as in the Ref. [1511, 1512, 1516, 1518, 1519, 1804]. The effect of these reaction parameters can be fortified by means of suitable reactor design (cf Sect. 1.6.1).

In the following examples it is useful to be able to obtain Oxo alcohols from the combined Oxo and hydrogenation steps: –
- When the Oxo products are to be further processed to alcohols (plasticizer or detergent alcohols). In these cases, as the initial product has already been largely hydrogenated only a post-hydrogenation is necessary. Thus a smaller reactor is adequate for the post-hydrogenation step.
- The olefin feedstock is contaminated with sulfur such that a separate heterogeneously catalyzed hydrogenation of the Oxo aldehydes would probably involve high consumption of catalyst. In this instance, the relative insensitivity of the homogeneous catalysts towards poisons can be exploited [1510, 1520, 1522], supported by their regeneration during recycling (cf Sects. 1.3.3.2 and 1.3.3.4).
- When only CO-containing hydrogen or syn gas is available for the hydrogenation [1519, 1523, 1528].

The economic impact of these Oxo variants has remained insignificant.

Exploiting the powerful hydrogenation activity of the ligand-modified Co catalysts (cf pp. 45/53) is of much more importance. In this way Oxo aldehydes can be converted into the corresponding alcohols using comparatively mild conditions to limit by-product formation. While there are a number of variants [1525–1527], the Shell process is the most well-known (Sect. 1.6.2.1.4) [1524]. Although around 400,000 t of Oxo alcohols are produced annually using the Shell process they represent less than 10% of the total world production (Sect. 1.6.4).

A summary of the advantages and disadvantages of converting olefins to Oxo alcohols in a single step is shown in Table 1.57.

Table 1.57. Advantages and disadvantages of the single-step process for Oxo alcohol production

Advantages:	• Oxo reaction and hydrogenation occur in *one* reactor (lowers investment costs) and uses only *one* catalytically active metal • sulfur-containing substrates can be reacted under certain conditions • it is possible to employ CO-containing hydrogen for the hydrogenation step • decobalting and hydrogenation can ensue in one step (Sect. 1.3.3.4 and Ref. [1511])
Disadvantages:	• the forced hydrogenation of the Oxo aldehydes also encourages other reactions • despite single stage hydroformylation/hydrogenation, a separate post-hydrogenation of the Oxo alcohols is almost always essential (invalidating the investment advantage) due to quality considerations • the partly severe conditions during alcohol formation lower the share of the unbranched alcohols and can cause difficulties with Co deposition in the reaction system

1.5.2.3 Synthesis of Isomeric Oxo Products

1.5.2.3.1 The n : iso Problem of the Oxo Synthesis

With Propylene or Butylenes

On reflecting that the maximum ratio of unbranched ('n') to branched ('iso') products is 80 : 20 for the Oxo reaction catalyzed by unmodified Co catalysts then the examples in Figs. 1.19 or 1.53 help underline the fact that the formation of isomeric Oxo aldehydes is usually more serious than the other parallel and secondary reactions. With propylene – depending on Oxo process – the following product compositions are obtained with the classical method:

Table 1.58. Composition of the products from the Oxo synthesis [1474]

Products	Content of crude Oxo product (%)	Yield in kg/100 kg propylene[a]
isobutyraldehyde	17.2	30.3
n-butyraldehyde	69.8	123.0
Σ butyl formates	5.2	–
isobutanol	0.9	3.4
n-butanol	2.5	9.2
heavy ends	4.4	6.9

[a] after total cracking of formates and partial cracking of heavy ends

Thus around 30 kg isobutyraldehyde (per 100 kg propylene) are formed compared to 20 kg of various by-products. As can be readily appreciated, this aldehyde has been the subject of considerable attention over a long period. Besides being employed as an intermediate for chemical syntheses (pp. 152 f), the cracking of isobutyraldehyde to propylene and syn gas (pp. 154 f) and its isomerization to an unbranched aldehyde (pp. 155/156) have also been studied. In addition, there was also intensive research to reduce by-product formation by varying reaction conditions (Sects. 1.3.1, 1.3.2, 1.3.4, and 1.3.5), the Oxo catalyst (Sect. 1.3.3) and via suitable reactor design (Sects. 1.3.6, 1.3.7 and 1.6).

Isobutyraldehyde – the "iso" product from the propylene hydroformylation – has undergone a change in demand as can be seen from the existing markets for this substance. Initially, it was mainly combusted, being regarded as an undesired by-product, then its cracking/isomerization became a topic of increasing interest. Currently, its application as an intermediate is of such growing interest that, already too little isobutyraldehyde would be available for several potential applications.

The isomeric distribution of the Oxo products of the olefin homologoues is of much lesser interest. With butenes for example, all the isomers – 1-butene and 2-butene-present as a mixture in the C_4 cut and isobutene-(2-methylpropene)–undergo industrial hydroformylation. Various products are obtained:

$$
\begin{array}{c}
CH_3-CH_2-CH=CH_2 \\
+ \\
CH_3-CH=CH-CH_3
\end{array}
\xrightarrow[\text{cat.}]{+CO/H_2}
\underset{\text{2-Methylbutanal}}{CH_3-CH_2-\overset{\overset{\displaystyle CH_3}{|}}{CH}-CHO} +
\underset{\text{n-Valeraldehyde}}{CH_3-(CH_2)_3-CHO}
\quad (159)
$$

$$
\underset{}{CH_3-\overset{\overset{\displaystyle CH_3}{|}}{C}=CH_2}
\xrightarrow[\text{cat.}]{+CO/H_2}
\underset{\text{3-Methylbutanal}}{CH_3-\overset{\overset{\displaystyle CH_3}{|}}{CH}-CH_2-CHO} +
\underset{\substack{\text{2,2-Dimethyl-}\\\text{propanal}}}{CH_3-\overset{\overset{\displaystyle CH_3}{|}}{\underset{\underset{\displaystyle CH_3}{|}}{C}}-CHO}
\quad (160)
$$

1-Butene or 2-butene generally yields mixtures of 50–80% n-valeraldehyde and 20–50% 2-methylbutanal [4, 1210] which, after hydrogenation, are marketed as amyl alcohol. The individual isomers – obtained via distillation of the aldehyde or alcohol product – can also be used as intermediates [4, 25, 35, 43, 57, 1210]. While the relative content of the individual isomers in the amyl alcohol mixture is unimportant for many uses sometimes a particularly high content of the 'iso' isomers are required for special applications on account of the desired viscosity of the secondary products e.g., processing of amyl alcohol to synthetic lubricants.

Besides containing a considerable amount of 3-methylbutanal (isovaleraldehyde), the crude Oxo product (from the isobutene hydroformylation) also has a little 2,2-dimethylpropanal present ('Keulemans' rule', cf Sect. 1.4.1).

In this case, the isomers are generally employed individually, both in the aldehyde and alcohol form.

With Higher Olefins as Feedstocks

Currently, the hydroformylation of pentenes or hexenes has made no industrial impact. However, the conversion of 1-hexene to n-heptylaldehyde could become significant.

$$CH_3-(CH_2)_3-CH=CH_2 \xrightarrow[\text{cat.}]{\text{+ CO/H}_2} CH_3-(CH_2)_4-CH_2-CHO \ + CH_3-(CH_2)_3-\underset{\underset{CH_3}{|}}{CH}-CHO$$

n-Heptanoic acid or n/isoheptanoic acid mixtures (from the oxidation of the aldehydes) are becoming increasingly important for the manufacture of synthetic lubricants. The present source (by-product from the production of ω-amino undecanoic acid, the nylon 11 precursor) will be exhausted in the not too distant future.

The C_9 Oxo products from diisobutylene — consisting mainly of 2,2,4-trimethyl-pentenes — must maintain a certain isomeric ratio for commercial reasons. The chief reaction product — the 3,5,5-trimethylhexyl isomer — must be present to more than 90% in the C_9 alcohol or in isononanoic acid [4]. These are the reasons for the special properties of the C_9 products from diisobutylene compared to those from the C_8 codimer.

The desired isomeric distribution of the Oxo products $> C_9$ (with mixtures: $> C_7$) is mainly determined by the application. Thus, isomeric mixtures of carboxylic acids of medium C number (7/8, 8/9, or 9/10) are particularly desired as the corresponding metal salts are highly oil-soluble, this property is important in their application as paint driers. With plasticizer alcohols of C number 8/9, 9/10, or 9/11, the volatility of the resultant esters (decreases with increasing C number and increasing share of unbranched compounds) must be balanced against PVC compatibility. In general, the compatibility is the more important aspect. Consequently, considerable amounts of branched plasticizer alcohols can be tolerated, e.g., 2-ethyl-hexanol — the most important branched plasticizer alcohol.

On the other hand, with detergent alcohols — manufactured via the hydroformylation of (preferably) α-olefins of C numbers 10–14, 12–14, 14–16 or 11–16 — minimum contents of the (more readily biologically degradable) unbranched secondary products are desired. According to present information — affected somewhat by the economic consequences of the oil crisis in 1973 — an n:iso ratio of $\geq 35:65$ is adequate. Such n:iso ratios can even be attained on hydroformylating olefins with internally located double bonds [1533, 1534]. However, high grade detergents based on Oxo alcohols contain, as before, over 65% unbranched derivatives.

On using Oxo alcohols or carboxylic acids based on Oxo aldehydes, the required n:iso ratio depends on the particular application. Frequently, minimum contents of branched isomers are required for reasons based on an acceptable viscosity index.

The Hydroformylation of Functionally Substituted Olefins

As several examples were already discussed in Sect. 1.4 only a few representative reactions will be mentioned here.

When hydroformylating allyl acetate, it is important that there is a high share of the unbranched 4-acetoxy isomers if they are to serve as precursors for 1,4-butanediol [244].

$$H_2C{=}CH{-}CH_2{-}O{-}CO{-}CH_3 + CO/H_2 \rightarrow OHC{-}CH_2{-}CH_2{-}CH_2{-}O{-}COCH_3 \qquad (161)$$

$$+ H_2 \qquad (162)$$

$$HOCH_2{-}(CH_2)_2{-}CH_2OH + CH_3COOH \xleftarrow{\ H_2O\ } HOCH_2{-}CH_2{-}CH_2{-}CH_2{-}O{-}COCH_3 \qquad (163)$$

With allyl alcohol there is a similar situation (cf Sect. 1.4), and on using acrylates, methacrylates or other unsaturated esters, the desired isomeric distribution is pre-determined by the properties of the potential secondary products.

This is also important in reactions with unsaturated nitriles. For example, with acrylonitrile the main Oxo product (β-cyanopropionaldehyde) is industrially manu-factured and converted to glutamic acid via secondary reactions [1112].

$$H_2C{=}CH{-}C{\equiv}N + CO/H_2 \xrightarrow[\text{cat.}]{} OHC{-}CH_2{-}CH_2{-}C{\equiv}N \qquad (164)$$

As discussed in Sect. 1.3, when hydroformylating functionally substituted olefins, the desired isomeric ratio — which should be better termed β:α and not as the n:iso ratio — must be considered in connection with steric *and* noticeable electronic effects. Besides the reaction parameters, the direction of the CO/H_2 addition can also be influenced by the choice of catalyst (Co or Rh, modified or unmodified). Very little has been published about the special know-how used in this sphere.

1.5.2.3.2 Isobutyraldehyde as the 'main by-product' of the Propylene Hydroformylation

Due to the great significance of the propylene hydroformylation (cf Sect. 1.6), iso-butyraldehyde represents — also in a quantitative sense — the 'main by-product' of the Oxo synthesis. Consequently, its further processing or up-grading has been the subject of intensive research.

Secondary Reactions with Isobutyraldehyde

n-Butyraldehyde is largely only further processed to one secondary product — 2-ethyl-hexanol, all other secondary reactions to n-butanol, n-butyric acid, n-butylamine, n-butyronitrile etc. being insignificant in comparison. The intensive research on possible useful secondary reactions of isobutyraldehyde has led to a greater number of reaction variants with this isomer [1631].

The conversion of isobutyraldehyde to isobutanol or isobutyric acid [1535–1541, 1633, 1634] is closely associated to the analogous reactions with n-butyraldehyde.

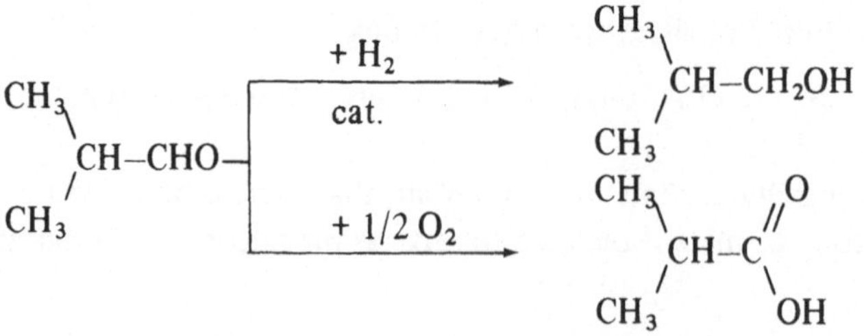

$$(165)$$

$$(166)$$

The Tishchenko reaction yields isobutyl isobutyrate [1542, 1543].

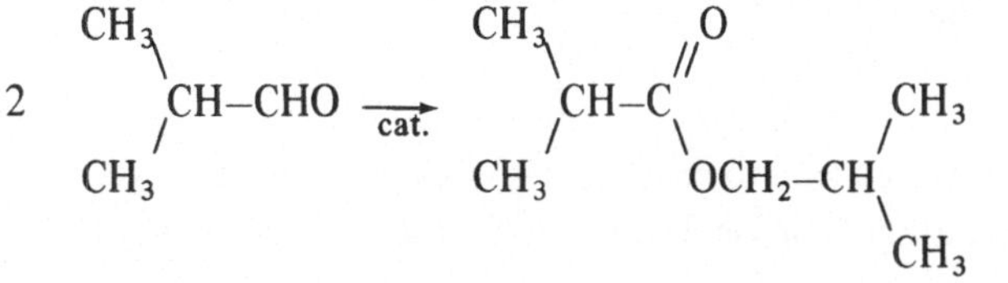

$$2\ \underset{CH_3}{\overset{CH_3}{\diagdown}}CH-CHO \xrightarrow{cat.} \underset{CH_3}{\overset{CH_3}{\diagdown}}CH-\overset{\overset{O}{\parallel}}{C}\underset{OCH_2-CH\diagdown CH_3}{\diagup}CH_3 \tag{167}$$

Depending on the secondary reaction, the aldol addition yields 2,2,4-trimethyl-1,3-pentanediol or 2,2,4-trimethylpentanol [1544–1548, 1620].

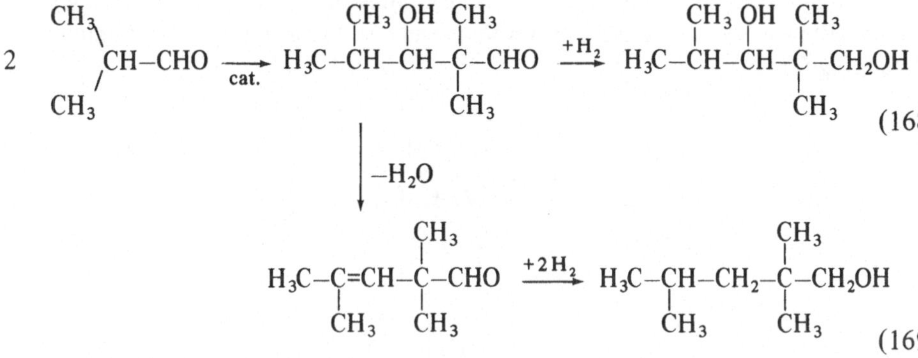

$$2\ \underset{CH_3}{\overset{CH_3}{\diagdown}}CH-CHO \xrightarrow{cat.} H_3C-\overset{CH_3}{\underset{}{CH}}-\overset{OH}{\underset{}{CH}}-\overset{CH_3}{\underset{CH_3}{C}}-CHO \xrightarrow{+H_2} H_3C-\overset{CH_3}{\underset{}{CH}}-\overset{OH}{\underset{}{CH}}-\overset{CH_3}{\underset{CH_3}{C}}-CH_2OH \tag{168}$$

$$\downarrow -H_2O$$

$$H_3C-\overset{CH_3}{\underset{}{C}}=CH-\overset{CH_3}{\underset{CH_3}{C}}-CHO \xrightarrow{+2H_2} H_3C-\overset{CH_3}{\underset{}{CH}}-CH_2-\overset{CH_3}{\underset{CH_3}{C}}-CH_2OH \tag{169}$$

The mixed aldolization of isobutyraldehyde yields – with formaldehyde – hydroxy-pivalaldehyde, which on being hydrogenated opens a route to the industrially important neopentyl glycol [1549–1568].

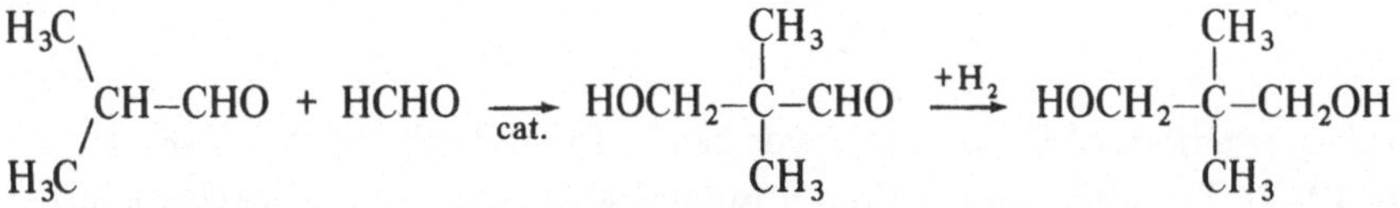

$$\underset{H_3C}{\overset{H_3C}{\diagdown}}CH-CHO + HCHO \xrightarrow{cat.} HOCH_2-\overset{CH_3}{\underset{CH_3}{C}}-CHO \xrightarrow{+H_2} HOCH_2-\overset{CH_3}{\underset{CH_3}{C}}-CH_2OH \tag{170}$$

Hydroxypivalaldehyde also serves as a precursor for hydroxypivalic acid or its amide, 2,3-dimethyl-1,3-propylenediamine, pantolactone and other compounds [1556, 1564, 1569–1575, 1631].

Acetone [1576–1580, 1918] can be obtained from isobutyraldehyde via oxidative decarbonylation and methyl ethyl ketone (2-butanone) results via rearrangement [1581–1583, 1641]. Diisopropyl ketone is yielded on subjecting 2,2,4-trimethylpentanediol to catalytic cleavage [1584].

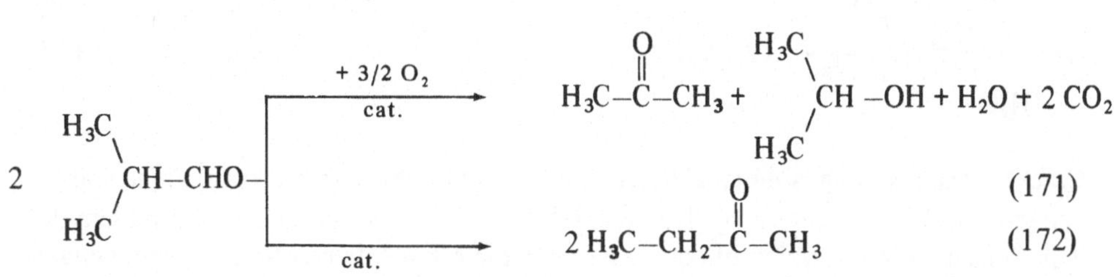

$$2\ \underset{H_3C}{\overset{H_3C}{\diagdown}}CH-CHO$$

$$\xrightarrow[cat.]{+3/2\ O_2} H_3C-\overset{\overset{O}{\parallel}}{C}-CH_3 + \underset{H_3C}{\overset{H_3C}{\diagdown}}CH-OH + H_2O + 2\ CO_2 \tag{171}$$

$$\xrightarrow{cat.} 2\ H_3C-CH_2-\overset{\overset{O}{\parallel}}{C}-CH_3 \tag{172}$$

153

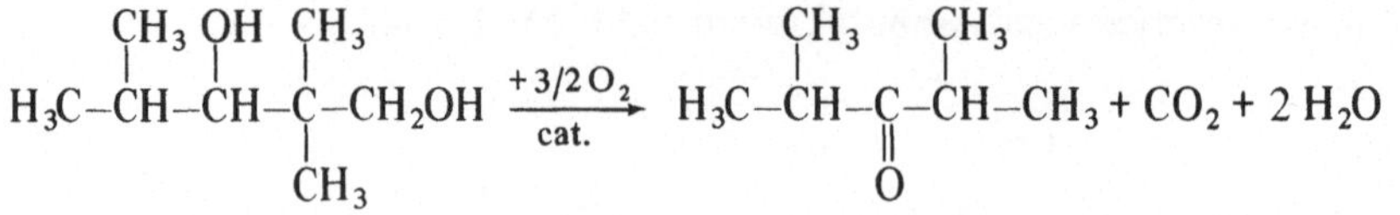

$$\underset{\underset{CH_3}{|}}{\overset{\overset{CH_3\ OH\ CH_3}{|\ \ \ |\ \ \ |}}{H_3C-CH-CH-C-CH_2OH}} \xrightarrow[\text{cat.}]{+3/2\,O_2} \underset{\overset{||}{O}}{\overset{\overset{CH_3\ \ \ CH_3}{|\ \ \ \ \ |}}{H_3C-CH-C-CH-CH_3}} + CO_2 + 2\,H_2O \quad (173)$$

The oxidative gas phase decarbonylation to acetone has already been tested on pilot plant scale. Acetone can be obtained in yields up to 93%, the isobutyraldehyde conversion being almost complete.

After dehydrogenating, isobutyraldehyde, isobutyric acid and methyl isobutyrate yield methacrolein, methacrylic acid and methyl methacrylate (MMA), respectively [1585–1597]. In particular, the reaction sequence isobutyraldehyde → isobutyric acid → methyl isobutyrate → MMA could become industrially important, replacing the previous route via acetone cyanohydrin which results in a large quantity of ammonium sulfate being formed.

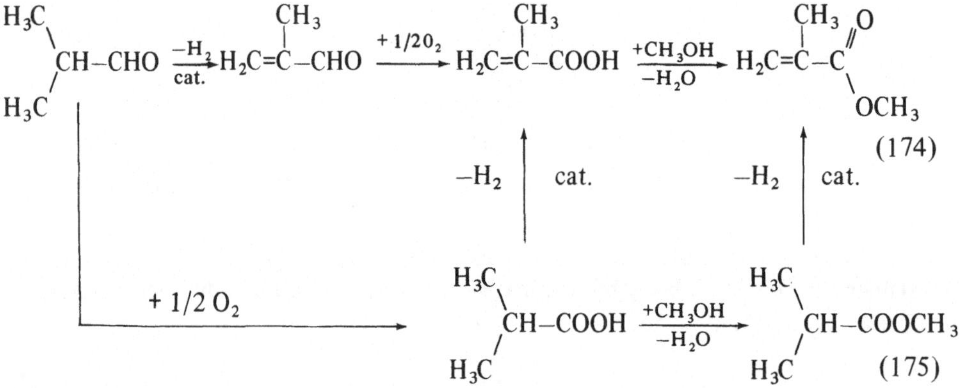

$$\underset{H_3C}{\overset{H_3C}{\diagdown}}CH-CHO \xrightarrow[\text{cat.}]{-H_2} \underset{}{\overset{\overset{CH_3}{|}}{H_2C=C-CHO}} \xrightarrow{+1/2O_2} \underset{}{\overset{\overset{CH_3}{|}}{H_2C=C-COOH}} \xrightarrow[-H_2O]{+CH_3OH} \underset{OCH_3}{\overset{\overset{CH_3\ \ \ O}{|\ \ \ \ ||}}{H_2C=C-C}}$$

$$\text{(174)}$$

$$-H_2 \Big| \text{cat.} \qquad -H_2 \Big| \text{cat.}$$

$$+1/2\,O_2 \longrightarrow \underset{H_3C}{\overset{H_3C}{\diagdown}}CH-COOH \xrightarrow[-H_2O]{+CH_3OH} \underset{H_3C}{\overset{H_3C}{\diagdown}}CH-COOCH_3$$

$$\text{(175)}$$

Other secondary reactions of isobutyraldehyde can be found in the Refs. [1598–1614, 1631, 1936, 1937]. The direct application of isobutyraldehyde – as a slow-release fertilizer – has been treated in several reports [1615–1617].

Cracking of Isobutyraldehyde

Besides converting isobutyraldehyde to oxygen-containing compounds such as alcohols, acids or ketones, it can also be cracked yielding its precursors (cf Falbe et al. [1636–1640] and Orlicek [1635]:

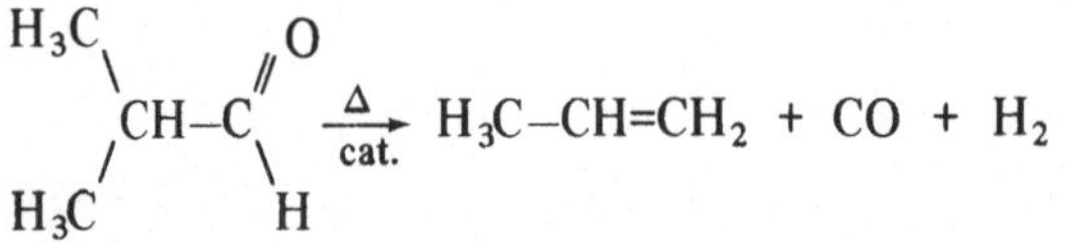

$$\underset{H_3C}{\overset{H_3C}{\diagdown}}CH-\underset{H}{\overset{\overset{O}{||}}{C}} \xrightarrow[\text{cat.}]{\Delta} H_3C-CH=CH_2 + CO + H_2 \quad (176)$$

Cracking the resulting isobutyraldehyde and recycling the reactants propylene and syn gas, would mean that propylene could be totally converted into n-butyraldehyde. This consideration – which of course neglects the other side reactions – would enable the n : iso problem to be overcome (Fig. 1.62):

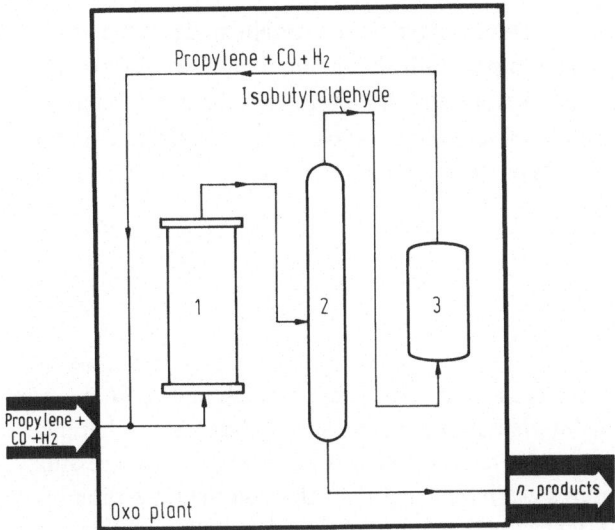

Fig. 1.62. Propylene hydroformylation coupled with isobutyraldehyde cracking [1642, 1643] (1-Oxo reactor, 2-aldehyde distillation, 3-isobutyraldehyde cracking)

Supported catalysts are favored, elements of the 8th group of the Period System being employed as the catalytically active metals, in particular nickel [1636] or noble metals such as palladium, platinum or rhodium [1635, 1637, 1639, 1642, 1645].

Introducing steam to the gaseous isobutyraldehyde improves not only reaction control but also the service life of the catalysts. The isobutyraldehyde cracking occurs – depending on catalyst – between 250° and 600 °C, pressures of 2 to 30 bar being applied to improve reactant separation.

While this additional variant of the Oxo synthesis has been industrially tested, the cracking of isobutyraldehyde to isobutene has aroused little interest [1644]. The cracking of acetoxybutanals – the hydroformylation products of allyl acetate – has been treated in Sect. 1.4.4.3.

The re-formation of olefins from acylcobalt complexes during the hydroformylation has already been dealt with in Sect. 1.2.2 [Eq. (26)] [152, 154].

Isomerization of Isobutylraldehyde

The earlier work of Eastman Kodak [1125] was based on the following isomerization equilibrium:

$$H_3C-CH_2-CH_2-C\overset{O}{\underset{H}{\diagdown}} \rightleftharpoons H_3C-\overset{CH_3}{\underset{|}{CH}}-C\overset{O}{\underset{H}{\diagdown}} \tag{177}$$

Eastman Kodak conducted the propylene hydroformylation (177) in dilute isobutyraldehyde, thereby apparently suppressing isobutyraldehyde formation. However, the claims made in the above patent could not be confirmed. Moreover, aldehydes are particularly unsuitable solvents for the Oxo reaction on account of their reactivity.

155

A more realistic possibility would be to isomerize isobutyraldehyde to the un-branched aldehyde in a separate reaction step with aluminium oxides. At 360 to 600 °C besides n-butanal mainly methyl ethyl ketone and diisopropyl ketone result, some-times along with methacrolein as by-product [1641, 1646]. Currently, the maximum selectivity to n-butyraldehyde lies around 16%.

1.5.3 Various Side Reactions

The basic component of each Oxo process is the Oxo reactor which must ensure – via constructive measures facilitating internal or external cooling – that the exothermic heat of reaction (ca. 125 kJ/mol) is removed (cf Sect. 1.6.1). If this is not successful then very high reaction temperatures will have a negative effect on the formation of unbranched compounds (Sect. 1.3.1) as well as encouraging side reactions (Sects. 1.5.1 and 1.5.2) or lead to cobalt deposition in the reaction system (Fig. 1.9).

When considering Oxo reactor design, besides taking efficient cooling and reaction parameters into account, it must also be guaranteed that the p_{CO} is sufficiently high in all sections of the reactor to prevent decomposition of the cobalt carbonyls leading to deposition of metallic cobalt.

Such deposits coupled with inadequate cooling could be the starting point for a disastrous sequence of reactions: If the olefin feed is disrupted this would lead to elevated partial pressures of CO and H_2. The syn gas could then react exothermally with the deposits forming $HCo(CO)_4 (\Delta H = -120 kJ/mol)$. This reaction could then initiate a subsequent methanation (also exothermic) of syn gas ($\Delta H = -252$ kJ/mol) or Fischer-Tropsch reaction ($\Delta H = -655$ kJ/mol for C_4H_{10}). "Self catalysis" can also ensue, whereby the exothermic hydrocarbonyl formation causes the liquid surround-ing the active cobalt to evaporate, releasing the dissolved CO and H_2. In this way CO and H_2, the reactants of the methanation or the Fischer-Tropsch synthesis, are released at exactly the point where catalytically active cobalt is present and where local super-heating has occurred (consequence of hydrocarbonyl formation). This supports the fact, that hydrocarbonyl formation ensues with consumption of much CO initially leading to a depletion of CO in the surrounding medium, making the remaining H_2-rich gas particularly suitable for the methanation. This interpretation can be experimentally verified on using autoclaves with excess cobalt metal. While the simulated reaction sequence hydrocarbonyl formation → methanation does not take place in the presence of CO or H_2 alone it ensues at lower temperatures the higher the H_2:CO ratio is [1414, 1647].

Other reactions, frequently denoted as side reactions, are actually secondary reactions involving the hydroformylation products e.g., decarbonylation of the alde-hydes during their hydrogenation to alcohols [1648], dehydration of the alcohols to ethers etc.

1.5.4 Work up of By-products of the Oxo Synthesis to Value Products

Part of the Oxo by-products result via an irreversible reaction and can neither be cracked to the feedstocks nor further processed to higher grade products. The paraffins, which are formed via olefin hydrogenation, are a typical example. They can generally be readily separated from the Oxo products, the exception being solely with mixtures of higher alcohols where the boiling range of the hydrocarbons C_{n+2} or C_{n+3} may interfere with that of the C_n alcohol e.g., detergent alcohols C_{12} – C_{15}.

Other by-products disturb the distillative purification of the commercially interesting Oxo products, necessitating either their previous chemical conversion (e.g., hydrogenation of ketones to sec.alcohol) or a more involved distillation (e.g., azeotropic removal of dibutyl ethers from butanols). In both cases, these special purification techniques must take place due to quality demands.

A third class of by-products can – via various process steps – be totally or partially converted into Oxo value products e.g., alkyl formates, acetals and trimers present in the heavy ends.

The alkyl formates can be introduced to the series of reactions shown below either as a distillatively-enriched "ester fraction" mixed with alcohols or mixed with alcohols and heavy ends.

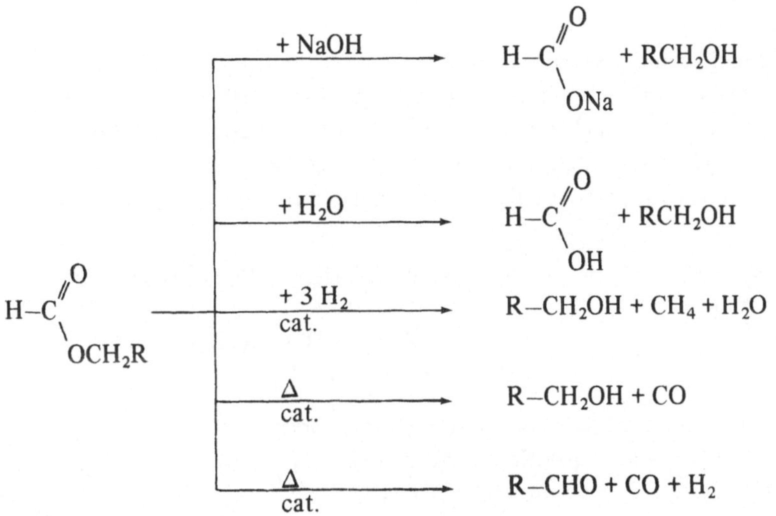

$$
\begin{array}{ll}
\xrightarrow{\ +\ NaOH\ } & H-C{\overset{O}{\underset{ONa}{\big\backslash}}} + RCH_2OH \qquad (178) \\[2em]
\xrightarrow{\ +\ H_2O\ } & H-C{\overset{O}{\underset{OH}{\big\backslash}}} + RCH_2OH \qquad (179) \\[2em]
\xrightarrow[\text{cat.}]{\ +\ 3\ H_2\ } & R-CH_2OH + CH_4 + H_2O \qquad (180) \\[1em]
\xrightarrow[\text{cat.}]{\ \Delta\ } & R-CH_2OH + CO \qquad (181) \\[1em]
\xrightarrow[\text{cat.}]{\ \Delta\ } & R-CHO + CO + H_2 \qquad (182)
\end{array}
$$

Starting material: $H-C{\overset{O}{\underset{OCH_2R}{\big\backslash}}}$

The most frequently used method is the acidic or alkaline saponification of the alkyl formates [995, 1235, 1236, 1649–1652]. Resulting formic acid/Na formate can then be converted into crack products in subsequent reaction steps. This is frequently necessary from pollution considerations and can have a negative effect on the application of this variant [1653–1655].

Hydrogenative cracking of alkyl formates [Eq. (180)] is also a standard method [1656–1658]. However, this reaction is hydrogen-consuming and the catalyst consumption is generally considerable on account of the drastic conditions. Supported nickel catalysts are usually employed. The formates are therefore subjected to alkaline cleavage before the hydrogenation [1649]. Moreover, distillates free of heavy ends should be employed.

In the thermal or thermal/catalytic cracking processes, the cleavage of the alkyl esters is combined in one step with the cleavage of the formate residue, Eq. (181) [1237, 1238, 1239, 1659]. The advantage is that neither the hydrogenating catalyst nor hydrogen are consumed and, in addition, no formate-containing effluents result. Under suitable reaction conditions, besides the esters, considerable amounts of heavy ends can be converted into value products – in particular aldehydes [1237, 1238]. There is a similar situation in cracking units according to Eq. (182) [1233].

The alkyl carboxylates (e.g., butyl butyrate) in the high boiling heavy ends are mostly cleaved via alkaline saponification. The resulting carboxylic acid can then be isolated [1660].

$$
R-C{\overset{\displaystyle O}{\underset{\displaystyle OCH_2-R}{}}} + NaOH \rightarrow
\begin{cases}
\xrightarrow{+ H_2SO_4} R-C{\overset{\displaystyle O}{\underset{\displaystyle OH}{}}} + NaHSO_4 & (183) \\
R-C{\overset{\displaystyle O}{\underset{\displaystyle ONa}{}}} + RCH_2OH & (184)
\end{cases}
$$

The following techniques have been used to crack the trimers, aldols and/or acetals present in the higher boiling fractions of the Oxo products: treatment with alkali [1661], acids [1954], hydrogenation [1662, 1663, 1667, 1912] or thermal cracking [1233, 1664]. Treatment with acidic ion-exchangers [1665], thermal cracking [1666] or reaction with steam [1240, 1668, 1669] is particularly recommended for converting acetals:

$$
R-CH_2-C{\overset{\displaystyle O-CH_2-CH_2-R}{\underset{\displaystyle O-CH_2-CH_2-R}{\big|}}}-H
\begin{cases}
\xrightarrow{+ H_2O} RCH_2CHO + 2\,RCH_2CH_2OH & (185) \\
\xrightarrow{\Delta} \begin{cases} R-CH_2CHO + RCH_2CH_2OH \\ + R-CH=CH_2 \end{cases} & (186)
\end{cases}
$$

It should be noted that the process variants for working up the by-products of the Oxo synthesis are frequently specific for the particular Oxo stage, exploiting the possibility of utilizing/recycling product streams.

1.6 The Industrial Oxo Synthesis: Process Variants and Economic Background

1.6.1 Industrial Aspects of the Oxo Synthesis

Fig. 1.63 is a schematic representation of an Oxo plant which is typical for the various modifications.

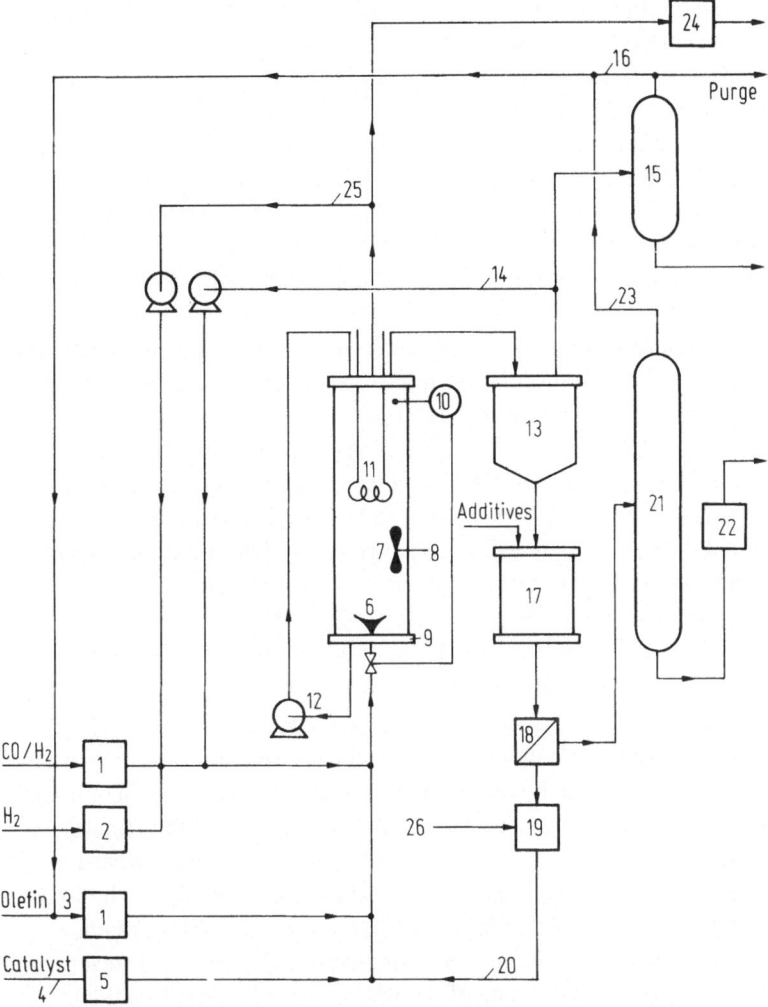

Fig. 1.63. Diagram of an industrial Oxo plant (for description see text)

The potential and actual industrial variants involving design and reaction control will now be discussed using the above diagram.

Sensitive catalyst systems, especially those based on rhodium, which is utilized at a relatively low concentration (cf Sect. 1.3.3.1.1), demand expensive prepurification of the reactants — syn gas and olefin (*1* in Fig. 1.63). This purification can ensue via direct measures e.g., adsorption [1148–1151] or via reactions preceding the hydroformylation (e.g., heterogeneous hydrogenation) which also act as a poison trap [324]. Product recycles represent a transition to the regeneration or make-up stages. In these recycles, as well as the cooling effect and/or an improvement in intermixing (see below), there is also a purification effect [628, 632, 878, 1059].

Certain processes, in particular those with ligand-modified Rh catalysts, demand extreme $CO:H_2$ ratios (up to 1:20). In these cases, the H_2 source, H_2 feed and its control *2* all require special equipment [935, 1152, 1153]. The question as to the type of olefin feed – gaseous or liquid *3* – can only be unequivocally answered for ethylene (gaseous) and the higher olefins > C_5 (liquid). Both propylene and butylene can be metered and transported in the liquid and gaseous (after vaporizing) states [35, 194, 263, 309, 437, 449, 1152, 1169].

Although the active catalytic species (e.g., $HCo(CO)_4$) can be formed via various routes (cf Sect. 1.3.3.4), the type of addition of fresh catalyst *4* and the catalyst recycle *20* are essentially determined by the catalyst system [1154, 1155]. In many cases – particularly with less active systems, e.g., water-soluble catalysts – it is worthwhile to connect a "carbonyl generator" *5* where the $HCo(CO)_4$ is preformed [35, 282, 1101, 1155–1158, 1167, 1346].

Particular attention must be paid to the introduction *6* of the reactants. The latter can be influenced by the type of transport, the reactor construction and requirements stemming from parameter profiles within the Oxo reactor. Special devices for pre-mixing [295, 328, 1153, 1159–1162, 1166, 1271, 1866] or preheating the reactants [295, 1160, 1161, 1162] along with suitable parameter arrangements are recommended for achieving optimal effects [252, 275, 1163, 1164]. This includes proposals for the most favorable relative gas content of Oxo reactors as this determines the rate of mass transfer via the interphases and the mean diameter of the bubbles. Under process conditions, the gas content should lie between 12 to 18% [1165].

Backmixed vessels [265, 273, 623, 1084, 1145, 1154, 1169, 1170, 1623, 1624, 1866] cascades (reactors in series) with a limited number of steps [1084, 1165, 1191] –independent of a multi-stage reaction control–or plug flow reactors [275, 295, 1152, 1158, 1171–1173] have not only been described but also employed industrially as reactors, *7*. In addition, stirred tank reactors *8* [623] and modifications thereof have also been reported [224, 1084, 1166, 1174]. After undergoing extensive tests (including, for example, turbulent flow [295, 1171, 1172]) interest in the flow tube has waned on account of disadvantages encountered in industrial operation i.e., mainly the limited throughputs. Currently, the standard reactor is a backmixed reactor. This is a consequence of the various demands on the mass transfer gas (CO/H_2) → liquid (reactor contents), the specific interface, the mixing of the gas and liquid phases (including macro- and micro-mixing), diffusion processes, heat transport problems (see below) etc. Besides the attention paid to the choice of suitable reaction parameters outlined in Sect. 1.3, process control has been described which is appropriate for particular types of reactor e.g., certain axial CO/H_2–[327] or temperature profiles (incl. adiabatic operation) [252, 253, 275, 283, 1168], reactor coupling without release of pressure [1175] etc. [252, 1176–1179]. Safety aspects must be taken into account, for example by preventing the deposition of cobalt in noncooled dead waters [1181].

Usually conventional stainless steel (e.g., DIN 4541 or 4571) is recommended as construction material *9* for the reactor when noncorrosive catalyst systems are employed e.g., cobalt metal, cobalt soaps or cobalt compounds suspended in organic liquids. Corrosive catalysts, particularly aqueous or acid-containing systems, necessitate high-alloy steels. This is underlined by the "apparently surprising fact" that on changing the catalyst, low-alloy steels can be utilized [328, 1149, 1151].

As the reaction parameters – pressure, temperature and reactant (catalyst) concentration – have a distinct effect on conversion and selectivity, the *control 10* of the Oxo reactor is particularly important. After the appearance of early publications dealing with optimal control of Oxo plants using process control computers thereby suggesting an open forum for exchange of experience [206, 1182] the subsequent Oxo literature disclosed few details and new developments fall under the know-how category [1183].

Fairly recently, Dubil and Gaube [194] among others studied the dynamics and control of the Oxo reaction in backmixed reactors. In this publication, stable, oscillating stable and unstable reactor behavior were presented using stability diagrams (as function of the operating parameters). Special control systems [1184] have been developed for unusual operating conditions, particularly in connection with unstable behavior (cf oscillation processes as a consequence of substrate inhibitors, Sect. 1.2.4).

The removal of the exothermic heat of reaction of the Oxo synthesis, *11* – approx. 30 kcal/mol (125,7 kJ) – represents part of the most important know-how of each Oxo process. For example, in average size Oxo reactors with an LHSV of 0.8 up to 3.1 million kcal (13 million kJ) heat of reaction are released per hour (incl. hydrogenation to alcohols and olefin hydrogenation). As only 0.6 million kcal (2.5 million kJ) are required for heating the product, 2.5 million kcal (10.5 million kJ) must be removed by cooling. If average cooling surfaces of 430 m^2 per reactor are assumed, then, with a temperature difference of approx. 35 $^\circ$C, the mean coefficient of heat transmission is 230 kcal (963,7 kJ)/m$^2 \cdot$ h $\cdot$ $^\circ$C [1185]. Indirect cooling has been particularly recommended for removing this heat of reaction. Internally located cooling coils or tubular bundle coolers [35, 623, 1163, 1171, 1186, 1192] (in particular, the field tube type [1187, 1188]) have been proposed for this application. Besides water lower alcohols are recommended as coolants [1189].

Besides adding water to the reaction mixture [1806], product recycles *12* can be used for direct cooling [253, 295, 1138, 1161, 1163] (internal [1174] or external [309, 1138, 1163, 1166, 1192]).

The reaction vessel *13* can be employed as a high pressure separator [449, 1152] (with possibility of a high pressure gas recycle *14* [327, 1141, 1152, 1153, 1160, 1164, 1175, 1190, 1273] or as a second reactor in the two-stage operation with a cascade arrangement [1186, 1191–1193]. The advantages of an operation of this type were already discussed in Sects. 1.3.5 and 1.3.6. The particularly favorable combination of a backmixed vessel (as the first reaction stage) and a plug flow reactor (as the second) is described in a patent filed by the Chemische Werke Hüls [1084]. In this instance, the comparatively high conversion rate in the second reactor – the flow tube – is exploited for the hydroformylation of less active internal olefins. Other publications report using a higher temperature [226, 294, 1192, 1194] in the second reactor which ensures residual conversion. In addition, a different CO:H$_2$ ratio to that in the initial reactor can be employed [309, 325, 327], in some cases combined with a counter-flow of olefin and syn gas in both reactors [1162].

If the reaction vessel *13* functions as a separator, then the off-gas can be either recycled or, after passing through a scrubber *15*, be freed of olefin which is then fed as the olefin recycle *16* [449, 1195, 1960].

The decobalting process *17* and the apparatus *18* for separating the catalyst from the crude Oxo product should be incorporated in a *single* unit. Generally, phase separators are employed where HCo(CO)$_4$ is oxidized to the water soluble Co^{2+} salts.

Filters or centrifuges are utilized with a solid recycle. The make-up stage *19* must be included as, in this instance, the cobalt compounds are reintroduced to the process via the catalyst recycle *20* — sometimes after chemical treatment *26* (cf Sect. 1.3.3.4).

After the crude Oxo product leaves the separator, the residual olefin content can be removed via a distillation column *21* before entering the subsequent separation or processing stages *22*. This olefin is then fed to the first reactor via the olefin recycle *23* [449, 623, 1162, 1196, 1197].

Off-gas from the Oxo reactor *7* or the separator *13* contains Oxo products in accordance with their vapor pressure. The main components are aldehydes which, after hydrogenation *24* to higher boiling alcohols, can be recovered as value products [1198, 1199]. Otherwise, the gas can be recycled, *25*, due to its useful olefin and syn gas content [273, 623, 449, 1149, 1160, 1190].

Alekseeva, Rudkovskij, and Trifel [1164] have published data pertaining to the various solubilities of the olefins and syn gas components in the Oxo products at different pressures. From these values, it can be concluded that the various $CO:H_2$ and olefin ratios in the gas recycles *14, 16, 23,* and *25* must be finely adjusted at the inlet to the Oxo reactor as the $CO:H_2$ ratio of ca. 1:1 in the syn gas from the generating plant does not correspond to the consumption ratio because the side reaction consume more H_2 than CO due to the hydrogenation of the aldehydes and olefins.

1.6.2 Process Variants of the Industrial Oxo Syntheses

The most important industrial Oxo processes will now be briefly discussed, particular attention being paid to their characteristic features (as far as is known). In addition, the most significant Oxo variants will also be shown schematically. The expiration of the basic Oxo patents during the first industrial realization phase encouraged the development of know-how pertinent to a particular Oxo variant. This readily explains the tactics behind the appearance of relatively superficial publications of restricted content. The following section will be arranged according to the central atom of the Oxo catalysts although classification according to the type of catalyst recycle — the most characteristic part of an Oxo process — would have been equally feasible.

1.6.2.1 Cobalt (Compounds) as Catalysts

1.6.2.1.1 Ruhrchemie Process

The Ruhrchemie process which has been frequently licensed encompasses a series of catalyst recycle variants enabling ready adaption to the quality and sulfur content of the syn gas feedstock [44, 289, 907, 909, 1246—1250]. Thus, all these process variants can dispense with a pre-purification of the reactants and a fine purification of the syn gas.

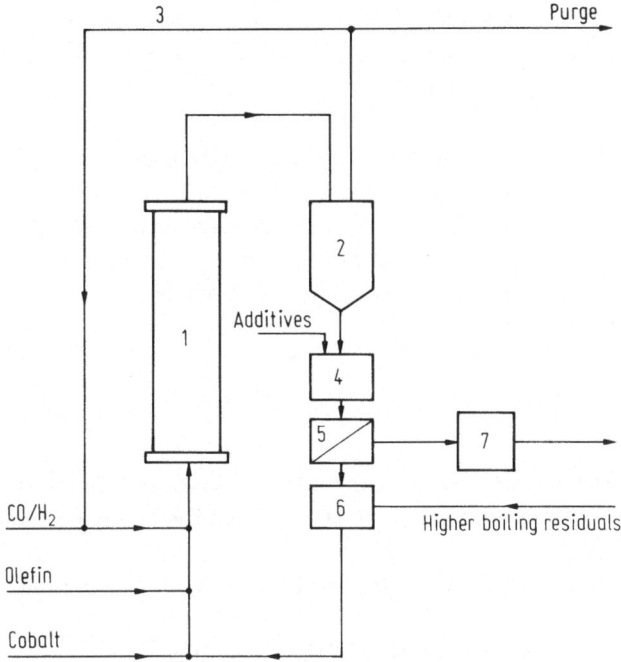

Fig. 1.64. Diagram of Ruhrchemie process

Reactor *1*, a tank reactor with highly effective cooling equipment [1251], is charged with commercial grade olefins. The reactor may contain devices for correcting the flow of current [224, 1139]. The cobalt supply can be supplemented with metallic cobalt or cobalt compounds from which cobalt hydridocarbonyl is formed in situ in the reactor. Larger quantities of sulfur may be present in the syn gas.

After passing the separating stage *2* – the off-gas being recycled, *3* – the Oxo products proceed to the decobalting step *4*. At this point, the Co^{1-} of $HCo(CO)_4$ is converted into $Co^{\pm 0}$ (metallic cobalt) or Co^{2+} [906, 995, 1039, 1044, 1048, 1049]. The conversion takes place by means of treatment with water, oxidizing agents, acidic species and/or other additives. When sulfur or sulfur compounds are present, metallic cobalt is usually obtained which can then be recycled to the reactor via the separator *5* and the make-up stage *6* where it forms a slurry with the heavy ends from the Oxo synthesis.

With syn gas of low sulfur content, the Co^{2+} compounds may be obtained in the solid state (Co formate [995]) and can then be recycled in the same way (after suspending in the heavy ends from the process). Cobalt compounds can be precipitated from aqueous cobalt salt solutions and, after mixing with the heavy ends, transported as a solid suspension. With these catalysts, throughputs as high as 2.0 LHSV (two parts by vol. olefin per part by volume reactor per hour) can be attained. These productivity figures mean, in actual fact, that the Ruhrchemie process has a predominant position amongst the Oxo processes. This is also due to constructive features of the reactors which enable the considerable amounts of heat from the exothermic Oxo synthesis to be removed – even at higher throughputs – and utilized as steam. Water is an adequate coolant, coolant mixtures as used in other processes being unnecessary.

The almost metal-free crude Oxo product which leaves the separator *5* can subsequently pass through various distillation, bottom product treatment and hydrogenation/fractionation stages *7* [1231, 1232, 1234—1245].

A particular attractive feature of the Ruhrchemie Oxo process is its high productivity which is a consequence of the high development level of its reactor and heat transfer technology along with the high catalytic activity and the large n:iso-ratio (for cobalt catalysts) of around 80:20. These factors facilitate lower specific investment costs. Moreover, the reactor used in the Ruhrchemie process can be employed for all olefins from ethylene up to C_{16}-C_{20}-α-olefins and thus forms the basis of a whole spectrum of products (cf Fig. 1.1).

1.6.2.1.2 BASF Process

Along with the Ruhrchemie and Kuhlmann processes, the BASF technology belongs to the prototypes of the cobalt-catalyzed Oxo process and has also been frequently licensed throughout the world [1200—1206].

The typical features of the BASF process are as follows [290, 393, 860, 862, 1154, 1155, 1159, 1207—1210, 1252—1254]:

- application of aqueous cobalt salt solution as catalyst precursor, if necessary
- after pre-carbonylation i.e., preforming of $HCo(CO)_4$ from the preliminary stages, as well as,
- decobalting via oxidation of Co^{1-} to watersoluble Co^{2+} in acidic solution.

In accordance with the above, the reactants — above all syn gas — must be pre-purified (*1* in Fig. 1.65) to prevent the formation of cobalt metal in the reaction and work-up stages [907]. In the improved version of the BASF process, the Co^{2+} salt (in aqueous solution) is converted into $HCo(CO)_4$ in the carbonyl generator *2* [393, 524, 860, 861, 862, 1032, 1155, 1156, 1208, 1211, 1212, 1254]. If necessary, extraction can then ensue with the olefin feedstock or oxygen-containing solvents from the reaction product [1155, 1254]. The aqueous catalyst solution introduces not only corrosion problems to the Oxo reactor *3* stemming from the anion of the catalyst precursor (cf Sect. 1.3.3.4) but demands extremely intensive mixing to ensure a homogeneous distribution of the aqueous catalyst solution and the organic phase containing the products [226, 968, 1000, 1159, 1166, 1174, 1189, 1194, 1208, 1213]. Thus, one modification of the BASF process involves directly employing Co-deficient water from the carbonyl generator as source of acid for the decobalting stage [1155, 393].

The actual Oxo reactor is termed an internal loop reactor with a connected post-reaction zone [1160, 1210]. The cooling is accomplished via the tube bundle cooler which is immersed in the Oxo reaction mixture. Water or lower alcohols can be employed as coolant (cf Sect. 1.6.1 [35, 1189]).

The off-gases from the high pressure separator *4* are still charged with Oxo products (in accordance with their vapor pressures). The aldehydes can be recovered on scrubbing the off-gases with higher boiling substances from the process, *5* [1195, 1210, 1214]. The off-gases can, of course, also be recycled.

BASF recommends decobalting (*6*) the catalyst-containing crude Oxo product with oxidizing gases or liquids (O_2, air, H_2O_2 etc.) in (formic) acid solution, thereby exploit-

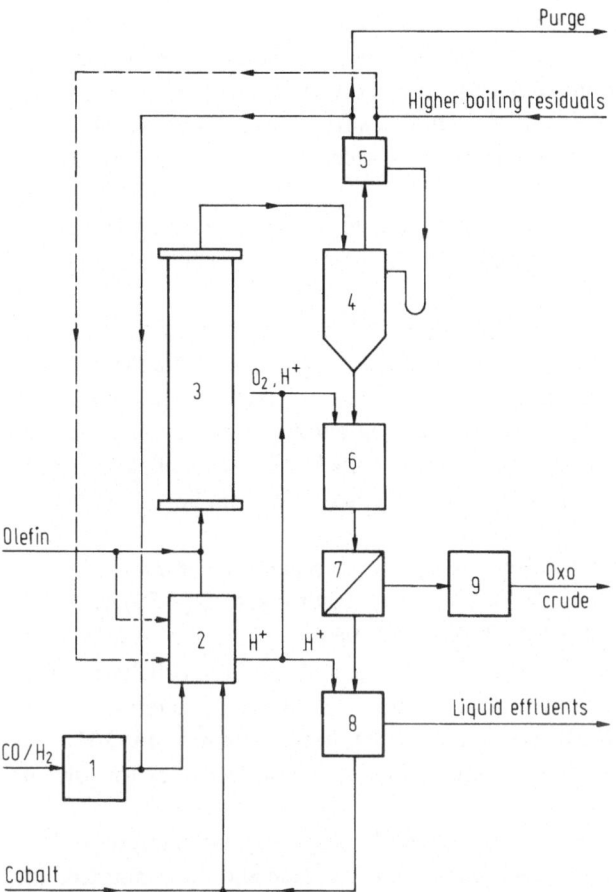

Fig. 1.65. Diagram of BASF process

ing the oxidation step $Co^{1-} \rightarrow Co^{2+}$ [860, 967, 968, 997, 998, 1154, 1155, 1208, 1215]. The separator *7* which consists of a phase separator or decanter facilitates the separation of the virtually cobalt-free organic phase (containing the Oxo products) and aqueous acidic Co^{2+} catalyst solution [997, 1210].

The object of the make-up stage *8* is to concentrate the catalyst solution which is dilute as a consequence of the decobalting step. If necessary, suitable promoters can be added and, on converting Co formate into Co acetate, the cobalt cycle to the reactor *3* or carbonyl generator *2* can be completed [999, 1020, 1035, 1057, 1212, 1228]. Decobalting and make-up stages are thus chemical consuming and produce effluent.

The after-treatment steps *9* (stabilization, aldehyde distillation, hydrogenation, alcohol distillation, aldolization etc.) are discussed in Sect. 1.5 as well as in various publications [290, 1017, 1056, 1154, 1209, 1210, 1216−1218].

1.6.2.1.3 Kuhlmann (PCUK) Process

The Kuhlmann process (PCUK, Produits Chimiques Ugine Kuhlmann) is a typical representative of Oxo processes in which the catalyst recycle involves no change in oxidation level. This process has also been frequently licensed [44, 1180, 1219–1222]. In addition the PCUK process basically requires that the formation of heterogeneous cobalt be prevented. Thus, the reactants 1 must be subject to extensive purification – this particularly applies to the removal of sulfur compounds from syn gas. One proposed Kuhlmann variant exploits the absorption capacity of nickel-containing hydrogenation catalysts [324].

As $HCo(CO)_4$ is recycled – using either the olefinic feedstock or other compounds as solvents – a carbonyl generator 2 is imperative to help introduce fresh cobalt to compensate for catalyst losses.

The hydroformylation in the Oxo reactor 3 ([1970]: an external loop reactor takes place under the usual conditions between the olefin feedstock – which previously served to extract $HCo(CO)_4$ in 8 – and the syn gas, part of which was used to preform the recycled catalyst in 7 [291, 962, 963, 966]. In older variants of the propylene hydroformylation, the recycled catalyst $HCo(CO)_4$ released in 7 (via acidification) was extracted not by the olefin feedstock but by dibutyl ether [523, 963, 964, 1080] which stemmed from further processing steps [1080, 1233].

The crude Oxo product obtained after passing the gas separator 4 is treated in 5 in countercurrent flow with aqueous Na_2CO_3 solution. According to Eq. (58), $HCo(CO)_4$ forms water-soluble $Na[Co(CO)_4]$ [291, 523, 962, 963, 966].

A virtually cobalt-free organic phase (Oxo products) and a sodium cobaltate-containing aqueous solution results after scrubbing with water in 6. $HCo(CO)_4$ is released in apparatus 7 (Eq. 59 [291, 523, 962]) after adding H_2SO_4 in the presence of CO/H_2. This "pre-formed" cobalt carbonyl is extracted in 8 by the olefin feedstock (or solvent, cf above).

The attractive features of the PCUK process are the recycle of the catalyst as $HCo(CO)_4$, the absence of heterogeneous phases in the Oxo step and the guaranteed introduction of the catalyst to the Oxo reactor in its most active form [523]. A disadvantage is that the catalyst separation and make-up stages must ensue in pressurized reactors.

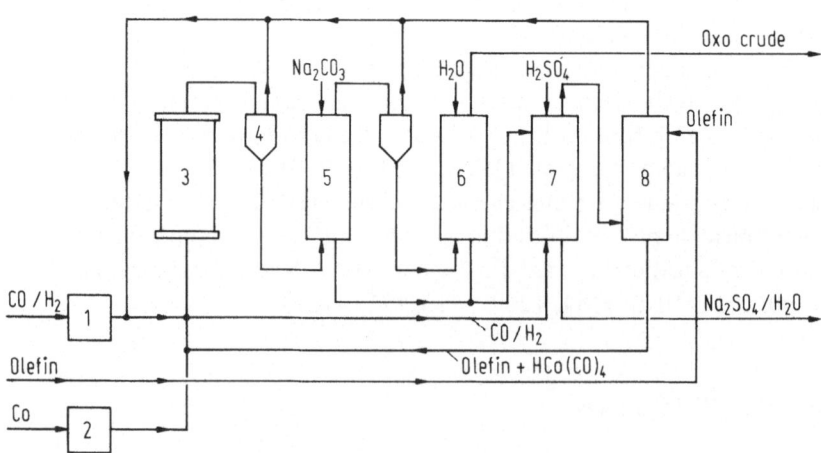

Fig. 1.66. Diagram of Kuhlmann Process [966]

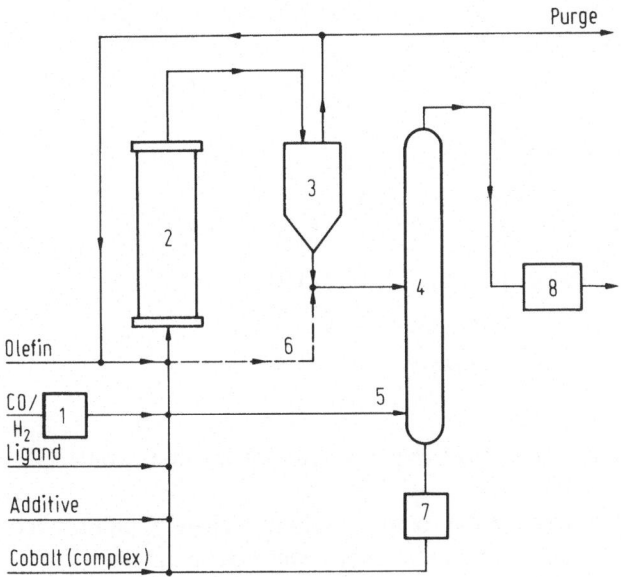

Fig. 1.67: Flow diagram of Shell process

1.6.2.1.4 Shell Process

The Shell process is the sole commercially operated Oxo process employing ligand-modified cobalt catalysts. There is only one smaller licensed plant known besides the Shell plants [1255]. Figure 1.67 illustrates the main steps in this process which serves mainly to manufacture higher alcohols from α-olefins.

The meticulous syn gas purification *1* and the comparatively large reactors *2* are particularly noteworthy. The purification step is intended to prevent the introduction of the catalyst poisons sulfur (affects the central atom) and/or O_2 (reacts with ligands forming phosphine oxides).

The large reactors are a result of the drop in activity arising from ligand modification which requires reactors five to six times larger than the conventional Oxo reactors (with unmodified cobalt) for the same productivity. The pressure distillation step *4* is also typical for this process. In this stage, the Oxo products are separated from the ligand-modified Co carbonyl. Adding olefins *6* and/or maintaining a CO partial pressure *5* prevents the thermal decomposition of the modified Co carbonyls which would lead to the deposition of metallic cobalt [592, 702, 970].

To avoid thermal damage, a flash distillation can be used for the pressure distillation step [1256]. As the bottoms of this column are Oxo active, they can be fed as catalyst recycle — if necessary after passing a make-up stage *7* — to the Oxo reactor [592, 1053, 1257].

Special measures must be taken for the precipitation, separation and make-up stages of the cobalt recycle if additives such as KOH are present along with the ligand-modified cobalt carbonyl $HCo(CO)_3PR_3$. These measures serve to encourage the in situ aldolization of (initially formed) aldehydes. Thus, alcohols with (n + 1) and

(2n + 2) carbon atoms can be obtained from olefins with n carbon atoms, e.g., butanol and 2-ethylhexanol from propylene [710, 899, 1258].

$$3\ R-CH=CH_2 \xrightarrow[\text{cat.}]{CO/H_2} [3\ R-CH_2-CH_2-CHO] \xrightarrow[-H_2O]{3H_2}$$

$$\begin{matrix} \text{Olefin} & & \text{Aldehyde} & \text{cat.} \\ C_n & & C_{n+1} \end{matrix}$$

$$\text{(187)}$$

$$R-CH_2-CH_2-CH_2OH\ +\ R-CH_2-CH_2-CH_2-\underset{|}{CH}-CH_2OH$$

$$\begin{matrix} \text{Alcohol} & & & \text{Alcohol } CH_2-R \\ C_{n+1} & & & C_{2n+2} \end{matrix}$$

In these cases it must be ensured that the cobalt complex and additive are recovered at the same point and recycled.

The advantages of the Shell process are the comparatively high share of unbranched alcohols (up to 88%, besides 12% branched products), the lower reaction pressures of 50 to 100 bar and the possibility of separating the active catalyst via a conventional pressure distillation. On the other hand, the disadvantages are the low catalytic activity (permitting LHSV values only max. 0.2), the product spectrum is limited to alcohols and the strong hydrogenation tendency of the catalyst system which also hydrogenates a considerable portion of the olefin feed to paraffins.

1.6.2.1.5 Various Processes

Important contributions to Oxo technology have been made by a number of processes which have not been licensed on the same scale as the previous ones.

The *ICI process,* which employs relatively high pressures [229, 1259] and hydrogen-rich syn gas [286, 299, 309] involves an ingenious system of separators and strippers (Fig. 1.68 [1259]).

The crude product of the Oxo reactor passes to a phase separator *1* where CO, H_2 and unreacted propylene are removed. The remaining liquid passes through a stripper *2* in countercurrent flow to a 2:1 mol ratio of syn gas. Both, the gases from *1* and *2* return as a gas recycle to the Oxo reactor. The liquid from stripper *2* passes a stripper *3* at 160 °C where it is fed in countercurrent to a large volume of syn gas. The gaseous stripper effluents are then passed via a heat exchanger/cooler *4* to a second phase separator *5* from which the reaction products reach the storage facilities. The liquid from the stripper *3* returns as a cobalt catalyst recycle to the Oxo reactor.

Other features of the ICI process are the cobalt carbonyl generation [1030, 1260] (for other purification methods cf [1261]), the application of solvents [218, 287, 948, 1031, 1083, 1106, 1108, 1332], the addition of fresh catalyst via oil soluble Co salts [271, 1262, 1331] etc. [299, 1263, 1264].

Chemische Werke Hüls have made several modifications to the BASF process which they took over [1265, 1267]. These concern in particular the preformation of the Oxo

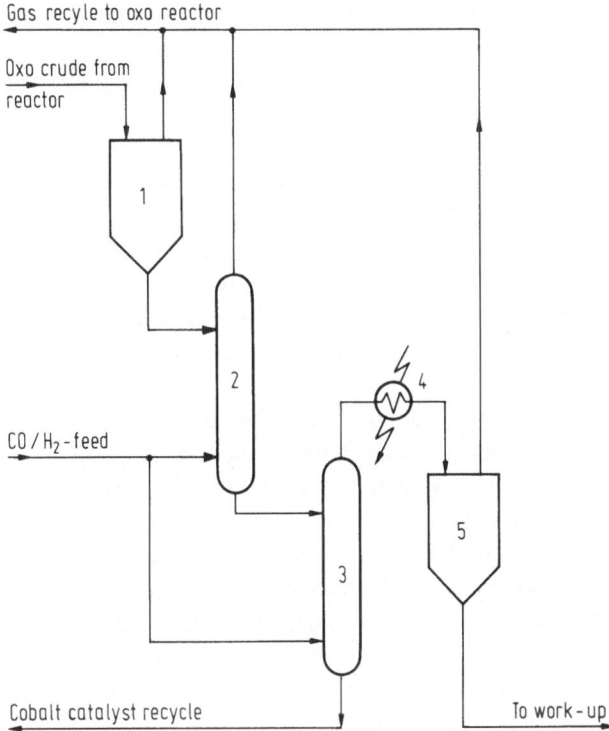

Gas recyle to oxo reactor

Oxo crude from reactor

CO / H$_2$ -feed

Cobalt catalyst recycle

To work-up

Fig. 1.68. Cobalt separator system of ICI process [1259]

catalyst HCo(CO)$_4$ [1266] before the reactor, the make-up stage [1266, 1267, 1954, 1960], process technology or engineering control techniques [184, 1084] and secondary steps [572, 1021, 1198, 1199, 1268].

With the *Gulf* [1001, 1009, 1163, 1171, 1314, 1328] and *Eastman-Kodak* processes [923, 978, 991, 1029, 1086, 1142, 1162, 1270–1273] the development work was concentrated on the Oxo reactor and reaction technology.

Besides incorporating many process steps typical for modern Oxo plants, *Esso's (Standard Oil Dev.)* process [1334, 1865] involves special process technology to operate and control the Aldox variant. The modern steps include reactant purification, use of H$_2$-rich syn gas, carbonyl generator [33, 277, 522, 523, 883, 1117, 1308, 1309, 1312], special reactor design [253, 277, 1138, 1190], gas or product recycles [253, 1004, 1141, 1161, 1190], the application of solvents or heavy ends from the process for catalyst make-up [277, 310, 1012, 1309. 1312] or special techniques for the secondary steps [279, 927, 958, 1308, 1310, 1333]. Besides the Oxo catalyst HCo(CO)$_4$, the Aldox variant includes additives (especially Zn salts) for the in situ aldolization of the resulting aldehydes (cf. Sect. 1.3.3.1.4 and Refs. [33, 296, 897, 898, 956, 1311, 1313, 1865]. *Union Carbide's* former cobalt catalyzed Oxo process was particularly characterized by the application of corrosive cobalt solutions [275, 295, 1005, 1102, 1173]. This was the background to UCC's development of a new Rh process with non-corrosive catalysts [1149] and higher on-stream factors. UCC's Co Oxo process had on stream factors of 90% which are well below the standard level [755].

Mitsubishi Chemical's process [1051, 1136, 1274, 1275, 1320] involves a low temperature hydroformylation which apparently guarantees a n:iso ratio of 4:1 and low formation of by-products [227, 1275, 1338, 1339], specially designed reactors are thought to be used in this variant. The process requires virtually sulfur-free syn gas (< 5 ppm S). It is also possible to form a slurry with the cobalt catalyst and the bottom products from the aldehyde distillation [1338]. This process has been frequently licensed for smaller plants [1335−1337].

There is little available information about the processes developed by the *Hungarian Petroleum Research Institute (MAFKI)* [293, 327, 523]. These Oxo processes incorporate a multi-stage conversion using sulfur-containing syn gas including an in situ hydrogenation of the resulting Oxo aldehydes with $HCo(CO)_4$ followed by special processing steps [1345].

Polish developments, which can only be gleaned indirectly from the literature [929, 1276, 1326] apparently did not prevent a foreign Oxo process being licensed in Poland [1277]. Russo-German developments were executed by Leuna/Neftechim [1278, 1279]. These processes features a two-stage hydroformylation with cobalt catalysts suspended in heavy ends which were fed to a carbonyl generator, a special decobalting step resulting in the formation of oil-soluble recyclable Co soaps and the necessity of using virtually sulfur-free syn gas. This joint project is supported by a number of Soviet publications [1157, 1280−1283, 1621−1627] dealing with problems associated with preforming cobalt hydridocarbonyls from cobalt compounds and syn gas [252, 521, 977, 980, 1107, 1284−1286], the adjustment of reaction parameters [944, 1197, 1287−1291], the precipitation/pressure distillation of the (cobalt-containing) crude Oxo product [971, 976, 977, 984, 990, 1281, 1282, 1285, 1292−1296] the make-up stage [975, 976, 977, 980, 1284] as well as processing and secondary steps [1297−1307].

1.6.2.2 Rhodium (Compounds) as Catalysts

1.6.2.2.1 Ruhrchemie Process

Olefins can be hydroformylated on a commercial scale with unmodified rhodium hydrocarbonyls using the well-tried reaction control techniques as well as reactors and ancillary equipment of the Ruhrchemie Oxo process (cobalt process). This method makes it possible to hydroformylate olefinic substrates which are usually hydrogenated (e.g., styrene or substituted styrenes) − a consequence of the low effective catalyst concentration. A series of other olefinic substrates exhibit high selectivity to aldehydes and low by-product formation, this is particularly important if the aldehydes are the desired intermediates (cf Fig. 1.1). These features of this Rh catalyzed process are particularly attractive when the formation of isomeric aldehydes can be ignored, e.g., ethylene, where there is no n:iso ratio, or with higher olefins which also previously yielded mixtures of isomeric compounds [26, 47].

Unmodified rhodium carbonyls can be employed in such low concentrations that their recycling does not introduce any problems. However, special process modifications are essential for the recycling, separation and recovery of the rhodium from the organic Oxo raw product or distillation residues [1039, 1044, 1048, 1049].

1.6.2.2.2 Process Developed by the Group — Union Carbide/Davy Powergas/Johnson, Matthey & Co. (Low Pressure Oxo Process — LPO)

The UCC/DP/JM process was the first commercial Oxo process to employ modified Rh catalysts, $HRh(CO)[P(C_6H_5)_3]_3$. Process development was initiated at Union Carbide; later Johnson, Matthey and Davy Powergas started work on a similar joint project in the U.K. In 1971 this culminated in the three companies combining their efforts [745]. UCC's endeavors must be seen as a consequence of their own Co Oxo process (cf Sect. 1.6.2.1.5) while Johnson, Matthey & Co. and Davy Powergas acquired Geoffrey Wilkinson's patents at an early stage [437, 438, 588, 590, 591, 741, 803]. Union Carbide's path thus went from ligand-modified cobalt catalysts [723] via phosphite-substituted rhodium carbonyls [636, 780, 1072, 1315] to phosphine-modified rhodium catalysts. This work was partially stimulated by the initial results of Booth et al. of Union Oil of California (see below) [589, 730, 780].

The application of $HRh(CO)[P(Ph)_3]_3$ dissolved in excess ligand as solvent at relatively low temperatures (60 to 120 °C favored) and comparatively low pressures (below 400–450 psia, ca. 28–31 bar) in the presence of hydrogen-rich syn gas demands a process operation which differs from that of the conventional Oxo process [623, 634, 635, 744, 745, 832, 878, 935, 1148, 1315–1317].

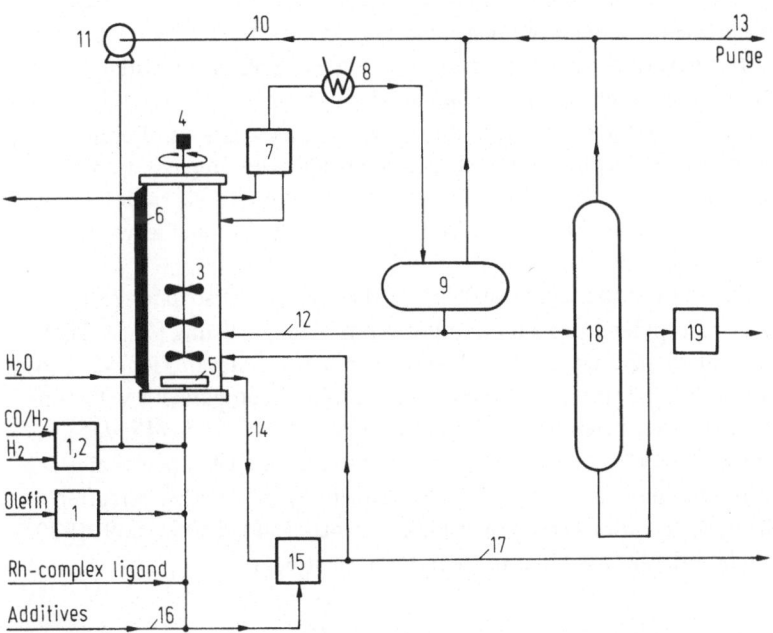

Fig. 1.69. Diagram of UCC/DP/JM process for the hydroformylation of propylene

171

In the LPO process the purification of reactants *1* via adsorption techniques is particularly important [628, 1149] due to the low catalyst concentration and its susceptibility to poisoning. Maintainance of high H_2 partial pressures necessitates that an external H_2 *2* source be available.

The stainless steel Oxo reactor *3* [1149, 1151] contains the total Rh feed. Initially, the rhodium species is dissolved in excess triphenylphosphine, then in a mixture of ligands and heavy ends from the process [623, 832, 878, 935]. The Rh concentration of this liquid charge lies between 25 and 1200 ppm, 250–400 ppm being favored and 0.5 to 30% triphenylphosphine. The ligand excess corresponds to > 10 mol PR_3 per mol Rh [628, 755]. The reactor content is stirred, *4*, the gaseous reactants being finely distributed throughout the reactor medium via a sprinkler system *5* [605, 1151]. The reactor is cooled externally, *6*, however, the exothermic heat of reaction cannot be utilized as it is at a low level (preferred reaction temperature is 100 °C) [1149]. The reaction temperature and pressure must be kept constant within very narrow limits (± 0.5 °C, ± 0.8 psi) [755].

After leaving the reactor, the products along with the unconverted reactants are fed to a demister *7*, thereafter they are condensed by means of the cooler *8* in the separator *9* [605]. Unconverted propylene (partial conversion 30% per pass) and propane, the level being regulated by the purge *13*, return via the gas recycle *10* (incl. the compressor *11*) to the reactor. If necessary, liquid products from the separator *9* can be returned via the product recycle *12* should the liquid level in the reactor *3* become too low.

Special precautions must be taken to counteract poisoning of the Rh charge via thermal effects, or – despite purification – via iron compounds or other phenomena (phosphine oxide formation etc. cf Sect. 1.3.3.2). These include a lowering of temperature. However, consequences of the limited reaction temperature [634] are a drop in olefin conversion and the (energy consuming) recompression of the reactants. Continuous removal *14* of part of the catalyst solution is another precaution which is necessary to facilitate its regeneration in a make-up stage *15*.

This regeneration can ensue either via addition of aqueous solutions of complexing agents (to remove Fe) or by introducing suitable additives *16* [628, 633, 634, 878, 1149]. At various intervals, the total catalyst charge is removed via outlet *17* and sent for reprocessing to Johnson, Matthey & Co. in the U.K. i.e., it must leave the Oxo plant site [1149].

The liquid products from the separator *9* are worked up by the stripper *18* thus enabling recovery of a propylene/propane fraction suitable for recycling [755]. Thereafter, a conventional distillative work-up of the Oxo products ensues in *19* [755, 1149].

The advantages and disadvantages of the LPO process and comparisons with conventional Oxo processes have often been topics of reports [26, 47, 527, 745, 755, 1148, 1149, 1151]. Since 1976 the process has been tested in the UCC plants at Texas City, Tex. (for ethylene) and Ponce, Puerto Rico (for propylene). Thereafter the process was licensed to various companies [1318, 1340–1344, 1805, 1809, 1939, 1940], however to date the licensed plants are not yet on stream.

1.6.2.2.3 Process Developed by Union Oil of California

Booth et al.'s (Union Oil of California) earlier work [187, 209, 273, 728, 729] supplied major impulses for the realization of the process variants utilizing ligand-modified Rh-catalysts (see fig. 1.70).

According to this scheme, the reactants are fed to the Oxo reactor 1 and the reaction products are withdrawn at the gas separator 2. The distillation tower 3 separates the lower boiling Oxo aldehydes (as distillate) from the Rh complex and the heavy ends. The bottom of this column contains the active $HRh(CO)[P(R)_3]_3$, dissolved in excess PR_3, and the heavy ends of Oxo synthesis. This Rh solution is returned to the Oxo reactor as an active "primary" Rh recycle 9. The overhead product of column 3 (isobutyraldehyde) may be hydrogenated in 4 and the isobutanol fractioned in column 5. With incomplete propylene conversion, a stabilizer column is necessary between separator 2 and column 3. Because of the low but noticeable formation of heavy ends, complete recycling of Rh via the recycle 9 would cause an enrichment of higher boiling components. To avoid this, part of the bottoms of column 3 is withdrawn continuously as a slipstream and distilled in column 6 to give "light aldols" overhead. Provided they possess sufficient activity, the bottoms which are thermally treated twice at this point, are recycled to the Oxo reactor as a 'secondary' Rh cycle 10 (similarly a "tertiary" Rh recycle via a slipstream of 10 and subject to fractionation in tower 7 gives 'heavy aldols' [26, 47, 729], thereby necessitating an external make-up step 8).

The significance of this process principle — which has not in fact been realized for lower olefins — is that while the Rh-containing product stream may be recycled (and thus more susceptible to Rh losses), the inherent limitation of the LPO process to low boiling substrates may be overcome [47].

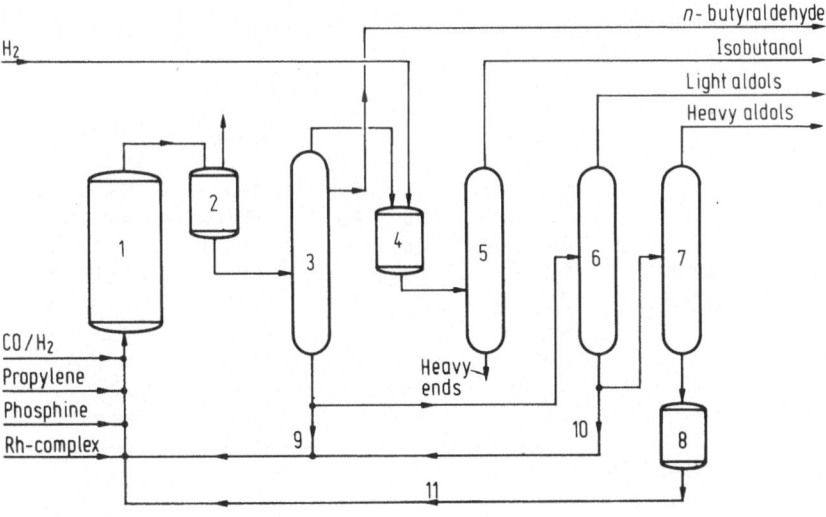

Fig. 1.70. Diagram of Union Oil's Oxo process [273, 729]

1.6.2.2.4 Various Processes

While several firms have made proposals or plans for an Oxo process with ligand-modified catalysts no industrial realization has resulted — the exceptions being Mitsubishi, Celanese and test runs by BASF.

The *Mitsubishi Chem. Ind. process* is catalyzed by HRh(CO)[P(C$_6$H$_5$)$_3$]$_3$ just as in the LPO process. However, to avoid patent infringements a lower excess of triphenylphosphine has been employed leading to the formation of greater amounts of isobutyraldehyde [307, 557, 627, 735, 736, 930, 932, 1040, 1069, 1104, 1113, 1320–1322, 1327, 1400, 1405, 1926]. This process is operated in Mizushima (Japan) [1322].

It has been known for some time that *BASF* has studied the applications of ligand-modified Rh carbonyls [314, 566, 553, 573, 784] and has even tested the hydroformylation of propylene with HRh(CO)[P(C$_6$H$_5$)$_3$]$_3$ on a larger scale. Their patents and patent applications do not suggest a specific process operation apart from the application of dibutyl phthalate as a high boiling solvent [1323].

From the very outset, *Celanese Corp.* has apparently been involved with plans for the application of ligand-modified Rh-carbonyls and even used them on a commercial scale before UCC/DP/JM [631, 726, 737, 818, 928, 945, 1103, 1329]. All known details bear resemblance to the LPO process.

While other proposals have been made by *Gulf Oil* [606, 734] and *Monsanto* [265, 409, 444, 449, 738, 771, 796, 1019, 1324] they have not led to industrial realization.

1.6.3 Comparative Considerations

The discussions about the advantages and disadvantages of the various Oxo processes have still to be settled. Table 1.59 presents a comparative summary of the current Oxo processes.

Table 1.59. Comparison of the various Oxo processes on employing propylene, L = Ligand

Metal	Cobalt		Rhodium	
Variant	Classical[a]	Modified[b]	Classical[c]	Modified[d]
Catalyst	HCo(CO)$_4$	HCo(CO)$_3$L	HRh(CO)$_4$	HRh(CO)L$_3$
Temperature (°C)	110–180	160–200	100–140	60–120
Pressure (bar)	200–300	50–100	200–300	1–50
Metal conc. (%) (rel. to olefin)	0.1–1	0.6	10^{-4}–0.01	0.01–0.1
Space time throughput (LHSV)	0.5–2.0	0.1–0.2	0.3–0.6	0.1–0.25
Hydrocarbon formation	low	high	low	high
Reaction products	aldehydes	alcohols	aldehydes	aldehydes
n : iso ratio	80 : 20	88 : 12	50 : 50	92 : 8
Sensitivity towards poisons	low	low	high	high

[a] e.g., Ruhrchemie or BASF process
[b] Shell process
[c] Ruhrchemie process
[d] According to Wilkinson/UCC/Davy Powergas/Johnson, Matthey & Co.

A basic appraisal of the cobalt and rhodium systems can be made when the various points in Table 1.60 are considered [26, 47, 527, 1732].

Table 1.60. Comparison of Oxo processes with cobalt and ligand-modified Rh catalysts

Process	Cobalt	Rhodium
Total yield	———— same ————	
Yield of unbranched compounds	lower	higher
Yield of branched compounds	high	low
Syn gas consumption	———— same ————	
Temperature	high	low
Pressure	high	low
Steam consumption	lower	higher
Electricity consumption	higher	lower
Cooling water requirement	low	high
Cost of purifying reactants	not necessary	high
Investment costs	similar, possibly advantage for Rh	
Availability of catalyst	high	lower
Flexibility	high	low

The similar total yield results from a somewhat higher formation of heavy ends in the cobalt process (consequence of higher concentration of catalyst) which is balanced out by the greater amount of propane formed (via hydrogenation of propylene) in the Rh process. The yield of unbranched compounds is greater in the Rh process due to the higher n:iso ratio, the converse, of course, applies to the iso compounds. The syn gas consumption is roughly the same for both processes. While the cobalt process requires higher reaction temperatures, this facilitates the exploitation of the exothermic heat from the Oxo reaction via steam generation thereby lowering the quantity of steam to be supplied from external sources.

As the reaction pressure is also lower with the modified process this enables less expensive reactors etc. to be employed. In addition, as no syn gas compression is necessary, the electricity consumption is also reduced.

The partial conversions of the reactants in the Rh process makes it necessary to repeat the cooling, condensation, and recycling steps thus cooling water consumption is higher than in the Co process. The purification of reactants also involves considerable costs in the Rh process. This step is not part of the classical Oxo synthesis due to the higher content of active metal which functions as a "filter" and is continuously regenerated in the catalyst recycle.

The investments for the Co and Rh process are either at the same level or are advantageous for the rhodium technology depending on the available infra-structure of existing plant and whether down-stream facilities are to be included. In the classical process, additional costs stem from the higher pressures and the compression of the syn gas to this level coupled with the costs for isolating the alcohols present in the Oxo products. On the other hand, while the lower pressure prevalent in the Rh process gives rise to lower costs this advantage is compensated by the investments for purification steps, recycling and, above all, the lower space-time yields. Reactors up to 86 m^3 are required for the production of 40000 t/a n-butyraldehyde using the Rh process while with the classical process the reactor volume is less than 10% of this figure. In any case, the decision as to which process is more suitable will be determined by market forces. On the one hand, the relative availability of the metals for the Oxo catalysts must be considered, the annual world production of cobalt and rhodium being around 22 000 t and slightly over 6 t respectively.

The price levels must also be taken into account on comparing these figures. With cobalt, the price tendency has been comparatively stable, the marked rise in 1978 [1938] was a consequence of the situation in Zaire where the cobalt and copper production were temporarily shut down.

It is a reasonable assumption that the Co supply will always be considerably better than that of Rh, particularly as South Africa and the Soviet Union together produce over 85% of the world rhodium supply. This is complemented by the strong speculative tendency of the rhodium market. If a mere quantitative comparison is made between Co and Rh then it becomes clear that there is absolutely no prospect of converting all present Oxo plants to rhodium as then, 2/3 of the world Rh production would have to be used to merely cover process losses. While Rh losses less than 1 mg/kg Oxo product would lower this figure, it must be mentioned that rhodium is recovered as a by-product from Pt production and the Rh consumption in other sectors is also undergoing expansion — as can be appreciated from the price tendency.

The relative flexibility of the cobalt — and rhodium — based Oxo plants are also important marketing considerations. The flexibility of plants using modified Rh catalysts for the hydroformylation is comparatively low as various olefinic feedstocks or other substrates cannot be readily employed (cf previous discussions). Moreover, plants built for the low pressure hydroformylation with modified Rh catalysts cannot be converted to cobalt should rhodium be in short supply or there be delays at the distant rhodium reprocessing plant. The Oxo plant would therefore have to be shut down under these circumstances.

Another important factor to be weighed in the pro and contra cobalt/rhodium argument will also be decided by the market i.e., the yield of isobutyraldehyde from the cobalt catalyzed hydroformylation of propylene. Although isobutyraldehyde is still a "surplus product" there are — due to its growing application as an intermediate — already periodical as well as regional shortages. Its further processing to neopentyl glycol or methyl methacrylate are particularly important and once this market experiences further growth, the by-product problem of the Oxo synthesis will appear in a quite different light (cf Sect. 1.5.2.3).

A universally applicable decision in favor of one or the other process cannot be made. In each case however, the choice must be made between a single product base and the possibility of high flexibility of product and process. In addition, external reprocessing and limited availability of the Rh catalyst is worth comparing with convenient on-site reprocessing.

1.6.4 Economic Aspects of Industrial Oxo Processes

As of 1st July 1978 the world Oxo capacities for the hydroformylation of propylene and other olefins were distributed as follows:

Table 1.61. World Oxo capacities

Country	Company	Capacity (metric t/year $\cdot$ 10^3)	Product range[a]	Process[b]
Australia	CSR Chemicals	24	C_4, C_8	1
Austria	Chemie Linz	50	C_4	2
Brazil	Ciquine	34[c]	C_4	3
Bulgaria	Technoimport	17	C_4	8
Canada	BASF	75	C_4	2
Chile	Petroquimica Chilena	66[d]	C_4	1
China	Taching	89[d]	C_4	2
	Taching	120[d]	C_4	4
Czechoslovakia	Chempol	25	C_4	3
Federal Republic	BASF	490	C_3-C_{13}	2
of Germany	Chemische Werke Hüls	320	C_4, C_8, C_9	2[f]
	Ruhrchemie	335	C_3-C_{18}	1
France	Oxochimie	200	C_4	1
	Produits Chimiques Ugine Kuhlmann	165	C_4-C_{10}	5
Hungary	State owned	165[d]	C_8, C_9	open
India	Indo Nippon	10[d]	C_8, C_9	5
	Nocil	24[c]	C_4, C_8, C_9	6
Iran	Iran Nippon Petrochem.	47	C_8, C_9	3
Italy	Liquichimica	100[e]	C_{8+}	5,6
	Montedison	84	C_4	8
	SIR	85[d]	C_4	5
Japan	Chisso	52	C_4	1
	Kyowa Yuka	175	C_4	1
	Mitsubishi Chemical	96[c]	C_4	3
	Mitsubishi Petrochemical	20[c]	C_8, C_9	3
	Nissan	80	C_{8+}	5
	Tonen	33	C_4	7
Netherlands	Konam	50	C_4	5
Poland	Chempol	180[d]	C_4	4
Rumania	Rimnicu Vilcea	40	C_4	1
	Rimnicu Vilcea	42	C_4	2
Sweden	Berol Kemi (Beroxo AB)	83[d]	C_4	4
Spain	BASF	38	C_4, C_8	2
United Kingdom	ICI	220	C_4, C_8, C_{9+}	8
	Shell	125	C_4, C_8, C_{9+}	6
United States	Celanese	68	C_3, C_4	8
	Dow Badische	95	C_4	2
	Exxon	136	C_8, C_{9+}	7
	Getty-Air Products	20	C_{8+}	5
	Monsanto	102	C_7, C_8, C_9	5
	Oxochem	215	C_4	2
	Shell	200	C_4, C_8, C_{9+}	6
	Texas Eastman	135	C_4, C_{8+}	8
	Union Carbide	300	C_4, C_{8+}	4,8
	USS	50	C_{8+}	7
USSR	Donieck (?)	76	C_4	8
Yugoslavia	INA	84[d]	C_4	open

$$\Sigma \sim 5\,200$$

Notes of Table 1.61 cf page 178

Notes of Table 1.61

[a] e.g., C_4 = n-butyraldehyde, n-butanol and corresponding
   iso-products

   C_{10} = isodecanol (from tripropylene)

[b] 1 = Ruhrchemie Process

2 = BASF Process

3 = Mitsubishi Process

4 = LPO-Process of UCC/JM/DP

5 = Kuhlmann Process

6 = Shell Process

7 = Enjay, Exxon, Standard Oil Process

8 = others

[c] other capacities planned

[d] planned/under construction

[e] production discontinued

[f] scheduled for LPO-process

The plants for the production of 2–EH are summarized in Table 1.62.

Table 1.62. World capacities for the manufacture of 2–EH

Process/Company	Capacities ($\cdot$ 10^3 metric t/year)		
	existing	planned/ under construction	total
BASF	462	50	512
Ruhrchemie/Rhone-Poulenc/ Melle Bezons	265	56	321
Mitsubishi	163	28	191
Chemische Werke Hüls	183	–	183
UCC	89	40	129
Shell	62	20	82
Kuhlmann	80	–	80
Kyowa Yuka	70	–	70
Texas Eastman	50	–	50
Berol Kemi (Beroxo AB)	–	35	35
Chisso Corp.	33	–	33
Exxon/Enjay	15	–	15
unknown	144	551	695
Total	1616	780	2396

The total Oxo capacities amount to 5.2 million t/a, the figure for 2–EH alone being around 2.4 million t/a. The installed capacities for the hydroformylation of propylene are ca. 5% higher than in the previous year while those for other Oxo alcohols are approx. 10% and for 2–EH the figure is around 6% higher.

Additional information can be found in the Refs. [33, 44, 289, 291, 292, 951, 1151, 1180, 1200–1205, 1219–1222, 1246–1250, 1255, 1276, 1316, 1318, 1320–1322, 1329, 1335–1337, 1340–1344, 1347–1358, 1810].

As can be expected, very little has been published about the manufacturing costs of the Oxo products using the various processes [1154, 1359–1362]. The comparative reports compiled by various market research organizations are only of limited value as technology is a continuously progressing – making corrections and amendments

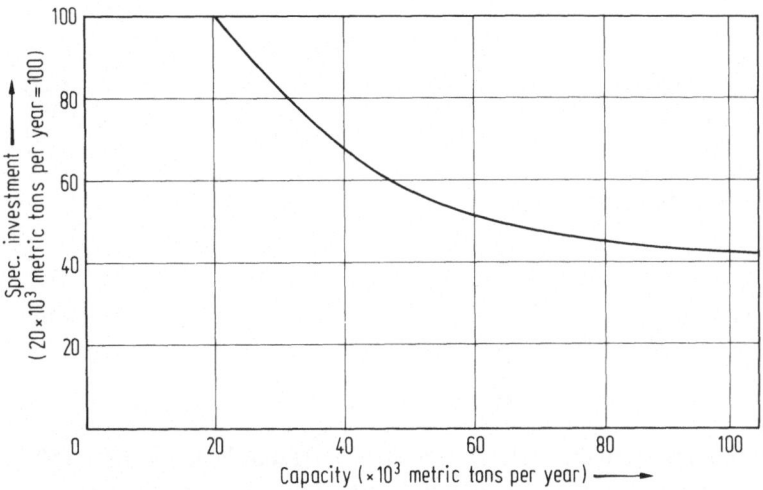

Fig. 1.71. Relation between the specific investment and capacity [1154]

necessary [1363–1366, 1399]. On account of the differing basis (location, extent of available infra-structure etc.) a detailed comparison is almost always indispensible for each projected plant. Fig. 1.71. gives an indication of the potential decreasing trend in costs on transferring to larger single stream plants.

1.6.5 Oxo-analogue Reactions

Some Oxo-analogue reactions will now be discussed. These conversions involve the introduction of carbon monoxide or syn gas components using transition metal complexes (carbonyls) under conditions similar to those prevailing in the hydroformylation. So far there have been no industrial applications. The following reaction can be regarded as representing a series of conversions.

Ketones, or more general carbonyl compounds, can be prepared on lab scale on reacting the alkyl transition metal carbonyls with carbon monoxide e.g., organomercury compounds [68, 69], zirconium [63], copper [1367], iron [1368, 1369], selenium [1370], nickel or π-allylnickel complexes [1371–1373] [Eq. (188)]:

$$
\begin{array}{c}
(R)_3P \diagdown \diagup X \\
Ni \\
\diagup \diagdown \\
H_3C \quad PR_3
\end{array}
\; + CO \; \rightarrow \;
\begin{array}{c}
(R)_3P \diagdown \diagup X \\
Ni \\
\diagup \diagdown \\
C \quad PR_3 \\
\diagup \; \|\\
H_3C \quad O
\end{array}
\qquad (188)
$$

The water gas shift reaction may be involved in the $Ru_3(CO)_{12}$-catalyzed carbonylation of acetylene to hydroquinone with H_2O as a hydrogen source [60–62, 1374–1376].

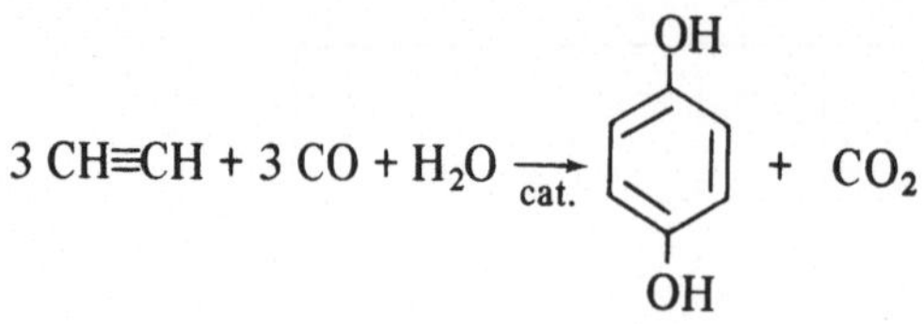

$$3 \; CH\equiv CH + 3 \; CO + H_2O \xrightarrow[\text{cat.}]{} \begin{array}{c} OH \\ \hline \\ OH \end{array} + CO_2 \qquad (189)$$

or to anthraquinone from benzene [62]

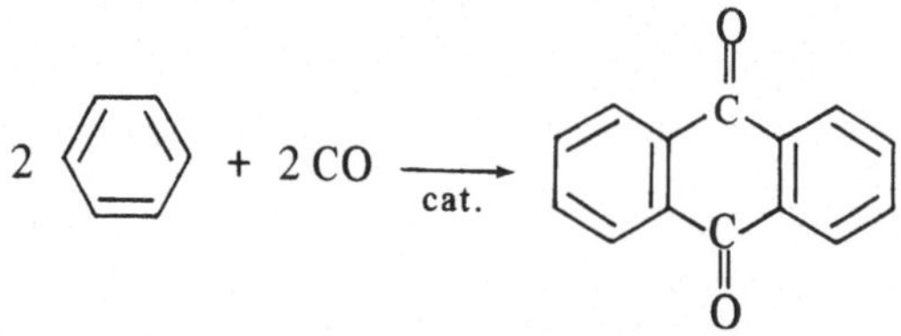

$$2 \; \bigcirc + 2 \; CO \xrightarrow[\text{cat.}]{} \qquad \qquad (190)$$

Ethers can be obtained via hydrosilylation and similar organosilicon reactions [53, 54, 1377–1379], alcohols and aldehydes via hydroboration and associated conversions [55–59, 1380, 1381, 1396], and ketones can be prepared via hydroziconation [63, 1382]. In addition, there is a series of other reactions: hydrocarboxylation of olefins with CO [1383], the dicarboalkoxylation of olefins or alkynes with CO and/or the reaction of organic halides with CO, olefins and acetylenes [1384–1386], the reaction of aromatic nitro compounds with CO to aryl-isocyanates [1387, 1388], the oxidative carbonylation [14] and other conversions [1389–1395, 1397, 1398, 1410–1412] (cf. Chap. 3 pp. 243).

An Oxo-analogue pathway to unbranched alcohols starting from internal olefins via organoaluminium intermediates might well prove to be significant in the future [64].

$$3 \; H_3C-(CH_2)_3-CH=CH-CH_2-CH_3 + Al \xrightarrow[\text{AlR}_3]{H_2} [H_3C-(CH_2)_6-CH_2]_3Al \qquad (191)$$

$$RhCl[P(Ph)_3]_3 \Big| + CO$$

$$CH_3-(CH_2)_7-CH_2OH$$

$$+$$

$$CH_3-(CH_2)_7-C\overset{\displaystyle O}{\underset{\displaystyle H}{\big\Vert}} \qquad (192)$$

1.7 References

Main Abbreviations used in Patent Citations

AT	Austria	HU	Hungary
AU	Australia	IN	India
BE	Belgium	IT	Italy
BG	Bulgaria	JP	Japan
BR	Brazil	NL	Netherlands
CA	Canada	NO	Norway
CH	Switzerland	PL	Poland
DD	German Democratic Republic (East Germany)	RO	Rumania
		SU	Soviet Union
DE	German Federal Republic (West Germany)	US	United States
		YU	Jugoslavia
ES	Spain	ZA	South Africa
FR	France		
GB	Great Britain		

1. Ruhrchemie A.G. (O. Roelen), DE 849.548 (1938), Roelen, O.: Chem. Exp. Didakt. *3*, 119 (1977), Roelen, O.: u. Ziegler, K.: Naturforschung u. Medizin in Deutschland 1939–1946, Bd. 36, part I, Wiesbaden: E. Dietrich-Verlag 1948
2. Chem. Ind. *20*, Nr. 11, 822 (1968)
3. Chem.-Ing.-Techn. *40*, A 1666a (1968)
4. „Produkte der Oxo-Synthese", Techn. Information Farbwerke Hoechst AG/Ruhrchemie AG, Frankfurt 1969, 2. Aufl. 1971, engl. Edition 1971
5. Andreas, F., Gröbe, K.: Propylenchemie. Berlin: Akademie-Verlag 1969, S. 202 f
6. Falbe, J.: Synthesen mit Kohlenmonoxid. Berlin-Heidelberg-New York: Springer-Verlag 1967, engl. Version: Carbon Monoxide in Organic Synthesis, Springer-Verlag 1970
7. Wender, I., Pino, P.: Organic Syntheses via Metal Carbonyls. Vol. 2. New York, London, Sydney, Toronto: Wiley-Interscience Publ. 1968/1976
8. Bond, G.C.: Homogeneous and Heterogeneous Catalysis by Noble Metals. In: Luberoff, B.J. (ed.): Homogeneous Catalysis, Amer. Chem. Soc., Adv. Chem. Ser. 70, Washington D.C., 1968
9. Bird, C.W.: Transition Metal Intermediates in Organic Synthesis. p. 117 f. New York: Logos Press/Academic Press 1967
10. Thompson, D.T., Whyman, R.: In: Schrauzer, G.N.: Transition Metals in Homogeneous Catalysis. New York: Marcel Dekker 1971
11. Kozikowski, A.P.: Transition Metals in Organic Synthesis. *1976*, 561
12. Tsuji, J.: Organic Synthesis by Means of Transition Metal Complexes. Vol. 28, p. 41, Berlin-New York: Springer-Verlag 1972
13. Taqui-Khan, M.M., Martell, A.E.: Homogeneous Catalysis by Metal Complexes. Vol. I 1971, Vol. II Academic Press. Inc. 1974
14. Rylander, P.N.: Organic Syntheses with Noble Metal catalysts, p. 215. New York, London: Academic Press, 1973
15. Paulik, F.E.: Catal. Rev. *6* (1), 49 (1972)
16. Falbe, J.: J. Organomet. Chem. *94*, 213 (1975)
17. Hegedus, L.S.: J. Organomet. Chem. *103*, 421 (1975)
    Bratermann, P.S.: J. Organomet. Chem. *103*, 307 (1975)
18. Benes, J., Hetflejs. J.: Chem. Listy *68*, 916 (1974)
19. Tolman, C.A.: In: Muetterties, E.L. (ed.): Transition Metal Hydrides. *1971*, 271–312; Chem. Reviews *77*, 313 (1977)
20. Heinemann, H.: Catalysis Reviews. Vol. 6, New York: M. Dekker 1972

21. Van der Kerk, G.J.M.: Organo-Metallverbindungen als homogene Katalysatoren. Dechema-Monographie *68*, 75 (1971)
22. Sittig, M.: Catalyst Manufacture, Recovery and Use, Noyes Data Corporation. Park Ridge N.J., 1967 and 1972
23. Falbe, J.: Chemie-Anlagen und Verfahren *1972*, Nr. 9, 50
24. Cornils, B. in: Falbe, J./Hasserodt, U.: Katalysatoren, Tenside und Mineralöladditive. p. 110. G. Thieme-Verlag 1978
25. Asinger, F.: Die petrolchemische Industrie. Vol. II, p. 1122. Berlin: Akademie-Verlag 1971
26. Cornils, B., Payer, R., Traenckner, K.-C.: Hydrocarbon Process. *1975*, Nr. 6, 83
27. Heil, B., Marko, L.: Magyar Kem. Lapja *23*, Nr. 12, 669 (1968)
    Chem. Titels, Nr. 1, 230 (1969)
28. Macho, V., Mistrik, J.: Ropa Uhlie *1975*, 17, 1; C.A. *82*, 173260 f (1975)
    Erdöl und Kohle *27*, Nr. 2, 98 (1974)
29. Miller, S.A.: Ethylene and its Industrial Derivatives. London: Ernest Benn 1969
30. Falbe, J. in: Hancock, E.G. (ed.): Propylene and its Industrial Derivatives, p. 333, New York: J. Wiley & Sons 1973
31. Söll, H. in: Houben-Weyl, Methoden der organ. Chemie, 4. Aufl., Bd. V/Teil 1 b, p. 1033. Stuttgart: G. Thieme-Verlag 1972
32. Kyle, H.E. in: Kirk-Othmer: Encyclopedia of Chem. Technology, Vol. XIV, 2nd (ed.), p. 373, New York: John Wiley & Sons 1967
33. Wickson, E.J., Dengler, H.P.: Hydrocarbon Process. *1972*, Nr. 11, 69
34. Hahn, V.G.: The Petrochemical Industrie, Markets and Economics. New York: McGraw-Hill 1970
35. Falbe, J. Weber, H.: Oil and Gas International *10*, Nr. 5, 90 (1970)
    Falbe, J., Payer, W. in: Ullmanns Encyklopädie d. Techn. Chem.,
    4. Aufl., Bd. 7, p. 118 and 203. Weinheim/Bergstr.: Verlag Chemie 1974
36. Marko, L.: Acta Cient. Venez., Supl. *1973*, 24 (2), 49
37. Bogdan, M.: Ing. Prelucrarii Hidrocarburilor *1974*, 2, 488; C.A. *83*, 100 363 b (1975)
38. Alekseeva, K.A. et al.: Zh. Vses. Khim. obsc. *1977*, 22 (1), 45; C.A. *86*, 120 712 f (1977)
39. Stern, E.W.: Advan. Chem. Ser. *1971*, 103
40. Falbe, J. in: Ullmanns Encyklopädie der Technischen Chemie, 3. Aufl., Erg. Band, p. 87, München-Berlin-Wien: Urban & Schwarzenberg 1970
41. Rudkovskij, D.M.: Z. Vses. Khim. Obsc. *14*, Nr. 3, 297 (1969)
42. Proc. Symposium on Chem. Hydroformylation and Related Reactions, Veszprem 1972
43. Falbe, J., Cornils, B.: Topics Curr. Chem. *11*, Nr. 1, 101 (1968)
44. Aleksejewa, K.A., Rudkovskij, D.M., Trifel, A.G.: Karbanilirovanie Nenassyshchennykh Uglevod. *1968*, 92; Anonym, Chem. Engng. News *48*, (7) 21 (1970); Anonym, Chem. Engng. News *1970*, (9), 18, Weber, H., Dimmling, W., Desai, A.M.: Hydrocarbon Proc. *1976*, (4), 127
45. Shell Oil Co. (Falbe, J., Korte, F.), US 3.159.653 (1962/1964); Esso Res. and Engng. Co. (Cull, L., Mertzweiller, J.K.), US 3.337.489 (1963/1967)
46. Bott, K., Pirke, G. (Chemische Werke Hüls A.G.): DE-OS 2.306.668 (1973)
    Bott, K.: Fette, Seifen, Anstrichmittel *75*, 629 (1973); *76*, 443 (1974)
47. Cornils, B., Kirchhof. D.: DGMK-Vortrag, Berlin 1978; ref. Erdöl & Kohle, Compendium *1978*, 463; Pryde, E.H., Frankel, E.N., Cowan, J.C.: Amer. J. Oil. Chem. Soc. *49*, 451 (1972)
48. Cornils, B., Feichtinger, H. in: Falbe, J./Hasserodt, U.: „Katalysatoren, Tenside und Mineralöladditive", p. 45. Stuttgart, G. Thieme-Verlag 1978
49. Wilkinson, G.: Angew. Chem. *86*, 664 (1974)
50. Chem. Labor Betrieb *25*, 1 (1974); Lion Fat Oil Co. Ltd. (Okamura, O., Hashiwa, I., Nagayama, M.), JP 74.53.610 (1972/1974) C.A *81*, 137.855 z (1974); Kolesov, M.L., El'man, I.V., Posysh. Effekt. Neftekhim Prom. *1977*, 30; C.A. 89.26425e (1978); Mitsubishi Chem. Ind., DE-OS 1.543.232 (1965/1972)
51. Falbe, J. (ed.): Chemierohstoffe aus Kohle, Stuttgart: G. Thieme-Verlag 1977
52. Cornils, B., Rottig, W. in: [51], p. 323

53. Seki, Y. et al.: Angew. Chem. *89*, 196, 818, 919 (1977); *90*, 139 (1978); J. Organomet. Chem. *140*, 361 (1977)

54. Chalk, A.J., Harrod, J.F.: Amer. J. Chem. Soc. *89*, 1640 (1967)
    Adv. Organomet. Chem. *6*, 119 (1968)

55. Brown, H.C.: Acc. Chem. Res. *2*, Nr. 3, 65(1969),
    Hydroboration, W.A., Benjamin Inc., New York 1962
    Boranes in Organic Chem. Ithaca/London: Cornell Univ. Press 1972

56. Brown, H.C. et al.: J. Amer. Chem. Soc. *91*, 2144, 4606 (1969), *89*, 4530 (1967);
    Brown, H.C. in: Stone, F.G.A., West, R.: Advances in Organometallic Chemistry, p.1. New York-London: Academic Press 1973

57. Cornils, B. in: Falbe, J. (ed.): Methodicum Chimicum, Vol. 5 p. 24 f. Stuttgart: G. Thieme-Verlag 1975

58. Brown, H.C., Coleman, R.A., Rathke, M.W.: J. Amer. Chem. Soc. *90*, 499 (1968)

59. Pusitskii, K.V. et al.: Izv. Akad. Nauk. SSSR, Ser. Khim. *1973*, (8), 1817; *1972*, (9), 1998,1

60. Hoogzand, C., Hübel, W. in: [7], Vol. 1, p. 343; Lonza AG (Pino, P., Piacenti, F., Bianchi, M.) DE 1.767.398 (1968/1977)

61. Weil, T.A.: J. Org. Chem. *39*, 48 (1974)

62. American Cyanamid Co. (Arzoumanidis, G.G., Rauch, F.C.), US 3.932.474 (1976)

63. Fachinetti, G., Fochi, G., Floriani, C.: J. Chem. Soc. Dalton *1976*, 1946, Hart, D.W., Schwartz, J.: J. Amer. Chem. Soc. *96*, 8115 (1974)

64. Universal Oil Products Co, US 3.959.386 (1974)

65. Behrens, H. et al.: Z. Naturforschg. *B32*, (11), 1217 (1977)

66. Sasaki, T., Yuki Gosei Kagaku Kyokai Shi *31*, (10), 850 (1973), Chem. Inform. 16-251 (1974)

67. Wakamatsu, H., Sakamaki, K.: Chem. Commun. *21*, 1140 (1967), C. *1968*, 56–0281

68. Larock, R.C.: Angew. Chem. *90*, 28 (1978)

69. Seyferth, D., Spohn, R.J.: J. Amer. Chem. Soc. *91*, 6192 (1969)

70. Piacenti, F., Cioni, C., Pino, P.: Chem. Ind. (London) *1961*, 1240

71. Piacenti, F.: Gazz. Chim. Ind. *92*, 225 (1962)

72. Piacenti, F., Bianchi, M., Pino, P.: J. Org. Chem. *33*, 3653 (1968)

73. Orchin, M., Rupilius, W.: Catal. Rev. *6* (1), 85 (1972)

74. Friedman, L: Chem. Eng. Progr., Symp. Ser. *63*, (76), 41 (1967)

75. Marko, L. in Ugo, R.: Aspects of Homogeneous Catalysis, D. Reidel Publ. Comp., Dordrecht 1974

76. Asinger, F. et al.: Forschungsberichte des Landes NRW, Vol. 1638, Köln/Opladen: Westdeutscher Verlag 1966

77. Alemdaroglu, N.H.: Dissertation Technische Hogeschool Twente, 1974

78. Henrici-Olivé, G., Olivé, S.: Coordination and Catalysis. Weinheim-New York: Verlag Chemie, 1977

79. Chalk, A.J., Harrod, J.F.: Adv. Organomet. Chem.,Vol. 6, p. 119. New York: Academic Press 1968

80. Heck, R.F., Breslow, D.S.: Actes congr. int. catalyse, 2e, Paris *1960*, p. 671; C.A. *55*, 24 545 (1961); J. Amer. Chem. Soc. *83*, 4023 (1961); Heck, R.F.: J. Amer. Chem. Soc. *85*, 651 (1963)

81. McQuillin, F.J.: Tetrahedron *30*, 1661 (1974)

82. Henrici-Olivé, G., Olivé, S.: Transition Met. Chem. *1*, 77 (1976)

83. Clark, A.C., Terapane, J.F., Orchin, M.: J. Org. Chem. *39*, 2405 (1974)

84. Gankin, W.Ju. et al.: in [179], p. 40

85. Falbe, J.: Angew. Chem. *80*, 568 (1968)

86. Whyman, R.: J. Organomet. Chem. *66*, C23–C25 (1974), *81*, 97 (1974), *94*, 303 (1975), J. Chem. Soc. *1972*, 1375, Nature, Phys. Sci. *1971*, (230) 13; A.J. Drakesmith, Whyman, R.: J. Chem. Soc. *1973*, 362

87. Vysokinskii, G.P., Gankin, W.Ju., Rudkowskii, D.M.: Katal. Reakts. Zhidk. Faze, Tr. Vses. Konf. 2nd, *1966*, 445; C.A. *69*, 18 367 j (1968)

88. Heil, B., Csontos, G., Marko, L.: Wiss. Z. Techn. Hochsch. „Carl Schorlemmer" Leuna-M. *1973*, 15 (1), 72; C.A. *79*, 4692 f (1973); Ann. N. Y. Acad. Sci. *1974*, 239; C.A. *82*, 97 411 p (1975)

89. Gankin, W. Ju., Novikov, V.P., Rybakow, V.A.: Kratk. Tezisy. Vses. Sov. Probl. Mekh. Geterol. Reakts. *1974*, 22; C.A. *85*, 77264b (1976); Zh. Org. Khim *1972*, 8 (2), 424, Kinet. Katal. *8*, (4), 908 (1967); C.A. *68*, 7735e (1968) und *13*, 14 (1), 101; C.A. *78*, 135274y (1973)

90. Osborn, J.A.: Endeavour *26*, 144 (1967)

91. Hagen, J.: Chem Lab. Betrieb *28*, (4) 125 (1977)

92. Moris, D.E., Tinker, H.B.: Chemtech *2*, 554 (1972)

93. Van Boven, M., Alemdaroglu, N., Penninger, J.M.L.: J. Organomet. Chem., *1975*, 84 (1), 65; Ind. Eng. Chem., Prod. Res. Dev. *14*, 259 (1975), Monatsh. Chem. *1976*, 1043, 1153

94. Baranova, G.I., Shmulyakovski, Y.E., in: Imyanitov, N.S. (ed.): Gidroformilirovanie, UdSSR *1972*, p. 229

95. Pergot, C.: C.R. Acad. Sci. Paris, Ser. C *268*, (10) 955 (1969)

96. Bor, G. et al.: J. Organomet. Chem. *64*, 367 (1974), Chem. Commun *1976*, 914

97. Imyanitov, N.S., Rudkowskii, D.M.: J. prakt. Chem. *311*, Nr. 5, 712 (1969) Kinet. Katal. *1968*, 9 (5), 1042; C.A. *70*, 57019 n (1969)

98. Gankin, W.Ju. Krinkin, D.P., Rudkovskii, D.M.: Karbonil. Nenasysh. Uglevod *1968*, 36, 45; C.A. *71*, 74 823 h, 13 196 j (1969)

99. Ungvary, F., Marko, L.: J. Organomet. Chem. *1969*, Nr. 1, 205 Kem. Kozlem. *1972*, 37 (1), 17; C.A. *77*, 39 671 j (1972)

100. Czizmadia, J., Ungvary, F., Marko, L.: Trans. Met. Chem. *1*, 170 (1976)

101. Heck, R.F. in: [7], Vol. 1, 1968, p. 373

102. Schurig, V.: Chem.-Ztg. *101*, 173 (1977)

103. Kitamura, T., Joh, T.: J. Organomet. Chem. *65*, 235 (1974)

104. Pino, P. et al.: J. Chem. Soc. *1971*, 1640

105. Gankin, W.Ju. et al.: Kinet. Katal. *1973*, 14 (5), 1149; C.A. *80*, 81 937 n (1974)

106. Heil, B., Marko, L. et al.: Chem. Ber. *102*, 2238 (1969), *104*, 3418 (1971)

107. Heck, R.F., Breslow, D.S.: J. Amer. Chem. Soc. *84*, 2499 (1962), Heck, R.F.: J. Amer. Chem. Soc. *85*, 3116 (1963)

108. Calderazzo, F.: Angew. Chem. *89*, 305 (1977)

109. Rupilius, W., Orchin, M. in: [42], p. 59

110. Piacenti, F. et al.: Coordination Chem. Rev. *16*, 9 (1975)

111. Costa, G., Mestroni, G.: Tetrahedron Lett. *19*, 1781, 1783 (1967)

112. Nagy-Magos, K., Bor, G., Marko, L.: J. Organomet. Chem. *14*, 205 (1968)

113. Bianchi, M. et al.: J. Organomet. Chem. *135*, 387 (1977)

114. Ungvary, F., Marko, L.: J. Organomet. Chem. *71*, 283 (1974) Ungvary, F.: J. Organomet. Chem. *36*, 363 (1972)

115. Pino, P., Piacenti, F., Bianchi, M. in: [7], p. 120

116. Pino, P., Piacenti, F., Bianchi, M. in: [7], p. 119

117. Bianchi, M. et al.: J. Organomet. Chem. *135*, 387 (1977); Bianchi, M., Piacenti, F.: J. Organomet. Chem. *137*, 361 (1977)

118. McCabe, M.V. et al.: Ind. Eng. Chem., Prod. Res. Dev. *14*, 4, 281 (1975)

119. Taylor, P., Orchin, M.: J. Amer. Chem. Soc. *93*, 6504 (1971)

120. Orchin, M.: Advan. Catal. *16*, 1 (1966), Wender, I., Sternberg, H.W., Orchin, M.: J. Amer. Chem. Soc. *75*, 3041 (1953), Karapinka, G., Orchin, M.: J. Org. Chem. *26*, 4187 (1961)

121. Pino, P., Piacenti, F. in: [7], p. 47

122. Piacenti, F. et al.: J. Chem. Soc. Chem. Commun. *1976*, 789; Chim. Ind. (Milano) *58*, (11), 759 (1976)

123. Vysokinskii, G.P. et al.: Gidroformilirovanie *1972*, 54; C.A. *77*, 22, 151205 (1972)

124. Goldfarb, I.J., Orchin, M.: Advan. Catal. *9*, 607 (1957)

125. Gankin, V.Yu., Dvinin, V.A.: Kinet. Katal. *1973*, 14 (1), 191 C.A. *78*, 135274 y

126. Gankin, V.Yu., Dvinin V.A., Rybakov, V.A.: Gidroformilirovanie *1972*, 57
127. Bezard, D.A., Consiglio, G., Pino.: Chimia *1974*, 28 (10), 610
128. Piacenti, F., Bianchi, M., Frediani, P.: Chim. Ind. (Milano) *55*, 262 (1973); Adv. Chem. Ser. *1974*, 132
129. Piacenti, F. et al.: J. Amer. Chem. Soc. *90*, 6847 (1968)
130. Fell, B., Barl, M.: Chem. Ztg. *101*, 343 (1977) Fell, B., Rupilius, W.: Angew. Chem. *81*, 916 (1969)
131. McCormack, W.E., Orchin, M.: J. Organomet. Chem. *129*, 127 (1977)
132. Lai, R. et al.: Bull. Soc. Chim. France *1969*, 793,
133. Casey, C.P., Cyr, C..R.: J. Amer. Chem. Soc. *95*, 2240 (1973)
134. Asinger, F., Berg, O.: Chem. Ber. *88*, 445 (1955)
135. Johnston, M.: J. Chem. Soc. *1963*, 4859
136. Rossi, R. et al.: J. Org. Chem. *32*, 842 (1967)
137. Piacenti, F., Bianchi, M., Pino, P.: J. Org. Chem. *33*, 3653 (1968)
138. Stefani, A. et al.: J. Amer. Chem. Soc. *95*, 6504 (1973)
139. Orchin, M., Rupilius, W., [73], p. 104 ff.
140. Wender, I., Pino, P. in [7], p. 118 ff.
141. Casey, C.P., Cyr, C.R.: J. Amer. Chem. Soc. *93*, 1280 (1971)
142. Ungvary, F., Marko, L.: Acta Chim. Hung. *62*, 425 (1969)
143. Piacenti, F. et al.: J. Amer. Chem. Soc. *90*, 6847 (1968)
144. Stefani, A. et al.: J. Amer. Chem. Soc. *99*, 1058 (1977)
145. Pino, P., Piacenti, F., Bianchi, M. in: [7], p. 121, 122
146. Falbe, J., Feichtinger, H., Schneller, P.: Chem.-Ztg. *95*, 644 (1971)
147. Piacenti, F. et al.: J. Organomet. Chem. *87*, C 54–C 55 (1975); *120*, 97 (1976) *135*, 387 (1977); Chim. Ind. (Milano) *58*, 223 (1976)
148. Takegami, Y. et al.: Bull. Chem. Soc. Japan *37*, 1190 (1964), *37*, 181 (1964); *38*, 787 (1965); *39*, 1495, 2430 (1966); *42*, 206 (1969)
149. Watanabe, Y. et al.: Yukagaku *1974*, 23 (5), 304; C.A. *81*, 24995b (1974)
150. Gankin, V.Y.: Catal. Proc. Int. Congr. 5th *1972*, (Publ. 1973), 1, 421; C.A. *80*, 107688j (1974)
151. Gankin, V.Y. Dvinin, V.A., Rybakov, V.A.: Gidroformilirovanie *1972*, 50
152. Rupilius. W., Orchin, M.: Proc. Symp. Chemistry of Hydroformylation and Related Reactions, Veszprem 1972, p. 59; C.A. *77*, 87513 (1972)
153. Piacenti, F. et al.: Coord. Chem. Rev. *1975*, 16 (1–2), 9
154. Rupilius, W., Orchin, M.: J. Org. Chem. *37*, 936 (1972)
155. Pino, P., Piacenti, F. and Bianchi, M. in: [7], p. 125
156. Taylor, P., Orchin, M.: J. Amer. Chem. Soc. *93*, 6504 (1971)
157. Orchin, M., Rupilius, W. in: [73], p. 108 f.
158. Imjanitov, N.S., Rudkovskij, D.M.: J. Prakt. Chem. *311*, 712 (1969), Kinet. Katal. *1968*, 9 (5), 1042
159. Cornils, B., Wiebus, E., Diekhaus, G.: Ruhrchemie A.G., unpublished
160. Grima, J.Ph., Choplin, F., Kaufmann, G.: J. Organomet. Chem. *129*, 221 (1977)
161. Johnson, Matthey & Co Ltd. (R. R. Hignett, Davidson, P.J.), DE-OS 2.827.300 and 2.827.301 (1978)
162. Angelici, R. J.: Organomet. Chem. Rev. *3*, 173 (1968)
163. Venanzi, L.M., Chem. in Britain *4*, 162 (1968)
164. Rupilius, W., McCoy, J., Orchin, M.: Ind. Eng. Chem., Prod. Res. Dev. *10*, 142 (1971)
165. Fichteman, W.L.: Univ. of Cincinnati, Thesis 1968, C.A. *71*, 80538 k (1969); Fichteman, W.L., Orchin, M.: J. Org. Chem. *34*, 2790 (1969)
166. Strohmeier, W., Fleischmann, R., Rehder-Stirnweiss, W.: J. Organomet. Chem. *47*, C37 (1973)
167. Strohmeier, W., Rehder-Stirnweiss, W.: J. Organomet. Chem. *22*, C27 (1970)
168. Yamaguchi, M.: Kogyo Kagaku Zasshi *72*, (3), 671 (1969), Chem. Titles *1969*, Nr. 10–173

169. Wilkinson G. et al.: Tetrahedron Lett. *1969*, 1725; J. Chem. Soc. (A) *1966*, 1711; *1969*, 725; *1968*, 2516

170. Brown, C.K., Wilkinson, G.: J. Chem. Soc. (A) *1970*, 2753

171. Davidson, P.J., Hignett, R.R., Thompson D.T. in: Kemball, C., Catalysis, Vol. I, p. 369 f, Chem. Soc. London 1977

172. Chini, P., Martinengo, S., Garlaschelli, G: J. Chem. Soc. Chem. Commun. *1972*, 709

173. Strohmeier, W., Kühn, A.: J. Organomet. Chem. *110*, 265 (1976)

174. Yagupsky, G., Brown, C.K., Wilkinson, G.: J. Chem. Soc., Chem. Commun. *1969*, 1244

175. Pino, P., Piacenti, F., Bianchi, M. in: [7], p. 190

176. Natta, G., Ercoli, R. Castellano, S.: Chim. Ind. *34*, 503 (1952), *37*, 6 (1955), Brennstoff-Chem. *36*, 176 (1955)., J. Amer. Chem. Soc. *76*, 4049 (1954)

177. Martin, R., Chem. Ind. *1954*, 1536

178. Iwanaga, R.: Bull. Chem. Soc. Japan *35*, 778 (1962)

179. Wysokinskii, G.P., Gankin, W.Yu., Rudkowski, D.M. in: Rudkowskii, D.M (ed.): Carbonylation of Unsaturated Hydrocarbons, Allunions Sci. Res., Inst. for Petrochem. Proc. Leningrad, Chem. Dep. 1968, p. 27 f

180. Ungvary, F.: J. Organomet. Chem. *36*, 363 (1972)

181. Heil, B., Marko, L.: Chem. Ber. *101*, 2209 (1968)

182. Yamaguchi, M.: Kogyo Kagaku Zasshi *72*, 671 (1969)

183. Paper presented by Pino, P. on 19.1.1978 at the Techn. Hochschule Aachen; s. Symp. on Rhodium in Homogeneous Catalysis, Sept. 1978 in Veszprem, Procedings p. 98 f

184. Pino, P., Piacenti, F. and Bianchi, M. in: [7], p. 105 f, p. 180 f

185. Niwa, M., Yamaguchi, M.: Shokubai *3*, 264 (1961)

186. Gankin, V. Yu., Genender, L.S., Rudkovskii, D.M. in: [179], p. 61; Zh. Prikl. Khim. *40*, 2029 (1967), CA. *68*, 77592 (1968)

187. Oliver, K.L., Booth, F.B.: Erdöl und Kohle *24*, 346 (1971)

188. Kuvaev, B.E., Imyanitov, N.S., Rudkovskii, D.M. in: [179], p. 222; C.A. *71*, 2708 b (1969)

189. Cappelli, A. et al.: Chem. React. Eng. Proc. Int. Symp. 4th, *1976*, 186, C.A. *86*, 139343 h (1977)

190. Polyakov, A.A., Gankin, V.Yu., Rybakov, V.A., Fuks, I.S.: Int. Chem. Eng. *16*, 518 (1976)

191. Seelig, F.F.: Z. Naturforschg. B. *1976*, 31 B (7), 929; 31 B (3), 336

192. Alekseeva, K.A. et al.: Poluchenie Maslyanykh Al'degidow i Butil. Spirtov Oks. *1977*, 29; C.A. *88*, 5850 d (1978)

193. Tjan, P.W.H.L., Scholten, J.F.: Proc. Int. Congr. Catal. 6th *1976*, I, 48898; C.A. *88*, 5823 x (1978)

194. Dubil, H., Gaube, J.: Chem. Ing. Techn. *44*, 950 (1972); *45*, 529 (1973)

195. Cornils, B., Ruprecht, P. and Koschnitzke, W.: unpublished

196. Toros, S.: Magyar Kem. Lapja *1974*, 29 (10), 543

197. Matsui, Y. et al.: Bull. Jap. Petr. Instr. *1977*, 19 (1), 62; C.A. *87*, 151. 639n (1977)

198. Imyanitov, N.S., Volkov, V.A.: Zh. Prikl. Khim. *46*, 2683 (1973)

199. Csontos, G. et al.: Hung. J. Ind. Chem. *1973*, 53; C.A. *79*, 41608 d (1973)

200. Macho, V. et al.: Chem. Prum. *1974*, 24 (5), 237; C.A. *81*, 36940t (1974), Hung. J. Ind. Chem. *1974* (2), 147

201. Macho, V. Mistrik, J., Ciha, M.: Chem. Commun. *29*, 826 (1964)

202. Gankin, W.Ju., Rudkovskij, D.M.: Karbonilirovanie Nenassys. Uglevod. *1968*, 17, 25

203. Fell, B., Asinger, F.: Tetrahedron Lett. *29*, 3261 (1968)

204. Deczy, Z., Belafi, K., Heil, B.: Proc. Conf. Appl. Phys. Chem. 2nd *1971*, I, 17; C.A. *76*, 87600 m (1972)

205. Kashina, V.V., Katsnel'son, M.G., Mishenkova, G.N.: Zh. Org. Khim. *1978*, 877; C.A. *89*, 23 741 u (1978)

206. Harders, H., Heller, G., Laurer, P.R.: Dechema-Monographie *50*, 19 (1963)

207. Cavalieri d'Oro, P. et al.: Paper presented at the Symp. on Rhodium in Homog. Catalysis, Veszprem, Sept. 1978, Proc. p. 76 f
208. Happel, J., Csuha, R.S.: J. Catalysis *20*, 132 (1971), A.I. Ch.E.J. *17*, 927 (1971)
209. Oliver, D.L., Booth, F.B.: Amer. Chem. Soc. Div. Petr. Chem. Preprints *14*, (3), A7 (1969)
210. Kaschina, W.W., Kaznjelson, M.G., Mischenkowa, G.N.: Z. Org. Chim. *14*, (4) 877 (1978)
211. Pino, P., Ercoli, R.: Chim. Ind. *37*, 782 (1955)
212. E.I. DuPont de Nemours & Co. (Barrick, P.L.), US 2.542.747 (1946/1951)
213. Hughes, V.L., Kirshenbaum, I.: Ind. Eng. Chem. *49*, 1999 (1957)
214. Hurd, V.N. et al.: 5. Welterdölkongreß, Sect. IV, Paper 14, 1–10 (1959) Oil Gas J. *57*, 199 (1959)
215. Matsubara, M. et al.: Hokkaido Daigaku Kogakubu Kenk. Hok. *1967*, (44), 157, 167; C.A. *68*, 86800z (1968) 8
216. Macho, V.: Chem. prumysl. *12*, 240 (1962)
217. Cornils, B., Ruprecht, P.: Ruhrchemie AG, unpublished
218. ICI Ltd. (Goddard, R.E., Mansfield, G.H.), GB 903.589 (1959/1962)
219. Alekseeva, K.A. et al.: Neftekhimija *6*, 276 (1966; C.A. *65*, 3731 (1966)
220. Pino, P., Piacenti, F., Bianchi, M. in: [7], p. 88 f, 221, Montecatini SpA (Pino, P., Piacenti, F.), FR 1.315.336 (1962/1963)
221. Consiglio, G et. al.: Helv. Chim. Acta *1978*, 1703
222. Kurokawa, K. et al.: J. Fuel Soc. Japan *41*, 539 (1962), C.*1963*, 10 669, Shigen Gijutsu Shikenjo Hokoku *64*, 67 (1966), C.A. *68*, 95 304 t (1968)
223. Toa Nenryo Kogyo K.K., JP 26 603/68 (1966/1968)
224. Ruhrchemie AG (Schnur, F., Cornils, B., Hibbel, J.) DE-AS 2 263.498 (1974), US 3.929.900 (1975)
225. Ajinomoto K.G. (Iwanaga, R., Fujii, T.): Kogyo Kagaku Zasshi *63*, 960 (1960)
226. BASF AG, BE 700 142 (1967), GB 1.180.433 (1967/1970), BE 727.015 (1969)
227. Ishikawa, Y.: Chem. Econ. Engng. Rev. (Tokyo) *2*, 31 (1970)
228. Falbe, J., Tummes, H., Weber, J.: Brennstoff-Chem. *50*, (2), 46 (1969)
229. Matsubara, M. et al.: Hokkaido Daigaku Kogakubu Kenkyo Hokuku *1967*, (44), 157; C.A. *68*, 86800 z (1968)
230. Pino, P., Piacenti, F., Bianchi, M. in: [7], p.93
231. Pino, P. et al.: Chim. Ind. (Milano) *50*, 106 (1968)
232. Zhesko, T.E., Polyakov, A.A.: Reakts. Sposobn. Org. Soedin. *1974*, 11 (1), 257; C.A. *82*, 42 584s (1975)
233. Pino, P., Pucci, S., Piacenti, F.: Chem. Ind. (London) *1963*, 294
234. Piacenti, F. et al.: J. Chem. Soc. C. *1966*, 488
235. Johnston, M.: J. Chem. Soc. *1963*, 4859
236. Chem. Verwertungsges. Oberhausen mbH (Nienburg, H.J., Gemassmer, A., Eckard, H.) DE 888.687 (1942)
237. Knap. J.E., Cox, N.R., Privette, W.R.: Chem. Engng. Progr. *62*, 4 (1966)
238. Wender, I. et al.: J. Amer. Chem. Soc. *77*, 5760 (1955)
239. Keulemans, A.J.M., Kwantes, A., van Bavel, T.: Rec. Trav. Chim. *67*, 298 (1948)
240. Falbe, J., Huppes, N.: Brennstoff-Chem. *48*, 46 (1967)
241. Falbe, J., Huppes, N., Korte, F.: Chem. Ber. *97*, 863 (1964)
242. Tanaka, M., Hayashi, T., Ogata, I.: Bull. Chem. Soc. Japan *50*, 2351 (1977)
243. Noguchi Res. Foundation, FR 1.370.004 (1964); C.A. *63*, 9823 (1965)
244. Theyßen, J. et al.: Erdöl u. Kohle *29*, 260 (1976)
245. Kurokawa, K. et al.: J. Fuel. Soc. Japan *41*, 860 (1962), C.*1964*, 1-2391
246. Macho, V., Mistrik, J.: Chem. Zvesti *18*, (10) 732 (1964); C.A. *62*, 414 (1965)

247. Rudkowskij, D.M., Imyanitov, N.J.: Zh. Prikl. Khim *35*, 2719 (1962)
248. Esso Res. Engng. Co (Staib, J.H., Guyer, W.R.F., Slotterbeck, O.C.) US 2.864.864 (1958)
249. Sittig, M.: Chem. Proc. Monogr. *1965*, (8), 69
250. Falbe, J., Korte, F.: Chem. Ing. Techn. *36*, 158 (1964)
251. Shell Dev. Co. (Falbe, J.), DE-AS 1.186.041 (1961/1965)
252. All-Unions Sci. Res. Inst. (Gankin, W.Ju., Rudkowskij, D.M., Krinkin, D.P.), DE-AS 1.240.064 (1964/1967), SU 180.581 (1965), 181,077 (1964) and 250.119 (1965)
253. Standard Oil Dev. Co. (Smith, W.M.), US 2.557.701 (1948), GB-PS 647.363 (1949); FR 979.283 (1950)
254. Cornils, B. in: [48], p. 108/109
255. Schweckendiek, L.: DE 841.589 (1952), DE 834.991 (1952)
256. BASF A.G. (von Kutepow, N., Bille, H.) DE 921.988 (1955)
257. Lautenschlager, H., Friedrich, H.: DE 1.046.030 (1959)
258. Slaugh, L.H., Mullineaux, R.D.: J. Organomet. Chem. *13*, 469 (1968)
259. Shell Dev. Co. (Canell, L.G., Slaugh, L.H., Mullineaux, R.D.), DE 1.186.455 (1965)
260. Shell Dev. Co (Slaugh, L.H., Mullineaux, R.D.) BE 603.820 (1961) 606.408 (1961), US 3.239.569 (1966), BE 619.344 (1962), US 3.239.566 (1966)
261. Shell Dev. Co. (Greene, C.R., Meeker, R.E.) BE 621.833 (1962), US 3.274.263 (1966), BE 623.213 (1962), 627.365 (1963), 627.371 (1963)
262. Tucci, E.R.: Ind. Engng. Chem., Prod. Res. Dev. *7*, 32, 125 (1968)
263. Cornils, B., Falbe, J., Tummes, H.: Chem.-Ztg. *97*, 368 (1973)
264. Tucci, E.R.: Chem. Ing. Techn. *41*, 883 (1969)
265. Hershman, A. et al.: Ind. Eng. Chem., Prod. Res. Dev. *8*, 291, 372 (1969)
266. Bianchi, M., Frediani, P., Piacenti, F.: Chim. Ind. (Milano), *55*, 798 (1973)
267. DuPont, GB 638.754 (1947/1950)
268. BASF (Reppe, W., Kröper, H.), DE 860.350 (1952)
269. Esso Res. Engng. Co (Wanless, G.G., Morway, A.J.), US 3.000.825 (1961)
270. Pino, P., Piacenti, F., Neggiani, P.P.: Chem. Ind. (London) *35*, 1400 (1961); Piacenti, F., Pino, P.: J. Chem. Soc. C *1966*, 488
271. Brewis, S.: J. Chem. Soc. *1964*, 5014
272. Piacenti, F. et al.: J. Organomet. Chem. *23*, 257 (1970)
273. Oliver, K.L., Booth, F.B.: Hydrocarbon Processing *1970*, (4), 112
274. Toyo Rayon Co., JP 23 005/67 (1964/1967)
275. Monsanto Comp. (Heimsch, R.A., Andersen, J.W., Weesner, W.E.) DE-AS 1.256.206 (1964)
276. Pruett, R.L., Smith, J.A.: J. Org. Chem. *34*, 327 (1969)
277. Esso Engng. Comp., DE-AS 1.218.428 (1956/1966)
278. DuPont (Gresham, W.F.), US 2.497.303 (1945/1950)
279. Esso Res. Co. (Mertzweiller, J.K.), US 3.182.090 (1961/1965)
280. Bor, G. et al.: J. Organomet. Chem. *154*, 301 (1978)
281. Marko, L.: Ber. Ungar. Mineralöl- und Erdgasversuchsanstalt *2*, 228 (1961)
282. Esso Res. Dev. Co. (Mertzweiller, J.K., Tenney, H.M.), US 3.255.259 (1961/1966), DE-AS 1.076.659 (1957/1960)
283. Monsanto Co. (Heimsch, R.A.), GB 1.041.101 (1964)
284. Montecatini SpA (Piacenti, F., Pino, P.), DE-AS 1.300.543 (1969)
285. All-Unions Sci. Res. Inst. (Gankin, W.Ju. et al.), SU 191.520 (1965) 191.520 (1965)
286. ICI Ltd. (Taylor, A.W.C., Harvey, P.G.) US 2.752.397 (1956), FR 1.024.451 (1953)
287. ICI Ltd. (Goddard, R.E., Mansfield, G.H.), CA 629.025 (1960/1971)
288. Studienges. Kohle mbH, BE 572.130 (1966)
289. Ruhrchemie AG, Hydrocarbon Proc. *1977*, (11), 134, 163

290. BASF A.G. Hydrocarbon Proc. *1977*, (11), 135, 172
291. Prod. Chim. Ugine-Kuhlmann, Inf. Chim. Ed. Special Export 1973, p. 105
     Nr. 121, 217 and Nr. 139, 165
292. Weber, H., Falbe, J.: Ind. Engng. Chem. *62* (4), 33 (1970)
293. Marko, L., Bathory, J.: Chem. Anlagen und Verfahren *1970*, (10), 65
294. DuPont (Gresham, W.F., McAlevy, A., Brooks, R.E. u.a.)
     GB 638 754 (1947/1950); 662 706 (1949/1952),
     CA 491 136 (1946/1953) and 491.137 (1946/1953)
     US 2 437 600 (1948), 2.564 130 (1951), 2 497 303 (1950),
     2 517 383 (1950), 2 549 454 (1951), 2 564 104 (1951) and
     3 081 357 (1960/1963)
295. Union Carbide Corp. (Schulz, H.W. et. al.), GB 815.566 (1959), CA 593 954 (1960)
     DE-AS 1 109 159 (1957/1961), FR 1.188 061 (1959) and 1 192 296 (1957/1959),
     US 3 014 970 (1957/1961)
296. Standard Oil Co. (Field, E., Hill, B.L.), US 2.587 576 (1950)
297. Mitsubishi Corp., US 2 992 275 (1957/1961)
298. Magyar Asvanyolaj (Freund, M. et al.: HU 145 570 (1956/1959); C.*1963*, Nr. 21, 8733
299. ICI Ltd. (Harvey, P.G., Taylor, A.W.C., Brewis, S.)
     FR 1 374 941 (1964), US 2 752 395 and 2 752 396 (1951/1956)
     DE-AS 1 273 518 (1963)
300. Natta, G. et al.: Chim. Ind. Milano *37*, 6, 865 (1955) and *42*, 587 (1960)
301. Marko, L.: Proc. Chem. Soc. London *1962*, 67
302. Matsubara, M., Hokaido Daigaku Kogakubu Kenkyu Hokuku *1966*, (40), 139; C.A. *68*,
     39 003 g (1968)
303. Wanaga, R.F.: Bull. Chem. Soc. Japan *35*, 778 (1962)
304. Esso Res. Dev. Co. (Robinson, J.W.), US 3 284 510 (1963/1966)
     GB 827.350 (1958/1960)
305. Alekseeva, K.A. et al.: Neftechimija *6*, 276 (1966)
306. Meis, J., Ruhrchemie AG, unpublished
307. Onoda, T. et al.: DE-OS 2.445.119 (1975)
308. Chemische Verwertungsges. mbH (Rottig, W.), DE 952 440 (1956)
309. ICI Ltd. (Harvey, P.G., Ackroyd, N.), FR 1.024.449 (1950),
     DE 860 199 (1952)
310. Standard Oil Dev. Co., GB 728 913 (1955)
311. Montecatini SpA (Natta, G., Ercoli, R. Castellano, S.),
     DE 1 079 025 (1969), GB 782 459 (1955),
     US 3 008 996 (1961)
312. Takegami, Y., Watanabe, A., Masada, H.: Bull. Chem. Soc. Japan *40*,
     1459 (1967)
313. Ube Ind. Ltd. (Umemura, S., Ikeda, Y.), JP 75 101 320 (1975)
     C.A. *84*, 43 342 q (1976)
314. BASF (Himmele, W.), DE-OS 2 106 243 (1971/1972)
315. Ajinomoto Co. Inc. (Kato, J., Wakamatsu, H., Ishiwara, H).
     US 2978 481 (1957/1961); C.*1963*, Nr. 3, 1067,
     JP 2780/66 (1963/1966)
316. Esso Res. Dev. Co. (Kirshenbaum, J.) US 2 909 538 (1959)
317. Shell Dev. Co. (Falbe, J., Korte, F.), AU 265860 (1963)
318. Rohm & Haas Comp. (Niederhauser, W.D.), DE 1 151 497 (1964)
319. Gankin, W.Ju.: Karbonilirovanie Nenasysh. Uglevod. *1968*, 61
320. Falbe, J., Huppes, N.: Brennstoffchem. *47*, 314 (1966); US 3.446.839 (1969)
321. Falbe, J., FR 1 576 058 (1969)
322. Inventa A.G. (Berther, C., Sailer, R., Giesen, J.), DE-OS 1 468 098
     (1964/1970), GB 1 161 147 (1969), DE-OS 1 468 065 (1963/1973)
323. Himmele, W., Siegel, H.: Tetrahedron Lett. *1976*, 907
324. Prod. Chim. Ugine Kuhlmann (Gueant, A., Mercier, S.) DE-AS 2 519 011
     (1975/1978)

325. Shell Dev. Co. (Shiras, R.N.) US 2.490.283 (1947)
326. BASF A.G. (Diewald, J., Bossert, C.), DE-AS 1.257.762 (1963/1968)
327. Magyar Asvanyolaj F.I. (Marko, L. et al.), HU 1304 (1970); C.A. *74*, 87 359 y (1971)
328. Eastman Kodak Co. (Magness, J.F., Morris, P.M.) GB 1 086 100 (1967); C.A. *68*, 59 116 z (1968)
329. Texaco Inc. (Macaluso, A., Rigdon, O.W.), US 3 907 909 (1975)
330. Tucci, E.R.: Ind. Engng. Chem. Prod. Res. Dev. *9*, 516 (1970)
331. Schwager, I., Knifton, J.F.: Catalysis *45*, 256 (1976)
332. BP Co., NL 67-14072 (1967), 70-08999 (1969)
333. BASF AG, NL 69-13 649 (1969), BE 738 676 (1969)
334. vergl. z. B. Chini, P.: Inorg. Chem. *8*, 1206 (1969); Chem. Commun. *1967*, 440; Gambino, O., Rossetti, R., Stanghellini, P.L., Cetini, G.: Inorg. Chem. *7*, 609 (1968)
335. Knap, J.E.: Chem. Engng. Progr. Symp. Ser. *63*, *76*, 47 (1967)
336. Chini, P. et al.: Inorg. Chim. Acta *3*, 21, 299, 315 (1969)
337. Chemische Verwertungsges. Oberhausen mbH (Ebel, A., Gemassmer, A., Wenzel, W.), DE 896 341 (1953)
338. Chemische Verwertungsges. Oberhausen mbH (Nienburg, H.J., Gemassmer, A.), DE 902 491 (1954)
339. See in Blossey, E.C., Neckers, D.C.; Solid Phase Synthesis. Dowden: Hutchinson & Ross, 1975
340. Neckers, D.C.: Chemtech *1978*, (2), 108
341. Mathur, N.K., Williams, R.E.: J. Macromol. Sci., Rev. Macromol. Chem. *C15* (1), 117 (1976)
342. Kohler, N., Dawans, F.: Rev. Inst. Franc. Petrole, Ann. Combust. Lig. *27*, 105 (1972)
343. Grubbs, R.H.: Chemtech *1977*, (8), 512
344. Michalska, Z.M., Webster, D.E.: Chemtech *1975*, (2), 117
345. Macho, V.: Ropa Uhlie *1977*, 19(1), 4; C.A. *87*, 55 302 b (1977)
346. Manassen, J. et al.: Catal.: Prog. Res., Proc., Sci. Comm. Conf. *1972*, 177; C.A. *82*, 77 513 v (1975)
347. Dawydoff, W.: Faserforsch. Textiltechn. *1976*, (27), 189; C.A. *85*, 99 713 z (1976)
348. Tarama, K.: Shokubai *1973*, 15 (2), 1
349. Grubbs, R.H. et al.: J. Amer. Chem. Soc. *95*, 2373 (1973)
350. Tjan, P.W.H.L.: Chem. Weekbl. *1973*, 69 (28) K11
351. Hares, R.M.: Dissertation 1976, Diss. Abstr. Int. B *1977*, 37 (12), 6130
352. Lang, W.H. et al.: Organomet. Polym. Symp. *1977* (Publ. 1978), 145; C.A. *89*, 23 586 x (1978); J. Organomet. Chem. *134*, 85 (1977)
353. Monsanto Co.: GB 1 185 453 (1968/1970)
354. Haag, W.O., Whitehurst, D.D.: Catal. Proc. Int. Congr., 5th 1972 (Publ. 1973), 1, 465, C.A. *80*, 107 936 p (1974)
355. BP Co. Ltd. (Hancock, R.D., Howell, I.V., Pitkethly, R.C.), GB 1 426 881 (1976)
356. Allum, K.G. et al.: J. Catalysis *43*, 322 (1976); Catal. Proc. Int. Congr. 5th 1972 (Publ. *1973*), I. 477; C.A. *80*, 125 325 t (1974)
357. Toa Nenryo Kogyo KK (Usami, S., Nishimura, K., Fukushi, S.), US 3 378 590 (1968); C.A. *69*, 35 434 w (1968) and JP *69*, 12 403 (1969); C.A. *71*, 90 820 k (1969)
358. Moffat, A.J.: J. Catalysis *18*, 193 (1970) and *19*, 322 (1970)
359. Shell Dev. Co. (Waterman, H.I., Keulemans, A.I.M.), CA 508 017 (1954)
360. Phillips Petr. Co. (Allen, J.D.), US 3 998 887 (1976)
361. Eastman Kodak Co., FR 1 457 354 (1965/1966)
362. All-Unions Oil Chem. Proc. Res. Inst. (Rudkovski, D.M., Trifel, A.G., Gankin, V. Yu, Krinkin, D.P.) SU 189 399 (1966); C.A. *68*, 77 689 t (1968)
363. Phillips Petrol. Co. (Solomon, P.W.), US 3 636 159 (1972)
364. Mantovani, E., Palladino, N., Zanobi, A.: J. Mol. Catal. *3*, 285 (1978) DE-OS 2 804 307 (1978)
365. Esso Res. Co. (Cull, N.L.), US 3 383 426 (1968), (Mertzweiller, J.K., Cull, N.L., Hawley, R.S.) US 3 425 895 (1965/1969); C.A. *70*, 69 099 b (1969)

366. Sanui, K., McKnight, W.J., Lanz, R.W.: Macromol. *7*, 952 (1974)
367. ICI Ltd. (Ragg, P.L.), DE-AS 2 000 829 (1972)
368. Arai, H.: J. Catalysis *51*, 135 (1978); Arai, H., Kaneko, T., Kunugi, T.: Chem. Lett. *1975*, 265
369. Bayer, E., Schurig, V.: Angew. Chem. *87*, 484 (1975)
370. Grubb, R.H., Sweet, E.M., Phisanbut, S.: in Rylander, P.N., Greenfield, H.: Catalysis in Organic Syntheses. p. 153. New York/London: Academic Press. Inc. (1976)
371. Atlantic Richfield Co., US 3 940 447 (1970)
372. Capka, M. et al.: Tetrahedron Lett. *50*, 4787 (1971)
373. Bilhou, J.L. et al.: J. Organomet. Chem. *153*, 73 (1978)
374. Esso Res. Co., NL 73-08749 (1973)
375. Mobil Oil Corp. (Mitchell, T.O., Whitehurst, D.D.), US 4 053 534 (1977)
376. BASF, GB 1 254 182 (1968)
377. Bailar, J.C.: Catal. Rev. *1974*, 17
378. Gankin, V.Yu., Rudkowskij, D.M.: SU 384 536 (1973)
379. Atlantic Richfield Co. (Yoo, J.S.), US 3 989 759 (1976), 3 991 119 (1976), 3 937 742 (1976)
380. The Distillers Co., JP 6 604/68 (1968), DE-AS 1 257 764 (1968)
381. Cornils, B., Frohning, C.D., Liebern, H. (Ruhrchemie AG), unpublished
382. Rony, P.R.: J. Catalysis *14*, 142 (1969) und US 3 855 307 (1974)
383. Stamicarbon B.V., (Gerritsen, L.A., Ambacht, H.I., Scholten, J.J.F.), DE-OS 2 802 276 (1978)
384. Mistrik, J.: Conference on the Chemistry and the Processing of Mineral Oil and Natural Gas *1965* (Budapest), Proceedings 358
385. Esso Res. Eng. Co. (Gladrow, E.M., Mattox, W.J.), US 3 352 924 (1963/1967)
386. Kraus, M.: Chem. Listy *1972*, 1281
387. Lang, W.H. et al.: ACS Div. Org. Coatings Plast., Chem. Pap. *37* (1), 304 (1977)
388. Pittman, C.U., Hirao, A.: J. Org. Chem. *43*, 640 (1978)
389. Phillips Petr. Co. (Allen, J.D.), US 3 847 997 (1974)
390. BP (Allum, K.G.), DE-OS 2 022 710 (1970)
391. BASF AG. (Kniese, W., Nienburg, H.J.), DE-OS 1 768 206 (1971)
392. Pittman, C.U. et al.: Ann. N.Y. Acad. Sci. *1977*, 295, 15; C.A. *88*, 135 911 k (1978)
393. BASF A.G. (Nienburg, H.J. et al.: DE-OS 2 206 252 (1973)
394. Pittman, C.U., Evans, G.O.: Chemtech. *1973*, (9), 560
395. Allum, K.G. et al.: J. Organomet. Chem. *87*, 189 (1975)
396. Pittman, C.U. et al.: J. Amer. Chem. Soc. 97, 1749, 4774 (1975)
397. Boucher, L.J., Oswald, A.A., Murrell, L.L.: Am. Chem. Soc. Div. Pet. Chem. Prepr. *19*, 162 (1974)
398. Mobil Oil Corp. (Mitchell, T.O., Whitehurst, D.D.), US 3 980 583 (1976)
399. UOP Inc. (Homeier, E.H.), US 4 070 403 (1978)
400. Esso Corp. (Oswald, A.A., Murrell, L.L.), DE-OS 2 332 167 (1974)
401. BP Ltd. (Hancock, R.D. et al.:) Catal., Proc. Ind. Symp. 1974 (Publ. 1975), 361; C.A. *84*, 121 050 j (1976); BP Ltd. (Allison, K., Foster, G., Lawrenson, M.J.), GB 1 275 733 (1972)
402. Boucher, L.J., Murrell, L.L., Oswald, A.A.: Erdöl & Kohle *28*, 253 (1975)
403. Esso Co. (Oswald, A.A., Murrell, L.L.), NL 7 308 749 (1976)
404. Mobil Oil Corp. (Haag, W.O., Whitehurst, D.D.), US 3 578 609 (1971); BE 721 686 (1968), DE-OS 1 800 371 (1969), NL 68-13 999 (1968)
405. Kim, T.H., Rase, H.F.: Ind. Eng. Chem., Prod. Res. Dev. *15*, 249 (1976)
406. Inst. Chem. Phys. Academ. Sci. USSR, SU 363 686 (1972)
407. Standard Oil Co. (Trevillyan, A.E.), US 3 998 864 (1976)
408. UOP Inc. (Massie, S.N.), US 3 880 938 (1975)
409. Monsanto Co. (Paulik, F.E.), US 3 487 112 (1969); C.A. *72*, 68 984 r (1970)
410. Phillips Petr. Co. (Kahle, G.R., Cleary, J.W.), US 3 652 676 (1972)
411. Shell Int. Res. Mat. (Neel, E.E.A., Gaucher, M., Clement, C.J.), DE-OS 2 127 624 (1971)

412. Johnson, Matthey & Co. Ltd. (Bond, G.C.), GB 1 332 894 (1973)
413. Atlantic Richfield Co. (Yoo, J.S.). US 3 940 447 (1976)
414. Eastman Kodak Co. (Hull, D.C., Hagemeyer, H.J.), GB 1 120 277 (1968), C.A. *69*, 76 644 h (1968)
415. All-Unions Sci. Res. Inst. Petr. Proc. (Rudkowskij, D.M.), SU 189 399 (1966)
416. VEB Leuna (Bemmann, R., Berndt, F.), DD 12 651 (1953)
417. Atlantic Richfield Co. (Yoo, J.S.), US 4 018 834 (1977)
418. The Distillers Co. Ltd., JP 6 604/68 (1968), GB 1 012 011 (1965)
419. Centola, P. et al.: Chim. Ind. (Milano) *54*, 775 (1972)
420. Matsubara, M.: Hokkaido Daigaku Kogakubu Kenkyu Hokoku *1967*, 167; C.A. *68*, 86 801 a (1968)
421. Gankin, V.Yu., Rudkowskij, D.M.: SU 384 536 (1973)
422. Gankin, V.Yu., Genender, L.S., Rudkowskij, D.M.: Zh. Prikl. Khim. *40*, 2029 (1967)
423. Gankin, V.Yu. et al.: Gidroformilirovanie *1972*, 90
424. Del'nik, V.B.: Karbonilirovanie Nenan. Ugle. *1968*, 161; C.A. *71*, 2903 m (1969)
425. Ethyl Corp. (Kehoe, L.J., Schell, R.A.), US 3 557 219 (1968/1971)
426. Toa Nenryo Kogyo K. K., JP 24 882/67
427. Rhône-Progil S. A. (Berthoux, J., Martinaud, J.P.), DE-OS 2 261 543 (1973)
428. Loktev, S.M. et al.: Neftepererab. Neftekhim *1972*, 34; C.A. *78*, 42 801 c (1973); Neftekhimiya *13*, 821 (1973)
429. Monsanto Co. (Rony, P.R., Roth, J.F.), Catal. Proc. Int. Symp. 1974 (Publ. 1975), 373; C.A. *84*, 30 289 c (1976)
430. Sanger, A.R., Schallig, L.R.: J. Mol. Catal. *3*, 101 (1977)
431. Gregorio, G. et al.: Symp. on Rhodium in Homogeneous Catalysis *1978*, (Veszprem), Proceedings p. 121
432. Evans, G.O. et al.: J. Organomet. Chem. *67*, 295 (1974)
433. Pichler, H., Firnhaber, B., Kioussis, D.: Brennstoff-Chem. *44*, 337 (1963)
434. Gankin, W.Yu.: Zh. Prikl. Khim. *40*, 1639, 1649 (1967); C.A. *67*, 118 616 s (1967)
435. Gankin, W.Yu.: Chem. Technol. Brennst. Öle (Moscow) *1966*, (4), 8
436. Gankin, W.Yu. et al.: Neftekhimiya 6, 271 (1966); C.A. *65*, 8704 (1966)
437. Johnson, Matthey & Co. Ltd. (Wilkinson, G.), GB 1 357 735 (1974)
438. Johnson, Matthey & Co. Ltd. (Wilkinson, G.), DE-OS 2 047 748 (1971) and 2 055 539 (1971)
439. Guyer, P., Guyer A.: DE-AS 1 229 060 (1966), NL 6412 249 (1965)
440. Toyo Gas K. K., JP 28 654/65 (1965)
441. Rudkowski, D.M. et al.: SU 177 871 (1966), C.A. *65*, 15 618 (1966)
442. Johnson, Matthey & Co., NL 70-16 532 (1970)
443. Gankin, W.Yu.: Chim. Technol. Topliv. Masel. *11*, 10 (1966); C. *1967*, 28-2590
444. Monsanto Co. (Paulik, F.E.), DE-OS 1 768 303 (1971)
445. Parshall, G.W.: J. Amer. Chem. Soc. *94*, 8716 (1972), and US 3 657 368 (1972) (to DuPont)
446. Alper, H., des Abbayes, H.: J. Organomet. Chem. *134*, C 11 (1977)
447. ICI Ltd. (Featherstone, W., Cox, T.), GB 1 432 561 (1972)
448. BP Ltd. (Ellis, J.E.), GB 1 312 076 (1973)
449. Robinson, K.K. et al.: J. Catalysis *15*, 245 (1969), Ind. Eng. Chem., Prod. Res. Dev. *8*, 372 (1969) US 3 487 112 (1969)
450. Esso Co., GB 801.734 (1958)
451. Chemische Verwertungsges. mbH (Schiller, G.), DE 953 605 (1956)
452. Stanolind Oil and Gas Co., US 2 679 378 (1954)
453. UOP In. (Homeier, E.H.), US 3 984 478 (1975), DE-OS 2 639 755 (1977)
454. BP Co. (Allum, K.G. et al.), DE-OS 2 062 352 (1970/1971)
455. California Inst. of Technol. (Gray, H.B., Rembaum, A., Gupta, A.), DE-OS 2 633 959 (1977)
456. Pittman, C.U., Ryan, R.C.: Chemtech *1978*, (3), 170
457. Basset, J.M., Smith, A.K.: XIX. Int. Conf. on Coord. Chemistry, Prague 1978, Proc. p. 161, and in [461], p. 69 f
458. Friedrich, J.P.: Ind. Eng. Chem., Prod. Res. Dev. *17*, 205 (1978)

459. Martinengo, S. et al.: J. Chem. Soc., Chem. Commun. *1977*, 39
460. Schmid, G.: Angew. Chem. *90*, 417 (1978)
461. Norton, J.R.: in Tsutsui, M., Ugo, R. (eds.): Fundamental Research in Homogeneous Catalysis, p. 99 f, New York/London: Plenum Press 1977
462. Imyanitov, N.S., Gogoradowskaya, N.M., Sjemyenova, T.A.: Kinetika i Katal. *19*, 573 (1978)
463. Cassar, L.: in [461], p. 115 f
464. Ryan, R.C., Pittman, C.U., O'Connor, J.P.: J. Amer. Chem. Soc. *99*, 1986 (1977); Abstracts of the 173rd Natl. Meeting of the A. C. S., New Orleans, La., 1977 p INOR 098 (ref. in [456])
465. Piacenti, F. et al.: Inorg. Chem. *10*, 2759 (1971)
466. Schulz, H.F., Bellstedt, F.: Ind. Eng. Chem., Prod. Res. Dev. *12*, 176 (1973)
467. Sanchez-Delgado, R.A., Bradley, J.S., Wilkinson, G.: J. Chem. Soc., Dalton Transactions *1976*, (5), 399
468. Csontos, G. et al.: Hung. J. Ind. Chem. Veszprem. *1*, 53 (1973)
469. Braca, G. et al.: Chim. Ind. (Milano) *52*, 1091 (1970)
470. BP Ltd. (Lawrenson, M.J., Green, M.), DE-OS 2 026 918 and 2 026 926 (1970/1971)
471. Ethyl Corp. (Kehoe, L.J.), US 3 534 103 (1970)
472. Mitsubishi Chem. Ind. Ltd., JP 7 102 244 (1971)
473. ICI Ltd. (Cox, G.F., Whitfield, G.H.), GB 999 461 (1962/1965); (Smith, P., Jager, H.H.), GB 966 482 (1960/1964); C. *1966*, N. 37-2617
474. King, R.B. et al.: J. Amer. Chem. Soc. *100*, 1687 (1978)
475. Weil, T.A., Metlin, S., Wender, I.: J. Organomet. Chem. *49*, 227 (1973)
476. Fell, B., Shanshool, J.: Chem.- Ztg. *99*, 231 (1975)
477. Sakakibara, Y., Nakamura, T.: J. Soc. Org. Synth. Chem. Japan *23*, 757 (1965)
478. Imjanitov, N.S., Rudkovskij, D.M.: Kinetika i Kataliz *9*, 1042 (1968), C. *1969*, 28-0603
479. Booth, B.L., Goldwhite, H., Haszeldine, R.N.: J. Chem. Soc. C. *1966*, 1447
480. Toyo Rayon K. K., JP 3361/68 (1968)
481. DuPont (Gresham, W.F., Brooks, R.E.), CA 543 582 (1953)
482. Kang, H.C. et al.: J. Amer. Chem. Soc. *99*, 8323 (1977)
483. Lapidus, A.L., Gildenberg, E.Z., Eidus, Y.I.: Kinet. i. Kataliz *16*, 252 (1975)
484. Bogoradovskaya, N.M., Imyanitov, N.S.: Gidroformilirovanie *1972*, 138
485. Shell Int. Res., NL 71-10255 (1971)
486. Shell Oil Co. (McClure, J.D.), US 3 875 240 (1975) and DE-OS 2 137 362 (1972)
487. UOP Inc. (Homeier, E.H.), DE-OS 2 639 755 (1977)
488. DuPont (Gresham, W.F., McAlery, A.), CA 491 136 (1953)
489. Toyo Rayon Co. Ltd., JP 23 005/67 and 23 007/67 (1967)
490. Zhir-Lebed, L.N. et al.: Kinet. i Katal. *15*, 537 (1974); C.A. *81*, 12 775 s (1974)
491. Consiglio, G.: Helv. Chim. Acta *1976*, 642
492. Fenton, D.M., Olivier, K.L.: Chemtech *2*, 220 (1972)
493. UOP Inc. (Homeier, E.H.), US 3 948 999 (1976)
494. Toyo Rayon Co. Ltd., JP 23 006/67 (1967)
495. BP Ltd. (Foster, G., Lawrenson, M.J.), DE-AS 1 911 631 (1969/1970); C.A. *72*, 21 322 h (1970)
496. Yamaguchi, M.: Shokubai (Catalyst, Tokyo) *9*, 160 (1967) C. *1969*, 39-1108
497. Ethyl Corp. (Kehoe, L.J., Schell, R.A.), US 3 544 635 (1968)
498. Yagupski, G., Brown, C.K., Wilkinson, G.: J. Chem. Soc., Chem. Commun. *21*, 1244/45 (1969)
499. BASF A. G. (Himmele, W., Hoffmann, W., Pasedach, H., Aquila, W.), DE-OS 1 964 962 (1969), NL 70-18 674 (1970)
500. BP Ltd. (Foster, G.), DE-OS 1 946 437 (1969/1971), FR 2 020 531 (1970)
501. Wilkinson, G.: FR 1 459 643 (1966)
502. King, R.B. et al.: J. Amer. Chem. Soc. *100*, 2925 (1978)
503. Progil S. A., NL 70-11 856 (1970)
504. Imyanitov, N.S. Rudkovskij, D.M.: Kinet. i Katal. *8*, 1051, 1240 (1967), C.A. *68*, 86 867 b (1968); Zh. Prikl. Khim. *40*, 2020 (1967)

505. UOP Ltd. (Homeier, E.H.), DE-OS 2 623 867 (1976)
506. Imyanitov, N.S.: Zh. Obsheh. Khim. *1975,* 1344
507. Shell Res. Dev. (Slaugh, L., Mullineaux, R.D.), US 3 239 571 (1966)
508. Texaco Inc. (Schwager, I., Knifton, J.F.), DE 2 322 751 (1973)
509. BASF A. G. (Kutepow, N. von, Mueller, F. J.), DE-OS 2 456 739 (1976)
510. Tummes, H., Bahrmann, H., Cornils, B.: Ruhrchemie AG, unpublished
511. Nakano, Y.: JP 16 726/68 (1965/1968)
512. Pino, P., von Bezard, D.: DE-OS 2 807 251 (1978) and CH-Appl. 2115/77 (1977)
513. Hsu, C. Y., Orchin, M.: J. Amer. Chem. Soc. *97,* 3553 (1975)
514. ICI Ltd., GB 1 185 156 (1966/1970); GB 1 368 434 (1974)
515. Texaco Inc. (Knifton, J.F., Schwager, I.), US 3 996 293 (1976)
516. Texaco Inc. (Knifton, J.F.), GB 1 391 395 (1975)
517. Dow Chemical Comp. (Tsai, J.H., Anderson, G.H.), US 3 679 722 (1970/1972)
518. Toa Nenryo Kogyo K. K., JP 12 403/69 (1966/1969)
519. Behrens, H. et al.: Z. Naturforsch. 32 b, 57 (1977)
520. Lion Fat and Oil Co. Ltd. (Inagaki, T., Kiyonaga, Y., Yoshimura, Y.), JP 7652, 390 (1976); C.A. *85,* 99 916 t (1976)
521. Gankin, V. Yu.: Zh. Prikl. Khim. *40,* 1788 (1967), C.A. *68,* 41 841 x (1968), SU 191 520 (1965/1967)
522. Esso Res. Eng. Co., NL 69-15 218 (1969)
523. Lemke, H.: Hydrocarbon Proc. *45,* (2), 148 (1966)
524. BASF A. G. (Strohmeyer, W. et al.), DE-OS 2 332 638 (1973)
525. Chem. Eng. News *55,* (13), 69 (1977)
526. Macho, V., Minarska, M.: Erdöl und Kohle *27,* 98 (1974)
527. Kuno, K.: Yuki Gosei Kagaku Kyokaishi *1977,* 35 (8), 683; C.A. *87,* 183 877 e (1977)
528. Chem. Verwertungsges. mbH Oberhausen (Schiller, G.), DE 953 605 (1952/1955)
529. Pino, P., Consiglio, G.: Sympos. Rhodium in Homogeneous Catalysis, Veszprem 1978, Proc. p. 98
530. Wakamatsu, H.: Nippon Kagaku Zasshi *85,* 227 (1964)
531. Pino, P., Piacenti, F., Bianchi, M. in [7], p. 171 f
532. Ethyl Corp. (Schnell, R.A.), US 3 752 859 (1967)
533. Mitsubishi Chem. Ind. Ltd. (Ohsumi, K.), JP 3733/1966 (1963/1966)
534. BP Ltd., NL 68-04417 (1968)
535. Esso Res. (Bartlett, J.H.), US 2 894 038 (1956/1959), GB 816 993 (1959), DE-OS 1 056 503 (1957)
536. Esso Res. Eng. Co. (Hughes, V.L.), US 2 880 241 (1959); GB 801 734 (1959)
537. Imyanitov, N.S., Rudkowskij, D.M.: Neftekhimiya *3,* 198 (1963), C.A. *59,* 7396 (1963)
538. Evans, D., Osborn, J.A., Wilkinson, G.: J. Chem. Soc. A *1968,* 3133 (1968)
539. Chini, P., Martinengo, S.: Inorg. Chim. Acta *3,* 299 (1969); Chem. Commun. *5,* 251 (1968)
540. BP Ltd. (Foster, G., Lawrenson, M.J.), DE-OS 1 901 144 (1969/1968); C.A. *71,* 83 136 p (1969)
541. Gankin, W. Yu. et al.: Zh. prikl. Chim. *40,* 2029 (1967), *41,* 209 (1968); Karbonilirovanie Nenas. Uglev. *1968,* 57, 60; Z. physik. Chem. (Frankfurt) *59,* 1577 (1968)
542. Yamaguchi, M.: Kogyo Kagaku Zasshi *1969,* 671
543. Marko, L.: Aspects of Homogeneous Catal. *1974,* 2, 3
544. Heil, B., Marko, L.: Magyar. Kem. Lapja *23,* 669 (1968), Chem. Ber. *104,* 3418 (1971)
545. Sittig, M.: Catalyst Manufacture, Recovery and Use. Neyes Dev. Corp., Park Ridge, N. J. 1972, p. 229
546. Cramer, R.: Acc. Chem. Res. *1,* (6), 186 (1968), C. *1969,* 15-757
547. Fenton, D.M., Oliver K.L.: Amer. Chem. Soc., Div. Petr. Chem. Prepr. *1971,* 16 (1), B5-B12; C.A. *78,* 29 128 u (1973)
548. Thomas, C.L.: Catalytic Processes and Proven Catalysts, p. 217, New York/London: Academic Press (1970)
549. Jefferson Chem. Co. Inc. (Gipson, R.M.), DE-OS 1 802 895 (1969); C.A. *71,* 70 067 s (1969)
550. Singer, H., Stein, W., Lepper, H.: Fette/Seifen/Anstrichm. *74,* 193 (1972)

551. Stefani, A. et al.: J. Amer. Chem. Soc. *95*, 6504 (1973)
552. Eidus, J.T., Lapidus, A.L.: Neftekhimiya *7*, 51 (1967)
553. BASF A. G. (Kummer, R., Platz, R.), DE-OS 2 354 217 (1973/1975)
554. Chini, P., Martinengo, S., Garlaschelli, G.: Symp. Chem. Hydroformylation and Related Reactions, Veszprem 1972, Proc. p. 68 f; Chem. Commun. *1972*, (12), 709
555. BP Ltd. (Lawrenson, M.J., Foster, G.), DE-OS 1 812 504 (1969); C.A. *71*, 101 313 a (1969), NL 70-09430 (1970)
556. Heil, B., Marko, L.: Chem. Ber. *102*, 2232 (1969)
557. Mitsubishi Chem. Ind. Co. Ltd. (Yamaguchi, M. et al.), JP 7 111 805 (1971), C.A. *74*, 140896 z (1971)
558. BASF A. G. (Schwirten, K., Disteldorf, W., Eisfeld, W.), DE-OS 2 604 545 (1977)
559. BP Ltd. (Lawrenson, M.J., Foster, G.), DE-OS 1 806 293 (1969), C.A. *71*, 70109 g (1969)
560. Törös, S.: Magyar Kem. Lapja *29*, 543 (1974), Chem. Inf. 12-122, 1975
561. Arakawa, M.: JP 7 797 930 (1977); C.A. *88*, 74 208 n (1978)
562. Lai, R., Ucciani, E.: C. R. Acad. Sci. Ser. C *273*, 1368 (1971)
563. Farbwerke Hoechst A. G. (Fischer, H., Röhrscheid, F.), DE-OS 2 163 753 (1971)
564. Chem. Techn. *26*, (6), 376 (1974)
565. Ruhrchemie AG (Falbe, J.), FR 1 576 057 (1969), BE 718 856 (1968), DE-OS 1 618 384 (1967), GB 1 179 226/1968
566. BASF A. G. (Nienburg, H.J.), DE-OS 2 317 625 (1973/1974); Eur. Chem. News *27*, (686), 36 (1973)
567. Falbe, J.: Brennstoff-Chem. *48*, (2) 46 (1967)
568. Ajinomoto K. K. (Takesada, M.), JP 1 575/1965; C. *1966*, 35-2629
569. Degussa AG, BE 674 150 (1965/1966)
570. Sato, S.: Nippon Kagaku Zasshi *90*, (6), 579 (1969)
571. Takesada, M., Wakamatsu, H.: Bull. Chem. Soc. Japan *43*, 2192 (1970)
572. Bott, K.: Symp. Rhodium in Homogeneous Catalysis, Veszprem 1978, Proc. p. 106; Chem. Ber. *108*, 997 (1975), Angew. Chem. *85*, 911 (1973)
573. BASF A. G. (Aquila, W., Himmele, W., Hoffmann, W.), DE-OS 2 050 677 (1970/1972)
574. Fell, B., Boll, W., Hagen, J.: Chem.-Ztg. *99*, 452, 485 (1975)
575. Fell, B., Beutler, M.: Erdöl und Kohle *29*, 149 (1976)
576. Chini, P.: Inorg. Chim. Acta *2*, 31 (1968)
577. Kang, H.C. et al.: J. Amer. Chem. Soc. *99*, 8323 (1977)
578. Imjanitov, N.S., Rudkovskij, D.M.: J. prakt. Chem. *311* (5), 712 (1969); Karbonilirovanie Nenassysh. Uglevod. *1968*, 28
579. Imyanitov, N.S.: in [179], p. 27
580. Roehm GmbH (Guenzler, W., Kabs, K., Schröder, G.), DE-OS 2 329 577 (1975)
581. Texaco Inc. (Bennett, R.H., Deever, W.R.), US 3 839 459 (1974)
582. Mitsubishi Chem. Ind. Co. Ltd., JP 688/68 (1965/1968)
583. Reppe, W. et al.: Liebigs Ann. Chem. *582*, 38 (1953)
584. Shell Res. Dev. (Slaugh, L.H., Mullineaux, R.D.), BE 606 408 (1962); J. Organomet. Chem. *13*, 469 (1968)
585. Chevron Res. Co. (Wilkes, J.B.), US 3 931 332 (1969/1976), US 3 976 703 (1970/1976), 3 839 471 (1969/1974)
586. Chevron Res. Co. (Wilkes, J.B.), US 3 647 842 (1969/1972) and 3 928 232 (1970/1975)
587. BASF A. G. (Himmele, W.), DE-OS 2 044 651 (1972)
588. Johnson, Matthey & Co. Ltd. (Wilkinson, G.), DE-OS 2 136 470 (1970/1972)
589. Union Carbide Corp. (Moyer, C.E. et al.), DE-AS 2 541 314 (1977), BE 833 454 (1974), FR 2 074 101 (1970)
590. Johnson, Matthey & Co. Ltd. (Wilkinson, G.), DE-OS 2 034 909 (1971)
591. Johnson, Matthey & Co. Ltd. (Wilkinson, G.), DE-OS 2 064 471 (1971)
592. Shell Oil Co. (Greene, C.R., Meeker, R.E.), US 3 274 263 (1966); DE-AS 1 212 953 (1967)
593. BASF A. G. (Kummer, R., Schwirten, K., Schindler, H.D.), DE-OS 2 448 005 (1976)
594. Texaco Inc. (Macalusco, A., Kuntschik, L.F.), US 4 060 557 (1977)
595. Sun Oil Comp. (Renick, L.E.), US 3 825 601 (1974)

596. Chisso Corp. (Tarao, R. et al.), JP 7 340 326 (1973)
597. Mitsubishi Chem. Ind. Ltd., JP 69-2683 (1969)
598. Pregaglia, G. et al.: Chim. Ind. (Milano), *55*, 203 (1973); C.A. *79*, 31 186 a (1973)
599. Imyanitov, N.S., Rudkowskij, D.M.: Zh. Prikl. Khim. *46*, 2683 (1973)
600. Piacenti, F. et al.: J. Organomet. Chem. *23*, 257 (1970)
601. BP Ltd., NL 70-08999 (1970)
602. Tucci, E.R.: Ind. Eng. Chem., Prod. Res. Dev. *8*, 215 (1969)
603. Asinger, F., Fell, B., Rupilius, W.: Ind. Eng. Chem., Prod. Res. Dev. *8*, 214 (1969)
604. Sanger, A.R.: J. Molec. Catal. *1978*, (4), 221
605. Union Carbide Corp. (Brewester, E.A., Pruett, R.L.), DE-OS 2 715 685 (1977)
606. Gulf Res. Dev. Co. (McCracken, J.H., Williamson, R.C.), CA 992 101 (1976)
607. Wilkinson, G.: DE-OS 1 816 063 (1969)
608. Frankel, E.N., Pryde, E.H.: J. Amer. Oil. Chemists Soc. *54*, 873 A (1977)
609. Fell, B., Boll, W.: Chem.-Ztg. *99*, 452 (1975)
610. Rupilius, W., McCoy, J.J., Orchin, M.: Ind. Engng. Chem., Prod. Res. Dev. *10*, (2), 142 (1971)
611. Gulf Res. Dev. Co. NL 67-11551 (1967); DE-OS 1 668 127 (1967)
612. Fell, B., Geurts, A.: Angew. Chem. (Nachr. Chem. Techn.) *20*, (3), 55 (1972), Chem. Ing. Techn. *44*, 434, 708 (1972), Angew. Chem. *83*, 901 (1971)
613. Imyanitov, N.S., Rudkowskij, D.M.: Zh. Prikl. Khim. *1974*, 2095; C.A. *82*, 3724 t (1975); Zh. Obshch. Khim. *44*, 2786 (1974)
614. Ethyl Corp. (Asinger, F.), US 3 657 354 (1972)
615. ICI Ltd. (Miles, D.H.), GB 1 090 993 (1967)
616. Mobil Oil Corp. (Rollmann, L.D., Whitehurst, D.D.), DE-OS 2 357 645 (1974), Adv. Chem. Ser. *1974*, 132
617. Feder, H.M., Halpern, J.: J. Amer. Chem. Soc. *97*, 7186 (1975)
618. Osborn, J.A. et al.: J. Chem. Soc. (A), *1966*, 1711
619. Yagupsky, G., Brown, C.K., Wilkinson, G.: J. Chem. Soc. (A) *1970*, 1392
620. Tucci, E.R.: Ind. Engng. Chem., Prod. Res. Dev. *7*, 32 (1968)
621. Bahrmann, H.: Dissertation Aachen 1975
622. Fell, B., Bahrmann, H.: J. Molec. Catalysis *2*, 211 (1977)
623. Union Carbide Corp. (Pruett, R.L., Smith, J.A.), US 3 917 661 (1975); 3 527 809 (1970); GB 1 197 804 (1970)
624. Ruhrchemie AG (Bahrmann, H. et al.), DE-OS 2 833 538 (1978)
625. Harnisch, H.: Angew. Chem. *88*, 517 (1976)
626. Mason, R., Meek, D.W.: Angew. Chem. *90*, 195 (1978)
627. Mitsubishi Chem. Ind. Co. Ltd. (Onada, T. et al.), JP 7549, 215 (1973), 7571, 610 (1973) and 74-1008, 207 (1974)
628. Union Carbide Corp. (Bryant, D.R.), DE-OS 2 736 278 (1978)
629. Maruzen Oil Co. Ltd. (Matsui, Y. et al.), Bull. Japan Petr. Inst. *1977*, (1), 68; C.A. *87*, 151 640 f (1977)
630. BASF A. G. (Kummer, R., Platz, R.), DE-OS 2 354 217 (1975)
631. Celanese Corp. (Stautzenberger, A.L., Paul, J.L.), US 4 009 003 (1977)
632. Texas Alkyls Inc., (Malpass, D.B., LaPonte, Y.), DE-OS 2 714 721 (1977)
633. Union Carbide Corp. (Bryant, D.R.), DE-OS 2 802 923 (1978)
634. Union Carbide Corp. (Morrell, D.G.), DE-OS 2 802 922 (1978)
635. Gregorio, G. et al.: Symposium on Rhodium in Homogeneous Catalysis, Veszprem 1978, Proceedings, p. 121
636. Pruett, R.L., Smith, J.A.: J. Org. Chem. *34*, 327 (1969)
637. Tucci, E.R.: Erdöl und Kohle *22*, 41 (1969)
638. Union Oil of California (Booth, F.B.), US 3 641 076 (1968)
639. Macho, V., Polievka, M.: Ropa Uhlie *1976*, 18 (1), 18; C.A. *85*, 108 240 x (1976)
640. Craddock, J.H. et al.: Ind. Eng. Chem., Prod. Res. Dev. *8*, (9) 291 (1969)
641. Gankin, V.Yu. et al.: Gidroformilirovanie *1972*, 66; C.A. *77*, 151 202 h (1972)
642. Tucci, E.R.: Ind. Eng. Chem., Prod. Res. Dev. *8*, 286 (1969)
643. Spooncer, W.W.: J. Organomet. Chem. *18*, 327 (1969)

644. Esso Res. Eng. Co. (Cull, N.L., Pine, L.A.), GB 1 097 364 (1968), C.A. *68*, 114 166 t (1968) DE-OS 1 518 765 (1969)
645. Van Boven, M., Alemdaroglu, N.H., Penninger, J.M.L.: Ind. Eng. Chem., Prod. Res. Dev. *14*, (4), 259 (1975)
646. Kniese, W., Nienburg, H.J., Fischer, R.: J. Organomet. Chem. *17*, 133 (1969)
647. Bianchi, M., Benedetti, E., Piacenti, F.: Chim. Ind. (Milano) *49*, 245 (1967); *51*, 613 (1969)
648. Imyanitov, N.S., Volkov, V.A.: Zh. Prikl. Khim. *46*, 2683 (1973)
649. BASF A. G. (Kummer, R.), DE-OS 2 401 553 (1975), FR 2 008 133 (1970)
650. Esso Res. Engng. Co. (Senn, W.L.), US 3 576 881 (1971)
651. BASF A. G. (Kniese, W., Nienburg, H.J.), DE-OS 2 026 163 (1971) and DE-AS 2 005 654 (1971)
652. Shell Int. Res. M., GB 1 191 815 (1970)
653. Dow. Chem. Corp. (Taylor, O.C., Lemaster, L.A.), US 3 989 675 (1976)
654. Inst. Francais du Petrole (Lassau, C.), GB 1 199 550 (1970), DE-OS 1 922 621 (1969), C.A. *72*, 45 755 c (1970)
655. Savitskii, A.V.: Zh. Obsch. Khim. *44*, 106 (1974)
656. Simon, A., Nagy-Magos, Z.: J. Organomet. Chem. *11*, 634 (1968)
657. Shell Oil Co. (Slaugh, L.H.), US 3 448 158 (1969)
658. Esso Res. Engng. Co. (Mertzweiller, J.K., Tenney, H.M.), DE-AS 1 293 735 (1969), US 3 351 666 (1967), BE 651 407 (1967)
659. Jefferson Chem. Comp. (Gipson, R.M.), DE-OS 1 817 700 (1969)
660. Inst. Francais du Petrole (Attali, S., Poilblanc, R. Lassau, C.), FR 1 603 462 (1971)
661. Diamond Alkali Co. (Eisenmann, J.L.), US 3 290 379 (1966)
662. Jefferson Chem. Co. (Brader, W.H.), DE-OS 1 668 244 (1967), FR 1 530 136 (1968), BE 700 691 (1967) and NL 6 709 039 (1968)
663. Shell Int. Res. M., NL 70-18 231 (1970), DE-OS 2 061 798 (1971)
664. BP Co., NL 69-13 756 (1969)
665. Iwanaga, R.: Bull. Chem. Soc. Japan *35*, 865 (1962), C. *1964*, 50-674
666. Chevron Res. Co. (Wilkes, J.G.), DE-OS 2 044 987 (1971) and 2 065 531 (1970)
667. UOP Co. (Massia, S.N.), US 3 969 413 (1975)
668. Montecatini SpA., BE 712 221 (1968); US 3 627 843 (1971)
669. Azote et Produits Chimiques (Prognon, P.), DE-OS 2 210 937 (1972)
670. Shell Oil Co. (Slaugh, L.H.), US 3 448 157 (1969)
671. BASF A. G. (Nienburg, H.J. et al.), DE-OS 1 955 828 (1969), 2 044 361 (1972) and 2 022 184 (1972)
672. ICI Ltd. (Jones, P.J.V.), DE-OS 2 242 646 (1973)
673. Pino, P. et al.: Adv. Chem. Ser. *1974*, 132; C.A. *82*, 85 634 a (1975)
674. BP Ltd. (Foster, G., Lawrenson, M.J.), DE-OS 1 946 437 (1971), GB 1 296 435 (1972)
675. Shell Int. Res. M. (Morris, R.C.), DE-OS 1 667 286 (1971)
676. Shell Int. Res. M. (van Winkle, J.L., Morris, R.C.), BE 728 951 (1969), 678 614 (1966), ZA 661 765 (1966)
677. BASF A. G. (Nienburg, H.J. et al.), DE-OS 2 056 342 (1972); DE-AS 1 230 010 (1966) and GB 1 273 042 (1972)
678. Gulf Res. Dev. Co. (Deffner, J.F., Tucci, E.R., Thayer, H.J.), US 3 725 483 (1973), FR 1 534 510 (1968)
679. Shell Int. Res. M. (Mason, R.F., van Winkle, J.L.), FR 1 502 250 (1967), C.A. *70*, 20 221 w (1969); DE-OS 1 593 466 (1970); NL 69-0123 (1966)
680. Shell Int. Res. M., NL 69-02960 (1969)
681. Shell Int. Res. M. (van Winkle, J.L.), DE-OS 1 909 614 and 1 909 620 (1969), BE 728 949 (1969), ZA 656 693 (1966)
682. Shell Int. Res. M. (Morris, R.C., van Winkle, J.L.), DE-OS 1 543 471 (1968)
683. Gulf Res. Dev. Co. (Deffner, J.F.), FR 1 534 510 (1968); GB 1 198 816 (1969)
684. Gulf Res. Dev. Co. (Deffner, J.F. et al.), US 3 725 483 (1973), 3 644 529 (1972)
685. Cornils, B., Wiebus, E., Dickhaus, G., Bahrmann, H.: unpublished
686. Montecatini SpA, DE-OS 2 313 102 (1973)

687. Falbe, J. et al.: Tetrahedron Lett. *27*, 3603 (1971); ZA 6806, 828 (1969)
688. Shell Res. Dev. Co. (Slaugh, L.H.), AU 256 468 (1965), US 3 231 621 (1966), CA 725 670 (1966), FR 1 472 711 (1967)
689. BASF A. G. (Tavs, P., Kniese, W., Nienburg, H.J.), DE-AS 1 768 441 (1968)
690. BASF A. G. (Nienburg, H.J.), DE-OS 2 056 342 (1972)
691. BASF A. G. (Tavs, P., Kniese, W., Nienburg, H.J.), DE-OS 1 793 439 (1972)
692. BASF A. G. BE 738 628 (1969)
693. Agency of Ind. Sci. Technol. (Matsuda, A. et al.), JP 7 479 991 (1974), 7 545 798 (1975); 7 546 589 (1975); 7 652 389 (1976) and 7 818 579 (1978); C.A. *85*, 99 917 u (1976), *89*, 43 121 v (1978)
694. Lion Fat and Oil Co. (Isa, H. et al.), DE-OS 2 447 068 and 2 447 069 (1975)
695. Lion Fat and Oil Co. (Isa, H. et al.), DE-OS 2 503 996 and 2 504 005 (1975)
696. Mitsubishi Gas Chem. Co. (Igasaki, Y.), JP 73-19 286 (1973)
697. Dokija, M., Bando, K.: Kogyo Kagaku Zasshi *71*, (11), 1866 (1968), C.A. *70*, 77 247 p (1969)
698. Goodman, R.J.: Pigment Resin Technol. *1972*, 1 (8), 4; C.A. *77*, 140 775 w (1972)
699. Jefferson Chem. Co. Inc. (Gipson, R.M.), DE-OS 1 817 700 (1969)
700. BASF A. G. (Kniese, W., Nienburg, H.J.), DE-OS 1 902 460 (1970)
701. Shell Int. Res. M., NL 65-16 164 (1966)
702. Shell Oil Co. (Slaugh, L.H., Mullineaux, R.D.), US 3 239 570 (1966); GB 1 002 428 (1965) and 988 944 (1965)
703. Jefferson Chem. Comp. (Gipson, R.M.), DE-OS 1 802 895 (1969), US 3 954 877 (1967)
704. Esso Res. Eng. Co. (Senn, W.L., Mertzweiller, J.K., Bentley, M.D.), US 3 488 296 (1970)
705. Palagyi, J., Marko, L.: J. Organomet. Chem. *17*, 453 (1969)
706. UOP Co., US 3 472 241 (1974)
707. Mathey, F. et al.: Inf. Chim. *179*, 191 (1978)
708. Shell Int. Res. M., GB 988 941 (1965)
709. Shell Int. Res. M. (Slaugh, L.H.), US 3 239 569 (1966)
710. Shell Int. Res. M. (Greene, C.R.), DE-OS 1 468 603 and 1 468 615 (1972)
711. Shell Int. Res. M. (van Winkle, J.L.), DE-AS 1 282 633 (1968), JP 24 643/68 (1968)
712. ICI Ltd., NL 72-06 256 (1972)
713. ICI Ltd. (Jones, P.J.V.), DE-OS 2 242 646 (1973), DE-OS 2 222 923 (1972)
714. BASF A. G. (Kruck, T., Lang, W.), DE-OS 1 592 215 (1970)
715. BASF A. G. (Kniese, W., Nienburg, H.J.), DE-OS 2 026 164 (1971), BE 732 866 (1969)
716. Standard Oil Co. (Indiana) (Trevillyan, A.E.), US 4 045 493 (1977)
717. Esso Res. Eng. Co., FR 1 438 811 (1966)
718. Gulf Res. Dev. Co., BE 702 934 (1967), GB 1 198 825 (1970)
719. Montecatini SpA., NL 68-03 440 (1968), IT 791 998 (1967), US 3 627 843 (1971)
720. Montecatini SpA. (Pregaglia, G., Andreetta, A., Benzoni, L.), DE-OS 1 668 484 (1971) and 1 804 518 (1969)
721. BP Ltd., FR 2 051 182 (1970)
722. Jefferson Chem. Co. Inc. (Gipson, R.M.), US 3 954 877 (1976)
723. Union Carbide Corp. (Pruett, R.L., Smith, J.A.), US 3 505 408 (1970)
724. Texaco Inc., US 3 950 439 (1972)
725. Chevron Res. Co., US 3 931 332 (1969)
726. Celanese Corp. (Stautzenberger, A.L., Paul, J.L.), BE 845 043 (1975)
727. Fell, B., Barl, M.: Chem.-Ztg. *101*, 343 (1977)
728. Union Oil of California (Booth, F.B.), US 3 560 539 (1971) and 3 965 192 (1976)
729. Union Oil of California (Olivier, K.L.), US 3 539 634 (1970) and 3 547 964 (1970)
730. Union Carbide Corp., NL 68-10 783 (1968)
731. Union Carbide Corp. (Pruett, R.L., Smith, J.A.), ZA 6 804 937 (1968)
732. BASF A. G. (Kummer, R.), DE-OS 2 414 253 (1975)
733. BASF A. G. (Himmele, W. et al.), DE-OS 2 410 156 (1975)
734. Gulf Res. Dev. Co. (McCracken, J.H., Williamson, R.C.), CA 992 101 (1976)
735. Mitsubishi Chem. Ind. Co. Ltd. (Onoda, T.), JP 7 627 649 (1976); C.A. *86*, 89 173 e (1977)

736. Mitsubishi Chem. Inc. Co., JP 76 006-124 (1976) and NL 7 508 829 (1976)
737. Celanese Corp. (Paul, J.L., Pieper, W.L., Wade, L.E. et al.), DE-OS 2 125 382 (1971), 2 552 351 (1976), 2 638 798 (1977) and 2 721 792 (1976)
738. Monsanto Co. (Paulik, F.E., Robinson, K.K., Roth, J.F.), AU 36 959/68 (1968), FR 1 560 961 (1969)
739. Shell Int. Res. M. (DeJong, A.J.), DE-OS 2 707 108 (1977)
740. BP Ltd. (Foster, G.), DE-OS 1 173 568 (1969), C.A. *72*, 663 889 (1970), DE-OS 1 801 145 (1969), C.A. *71*, 123 572 m (1969)
741. Johnson Matthey & Co. (Wilkinson, G.), DE-OS 1 518 236 (1965)
742. Johnson, Matthey & Co. (Bond, G.C.), DE-OS 2 053 218 (1970), DE-OS 2 055 539 (1971) and FR 2 069 322 (1971)
743. Johnson, Matthey & Co., JP 7 553 293 (1975)
744. Davy Powergas Ltd. (Fowler, R.), GB 1 387 657 (1975)
745. Fowler, R., Connor, H., Baehl, R.A.: Chemtech *1976*, 772
746. Wilkinson, G.: US 3 933 919 (1974), FR 1 601 798 (1970)
747. Wilkinson, G. et al.: J. Chem. Soc. (A) *1970*, 2753; Bull. Soc. Chim. France *1968*, 5055; Tetrahedron Lett. *1969* (22), 172
748. Ethyl Corp. (Sibert, J.W.), US 3 515 757 (1970)
749. Fell, B.: Chem.-Ztg. *101*, 343 (1977)
750. Fell, B., Hagen, J., Rupilius, W.: Chem.-Ztg. *100*, 308 (1976)
751. Fell, B., Beutler, M.: Erdöl & Kohle *29*, 49 (1976), Tetrahedron Lett. *1972*, (33), 3455
752. Fell, B., Müller, E.: Monatsh. Chem. *103*, (5), 1222 (1972)
753. Anonym, Kontakte (Merck AG) *1*, 24 (1972)
754. Tsuyi, J.: Kagaku (Kyoto) *1972*, 292, 382
755. Anonym, Chem. Eng. (N. Y.) *1977*, 84 (26)
756. Meakin, P., Jesson, J.P., Tolman, C.A.: J. Amer. Chem. Soc. *94*, 3240 (1972)
757. Frankel, E.N., Thomas, F.L., Rohwedder, W.K.: J. Amer. Oil Chem. Soc. *48*, 248 (1971), *49*, 10 (1972), Ind. Eng. Chem., Prod. Res. Dev. *12*, 47 (1973)
758. Botteghi, C., Consiglio, G., Pino, P.: Liebigs Ann. Chem. *1974*, Nr. 6, 864
759. Winzer, A., Griebel, R.: Z. Chem. *12*, 181 (1972)
760. Varshavsky, Yu.S., Cherkasowa, T.G., Buzina, N.A.: J. Organomet. Chem. *56*, 375 (1973)
761. Rosas, N., Gomez.Lara, J., Cabrera, A.: C. Alvarez Rev. Latinoam. Quim. *8*, 121 (1977); C.A. *88*, 6062 k (1978)
762. Svoboda, P., Hetflejs, J.: Collect. Czech. Chem. Commun. 40, 1746 (1975)
763. Fujikura, Y. et al.: Synth. Commun. *1976*, 6 (3), 199; C.A. *85*, 32 502 m (1976)
764. Kuraray K. K., DE-OS 2 538 364 (1976)
765. Exxon Res. Eng. Co., FR 2 275 434 (1976)
766. BP Ltd. (Foster, G., Lawrenson, M.J.), GB 1 207 561 (1970), FR 1 600 074 (1970), GB 1 284 615 (1972), BE 726 744 (1970)
767. Shell Res. Dev. Co. (Slaugh, L.H.), US 3 239 566 (1966)
768. Fell, B., Rupilius, W.: Angew. Chem. *81*, 916 (1969); Tetrahedron Lett. *1969*, (32), 2721
769. Union Oil Co. (Booth, F.B.), US 3 857 895 (1974)
770. Agency Ind. Sci. Technol., JA 7 399 131 (1973), C.A. *80*, 95 498 g (1974)
771. Monsanto Co. (Morris, D.E., Tinker, H.B.), DE-OS 2 359 377 (1974)
772. Shell Int. Res. Mat. (Downing, R.S., van Helden, R.), DE-OS 2 724 484 (1977)
773. Capka, M., Svoboda, P., Hetflejs, J.: CS 154 469 (1974)
774. Chisso Corp. (Tarao, R. et al.), JA 7323, 708 (1973)
775. Lednor, P.W. et al.: Chem. Ber. *111*, 615 (1978)
776. Tanaka, M. et al.: Bull. Chem. Soc. Japan *1974*, 1698 and *1977*, 2351 (C.A. *88*, 21 727 z (1978), Chem. Lett. *1972*, (6), 483; JA-PS 7783, 311 (1977); C.A. *87*, 200 792 t (1977))
777. Ruhrchemie A. G., NL 69-13 384 (1969)
778. Mitsubishi Chem. Ind. Co. Ltd. (Nakajima, C.), JP 7317, 251 (1973)
779. Mitsubishi Chem. Ind. Co. Ltd. (Yamaguchi, M., Onoda, T.), JP 7627, 649 (1976), JP 69-10,765 (1969)
780. Union Carbide Corp. (Pruett, R.L. et al.), J. Org. Chem. *34*, 327 (1969), DE-OS 1 793 069 (1972) and US 3 499 932 and 3 499 933 (1970)

781. Gulf Res. Dev. Co. (Deffner, J.F.), US 3 733 361 (1973)
782. Maruzen Oil Co. Ltd. (Iriuchijima, M. et al.), JP 7226, 762 (1972)
783. Ethyl Corp. (Keblys, K.A.), US 3 907 847 (1975)
784. BASF A. G. (Nienburg, H.J., Kummer, R.), DE-OS 2 317 625 (1975)
785. Mitsubishi Chem. Ind. Ltd., JP 63-653/1966 (1966)
786. Chevron Res. Co. (Suzuki, S.), DE-OS 2 456 056 (1975)
787. Mobil Oil Corp. (Rollmann, L.D., Whitehurst, D.D.), GB 1 448 255 (1976)
788. Fell, B.: Chem. Ing. Technol. *43*, 1271 (1971); Chem.-Ztg. *95*, 1017 (1971)
789. BP Ltd. (Lawrenson, M.J., Johnson, P.), GB 1 205 027 (1970), US 3 660 493 (1967), DE-OS 1 806 293 (1969), 1 768 160 (1972), 1 643 678 (1967), FR 1 558 222 (1970), 1 549 414 (1968) and NL 68-04 417 (1968), and 68-04 734 (1968)
790. BP Ltd. (Lawrenson, M.J.), DE-OS 2 031 380 (1971), FR 2 052 874 (1970), BE 724 877 (1970)
791. BP Ltd. (Foster, G., Lawrenson, M.J.), ZA 6905, 913 (1970), C.A. *73*, 87 430 u (1970); GB 1 263 720 (1972)
792. Pino, P. et al.: Adv. Chem. Ser. *1974*, 132, C.A. *82*, 85 634 a (1975)
793. Ogata, I., Ikeda, Y.: Tokyo Kogyo Shikensho Hokuku *1972*, 67(9), 340, Chem. Lett. *1972*, 487
794. BP Ltd. (Lawrenson, M.J., Foster, G.), NL 6817, 411 (1969), DE-OS 1 812 504 (1969)
795. Ethyl Corp. (Keblys, K.A.), US 3 859 359 (1975)
796. Monsanto Co. (Tinker, H.B.), CA 1 027 141 (1978); C.A. *89*, 42 440 m (1978), NL-P Appl. 75 00,745 (1975)
797. BP Ltd., NL 70-17 086 (1970)
798. Anonym, ECN *16*, Nr. 394 26 (1969)
799. BP Ltd. (Lawrenson, M.J.), GB 1 254 222 (1971), DE-OS 2 030 575 (1971) and NL 70-0899 (1969)
800. Phillips Petroleum Co. (Zuech, E.A.), US 3 956 177 (1976)
801. Yamada, A., Fukuda, T., Yanagita, M.: Bull. Chem. Soc. Japan *48*, 353 (1975)
802. BASF A. G. (Hoffmann, W., Siegel, H.), DE-OS 2 447 170 (1976); (Himmele, W., Siegel, H., Aquila, W.), DE-OS 2 219 168 (1973)
803. Johnson, Matthey & Co. (Wilkinson, G.), JP 0053-293 (1975), DE-OS 2 034 909 (1971)
804. UOP Inc. (Homeier, E.H.), US 3 984 478 (1976)
805. Rhône-Poulenc S. A. (Kuntz, E.), DE-OS 2 627 354 (1976), NL 7 606 634 (1976)
806. Shell Int. (van Winkle, J.L., Morris, R.C., Mason, R.F.), DE-OS 1 909 620 (1969)
807. BP Ltd., NL 71-04519 (1971)
808. Sun Oil Comp. (Renick, L.E.), US 3 825 601 (1974)
809. Salomon, C. et al.: Chimia *27*, 215 (1973)
810. BASF A. G. (Himmele, W., Aquila, W., Schecker, H.), DE-OS 2 101 489 (1972)
811. Exxon Res. Eng. Co. (McVicker, G.B.), DE-OS 2 424 526 (1975), US 3 939 188 (1976), 3 946 082 (1976)
812. Booth, B.L. et al.: J. Organomet. Chem. *27*, 119 (1971)
813. Lai, R., Ucciani, E.: C. R. Acad. Sci., Ser. C *1973*, 276 (5), 425
814. BASF A. G. (Tavs, P., Nienburg, H.), DE 1 768 391 (1978)
815. Spencer, A.: J. Organomet. Chem. *124*, 85 (1977)
816. Dow Corning Corp., GB 1 421 136 (1976)
817. Bayer A. G. (Braden, R.), DE-OS 2 453 229 (1976)
818. Celanese Corp. (Unruh, J.D., Wells, W.J.), DE-OS 2 617 306 (1976)
819. Gupta, A., Rembaum, A., Gray, H.B.: Organomet. Polym. (Symp.) *1977* (Publ. 1978), 155; C.A. *89*, 41 881 u (1978)
820. Ryan, R.C., Pittman, C.U.: J. Amer. Chem. Soc. *99*, 1986 (1977)
821. Chini, P., Heaton, B.T.: Tetranuclear Carbonyl Clusters. In: Dewar, M.J.S., Hafner, K. et al. (eds.): Inorganic Chemistry, Metal Carbonyl Chemistry. Berlin-Heidelberg-New York: Springer Verlag (1977)
822. Ethyl Corp. (Wilkinson, G.), US 3 501 531 (1970)
823. Mitsui Petr. Ind. Ltd. (Tanaka, M., Kiso, Y., Saeki, K.), JP 78.50,102 (1976/1978); C.A. *89*, 59 659 n (1978)

824. Texaco Dev. Corp. (Knifton, J.F., Moss, P.H.), DE-OS 2 322 751 (1973) and 2 811 403 (1978)
825. Spetz, T.G., Scholten, J.J.F.: J. Mol. Catal. *3*, 81 (1977)
826. Esso Res. Engng. Co., GB 801 734 (1958), US 2 880 241 (1959)
827. Esso Res. Eng. Co., JP 29 043/69 (1969)
828. Bychkova, A.Ya. et al.: Tr. Gos. Nauchno-Issled. Proektn. Inst. Azotn. *1974*, 27, 56, C.A. *85*, 93 738 q (1976)
829. BASF A. G., BE 744 531 (1970)
830. Azote et Produits chimiques S. A. (Pasquon, I., Albanesi, G., Prognon, P.), FR 2 031 648 (1970)
831. Union Oil of Calif. (Booth, F.B.), US 3 511 880 (1970)
832. Union Carbide Corp. (Pruett, R.L., Smith, J.A.), DE-OS 2 062 703 (1970)
833. BP Ltd. (Lawrenson, M.J.), DE-OS 2 058 814 (1971)
834. BASF A. G. (Kniese, W., Nienburg, H.J., Kummer, R.), DE-OS 2 026 164 (1971)
835. BASF A. G. (Platz, R. et al.), DE-OS 2 033 573 (1972)
836. ICI Ltd. (Keating, T., Jones, P.J.V.), DE-OS 2 222 923 (1972)
837. BASF A. G. (Himmele, W., Fliege, W., Aquila, W., Siegel, H.), DE-OS 2 404 312 (1975)
838. Monsanto Co. (Tinker, H.B.), DE-OS 2 623 673 (1976)
839. Evans, D., Yagupsky, G., Wilkinson, G.: J. Chem. Soc. (A) *1968*, 2660
840. Consiglio, G. et al.: Angew. Chem. *85*, 663 (1973)
841. Hershman, A., Craddock, J.H.: Ind. Engng. Chem., Prod. Res. Dev. *7*, 226 (1968)
842. Miller, W.W.: Diss. Abstr. B *1968*, 29(3), 915; C.A. *70*, 17 237 a (1969)
843. James, B.R.: Inorg. Chim. Acta Rev. *4*, 86 (1970)
844. Pino, P. et al.: Proceedings of New Aspects of the Chemistry of Metal Carbonyls and Derivatives, Venice 1968, paper E2
845. Pino, P. et al.: Chim. Ind. (Milano) *50*, 106 (1968)
846. Lapidus, A.L. et al.: Kinet. Katal. *17*, 1483 (1976); C.A. *86*, 139 301 t (1977)
847. Tsuji, J., Mori, J.: Bull. Chem. Soc. Japan *42*, 527 (1969), Chem. Titles *1969*, Nr. 7-151
848. Toyo Rayon K. K., JP 3 361/68 (1968)
849. DuPont (Mrowca, J.J.), DE-OS 2 507 447 (1977)
850. Wilkinson, G.: NL 68-01 410 (1968)
851. Mague, J.T., Mitchener, J.P.: Inorg. Chem. *11*, 2714 (1972)
852. BP Ltd., FR 2 049 773 (1970)
853. James, B.R., Markham, L.D.: Inorg. Nucl. Chem. Lett. *7*, 373 (1971)
854. ICI Ltd. (Rowe, G.A.), GB 1 368 802 (1974)
855. ICI Ltd. (Coffey, R.S.), GB 1 130 743 (1968), C.A. *70*, 59 417 q (1969)
856. DuPont (Mrowca, J.J.), US 3 876 672 (1975)
857. ICI Ltd. (Rowe, G.A.), GB 1 368 434 (1974)
858. Toa Nenryo Kogyo K. K., JP 23 443/68 (1964/1968); US 3 378 590 (1968)
859. Toa Nenryo Kogyo K. K. (Usami, S.), JP 69-12 403 (1966/1969); C.A. *71*, 90 820 k (1969)
860. BASF A. G. (Kummer, R. et al.), Adv. Chem. Ser. *1974*, 132
861. BASF A. G. (Nienburg, H.J. et al.), DE-OS 2 139 630 (1973) and 2 106 244 (1972)
862. BASF A. G. (Kniese, W. et al.), DE-OS 2 103 454 (1972), BE 778 177 (1971)
863. BASF A. G., BE 802 875 (1972)
864. The Distillers Co. Ltd., GB 1 012 011 (1963/1965)
865. Res. Institut Japan JP 14 608/61 (1958/1961)
866. Chevron Res. Co. (Wilkes, J.B.), US 3 928 232 (1975) and US 3 976 703 (1976)
867. Pino, P., Piacenti, F., Bianchi, M. in [7], p. 110
868. Eastman Kodak Co. (Hasek, R.H., Wayman, C.E.), US 2 820 059 (1958)
869. Roos, L., Orchin, M.: J. Org. Chem. *31*, 3015 (1966)
870. BASF A. G., DE-OS 2 106 243 (1971)
871. BASF A. G. (Nienburg, H.J. et al.), DE-OS 2 056 342 (1972)
872. BASF A. G. (Kummer, R., Nienburg, H.J., Kniese, W.), DE-OS 2 045 169 (1972)
873. Esso Res. Engng. Co., FR 2 022 880 (1970)
874. DuPont (Gresham, W.F., McAlevy, A.), US 2 564 104 (1951)

875. Showa Denko K. K. (Yoshinaga, Y., Onochi, T.), JP 7380,504 (1973/1976); C.A. *80*, 59 413 s (1977)
876. Ruhrchemie A. G. (Büchner, K., Kühnel, P.), DE 837 847 (1952)
877. Ajinomoto Co. Inc. (Kato, J., Iwanaga, R. et al.), US 3 337 603 (1967), C.A. *68*, 21 535 x (1968)
878. Union Carbide Corp. (Halstead, R.W., Chaty, J.C.), DE-OS 2 730 527 (1978)
879. Mitsubishi Chem. Ind., JP 76-23 212 (1974)
880. Celanese Corp. (Wu, A., Thigpen, H.H.), DE-OS 2 145 427 (1972)
881. Szabo, P., Marko, L.: Conf. on Min. Oil *1965*, Proc. p. 405; C. *1969*, 41-1705
882. Bressan, G., Broggi, R.: Chem. Ind. (Milano) *50*, 1194 (1968)
883. Esso Res. Engng. Co. (Rehner, J.), DE-AS 1 093 347 (1960)
884. Degussa A. G. (Weigert, W.), DE-AS 1 241 847 (1964)
885. BASF A. G. (Aquila, W., Himmele, W., Scheiper, H.J.), DE-OS 2 406 223 (1975)
886. Tokuyama Soda K. K., JP 7 690/69 (1964/1969)
887. Frankel, E.N.: J. Amer. Oil Chem. Soc. *53*, Nr. 4, 138 (1976)
888. Esso Res. Engng. Co., DE-AS 1 076 658 (1960)
889. Ruhrchemie A. G., NL 76 974 (1955)
890. UOP Co. (Massie, S.N.), US-Publ. Pat. Appl. B 472 241 (1976), C.A. *84*, 121 119 p (1976)
891. Standard Oil Co., FR 1 041 838 (1953)
892. Montecatini Edison SpA (Ferrari, G., Griselli, P.L.), DE-OS 2 119 334 (1971)
893. Macho, V.: Chem. Zvesti *25*, 49 (1971)
894. Cornils, B., Ruprecht, P.: Ruhrchemie A. G., unpublished
895. Monsanto Chem. Co. (Weesner, W.E., Heimsch, R.A.), US 2 949 486 (1960)
896. Nalco Chem. Co. (Hesler, J.C.), US 3 488 184 (1970), C.A. *72*, 71 143 w (1970)
897. Esso Res. Engng. Co. (Roming, C.), FR 1 324 873 (1962); C. *1967*, 36-2433
898. Esso Res. Engng. Co. (Mason, R.B.), GB 761 024 (1956) and 776 998 (1957)
899. Shell Int. Res. Mij., GB 1 002 429 (1965)
900. Shell Int. Res. Mij., BE 627 371 (1964)
901. Toa Gosei Chem. Ind. Co. Ltd., JA 54 448/66
902. Aleksejewa, K.A. et al.: Karbonilirovanie Nenass. Uglevod. *1968*, 100; Dokl. Akad. Nauk. Beloruss. SSR *13*, 914 (1969); C. A. *72*, 36 165 h (1970)
903. Inst. Gen. Inorg. Chem. (Ermolenko, N.F.), SU 251 738 (1969); C.A. *72*, 34 077 g (1970)
904. Komora, L., Macho, V., Lustik, O.: Ropa Uhlie *1972*, 281, C.A. *77*, 93 382 d (1972)
905. Ajinomoto Co. Ltd., JP 26 170/68 (1968)
906. Cornils, B., Förster, I.: Chem.-Ztg. *97*, 374 (1973)
907. Ruhrchemie A. G. (Tummes, H., Heim, W.), DE 1 937 662 (1971)
908. Aleksejewa, K.A. et al.: Karbonilirovanie Nenass. Uglevod. *1968*, 87, C.A. *71*, 12 457 h (1969)
909. Ruhrchemie A. G. (Tummes, H. et al.), DE-AS 2 245 565 (1974), US 3 993 695 (1976)
910. Montecatini SpA (Chiusoli, G.P., Mondelli, G.), FR 1 381 022 (1964); C. *1968*, 26-2312
911. Klumpp, E., Bor, G., Marko, L.: J. Organomet. Chem. *11*, 207 (1968)
912. Marko, L., Bor, G.: J. Organomet. Chem. *3*, 162 (1965)
913. Abel, E.W., Crosse, B.C.: Organomet. Chem. Rev. *2*, 443 (1963)
914. Vahrenkamp, H.: Angew. Chem. *87*, 363 (1975)
915. Marko, L.: Acta Chim. Acad. Sci. hung. *59*, 367, 389 (1969)
916. Field, D.S., Newlands, M.J.: J. Organomet. Chem. *27*, 221 (1971)
917. UOP Co. (Massie, S.N., Vesely, J.A.), DE-OS 2 415 902 (1974)
918. Macho, V.: Chem. Zvesti *20*, 870 (1966)
919. Esso Res. Eng. Co., DE 1 116 645 (1962)
920. Eastman Kodak Co. (Mooney, E.J., Hart, W.F.), US 3 903 172 (1975)
921. VEB Leuna-Werke (Blatz, H., Schröder, L.), DE-OS 2 149 120 (1972)
922. Alekseeva, K.A. et al.: DD 121 458 (1976)
923. Eastman Kodak Co., CA 761 426 (1967)
924. BASF A. G. (Elliehausen, H. et al.), DE-OS 2 650 829 (1978)
925. Macho, V., Komora, L.: Chem. Zvesti *21*, 164 (1967)

926. Union Oil Co. of California (Biale, G.), US 3 481 987 (1967), C.A. *72*, 45 625 k (1970)

927. Esso Res. Eng. Co., GB 814 706 (1969)

928. Celanese Corp. (Maddox, L.A.), DE-OS 2 638 798 (1977)

929. Kotowski, W.: Chem. Technol. *30*, 360 (1978)

930. Mitsubishi Chem. Ind., JP 41-805/73

931. Macho, V., Rihova, H., Polievka, M.: Chem. Zvesti *29*, 474 (1975)

932. Mitsubishi Chem. Ind. Co. Ltd. (Onoda, T. et al.), JP 7608, 207 (1976), C.A. *84*, 164 170 r (1976)

933. Montedison SpA (Tampieri, M., Gregorio, G., Andreetta, A.), DE-OS 2 526 129 (1976)

934. Institut Francais du Petrole (Lassau, I.), DE-OS 1 922 621 (1969)

935. Union Carbide Corp. (Brewester, E.A.V., Pruett R.L.), DE-OS 2 715 685 (1977)

936. Chini, P., Martinengo, S., Garlaschelli, G.: J. Chem. Soc., Chem. Commun. *1972*, 709

937. Tummes, H., Cornils, B.: Ruhrchemie A. G., unpublished

938. Erdölchemie GmbH (Marx, H.D., Scherb, H., Schleppinghoff, B.), DE-OS 2 141 469 (1973)

939. Morikawa, M.: Bull. Chem. Soc. Japan *37*, 430 (1964)

940. Cornils, B., Ruprecht, P.: Ruhrchemie A. G., unpublished

941. Esso Res. Eng. Co., NL 6 400 701/1964

942. Consiglio, G., Pino, P.: Isr. J. Chem. *1977*, 221; C.A. *88*, 88 960 p (1978)

943. Gulf Res. Dev. Co. (Tucci, E.R.), US 3 631 111 (1971)

944. All-Unions Petro. Chem. Proc. Inst. (Rudkowskii, D.M.), SU 193 480 (1967)

945. Celanese Corp. (Hughes, R.O., Hillman, M.E.D.), US 3 821 311 (1974)

946. BASF A. G. (Himmele, W., Müller, F.J., Aquila, W.), DE-OS 2 039 078 (1972)

947. Gavrilova, V.M., Delnik, V.B., Kagua, S.Sh.: Khim. Prom. *1974*, 739; Sov. Chem. Inc. *6*, (10), 617 (1974)

948. ICI Ltd. (Goddard, R.E., Mansfield, G.H.), GB 903 589 (1959)

949. Cornils, B., Ruprecht, P., Wiebus, E.: Ruhrchemie A. G., unpublished

950. Weber, J., Pluta, W.: Ruhrchemie A. G., unpublished

951. Anonym, Inf. Chim., Special Export *1973*, 105

952. Tca Gosei Chem. Ind. Co. Ltd., JP 30 283/68 (1968)

953. Union Oil Co. of California (Oliver, K.L., Booth, F.B., Mears, D.E.), US 3 555 098 (1971)

954. Stresinka, J.: CS 126 708 (1968), C.A. *70*, 46 860 d (1969)

955. Magyar Asvanyolaj es Foldgazkizerleti Intezel (Bathory, J. et al.), AT 256 799 (1967), HU 153 335 (1965)

956. Esso Res. Eng. Co. (Mertzweiler, J.K.), JP 3 530/66 (1966)

957. Ruhrchemie A. G. (Falbe, J., Weber, J.), DE 1 592 502 (1976)

958. Standard Oil Co. (Ruscilli, A.E.), US 3 451 943 (1966); C.A. *71*, 51 958 s (1969)

959. Gulf Res. Co., US 2 743 302 (1969)

960. ICI Ltd. (Lamb, S.A., Smith, H.O., Norcross, G.), GB 815 091 (1959)

961. Friedrich, J.F., List, G.R., Sohns, V.E.: J. Amer. Oil Chem. Soc. *50*, 455 (1973)

962. Ugine Kuhlmann (Lemke, H.), DE-OS 1 443 799 (1969), DE 1 206 419 (1968)

963. Ugine Kuhlmann (Lemke, H. et al.), FR 1 563 043 (1969), DE-OS 1 906 850 (1969)

964. Ugine Kuhlmann (Lemke, H.), BE 727 913 (1969)

965. Lemke, H.: Genie Chimique *89*, 118 (1963)

966. Ugine Kuhlmann, (Lemke, H.), FR 1 223 381 (1960), 1 223 381, Addition 84 178 (1964), US 3 188 351 (1965)

967. BASF A. G. (Nienburg, H.J., Appl, M.), BE 641 206 (1964)

968. BASF A. G. (Wiese, F.F. et al.), DE-AS 1 290 927 (1969)

969. ICI Ltd. (Hibbs, F.M.), GB 1 458 375 (1976)

970. Shell Oil Co. (Greene, C.R.), US 3 369 050 (1964), C.A. *68*, 104 553 c (1968)

971. Gankin, W.Yu. et al.: Khim i Tekhnol. Topliv i Masel *11*, 5 (1966)

972. General Electric Co. (Aycock, D.F., Sliva, D.E.), DE-OS 2 426 650 (1975)

973. Gronoglasov, Y.A. et al.: SU 451 682 (1974); C.A. *82*, 155 294 r (1975)

974. Arostovich, V.Yu. et al.: DD 126 445 (1977); C.A. *88*, 74 043 c (1978)

975. Gankin, V.Yu. et al.: SU 303 313 (1971)

976. Alekseeva, K.A. et al.: Neftepererab. Neftekhim *1971* (5), 27; DD 111 675 (1975), 108 731 (1974); SU 245 759 (1969) and SU 143 788 (1966)
977. Vsesoj. Naucho-Issled. Inst., GB 1 002 691 (1965), FR 1 315 589 (1962)
978. Eastman Kodak Co. (Morris, P.M.), DE 1 235 283 (1967), JP 13 487/69 (1969)
979. Stresinka, J., Mistrik, E.J., Macho, V.: CS 121 418 (1966); C.A. *68*, 31 787 v (1968); DD 55 650 (1967)
980. Alekseeva, K.A. et al.: SU 541 493 (1977); C.A. *86*, 128 088 c (1977)
981. Gankin, W.Yu. et al.: Karbonil. Nenasysh. Uglevod. *1968*, 51; C.A. *71*, 9268 s (1969); SU 184 257 (1965)
982. All-Unions Sci. Res. Inst. (Gankin, W.Yu.), SU 248 633 (1969); C.A. *72*, 89 785 s (1970)
983. Ungvary, F., Marko, L.: Inorg. Chim. Acta *4*, 324 (1970)
984. Rudkovskij, D.M. et al.: Khim. Prom. *42*, 801 (1966), C.A. *66*, 97 870 m (1967)
985. Mironyuk, O.I. et al.: SU 417 409 (1974), C.A. *80*, 145 422 a (1974)
986. Ajinomoto Co., JP 29 936/68 (1968)
987. Macho, V. et al.: Chem. Prumysl. *16*, 65 (1966), C. *1968*, 26-2611
988. Gankin, W.Yu. et al.: SU 247 247 (1969); C.A. *71*, 12 3097 k (1969)
989. Krinkin, D.P., Rudkovskij, D.M., Trifel, A.G.: Zh. prinkl. Khim. *40*, 614 (1967); C.A. *67*, 17 450 (1967); SU 205 004 (1964)
990. Baltz, H., Schröder, L.: DD 127 163 (1968)
991. Eastman Kodak Co. (Peacock, D.R., Magness, J.F., Morris, P.M.), DE-OS 1 568 478 (1970)
992. Hess, P.H., Parker, P.H.: Polymer Preprints *7*, 900 (1966)
993. Freund, M., Laky, J.: Magyar Asvanyolaj es Földgaz Kit. Intezet Publication
994. Cornils, B., Förster, I., Krüger, C., Tsay, Y.H.: Transition Met. Chem. *1*, 151 (1976)
995. Ruhrchemie A. G. (Tummes, H., Meis, J.), DE 1 235 285 (1967), GB 1 026 283 (1966)
996. Ichinokawa, H., Ogawa, K. et al.: Tokyo Kogo Shi. Hok. *61*, 418 (1966); C.A. *66*, 54 791 q (1967)
997. BASF A. G. (Moell, H. et al.), DE-AS 1 272 911 (1968)
998. BASF A. G., CA 796 771 (1968); BE 696 616 (1967)
999. BASF A. G. (Nienburg, H.J. et al.), DE-AS 1 642 944 (1975)
1000. BASF A. G. (Nienburg, H.J., Eckert, E., Goilav, M.), DE 1 106 307 (1961)
1001. Gulf Res. Dev. Co. (Gwynn, B.H., Pardee, W.A., Ward, J.V.), US 3 246 024 (1966)
1002. VEB Leuna-Werke (Haack, H. et al.), DD 106 819 (1974)
1003. Daicel Ltd. (Takasu, I., Higuchi, M.), DE-OS 2 145 687 (1972)
1004. Standard Oil Comp. (Gunter, C.G., Baldner, R.L., Wennerberg, A.N.), DE-AS 1 264 415 (1968)
1005. Union Carbide Corp. (Kyle, H.E. et al.), US 3 288 857 (1966) and 3 159 679 (1964)
1006. Gavrilova, V.M., Delnik, U.B., Kagna, S.Sh.: Khim. Prom. *1974*, (10), 739
1007. Sumitomo Ltd., JP 6168/67 (1966)
1008. Toa Nenryo K. K. (Usami, S. et al.), US 3 634 291 (1967)
1009. Gulf Res. Dev. Co. (Elder, H.G., Gwynn, B.H., Ward, J.V.), US 3 332 871 (1967), C.A. *67*, 108 234 b (1967)
1010. Montecatini SpA (Endler, H.), DE-AS 1 237 999 (1967); DE-OS 1 468 287 (1968) and GB 1 015 086 (1965)
1011. Monsanto Co. (Null, H.R., Bowe, L.E.), DE-OS 1 468 301 (1969)
1012. Standard Oil Co., CA 736 981 (1966)
1013. Rudkovskij, D.M., Trifel, A.G., Gankin, V.Yu.: SU 178 814 (1966); C.A. *65*, 5368 d (1966)
1014. Gankin, W.Yu., Delnik, V.B.: SU 370 196 (1973), C.A. *79*, 18 025 c (1973)
1015. Alekseeva, K.A. et al.: Khim. Prom. *1972*, (6), 410
1016. Teijin Ltd., DE-AS 1 245 918 (1967)
1017. BASF A. G. (Kniese, W. et al.), DE-OS 1 954 315 (1971)
1018. Snia Viscosa SpA (Notarbartoli, L., Morbidelli, G.), DE-AS 1 239 282 (1967); C. *1968*, 6-2155; AT 239 139 (1965)
1019. Monsanto Co. (Leach, H.S., Slate, J.L., Urrg, W.H.), DE-OS 1 793 126 (1971)
1020. BASF A. G. (Kniese, W.), DE-OS 2 045 415 (1972)
1021. Chem. Werke Hüls A. G., GB 1 132 666 (1968)

1022. Alekseeva, K.A.: Dokl. Akad. Nauk. Beloruss. SSR *13*, (10), 914 (1969), Chem. Titles *1969*, 24-176
1023. Daicel Ltd. (Takasu, I., Higuchi, M., Hijioka, Y.), JP 7414,634 (1970)
1024. Azote and Produits Chimiques S. A. (Prognon, P.), DE-OS 1 910 959 (1969)
1025. Toa Gosei Chem. Ind. Co. Ltd., JP 13 288/67 (1967)
1026. Mitsubishi Chem. Ind., JP 5224/56 and 9978/56 (1956), 15 164/61 (1961) and 29 567/68 (1968)
1027. The Noguchi Inst. (Noi, G. et al.), JP 6734/66 (1966)
1028. Daicel Ltd., DE-OS 2 145 688 (1972)
1029. Eastman Kodak Co., DE-AS 1 277 236 (1968), GB 1 100 422 (1968), JP 320/68 (1968)
1030. ICI Ltd. (Taylor, A.W.C.), US 2 815 387 (1957), FR 1 079 469 (1957)
1031. ICI Ltd. (Harrison, A.C., Hayes, R.), GB 1 201 083 (1970)
1032. BASF A. G. (Kummer, R. et al.), DE-OS 2 237 373 (1972); DE-OS 2 206 252 (1973) and DE-AS 1 642 944 (1975), NL 71-17 830 (1971)
1033. ICI Ltd., NL 68-05 466 (1968)
1034. Vysotskii, M.P. et al.: SU 598 635 (1978); C.A. *88*, 190 090 s (1978)
1035. BASF A. G. (Nienburg, H.J. et al.), US 3 607 786 (1971)
1036. Toa Nenryo Kogyo Co. Ltd. (Koyama, T., Nishimura, K., Usami, M.), JP 73 16.891 (1973)
1037. Macek, V. et al.: CS 157 298 (1975); C.A. *83*, 78 603 g (1975)
1038. Ajinomoto Co. Ind., FR 1518 305 (1968), C.A. *71*, 12 665 z (1969)
1039. Ruhrchemie AG (Falbe, J., Tummes, H., Meis, J.), DE 1 295 537 (1969), FR 1 588 014 (1970)
1040. Mitsubishi Co. Ltd. (Yamaguchi, M. et al.), DE-OS 1 945 574 (1970), JP 7 111 805 (1971)
1041. DuPont (Balmat, J.L.), US 3 998 622 (1976)
1042. Ajinomoto Co. Inc., DE-OS 1 558 395 (1970), JP 23 763/66 (1966)
1043. BASF A. G. (von Kutepow, N., Müller, F.J., Reuter, P.), DE-OS 2 262 852 (1972)
1044. Ruhrchemie A. G. (Tummes, H., Weber, J., Bexten, L.), DE-OS 2 406 323 (1975)
1045. Awl, R.A., Frankel, E.N., Pryde, E.H.: J. Amer. Oil Chem. Soc. *53*, 190 (1976)
1046. United States, Depl. of Agriculture (Friedrich, J.P.), US-Pat. Appl. 438 307 (1974); C.A. *83*, 153 206 k (1975)
1047. Dufek, E., List, G.R.: J. Amer. Oil Chem. Soc. *54*, (7) 276 (1977)
1048. Ruhrchemie A. G. (Falbe, J., Weber, J.), FR 1 598 768 (1970); DE-AS 1 290 535 (1969)
1049. Ruhrchemie A. G. (Falbe, J., Weber, J.), DE-OS 1 592 502 (1972), FR 1 590 393 (1968)
1050. BASF A. G. (Kummer, R., Schneider, H.W., Schwirten, K.), DE-OS 2 614 799 (1976), DE-OS 2 045 416 (1972)
1051. Mitsubishi Co. Ltd., US 2 992 275 (1961)
1052. Shell Int. Res., NL 6 606 606 (1966), C.A. *66*, 65 074 j (1967)
1053. Shell Int. Res., DE-AS 1 593 368 (1966), C.A. 728 676 (1966, AU 5586/66 (1966)
1054. BASF A. G. (Nienburg, H.J., Kummer, R.), DE-OS 2 045 745 (1972) and 2 145 532 (1973)
1055. BASF A. G. (Kummer, R.), DE-OS 2 045 910 (1972)
1056. BASF A. G. (Nienburg, H.J.), DE-OS 2 145 532 (1973)
1057. BASF A. G. (Nienburg, H.J. et al.), DE-AS 1 642 944 (1967)
1058. Union Oil of California (Booth, F.B.), US 3 857 895 (1974)
1059. Montedison SpA (Tampieri, M., Gregorio, G., Andreetta, A.), DE-OS 2 526 129 (1974) and 2 502 233 (1974)
1060. Hartwig, U.: Dissertation TH Aachen 1973, Nachr. Chem. Techn. *21*, (15), 366 (1973), DE-OS 2 311 388 (1974)
1061. Union Oil of California (Olivier, K.L., Snyder, L.R.), US 3 539 634 (1970)
1062. Montedison SpA (Gregorio, G., Montrasi, G.), IT 1 007 026 (1976)
1063. Gosser, L.W., Knoth, W.H., Parshall, G.W.: J. Molec. Catalysis *1977*, 2 (4), 253, C.A. *87*, 91 312 p (1977)
1064. BP Ltd. (Goldup, A., Westaway, M.T., Walker, G.), US 3 645 891 (1972)
1065. BASF A. G., BE 833 862 (1974)
1066. Union Oil of California (Olivier, K.L.), US 3 547 964 (1970), 3 560 539 (1971), 3 641 076 (1972)

1067. Mitsubishi Chem. Ind. Co., JP 7 486 301 (1974)

1068. Erdölchemie GmbH (Fell, B., Dolkemeyer, W.), DE-OS 2 637 262 (1978)

1069. Mitsubishi Chem. Ind. Co., DE-OS 2 533 320 (1975)

1070. Esso Res. Engng. Co. (Mertzweiller, J.K., McLean, L.W.), US 3 513 130 (1970) and 3 330 875 (1967)

1071. Gavrilova, V.M., Delnik V.B., Kagna, S.Sh.: Sov. Chem. Ind. 6 (10), 617 (1974)

1072. Union Carbide Corp. (Pruett, R.L., Smith, J.A.), ZA 6 804 937 (1968), C.A. 71, 90 819 s (1969)

1073. Union Carbide Corp. (Bowditsch, R.P.), DE-AS 1 065 829 (1959)

1074. Mitsubishi Chem. Ind. Co. Ltd. (Onoda, T. et al.), JP 7 541 805 (1975); C.A. 83, 78 606 k (1975)

1075. Gankin, W.Yu.: Zh. Prikl. Khim 40, 1862 (1967); C.A. 68, 12 411 b (1968)

1076. Kraftco Corp., DE-OS 2 205 899 (1972)

1077. Kashina, V.V., Katsnelson, M.G., Mishenkova, G.N.: Zh. Org. Khim 1978, 877; C.A. 89, 23 741 u (1978)

1078. Mitsubishi Chem. Ind. Co. Ltd. (Onoda, T., Tano, K.), JP 7 486 301 (1974)

1079. Mitsubishi Chem. Ind. Co. Ltd. (Onoda, T. et al.), DE-AS 2 438 847 (1975)

1080. Ugine Kuhlmann (Lemke, H., Duval, R.), DE-OS 1 966 388 (1972)

1081. Mistrik, E.J., Lustik, O.: CS 124 147 (1965/1967); C.A. 69, 95 963 c (1968)

1082. BP Ltd., NL 69-13 756 (1969)

1083. ICI Ltd. (Thornton, J.M.), DE-OS 2 224 089 (1972)

1084. Chemische Werke Hüls A. G. (Kaufhold, M., Gaube, J.), DE-OS 2 538 037 (1975/1977)

1085. Rybakov, V.A. et al.: SU 403 665 (1973); C.A. 80, 47 459 y (1974)

1086. Eastman Kodak Co., FR 1 430 719 (1966)

1087. Pino, P., Piacenti, F., Bianchi, M.: in [7], p. 112

1088. Kashina, V.V. et al.: Zh. Prikl. Khim 50, 598 (1977), C.A. 87, 22 101 g (1977)

1089. Wender, I.: Petroleum Refiner 35, 197 (1956)

1090. BP Ltd., BE 724 977 (1970)

1091. Imyanitov, N.S., Volkov, V.A.: Zh. Prikl. Khim 46, 2683 (1973)

1092. Koga, T., Noyori, G.: Kogyo Kagaku Zasshi 70, 1172 (1967), C.A. 68, 28 966 x (1968)

1093. Bortinger, A., Busse, P.J., Orchin, M.: J. Catalysis 52, 385 (1978)

1094. Takegami, Y. et al.: Bull. Chem. Soc. Japan 42, 206 (1969)

1095. Monsanto Chem. Co. (Andersen, J.W.), FR 1 242 088 (1960), C. 1964, 51-2427

1096. Clark, A.C., Orchin, M.: J. Org. Chem. 38, 4004 (1973)

1097. Union Carbide Corp. (Privette, W.R., Knap J.E.), US 3 404 188 (1968), C.A. 69, 106 011 u (1968)

1098. Cornils, B., Wiebus, E., Tummes, H.: Ruhrchemie AG, unpublished

1099. Ethyl Corp. (Kehoe, L.J., Schell, R.A.), US 3 544 635 (1968)

1100. Silich, M.I. et al.: SU 300 453 (1971)

1101. Esso Res. Eng. Co., BE 545 960 (1956)

1102. Union Carbide Corp. (Knap, J.E.), US 3 236 597 (1966)

1103. Celanese Corp. (Stautzenberger, A.L., Unruh, J.D., Paul, J.L.), DE-OS 2 552 351 (1975)

1104. Mitsubishi Chem. Ind. Co. Ltd., GB 1 202 507 (1970)

1105. Parshall, G.W.: J. Amer. Chem. Soc. 94, 8716 (1972)

1106. ICI Ltd., DE-OS 2 020 550 (1963)

1107. Gankin, W.Yu.: SU 213 795 (1968), C.A. 69, 43 394 y (1968)

1108. ICI Ltd., FR 2 040 373 (1970)

1109. Rudkovskii, D.M.: GB 1 099 712 (1968)

1110. BASF A. G. (von Kutepow, N.), DE-OS 2 037 782 (1972)

1111. Esso Res. Eng. Co., GB 1 072 796 (1967)

1112. Yoshida, T.: Chem. Ing. Techn. 42, 641 (1970)

1113. Mitsubishi Chem. Ind. Co. Ltd. (Yamaguchi, M.), DE-OS 1 920 960 (1969); C.A. 72, 31 230 r (1970)

1114. Alekseeva, K.A., Rudkowskij, D.M., Trifel, A.G.: Karbon. Nenas. Uglevod. 1968, 75; C.A. 70, 114 537 f (1969)

1115. Toyo Rayon Co., JP 21 976/65 (1965)
1116. Pino, P., Botteghi, C.: Org. Synth. *1977*, 11
1117. Esso Res. Eng. Co. (Mertzweiller, J.K., Tenney, H.M.), FR 1 315 275 (1966)
1118. Chisso Corp. (Tarao, R. et al.), JP 23 708/73 (1973), C.A. *79*, 41 932 e (1973)
1119. Hershman, A., Craddock, J.H.: Ind. Eng. Chem., Prod. Res. Dev. *7*, 226 (1968)
1120. Osborn, J.A., Wilkinson, G., Young, J.F.: Chem. Commun. *2*, 17 (1965)
1121. Gankin, W.Yu.: Karbonil. Nenas. Uglevod. *1968*, 61
1122. Institut Francais du Petrole (Lassau, C.), DE-OS 1 928 101 (1969)
1123. Esso Res. Eng. Comp. (Hughes, V.L., Kirshenbaum, I.), DE-AS 1 066 570 (1959)
1124. Macho, V., Minarsha, M.: Petrochemia *1973*, 149
1125. Eastman Kodak Comp. (Hagemeyer, H.J., Hull, D.C.), US 2 790 832 (1957)
1126. Lion Fat and Oil Co. Ltd. (Inagaki, T., Kiyonaga, Y., Yoshimura, Y.), JP 52 390/76 (1976)
1127. Montecatini Edison S. p. A., NL 68-03 440 (1968), BE 712 221 (1968)
1128. Japan Bureau of Ind. Technol. (Matsuda, A.), JP 17 900/69 (1969), C.A. *71*, 112 436 u (1969)
1129. Petrov, A.N. et al.: SU 485 104 (1975); C.A. *84*, 4480 p (1975)
1130. Borowski, A.F., Cole-Hamilton, D.J., Wilkinson, G.: Nouv. J. Chem. *2* (2), 137 (1978)
1131. Shell Int. Res. Mat. (Gray, R.T., DeJong, A.J.), DE-OS 2 753 644 (1978)
1132. Agency Ind. Sci. Technol. (Matsuda, A.), JP 14 190/78 and 16 385/78 (1978); C.A. *89*, 118 478 c, 118 479 d (1978)
1133. Mitsui Petrochem. Ind. Ltd. (Hayashi, T., Tanaka, M., Saeki, K.), JP 31 614/78 (1978); C.A. *89*, 108 170 g (1978)
1134. Consiglio, G., Pino, P.: Chim. Ind. (Milano) *1978*, 396
1135. Uson, R. et al.: J. Mol. Catal. *4*, 231 (1978)
1136. Mitsubishi Kasei Kogyo KK, JP 12 256/62 (1962)
1137. BP Ltd. (Lawrenson, M.J., Green, M.), DE-OS 2 026 926 (1971)
1138. Esso Res. Engng. Co., DE 975 040 (1961)
1139. Ruhrchemie AG (Hibbel, J., Kessen, G., Meis, J.), DE-OS 2 747 302 (1977)
1140. Farbenfabriken Bayer A. G. (Wegner, C., Wambach, R.), DE-AS 1 135 881 (1962)
1141. Esso Res. Eng. Co. (Ellis, W.J., Roming, C.), US 3 271 458 (1966); C.A. *66*, 10 589 f (1967)
1142. Eastman Kodak Co. (Hull, D.C., Hagemeyer, H.J.), DE-AS 1 255 097 (1967)
1143. Cornils, B., Ruprecht, P.: Ruhrchemie AG, unpublished
1144. Cornils, B., Ruprecht, P., Wiebus, E.: Ruhrchemie AG, unpublished
1145. Balint, T., Hanel, E., Laky, J.: Kem. Kozlem. *1969*, 32 (2), 151; C.A. *72*, 113 239 q (1970)
1146. Rudkowskii, D.M., Krinkin, D.P., Gankin, V.Yu.: SU 190 881 (1967), C.A. *68*, 29 257 d (1968)
1147. Levin, S.Z., Gurevich, G.S., Sedova, I.G.: SU 364 587 (1972), C.A. *78*, 147 342 g (1973)
1148. Fowler, R., Connor, H., Baehl, R.A.: Hydrocarbon Processing *1976*, (9), 247
1149. Brewester, E.A.V.: Chem. Engng. *1976*, (1), 90
1150. Fowler, R., Connor, H., Baehl, R.A.: Chemtech *1976*, (12), 773
1151. Anonym, Chem. Eng. News *1976*, (4), 25
1152. Sittig, M.: Catalyst Manufacture, Recovery and Use 1972, Noyes Data Corp., Park Ridge/N. J. 1972, p. 222
1153. Kogyo K. K. (Usami, S. et al.), US 3 378 590 (1968)
1154. Dümbgen, G., Neubauer, D.: Chem. Ing. Technik *41*, 974 (1969)
1155. BASF A. G. (Nienburg, H.J. et al.), DE-OS 2 139 630 (1973)
1156. BASF A. G., NL 72-01071/1972
1157. Gankin, V.Yu. et al: SU 239 310 (1969); C.A. *71*, 41 067 n (1969), ECN *16* (394), 26 (1969)
1158. Ajinomoto Co. (Kato, J., Iwanaga, R. et al.), DE-AS 1 259 324 (1968)
1159. BASF A. G. (Dümbgen, G. et al.), DE-OS 1 938 102 (1969)
1160. Chemo Co., DE-AS 1 002 308 (1957)
1161. Esso Res. Eng. Co. (Eliss, W.J. et al.), JP 3732/66 (1966)
1162. Eastman Kodak Co. (Hart, W.F., Hagemeyer, H.J.), US 3 868 422 (1975)
1163. Gulf Res. Dev. Co. (Gwynn, B.H.), GB 777 388 (1957)
1164. Alekseeva, K.A. et al.: Chimija. Technol. Topliv. i Masel *4*, 24 (1959)

1165. Cornils, B., Wiebus, E.: Ruhrchemie A. G., unpublished
1166. BASF A. G., FR 2 053 177 (1971)
1167. Showa Denko K. K. (Sugiyama, K., Yonemoto, S.), JP 11 961/63 (1963)
1168. Monsanto Co. (Heimsch, R.A.), FR 1 394 118 (1965), C. *1968*, 29-2706
1169. Gankin, V.Yu. et al.: Model. Khim. Reaktorov (1970) *2*, 128; C.A. *81*, 37 185 n (1974)
1170. Ruprecht, P., Cornils, B.: Ruhrchemie AG, unpublished
1171. Gulf Res. Dev. Co., GB 786 809 (1957)
1172. Gankin, V.Yu. et al.: SU 274 103 (1970); C.A. *74*, 12 613 t (1971)
1173. Union Carbide Corp. (Schulz, H.W., Voorhees, B.), CA 593 954 (1960)
1174. BASF A. G., BE 753 916 (1970)
1175. VEB Leuna-Werke (Smeykal, K., Walther, H.), DE 1 055 516 (1963)
1176. Anonym, Chem. Eng. *1977*, (12), 110
1177. Gankin, W.Yu. et al.: SU 251 564 (1969); C.A. *72*, 54 766 j (1970)
1178. Badikov, I.D., Til'dikov, Yu.V., Mandrusenko, G.I.: Visb. Automatiz. Khim. Proiz. M. *1975*, (4), 10; C.A. *85*, 142 546 m (1976)
1179. Berdnikov, V.V. et al.: Karbonil. Nenasysh. Uglevod. *1968*, 111; C.A. *71*, 101 217 (1969)
1180. Anonym, Europ. Chem. News *26*, (6), 655 (1974)
1181. Cornils, B.: in [24], p. 109/110
1182. Harders, H., Heller, G., Laurer, P.R.: Digital Computer Appl. Process Control, 1st Int. Conf., Stockholm *1964*, 89; C. *1967*, 46-2446
1183. Heller, G., Peinke, W.: Messen, Steuern, Regeln, Beilage Automat. Praxis *9*, (3), 51 (1966)
1184. Ruhrchemie A. G. (Cornils, B., Wiebus, E.), DE-Application
1185. Schnur, F., Cornils, B.: Ruhrchemie A. G., unpublished
1186. Kafarov, V.V. et al.: Tr. Mosk. Khim. Tekhnol. Inst. *72*, 111 (1973); C.A. *81*, 138 074 f (1974)
1187. Wdowin, A., Kriwonosov, A.I.: Chem. Stosow. *17*, (2), 229 (1973), C.A. *80*, 36 662 h (1974)
1188. Fal'kevich, G.S., Telkov, Yu.K., Solov'eva, S.S.: Sb. Tr. Vses. Ob'edin. Neftekhim. *1976*, (12), 134; C.A. *88*, 24 555 c (1978)
1189. BASF A. G. (Dümbgen, G. et al.), DE-OS 1 938 104 (1971)
1190. Esso Res. Eng. Co. (Ellis, W.J., Roming, C.), US 3 271 458 (1966); C.A. *66*, 10 589 (1967)
1191. Mandrusenko, G.I., Badikov, I.D., Til'dikov, Yu.V.: Khim. Tekhnol. (Kiev) *1973*, (1), 20; C.A. *79*, 4865 q (1973)
1192. All-Unions Sci. Res. Inst./VEB Leuna-Werke (Alekseeva, K.A. et al.), DD 122 761 (1976)
1193. BASF A. G., FR 2 000 423 (1969)
1194. BASF A. G., NL 6900 586 (1969); DE-AS 1 283 219 (1968)
1195. BASF A. G., GB 1 195 877 (1970)
1196. Lippert, B. et al.: Poluch. Maslyan. Al'degidov i Butilovykh Spirtov Oksosintezom *1977*, (Ch. 2), 68; C.A. *88*, 22 070 s (1978)
1197. Al-Perovich, V.Ya., Gankin, V.Yu.: Khim. Prom. *44*, 663 (1968)
1198. Reich, M.: Chem. Ing. Techn. *47*, 532 (1975)
1199. Chemische Werke Hüls A. G. (Reich, M., Müller, W., Zur Hausen, M.), DE 1 935 900 (1969)
1200. Anonym, Chem. Age. *99*, Nr. 2628, 28 (1969); Eur. Chem. *20*, 11 (1967); VWD-Chem. *141*, 2 (1969); *222*, 5 (1969), ECN *11*, Nr. 277, 18 (1967)
1201. Anonym, ECN *13*, Nr. 232, 25 (1968) *16*, Nr. 408, 15 (1969), *27*, Nr. 680, 12 (1975), Chem. Age. *110*, Nr. 2904, 14 (1975), Eur. Chem. *22*, 454 (1974)
1202. Eur. Chem. *1972*, Nr. 4, 68; *1976*, Nr. 6, 96; Erdöl und Kohle *29*, (8), 334 (1976); Chem. Age *104*, Nr. 2744, 15 (1972); VWD-Chem. *1973*, (32), 10; ECN *21*, Nr. 519, 14 (1972);
1203. Chem. Ind. *21*, (4), 257 (1969), *22*, (8), 523 (1970); ECN *16*, Nr. 398, 12 (1969); Eur. Chem. *1969*, (4), 12; Canad. Chem. Proc. *53*, (10), 67 (1969)
1204. ECN *11*, Nr. 277, 21 (1967), Chem. Mark. Abstr. *66*, (7), 125 (1974); Plast-Verarbeiter *19*, (1), 92 (1968)
1205. Chem. Ind. *20*, (8), 519 (1968); Chem. Engng. *75*, (16), 69 (1968)
1206. Chem. & Engng. News of March 29, 1976, p. 8; K.-Plasticzeitg. *1976*, No. 114, p. 1
1207. Anonym, Chem. Lab. Betr. *19*, (5), 231 (1968)
1208. Kummer, R. et al., 166th Nat. Meeting ACS, Chicago, Aug. *1973;* Kobalt *1974*, (1), A 17

1209. Neubauer, D., Dümbgen, G.: Erdöl und Kohle 22, (11), 664 (1969)

1210. Scheidmeier, W.J.: Chem.-Ztg. 96, 383 (1972)

1211. BASF A. G., NL 70-10934 (1970)

1212. BASF A. G., DE 2 451 473 (1974)

1213. BASF A. G. (Nienburg, H.J.), DE-OS 1 643 745 (1971)

1214. BASF A. G. (Hagen, W., Kerber, H., Schröder, W.), DE-AS 1 294 950 (1969); Brennstoff-Chem. 50, (6), 190 (1969)

1215. BASF A. G. (Moell, H. et al.), DE-OS 2 404 855 (1975); BE 700 142 (1967)

1216. BASF A. G. (Rauch, K., Scheidmeir, W.J.), DE-AS 1 283 216 (1968); C.A. 71, 12 531 c (1969)

1217. BASF A. G., GB 1 240 847 (1971)

1218. BASF A. G. (Kerber, H., Hohenschutz, H.), DE-AS 1 301 806 (1969)

1219. Inf. Chim. 97, 74 (1971), Chem. Ind. 20, (8), 540 (1968) and 25, (9), 596 (1973), Inf. Chim., Int. Ed. 1973, (12), 105

1220. Chem. Age 102, Nr. 2704, p. 26 (1971); 102, Nr. 2705, 20 (1971); Chem. Ind. 24, (4), 234 (1972); Chem. Mark. Abstr. 66, (7), 121 (1974); Eur. Chem. 1971, (10), 12

1221. Chem. Age 100, Nr. 2639, 8 (1970) and 102, Nr. 2704, p. 26 (1971); Inf. Chim. 1971, Nr. 103, 50; Chem. Eng. 77, (4), 45 (1970) and (23), 141 (1970); Chem. Ing. Techn. 43, 1239 (1971); Chem. Lab. Betrieb 22, 476 (1971)

1222. Chem. Mark. Abstr. 66, (11), 111 (1974)

1223. All-Unions Sci. Res. Inst. (Rudkovskij, D.M.), SU 193 480 (1967); C.A. 69, 10088 y (1968)

1224. Agency Ind. Sci. Technol. (Matsuda, A.), JP 16 385/78 (1978); C.A. 89, 118 478 c (1978)

1225. Nissan Chem. Ind. Ltd. (Fukushima, S., Hoshiyama, K., Ono, K.), JP 79 806/78; C.A. 89, 146 427 d (1978)

1226. Mitsubishi Petrochemical Co. Ltd. (Tanaka, M.), JP 65 810/78 (1978); C.A. 89, 146 420 w (1978)

1227. Polievka, M., Uhlar, L., Macho, V.: Petrochemie 1978, 9; C.A. 89, 146 328 x (1978)

1228. BASF A. G., DE-OS 2 460 784 (1974)

1229. Mason, T., Grote, D., Trevedi, B.: in Smith, G. V.: Catalysis in Organic Chemistry 1977, p. 165 (1978)

1230. Johnson, M.: J. Chem. Soc. 1963, 4859

1231. Ruhrchemie A. G. (Tummes, H., Falbe, J., Cornils, B.), DE 1 718 051 (1972)

1232. Ruhrchemie A. G. (Tummes, H., Cornils, B., Kascha, W.), DE-AS 2 621 405 (1977)

1233. Produits Chimiques Ugine Kuhlmann (Guerant, A., Mercier, S.), DE-OS 2 443 995 (1975)

1234. Ruhrchemie A. G. (Bexten, L., Noeske, H., Tummes, H., Cornils, B.), DE 2 459 152 (1976)

1235. Ruhrchemie A. G. (Meis, J., Tummes, H.), DE 1 258 855 (1967)

1236. Ruhrchemie A. G. (Büchner, K., Tummes, H.), DE 1 085 513 (1960)

1237. Ruhrchemie A. G. (Tummes, H., Falbe, J., Cornils, B.), DE 1 817 051 (1972)

1238. Ruhrchemie A. G. (Tummes, H., Falbe, J., Cornils, B.), DD 80 691 (1968)

1239. Ruhrchemie A. G. (Tummes, H., Noeske, H., Cornils, B., Kascha, W.), DE 2 713 434 (1979)

1240. Ruhrchemie A. G., NL 100 479 (1962)

1241. Ruhrchemie A. G. (Bexten, L., Cornils, B., Hahn, H.D., Tummes, H.), DE-AS 2 737 633 (1979)

1242. Ruhrchemie A. G. (Falbe, J., Tummes, H.), DE-AS 1 768 768 (1968)

1243. Ruhrchemie A. G. (Hibbel, J., Schmidt, V.), DE 1 432 864 (1964)

1244. Ruhrchemie A. G. (Tomuschat, H.J.), DE-OS 2 851 515 (1978)

1245. Ruhrchemie A. G. (Cornils, B., Bahrmann, H., Frohning, C.D., Weber, J.), DE-OS 2 837 480 (1975)

1246. Anonym, Eur. Chem. News 12, Nr. 300, 8 (1967); 14, No. 348, 6 (1968); 15, No. 367, 10 (1969); 15, No. 369, 16 (1969) and 15, No. 370, 8 (1969); 17, No. 418, 8 (1970) and 17, No. 423, 4 (1970); 20, No. 501, 20 (1971); 21, No. 515, 16 (1972); 22, No. 540, 34 (1972) as well as 26, No. 645, 25 and No. 649, 28 (1974)

1247. Chem. Age 99, No. 2630, 30 (1969); 102, No. 2709, 2 (1971) and 105, No. 2765, 2 (1972)

1248. Chem. Ind. 1977, No. 7, 240; Chem. Mark. Abstr. 66, (7) 126 (1974)

1249. Chem. Ind. 21 (8), 514, 552 (1969); 23, (11), 787 (1971) and 24, (8), 494 (1972); Inf. Chim. 1973, No. 120, 61; 1974, No. 132, 118; 1974, No. 139, 241

1250. Eur. Chem. *1971*, No. 13, 9; *1971*, No. 19, 12 and *1974*, No. 11, 227
1251. Eur. Chem. *1976*, No. 13, 226
1252. BASF A. G. (Häuber, H., Hagen, W.), DE 1 036 839 (1961)
1253. BASF A. G. (Neubauer, D., Bach, J.M., Garrido, R.), DE-OS 2 544 570 (1975)
1254. BASF A. G. (Kummer, R. et al.), DE-OS 2 244 373 (1974)
1255. Chem. Ind. *1972*, (6), 333; *1976*, (11), 675; Inf. Chim. *1976*, No. 160, p. 85; *1975*, No. 149, p. 95; ECN *17*, No. 418, p. 7 (1970); Chem. Age. *100*, No. 263; Chem. Engng. News *53*, No. 43, 11 (1975), Chemistry and Industry *1972*, (5)
1256. Cornils, B., Ruprecht, P.: Ruhrchemie A. G., unpublished
1257. Shell Int. Res. (Greene, C.R., Brown, W.A.), DE-AS 1 593 368 (1970)
1258. Shell Int. Res. M. (Greene, C.R.), DE-OS 1 468 615 (1972)
1259. ICI Ltd. (Hibbs, F.M.), GB 1 458 375 (1976)
1260. ICI Ltd., NL 68-05466 (1968)
1261. ICI Ltd. (Harvey, P.O., Lamb, S.A.), DE 843 849 (1952)
1262. ICI Ltd. (Brewis, S.), GB 1 045 679 (1966)
1263. Anonym, Chem. Age. *1974*, May, p. 7
1264. Murfitt, H.C.: ECN Polymer Intermed. *1970*, (10), p. 18
1265. Schneider, K., Wilck, I.: Lichtbogen (Chem. Werke Hüls) *16*, (3), 7 (1967)
1266. Chem. Werke Hüls A. G. (Kaufhold, M., Wulf, H.D.), DE-OS 2 165 515 (1973)
1267. Wilke, N.: Chem.-Ztg. *95*, 16 (1971)
1268. Chem. Werke Hüls A. G., GB 1 132 666 (1968)
1269. Chem. Werke Hüls A. G. (Broich, F. et al.), DE-AS 1 003 702 (1957)
1270. Eastman Kodak Co., JP 29 044/69 (1969)
1271. Eastman Kodak Co. (Hart, W.F., Hagemeyer, H.J., Park, W.P.), US 3 868 422 (1975)
1272. Eastman Kodak Co. (Hagemeyer, H.J.), US 2 748 167 (1956)
1273. Eastman Kodak Co. (Peacock, D.R., Magness, J.F., Morris, P.M.), DE-OS 1 568 478 (1970)
1274. Chem. Ind. *24*, (3), 118 (1972)
1275. Ishikawa, Y.: Chem. Econ. Engng. Rev. (Tokyo), *2*, (1), 31 (1970)
1276. Anonym, Eur. Chem. *1968*, (6), p. 12
1277. Chem. Ing. Techn. *51*, A 80 (1979); Chem. Engng. News, Nov. 27, 1978, p. 15
1278. Eur. Chem. *1974*, (21), p. 428
1279. VEB Leuna-Werke „Walter Ulbricht", Firmenschrift Juli *1973*
1280. All-Unions Sci. Res. Inst. Petrochem. (Rudkowskij, D.M., Krinkin, D.P., Gankin, V.Yu.), DE-AS 1 240 064 (1967); C.A. *67*, 63 796 t (1967)
1281. Knebel, H. et al.: Poliechenie Maslyan. Al'degidov i Butilov Spirtov. ok. *1977* (Ch. 2), 32, 51; C.A. *88*, 22 071 t (1978)
1282. Alekseeva, K.A. et al.: Karbonil. Nenass. Uglev. *1968*, 127
1283. Dawidoff, W.: Katal. Reakts. Zhid. Faze. Tr. Vses. Konf. *1966*, 441; C.A. *69*, 35 145 c (1968)
1284. Alekseeva, K.A. et al.: Nenas. Uglevod. *1968*, 140; C.A. *71*, 2914 r (1969)
1285. Eidus, Y.T., Bulanova, T.F.: Sci. Select. Catal. *1968*, 206; C.A. *72*, 32 333 p (1970)
1286. Alekseeva, K.A. et al.: Nefteper. Neftekhim. *1977*, (5), 28; C.A. *88*, 6267 f (1978)
1287. Krinkin, J.P., Rudkowskij, D.M.: Karbonsl. Nenasysh. Uglevod. *1968*, 147; C.A. *71*, 23 448 g (1969)
1288. All-Unions Sci. Res. Inst. (Gankin, W.Yu., Razumovskii, S.D.), SU 215 958 (1968); C.A. *69*, 66 891 x (1968)
1289. Petro-Chem. Proc. Inst. (Rudkowskij, D.M.), SU 193 480 (1967)
1290. Petrochem. Inst. (Gankin, W.Yu.), SU 191 520 (1967)
1291. Mandrusenko, G.I., Fuks, I.S., Goldshtein, E.V.: Automat. Khim. Proiz. *1970*, (4), 62; C.A. *75*, 50 745 r (1971)
1292. Baltz, H.: DD 63 477 (1968)
1293. Motzhukhin, A.S. et al.: Khim. i. Tekhnol. Topliv i Masel. *11*, (10), 5 (1966), C.A. *66*, 54 975 c (1967)
1294. Alekseeva, K.A., Trifel, A.G., Rudkovskij, D.M.: Karbonil. Nenass. Uglevod. *1968*, 134
1295. Gankin, W.Yu., Krinkin, D.P., Rudkovskij, D.M.: Zh. prikl. Chim. *40*, (8), 1788 (1967); C.A. *68*, 41 841 x (1968)

1296. Alekseeva, K.A., Kuvneva, M.M., Ermolenko, N.F. et al.: Dokl. Akad. Nauk. Beloruss. SSR *13*, 914 (1969); C.A. *72*, 36 165 h (1970)

1297. All-Unions Sci. Res. (Itsikson, L.B. et al.), SU 242 865 (1969); C.A. *71*, 90 795 f (1969)

1298. Gromoglasov, Yu.A. et al.: SU 451 682 (1974); C.A. *82*, 155 294 r (1975)

1299. Icikson, L.B. et al.: Chim. Technol. Topliv i Masel *10*, (8) 25 (1965); C. *1967*, 40-2478

1300. Barba, N.A., Shur, A.M.: SU 229 479 (1968); C.A. *70*, 57 147 c (1969)

1301. Zvezdkina, L.I., Grigor'eva, S.A., Muravlyanskaya, T.B.: SU 299 502 (1971); C.A. *75*, 63 113 h (1971)

1302. Borkowski, B. et al.: Przemysl. Chem. *46*, (7), 742 (1967); C. *1968*, 48-2097

1303. Aristovich, V.Yu. et al.: SU 245 063 (1969); C.A. *72*, 2740 b (1970)

1304. Klinova, L.L., Silich, M.I.: Khim. Prom. Moskau *1975*, (5), 341

1305. ECN *13*, No. 18, p. 32 (1968)

1306. Aleksandrova, M.V., Pavlenko, T.G.: Iz. Vses. Chimija i Chimiceskaja Technol. *10*, (3), 331 (1967)

1307. Timofejew, W.S. et al.: Chim. Technol. Topliv i. Magel *9*, (7), 18 (1964); C. *1966*, 33-2434

1308. Esso Res. Engng. Co. (Bearden, R., Mertzweiller, J.K.), US 3 409 648 (1968)

1309. Esso Res. Engng. Co. (Mertzweiller, J.K.), US 2 767 048 (1956)

1310. Standard Oil Dev. Co. (Hale, C.H., Starr, C.E.), US 2 595 785 and 2 595 786 (1952)

1311. Standard Oil Dev. Co., FR 1 041 838 (1953)

1312. Esso Res. Engng. Co. (Mertzweiller, J.K., Mason, R.B.), US 2 834 815 (1958)

1313. Esso Res. Engng. Co. (Cull, N.L.), US 3 454 649 (1963)

1314. Gulf Res. Dev. Co. (Gwynn, B.H.), US 3 246 024 (1966)

1315. Union Carbide Corp. (Pruett, R.L., Smith, J.A.), US 3 527 809 (1970), GB 1 197 847 (1967)

1316. Hydrocarbon Proc. *53*, (8), 33 (1974); *56* (11), 190 (1977)

1317. Fowler, R., Connor, H., Baehl, R.A.: Inf. Chim. *159* (10), 193 (1976)

1318. Chem. Engng. *83*, (3), 104 (1976), ECN *30*, No. 778, 44, 50 (1977) and *31*, No. 810, 39 (1977)

1319. Mitsubishi Chem. Ind. Co. (Onoda, T., Tano, K.), JP 86 301/74 (1974); C.A. *82*, 57 941 e (1975)

1320. Japan Chem. Ind. Ass. Monthly *27*, (7), (1974); Chem. Eng. *84*, (15), 104 (1977)

1321. Japan Chem. Rep. March 1974; Japan Chem. Week Sept. 25, 1975

1322. Nikhan Kogyo Shimbum *1977*, Nov. 25; Chem. Mark. Abstr. *66*, No. 11, 110 (1974)

1323. BASF A.G. (Kummer, R., Platz, R.), DE-OS 2 354 217 (1973)

1324. Rony, P.R., Roth, J.F.: Catal. Proc. Int. Symp. *1974* (Publ. 1975), 373; C.A. *84*, 30 289 c (1976)

1325. Union Oil of California (Booth, F.B.), US 3 801 646 (1974)

1326. Wdowin, A., Zielinski, K.: Wiad Chem. *21*, (6), 455 (1967); C.A. *67*, 63 417 v (1967)

1327. Mitsubishi Chem. Ind. Co. Ltd. (Onoda, T. et al.), JP 71 610/75 (1975); C.A. *83*, 178 314 g (1975)

1328. Gulf Res. Dev. Co. (Swift, H.E.), US 3 542 878 (1970)

1329. Chem. Eng. News *54*, No. 26, 19 (1976), ECN *28*, No. 741, 62 (1976)

1330. ICI Ltd. (Harvey, P.G.), US 2 752 395 (1956)

1331. ICI Ltd. (Brewis, S.), GB 1 045 679 (1966); C. *1968*, 46-2304

1332. ICI Ltd. (Goddart, R.E., Mansfield, G.H.), JP 7719/63 (1963)

1333. Standard Oil Dev. Co., GB 698 631 (1953)

1334. Standard Oil Dev. Co., Petroleum Refiner *44*, (11), 249 (1965)

1335. Chem. Ind. *21*, 429 (1969) and *22*, 617 (1970)

1336. Chem. Age. *100*, No. 2664, 16 (1970); *102*, No. 2707, 17 (1971), *105*, No. 2761, 20 (1972) and *115*, No. 3025, 8 (1977)

1337. ECN *18*, No. 454, 16 (1970), *19*, No. 473, 16 (1971), *21*, No. 515, 13 (1972) and *32*, No. 975, 30 (1977)

1338. Iwasaki, I., Yumura, S.: Japan Chem. Quart. *2*, 71 (1966)

1339. Hydrocarbon Proc. *46*, (11), 179, (1967)

1340. Chem. Age. *106*, No. 2806, 14 (1973); *114*, No. 3009, 1 (1977) and No. 2964, 13 (1976); Chem. Eng. News *54*, (18), 25 (1976)

1341. Inf. Chim. *149*, 101 (1975); *156*, 65 (1976); CRN *53*, (41) 6 (1975)
1342. Chem. Mark. Abstr. *66*, No. 8, 133, 256 (1974), *67*, (2), 125 (1975); Chem. Mark. Rep./ OPD April 16, 1973 and June 24, 1974
1343. Eur. Chem. *1975*, No. 6, 116 (1975) and *1977*, No. 7, 146
1344. Ahlstrom, L.: Kem. Tidskr. *89*, (5), 18 (1977), C.A. *87*, 203 721 e (1977)
1345. Eur. Chem. *1967*, No. 13, 10
1346. Montedison SpA (Pregaglia, G. et al.), IT 913 125 (1972)
1347. ECN *1970*, (3), 86; Chem. Eng. News *48*, No. 41, 18 (1970), Chem. Ind. *19*, (11), 744 (1967)
1348. Kotowski, W., Bazan, A., Dzwig, E.: Przemysl. Chem. *45*, 354 (1966); C.A. *66*, 10 503 a (1967)
1349. Hydrocarbon Proc. *47*, (6), CR-3 (1968); *49*, (6), 3-72 (1970); *49*, (2), Section 2 (1970); *49*, (10), 2, CR-5-75 (1970), *50*, (2), p. 2 f (1971)
1350. Chem. Week of May 20, 1970; Chem. Engng. *74*, (21), 221 (1967)
1351. Thompson, B.M.M.: Oel *8*, (2), 34 (1970)
1352. Thomas, C.L.: Catalytic Processes and Proven Catalysts, Academic Press, New York/London (1970)
1353. Ferguson, G.U.: in Long, R.: The Production of Polymer and Plastic Intermediates from Petroleum, p. 86, (1967)
1354. VDI-Nachr. *23*, (1), 4 (1969); Chem. Ing. Techn. *43*, 407 (1971); Chem. Ind. *20*, (10), 733 (1968); *20*, (12), 869 (1968) and *22*, (1), 25 (1970)
1355. ECN *17*, No. 432, 36 (1970); *18*, No. 458, 18 (1970); *30*, No. 791, 49 (1977); *32*, No. 795, 23 (1977); Chem. Eng. News *55*, (9), 14 (1977)
1356. Chem. Eng. *74* (21), 221 (1967); Japan Chem. News of May 31, 1974 and April 18, 1974; Chem. Mark. Abstr. *66* (7), 127, 251 (1974); Chem. Age *104*, No. 2759, 6 (1972), *108*, No. 2867, 10 (1974) and *114*, No. 3016, 16 (1977)
1357. Inf. Chim. *1973*, No. 124, 82
1358. Striebeck, P.: Der Lichtbogen (Chem. Werke Hüls A. G.), XIX, Nr. 156, 14 (1970)
1359. Eur. Chem. News *12*, No. 290, 18 (1967)
1360. Chem. Ind. *1976*, No. 12, 733
1361. Eur. Chem. News *18*, No. 458, 32 (1970)
1362. Eur. Chem. News *18*, No. 459/460, 50/60 (1970)
1363. ChemSystems Inc. (New York), PERP-Quarterly-Report, Dec. 1975
1364. ChemSystems Inc. (New York), PERP-Report 77-4 (May 1978)
1365. Stanford Res. Inst. (Menlo Park, Calif.), Rep. No. 21 (1966)
1366. Stanford Res. Inst. (Menlo Park, Calif.), Rep. No. 21 B (May 1978)
1367. Schwartz, J.: Tetrahedron Lett. *1972*, 2803, J. Amer. Chem. Soc. *96*, 4721 (1974) and *99*, 5831 (1977)
1368. Aumann, R., Knecht, J.: Chem. Ber. *109*, 174 (1976)
1369. Alper, H.: in [7], Vol. 2, p. 545 f
1370. Sonoda, N. et al.: J. Amer. Chem. Soc. *93*, 6344 (1971)
1371. Klein, H.F.: Angew. Chem. *85*, 403 (1973)
1372. Baker, R., Copeland, A.H.: Tetrahedron Lett. *1976*, 4535
1373. Chiusoli, G.P., Cassar L.: in [7], Vol. 2, p. 297 ff
1374. Ajinomoto Co. Inc., JP 18 535/68 (1968); 30 293/68 (1968) and JP 2 691/68 (1968)
1375. Lonza Ltd., GB 1 119 520 (1968), DE-OS 1 930 431 (1970)
1376. Pino, P.: Chem. Ind. *49*, 1732 (1968)
1377. Hosomi, A., Sakurai, H.: Tetrahedron Lett. *1976*, 1295
1378. Ojima, I., Kumagai, M.: J. Organomet. Chem. *111*, 43 (1976)
1379. Murai, S. et al.: Angew. Chemie *89*, 196, 818, 919 (1977), J. Organomet. Chem. *140*, 361 (1977)
1380. Magomedov, G.K.I. et al.: J. Organomet. Chem. *149*, 23 (1978)
1381. Puzitskii, K.V. et al.: Izv. Akad. Nauk. SSSR, Ser. Chim. *1972*, (9), 1998
1382. Bertelo, C.A., Schwartz, J.: J. Amer. Chem. Soc. *97*, 228 (1975)
1383. Bulanova, T.F., Eidus, Ya.T., Sergeeva, N.S.: Izv. Akad. Nauk. SSSR, Ser. Khim. *1966*, (10), 1814, C.A. *66*, 18 877 (1967)

1384. Heck, R.F.: J. Amer. Chem. Soc. *94*, 2712 (1972)
1385. Heck, R.F.: Transition-Metal-Catalyzed Reactions of Organic Halides with CO, Olefins and Acetylenes, Adv. Catal. *26*, 323 (1977)
1386. Weil, T.A., Cassar, L., Foa, M.: in [7], Vol. 2, p. 517 f
1387. Weigert, F.J.: J. Org. Chem. *38*, 1316 (1973)
1388. Mitsui Toatsu Chem. Inc., DE-OS 2 555 557 (1974)
1389. Farbenfabriken Bayer A. G. (Enders, E., Hüllstrung, D.), DE-OS 1 668 906 (1971)
1390. Shell Dev. Co., GB 1 005 493 (1965)
1391. Kagan, J.B. et al.: Neftechimija *4*, 106 (1964); C. *1967*, 33-1024
1392. Toyo Rayon K. K., JP 10 250/69 (1969)
1393. Marko, L., Bakos, J.: J. Organomet. Chem. *81*, 411 (1974)
1394. Takezaki, Y., Teraoka, T. et al.: J. Chem. Soc. Japan, Ind. Chem. Sect. *69*, 907 (1966); C. *1967*, 34-981
1395. Muetterties, E.L., Hirsekorn, F.J.: Chem. Commun. *1973* (18), 683
1396. Studiengesellschaft Kohle GmbH (Koester, R., Pourzal, A.A.), US 4 010 204 (1977)
1397. Universal Oil Prod., US 3 438 484 (1971)
1398. [12], p. 117 f
1399. Murfitt, H.C.: ECN (Chemscope) of Oct. 30, 1970, p. 14
1400. Mitsubishi Chem. Co. Ltd. (Osumi, Y.), JP 14 738/66 (1966)
1401. Snam Progetti SpA, GB 1 193 547 (1970)
1402. Montecatini Edison SpA, IT 802 536 (1968)
1403. ICI Ltd., GB 1 185 168 (1970)
1404. BP Ltd. (Gevers, J.), BE 759 729 (1970)
1405. Mitsubishi Chem. Co. Ltd. (Onoda, T. et al.), JP 24 928/78 (1978), C.A. *89*, 163 068 e (1978)
1406. Chevron Res. Co. (Wilkes, J.B.), US 4 096 188 (1978)
1407. Institut Francais du Petrole, FR 2 041 776 (1969)
1408. Monsanto Co. (Paulik, F.E.), DE-OS 1 768 306 (1968)
1409. Monsanto Co. (Ries, J.G., Van Wazer, J.R.), US 3 414 390 (1968); C.A. *70*, 49 123 b (1969)
1410. National Distillers and Chem. Corp. (Horvitz, D., Baugh, W.D.), DE-OS 2 657 346 (1977)
1411. General Electric Co. (Hallgren, J.E., Chalk, A.J.), DE-OS 2 738 437, 2 738 487, 2 738 488, 2 738 519 and 2 738 520 (1978)
1412. Fujiyama, S., Kasahara, T.: Hydrocarbon Proc. *1978*, (11), 149
1413. BASF A. G. (Nienburg, H.J. et al.), DE 2 037 783 (1978)
1414. Krinkin, D.P., Rudkowskij. D.M., Trifel, A.G.: Neftekhimija *4*, 373 (1956); C. *1966*, 24-0775
1415. Pino, P., Piacenti, F., Bianchi, M.: in [7], p. 74 f
1416. [6], p. 56, 58, 61 f
1417. Porter, R.A., Shriver, D.F.: J. Organomet. Chem. *90*, (1), 41 (1975)
1418. Feder, H.M.: J. Amer. Chem. Soc. *97*, 7186 (1975)
1419. Ungvary, F., Marko, L.: Acta Chim. Acad. Sci. Hung. *62*, 425 (1969)
1420. Dawydoff, W.: Katal. Reactions Zhidk. Faze, Tr. Vses. Konf. 2nd Alma Ata., Kaz. SSR *1966*, Publ. *1967*, 441; C.A. *69*, 35 145 c (1968)
1421. Aldridge, C.L., Jonassen, H.B.: J. Amer. Chem. Soc. *85*, 886 (1963)
1422. Shell Int. Res., GB 942 435 (1963)
1423. Brown, C.K., Wilkinson, G.: J. Chem. Soc. (A) *1970*, 2753
1424. Babos, B., Ungvary, F., Papp, L., Marko, L.: Mag. Kem. Folyoirat *75*, (3), 126 (1963)
1425. Litvin, E.F., Freidlin, L.KH., Karimov, K.G.: Neftekhimiya *12*, (3), 318 (1972)
1426. Pregaglia, G.F., Andreetta, A., Ferrari, G.F.: J. Organomet. Chem. *30*, 387 (1971)
1427. Kitamura, T., Sakamoto, N., Joh, T.: Chem. Lett. *1973*, (4), 379
1428. Ucciani, E., Lai, R., Tanguy, L.: C. R. Hebd. Séances, Acad. Sci., Ser. C *1976*, 283 (1), 17; C.A. *86*, 16 257 a (1977)
1429. Ucciani, E., Lai, R., Tanguy, L.: C. R. Hebd. Séances Acad. Sci., Ser. C *1975*, 281 (21), 877; C.A. *84*, 104 934 h (1976)
1430. Baird, M.C., Nyman, C.J., Wilkinson, G.: J. Chem. Soc. A *1968*, 348
1431. Marko, L.: Chem. Ind. *1962*, 260

1432. Marko, L., Szabo, P.: Chem. Technol. (Berlin) *13*, 482 (1961)

1433. Polievka, M., Mistrik, E.J.: Chem. Zvesti *1972*, 26 (2), 149; C.A. *77*, 113, 388 r (1972)

1434. Cornils, B., Förster, I.: Ruhrchemie A. G., unpublished

1435. ICI Ltd. (Smith, R.D., Wilson, R.A.L.), GB 970 072 (1964); C. *1966*, 23-2411

1436. Gavrilowa, V.M., Rudkowskij, D.M., Trifel, A.G.: Zh. Prikl. Khim. *45* (6), 1320 (1972)

1437. BP Ltd., NL 71-03178 (1971)

1438. Ruhrchemie A. G. (Cornils, B., Bahrmann, H., Tummes, H.), DE application

1439. Ruhrchemie A. G. (Tummes, H., Falbe, J., Cornils, B.), DE 1 817 051 (1972)

1440. Otto Roelen, private communication

1441. Bertrand, J.A. et al.: J. Org. Chem. *29*, 790 (1964)

1442. Polievka, M., Mistrik, J.E.: Ropa Uhlie *11*, (12) 665 (1969) and *13*, (4), 199 (1971)

1443. Dokiya, M., Bando, K.: Kogyo Kagaku Zasshi *1968*, 71 (11), 1866; C.A. *70*, 77 247 p (1969) and *1973*, (8) 1523; C.A. *79*, 125 365 k (1973)

1444. Natta, G., Pino, P., Ercoli, R.: J. Amer. Chem. Soc. *74*, 4496 (1952)

1445. Esso Res. Engng. Co. (Staib, J.H., Guyer, W.R.F., Slotterbeck, O.C.), US 2 864 864 (1958)

1446. Agency of Ind. Sci. Techn. (Tokiya, M., Bando, K.), JP 34 651/74 (1974); C.A. *82*, 139 355 f (1975)

1447. Karmilchik, A.Y., Stonkus, V.V. et al.: J. appl. Chem. USSR *48*, 1623 (1975)

1448. Damico, R., Logan, T.J.: J. Org. Chem. *32*, 2356 (1967)

1449. Minacev, C.M., Loginov, G.A., Markov, M.A.: Kinetika i Kataliz *7*, 904 (1966); C. 1948, 32-1005

1450. Sasson, Y., Blum, J.: J. Org. Chem. *40*, 1887 (1975)

1451. Jones, J.H., Ernst, W.R.: Amer. Chem. Soc. Div. Petr. Chem. Prepr. *16* (3), B5-B9 (1971)

1452. [6], p. 60

1453. Ogawa Co. Ltd. (Saeki, Y. et al.), DE-OS 2 424 128 (1974)

1454. Lion Fat Oil Co. Ltd. (Isa, L.H. et al.), DE-OS 2 504 005 (1975)

1455. Rudkovskij, D.M. et al: Khim. Prom. *42*, 95 (1966); C.A. *64*, 19 357 (1966); Neftekhimiya 6, 458 (1966)

1456. Alekseeva, K.A., Efimova, N.I.: Neftekhimiya *7*, 407 (1967); SU 258 296 (1969); C.A. *72*, 132 055 t (1970)

1457. Polievka, M.: Ropa Uhlie *1973*, 366; C.A. *79*, 125 500 a (1973)

1458. Kasano, K., Takahashi, T.: J. Soc. Org. Synth. Chem. Japan *23*, 144 (1965); C. *1967*, 3-2578

1459. Montecatini Edison SpA, BE 698 544 (1967)

1460. Bonner, T.G. et al.: J. Chem. Soc. (B) *1971*, 957

1461. Kutsenko, A.I., Lyubomilov, V.I.: Vyss. Zhirnye. Spirty *1970*, 175, 194; Plast. Massy *1970*, 41

1462. Nippon Synthetic Chem. Inc. Co. Ltd., JP 7 633/67 (1967)

1463. Japan Synthetic Chem. Ind. (Noro, K., Takida, H.), JP 18 149 (1968)

1464. Chen, W.: Hua Hsueh *1967*, 75; Chem. Titles *68*, 20-128

1465. Insue, H., Kunikawa, K., Imoto, E.: Bull. Chem. Soc. Japan *1973*, 518

1466. Pomazkov, B.K. et al.: Zaved. Khim. Tekhnol. *1971*, 14 (11), 1761; C.A. *76*, 58 905 j (1972)

1467. Aotani, K., Imoto, T.: J. Chem. Soc. Japan *89*, (3), 235 (1968)

1468. Wolf, F., Lesse, A., Mücke, J.: J. prakt. Chem. *313*, 137 (1971)

1469. BP Ltd. (Yeomans, B.), DE-OS 2 111 669 (1971)

1470. Gorgues, A.: Bull. Soc. Chim. France *1974*, (3, 4), 529

1471. Gregorio, G., Pregaglia, G.F., Ugo, R.: J. Organomet. Chem. *37*, 385 (1972)

1472. Henkel & Cie. GmbH, DE-OS 2 634 676 (1977)

1473. Cornils, B., Zilly, E. in [57], p. 84 f

1474. Cornils, B., Wiebus, E., Kascha, W.: unpublished

1475. Wolf, F., Losse, A., Mücke, J.: J. Prakt. Chem. *313*, 137, 145 (1971)

1476. Matsuda, A., Uchida, H.: Tokyo Kogyo Shikensho Hokoku *57*, (1), 50 (1962); C.A. *62*, 7625 (1965)

1477. Belafi-Rethy, K. et al.: Erdöl und Kohle *24*, (1), 19 (1971)

1478. Prokopenko, N.A. et al.: Zh. Anal. Khim. *1978*, 5, 1196, 1228; C.A. *89*, 52 926 n, 139 994 h and *89*, 139 997 m (1978)

1479. Lechenko, G.T. et al.: Gazor. Khromatogr. *4*, 110 (1966); C.A. *68*, 26 757 f (1968)
1480. Cornils, B., Ruprecht, P.: Ruhrchemie A. G., unpublished
1481. Toa Nenryo Kogyo K. K. (Wada, S., Takeuchi, J., Yoda, M.), JP 10 127/70 (1970); C.A. *73*, 44 899 m (1970)
1482. Cherntsov, O.M. et al.: Neftekhimiya *14*, (1), 56 (1974)
1483. Adkins, H., Krsek, G.: J. Amer. Chem. Soc. *71*, 3051 (1949)
1484. Alekseeva, K.A. et al.: SU 406 823 (1973); C.A. *75*, 82 068 s (1974)
1485. Koshechkina, L.P., Mel'nichenko, I.V.: Ukr. Khim. Zh. *1974*, 40 (2), 172; C.A. *80*, 132 739 j (1974)
1486. Eastman Kodak Co. (McCollum, A.W., Hagemeyer, H.J.), US 4 017 537 (1977)
1487. Standard Oil Co. of Indiana (Field, E., Hill, B.L.), US 2 587 576 (1952)
1488. Monsanto Co. (Null, H.R., Bowe, L.E.), FR 1 394 095 (1965); C.A. *63*, 5530 c (1965)
1489. Esso Res. Engng. Co. (Jaros, S.E., Roming, C.), US 3 119 876 (1964)
1490. Esso Res. Engng. Co. (Robbins, L.V., Mertzweiller, J.K.), US 3 118 954 (1964)
1491. Exxon Res. Engng. Co. (McVicker, G.B.), US 3 946 082 (1976)
1492. Toa Noryo Kogyo K. K. (Usami, M., Nishimura, K., Fukushi, S.), JP 7 526/72 (1972) and JA 11 410/72 (1972); C.A. *77*, 19 182 x (1972)
1493. Tonen Sekiya Kagaku K. K., JP 24 404/68 (1968)
1494. Esso Res. Engng. Co. (Mason, R.B.), US 2 811 567 (1957)
1495. Esso Res. Engng. Co. (Cull, N.L., Aldridge, C.L.), US 3 454 649 (1963)
1496. Agency of Ind. Sci. Tech., JP 19 347/64 (1964)
1497. Esso Res. Engng. Co. (Mertzweiller, J.K., Watts, R.N.), DE-AS 1 114 469 (1961)
1498. Macho, V.: Chem. Zvesti. *17*, (8), 525 (1963); C.A. *60*, 10 534 e (1964)
1499. Esso Res. Engng. Co., GB 867 799 (1961)
1500. Esso Res. Engng. Co. (Aldridge, C.L., Cull, N.L.), DE 1 153 006 (1963)
1501. Macho, V.: Ropa Uhlie *6* (10), 297 (1964); C.A. *62*, 8996 b (1965)
1502. Goetz, R.W., Orchin, M.: J. Org. Chem. *27*, 3698 (1962)
1503. Roelen, O., Ziegler, K. (ed.): Naturforschung und Medizin in Deutschland, Vol. 36/Part I, E. Dietrich-Verlag, Wiesbaden 1948
1504. Wender, I., Levine, R., Orchin, M.: J. Amer. Chem. Soc. *72*, 4375 (1950)
1505. Berty, I., Marko, L.: Acta Chimica *1952*, 177
1506. Heil, B., Marko, L.: Chem. Ber. *99*, 1086 (1966)
1507. Mitsubishi Chem. Ind. Co., JP 268/65 (1965), JP 2445/69 (1969)
1508. Heil, B., Marko, L.: Acta Chim. Acad. Sci. Hung. *55*, 107 (1968); C.A. *68*, 77 430 b (1968)
1509. Falbe, J., Huppes, N.: Brennstoff-Chem. *48*, (6), 182 (1967)
1510. Berty, J., Marko, L.: Acta Chim. Acad. Sci. Hung. *3*, 177 (1953), C.A. *48*, 11 294 b (1954)
1511. Esso Res. Eng. Co. (Mertzweiller, J.K., Watts, R.N.), US 3 182 090 (1965)
1512. California Res. Corp. (Bendit, P.C., Spence, J.A.), US 2 647 149 (1953)
1513. Hasegawa Co. Ltd. (Kumanodani, J. et al.), JP 1380/73 (1973)
1514. DuPont & Co. (Gresham, W.F.), GB 638 754 (1950)
1515. Anglo Iranian Oil Co. Ltd. (Birch, S.F., Habeshaw, J., Rae, R.W.), GB 702 191 (1954)
1516. [25], p. 1129
1517. Chen, W.: Hua Hsueh *1967*, (3), 75; Chem. Titles *1968* (20), 128
1518. Standard Oil Dev. Co., GB 668 963 (1952)
1519. Anglo Iranian Oil Co. Ltd., GB 702 217 (1954)
1520. Berty, J., Marko, L.: Chem. Tech. *9*, (5), 283 (1957)
1521. Marko, L., Szabo, P.: Chem. Techn. *13*, (7/8), 482 (1961)
1522. Dawidoff, W.: Chem. Tech. *11*, 431 (1959) and *12*, 414 (1960): C.A. *55*, 7337, 10 862 (1961)
1523. Ugine Kuhlmann (Guerant, A., Mercier, S.), DE-OS 2 519 011 (1975)
1524. Shell Dev. Co., GB 942 435 (1962)
1525. Imyanitov, N.S., Volkov, V.A.: Zh. Prikl. Khim. *47*, 2095 (1974); C.A. *82*, 3724 t (1975)
1526. BASF A. G. (Kniese, W.), DE-OS 1 793 398 (1972), NL 69-13 649 (1969)
1527. Strohmeier, W., Steigerwald, H.: J. Organomet. Chem. *129*, (2), C 43 (1977)
1528. Ucciani, E., Lai, R., Tanguy, L.: C. R. Hebd. Séances Acad. Sci., Ser. C *1975*, 281 (21), 877; *1976*, 283 (1), 17

1529. Cecchi, G., Ucciani, E.: Rev. Fr. Corps. Gras. *24*, (6), 321 (1977)
1530. Dawydoff, W.: Katal. Reakts. Zhidk. Faze, Tr. Vses. Konf., 2nd Alma-Ata Kaz. SSSR *1966*, 441; C.A. *69*, 35 145 c (1968)
1531. Wakamatsu, H., Furukawa, J., Yamakami, N.: Bull. Chem. Soc. Jap. *44*, (1), 288 (1971)
1532. Krinkin, D.P., Rudkowskij, D.M., Trifel, A.G.: Chim. Promysh. *1966*, (7), 501
1533. Liquichimica Italiana SpA, Data sheet SP 197-198 (1974)
1534. Anonym, Eur. Chem. News of Oct. 13, 1972
1535. Agency of Ind. Sci. Technol. (Imamura, J., Kubota, K., Kobiyama, T.), JP 29 709/73 (1973)
1536. Kryshchenko, K.I., Khim. Promysl. *1969*, (7), 496
1537. Cheryak, B.I., Andrianova, L.A.: Neftekhimiya *14*, 97 (1974); Ukr. Khim. Zh. *1973*, 199
1538. Toyo Rayon Ltd., JP 6201/68 (1968)
1539. Asahi Chem. Ind. Co. (Imamura, J., Wakasa, R., Kataoka, K.), JP 43 926/74 (1974); C.A. *82*, 171 682 w (1975)
1540. Weissermel, K., Arpe, H.J.: Industrielle Organische Chemie, Weinheim-New York: Verlag Chemie (1976), p. 109 f, 166 f
1541. Cornils, B., Zilly, E.: in [57], p. 49
1542. Gavrilova, V.M., Katsnel'son, M.G.: SU 566 826 (1977), C.A. *87*, 151 715 j (1977)
1543. Eastman Kodak Co. (Hagemeyer, H.J., Wright, H.N.), US 3 081 344 (1961)
1544. Gankin, V.Yu., Rudkowskij, D.M., Rybakov, V.A.: Zh. Prikl. Khim. *1971*, 1347
1545. BASF A.G. (Merger, F.), DE-OS 2 003 600 (1971), DE-OS 1 643 727 (1971)
1546. BP Ltd. (Yeomans, B.), GB 1 290 094 (1972)
1547. Texaco Co. (Chafetz, H.), US 3 201 478 (1964); C.A. *63*, 11 360 a (1965)
1548. Eastman Kodak Co. (Hagemeyer, H.J.), US 2 829 169 (1960); C. *1960*, 6321
1549. Cornils, B., Feichtinger, H.: Chem.-Ztg. *100*, 504 (1976)
1550. Ruhrchemie A.G. (Tummes, H. et al.) DE-OS 2 045 669 (1970), NL 71-12 570 (1971)
1551. Ruhrchemie A.G. (Rottig, W. et al.), DE-AS 2 054 601 (1972) and 1 793 512 (1972)
1552. BASF A.G. (Merger, F.), DE-OS 1 957 591 (1971), DE-OS 1 958 463 (1971), DE-OS 2 000 699 (1971); FR 2 019 028 (1970)
1553. Eastman Kodak Co. (Hagemeier, H.J.), US 2 811 562 (1957); (Wright, R.L.), US 4 021 496 (1977); (Palmer, B.W., Bondurant, D.L., Hagemeyer, H.J.), US 3 975 450 (1976) and (James, H.B.), US 3 920 760 (1975)
1554. Hoechst AG (Jacobsen, G., Freudenberger, D., Fernholz, H.), DE-OS 1 668 117 (1971) and 1 768 274 (1971)
1555. Nippon Gasu Kagaku Kogyo Co., JP 26 283/68 (1968)
1556. Arpe, H.J.: Chem.-Ztg. *97*, 53 (1973)
1557. Vysotskii, M.P. et al.: Khim. Promysl. *1977* (2), 98; C.A. *86*, 170 776 a (1977), Vses. Soveseh. Sic. Zhirozamen *1965*, 217; C.A. *67*, 99 567 h (1967)
1558. Mitsubishi Gas Chem. Co. Inc. (Ida, N., Okawa, T., Obata, Y.), JP 138 607/76 (1976); C.A. *87*, 5377 j (1977)
1559. Bathory, J., Repasy, O.: Petrochemia *16*, 94 (1976); C.A. *86*, 155 144 z (1977)
1560. Kyowa Petrochem Co. Ltd., JP 10 015/73 (1973)
1561. Kusy, V.: Sb. Pr. Vyzk. Chem., Vyuziti Uhli Dehtu Ropy *1972*, (12), 145; C.A. *78*, 136 858 d (1973)
1562. Jelinck, J.: CS 161 450 (1975); C.A. *85*, 93 805 j (1976)
1563. CdF Chimie (Couderc, P., Hilmoine, S.), DE-OS 2 547 540 (1976)
1564. VEB Jenapharm (Schmidt, J. et al.), GB 1 345 459 (1974)
1565. Chem. Age. *102*, No. 2709, 19 (1971); Eur. Chem. *1971*, No. 12, 5
1566. Chem. Mark. Rep. Sept. 11, 1972
1567. Chem. Age June 22, 1973
1568. Hoechst A.G., Technical Data Sheet "NPG"
1569. BASF A.G. (Berger, F., Segnitz, A.), DE-OS 2 722 957 (1978)
1570. Dynamit A.G. (Zoche, G., Richtzenhain, H.), DE-OS 2 003 238 (1971)
1571. BASF A.G., NL 68-12 005 (1968)
1572. Frank, H.P., Krzemicki, K.: Monatsh. Chem. *95*, 410 (1964)
1573. Japan Gas Chem. Co. (Huang, C.Y., Shimono, Y.), JP 24 888/68 (1966)

1574. Shell Int. Res. Mig. (Radder, J.), GB 1 140 928 (1967)

1575. Hoechst A. G. (Arpe, H. J.), DE-AS 1 941 184 (1977)

1576. Rhône-Progil S. A. (Gobron, G., Falize, C., Dufour, H.), US 3 987 103 (1968)

1577. Chemische Werke Hüls A. G. (Obenaus, F., Droste, W.), DE-OS 2 415 151 (1975)

1578. Droste, W., Obenaus, F.: Erdöl und Kohle-Compendium 1974/1975, Vol. I, p. 478; Erdöl und Kohle 28, 291 (1975)

1579. Chemtech 7 (6) 365 (1977)

1580. Daicel Ltd. (Sakakibara, K., Yasuda, K.), JP 6292/74 (1974); C.A. 82, 86 085 j (1975), NL 71-16 058 (1971)

1581. Akimov, S.A. et al.: SU 583 117 (1977), C.A. 88, 61 988 s (1978)

1582. Toa Nenryo Kogyo KK, JP 48 611/77 (1977); C.A. 87, 84 531 f (1977)

1583. Lange, S.A., Evdokimova, Z.A., Levchenko, N.G.: Neftekhimiya 16, 818 (1976); C.A. 86, 120 744 t (1977)

1584. Baltz, H. et al.: DD 76 491 (1970)

1585. Asahi Kasei Kogyo KK (Nakajima, H., Chono M.), DE-OS 2 129 920 (1971) and 2 208 580 (1972)

1586. Mitsubishi Chem. Co. Ltd. (Kita, T., Ishii, C.), JP 14 085/72 (1972); C.A. 77, 47 883 v (1972); (Otaki, T., Sakurada, H., Otake, M.), JP 78 120/73; C.A. 80, 59 478 s (1974); DE-OS 2 633 593 (1978)

1587. Toa Gosei Chem. Ind. Co. Ltd. (Inoue, H., Mizutani, K., Ito, H.), JP 19 614/73 (1973); C.A. 79, 91 606 x (1973)

1588. Szelejewski, W. et al.: DE-OS 2 432 527 (1975); C.A. 82, 156 983 b (1975)

1589. BASF A. G. (Bressel, U., Fuchs, W., Platz, R.), DE-OS 2 064 576 (1972) and 2 144 148 (1973)

1590. Ohtsuka, M., Kikuchi, Y., Ikawa, T.: Kogyo Kagaku Zasshi 74, (2), 191 (1971)

1591. Tihanyi, B.: DE-OS 2 504 510 (1976)

1592. Nissan Chem. Ind. Ltd. (Ikawa, T., Murata, A.), JP 1 644/73 (1973)

1593. Akimoto, M., Ichikawa, K., Echigoya, E.: Nippon Kagaku Kaishi 1977, (3), 320; C.A. 86, 178 070 m (1977)

1594. Gulf Research Co. (Coyne, D.M., Havens, R.H.), US 3 391 193 (1968); C.A. 69, 95 965 e (1968)

1595. Montecatini SpA (Petrini, G., Moreschini, L., Marciandi, F.), NL 72-08009 (1972), IT 930 657 (1972)

1596. Eastman Kodak Co. (Watkins, W.C.), DE-OS 2 624 555 (1976)

1597. Melle-Bezons S. A. (Bouniot, A.), DE-OS 2 222 669 (1973)

1598. Droste, W., Obenaus, F.: Monatsh. Chem. 104, (2), 485 (1973)

1599. Becker, B., Fritz, H.P.: Chem. Ber. 108, 3292 (1975)

1600. Moehrle, H., Schnaedelbach, D.: Pharmazie 1975, 352, 699

1601. Rohm and Haas Co. (Lewis, S.N., Levy, J.F.), US 3 937 716 (1976)

1602. Dilling, W.L.: J. Org. Chem. 37, 4159 (1972)

1603. Feichtinger, H., Payer, W., Cornils, B.: Chem. Ber. 111, 1721 (1978)

1604. Protekhin, A.A., Barkova, T.F.: Zh. Org. Khim. 8, 657 (1972);

1605. BASF A. G. (Merger, F.), DE-OS 2 000 511 (1971)

1606. Smeykal, K. et al.: J. Prakt. Chem. 4, (29), 250 (1965)

1607. Braune, W., Bartel, D., Schmidt, D.: DD 41 651 (1965), C. 1966, 35-2708

1608. Novikov, L.S., Tishchenko, I.G.: Vestsi Akad. Navuk-Belarus. SSR, Ser. Khim. Navuk 1970, 79; C.A. 74, 42 079 r (1971)

1609. Beal, R.E. et al.: J. Amer. Oil Chem. Soc. 44, 55 (1967)

1610. Komori, S., Hattori, S., Oshiro, Y.: Kogyo Kagaku Zasshi 70, 1332 (1967); C.A. 68, 48 978 f (1968)

1611. Ohshiro, Y., Tsuruda, S., Hattori, S., Komori, S.: Kogyo Kagaku Zasshi 70, 1329 (1967); C.A. 68, 48 981 b (1968)

1612. Dow Chem. Co., (Dalman, D.A.), US 3 801 645 (1974)

1613. Kiragawa, Y., Takata, Y., Kataoka, K.: Hokkaido Daigaku Kogakubu Kenkyu Hokuku 1973, 155; C.A. 80, 47 401 y (1974)

1614. Nippon Synthetic Chem. Ind. Co. Ltd. (Nakajima, K., Sato, T.), JP 39 966/74 (1974); C.A. *82*, 139 433 e (1975)
1615. Mitsubishi Chem. Ind. Co. Ltd. (Yoshida, S.), JP 18 567/71 (1971), 13 683/72 (1972)
1616. Mitsui Toatsu Chem. Co. Ltd. (Takahashi, T. et al.), JP 7 329/72 (1972)
1617. Farbwerke Hoechst A. G. (Schafer, H.K., Kohlhaas, R.), US 3 441 539 (1969); C.A. *71*, 21 711 p (1969)
1618. Alekseeva, K.A., Gordina, N.Y., Rudkowskij, D.M., Trifel, A.G.: in Rudkowskij, D.M.: Carbonylierung ungesättigter Kohlenwasserstoffe, Edition Chimia, Leningrad 1968, p. 80 f
1619. Matsuda, A., Uchida, H.: C. *1964*, 38-2369
1620. Baltz, H. et al.: J. prakt. Chem. 4, *29*, 250 (1965)
1621. Del'nik, V.B. et al.: Poluchemie Maslyan. Al'degidov i Butilovykh Spirtov Oksosintezom *1977*, 134; C.A. *89*, 179 479 a (1978)
1622. Alekseeva, K.A. et al.: Poluchemie Maslyan. Al'degidov i Butil. Spirtov Oksosint. *1977*, 66; C.A. *89*, 179 480 u (1978)
1623. Poredda, S. et al.: see [1621], p. 19; C.A. 179 482 w (1978)
1624. Kulik, V.G. et al.: see [1621], p. 150; C.A. 179 481 v (1978)
1625. Alekseeva, K.A. et al.: see [1621], p. 76; C.A. *89*, 179 483 x (1978)
1626. Alekseeva, K.A. et al.: see [1621], p. 119; C.A. *89*, 179 492 z (1978)
1627. Gankin, V.Yu., Rybakov, V.A., Novikov, V.P.: Kinet. Katal. *19*, 1085 (1978); C.A. *89*, 179 277 h (1978)
1628. Johnson, Matthey & Co., JP 53 293/75 (1975); C.A. *85*, 176 838 y (1976)
1629. Sanger, A.R.: Pepr. Canad. Symp. Catal. 5th, *1977*, 281; C.A. *89*, 179 250 u (1978)
1630. Parizek, M., Macek, V., Svajgl, O.: Sb. Pr. Vyzk. Chem. Vyuziti Uhli, Dehtu Ropy *1976*, (14), 231; C.A. *85*, 176 768 n (1976)
1631. Hagemeyer, H.J., de Croes, G.C.: The Chemistry of Isobutyraldehyde and its Derivatives, Tennessee Eastman Comp. Publ., Kingsport (Tenn.), 1953
1632. Gurevich, G.S., Levin, S.Z., Sedova, I.G.: Protsessy Katal. Gidrirovaniya, vses. Nauch. Issled. Inst. Neftkhim. Protsessov *1966*, 49, 61; C.A. *67*, 43 344, 53 345 (1967)
1633. The Distillers Co., FR 1 349 816 (1963)
1634. Mistrik, E.J.: Chem. Prumysl. *13*, 622 (1963)
1635. Orlicek, A.: AT 277 949 (1967), DE-OS 1 913 198 (1970)
1636. Ruhrchemie A. G. (Falbe, J., Hahn, H.D.), FR 2 023 625 (1969), DE 1 767 281 (1968)
1637. Ruhrchemie A. G. (Falbe, J., Hahn, H.D., Tummes, H.), DE 1 917 244 (1969)
1638. Falbe, J., Hahn, H.D.: Chem.-Ztg. *96*, 164 (1972)
1639. Anonym, Eur. Chem. News *19* of May 21, 1971
1640. Ruhrchemie A. G., NL 70-05825 (1969)
1641. Anonym, Inf. Chim. *1972*, No. 113, p. 47
1642. Ruhrchemie A. G., Hydrocarbon Proc. *50*, (11), 166 (1971)
1643. Falbe, J.: Angew. Chem. Int. Ed. Engl. *11*, (2), 155 (1972)
1644. Toa Nenryo Kogyo K. K., JP 48 603/77 (1977); C.A. *87*, 84 496 y (1977)
1645. Fujimoto, K., Kato, Y.: JP 122 308 (1977); C.A. *88*, 61 955 d (1978)
1646. Ploder, W.H.: J. Catalysis *23*, 358 (1971)
1647. Cornils, B., Tummes, H., Noeske, H.: Ruhrchemie A. G., unpublished
1648. Valitov, N.K., Nosal, G.I.: Neftepererab. Neftekhim *1969*, (8), 31
1649. BASF A. G. (Neubauer, D., Bach, J.M., Garrido, R.), DE 2 544 570 (1977)
1650. BASF A. G. (Hagen, W., Hohenschutz, H., Strohmeyer, M.), DE-AS 2 460 984 (1974); (Kerber, H., Hohenschutz, H.), DE-AS 1 301 806 (1969)
1651. BASF A. G., GB 1 240 847 (1971)
1652. Standard Oil Co. (Smith, W.M.), US 2 626 284 (1953)
1653. Mitsubishi Chem. Ind. Co. Ltd. (Suzuki, Y. et al.), JP 96 512/73 (1973); C.A. *75*, 95 278 k (1974)
1654. Gates, B.C., Schwab, G.M.: J. Catalysis. *15*, 430 (1969)
1655. Noto, Y. et al.: Bull. Chem. Soc. Japan *40*, (10), 2459 (1967); C. 1968, 44-2357: Trans. Faraday Soc. *63*, No. 540, 3081 (1967)
1656. VEB Leuna (Balts, H. et al.), DD 92 440 (1972), 109 372 (1974) and 122 180 (1976); DE-OS 2 246 718 (1973)

1657. Levin, S.Z., Gurevich, G.S., Sedova, I.G.: SU 364 587 (1972)
1658. Chemische Werke Hüls A. G. (Reich, M.), DE-AS 1 805 403 (1973)
1659. BP Ltd. (Habeshaw, J., Rae, R.W.), CA 607 753 (1960)
1660. BASF A. G., BE 820 311 (1975), DE-OS 2 460 784 (1974)
1661. Usines de Melle (Alheritiere, L.), FR 1 480 337 (1967); C.A. *68*, 12 481 z (1968)
1662. Oberender, H., Stoss, W., Voigt, D., Zimny, H.W.: DD 103 884 (1974)
1663. Eastman Kodak Co. (Johnson, S., Woude, J.C.), DE-AS 1 300 541 (1969)
1664. Petrochem. Process, SU 487 056 (1974)
1665. Usines de Melle (Mercier, J.), FR 1 524 289 (1968); C.A. *71*, 60 713 k (1969)
1666. Kasano, K., Hashimoto, T., Takahashi, T.: J. Soc. Org. Synth. Chem. Japan *23*, 883 (1965);
     C. *1968*, 8-0941
1667. Mistrik, E.J.: Chem. Technol. *19*, 154 (1967)
1668. Hoechst A. G. (Rehn, K., Theilig, G.), DE-AS 1 108 195 (1961)
1669. Montecatini Edison SpA (Ferrari, G., Simonazzi, T.), DE-OS 2 131 628 (1972), BE 769 317
     (1970)
1670. Esso Res. Eng. Co. (Cull, N.L., Mertzweiller, J.K., Tenney, H.M. et al.), US 3 284 510 (1966),
     3 314 911 (1967), 3 330 875 (1967), 3 334 076 (1967), 3 337 489 (1967), 3 346 664
     (1967), 3 365 411 (1968), 3 383 426 (1968), 3 409 648 (1968), 3 425 895 (1969),
     3 454 649 (1969) and 3 513 130 (1970); GB 1 072 796 (1967)
1671. Ramp, F.L., Dewitt, E.J., Trapaso, L.E.: J. Poly. Sci. A-1, *4* (9), 2267 (1966)
1672. Shell Int. Res. Mij., GB 1 035 011 (1966)
1673. Wender, I. et al.: J. Amer. Chem. Soc. *78*, 5401 (1956)
1674. Esso Res. Eng. Co., GB 1 072 796 (1967)
1675. Esso Res. Eng. Co. (Mertzweiller, J.K., Tenney, H.M.), US 3 334 076 (1967)
1676. Guyer, P., Bosshard, E.: Chimia *18*, 131 (1964)
1677. Bird, C.W.: Chem. Rev. *62*, 290 (1962)
1678. Asinger, F.: Chemie und Technologie der Monoolefine, Berlin: Akademie-Verlag 1957,
     p. 650 f
1679. Adkins, H., Krsek, G.: J. Amer. Chem. Soc. *70*, 383 (1948)
1680. Imyanitov, N.S., Rudkowskij, D.M.: Petrol. Chem. (USSR) *3*, (1), 91 (1964): C.A. *60*,
     9072 (1964)
1681. Pino, P.: Gazz. Chim. Ital. *81*, 625 (1951)
1682. DuPont, GB 614 010 (1948)
1683. Eastman Kodak Co. (Hagemeyer, H.J., Hull, D.C.), US 2 694 734 and 2 694 735 (1954)
1684. Montecatini SpA (Natta, G., Ercoli, R., Castellano, S.), IT 516 716 (1955)
1685. Mitsubishi Chem. Ind. (Niwa, A. et al.), JP 1107 (1957); C.A. *52*, 4680 (1958)
1686. Pino, P., Paleari, C.: Gazz. Chim. Ital. *81*, 646 (1951)
1687. BASF A. G. (Reppe, W., Friedrich, H.), DE 897 403 (1953)
1688. El Daoushy, M.A.F.: Dissertation, Technol. Hochsch. Aachen 1964
1689. ICI Ltd. (Taylor, A.W.C.), GB 798 541 (1958)
1690. Anglo-Iranian Oil Ltd. (Habeshaw, J., Thornes, L.S.), GB 702 195 (1954)
1691. ICI Ltd. (Taylor, A.W.C., Lamb, S.A.), GB 684 673 (1952)
1692. Mitsubishi Chem. Ind. (Osumi, Y. et al.), JP 22 735/65 (1965); C.A. *64*, 4943 (1966)
1693. Natta, G., Beati, F.: GB 646 424 (1950)
1694. Falbe, J., Huppes, N., Korte, F.: Brennstoff-Chem. *47*, 207 (1966); *48*, 24 (1967); BE 33 538
     and 33 539 (1966)
1695. Distillers Ltd. (Millidge, A.F.), FR 1 411 602 (1965)
1696. Shell Int. Res., GB 995 459 (1965)
1697. Diamond Alkali Co. (Eisenmann, J.L., Yamartino, R.L.), GB 941 996 (1963)
1698. Morikawa, M.: Bull. Chem. Soc. Japan *37*, 379 (1964)
1699. Toyo Rayon Co., JP 8929/66 (1966)
1700. BASF A. G. (Nienburg, H.J., Kummer, R.), DE 2 314 694 (1973)
1701. Kururay Co. (Tanomura, M.), JP 39 415/73 (1973); C.A. *79*, 321 (1973)
1702. Shell Int. Res. (DeJong, A.J.), DE-OS 2 719 735 (1977)

1703. Mitsubishi Chem. Ind. (Yamaguchi, M., Nakajima, C.), JP 17 251/73 (1973), C.A. *79*, 91 591 (1973)
1704. Inventa A. G. (Torre, H.D., Jäger, P.), DE-OS 1 668 255 (1966)
1705. Esso Res. Eng. Co. (Cull, N.L., Pino, L.A.), DE-OS 1 518 765 (1965)
1706. Inventa A. G., NL 298 834 (1964), FR 1 371 085 (1964)
1707. Kogami, K.: Japan Yakagaku *22*, (6), 316 (1973)
1708. Ruhrchemie A. G. (Büchner, K.), GB 765 742 (1957)
1709. Matsubara, M.: Hokkaido Kogyo Shik. Hok. *1968*, 74; C.A. *74*, 12 667 p (1971)
1710. Komora, L., Macho, V.: Chem. Prumysl. *19*, 359 (1969)
1711. Adkins, H., Williams, J.L.R.: J. Org. Chem. *17*, 980 (1952)
1712. BASF A. G. (Nienburg, H.J.), DE 800 400 (1950)
1713. Holm, M.M. et al.: FIAT-Report 1000, p. 31
1714. Inventa A. G., GB 1 007 627 (1965)
1715. Tummes, H., Meis, J.: Ruhrchemie AG, unpublished
1716. DuPont (Brooks, R.E.), US 2 517 383 (1950)
1717. Whitman, G.M.: US 2 462 448 (1946)
1718. Fell, B., Rupilius, W.: Tetrahedron Lett. *1968*, 3261 and *1969*, 2721
1719. Husebye, S., Jonassen, H.B.: Acta Chem. Scand. *18*, 1581 (1964)
1720. Fell, B., Boll, W., Hagen, J.: Chem.-Zeitg. *99*, 485 (1975)
1721. Greenfield, H., Wotiz, J.H., Wender, I.: J. Org. Chem. *22*, 542 (1957)
1722. Botteghi, C., Salomon, C.: Tetrahedron Lett. *1974*, 4285
1723. Jardine, F.H. et al.: Chem. Ind. *1965*, 560
1724. Orchin, M., Wender, I.: Catalysis, Reinhold Publ. Corp. 1957, Vol. I, p. 1
1725. Falbe, J., Schulze-Steinen, M.J., Korte, F.: Chem. Ber. *98*, 886 (1965)
1726. Macho, V.: Chem. Zvesti *21*, 170 (1967)
1727. Fell, B., private communication
1728. Botteghi, C. et al.: J. Org. Chem. *37*, 1835 (1972), J. prakt. Chem. *314*, 840, (1972)
1729. DuPont (Copelin, H.B.), CA 983 942 (1972)
1730. Shell Int. Res. (Falbe, J.), DE-AS 1 227 030 (1961)
1731. Goetz, R.W., Orchin, M.: J. Amer. Chem. Soc. *85*, 2782 (1963)
1732. Inf. Chim. No. *178*, 185 (1978), Spécial Juin 1978
1733. Boll, W.: Dissertation, Techn. Hochsch. Aachen, 1973
1734. Diels, O., Alder, K.: Liebig Ann. Chem. 460, 98 (1928)
1735. Stockhausen, F.: Dissertation, Köln 1959
1736. Degussa A. G. (Fahnenstiel, R., Weigert, W.), DE-AS 1 286 046 (1964)
1737. Anglo-Iranian Oil Co. (Habeshaw, J., Rae, R.W.), GB 702 201 (1954)
1738. Cornils, B., Weber, J., Bahrmann, H.: Ruhrchemie A. G., unpublished
1739. Wender, I., Sternberg, H.W., Orchin, M.: Catalysis, Reinhold Publ., New York 1957, Vol. V, p. 73
1740. Ajinomoto Co. (Iwanaga, R., Mori, Y., Yoslinde, T.), JP 8177/57 (1957); C.A. *52*, 14 661 (1958)
1741. Pino, P., Paleari, L.: in Ullmann, Encyklopädie der technischen Chemie, Urban & Schwarzenberg, München 1962, Vol. 13, p. 65
1742. Eastman Kodak Co. (Hagemeyer, H.J., Hull, D.C.), US 2 610 203 (1952)
1743. Stolle, M., Bolle, P.: Helv. Chim. Acta *21*, 1551 (1938)
1744. Toa Gosei, JP 21 603/66 (1966)
1745. Matsuda, A., JP 85 530/73 (1973); C.A. *80*, 59 502 v (1974)
1746. Bayer A. G. (Rasp, C., Grolig, J.), DE-AS 2 430 038 (1976)
1747. General Electric Co. (Corn, J.E., Swiger, R.T., Webb, J.L.), DE-OS 2 425 843 (1975)
1748. General Electric Co. (Merritt, W.D.), DE-OS 2 425 761 (1975)
1749. General Electric Co. (Smith, W.H.), DE-OS 2 425 654 (1975), US 3 941 851 (1976) and US 4 035 408 (1977)
1750. Laine, R.M.: J. Amer. Chem. Soc. *100*, 6451 (1978)
1751. Ziegenhagen, J.: Diplomarbeit, Techn. Hochschule, Aachen 1972
1752. Ajinomoto Co. (Takesada, M.), DE-AS 1 225 627 (1963)

1753. Fell, B., Barl, M.: J. Mol. Catalysis 2, 301 (1977)
1754. General Electric Co. (Smith, W.E.), DE-OS 2 425 653 (1974), 2 425 844 (1974) and US 4 039 592 (1977)
1755. Bayer A. G. (Scharfe, G., Grolig, J., Rasp, C.), DE-AS 2 430 022 (1975)
1756. General Electric Co. (Aycock, D.F., Sliva, D.E.), DE-OS 2 426 650 and 2 426 684 (1974)
1757. BASF A. G. (Kummer, R.), DE-OS 2 451 473 (1976)
1758. Bayer A. G. (Rasp, C., Massoubra, P., Scharfe, G.), DE-OS 2 430 037 (1976)
1759. Falbe, J., Korte, F.: Chem. Ber. 97, 1104 (1964)
1760. Rosenthal, A., Abson, D.: Can. J. Chem. 42, 1811 (1964)
1761. Rosenthal, A., Koch, J.: Can. J. Chem. 42, 2025 (1964)
1762. Rosenthal, A., Abson, D.: J. Amer. Chem. Soc. 86, 5356 (1964)
1763. Shell Int. Res. (Falbe, J., Korte, F.), DE-AS 1 227 030 (1961)
1764. Stresinka, J., Marko, M., Macho, V.: Chem. Zvesti 22, 844 (1968); C.A. 70, 77 517 b (1969) Chem. Zvesti 22, 263 (1968) and 22, 656 (1968)
1765. Anglo- Iranian Oil Co. (Habeshaw, J., Geach, C.J.), GB 702 206 (1954)
1766. BASF A. G. (Himmele, W., Aquila, W.), DE-OS 2 111 116 (1971)
1767. Eastman Kodak Co. (Snapp, T.C., Blood, A.E.), US 3 888 880 (1973)
1768. DuPont (Bhatia, K.K., Cumbo, C.C.), DE-OS 2 714 237 (1977)
1769. DuPont (Bhatia, K.K., Cumbo, C.C.), US 4 024 197 (1977)
1770. DuPont (Bhatia, K.K., Cumbo, C.C.), DE-OS 2 523 889 (1976)
1771. Eastman Kodak Co. (Hagemeyer, H.J.), DE-OS 1 443 807 (1964)
1772. Rosenthal, A.: Can. J. Chem. 45, 1525 (1967)
1773. Malda, J., Yoshida, R.: Bull. Chem. Soc. Japan 41, 2969 (1968)
1774. Marko, M. et al.: CS 110 106 (1964), C.A. 61, 3025 (1964)
1775. Marko, M.: CS 109 561 (1964), C.A. 61, 13 138 (1964)
1776. Lonza AG, NL 6 406 299 (1964), DE-AS 1 220 405 (1966)
1777. Gut, G., Makhzangi, M.H., Guyer, A.: Helv. Chim. Acta 48, 1151 (1965)
1778. Rudkowskij, D.M., Imyanitov, N.S., Gankin, V.Yu.: Tr., Vses. Nauchn.-Issled. Inst. Neftekhim-Protessov 1960, (2), 121; C.A. 57, 10 989 (1962)
1779. Hoechst A. G. (Bestian, H., Rehn, K.), NL 1 007 771 (1971)
1780. Ajinomoto Co., GB 828 946 (1960)
1781. Ono, Y., Sato, S., Takesada, M., Wakamatsu, H.: Chem. Commun. 1970, 1255
1782. Sato, S., Ono, Y., Tatsuma, S., Wakamatsu, H.: Nippon Kagaku Zasshi 92, 178 (1971)
1783. Kurokawa, H.: Kogyo Kag. Zasshi 70, 1168 (1967)
1784. Kato, J.: US 2 978 481 (1961)
1785. Ajinomoto Co., JP 58 677/67 (1965) and 29 936/68 (1968)
1786. Sato, S.: Kogyo Kagaku Zasshi 74, 1830 (1971)
1787. Kodoma, Sh., Taniguchi, J., Takegami, Y.: JP 7 770 (1957); C.A. 52, 13 777 (1958)
1788. Kato, J., Wakamatsu, H., Komatsu, T.: Kogyo Kagaku Zasshi 64, 2139 (1961); C.A. 57, 2064 (1962)
1789. Ajinomoto Co. (Kato, J., Wakamatsu, H., Ishihara, H.), JP 2 574 (1961); C.A. 56, 9977 (1962), GB 838 737 (1960), US 2 978 481 (1961)
1790. Kato, J., Ito, T., Yabo, Y.: Kogyo Kagaku Zasshi 65, 184 (1962), C.A. 58, 4420 (1963)
1791. Houda, M., Koga, T., Noyori, G.: Kogyo Kagaku Zasshi 70, 1346 (1967)
1792. Noguchi Res. Foundation, JP 21 287/68 (1968)
1793. Koga, T., Noyori, G.: Kogyo Kagaku Zasshi 70, 1172 (1967)
1794. Ajinomoto Co., FR 1 568 444 (1969)
1795. Ajinomoto Co., GB 1 198 028 (1968)
1796. Phillips Petroleum Co. (Kuper, D.G.), US 3 520 914 (1970)
1797. Ajinomoto Co., DE 1 196 630 (1960)
1798. Houda, M. et al.: Kogyo Kagaki Zasshi 70, 1350 (1967)
1799. DuPont (Schreyer, R.C.), US 2 564 131 (1951)
1800. BASF A. G. (Nienburg, H.J., Kummer, R.), DE 2 132 548 (1971)
1801. Sato, S.: J. Chem. Soc. Japan 90, 404 (1969)
1802. Kurokawa, H., Koga, T., Noyori, G.: Kogyo Kagaku Zasshi 70, 1355 (1967)

1803. Shell Oil Co. (Dewhirst, K.C.), US 3 489 786 (1970); C.A. *72*, 89 833 f (1970)
1804. Sumitomo Co., JP 6 168/67 (1966)
1805. Eur. Chem. News of August 11, 1978 and of Jan. 15, 1979
1806. DuPont, GB 665 705 (1952)
1807. Anglo Iranian Oil Co. (Habeshaw, J., Geach, C.J.), GB 702 192 (1954)
1808. Anglo Iranian Oil Co. (Habeshaw, J., Thornes, L.S.), GB 702 204 (1954)
1809. Anonym, Hydrocarbon Processing, Dez. 1978, p. 56-C
1810. Anonym, Chem. Mark. Rep. *214*, (17), Oct. 23, 1978, p. 9
1811. Oro, L.A., Manrique, A., Royo, M.: Transit. Met. Chem. *3*, 383 (1978)
1812. Mitsubishi Chem. Ind. Co. Ltd., JP 22 735/65 (1965)
1813. Noguchi Res. Foundation, JP 5853/67 (1967)
1814. Noguchi Res. Foundation, JP 21 287/68 (1968)
1815. BASF A. G., GB 1 254 182 (1968)
1816. Ajinomoto Co., JP 17 209/62 (1962)
1817. Noguchi Res. Foundation, FR 1 355 247 (1963)
1818. Toa Gosei Chem. Ind. Co. Ltd., JP 9 730/68
1819. Katsnelson, M.G.: SU 558 906 (1975)
1820. Karapinka, G., Orchin, M.: Abstracts 137th A. C. S. Meeting, Cleveland/Ohio, Apr. 5, 14 (1960), p. 92-100
1821. BASF A. G. (Himmele, W., Aquila, W.), DE 2 050 679 (1970)
1822. BASF A. G. (Himmele, W., Aquila, W.), DE 2 053 736 (1970)
1823. Orchin, M: in Advances in Catalysis, Vol. V, p. 401, Academic Press New York: 1953
1824. Shaw, J.T., Tyson, F.T.: J. Amer. Chem. Soc. *78*, 2538 (1956)
1825. Clement, W.H., Orchin, M.: Ind. Eng. Chem., Prod. Res. Dev. *4*, 283 (1965)
1826. Locicero, J.C., Johnson, R.F.: J. Amer. Chem. Soc. *74*, 2094 (1952)
1827. Newport Ind., US 2 790 006 (1957)
1828. Bahrmann, H., Tummes, H.: Ruhrchemie A. G., unpublished
1829. Levering, D.R., Glasebrook, A.L.: Ind. Eng. Chem. *50*, 317 (1958)
1830. Nussbaum, A.L. et al.: J. Amer. Chem. Soc. *81*, 1228 (1959)
1831. Beal, P.F., Rebenstorf, M.A.: J. Amer. Chem. Soc. *81*, 1231 (1959)
1832. Burkhard, C.A., Hurd, D.T.: J. Org. Chem. *17*, 1107 (1952)
1833. Upjohn Co. (Beal, P.F., Rebenstorf, M.A.), DE 1 124 942 (1962)
1834. Wender, I., Sternberg, H.W., Orchin, M.: J. Amer. Chem. Soc. *75*, 3041 (1953)
1835. Heck, R.F., Breslow, D.S.: Chem. Ind. *1960*, 467
1836. Takegami, Y. et al.: Bull. Chem. Soc. Japan *37*, 672 (1964)
1837. Yokokawa, C., Watanabe, Y., Takegami, Y.: Bull. Chem. Soc., Japan *37*, 677 (1964)
1838. Heck, R.F.: J. Amer. Chem. Soc. *85*, 1460 (1963)
1839. Takegami, Y. et al.: Bull. Chem. Soc. Japan *38*, 1649 (1965)
1840. Rohm & Haas Co. (Niederhauser, W.D.), US 3 054 813 (1962)
1841. Roos, L., Goetz, R.E., Orchin, M.: J. Amer. Chem. Soc. *30*, 3023 (1965)
1842. Shell Res. Co. (Smith, C.W.), US 3 463 819 (1969) and 3 456 017 (1969)
1843. DuPont (Lawrence, F.R., Sullivan, R.H.), US 3 687 981 (1972)
1844. Bogdanovic, B.: Angew. Chem. *85*, 1013 (1973)
1845. Izumi, Y.: Angew. Chem. *83*, 956 (1971)
1846. Botteghi, C., Consiglio, G., Pino, P.: Chimia *26*, 141 (1972)
1847. Stern, R., Hirschauer, A., Sajus, L.: Tetrahedron Lett. *35*, 3247 (1973)
1848. Botteghi, C., Salomon, Ch.: Tetrahedron Lett. *49/50*, 4285 (1974)
1849. Botteghi, C.: Gazz. Chim. Ital. *1975*, 105, 233
1850. Tanaka, M., Ikeda, Y., Ogata, J.: Chem. Lett. *1975*, 1115
1851. Agency of Ind. Sci. and Technol. JP 99 131/73 (1973)
1852. Inst. Francais du Petrole (Stern, R., Commerenc, D.), FR 2 208 872 (1977)
1853. BASF A. G. (Himmele, W., Siegel, H., Aquila, W.), DE-AS 2 132 414 (1971)
1854. Pino, P. et al., DE-OS 2 359 101 (1972)
1855. Kagan, H.B., Dang, T.P.: J. Amer. Chem. Soc. *94*, 6429 (1972)
1856. BASF A. G., NL 71-14 729 (1971)

1857. Dynamit Nobel A. G. (El Chahawi, M., Prange, U., Richtzenhain, H.), DE-OS 2 404 776 (1975)
1858. Farbwerke Hoechst A. G. (Röhrscheid, F.), DE-OS 2 140 644 (1973) and 2 163 752 (1973)
1859. DuPont (Copelin, H.B.), Def. Publ. US Pat. 904 021 (1972)
1860. Kogyo Gizyutsu In., JP 17 441/66 (1969)
1861. Kururay Co. Ltd. (Shimizu, T.), DE-OS 2 538 364 (1976)
1862. Toray Ind. Inc. (Aya, T., Sasagawa, T., Izumi, Z.), JP 104 524/78 (1978); C.A. *89*, 201 050 c (1978)
1863. Fonnesbach, N., Hjortkjaer, J., Johansen, H.: Int. J. Quantum Chem. *1977*, 12 (Suppl. 2), 95; C.A. *89*, 214 751 g (1978)
1864. Nahum, L.S.: J. Org. Chem. *33*, 3601 (1968)
1865. Chem. Mark. Rep./OPD of Nov. 27, 1978
1866. Sittig, M.: Handbook of Catalyst Manufacture, Noyes Data Corp., Park Ridge, N. J., USA, 1978
1867. Nakum, L.S.: Tappi *52*, (4), 712 (1969)
1868. Ucciani, E., Bonfand, A., Lai, R., Naudet, M.: Bull. Soc. Chim. France *1969*, (8), 2826
1869. Lai, R., Ucciani, E.: C. R. Acad. Sci. Ser. C *275*, (18), 1033 (1972)
1870. BASF A. G. (Aquila, W., Hoffmann, W., Himmele, W., Siegel, H.), DE-OS 2 340 812 (1975)
1871. Cornils, B., Payer, R.: Chem.-Ztg. *98*, 596 (1974)
1872. Mead, C.A.: Topics Curr. Chem. *49*, 1974
1873. Pracejus, H.: Topics Curr. Chem. *8*, No. 4, 493 (1967)
1874. Stefani, A., Tatone, D.: Helv. Chimia Acta *60*, 518 (1977)
1875. Rossi, R. et al.: Gazz. Chim. Ital. *97*, 1194 (1967), C. *1968*, 54-218
1876. Llinas, J.R., Lai, R.: C. R. Hebd. Séances Acad. Sci., Ser. C *280*, (4), 201 (1975)
1877. Khoe, E.: J. Amer. Oil Chem. Soc. *49*, 134 (1972)
1878. Dufek, E.J., List, G.R.: J. Amer. Oil Chem. Soc. *54*, 271 (1977)
1879. General Electric Co. (Smith, W.E.), DE-OS 2 758 475 (1978)
1880. Mistrik, E.J., Komora, L.: CS 124 842 (1967); C.A. *69*, 86 397 u (1968)
1881. Ajinomoto Co. Inc., JP 487/68 (1968)
1882. Shell Int. Res., NL 6 613 226 (1967); C.A. *67*, 43 433 u (1967)
1883. Shell Oil Co. (Falbe, J., Huppes, N.), US 3 318 913 (1967)
1884. Brownstein, A., List, H.L.: Hydrocarbon Proc. *1977*, (9), 159
1885. Ogata, I., Asakawa, T.: Kogyo Kagaku Zasshi *74*, 1640 (1971)
1886. Toyo Soda Mfg. Co. Ltd. (Tsutsumi, Y., Ohshio, M.), JP 30 822/75 (1975); C.A. *83*, 42 828 v (1975)
1887. Ohshio, M., Tsutsumi, Y.: Toyo Soda Kenkyu Hokuku *1975*, 19(2), 60; C.A. *85*, 142 568 v (1976)
1888. Friedrich, J.P. et al.: J. Amer. Oil Chem. Soc. *50*, 455 (1973)
1889. Mistrik, E.J., Durmis, J.: Chem. Zvesti *23*, 286 (1969)
1890. Schwab, A.W. et al.: J. Amer. Oil Chem. Soc. *49*, 75 (1972)
1891. Frankel, E.N.: J. Amer. Oil Chem. Soc. *46*, 133 (1969); Ann. N. Y. Acad. Sci. *1973*, (214), 79
1892. Lai, R. et al.: Rev. Francais Corps. Grai *15*, 15 (1968); Bull. Soc. Chim. France *1969*, 793, C. R. Hebd. Séances, Acad. Sci., Ser. C *271*, 1588 (1970)
1893. Ajinomoto Co. (Yamagami, N. et al.), DE-OS 2 016 061 (1970)
1894. Fell, B.: Chem.-Ztg. *100*, 308 (1976)
1895. Texaco Inc. (Macaluso, A., Rigdon, O.W.), US 3 984 486 (1973)
1896. Falbe, J., GB 1 170 025 (1969)
1897. Nakai, S.: Tetrahedron Lett. *28*, 2425 (1977)
1898. Maeda, J., Yoshida, R.: Bull. Chem. Soc. Japan *41*, 2969 (1968) Chem. Titles, *1969*, Nr. 6, 123
1899. Del'nik, V.B. et al.: Zh. Prikl. Khim *51*, 1912 (1978)
1900. Shell Oil Co (Falbe, J., Korte, F.), US 3 159 653 (1964)
1901. Shell Int. Res. Mij., GB 960 330 (1961)
1902. DuPont (Lawrence, F.R., Sullivan, R.H.), US 3 687 981 (1972)
1903. Rohm & Haas Co. (Niederhauser, W.D.), US 3 130 233 (1964)

1904. Falbe, J.: GB 1 170 226 (1969), ZA 68-04 722 (1968)
1905. Shell Oil Co. (Falbe, J., Huppes, N.), US 3 509 221 (1970)
1906. Dokiya, M., Bando, K.: Bull. Chem. Soc. Japan, *41*, 1741 (1968)
1907. Union Carbide Corp. (Pruett, R.L.), US 3 499 932 (1970)
1908. Imyanitov, N.S., Rudkowskij, D.M.: Karbon Nenasyshch. Uglev. *1968*, 206; C.A. *71*, 29 961 r (1969)
1909. Cornils, B., Payer, R., Tummes, H., Weber, J.: Eur. Chem. News *27*, No. 712, 36 (1975)
1910. Institut Francais du Petrole (Lassau, I.), DE-OS 1 928 101 (1969)
1911. Rybakov, V.A. et al.: SU 403 665 (1973); C.A. *80*, 47 459 y (1974)
1912. BASF A. G. (Nienburg, H.J.), DE 891 254 (1953)
1913. Andreetta, A. et al.: Nouv. J. Chim. *1978*, 2 (5), 463
1914. Celanese Corp. (Poist, J.E.), US 4 101 565 (1978)
1915. SNAM-Progetti SpA, JP 96 999/78 (1978); C.A. *90*, 12 907 x (1979)
1916. Inst. Gen. Inorg. Chem., SU 251 738 (1969); C.A. *72*, 34 077 g (1970)
1917. Macho, V., Polievka, M., Komora, L.: Chem. Zvesti *21*, 170 (1967)
1918. Chemische Werke Hüls AG (Scharf, H.), DE-OS 2 802 672 (1978)
1919. Pruett R.L. in Stone, F.G.A., West, R.: Advances in Organometallic Chemistry, Vol. 17, p. 1–57, Academic Press, New York, San Francisco, London 1979
1920. Murai, S., Kato, T., Sonoda, N., Seki, Y., Kawamoto, K.: Angew.-Chem. *91*, 421 (1979)
1921. Werner, P., Ault, B., Orchin, M.: J. Organometall. Chem. *162*, 189 (1978)
1922. King, R.B., King, A.D., Iqbal, M.Z.: J. Amer. Chem. Soc. *101*, 4893 (1979)
1923. Laine, R.M.: J. Amer. Chem. Soc. *100*, 6452 (1977)
1924. Murata, K., Matsuda, A., Bando, K., Sugi, Y.: J. Chem. Soc./Chem. Commun. *18*, 785 (1979)
1925. des Abbayes, H., Buloup, A.: J. Organomet. Chem. *179*, C 21 (1979)
1926. Yamaguchi, M.: Japan Chem. Ind. Assoc. Monthly Rep. *1979*, 14
1927. Pino, P., Consiglio, G.: Symp. Rhodium Hom. Catal. (Proc.) *1978*, 98; C.A. *90*, 21 893 s (1979)
1928. Kawabata, Y., Hayashi, T., Ogata, I.: J. Chem. Soc./Chem. Commun. *1979*, 462
1929. Consiglio, G., Arber, W., Pino, P.: Chim. Ind. (Milano), *60*, 396 (1978)
1930. Celanese Corp. (Unruh, J.D.), DE-OS 2 834 742 (1979)
1931. Johnson, Matthey & Co. Ltd. (Hignett, R.R., Davidson, P.J.), BE 868 278 (1978)
1932. Hjortkjaer, J.: J. Mol. Catalysis *5*, 377 (1979)
1933. Borisov, R.B., Gankin, W.Yu., Rybakov, V.A.: Zh. Prikl. Khim. *1979*, 625; C.A. *91*, 4930 e (1979)
1934. DuPont (Peterson, M.L.), US 4 137 240 (1979)
1935. Ushakow, V.M. et al.: Neftekhimiya *1979*, 62; C.A. *90*, 186 438 k (1979)
1936. Chisso Corp. (Ogawa, T.), JP 78 101 386 (1978); C.A. *90*, 152 247 g (1979)
1937. Uglea, C., Georgescu, F.: RO 63 405 (1977); C.A. *90*, 103 968 j (1979)
1938. Anonym, Chem. Marketing Reporter, Feb. 29, 1979, p. 31; May 7, 1979, p. 37; Europa-Chemie *5*, 81 (1979); Chem. Ind. *31*, 52 (1979)
1939. Anonym, Chem. Ing. Techn. *51*, 80 a (1979)
1940. Anonym, Europa-Chemie *12*, April, 217 (1979); Chem. Engng. *86*, (10), 31 (1979)
1941. DuPont, JP 79 26 218 (1979); C.A. *91*, 57 027 x (1979)
1942. Nissan Chem. Ind. K. K., JP 108 589 (9.9.1977)
1943. Exxon Res. and Engng. Co. (Oswald, A.A.), US 4 151 114 (1979)
1944. Farrell, M.O., van Dyke, C.H., Boucher, L.J., Metlin, S.J.: J. Organomet. Chem. *169*, 199 (1979)
1945. Dow Chemical Co. (Hartwell, G.E., Garrou, P.E.), US 4 141 191 (1979)
1946. Nissan Chem. Ind. K. K. (Ichikawa, M., Kido, Y.), JP 79 41 293 (1979); C.A. *91*, 56 355 r (1979)
1947. Penninger, J.M.L.: J. Catal. *1979*, 287
1948. Mathey, F., Muller, G., Demay, C., Lemke, H.: Informations Chimie *179*, 191 (1978)
1949. Texaco Inc. (Love, R.F., Kerr, E.R., Knifton, J.F.), US 4 147 730 (1979)
1950. Mobil Oil Corp. (Haag, W.O.), US 4 098 727 (1978)
1951. Celanese Corp., US 4 151 209 (1976)

1952. Kuzmina, L.S. et al.: Khim. Tekhnol. Topl. Masel *1979*, (2), 30; C.A. *90*, 167 973 q (1979)
1953. Kagna, S.S., Gankin, V.J. et al.: GB 1 524 775 (1978), SU 518 931 (1979); C.A. *90*, 186 388 u (1979)
1954. Chemische Werke Hüls A. G. (Nehring, R., Zur Hausen, M., Neumann, W.), DE 2 710 216 (1979)
1955. Chevron Res. Co. (Wilkes, J.B.), US 4 142 061 (1979)
1956. Crabtree, R.H., Felkin, H.: J. Mol. Catal. *5*, 75 (1979)
1957. Matsuda, A. et al.: Bull. Chem. Soc. Japan *51*, 3016 (1978); C.A. *90*, 22 219 p (1979)
1958. Ugine Kuhlmann S. A., BE 871 814 (1977)
1959. Johnson, Matthey & Co. Ltd. (Hignett, R.R., Davidson, P.J.), BE 868 279 (1978)
1960. Chemische Werke Hüls A. G. (Sridhar, S.), DE-OS 2 749 890 (1979)
1961. Lennertz, A.M., Laege, J., Mirbach, M.J.: J. Organomet. Chem. *171*, 203 (1979)
1962. Celanese Corp. (Poist, J.E.), US 4 101 565 (1978)
1963. General Electric Co. (Smith, W.E.), US 4 139 542 (1979)
1964. Lai, R., Ucciani, E.: J. Mol. Catal. *4*, 401 (1978)
1965. Mitsubishi Petr. Co. Ltd. (Takeda, M. et al.), JP 79 24 843 (1979); C.A. *90*, 203 686 j (1979)
1966. Botteghi, C., Branca, M., Marchetti, M., Saba, A.: J. Organomet. Chem. *161*, 197 (1978)
1967. Celanese Corp. (Taylor, P.D.), US 4 105 677 (1978)
1968. DuPont, BE 831 988 (1974)
1969. Siegel, H., Himmele, W., Angew.-Chem. *92*, 182 (1980)
1970. Produits Chimiques Usine Kuhlmann (Papp, R., Mongenet, F.), DE-OS 2 927 979 (1980)

2. Homologation of Alcohols

H. Bahrmann and B. Cornils

2.1 Introduction

During the years 1941 [1]–1943 [2], Wietzel, Eder, Vorbach and Scheuermann discovered that aliphatic primary alcohols can be converted into the next higher alcohols on reacting with syn gas at raised temperatures and pressures:

$$ROH + CO + 2\,H_2 \xrightarrow{\text{metal carbonyls}} RCH_2OH + H_2O$$

The by-products consisted of higher acids, their esters and hydroxyethers.

In 1949, Wender [3] applied this reaction to secondary and tertiary aliphatic and aromatic alcohols and introduced the term 'homologation' [1, 3] which has widespread use. However, the most industrially interesting conversion – methanol → ethanol – has satisfactory conversion rates only when drastic conditions are applied (extreme high pressure [4, 5, 11–13] or long reaction periods [6, 7, 14]). The latter have however an unfavorable effect on the selectivity.

A noticeable increase in the reaction rate was made possible in 1956 when I. Berty [8] introduced iodine promoters. This represented the transition from the high pressure synthesis (400–1000 bar) to the medium pressure synthesis (180–400 bar).

In recent years, the selectivity has gradually been improved via utilization of phosphine-modified transition metal catalysts [9, 10], heterogeneous catalysts as well as by continuous operation [70]. Nevertheless, unsatisfactory conversion and selectivity has still prevented commercial realization of the homologation process.

2.2 Reaction Mechanism

The material available does not allow clear deductions to be made about the mechanism. There have however, been two main approaches – Wender's carbenium ion mechanism [3] and Ziesecke's dehydration mechanism [4]. A new insertion mechanism is also presented by the authors.

Based on his work with substituted benzyl alcohols, Wender [17] postulated the formation of the benzyl carbenium ion from the alcohol and $HCo(CO)_4$ [18], the latter functioning as an acid under the prevailing reaction conditions. The tendency of the carbenium ion to form is particularly great when its degree of resonance can be fortified via electron donating substituents.

The close relationship between the tendency of carbenium ions to form and the relative reaction rate was later confirmed using aliphatic alcohols [8]. Apparently, the carbenium ion initially forms an aldehyde via addition to the hydrido metal complex followed by reaction with CO and H_2. Thereafter, the alcohol results from the aldehyde.

Ziesecke assumes that the synthesis largely occurs via an intermediate dehydration of the alcoholic starting material to the olefin, followed by hydroformylation and hydrogenation of the aldehyde to the alcohol. This explains the resulting product spectrum from the tert-butanol feedstock, 3-methylbutanol(-al) being almost exclusively obtained and *not* 2,2-dimethylpropanol(-al) [8]. This dehydration variant is also supported by Burn's experiments [20] with ^{14}C-labelled methanol. While the activity of the ethanol product is concentrated in the CH_3 group, the activity of the n-propanol is equally distributed between C_2 and C_3 (i.e., only half of the activity relative to the methanol feed). Thus, the reaction must ensue via a symmetric intermediate ethylene.

Both interpretations must be modified especially in the case of methanol as dehydration would produce a fairly improbable carbene complex. The usually observed relationship between the tendency of carbenium ion formation and reactivity is not evident with methanol.

Thus at an early stage, Hecht and Kröper [67] and others [10, 49] interpretated alkyl complex formation according to Eq. (1) as being an esterification in which the very acidic cobalt hydridocarbonyl acts both as catalyst and reactant.

$$CH_3OH + HCo(CO)_4 \rightarrow CH_3-Co(CO)_4 + H_2O \tag{1}$$

The function of the iodine (or hydrogen iodide which results under reaction conditions) as promoter is interesting in this connection. In the iodine-activated carbonylation of methanol to acetic acid in the presence of Rh catalysts (Monsanto process), the rate determining step was the esterification to methyl iodide.

$$CH_3OH + HI \rightarrow CH_3I + H_2O \tag{2}$$

It can however be assumed — based on more recent studies by Mizoroki [49] on the cobalt catalyzed carbonylation of methanol to acetic acid — that the activating effect of iodine (in the homologation) does not ensue via methyl iodide formation [48] but via the effect of the iodine ligands of the cobalt carbonyls which cause lability to occur in the coordination sphere.

As a Co:I ratio of approx. 1:1 has been found to be suitable, the following reaction sequence can be postulated (Oxo synthesis analogue):

$$CH_3OH + HCo(CO)_3I \rightleftharpoons CH_3-Co(CO)_3I + H_2O \tag{3}$$

$$CH_3-Co(CO)_3I \xrightarrow{CO} CH_3COCo(CO)_3I \tag{4}$$

$$CH_3COCo(CO)_3I + HCo(CO)_3I \rightarrow CH_3CHO + Co_2(CO)_6I_2 \tag{5}$$

The further hydrogenation of the aldehyde is discussed on pp. 5 and 147.

As it has not been established whether a five- or sixfold coordinated complex is formed in the homogeneously catalyzed reaction with cobalt (cf Marko's work on hydrogenation [68]) in analogy to Calderazzo [19] an "insertion mechanism" is proposed which ensues via sixfold coordination (depending on CO partial pressure), carbonylation and homologation occurring simultaneously (cf Fig. 2.1.)

After the 'oxidative addition' of methanol to the cobalt hydrido complex 1, at high CO partial pressure CO addition occurs with subsequent insertion leading to the acyl complex 6 and then to formation of acetic acid. The alkyl group migration (CO insertion) is activated via the I-ligands in the coordination sphere of the central atom. These ligands effect a loosening of the metal-alkyl bond [49].

At higher H_2 partial pressures the complex 2 is dehydrated to the alkyl complex 3 before CO insertion takes place. After complex 4 results via oxidative addition of hydrogen, it is transformed into the acyl complex 5 via CO insertion. At this point, acetaldehyde is produced with simultaneous reformation of the active catalyst 1. Figure 2.2 summarizes the primary products according to the insertion mechanism.

In addition, the following products are mentioned in the Ref. [70]: 2-propanol, 1-butanol, 2-butanol, 2-methyl-2-butanol, 1-methoxy-2-propanol, 2-methoxy-1-propanol, methyl propyl ether.

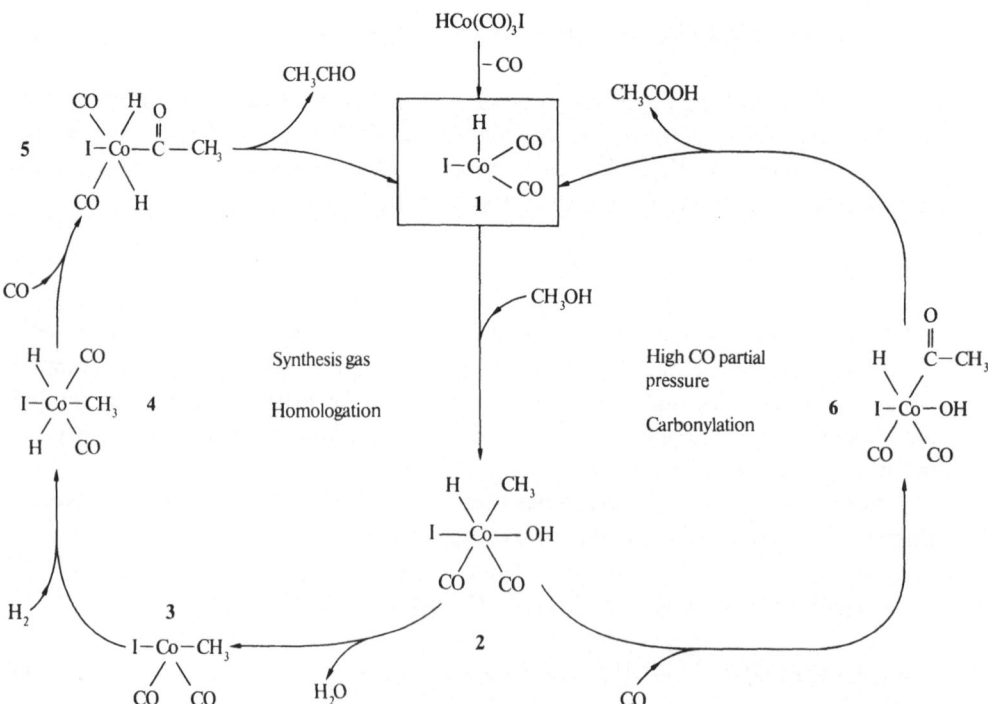

Fig. 2.1. Insertion mechanism for the homologation and carbonylation of methanol with cobalt carbonyl/iodine catalysts

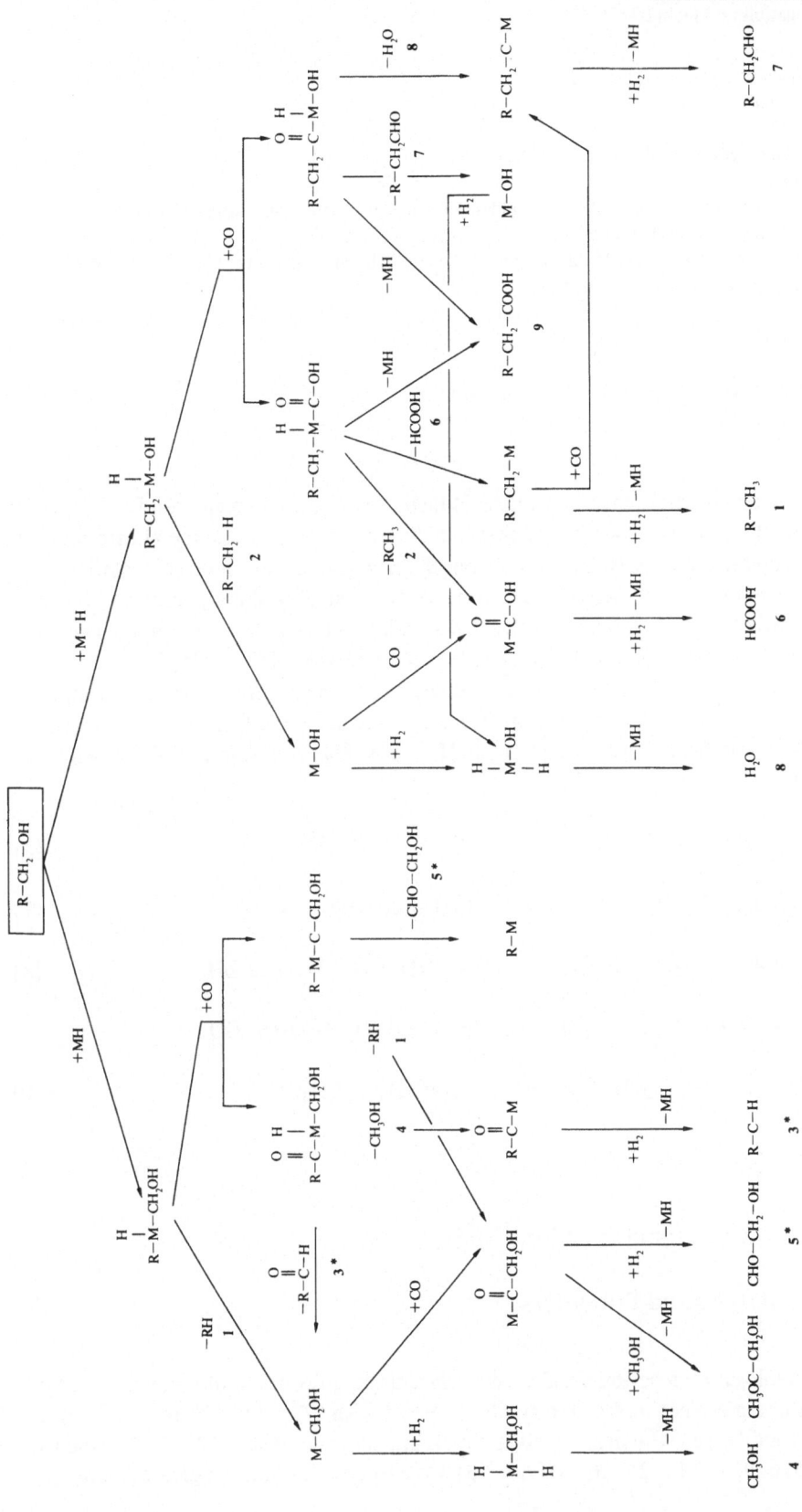

Fig. 2.2. Homologation products according to the insertion mechanism

Explanation of symbols in Fig. 2.2:
1 Hydrogen
2 Methan
3* Formaldehyde/formaldehyde dimethyl acetal
4 Methanol
5* Glycolaldehyde/glycol, methoxyacetaldehyde-dimethyl acetal, 2-methoxyethanol
6 Formic acid/methyl formate
7 Acetaldehyde/acetaldehyde dimethyl acetal, acetaldehyde methyl ethyl acetal, acetaldehyde
   diethyl acetal, ethanol, propanol
8 Water
9 Acetic acid/methyl acetate/ethyl acetate

* so far not directly identified

The insertion mechanism (cf Fig. 2.2) facilitates a ready explanation for the occurrence of all primary and secondary by-products stemming from the metal carbonyl catalyzed homologation (as well as those from the new glycol and acetic acid syntheses). This does not of course include the thermal or H-ion catalyzed dehydration of higher alcohols which yields olefins and ethers. According to recent work, methanol can also result from the homogeneously catalyzed reaction between CO and H_2 [69].

There are indications [4, 21] that, besides the ion- and metal-catalyzed constructive reactions, a degradation reaction also occurs involving the formaldehyde intermediate (from methanol) producing CO and H_2 — possibly according to the following scheme:

$$CH_2O + HCo(CO)_3 \qquad \rightarrow \qquad HCO-Co(CO)_3(H)_2 \qquad\qquad (6)$$

$$HCO-Co(CO)_3H_2 \qquad \rightleftharpoons \qquad HCO-Co(CO)_3 + H_2 \qquad\qquad (7)$$

$$HCO-Co(CO)_3 + CH_3OH \quad \rightleftharpoons \quad HCO(H)-Co(CO)_3-CH_2OH \qquad\qquad (8)$$

$$HCO(H)-Co(CO)_3-CH_2OH \rightleftharpoons \quad HCo(CO)_3(H)CH_2OH + CO \qquad\qquad (9)$$

$$HCo(CO)_3(H)CH_2OH \qquad \rightleftharpoons \qquad HCo(CO)_3 + CH_3OH \qquad\qquad (10)$$

2.3 Effect of Reaction Conditions

2.3.1 Catalysts and Promoters

The homologation reaction usually takes place under pressures and temperatures similar to those prevalent in the Oxo synthesis. Besides Rh [22], Pd [22], Ir [22], Mn [22], Cr [22] and Os [25] the main catalyst metals examined were Ru [13, 22–25] and Co [1, 2, 5, 6, 9, 11–14, 24, 26–35, 64–66]. Co clearly exhibited the greatest activity.

Salts such as $KHCO_3$ [13], $(NH_4)_2HPO_4$ [29], HBO_3 [32], bromides [6, 9, 35] and above all iodides [5, 6, 8, 9, 24, 27, 29, 30, 32, 34, 35, 65] were suggested as promoters (cf Sect. 2.2). In addition, amino [64], antimony [64] and phosphor compounds [29, 65, 66] or phosphines were proposed as modifiers [9]. However, frequently the increase in selectivity is at the expense of the conversion rate. Recent studies claim a molecular selectivity of 82% to ethanol on combining various catalyst metals (cobalt with traces of ruthenium) with various promoters (iodine and a small quantity of chloride) [23]. Generally speaking however, the available experimental data does not permit basic relationships to be derived.

2.3.2 Temperature

The highest conversion rates of methanol to ethanol or acetaldehyde were achieved using higher pressures (900 bar) at 225 °C [4], and at 300–400 bar at approx. 190–200 °C [8, 22, 36]. Apparently, the reaction step leading to the intermediate stage — acetaldehyde or 1,1-dimethoxyethane — occurs at 110–130 °C [6, 37, 70], however at an insufficient rate. An increase in temperature above 200 °C mainly benefits the side reactions such as the hydrogenation to methane [36] and the thermally induced acid (e.g., $HCo(CO)_4$) catalyzed ether formation.

2.3.3 Pressure

While an increase in pressure generally leads to higher yields, there is no essential change in the conversion rate on employing unmodified cobalt catalysts at pressures above 900 bar and with iodine-modified catalysts at above 400 bar. There are high conversion rates (90%) at pressures up to 3000 bar and the product distribution becomes increasingly displaced towards higher alcohols ($\leqslant C_5$) [21] and glycol ethers [29].

2.3.4 CO:H_2 Ratio

Although the stoichiometry of the homologation demands a CO:H_2 ratio of 1:2, the highest conversion rates were found using a 1:1 ratio [22, 29, 36]. Hydrogen-rich syn gas markedly suppresses the carbonylation to acetic acid and its derivatives. Consequently, the selectivity to ethanol [70] or its precursors — acetaldehyde or acetaldehyde dimethyl acetal — clearly increases [37].

2.3.5 Catalyst Concentration

Initially, the conversion rate increases with rising catalyst concentration [3, 70], however a maximum is reached at 5% cobalt (II) acetate and 2–4% iodine (relative to methanol). Generally, however lower concentrations (> 0.1 mol% relative to methanol) are recommended [26].

2.4 Homologation of Particular Structures

With alcohols capable of being dehydrated, the homologation ensues according to the carbenium ion mechanism (cf Sect. 2.2) i.e., most rapidly with tertiary alcohols, then secondary followed by primary. With homologues of aliphatic alcohols, the branched alcohols are more reactive than the straight chained [3] and the lower alcohols are [4, 38] more readily converted than the higher. Relatively high reaction rates were achieved with the nondehydratable primary alcohols – methanol and benzyl alcohol:

Table 2.1. Synthesis gas consumption during homologation of various alcohols as measure of the reaction rate (catalyst: 0.025 mol = 6.2 CoAc$_2$, 0.006 mol = 1.5 g iodine: pressure 200 to 250 bar; temp. 195–205 °C) [8]

Alcohol	'Reaction rate' (mmol gas/min)
methanol	58
ethanol	1.4
isopropanol	1.2
tert-butanol	140
glycol	5
benzyl alcohol	20
p–OCH$_3$–benzyl alcohol	very high

Acetals [14, 39, 40], ethers [21] and aromatic [45] and aliphatic [5] ketones undergo this reaction, the latter via their corresponding alcohols.

Wender et al. [17] made semi-quantitative studies of the relative reaction rates of substituted benzyl alcohols with CO and H$_2$:

The experimental data show that electron-donating substituents in the m- and p-positions markedly increase the reaction rate.

Tables 2.3 and 2.4 give a summary of currently available experimental data relating to the homologation of various structures.

Table 2.2. Relative reaction rates of substituted benzyl alcohols [17]

Substituent[a]	Relative rate
p-methoxy	10,000
p-methyl	200
m-methyl	50
p-tert-butyl	50
hydrogen	1
p-chloro	0.8
p-carbethoxy	0.4
m-methoxy	0.3
m-trifluoromethyl	0.01

[a] Temperature was $188-190\,°C$ in all cases, except p-methoxy ($92\,°C$) and p-methyl ($166\,°C$)

Table 2.3. Homologation of alcohols

Alcohol	Press. (bar)	Temp. (°C)	Catalyst	Conv.	Products/yields (%)	Ref.
methanol	1000	180	$Co/AgJ/Bi$	71	ethanol 67%	[5]
	200–1500	200–300	Co	15	methyl acetate, acetaldehyde	[11]
	200–340	180	$Co_2(CO)_8$	76	ethanol 39%, propanol 5% butanol 1%, methyl formate 2%, methyl acetate 9%, ethyl acetate 6%, methane 8%	[7]
	180–250	195–205	$CoAc_2/J_2$		ethanol 43%, acetic acid 32% higher alcohols 9%	[8]
	300	185	$Co_2(CO)_8RuCl_3$		ethanol 82%	[23]
	140	200	CoJ_2, PR_3	10	ethanol 90%, methyl acetate 5%, methane 5%	[9]
	260	100–110	$HCo(CO)_4$	14	dimethyl acetal 75%,	[37]
	300	190	$HCo(CO)_4$	25	ethanol 57%, dimethyl acetal 12%	[70]
	300	200	$Ni(CO)_4/J_2$		ethanol (28%) after hydrogenation	[76]
ethanol	900	225	$Co_2(CO)_8$	60	n-propanol 20%, n-butanol 4% 2-methylbutanol(4) 5%, diethyl ether 16%, various 17%	[4]
	1000	190	$Co/AgJ/Bi$	53	n-propanol 48%	[5]
n-propanol	1000	225	$Co_2(CO)_8$	50	n-butanol 11%, isobutanol 4%, n-pentanol 3%, di-n-propyl ether 9%, ethylene glycol monopropyl ether 8%	[4]
n-butanol	1000	225	Co	38	ether 4%, $n-C_5$-alcohols 11%	[4]
2-propanol	1000	225	Co	66	ethers 25%, isobutanol 16% n-butanol 18%	[4]
	220	180	$Co(OAc)_2$		n-butanol/isobutanol 11%	[3]
	200	130	$Co(OAc)_2$		3-methyl-1-butanol 51%, 2,2-dimethyl-1-propanol 10%	[41] [43]
tert-butyl alcohol	265	200	$Co_2(CO)_8$		3-methyl-1-butanol 60%, 2,2-dimethyl-1-propanol 4% isobutane 3%, isobutene 3%	[42]

Table 2.3 (continued)

Alcohol	Press. (bar)	Temp. (°C)	Catalyst	Conv.	Products/yields (%)	Ref.
pinacol	225	185	$Co_2(CO)_8$		pinacolone 17%, pinacolyl alcohol 4%, 3,4-dimethyl-1-pentanol 26%	[44]
2-butanol	1000	225	Co	70	first runnings 7%, 2-methyl-1-butanol 18%, n-pentanol 33%, C_6-alcohols 4%	[4]
cyclohexanol	1000	225	Co/Fe	68	cyclohexylcarbinol 44%, higher boiling substances	[4]
benzyl alcohol	250	185	$Co_2(CO)_8$	100	2-phenylethanol 32%, toluene 63%	[45]
benzyl alcohol	270	130	$Co_2(CO)_8$ + $RuCl_3$	47	2-phenylethanol 79%, toluene 16%	[77]
p-methylbenzyl alcohol	238	185	$Co_2(CO)_8$		2 (p-tolyl) ethanol 24% p-xylene 58%	[17]
m-methylbenzyl alcohol	238	185	$Co_2(CO)_8$		2-(m-tolyl)ethanol 36%, m-xylene 52%	[17]
p-(tert-butyl) benzyl alcohol	238	185	$Co_2(CO)_8$		2-(p-tert-butylphenyl)ethanol 28%, 4-tert-butyltoluene 54%	[17]
2,4,6-trimethyl- benzyl alcohol	238	185	$Co_2(CO)_8$		2-(2,4,6-trimethylphenyl)-ethanol 18%, 1,2,3,5-tetramethyl-benzene 58%	[17]
p-hydroxy-methylbenzyl alcohol	238	185	$Co_2(CO)_8$		2-(p-tolyl)ethanol 39% 1,4-bis(hydroxyethyl)-benzene 12%, p-xylene 27%	[17]
p-methoxy-benzyl alcohol	238	185	$Co_2(CO)_8$		2-(p-anisyl)ethanol 44%, p-methoxytoluene 16%	[17]
m-methoxy-benzyl alcohol	238	185	$Co_2(CO)_8$		m-methoxybenzaldehyde 4%, m-methoxytoluene 23%	[17]
p-chlorobenzyl-alcohol	238	185	$Co_2(CO)_8$		2(p-chlorophenyl)ethanol 16% p-chlorotoluene 41%	[17]
m-trifluoro-methylbenzyl alcohol	238	185	$Co_2(CO)_8$		ethyl p-toluate 27%	[17]
p-nitrobenzyl alcohol	238	185	$Co_2(CO)_8$		polymeric p-aminobenzyl alcohol	[17]

Table 2.4. Homologation of other structures

Compound	Press. (bar)	Temp. (°C)	Catalyst	Conv.	Products/yields (%)	Ref.
butanone	1100	225	Co	$\sim$50	2-methyl-1-butanol 19%, 1-pentanol 20%	[4]
benzophenone	210–245	180–185	$Co_2(CO)_8$	92	diphenylmethane 86%	[45]
formaldehyde dimethyl acetal	180	115	Co		methanol, dimethyl ether glycol methyl ether, glycols	[14(50)]
	100–1000	150–250	$CoI_2/RuCl_3$	$\sim$100	ethanol 30%	[40(50)]
	200	120	$Co_2(CO)_8$		glycolaldehyde 50%	[31]
	300	150	$Rh(CO)_2(C_2H_7O_2)$			
				$\sim$80	methanol, ethylene glycol, glycolaldehyde	[46]
formaldehyde	70–700	100–200	$Rh + Co_2(CO)_8$		ethylmylycol ethylenglycolmono- methyl ether	[74]
tetrahydro-furan	900–1500	200	$Co_2(CO)_8/I_2$		2-hydroxymethyltetra- methyltetrahydrofuran 15%	[73]
methylacetate	550	180	$Co_2(CO)_8/I_2$		ethylacetate 78%	[75]

2.5 Parallel and Secondary Reactions of the Homologation

2.5.1 Hydrogenation

While only 5–10% methane is formed with a methanol feedstock, the share of the corresponding hydrocarbon increases to 60–70% with aromatic primary alcohols such as benzyl alcohol. This figure even reaches over 90% with aromatic ketones which are reduced to the corresponding secondary alcohols under reaction conditions [45]. With substituted benzyl alcohols, electron-attracting substituents favor the hydrogenation at the cost of the homologation [17].

Wender [45] proposed the following radical mechanism for the hydrogenation which takes place under homologation conditions:

$$Co_2(CO)_8 \quad\rightleftharpoons\quad 2 \cdot Co(CO)_4 \tag{11}$$

$$\cdot Co(CO)_4 + H_2 \quad\rightleftharpoons\quad H\cdot + HCo(CO)_4 \tag{12}$$

$$H\cdot + R-OH \quad\rightarrow\quad R\cdot + H_2O \tag{13}$$

$$R\cdot + HCo(CO)_4 \quad\rightarrow\quad RH + \cdot Co(CO)_4 \tag{14}$$

235

Table 2.5. Reaction of substituted benzyl alcohols with synthesis gas (238 bar initial pressure, $CO:H_2 = 1:2$) [17]

Substituent	Hydro-carbon %	Homolo-gated alcohol %	% Homolo-gated % hydro-carbon	Recovered alcoholic feedstock %	Other products
p-methoxy	16	44	4.9	0	approximately 34% high-boiling polymer, probably aldol of p-methoxyphenylacetaldehyde
p-hydroxymethyl	27	39 (12[a])	1.9	0	
p-methyl	58	24	0.6	0	8% of high-boiling polymer, probably aldol
m-methyl	52	36	7	0	1–2% m-methylbenzaldehyde
p-tert-butyl	54	28	5	0	1% p-tert-butylbenzaldehyde
benzyl	63	31	5	0	—
2,4,6-trimethyl	58	18	3	0	—
p-chloro	41	16	4	31	4% p-chlorobenzaldehyde
m-methoxy	23	ca. 2	0.1	56	4% m-methoxybenzaldehyde
m-trifluoromethyl	5			78	4% m-trifluoromethyl benz-aldehyde
p-carbethoxy	27				high-boiling products
p-nitro				44	polymer of p-aminobenzyl alcohol

[a] This yield refers to the dihomologated alcohol, p-phenylene-β,β'-diethanol

However, the following reaction sequence is also plausible for the benzyl alcohol hydrogenation on combining the carbenium ion and radical mechanisms:

$$C_6H_5CH_2OH + H^+Co(CO)_4^- \rightarrow C_6H_5CH_2^+ + H_2O + Co(CO_4)^- \qquad (15)$$

$$C_6H_5CH_2^+ + Co(CO)_4^- \rightarrow C_6H_5CH_2Co(CO)_4 \qquad (16)$$

$$C_6H_5CH_2Co(CO)_4 \rightarrow C_6H_5CH_2\cdot + \cdot Co(CO)_4 \qquad (17)$$

$$C_6H_5CH_2\cdot + H_2 \rightarrow C_6H_5CH_3 + H\cdot \qquad (18)$$

$$H\cdot + \cdot Co(CO)_4 \rightarrow HCo(CO)_4 \qquad (19)$$

Recent high pressure IR studies on arylcarbonyl metal complexes confirmed the ready homolytic cleavage of the arylcarbonyl complex in Eq. (17) [47]. This step is apparently encouraged by resonance stabilization via the benzene ring:

$$RMo(CO)_3C_5H_5 \rightarrow R\cdot + \cdot C_5H_5-Mo(CO)_3 \qquad (20)$$

A reaction sequence via a sixfold coordinated cobalt complex is also feasible:

$$R-OH + HCo(CO)_4 \rightarrow R-Co(CO)_4 + H_2O \qquad (21)$$

$$R-Co(CO)_4 \qquad \rightleftharpoons \quad R-Co(CO)_3 + CO \tag{22}$$

$$RCo(CO)_3 + H_2 \qquad \rightleftharpoons \quad R(H)_2Co(CO)_3 \tag{23}$$

$$R(H_2)Co(CO)_3 \qquad \rightarrow \quad \textbf{RH} + HCo(CO)_3 \tag{24}$$

$$HCo(CO)_3 + CO \qquad \rightleftharpoons \quad HCo(CO)_4 \tag{25}$$

2.5.2 Carbonylation to Acids

The reaction mechanism of the cobalt catalyzed carbonylation has not been completely established (cf Chap. 3.6.3 for more detailed treatment of alcohol carbonylations).

A possible insertion mechanism for the formation of acetic acid from methanol with Co-catalyst has already been presented in Fig. 2.1. In addition, the significance of the halogen promoters was also discussed [6, 8, 36, 49].

A reaction sequence was proposed [49] in which the intermediate acyl complex is hydrolytically cleaved yielding acetic acid or the corresponding ester. This reaction is exploited in the commercial production of acetic acid (BASF process cf Chap. 3.7.2).

$$
CH_3COCo(CO)_n \rightarrow
\begin{cases}
\xrightarrow{\;H_2O\;} CH_3COOH + HCo(CO)_n & (26) \\[2ex]
\xrightarrow{\;CH_3OH\;} CH_3COOCH_3 + HCo(CO)_n & (27)
\end{cases}
$$

In the hydroformylation of ethylene for example, which takes place under the same conditions as the homologation, the reaction, which is catalyzed by 1% cobalt (doped with 0.01% iodine dissolved in 20% H_2O), must give rise to a certain amount of propionic acid from the acyl complex. In fact, only traces of a decomposition of this type were found [21], indicating that the route involving the hydrolytic cleavage of the acyl complex is not predominant with a C_2 moiety.

2.5.3 Acetal Formation

At low temperatures or minor conversion rates, the aldehyde intermediates are not reduced to the corresponding alcohols but react with excess alcohols to form acetals which can then be isolated [37]. Under drastic reaction conditions, the acetals react — via a series of cleavage and constructive reactions — to yield higher condensation products which, along with the aldol products, probably account for the largest share of the heavy ends.

Formaldehyde dimethyl acetal [14, 46, 50] which results in small quantities with a methanol feedstock [Eqs. (28–30)] – is particularly reactive [37]:

$$Co_2(CO)_7 + CH_3OH \rightarrow HCo(CO)_4 + CH_3-OCo(CO)_3 \tag{28}$$

$$CH_3OCo(CO)_3 \rightarrow CH_2O + HCo(CO)_3 \tag{29}$$

$$CH_2O + 2\,CH_3OH \rightleftharpoons CH_2(OCH_3)_2 + H_2O \tag{30}$$

It was shown [21, 50, 51] that methylal is readily convertible into 1,1,2-trimethoxy-ethane or tetraalkoxypropane:

$$CH_2(OCH_3)_2 + HCo(CO)_4 \qquad \rightarrow \quad CH_3OH + CH_3-O-CH_2-Co(CO)_4 \tag{31}$$

$$CH_3-O-CH_2-Co(CO)_4 \qquad \rightarrow \quad CH_3O-CH_2-COCo(CO)_3 \tag{32}$$

$$CH_3O-CH_2-COCo(CO)_3 + CO \quad \rightarrow \quad CH_3O-CH_2CO\,Co(CO)_4 \tag{33}$$

$$CH_3O-CH_2-COCo(CO)_4 + HCo(CO)_4 \rightarrow CH_3O-CH_2-CHO + Co_2(CO)_8 \tag{34}$$

$$CH_3O-CH_2-CHO + 2\,CH_3OH \qquad \rightleftharpoons \quad CH_3O-CH_2-CH(OCH_3)_2 \tag{35}$$

and

$$CH_3OCH_2\,CH(OCH_3)_2 \xrightarrow{\ Co;\,CO/H_2\ } CH_3OCH_2CH(OCH_3)CH(OCH_3)_2 \tag{36}$$

etc.

2.5.4 Ether Formation

At temperatures above 200 °C, dimethyl ether and mixed ethers are formed from the resulting higher alcohols under the acidic conditions of the homologation reaction. They then in turn undergo secondary reactions e.g., carbonylation to esters.

2.6 Heterogeneously Catalyzed Homologation

Besides the homogeneously catalyzed homologation, processes have been recently patented which permit the homologation to ensue with fixed-bed catalysts [57, 58]. In the majority of cases, the feedstock is syn gas which probably forms the higher oxygen-containing compounds [52–55] via a methanol intermediate or is directly converted into ethylene via dehydration [56]. So far, conversion rates have been unsatisfactory and, furthermore, the high enthalpy of reaction (homologation – approx. 51 kcal/mol) [59] indicates that heat-removal difficulties are to be expected in the commercial process.

2.7 Future Prospects of the Homologation

2.7.1 As an Alternative Source of Ethylene

A series of alternatives are currently being developed to replace oil as the traditional ethylene feedstock (cf Fig. 2.3). The intensity of this development work and its realization will depend not only on process improvements but mainly on the future oil supply situation:

The economics of all alternative routes to ethylene are impaired by one decisive factor – over 50% of the oxygen-containing compounds used as feedstocks (to supply carbon) are lost as (worthless) carbon dioxide or water during the process.

Other processes based on syn gas derived from coal are competing with the homologation e.g., modern selective variants of FT process [15, 71, 72] as well as the UCC [52–54] or Mobil-Oil [56] process. To date, none of these variants has exhibited sufficiently high conversion rates and selectivities. Thus, in the long term the homologation has good prospects of economic realization – assuming an increase in selectivity and yield along with milder reaction conditions – even although the homologation represents only one step in a three-stage process. In the medium-term, the economic barrier can probably breached when – as outlined above – the highest possible oxygen content remains in the molecule. This is the case when the final product is either ethanol or acetaldehyde [61].

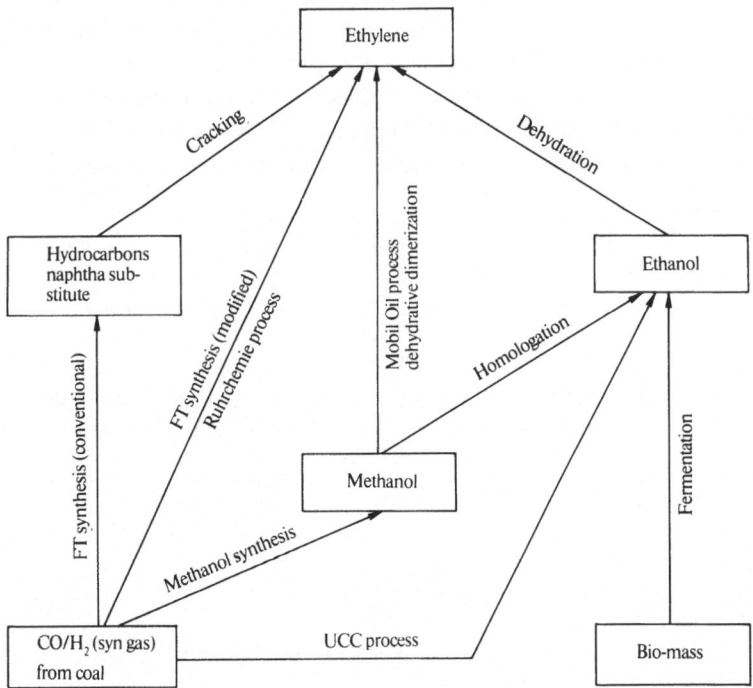

Fig. 2.3. Alternative paths to the chemical feedstock ethylene starting from synthesis gas

2.7.2 Production of Styrene

Toluene can be oxidized to benzyl alcohol via a new single step process [62]. After homologation to 2-phenylethanol, the latter can be dehydrated to styrene. Thus, these reactions present the possibility of producing styrene from toluene:

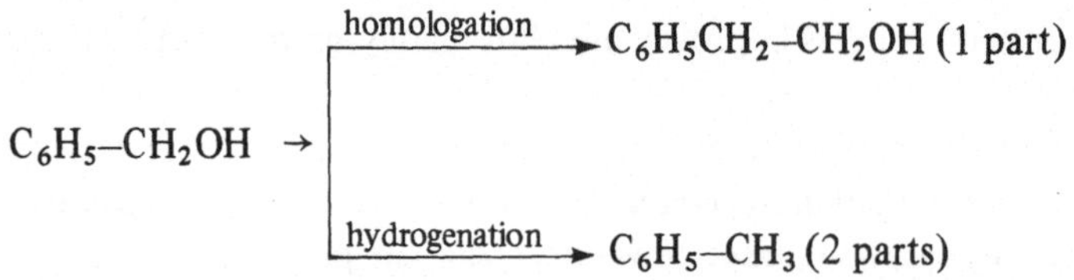

$$
C_6H_5-CH_2OH \rightarrow
\begin{cases}
\xrightarrow{\text{homologation}} C_6H_5CH_2-CH_2OH \text{ (1 part)} \\
\\
\xrightarrow{\text{hydrogenation}} C_6H_5-CH_3 \text{ (2 parts)}
\end{cases}
$$

The main aim of the development work will be to improve the selectivity of the homologation step to 2-phenylethanol. The toluene feedstock can either be obtained from the oil product from the coal liquefaction or from syn gas (via bifunctional catalysts) or from the Mobil-Oil process [63].

2.7.3 The Production of Methyl Fuels

Besides extending the life of the oil reserves, the partial or complete replacement of motor-fuels (based on hydrocarbons) by methanol or ethanol also introduces a series of advantages such as more effective usage of motor fuels and combustion with low environmental impact. However, the intermediate solution planned in several countries i.e., mixing alcohols with gasoline (10–15%) presents the danger of phase separation. This problem can be solved when isobutanol (from the Oxo synthesis) is used as solubizer. The application of higher oxygen-containing compounds from the homologation is currently being studied.

2.8 References

1. Wietzel, G., et al.: (BASF A.G.), 489 764 (1941) or DE-Appl. 442 318 and DE 867849 (1941)
2. Wietzel, G., Vorbach, O., Scheuermann, A., (BASF A.G.), 875 346 (1953); Supplement to 867 849 (1941)
3. Wender, I., Levine, R., Milton, M.: J. Amer. Chem. Soc. *71* 4160 (1949)
4. Ziesecke, K.H.: Brennstoff-Chem. *33*, 385 (1952)
5. Vorbach, O., Wietzel, G., Scheuermann, A. (BASF A.G.), 877 598 (1953)
6. Reppe, W., Friedrich, H. (BASF A.G.), 897 403 (1953)
7. Wender, I., Friedel, R.A., Orchin, M.: Science *113*, 206 (1951)
8. Berty, J., Markó, L., Kollo, D.: Chem. Techn. *8*, 260 (1956)
9. Slaugh, L.H. (Shell Intern. Res. Maatschappij B.V.), DE-OS 2 625 627 (1976)
10. Wender, I., Catal. Rev. Sci. Eng. 14 (1) 112 (1976)
11. Brooks, R.E. (DuPont), U.S. 2 457 204 (1948)
12. Gresham, W.F. (DuPont), U.S. 2 623 906 (1952)

13. Gresham, W.F. (DuPont), U.S. 2 535 060 (1950)
14. Pieroh, K. (BASF A.G.), 875 802 (1941)
15. Status and potential developments in the Fischer-Tropsch Synthesis, Study conducted by Ruhrchemie AG on behalf of Federal Minister for Research and Technology, Bonn, 1976
16. Comprehensive reviews by:
    − Asinger, F., Die petrolchemische Industrie. Vol. II, p. 1183 ff. Berlin: Akademie-Verlag 1971
    − Cornils, B., in Falbe, J.: Chemeierohstoffe aus Kohle. p. 329 ff. G. Thieme Verlag 1977
    − Falbe, J., Carbon Monoxide in Organic Syntheses. p. 59 ff. Springer Verlag 1970
    − Wender, I., Pino, P., Organic Synthesis via Metal Carbonyls. Vol.2. New York: Wiley 1977
17. Wender, I., et al.: J. Amer. Chem. Soc. 74, 4079 (1952)
18. Sternberg, H.W., et al.: J. Amer. Chem. Soc. 75, 2717 (1953)
19. Calderazzo, F., Angew. Chem. 89, 305 (1977)
20. Burns, G.R., J. Amer. Chem. Soc. 77, 6615 (1955)
21. Bahrmann, H., Tummes, H., Cornils, B., Ruhrchemie A.G. unpublished
22. Fell, B., Beutler, M., Thesis, RWTH Aachen, 1971
23. Wender, I. at CHEMRAWN, Proc. 1st World Conf., Ontario, Canada 10−13 July 1978 Pergamon Press
24. Butter, G.N. (Commercial Solvents Comp.), U.S. 3 285 949 (1966)
25. Mitsubishi Gas Chem. Ind., Japan. 5 207 804 (1977)
26. Kuraishi, M., U.S. 3 356 734 (1967)
27. Kuraishi, M., U.S. 3 387 043 (1968)
28. Kölbel, H., Engelhardt, F. (Rheinpreussen A.G.), DE 733 792 (1952)
29. Riley, D., Bell, W.O. (Commercial Solvents Comp.), DE 1 173 075 (1965)
30. Butter, N. (Commercial Solvents Comp.), DE 1 215 673 (1966)
31. Toshihide, Y. (Ajinomoto Co.), DE-OS 2 427 954 (1975)
32. Gijutsuin, K., JP 18 682/68 (1968)
33. Mitsubishi Gas Chem. Ind., JP 73.02.525 (1973)
34. Japan Gas Chem. Ind., JP 15 692/66 (1964)
35. Nippon Gas Chem. Ind., JP 12 802/68 (1963)
36. Mizoroki, T., Nakayama, M., Bull Chem. Soc., Japan 37, (2), 263 (1964)
37. Albanesi, G., Chim. Ind. (Milano), 55, 319 (1973)
38. Guyer, P., Friedli, H.R., Guyer, A., Chimia 13, 331 (1959)
39. Müller-Cunradi, M., et al.: (BASF A.G.) DE 890 935 (1942)
40. Dubeck, M., Knapp, G.G. (Ethyl Corp.), U.S. 4 062 898 (1975)
41. Kröper, H., Häuber, H., Hagen, W. (BASF A.G.), DE 921 936 (1955) C.A. 53, 222 (1959)
42. Wender, I., et al.: J. Amer. Chem. Soc. 77, 5760 (1955)
43. Mönkemeyer, K. (Chem. Werke Hüls), U.S. 2 770 655 (1956), C.A. 51, 5817 (1957)
44. Wender, I., Metlin, S., Orchin, M., J. Amer. Chem. Soc. 73, 5704 (1951)
45. Wender, I., Greenfield, H., Orchin, M., J. Amer. Chem. Soc. 73, 2656 (1951)
46. Goetz, R.W. (National Distillers and Chemical Corp.), DE-OS 2741 589 (1977)
47. King, R.B., King, A.D., et al., J. Amer. Chem. Soc. 100, 1687 (1978)
48. von Kutepow, N., Himmele, W., Hohenschutz, H., Chem.-Ing. Techn. 37, (4), 383 (1965)
49. Mizoroki, T., Nakayama, M., Bull. Chem. Soc. Japan 41, 1628 (1968)
50. Müller-Cunradi, M., Pieroh, K., Lorenz, L. (BASF A.G.), DE 890 945 (1953)
51. Gresham, W.F., Brooks, R.E. (DuPont), U.S. 2 525 793 (1950)
52. Cropley, J.B. (Union Carbide Co.), DE-OS 2 628 576 (1976)
53. Bhasin, M.M. (Union Carbide Co.), DE-AS 2 503 204 (1975)
54. Bhasin, M.M., Connor, O., Lawrence, G. (Union Carbide Corp.), DE-AS 2 503 233 (1975)
55. Sugier, A., Freund, E. (Inst. Francais du Petrole), DE-OS 2 748 097 (1978)
56. Chang, C.D., et al.: (Mobil Oil Corp.), DE-OS 2 615 150 (1976)
57. Marullo, G., Baroni, A. (Montecatini SpA), IT 484 182 (1953)
58. Kölbel, H., Engelhardt, F. (Rheinpreussen AG), GB 733 792 (1955)
59. Mizoroki, T., Nakayama, M., Bull. Chem. Soc. Japan 36, 1876 (1965)
60. Ichikawa, M., Chem. Comm. 1978, 566
61. Spitz, P.H., Chemtech 1977, 295

62. Anon., Chem. Engng. News, July 4, 1977, p 13
63. Meisel, S.L. et al.: Chemtech *1976*, (2) 86
64. Mitsubishi Gas Chem. Ind., JP 52 133 914 (1976)
65. Mitsubishi Gas Chem. Ind., JP 52 136 110 (1976)
66. Mitsubishi Gas Chem. Ind., JP 52 136 111 (1976)
67. Hecht, O., Kröper, H., Naturforsch. u. Medizin in Deutschland, 1939–1947, Vol. *36*, Part I, p. 143
68. Marko, L., Proc. Chem. Soc. *1962*, 67
69. Rathke, J.W., Feder, H.M., J. Amer. Chem. Soc. *100*, 3623 (1978)
70. Koermer, G.S., Slinkard, W.E., Ind. Engng. Chem., Prod. Res. Dev. *17* (3), 231 (1978)
71. Büssemeier, B., Frohning, C.D., Cornils, B., Hydrocarbon Proc. *55* 105 (1976)
72. Cornils, B., Büssemeier, B., Frohning, C.D., Erdöl/Kohle/Erdgas, Petrochem. *30*, 137 (1977)
73. Wilson, J.D.C (Du Pont de Nemours Co) US 2 524 503 (1950)
74. Wall, R.G. (Chevron Research Co) BE 85 8 628 (1978)
75. Imhausen Chemie GmbH DE-OS 2 731 962 (1977)
76. Fremery, M. et al. (Union Rheinische Braunkohlen Kraftstoff AG) DE-OS 2 726 978 (1979)
77. Sherwin, M.B (Chem. Systems Inc.) DE-OS 2 848 665 (1979)

3. Carbonylations Catalyzed by Metal Carbonyls–Reppe Reactions

A. Mullen

3.1 Introduction

The large number of reviews dealing with various aspects of metal carbonyl catalyzed carbonylations i.e. Reppe reactions, is indicative of the interest as well as the level of activity in this industrially significant field [1–12, 51, 242, 272].

The initial work on metal carbonyl catalyzed conversions – involving the introduction (insertion/addition) of CO into various substrates – was conducted by W. Reppe et al. (BASF) during the thirties and forties. The term 'carbonylation' was coined by Reppe. Carbonylations are those reactions in which CO, alone or together with other compounds, is introduced into a particular substrate. With the exception of hydroformylation this Chapter is devoted to carbonylations with metal carbonyls. The carbonylations, which usually ensue in the presence of a nucleophile with a mobile H-atom may occur either via addition of CO to unsaturated compounds (such as alkynes or alkenes, cf Sect. 6.1 and 3.6.2) or via insertion of CO into existing bonds, (e.g., alcohols etc., cf. Sects. 3.6.3–3.6.8). Some of the more typical reactions are (1–3):

$$HC\equiv HC + CO + H_2O \rightarrow H_2C=CH-COOH \qquad (1)$$

$$H_2C=CH_2 + CO + H_2O \rightarrow H_3C-CH_2-COOH \qquad (2)$$

$$RCH_2OH + CO + (H_2O) \rightarrow RCH_2-COOH \qquad (3)$$

The carbonylation of alkenes and alkynes with carbon monoxide and water is also known as the 'hydrocarboxylation reaction' (with alcohols instead of water 'hydroesterification').

The Reppe reactions thus represent routes for the manufacture of unsaturated and saturated acids, anhydrides, esters, amides etc. from a wide variety of feedstocks. Due to the industrial significance of the addition of CO/H_2 (syn gas) to olefins (cf Fig. 3.1) the 'hydroformylation' or 'Oxo synthesis' will be treated separately (cf Chap. 1). Figure 3.1 gives an overall impression of the relative industrial impact of the various carbonylation processes (cf Sect. 3.7).

The picture will shift in favor of the Reppe reactions (methanol carbonylation) when projected plants go into production in a few years time (cf Sect. 3.7.2). Acrylic acid [Eq. (1)] is an important feedstock for acrylic resins, the world production of acrylic acid and its esters being around 550,000 t/a (in 1973) [15]. Besides the acetylene-

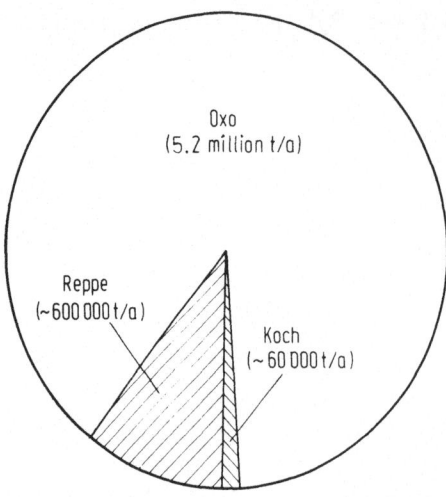

Fig. 3.1. Relative capacities of the various carbonylation processes

based Reppe reaction there are other commercially important routes, the cyanohydrin, propiolactone and propylene oxidation processes [15]. The acetylene route is the most important in terms of the capacity of existing plants.

$$H_2C=CH_2 + CO + 1/2\, O_2 \xrightarrow[Ac_2O]{cat.} H_2C=CH-COOH \qquad (4)$$

Due to the high cost of acetylene, attention is being devoted to the production of acrylates and their homologues from olefins, e.g., via oxidative carbonylation [Eq. (4)], or carbonylation followed by dehydrogenation (cf Sects. 3.6.2.1.1 and 3.6.2.2). Large acetylene-based plants which have already been written off against tax are of course commercially viable. The BASF process is used for the production of over 160,000 t/a acrylic acid (cf Sect. 3.7.1).

Recently, there has been considerable commercial interest in the carbonylation of alcohols, particularly methanol (cf. Sects. 3.6.3 and 3.7.2).

Monsanto's low pressure liquid-phase process for acetic acid production via the Rh catalyzed carbonylation of methanol is currently being licensed world-wide. If the planned plants go into production, their combined capacities (Rh- + Co-based) will exceed 1,500,000 t/a in a few years time [279–281]. Alcohol carbonylation is – after hydroformylation (cf Chap. 1) – the most significant industrial branch of carbonylation chemistry [279].

Various transition metal compounds can be employed as catalysts for carbonylation reactions, their carbonyls or hydrocarbonyls being the actual catalytic species (cf Sect. 3.3). The reactions take place under CO pressure in the presence of the metal carbonyls or carbonyl-forming metals as well as stoichiometrically at atmospheric or slightly raised pressures employing metal carbonyls as CO source. The stoichiometric method [10], which generally employs Ni(CO)$_4$, has the advantage that no high pressure reactors and elevated temperatures are necessary. There are thus three main types of carbonylation reactions differing in the amount of metal carbonyl employed:

244

- stoichiometric
- modified
- catalytic

The Reppe [14] hydrocarboxylation of acetylene utilizes the first method with $Ni(CO)_4$. The second method involves a less-than-stoichiometric amount of $Ni(CO)_4$, whereas the third concerns the use of catalytic amounts of, for example, $NiBr_2$ and a promoter such as a copper halide. The advantage of the latter method is that there is no economic requirement that the catalyst be recovered. However, the catalytic method demands conditions of 40–55 bar and 160–200 °C. Industrially, only the modified and catalytic processes have been of interest [15].

Rh catalyzed carbonylation processes have made a considerable commercial impact as they often operate at ambient pressures and temperatures (cf Sects. 3.6.3 and 3.7.2).

A positive aspect of the Reppe reactions is that relatively little structural isomerization takes place — quite different from the spectrum of products stemming from the Koch, acid-catalyzed, carbonylations (cf Chap. 5).

The rising price of oil and the necessity of utilizing the native coal reserves in Western countries will give even greater impetus to activities in the carbon monoxide field, i.e., carbonylation chemistry [279]. Interesting developments can thus be expected over coming decades.

3.2 Reaction Mechanism

The Reppe carbonylations are catalyzed by transition metal complexes with carbon monoxide, the latter stabilizing the lower oxidation states of these metals. The mechanisms of the Pd and $Ni(CO)_4$ catalyzed hydroesterification of alkynes will be discussed in Sect. 3.6.1.2. The cobalt catalyzed hydrocarboxylation of alkenes was studied by Heck, who proposed for α-olefins a course of reaction [1] [Eq. (5)]. This mechanism is analog to Heck's work in the Oxo field (cf Chap. 1).

As the cobalt catalyst is capable of bonding to either of the olefinic carbon atoms, two isomeric carboxylic acids can result. Moreover, as this catalyst can induce a displacement of the double bond, further isomers can be expected. However, the linear carboxylic acid usually dominates in the product. Even with non-terminal olefins, the non-branched carboxylic acids generally account for the major part of the product (cf Sect. 3.3.2). The presence of pyridine can further increase the linear acid content when using α- or β-olefins as precursor [4].

On employing $Ni(CO)_4$ with traces of hydrogen halides in the olefin carbonylation, Heck proposed the mechanism shown in [Eq. (6)], [16].

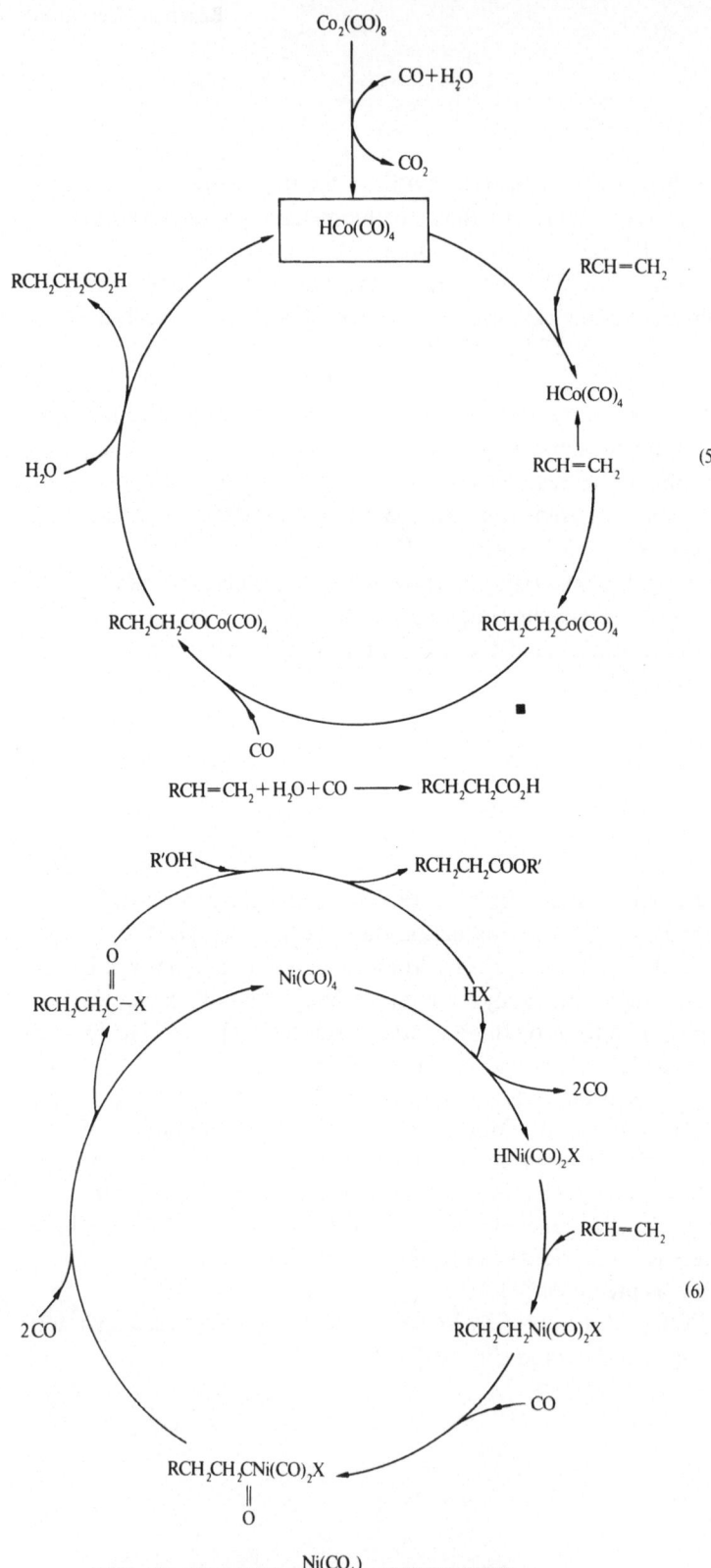

Co$_2$(CO)$_8$

CO+H$_2$O

CO$_2$

HCo(CO)$_4$

RCH=CH$_2$

RCH$_2$CH$_2$CO$_2$H

HCo(CO)$_4$

RCH=CH$_2$

H$_2$O

(5)

RCH$_2$CH$_2$COCo(CO)$_4$

RCH$_2$CH$_2$Co(CO)$_4$

CO

RCH=CH$_2$+H$_2$O+CO $\longrightarrow$ RCH$_2$CH$_2$CO$_2$H

R'OH

RCH$_2$CH$_2$COOR'

$$RCH_2CH_2\overset{O}{\overset{\|}{C}}-X$$

Ni(CO)$_4$

HX

2CO

HNi(CO)$_2$X

RCH=CH$_2$

(6)

2CO

RCH$_2$CH$_2$Ni(CO)$_2$X

CO

$$RCH_2CH_2\underset{\|}{\overset{}{C}}Ni(CO)_2X$$
O

RCH=CH$_2$+CO+R'OH $\xrightarrow[HX]{Ni(CO_4)}$ RCH$_2$CH$_2$COOR'

246

The mechanism of the Pd catalyzed oxidative carbonylation of olefins has also been given special attention due to its industrial potential [Eq. (7) [13]:

$$H_2C{=}CH_2 + CO + 1/2\,O_2 \xrightarrow[Ac_2O]{PdCl_2} \begin{array}{l} H_2C{=}CHCOOH \\ AcO{-}CH_2{-}CH_2{-}COOH \end{array} \tag{7}$$

The following reaction steps have been proposed [Eq. (8)] [13]:

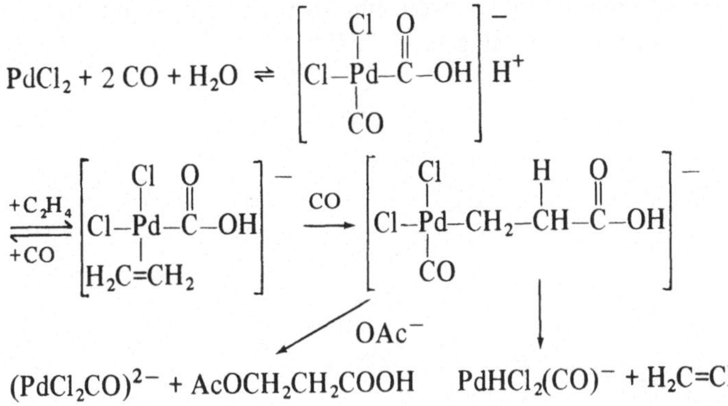

$$PdCl_2 + 2\,CO + H_2O \rightleftharpoons \left[\begin{array}{c} Cl\quad O \\ | \quad\; \| \\ Cl{-}Pd{-}C{-}OH \\ | \\ CO \end{array}\right]^{-} H^{+}$$

$$\underset{+CO}{\overset{+C_2H_4}{\rightleftharpoons}} \left[\begin{array}{c} Cl\quad O \\ | \quad\; \| \\ Cl{-}Pd{-}C{-}OH \\ | \\ H_2C{=}CH_2 \end{array}\right]^{-} \xrightarrow{CO} \left[\begin{array}{c} Cl\qquad\;\; H\;\; O \\ | \qquad\quad | \;\;\; \| \\ Cl{-}Pd{-}CH_2{-}CH{-}C{-}OH \\ | \\ CO \end{array}\right]^{-} \tag{8}$$

$$\swarrow OAc^{-} \qquad\qquad \downarrow$$

$$(PdCl_2CO)^{2-} + AcOCH_2CH_2COOH \qquad PdHCl_2(CO)^{-} + H_2C{=}CHCOOH$$

Heck et al. also proposed a mechanism for the palladium triphenylphosphine complex catalyzed carbalkoxylation of organic halides, in the presence of an alcohol and an amine [Eq. (9)] [17].

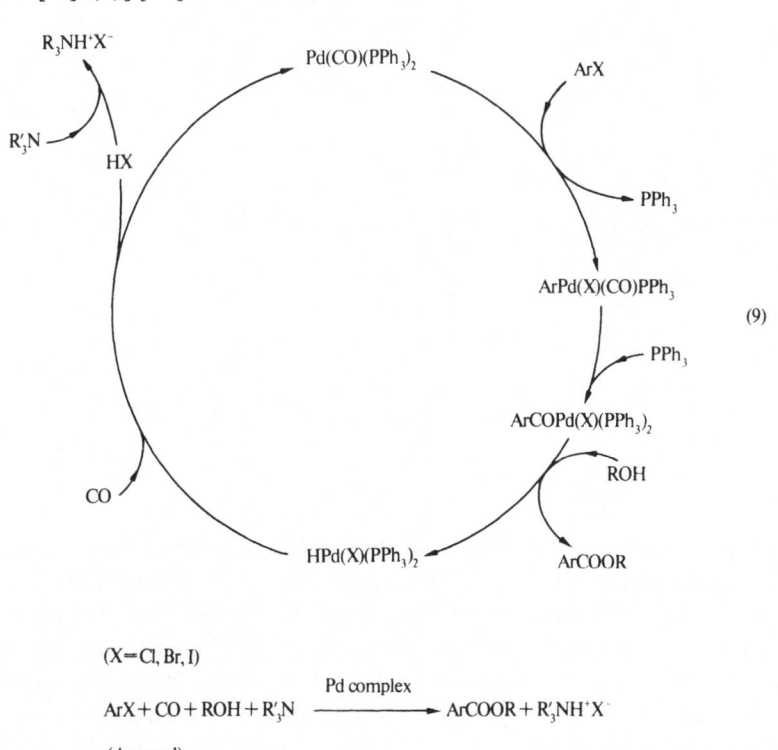

(9)

$$(X{=}Cl, Br, I)$$

$$ArX + CO + ROH + R'_3N \xrightarrow{\text{Pd complex}} ArCOOR + R'_3NH^{+}X^{-}$$

$$(Ar = aryl)$$

247

The reaction occurs under mild conditions [17] (cf Sect. 3.6.6 for mechanism of the Pd catalyzed carbonylation of vinyl halides).

The mechanism of the rhodium catalyzed carbonylation of methanol to acetic acid was recently clarified by Forster [18]. Based on catalytic studies with various Rh compounds, it was suggested that the same moiety was responsible for catalyzing the carbonylation [18].

The scheme shown in Eq. (10) is in agreement with the independence of the overall reaction rate on the methanol concentration and CO pressure [19]. In this case, Rh(III) halides, together with methyl iodide, were employed as catalyst precursors. $[Rh(CO)_2X_2]^-$ [18, 20] was formed on using excess halide

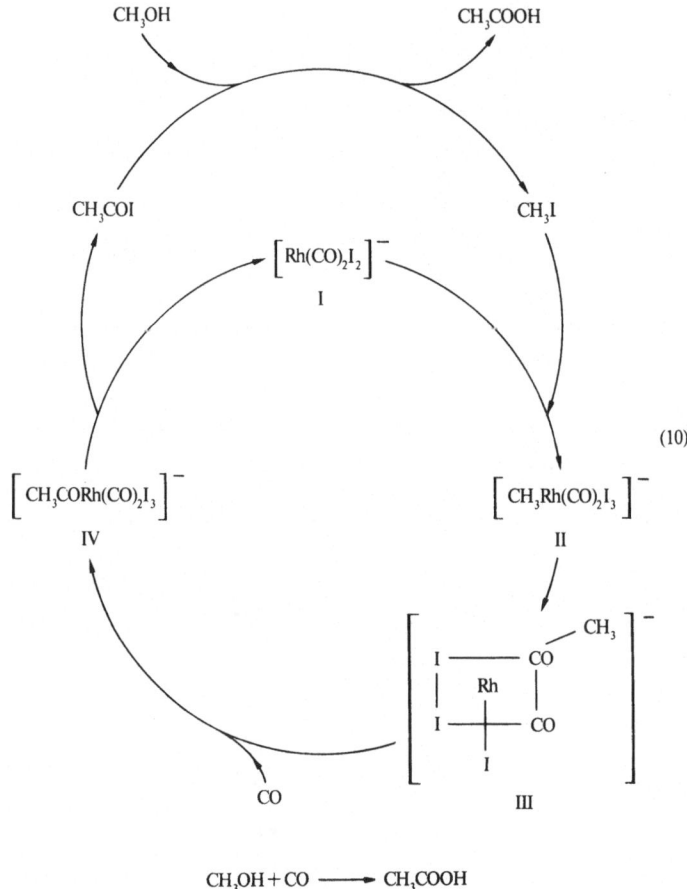

$$(10)$$

$$CH_3OH + CO \longrightarrow CH_3COOH$$

The fact that the initial formation of an acetyl complex (e.g. III) ensues in the absence of CO and that III reacts with CO at 1 bar pressure provide an explanation for the lack of rate dependence on the CO partial pressure [18]. The rate determining step is the oxidative addition of CH_3I to $[Rh(CO)_2I_2]^-$ which, as already mentioned, can be formed from a variety of Rh compounds [21]. The steps in the mechanism are: insertion of CO between Rh and CH_3, reductive elimination and exchange between acetyl iodide and methanol to yield acetic acid whilst regenerating the promoter (HI or MeI). Recently, work was carried out to find a suitable promoter possessing the activity of HI but not its corrosive properties [21]. It was assumed that the effectiveness of a promoter CH$_3$X is based on the leaving group ability of X. Several substances were examined, the results are presented in Fig. 3.2 [21].

This comparison (Fig. 3.2) clearly indicates the superior leaving group ability of pentachlorobenzenethiol although is it not equal to that of MeI. The rate of methanol carbonylation with pentachlorobenzenethiol at 156 °C is about 4% of the CH$_3$I rate at that temperature [21]. To compensate for this drop in activity, higher concentrations of promoter could be employed at higher temperatures. Pentachlorobenzenethiol is relatively non-corrosive and enables carbonylation to occur under mild reaction conditions. Furthermore, this promoter has a higher boiling point than the products of the carbonylation, facilitating a ready distillative separation. In addition, both chloro- and fluoro-derivatives exhibit the same high degree of selectivity as MeI [21].

The mechanism has also been reviewed by Hjartkjaer [22]. The considerable industrial significance (see Sect. 3.7) of this reaction has led to a great deal of effort being devoted to the study of its mechanism [18–22]. The mechanisms of the Ni and Co catalyzed carbonylations of methanol are discussed in Sect. 3.6.3.

Although the Rh catalyzed hydrocarboxylation of alkenes has not achieved any notable significance, several firms (Monsanto in particular) have been working in this field [271–278]. A reaction mechanism [Eq. (10a)] based on radio-tracer experiments, was proposed by Morris and Johnson [278].

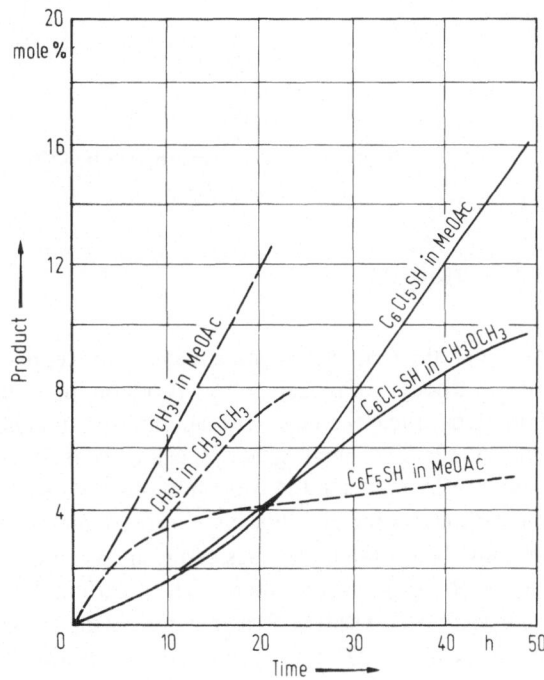

Fig. 3.2. Effectiveness of CH$_3$I, C$_6$Cl$_5$SH and C$_6$F$_5$SH as promoter in Rh catalyzed carbonylation of methanol [21].

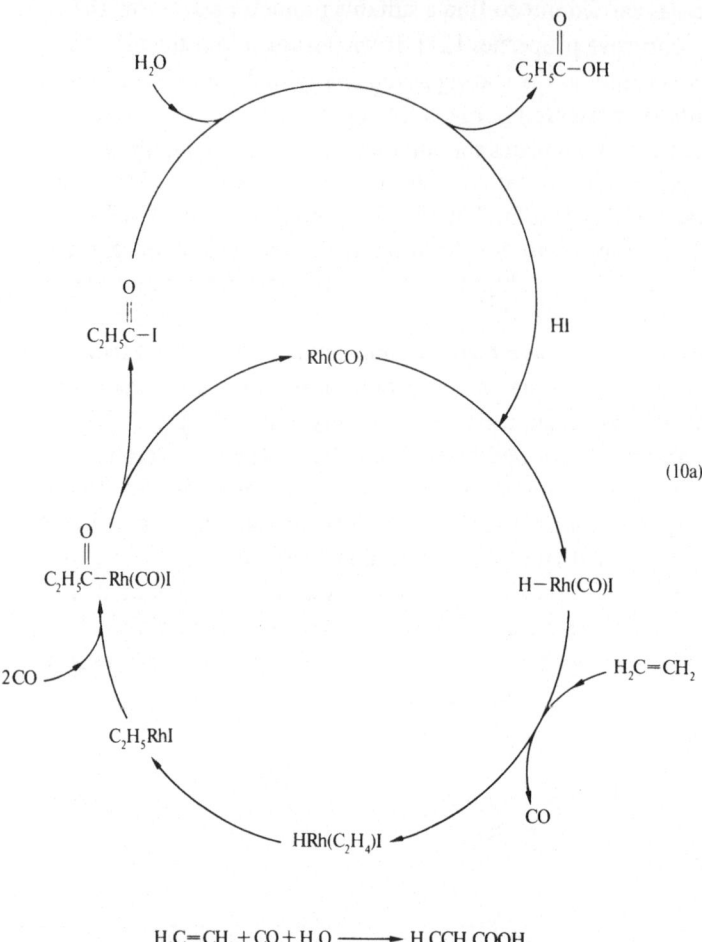

$$H_2C=CH_2 + CO + H_2O \longrightarrow H_3CCH_2COOH$$

3.3 Catalysts

Although the most significant catalysts for Reppe carbonylations are based on the metal carbonyls of nickel, cobalt, rhodium, platinum, palladium and iron, other transition metals such as copper, ruthenium, osmium and manganese have also found application [5, 12]. The metal carbonyls can be generated by reacting either the finely divided metals or by reducing the corresponding salts with CO. Above 100 °C these volatile carbonyls generally dissociate into the metal and CO unless a high CO pressure prevails. Transition metals possessing an even number of valence electrons (Cr, Mo, W, Ni, Fe, Ru, Os) can form mononuclear carbonyls, whereas polynuclear carbonyls or hydrocarbonyls result with metals possessing an odd number of valence electrons (Mn, Re, Co, Rh, Ir) [23]. Mixed carbonyls are readily formed by ligand exchange reactions. The steric effects of phosphorus ligands were recently reviewed [24] and a recent monograph [289] discusses model systems for homogeneous catalysis with CO and metal carbonyls.

With an appropriate choice of metal and ligand, catalysts can be obtained capable of carbonylating under very mild conditions. There has been a great deal of industrial activity in this field i.e. to produce catalysts for commercial use which enable carbonylation processes to ensue at ambient temperatures and pressures.

3.3.1 Nickel Catalysts

Nickel can be employed in one of three ways when carbonylating alkynes: stoichiometrically, stoichiometric-catalytically (modified) and catalytically. When $Ni(CO)_4$ is used in stoichiometric amounts, the carbonylation requires only mild reaction conditions, as the former serves both as CO source and catalyst in accordance with Eq. (11):

$$4\ HC\equiv CH + 4\ H_2O + Ni(CO)_4 + 2\ HCl \xrightarrow[\text{1 bar}]{40\,^{\circ}C} 4\ H_2C=CH-COOH + NiCl_2 + [2\ H]$$

$$(11)$$

The $NiCl_2$ can be recycled [Eq. (12)]:

$$[Ni(NH_3)_6]\,Cl_2 + 5\ CO + 2\ H_2O \rightarrow Ni(CO)_4 + 2\ NH_4Cl + (NH_4)_2CO_3 + 2\ NH_3 \quad (12)$$

The stoichiometric method has found no widespread industrial application on account of the involved recovery, work-up and toxicity of $Ni(CO)_4$ [4, 11].

The stoichiometric-catalytic or modified process, which under mild conditions employs a less-than-stoichiometric amount of $Ni(CO)_4$, has been described [4].

The catalytic processes generally take place below 100 bar and above 250 °C using Ni salts along with a halide etc. as promoter. The effectiveness of the halides increases in the following sequence:

fluoride < chloride < bromide < iodide [16].

Other substances such as pyridine, alkyl- or aryl-phosphines, hydrogen or copper have also a positive effect on the course of the reaction [1]. Nickel catalysts are very suitable for the carbonylation of alkynes, whereas for olefins, Co, Rh, Pd, Pt, and Ru are equally good if not better [4]. In the nickel catalyzed hydrocarboxylation of α-olefins, the main product (60–70%) is usually the branched carboxylic acid [25]. With internal olefins, the branched acid is almost exclusively formed.

The hydrocarboxylation of alkenes with nickel tetracarbonyl and water has recently been carried out at atmospheric pressure and ambient temperatures on introducing an organic acid [6, 26]. In the presence of the latter, the ratio of branched to linear acid in the product is 19:1 with an olefin precursor [26].

Recent work [27] on the hydrocarboxylation of propene [Eq. (13)] with $Ni(CO)_4$ as catalyst showed that with excess PPh_3 the reaction took place under milder conditions than with $Ni(CO)_4$ alone. Furthermore, the selectivity to isobutyric acid fell from 70 to 50%. There are certain analogies with the hydroformylation reaction. The catalytically active species is thought to be $NiHX(CO)\,(PPh_3)$ [27].

$$CH_3-CH=CH_2 + CO + H_2O \xrightarrow{\text{cat.}} \begin{array}{l} CH_3CH_2CH_2CO_2H \\ (CH_3)_2CHCO_2H \end{array} \qquad (13)$$

3.3.2 Cobalt catalysts

The $Co_2(CO)_8$ catalyzed hydrocarboxylation of linear α-olefins usually gives rise to ~55–60% linear carboxylic acids [28], the total carboxylic acid yield being 80–90%. The conversion takes place at 150–200 °C and 150–250 bar [28]. As cobalt carbonyls readily isomerize alkenes at ambient temperatures, when β-alkenes are hydrocarboxylated the main product is the linear acid [cf Eq. (14) and Sect. 3.2] [16]:

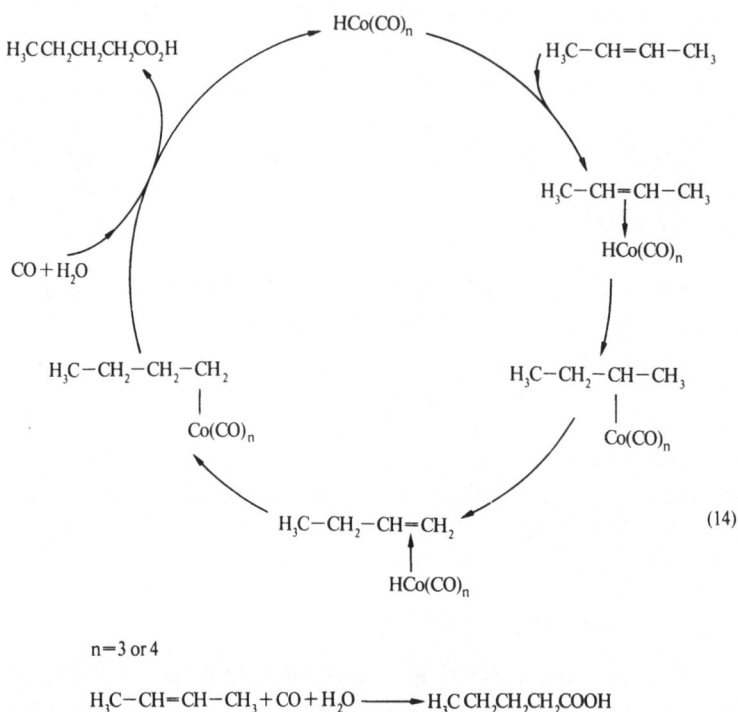

$$n = 3 \text{ or } 4$$

$$H_3C-CH=CH-CH_3 + CO + H_2O \longrightarrow H_3C\,CH_2CH_2CH_2COOH$$

(14)

When the cobalt catalyzed hydrocarboxylation of α-olefins is conducted in the presence of pyridine, the content of linear acid in the product can be increased to ~80% [29]. This catalyst system has recently been reviewed [30]. A similar effect is observed with internal olefins although to a more limited extent and necessitating longer reaction times. While the addition of hydrogen can increase the reaction rate it has the disadvantage that hydroformylation can ensue [4]. Both of these additives encourage cobalt hydridocarbonyl formation [31]. Butadiene can be converted under relatively mild conditions into methyl 3-pentenoate when carbonylated with $Co_2(CO)_8$ in the presence of methanol and a base such as pyridine or isoquinoline [32]. The cobalt catalyst can be recovered on distilling over the products of the reaction. Satisfactory yields were reported using this residue as hydrocarboxylation catalyst [33].

In the hydrocarboxylation of ethylene with CoI_2–EtI as catalyst, it was found that the Ir, Ni and Rh salts were much less effective than cobalt [34], which gave > 99% selectivity (based on $EtCO_2H$).

BASF [36] and Shell [36] have disclosed using Co/I_2 catalysts with Pd, Pt, Ir, Ru or Cu salts or complexes, thereby facilitating the carbonylation of methanol at lower temperatures (80–200 °C) and pressures (70–300 bar) than the conventional process (CoI_2 carbonylation at 250 °C and 680 bar).

3.3.3 Rhodium Catalysts

The most widely used catalysts for the carbonylation of alcohols are rhodium or cobalt [3], the former possessing the advantage that the reaction can ensue under milder conditions than with cobalt or nickel [22]. The best catalysts for the carbonylation are based on Rh with halide promoters (cf Sects. 3.6.3 and 3.7.2). To combat losses of this precious metal, in addition to low-loss recycling techniques, processes have been developed involving anchored catalysts [37, 38] which – coupled with virtually the same activity – give rise to 85% less Rh in the effluent [38].

In the Rh catalyzed carbonylation of methanol, the acetic acid selectivity is around 99% at a CO pressure as low as 1 bar [19]. The reaction rate is almost constant for a variety of rhodium complexes, suggesting the same catalytic moiety [19–22]. $[Rh(CO)Cl(PPh_3)]_2$ and $[Rh(CO)_2Cl_2]$ are useful for the carbonylation of alcohols.

Several firms have been working on the rhodium catalyzed hydrocarboxylation of alkenes [271–278], $RhCl_3$ or $Rh(CO)_2Cl_2$ being often used as catalyst. This type of catalysis apparently gives rise to fewer by-products [272]. The reaction conditions were generally 25–50 bar and < 200 °C.

Rhodium complexes, e.g., chlorocarbonyl-bis(triphenylphosphine)rhodium, have been employed in the carbonylation of alkyl halides, necessitating, however, high temperatures and pressures [7, 39].

Recently, 100% Rh recovery was reported using a process for the regeneration of spent complex catalysts from the methanol carbonylation [40]. The economic viability of the attractive Rh process depends on an extremely efficient catalyst recovery.

For the mechanisms of the Rh catalyzed alkene and methanol carbonylations, see Sect. 3.2.

3.3.4 Palladium and Platinum Catalysts

In terms of catalytic activity, these metals generally fall into fourth place behind nickel, cobalt, and rhodium although the actual sequence depends on the ligand involved as well as on the substrate and promoter [5].

Platinum catalysts can facilitate mild hydrocarboxylation of olefins and dienes (50–100 °C/atmos. pressure) [23, 41].

With H_2PtCl_6–$SnCl_2$, a series of α-olefins were found to react with carbon monoxide and methanol producing ~85% yields of the corresponding linear methyl ester, branched esters being formed in ~15% yield [41].

$$RCH{=}CH_2 + CO + CH_3OH \xrightarrow{\text{Pt cat.}} RCH_2{-}CH_2{-}CO_2CH_3 \qquad (15)$$

The reaction takes place under relatively mild conditions, the acid being produced on substituting CH_3OH with H_2O [41].

In palladium catalyzed carbonylations [4, 5, 7, 8, 42, 49] no simple carbonyls are formed. The reaction generally proceeds via formation of an olefin-palladium (π) complex followed by CO attack in absence or presence of an alcohol to produce an acyl chloride or ester [8, 42].

The carbonylation of styrene with palladium-phosphine complexes [Eq. (16)] gives a product specificity depending on the phosphine ligand used [43].

$$
\begin{array}{c}
[PdCl_2(PPh_3)_2] \\
\end{array}
$$

$$
\bigcirc\!\!\!\!\!\nearrow + CO + EtOH
\begin{cases}
[PdCl_2(PPh_3)_2] \longrightarrow H_3C\text{–}CH\text{–}CO_2Et \;\; (Ph) \\
[PdCl_2\{Ph_2P(CH_2)_4PPh_2\}] \longrightarrow Ph\text{–}CH_2\text{–}CH_2\text{–}CO_2Et \\
[PdCl_2\{Ph_2P(CH_2)_nPPh_2\}] \longrightarrow \text{no reaction}
\end{cases}
$$

$$n = 1, 2 \text{ or } 3 \tag{16}$$

Much work has been done on the Pd catalyzed carbonylation of aryl and benzyl halides under mild conditions ($> 100\,°C$, 1 bar), the best yields being obtained with iodides [44] (cf Sect. 3.2 for mechanism). When the reaction is conducted in the presence of an amine instead of an alcohol, the corresponding amide is obtained [45]. The carbonylation of olefins with $PdCl_2$-phosphine complexes can also be carried out under mild conditions ($< 100\,°C$) which help to ensure a lower degree of by-product formation [46]. In the hydrocarboxylation of ethylene with $Pd(PPh_3)_4$ no propionic acid was formed in the absence of PPh_3 [Eq. (17)] [47].

$$PPh_3 + H_2C=CH_2 + CO + H_2O \xrightarrow[135\,°C/135\,bar]{\text{cat.}} H_3C\text{–}CH_2\text{–}COOH + PPh_3 \tag{17}$$

The $PdCl_2(PPh_3)_2$ catalyst from the hydrocarboxylation of propene can be reactivated using HCl [48].

In the carbonylation of butadiene different products are obtained depending on the Pd catalyst employed [Eq. (18)] [49].

$$\bigwedge\!\!\bigwedge + CO + ROH \xrightarrow{PdCl_2} \bigwedge\!\!\bigwedge_{COOR} \tag{18}$$

$$2\;\bigwedge\!\!\bigwedge + CO + ROH \xrightarrow{Pd/PPh_3} H_2C=CH\text{–}CH_2\text{–}CH_2\text{–}CH_2\text{–}CH=CH\text{–}CH_2\text{–}COOR$$

Whether a monomeric or dimeric carbonylation takes place is determined by the presence or absence of a halide ion coordinated to Pd [49]. In the carbonylation reaction, 2 moles of butadiene form the diallylic palladium complex which is then followed by CO insertion producing the 3,8-nonadienoate. This step is prevented when the chloride ion coordinates with the palladium resulting in monomeric complex formation and CO insertion to give 3-pentenoate [49].

The reactivation of spent $PdCl_2/PPh_3$ hydrocarboxylation catalysts can be carried out by distilling over the product, oxidizing the pot residues with HNO_3, extracting with ether then treating the residue with PPh_3 and HCl. The regenerated catalyst was reported to possess activity comparable to that of the fresh catalyst [50].

3.3.5 Iron Catalysts

$Fe(CO)_5$ may be used in the hydrocarboxylation of acetylenes [10], resulting in lower yields ($\sim28\%$) compared to reactions catalyzed by other metal carbonyls (e.g. Ni carbonyl $\sim50\%$) [12, 51].

In alcohol carbonylations, iron carbonyls have been used successfully in conjunction with promoters (halides) although yields do not approach those obtained with rhodium [22]. Iron hydridocarbonyl can isomerize olefins at room temperature – a property it shares with the corresponding Co compounds [23].

When $Fe_2(CO)_9$ was employed together with $PdCl_2$ in the hydrocarboxylation of 1-octene, the acids were obtained in 50% yield, the ratio of linear to branched being $3:1$ [52]. When $Fe_2(CO)_9$ was replaced by $FeCl_2$, the total yield increased to 63%, but the ratio of nonanoic to α-methyloctanoic acid was $4:5$ [52].

Stoichiometric amounts of iron carbonyl lead to the remarkable formation of hydroquinone and its derivatives from substituted alkynes, carbon monoxide and water [1, 4, 12]. Aryl carboxylic acids may be obtained via the stoichiometric reaction of aryldiazonium salts and $Fe(CO)_5$ at ambient pressure and temperature, using ethanol, acetone or acetic acid as solvent.

The reaction of N,N-dichloroamines or -amides with $Fe_2(CO)_9$ generates the corresponding isocyanate in reasonable yield [55].

With iron hydridocarbonyl, propanol and not propionic acid is obtained from the reaction with ethylene, CO and H_2O [56]. In the presence of Pd catalysts, iron carbonyl has been prepared in $\sim22\%$ yield from Fe compounds and synthesis gas [57].

There is growing interest in the cluster complexes of iron which exhibit greater activity than $PdCl_2(PPh_3)_2$ in the carbonylation of 1-octene [58].

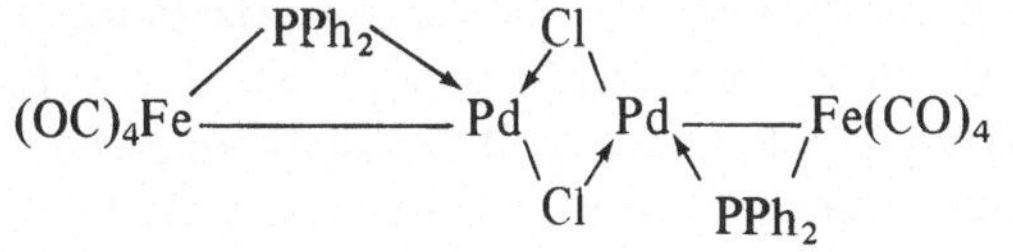

$$(OC)_4Fe \overset{PPh_2}{\underset{}{\diagup}} Pd \overset{Cl}{\underset{Cl}{}} Pd \overset{}{\underset{PPh_2}{}} Fe(CO)_4 \tag{19}$$

3.3.6 Copper Catalysts

Copper (I) carbonyl (in strong acid) catalyzes the hydroesterification of olefins and alcohols under very mild conditions (25 °C, 1 bar), but branched (tertiary) carboxylic acids predominate in the reaction product [59–61]. The reaction is a modified Reppe-Koch conversion (see Chap. 5).

Dienes and diols, between C_6 and C_{12}, give with copper carbonyls, mixtures of mono- and dicarboxylic acids at ambient temperature and pressure [61]. Using $Co_2(CO)_8$, Rh or Ir complexes as catalyst necessitate higher temperatures and pressures.

$$Cu^+ + CO \rightarrow Cu(CO)^+ \xrightarrow{n\,CO} Cu(CO)_n^+ \qquad\qquad (20)$$

$$Cu(CO)_n^+ \rightleftharpoons Cu(CO)^+ + (n-1)CO \qquad n = 3,4$$

The copper carbonyl ion [Eq. (20)] functions as a CO supplier, transporting the CO from the gas phase to the H_2SO_4 solution. Although diene carbonylation in acidic solution is usually difficult (due to the danger of polymerization) copper (I) carbonyl facilitates the conductance of the reaction in high yields — even in strong acid solution [61].

Saturated hydrocarbons [Eq. (21)] react with CO in the presence of an alcohol or olefin, using Cu (I) carbonyl generated in conc. H_2SO_4 (> 80%). Tertiary carboxylic acids [Eq. (21)] are formed in yields between 20–70% at ambient pressure and temperature, the unstable $Cu(CO)_n^+$ ion acting as a catalyst by releasing CO in the presence of acceptors [62]. The saturated hydrocarbon undergoes a hydride transfer reaction with the carbenium ion derived from the olefin or alcohol, resulting in the formation of a new carbenium ion which participates in the carbonylation reaction. Only saturated compounds possessing a tertiary C atom react. Octane and cyclohexane are unsuitable and remain unchanged [62].

$$\text{(cyclohexyl)}\overset{Me}{\underset{}{}}H + Me(CH_2)_3 CH=CH_2 + CO \xrightarrow{Cu\ catalyst} \text{(cyclohexyl)}\overset{Me}{\underset{}{COOH}} + \text{tert-}C_7\ acids \qquad (21)$$

3.4 Effect of Temperature and Pressure

With an appropriate choice of metal and ligand, catalysts can be obtained capable of carbonylating under very mild conditions [6]. High temperatures and pressures not only encourage side reactions such as decomposition, polymerization or isomerization and thus by-product formation but also can be a source of severe corrosion problems, e.g., BASF had to develop a special Mo-Ni alloy (Hastelloy B, C) for their Co catalyzed carbonylation of methanol [11]. When high temperatures are necessary they must often be accompanied by high pressures as the metal carbonyls otherwise dissociate into the metal and CO [23].

When hydrocarboxylating alkynes, a wide range of temperatures and pressures (Table 3.1) can be employed depending on the catalyst:
It was found that the rate of acetylene carbonylation with $Co_2(CO)_8$ was a function of the pressure, whereas the selectivity of the process was independent of the pressure [65].

As the result of an appropriate reactor design, the catalytic hydrocarboxylation of HC≡CH under pressure has become a virtually risk-free operation [11].

Table 3.1. Carbonylation of alkynes with various catalysts

Catalyst	Temp. [°C]	Pressure [bar]	Ref.
$Ni(CO)_4$	180–205	10–55	[4]
$Ni(CO)_4$ (stoichiometric method)	25– 70	1	[4]
$PdCl_2/HgCl_2$	25	2	[63]
$Co_2(CO)_8$	110	210	[64]

Table 3.2. Carbonylation of alkenes with various catalysts

Catalyst	Temp. [°C]	Pressure [bar]	Ref.
Ni	250	200	[1, 9]
Ni (stoichiometric)	150	50	[1, 9]
Co	100–260	30–900	[1, 9]
(with higher alkenes)	280	200	[1, 9]
Rh/Ir complexes	50–300	slightly raised pressure	[9]
$Ru(CO)_{12}$	190	20	[66]
Pd	80–150	50	[6]

The catalytic carbonylation of the olefinic double bond requires higher temperatures and pressures (Table 3.2) than the corresponding reaction with the alkynes [1, 4, 6].

Due to the mild reaction conditions, reactive olefinic compounds such as butadiene and styrene can be carbonylated without polymerization taking place [4].

The hydrocarboxylation reaction generally occurs at temperatures ~50 °C higher than the corresponding hydroesterification reaction (with ROH instead of H_2O).

In the oxidative carbonylation of ethylene [Eq. (22)], the effect of temperature and pressure on the conversion to acrylic and β-acetoxypropionic acid is shown in Fig. 3.3 and 3.4 [13].

$$H_2C=CH_2 + CO + 1/2 \ O_2 \xrightarrow[Ac_2O]{PdCl_2} \begin{array}{l} H_2C=CHCOOH \\ AcO-CH_2-CH_2-COOH \end{array} \qquad (22)$$

The conversion is at a maximum around 138 °C, while the selectivity is reported as being at an optimum around 127 °C [13]. With increasing pressure the β-acetoxypropionic acid content increases [13] (for mechanism cf Sect. 3.2).

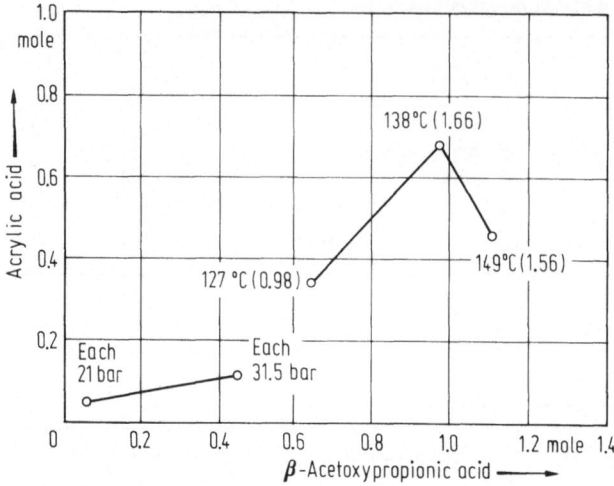

Fig. 3.3. Effect of temperature and pressure on conversion of ethylene [15] (figures in parenthesis represent total acid yield)

Figure 3.4 shows that high ethylene pressure favors the formation of acrylic acid, whilst high CO pressure enhances β-acetoxypropionic acid formation [13].

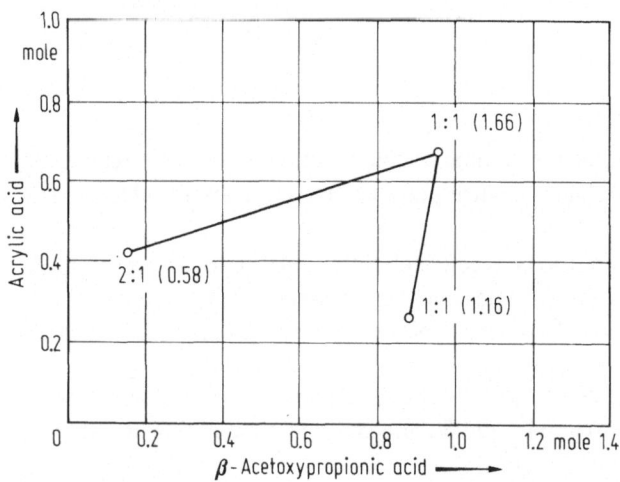

Fig. 3.4. Effect of partial pressures (C_2H_4:CO) on selectivity [13] (figures in parenthesis represent total yield)

Typical conditions for alcohol carbonylation are presented in Table 3.3

Table 3.3. Reaction conditions for alcohol carbonylation

Catalyst	Temp. [°C]	Pressure [bar]	Ref.
Ni/I	150–280	40–300	[14]
Co/I	180	70–700	[67]
Rh/I	⩾ 50	1–15	[68]

The BASF Co catalyzed process for acetic acid manufacture from methanol requires temperatures of ~250 °C and pressures of 680 bar. Recent work conducted by BASF and Shell using noble metal additives has effected a lowering of pressure (70–300 bar) and temperature (80–200 °C) [36]. The cobalt catalyzed CH_3OH carbonylation is markedly dependent on CO concentration.

The rhodium catalyzed carbonylation of methanol showed the following dependency on reaction temperature (Table 3.4) [69].

Table 3.4. Effect of temperature on Rh catalyzed methanol conversion

Temp.	Pressure	Feed rate (moles/hr)				CH_3OH conversion (mol %)
		CH_3OH	CH_3I	CO	LHSV[a]	
175 °C	1.05 bar	0.25	0.02	0.50	1.0	9.5
210 °C	1.05 bar	0.25	0.02	0.50	1.0	14.8
245 °C	1.05 bar	0.25	0.02	0.50	1.0	21.5

[a] LHSV = *L*iquid *H*ourly *S*pace *V*elocities

The share of CH_3COOH in the product increased in the same manner (Table 3.4). On raising the reaction pressure at 200 °C, the CH_3COOH product share also increased accordingly (Table 3.5). Generally, the process is conducted at 150–250 °C and 0.35 bar [69].

Table 3.5. Effect of reaction pressure on Rh catalyzed methanol conversion

Pressure	Feed rate (moles/hr)				CH_3OH conversion (mol %)
	CH_3OH	CH_3I	CO	LHSV	
1.05 bar	0.25	0.02	0.50	1.0	20.0
7 bar	0.25	0.002	0.50	1.0	70.3
14 bar	0.25	0.002	0.50	1.0	85.9
15.05 bar	0.25	0.02	0.50	1.0	98.7

The carbonylation of organic halides can be carried out over a wide range of temperatures and pressures (Table 3.6) [4].

Table 3.6. Conditions for carbonylation of organic halides

Catalyst	Temperature [°C]	Pressure [bar]
$NaCo(CO)_4$	0–100	1
Ni Co Fe Pd	200–300	200–700

Allylic halides are oxidatively carbonylated with $Ni(CO)_4$ thereby generating unsaturated acids. The reaction starts when the temperature is $> 100\,°C$ and a high pressure prevails. The latter is necessary in order that coordinative unsaturation is present in the complex to facilitate oxidative addition to the allylic halide, CO has to be removed to attain this state. Since high CO pressure hinders the evolution of CO from $Ni(CO)_4$, an appropriate temperature has to be employed [6, 9].

$$R-CH=CH_2 + CCl_4 + CO \xrightarrow{Co_2(CO)_8} \begin{array}{l} R-CH-CH_2-CCl_3 \\ \quad\quad | \\ \quad\quad COCl \\[1em] RCH-CH_2-CH_3 \\ \ | \\ \ Cl \end{array} \tag{23}$$

In reaction (23), the acid chloride selectivity is favored on increasing the CO pressure and lowering the temperature [70].

3.5 Solvents

As alkenes and alkynes form a heterogeneous system with water, the hydrocarboxylation must be conducted in a solvent to ensue homogeneity, e.g., tetrahydrofuran, 1,4-dioxane, acetone, acetonitrile, primary, secondary and tertiary alcohols, methyl ethyl ketone, dimethylformamide, ethyl acetate, pyridine, acetic acid or anisole [1, 3, 51]. Furthermore, the solvent can influence the induction period of the reaction [1]. Side reactions are reported to be favored by non-hydroxylic media such as dioxane [71].

A wide variety of esters can be synthesized by conducting the reaction in the presence of an appropriate alcohol or thiol. Amides result when an amine is employed as solvent [14].

Depending on the solvent system used, widely varying products can be obtained on carbonylating acetylene. When an iodotricarbonylnickel catalyst is employed along with a mixture of alcoholic and aprotic solvents (acetone, dimethylformamide), the main product is 2,4-pentadienoic acid ester [Eq. (24)].

$$2\,HC≡CH + CO + ROH \xrightarrow{[Ni(CO)_3I]^-} H_2C=CH-CH=CH-COOR \tag{24}$$

(R = Alkyl)

The aprotic solvent apparently facilitates a halotricarbonylnickel formation [72]. However, when acetone and a small quantity of water ($\sim 1\%$) are employed as reaction medium, the spectrum of products increases and 2,4-pentadienoic acid is only obtained in very limited amounts [Eq. (25) [72].

$$CO + HC\equiv CH \xrightarrow{[Ni(CO)_3]I^-} \underset{HO}{\bigcirc}\text{--CH=CH--CO}_2\text{H} + H_2C=CH\text{--CH=CH} \atop \qquad\qquad\qquad\qquad\qquad\qquad\qquad\qquad CO_2H \qquad (25)$$

$$+$$

$$HOOCCH_2\text{--CH}_2\text{--C}\underset{H}{\diagup}\diagdown O \diagdown O$$

In the oxidative carbonylation of ethylene with CO in the presence of $PdCl_2$ [Eq. (22)] it is advantageous to employ a solvent with a boiling point exceeding that of the acrylic acid product.

$$H_2C=CH_2 + CO + 1/2\,O_2 \underset{Ac_2O}{\overset{PdCl_2}{\longrightarrow}} \begin{array}{l} H_2C=CH\text{--COOH} \\[2mm] \underset{OAc}{CH_2}\text{--CH}_2\text{--COOH} \end{array} \qquad (22)$$

As the reaction rate decreases in acidic solvents above C_3, mixtures of acetic acid and higher boiling solvents are often used, e.g., acetic acid and β-acetoxypropionic acid [13]. The effect of various solvent systems on the selectivity is shown in Fig. 3.5 [13]. In the rhodium catalyzed carbonylation of methanol [73], the solvent medium chosen is determined by the solubility and compatibility of the species in the reacting system. Studies show that faster reaction rates are achieved with more polar solvents (Table 3.7) [73], indicating that the active catalytic species is probably ionic [74]. The best medium had a dielectric constant between 11 and 23.

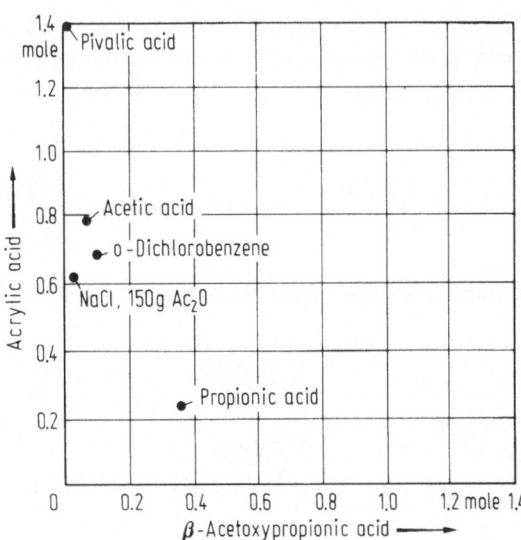

Fig. 3.5. Effect of solvent on selectivity of oxidative carbonylation of ethylene [13]

Table 3.7. Effect of solvent on Rh catalyzed carbonylation of methanol

Solvent	Rate mol/l · s (x10^3)
Benzene	0.14
Methyl ethyl ketone	0.30
Dioxane	0.40
Acetic acid	0.64

However, with dioxane (Fig. 3.6) the polarity can be adjusted by adding H_2O [73].

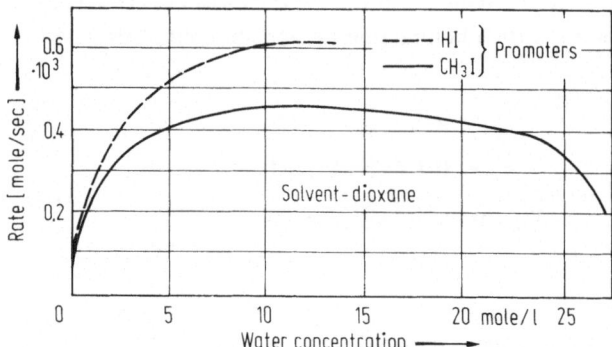

Fig. 3.6. Effect of addition of water to dioxane solvent in Rh catalyzed carbonylation of methanol [73].

On carbonylating halides [Eq. (26)] with Ni(CO)$_4$ there are also interesting solvent effects.

$$PhI + Ni(CO)_4 \rightarrow Ph-\underset{O}{\overset{\parallel}{C}}-\underset{O}{\overset{\parallel}{C}}-Ph \qquad (26)$$

The carbonylation of aryl halides even ensues in non-polar solvents such as benzene, yielding α-diketones, whereas the carbonylation of benzyl halides [Eq. (27)] is only feasible in solvents such as DMF, THF or ethanol [75].

$$2 \quad \text{(benzene)}-CH_2X + Ni(CO)_4 \longrightarrow \text{(benzene)}-CH_2-\overset{O}{\overset{\parallel}{C}}-CH_2-\text{(benzene)} \qquad (27)$$

A marked solvent effect was also found with acyl halides [Eq. (28)] [75].

$$2\,Ph-\underset{O}{\overset{\parallel}{C}}-CH_2Br \quad \overset{DMF/Ni(CO)_4}{\underset{THF/Ni(CO)_4}{\underset{50-60°C}{\xrightarrow{\hspace{2cm}}}}} \quad \begin{array}{c} Ph\diagdown_{O}\diagup Ph \\[4pt] Ph-\underset{O}{\overset{\parallel}{C}}-CH_2-CH_2-\underset{O}{\overset{\parallel}{C}}-Ph \end{array} \qquad (28)$$

Cyclization readily took place with $Ni(CO)_4$ at 30 °C with DMF as solvent, whereas in THF the diketone was obtained [Eq. (28)] [75].

3.6 Carbonylation of Various Structures

3.6.1 Carbonylation of Alkynes

3.6.1.1 Alkynes and Functional Derivatives in the Presence of Water

The hydrocarboxylation of acetylenes yields substituted unsaturated acids, i.e., acrylic acid and its homologues [Eq. (29)].

$$HC{\equiv}CH + CO + H_2O \xrightarrow{\text{cat.}} H_2C{=}CH{-}CO_2H \tag{29}$$

The product yield and distribution can be influenced by:

- catalyst metal (cf Sect. 3.3)
- ligands (cf Sect. 3.3)
- pH of reaction medium/solvent (cf Sect. 3.5)

In contrast to hydroformylation reactions with alkenes, there is generally no isomerisation of triple bonds during carbonylation. Some typical examples of the hydrocarboxylation of alkynes are shown in Table 3.8.

Table 3.8. Hydrocarboxylation of alkynes with carbon monoxide and water

Substrate	Product	Yield (%)	Ref.
Acetylene	Acrylic acid	95	[14, 78, 245]
Propyne	Methacrylic acid	50	[78]
1-Butyne	2-Ethylacrylic acid	45	[78]
1-Hexyne	2-Butylacrylic acid	35	[79]
1-Octyne	2-Hexylacrylic acid	20	[14]
2-Nonyne	2-Nonene-2-carboxylic acid 2-Nonene-3-carboxylic acid	32	[14]
5-Decyne	2-Butyl-2-heptenoic acid	52	[249]
Phenylacetylene	Atropic acid and traces of cinnamic acid	48	[14, 79]
1-Phenyl-1-propyne	α-Methylcinnamic acid and α-phenylcrotonic acid	54	[14]
Diphenylacetylene	α-Phenyl-trans-cinnamic acid	48	[14, 248–249]
3-Acetoxypropyne	3-Acetoxy-1-propene-2-carboxylic acid	32	[79]
3-Hydroxy-1-butyne	3-Hydroxy-1-butene-2-carboxylic acid	–	[14]

Table 3.8 (continued)

Substrate	Product	Yield (%)	Ref.
1-Hydroxy-3-butyne	α-Methylene-γ-butyrolactone	23	[79]
1-Acetoxy-3-butyne	4-Acetoxy-1-butene-2-carboxylic acid	32	[79]
4-(2-Hydroxytetrahydro-pyranyl)-1-butyne	4-(2-Hydroxytetrahydropranyl)-1-butene-2-carboxylic acid	20	[79]
1,4-Diacetoxy-2-butyne	1,4-Diacetoxy-2-butene-2-carboxylic acid	58	[79]
2-Hydroxy-4-pentyne	α-Methylene-γ-methyl-γ-butyrolactone	30	[79]
2-Hydroxy-3-pentyne	4-Hydroxy-2-pentene-2-carboxylic acid	60	[249]
3-Acetoxy-1-hexyne	3-Acetoxy-1-hexene-2-carboxylic acid	48	[79]
2-Hydroxy-2-methyl-4-pentyne	α-Methylene-γ-dimethyl-γ-butyrolactone	7.4	[79]
1-Acetoxy-1-ethynecyclohexane	α-(1-Acetoxycyclohexyl)-acrylic acid	3.5	[79]
3-Acetoxy-3-phenylpropyne	3-Acetoxy-3-phenyl-1-propene-2-carboxylic acid	50	[79]
1-Hydroxy-1-phenylpropyne	3-Hydroxy-3-phenyl-1-propene-2-carboxylic acid	12	[79]
5-Bromopentyne	5-Bromo-1-pentene-2-carboxylic acid	40	[244]
3-Pentyne-2-one	cis- and trans-4-Oxo-2-pentene-2-carboxylic acid	30	[249]
3-Octyne-2-one	2-Oxo-3-octene-4-carboxylic acid	40	[249]
1-Hydroxy-4-pentyne	α-Methylene-δ-valerolactone	21	[79]
Ethyl 3-butynoate	Ethyl 3-carboxy-3-butenoate	28	[250]
Ethyl 4-pentynoate	Ethyl 4-carboxy-4-pentenoate	46	[250]
Ethyl 5-hexynoate	Ethyl 5-carboxy-5-hexenoate	40	[250]
Ethyl 6-heptynoate	Ethyl 6-carboxy-6-heptenoate	37	[249, 250]
5-Cyano-1-pentyne	5-Cyano-1-pentene-2-carboxylic acid	39	[250]

The addition of the H—COOH to the acetylenic linkage generally takes place in a cis manner with Ni catalysts and according to Markovnikoff, although there are exceptions [1, 14, 78, 245]. With substituted acetylenes (XC≡CY) the substituents play an important role in determining which acetylenic carbon atom bonds to the carboxyl group [1, 5, 51].

$$A–C≡C–H \tag{30}$$

For example, when A is an alkyl, aryl, $–CH_2OH$, $–CH_2–CH_2OH$, $–CH_2–CH_2–CH_2OH$ or a $CH_3–COO–CHR$ group, then CO bonds to the acetylenic carbon atom attached to A [Eq. (31)] [51].

$$A–C≡C–H \; + \; CO \; + \; H_2O \; \xrightarrow{\text{cat.}} \; \underset{CO_2H}{\overset{A}{\underset{\diagup}{\overset{\diagdown}{C}}}}=CH_2 \tag{31}$$

The converse is the case when a substituent of type B is present (H, CHROH, CR_2OH, CO_2H, $COOCH_3$, $COCH_3$, $CH_2C(CH_3)_2OH$) [Eq. (32)].

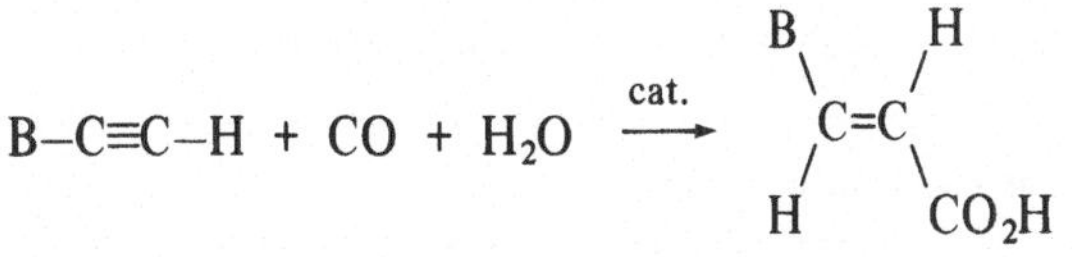

$$B-C\equiv C-H + CO + H_2O \xrightarrow{\text{cat.}} \begin{matrix} B \\ \diagdown \\ \diagup \\ H \end{matrix} C=C \begin{matrix} H \\ \diagup \\ \diagdown \\ CO_2H \end{matrix} \tag{32}$$

If in a disubstituted alkyne $XC\equiv CY$, both X and Y are members of group A, then carbonylation takes place rapidly. When X = A and Y = B, it is still a reasonably fast conversion. However, when X = Y = B, the reaction is no longer exothermic and low yields are obtained even after prolonged heating [51]. One of the exceptions to this rule is when X = Y = H, i.e. acetylene, which is very reactive.

In the hydrocarboxylation of diarylacetylenes [Eq. (33)] a similar effect is observed.

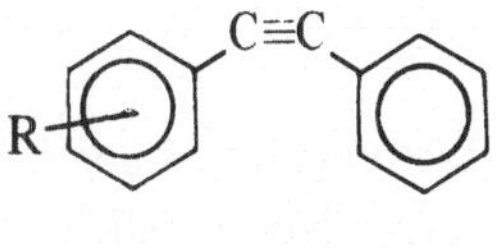

(33)

When R is a para substituted group with +I or +M properties (electron donating) then CO addition ensues at the acetylenic carbon adjacent to the aryl group [Eq. (34)] [9, 76].

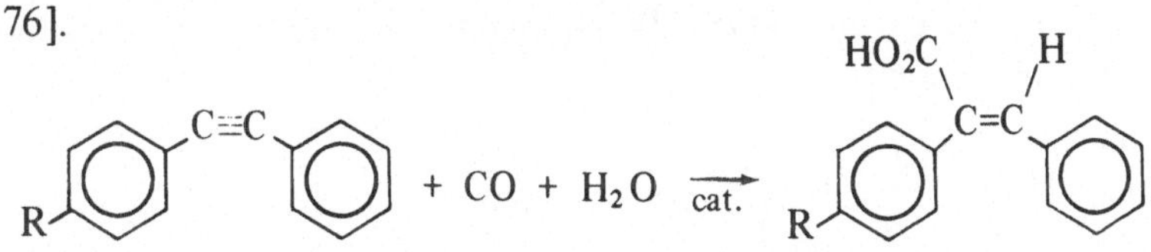

(34)

Conversely, when electron withdrawing groups are present at the para position they encourage substitution at the acetylenic carbon atom next to the unsubstituted phenyl group. However, when either an electron releasing or donating group is at the ortho position to the acetylenic bond, then CO addition occurs at the carbon atom adjoining the substituted ring [Eq. (35)] [76].

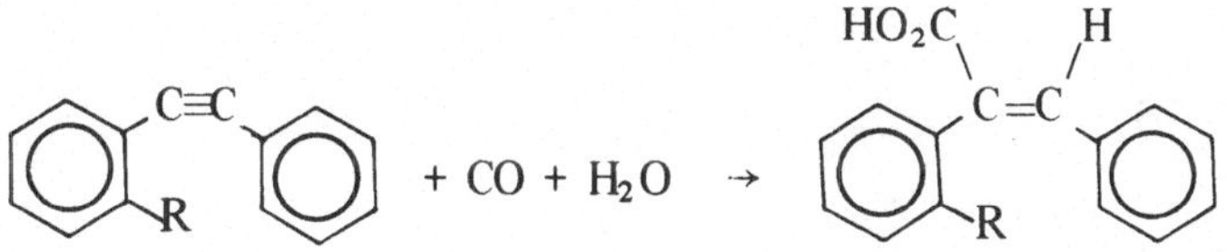

(35)

A few examples with product ratios are shown in Table 3.9 [9, 76].

3.6.1.1.1 Various Catalysts

The most important substance for catalyzing the hydrocarboxylation of alkynes is $Ni(CO)_4$. Originally, stoichiometric amounts were employed, serving as CO source thereby facilitating pressure-free operation at ambient temperatures [14].

$$4 HC\equiv CH + 4 H_2O + Ni(CO)_4 + 2 HCl \rightarrow 4 H_2C=CHCO_2H + NiCl_2 + [2 H] \tag{36}$$

265

Table 3.9. Effect of substituents in diarylacetylenes on product distribution [9, 76]

Precursor $\underset{R}{\overset{C\equiv C}{\diagup}} \bigcirc$ where R is:	Product ratio in alkene		
	COOH adjacent to phenyl group		COOH adjacent to substituted phenyl group [Eq. 34, 35]
p–CH$_3$OPh	1	:	2
p–CH$_3$Ph	1	:	1
p–ClPh	2	:	1
p–NO$_2$Ph	6	:	1
o–CH$_3$Ph	1	:	3
o–ClPh	1	:	3

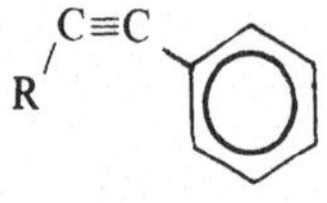

The hydrogen is not released [77]. It reacts with the precursors producing ethane and ethylene as well as reducing the product to propionic acid [14]. The Ni salt is reconverted into nickel carbonyl.

The rate of hydrocarboxylation of mono-substituted alkyl-alkynes decreases with the steric hindrance of the alkyl group [78]. Two products can arise from the addition of H–COOH to the triple bond [Eq. (37)]:

$$4\ R-C\equiv C-R' + H_2O \xrightarrow{Ni(CO)_4} \underset{HO_2C}{R-C=C-R'} + \underset{H\ \ CO_2H}{R-C=C-R'} \tag{37}$$

In the Ni(CO)$_4$-pyridine catalyzed stoichiometric hydrocarboxylation of vinylacetylene [Eq. (38)] a diacid product is formed via a Diels-Alder reaction [14, 78, 84, 246].

$$2\ HC\equiv C-CH=CH_2 + 2\ CO + 2\ H_2O \xrightarrow{Ni(CO)_4} 2\ \underset{CO_2H}{H_2C=C-CH=CH_2} \tag{38}$$

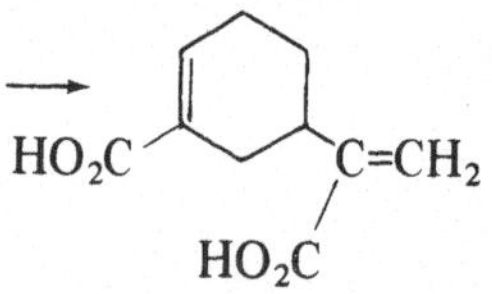

The hydrocarboxylation of alkynes using a stoichiometric method with HCo(CO)$_4$ or Co$_2$(CO)$_8$ was not particularly successful [Eq. (39)]. It occurs only above 70 °C with poor yields, probably on account of the high stability of the intermediate cobalt complex [81].

$$HC\equiv CH + Co_2(CO)_8 + H_2O + CO \xrightarrow[70°C]{THF} \underset{CH_2-COOH}{CH_2-COOH} \tag{39}$$

The stoichiometric-catalytic or modified carbonylation with Ni(CO)$_4$ has been used industrially because of the quantitatively lower Ni requirement compared to the corresponding stoichiometric reaction. The process is conducted at slightly raised pressure and 40–50 °C [82].

With the catalytic method, the low rate of reaction makes temperatures of 180–200 °C and pressures of 40–50 bar necessary before reaction ensues with, for example, $Ni(CO)_4/CuBr_2/HBr$. When the reaction is conducted in THF, which has good solvent properties for C_2H_2, the acrylic acid yield is 90% (based on C_2H_2) [83].

Unsaturated acids with two C=C bonds in the α and δ positions adjacent to the carboxyl group can be prepared in 80–85% yield on carbonylating acetylene in the presence of an allyl halide with a $Ni(CO)_4$ catalyst at ambient temperatures and pressures [12, 88] [Eq. (40)]

$$RCH=CHCH_2X + HC\equiv CH + H_2O + CO \xrightarrow{Ni(CO)_4} RCH=CH-CH_2-CH=CHCO_2H + HX \quad (40)$$

Instead of allyl halides, allyl alcohols can also be employed with good results [88].

Various acetylenic halides which can be used for the preparation of allenic acids are summarized in Table 3.10.

With $Co_2(CO)_8$ as catalyst in the acetylene hydrocarboxylation, the reaction ensues at 80–100 °C/100–200 bar, in the absence of HX, alkyl halides etc., producing succinic acid in 80% yield [85]. The catalyst is so active that double carbonylation readily occurs, however, at lower CO pressure (50 bar) using $CoBr_2/CuBr_2$ as catalyst, acetylene was converted into acrylic acid.

At higher pressures, along with the main product (succinic acid), small quantities of propionic and acrylic acids as well as cyclopentanone were also formed [86], the hydrogen for these conversions being supplied by the water gas shift reaction [Eq. (41)].

$$CO + H_2O \rightarrow CO_2 + H_2 \quad (41)$$

It was established that there was a correlation between the quantity of CO_2 formed and the amount of reduced substances.

The choice of solvent has a powerful effect on the product distribution, e.g., on catalyzing the hydrocarboxylation with $Co_2(CO)_8$ at 110–140 °C/~180 bar, succinic

Table 3.10. Allenic acids via hydrocarboxylation of α-halogen substituted alkynes

Precursor	Allenic acid[a]	Yield (%)[b]	Ref.
3-Chloro-1-propyne	2,3-Butadiene-1-carboxylic acid	6	[247, 251]
3-Chloro-1-butyne	2,3-Pentadiene-1-carboxylic acid	10	[247]
1-Chloro-2-butyne	2-Methyl-2,3-butadiene-1-carboxylic acid	15	[252]
3-Chloro-3-methyl-1-butyne	4-Methyl-2,3-pentadiene-1-carboxylic acid	45	[247]
3-Chloro-2-heptyne	2-Butyl-2,3-butadiene-1-carboxylic acid	15	[247, 252]
1-Bromo-2-heptyne	2-Butyl-2,3-butadiene-1-carboxylic acid	11.5	[252]
1-Iodo-2-heptyne	2-Butyl-2,3-butadiene-1-carboxylic acid	22	[252]
p-Toluenesulfonic acid ester of 2-heptyn-1-ol	2-Butyl-2,3-butadiene-1-carboxylic acid	31	[252]
2-Chloro-2-methyl-3-octyne	2-Butyl-4-methyl-2,3-pentadiene-1-carboxylic acid	13	[247]
3-Chloro-3-phenyl-1-propyne	4-Phenyl-2,3-butadiene-1-carboxylic acid	12	[247]

[a] acid + ethyl ester   [b] total yield of acid + ester

acid was obtained in 8% yield with toluene [87], 25% yield with ethyl acetate [81] and 85% with acetone [65] as solvent (cf Sect. 3.5).

With palladium catalysts, the hydrocarboxylation of acetylene led to dicarboxyl-ation i.e., the formation of fumaric and maleic acids [89–92].

Recently, there has been growing interest in polynuclear complexes of palladium and iron (19) for the hydrocarboxylation of acetylene [93]. The hydrocarboxylation reactions catalyzed by $Fe(CO)_5$ have been reviewed [9]. An 80% yield of acrylic acid was obtained [94] on carbonylating acetylene in the presence of the system $Fe(CO)_5/CuBr_2/Br_2$ in tetrahydrofuran at 208 °C, whereas with acetone as solvent, at 180 °C/200 bar, succinic acid resulted in 34% yield [87]. Iron would thus appear to have more similarities to cobalt than to nickel although there has been little systematic work done on the Fe system.

Ruthenium and rhodium have also been employed as hydrocarboxylation catalysts, although ruthenium carbonyl possesses a greater tendency to catalyze ring-forming. reactions [81].

3.6.1.2 Alkynes and Derivatives in the Presence of Alcohols

When the hydrocarboxylation is conducted in the presence of an alcohol, the corres-ponding ester is formed. This reaction is also known as hydroesterification. However, depending on the catalyst system, solvent and reaction conditions, a whole series of different products can arise [Eq. (42)].

$$x\,HC{\equiv}CH + y\,CO + z\,ROH \rightleftharpoons \begin{array}{l} CH_3{-}CH_2{-}CO_2R \\ CH_3{-}CH(CO_2R)_2 \\ RO_2C{-}CH{=}CH{-}CO_2R \\ RO_2C{-}CH_2{-}CH_2{-}CO_2R \\ H_2C{=}CH{-}CO_2R \end{array} \qquad (42)$$

Water is essential for the reaction. Under anhydrous conditions there was an induction period lasting until water resulted from the ether synthesis [Eq. (43)] at high tempera-tures [1].

$$2\,R{-}OH \rightarrow R{-}O{-}R + H_2O \qquad (43)$$

The rate determining step of the subsequent reaction is thus ester formation which releases the water for the hydrocarboxylation [Eq. (44)].

$$HC{\equiv}CH + CO + H_2O \rightleftharpoons H_2C{=}CH{-}COOH$$

$$ \qquad (44)$$

$$H_2C{=}CH{-}COOH + ROH \rightleftharpoons H_2C{=}CH{-}CO_2R + H_2O$$

The most commonly used Group VIII metals for catalyst purposes are Ni, Co, Fe, Rh and Pd. The conditions of reaction are similar to those outlined in the previous Section (hydrocarboxylation of acetylenes). It is necessary to have an acid present in the $Ni(CO)_4$ catalyzed stoichiometric synthesis of acrylates from acetylene. In this connection, the effect of water with weak acids was studied [79]. With hydrochloric acid, methyl acrylate was obtained in ~80% yield whereas with acetic acid, the yield fell below 45% [4, 14]. The type of acid has thus a powerful effect on the product distribution.

The ester yield was also found to be inversely proportional to the molecular weight of the alcohol [78]. Some results of the stoichiometric reaction at ambient pressures and temperatures with $Ni(CO)_4$ are shown in Table 3.11.

Table 3.11. Stoichiometric hydroesterification reaction with $Ni(CO)_4$

Reactants	Acid	Product	Yield	Ref.
Acetylene/methanol	HCl	Methyl acrylate	80%	[4, 14]
Propyne/methanol	HCl	Methyl methacrylate	50%	[78]
1-Butyne/ethanol	HCl	Ethyl 2-ethylacrylate	45%	[78]
1-Hexyne/ethanol	HCl	Ethyl 2-butylacrylate	35–50%	[78]
1-Octyne/ethanol	HOAc	Ethyl 2-hexylacrylate	60%	[4, 14]

With propagryl alcohol using $Ni(CO)_4$ as stoichiometric reagent the following reaction takes place:

$$HC\equiv C-CH_2OH + CO + CH_3CH_2OH + Ni(CO)_4$$

$$\xrightarrow[1\ bar]{55\,^\circ C} \ \overset{\displaystyle CH_2OH}{\underset{\displaystyle |}{H_2C=C}}-CO_2C_2H_5 + HOCH_2-CH=CH-CO_2C_2H_5 \qquad (45)$$

$$58\% 11\%$$

The substitution pattern is in accordance with the effect outlined in the previous Sect. [9]. The corresponding methyl, isobutyl, n-butyl esters can be prepared similarly. Reactions of other substituted acetylenes are also well documented [5, 8].

Using Ni catalysts, alkyl acrylate can be prepared at higher pressure and temperature, the activity of the Ni halides being as follows [23]:

$$NiCl_2 < NiBr_2 < NiI_2 \qquad\qquad (46)$$
$$200\,^\circ C \ \ 180\,^\circ C \ \ 150\,^\circ C$$

As in the hydrocarboxylation, $Ni(CO)_4$ can also catalyze the formation of unsaturated esters. These reactions have been described by many authors [1, 8, 9]. Good yields can be obtained [8].

When acetylene is reacted with carbon monoxide and methanol using a $[Ni(CO)_3I]^-$ catalyst, methyl 2,4-pentadienoate is the main product [95]. This reaction is usually carried out in a medium consisting of aprotic and alcoholic solvents which facilitate catalyst formation [95].

When $CuCl_2/RNH_4Cl$ is added to the $Ni(CO)_4$ catalyst, double carbonylation takes place leading to unsaturated dicarboxylic acid esters in good yield [72].

A mechanism [Eq. (47)] for the $Ni(CO)_4$ catalyzed conversion of methylacetylenes into methacrylates has been proposed by Japanese workers [96].

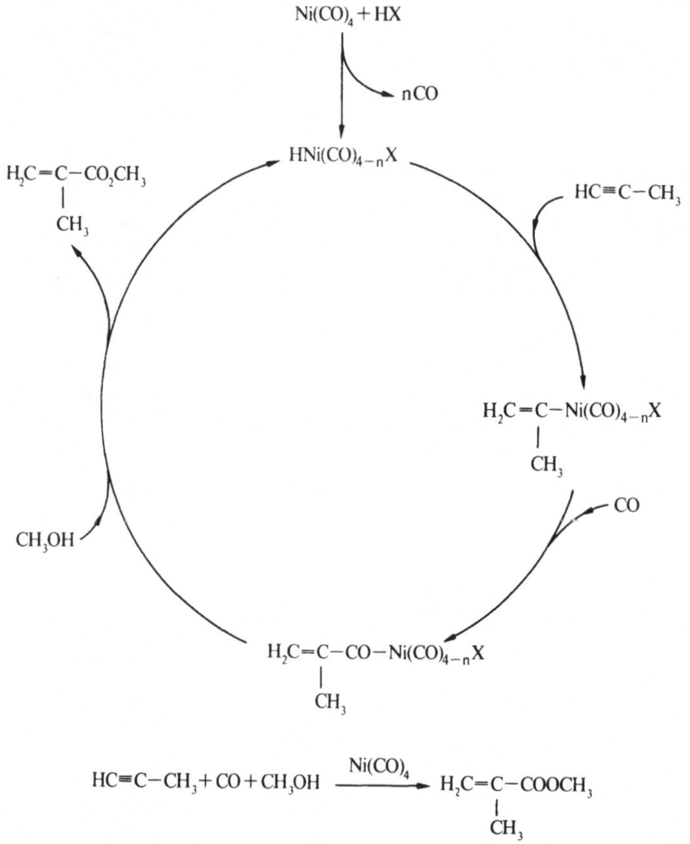

$Ni(CO)_4 + HX$

nCO

$HNi(CO)_{4-n}X$

$H_2C=C-CO_2CH_3$
$|$
CH_3

$HC\equiv C-CH_3$

$H_2C=C-Ni(CO)_{4-n}X$
$|$
CH_3

CO

CH_3OH

$H_2C=C-CO-Ni(CO)_{4-n}X$
$|$
CH_3

(47)

$$HC\equiv C-CH_3 + CO + CH_3OH \xrightarrow{Ni(CO)_4} H_2C=C-COOCH_3$$
$$|$$
$$CH_3$$

In the hydroesterification of propyne at $130\,°C/13-5$ bar in the presence of a $Ni(CO)_4$ catalyst, methyl methacrylate was obtained in $> 80\%$ yield, the major by-product being methyl crotonate [97].

With a stoichiometric amount of $Co_2(CO)_8$ more severe reaction conditions are necessary than with Ni. The desired product is generally obtained in lower yield [9]. Cobalt catalysis has a tendency to promote double carbonylation compared to nickel catalysis. However, alkyl acrylate formation is encouraged with a lower initial acetylene concentration and a low catalyst concentration at ca. $110\,°C/210$ bar using the alcohol as reaction medium [9]. An increase in the CO pressure can improve the yield. Alterations in temperature only affect the product distribution of the diesters [98].

There has been considerable activity in the use of Pd catalysts in the hydrocarboxylation of alkynes during the last few years [7, 9]. Pd generally tends to favor dicarbonylation. As Pd forms a very active catalyst, conversions can be conducted under very mild conditions. Pd catalysis in carbonylation reactions has been reviewed by

Tsuji [7]. Using Pd/C under mild conditions substituted acrylates were obtained in high yields from alkynes [Eq. (48)] [72]:

$$RC\equiv CH + CO + R'OH \xrightarrow{\text{Pd/C/R''X}} H_2C=CR-CO_2R' \qquad (48)$$

X = halogen
R'' = H, alkyl

In the presence of a palladium complex such as $PdBr_2[P(OPh)_3]_2$ and perchloric acid, acetylene readily reacts to produce the corresponding acrylate in 95% yield [99].

Using palladium halides, the most effective being PdI_2, acetylene was converted into a number of products [Eq. (49)] [100]:

$$HC\equiv CH + CO + CH_3-(CH_2)_2-CH_2OH \rightarrow H_2C=CH-CO_2Bu + CH_3-CH_2-CO_2Bu$$

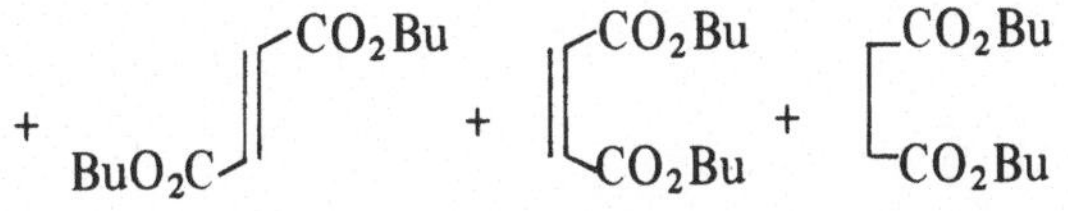

$$+ \quad \underset{BuO_2C}{\overset{CO_2Bu}{\diagup}} \quad + \quad \underset{CO_2Bu}{\overset{CO_2Bu}{\diagup}} \quad + \quad \underset{CO_2Bu}{\overset{CO_2Bu}{\diagup}} \qquad (49)$$

Bu = butyl

The main product was dibutyl succinate. However, on adding minor amounts of HX (X = halogen) the product distribution was altered in favor of butyl acrylate. When the $CO/HC\equiv CH$ ratio was increased, dibutyl fumarate and succinate formation were encouraged [100].

With a thiourea additive, the $PdCl_2$ catalyzed hydrocarboxylation [Eq. (50)] of $HC\equiv CH$, in the presence of CH_3OH, produced dimethyl maleate in 90% yield [6].

$$HC\equiv CH + 2\,CO + 2\,CH_3OH \xrightarrow[\text{thiourea}]{PdCl_2} \underset{HC}{\overset{HC}{\underset{\diagdown}{\overset{\diagup}{\parallel}}}} \overset{CO_2CH_3}{\underset{CO_2CH_3}{}} + 2\,[H] \qquad (50)$$

The reaction occurred under ambient conditions, a minor amount of oxygen being necessary to remove the resulting hydrogen which would otherwise reduce the palladium ion to the metal [6]. A mechanism [Eq. (51)] was proposed for the reaction [6].

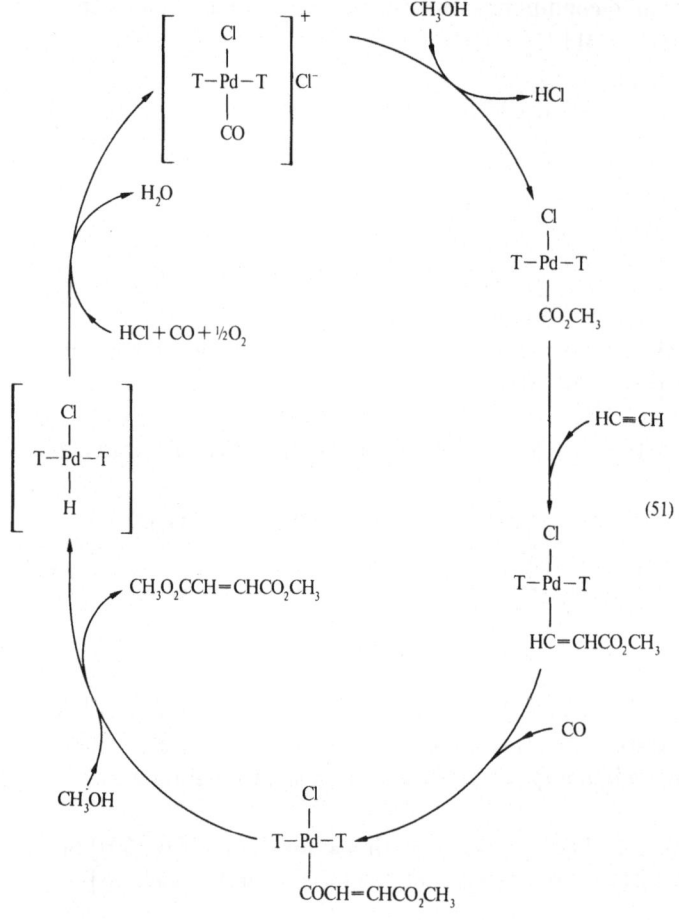

(51)

T = thiourea

$$HC\equiv CH + 2CH_3OH + 2CO \longrightarrow CH_3O_2CCH=CHCO_2CH_3 + 2[H]$$

In absence of thiourea, Heck proposed the following mechanism for the formation of diesters from acetylenes [Eq. (52)] [101].

$$HC\equiv CR + PdCl_2 + CO \rightleftharpoons Cl-\overset{\overset{\displaystyle HC\equiv CR}{\downarrow}}{\underset{\underset{\displaystyle CO}{|}}{Pd}}-Cl \xrightarrow{CH_3OH} \left[Cl-\overset{\overset{\displaystyle HC\equiv CR}{\downarrow}}{\underset{\underset{\displaystyle COOCH_3}{|}}{Pd}}-Cl \right]^- H^+$$

$$\xrightarrow{-HCl} [CH_3O_2C-CH=CR-PdCl]$$

(52)

$$\xrightarrow{CO} CH_3O_2C-CH=CR-COPdCl \xrightarrow{CH_3OH} CH_3O_2CCH=CRCO_2CH_3 + HCl + Pd^\circ.$$

The hydroesterification is facilitated on using acetylenic diesters as substrate. Reaction (53) readily took place at ambient pressure and temperature in the presence of $PdCl_2$ [7].

$$RO_2C-C\equiv C-CO_2R + CO + ROH \xrightarrow{PdCl_2/HCl} \underline{(RO_2C)_2C=CH(CO_2R) + (RO_2C)_2CH-CH_2-CO_2R}$$

$$\underbrace{\qquad\qquad}_{42.5\%}$$

$$+ (RO_2C)_2CHCH(CO_2R)_2 \qquad (53)$$

$$6.5\%$$

The product distribution depends on the HCl concentration in the reaction medium. Higher concentrations favor dicarbonylations and saturated esters [7]. A 44% yield of the saturated tetracarboxylate of a diester of the above type was reported on using PdCl$_2$ at 100 bar and 25 °C [102].

With diphenylacetylene, carbon monoxide, and methanol, the following products arose when the hydrocarboxylation was catalyzed by Pd/HCl at 100 °C under raised pressure [7].

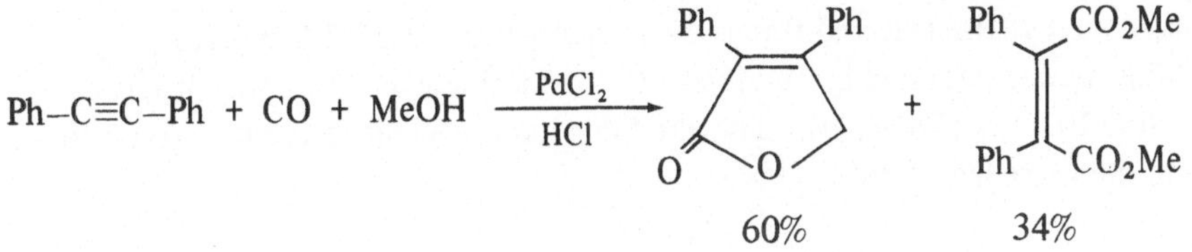

$$Ph-C\equiv C-Ph + CO + MeOH \xrightarrow[HCl]{PdCl_2}$$

60%   34%   (54)

$$+ \; PhCH=CPh(CO_2Me)$$

Lower total yields were obtained at lower concentrations of hydrochloric acid [7]. Iron and rhodium catalysts have both led to dicarbonylation on hydrocarboxylating acetylene [9, 103].

3.6.1.3 Alkynes in Presence of Carboxylic Acids, Hydrogen Halides, Mercaptans or Amines

This type of carbonylation reaction (see Reppe [14]), permits the formation of various carboxylic acid derivatives directly from the alkyne substrate. The potential routes are shown in Eq. (55) [1, 9, 51]:

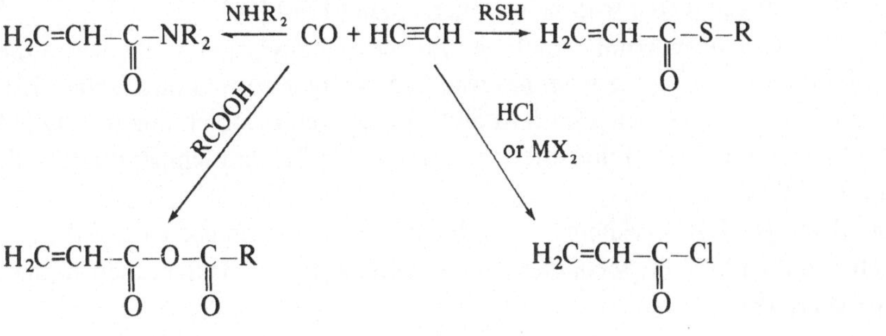

(55)

273

The acid chloride of acrylic acid (acroyl chloride) was synthesized:

$$HC \equiv CH + CO + HCl \xrightarrow[120-200\,°C/500-1000\,bar]{Ru\ or\ Rh\ compounds} H_2C=CH-COCl \tag{56}$$

In toluene, succinoyl chloride was also formed [104]. Hydrogen fluoride gives rise to the acid fluoride. With a stoichiometric quantity of $PdCl_2$, Tsuiji [105] managed to dicarbonylate acetylene [Eq. (57)] whilst simultaneously producing the acid chlorides:

$$4\ HC \equiv CH + 6\ CO + 3\ PdCl_2 \xrightarrow[100\,bar/100°C]{C_6H_6}
\begin{array}{c} CH=CH-COCl \\ | \\ CH=CH-COCl \end{array}
+
\begin{array}{c} CH-COCl \\ || \\ CH-COCl \end{array}
\tag{57}$$

$$+ \quad \begin{array}{c} CH-COCl \\ || \\ ClOC-CH \end{array} \qquad + 3\ Pd$$

The combined maleyl chloride and fumaryl chloride yield was 31.7%, whereas muconyl chloride was obtained in 38.5% yield. The mechanism which was proposed for the reaction [9] [Eq. (58)] is based on the fact that acetylene is cyclized to cyclo-butadiene in the presence of $PdCl_2$ [106].

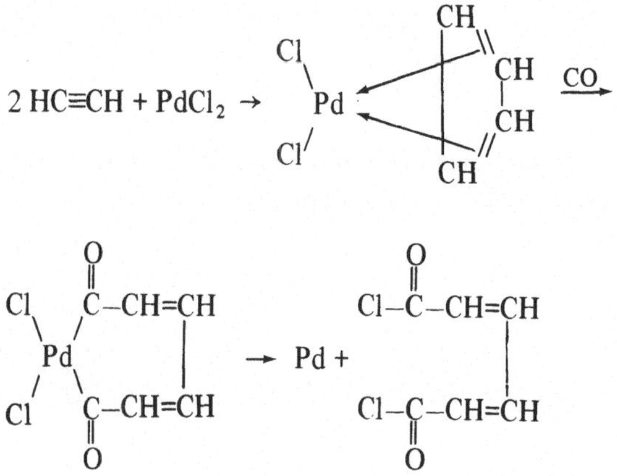

$$\tag{58}$$

Many thioesters were prepared on carbonylating alkynes with carbon monoxide in the presence of thiols. Stoichiometric quantities of $Ni(CO)_4$ were employed along with polymerization inhibitors such as hydroquinone. Some examples are presented in Table 3.12, the best yields being reported with benzyl mercaptan (77%) [14].

Nickel catalysts have been mainly employed for the carbonylation of alkynes in the presence of primary or secondary amines resulting in amide formation (Table 3.13). The reaction has been conducted both catalytically as well as stoichiometrically [145]. The catalytic method requires pressures between 30–35 bar and temperatures in the range of 80–180 °C.

The N-butyl and N-ethyl acrylamide were obtained as dimers on account of their high reactivity. As a rule, the secondary amines gave better yields than their primary counterparts [1, 14].

Table 3.12. Thioester formation via carbonylation of alkynes with carbon monoxide in the presence of thiols [14]

Alkyne	Thiol	Reaction product
Acetylene	Hydrogen sulfide	Thioacrylic acid
Acetylene	Dodecyl mercaptan	Thiododecyl acrylate
Acetylene	Benzyl mercaptan	Thiobenzyl acrylate
Acetylene	Thiophenol	Thiophenyl acrylate
Acetylene	p-Thiocresol	Thio-p-tolyl acrylate
Acetylene	Thioglycolic acid	S-Acryl thioglycolic acid
Phenylacetylene	Ethyl mercaptan	Thioethyl-α-phenylacrylate

Table 3.13. Formation of unsaturated acid amides via carbonylation of alkynes with CO in presence of amines or amides [14]

Alkyne	Amine	Reaction product
Acetylene	Ethylamine	N-Ethylacrylamide (dimer)
Acetylene	Butylamine	N-Butylacrylamide (dimer)
Acetylene	Aniline	Acrylanilide
Acetylene	Pyrrolidone	N-Acrylpyrrolidone
Acetylene	Bicyclohexylamine	N-Bicyclohexylacrylamide
Acetylene	Diphenylamine	N-Diphenylacrylamide (polymer)
Acetylene	Urea	N-Acryl urea (polymer)
Phenylacetylene	Aniline	α-Phenylacrylanilide
Acetylene	Acetamide	N-Acetylacrylamide

3.6.2 Carbonylation of Alkenes

3.6.2.1 Alkenes and Functional Derivatives in the Presence of Water

Olefins can be hydrocarboxylated with carbon monoxide and water using transition metal compounds as catalysts or as stoichiometric reagents:

$$RCH{=}CH_2 + CO + H_2O \xrightarrow{\text{cat.}} \begin{cases} RCH_2CH_2COOH \\ \\ RCH{-}COOH \\ \quad | \\ \quad CH_3 \end{cases} \tag{59}$$

This type of carbonylation has been reviewed several times [1, 5, 7–9, 12]

Compared to the alkynes (Sect. 3.6.1) more severe conditions are necessary when hydrocarboxylating alkenes [4]. The activity of Co, Rh, Ru, Pd, Pt-based catalysts is equal to or better than Ni [107]. Analogous to the alkyne hydrocarboxylation – the yield, selectivity and product distribution can be influenced by a suitable choice of ligand and central atom (of the catalyst), phosphine ligands enhancing the activity as they increase the solubility of the less soluble metal salts. The resulting carboxylic acids are always obtained as an isomeric mixture. The ratio of branched to linear acids can be influenced by the catalyst and solvent system as well as by additives. While isomerization of the triple bond in alkynes does not occur during hydrocarboxylation, the double bond in the alkenes is subject to a varying degree of isomerization depending on the catalyst system.

The reaction mechanism with Co, Ni, Pd and Rh catalysts has been treated in Sect. 3.2.

The reactivity of the olefin substrate is directly related to its degree of branching – the more substitution present, the more difficult it is to hydrocarboxylate. However, highly reactive olefins react readily under 'acetylenic' conditions.

Table 3.14. Hydrocarboxylation of olefins

Olefin	Catalyst	Products	Yield (%)	Ref.
Ethylene	$NiCl_2$, Co $Pd[P(C_6H_5)_3]/HCl$ $Pd/HI/I_2$	Propionic acid	90	[14, 253–60]
Propene	$Ni(CO)_4$	Butyric acid + isobutyric acid (2 : 1)	60	[14, 255, 257]
Propene	$Co_2(CO)_8$	Butyric acid	88	[261]
1-Butene	$Ni(CO)_4$	n- and iso-Valeric acid	–	[14]
Isobutene	$Ni(CO)_4/NiI_2$	Isovaleric acid + trimethylacetic acid (6 : 1)	100	[255, 14] 255]
1-Hexene		2-Methylhexanoic acid + heptanoic acid	70	[245]
1-Octene	Ni/silica gel	2-Methyloctanoic acid + nonanoic acids	84	[14, 262]
2-Ethyl-1-hexene	Raney-Ni, NiI_2, CuI	2-Ethylheptanoic acid	60	[14]
2-Ethyl-1-hexene	$Co_2(CO)_8$	Nonanoic acids	57	[261]
1-Vinyl-3-cyclo-hexene	$Pd(PPh_3)_2Cl_2/HCl$	Isomeric (carboxy-cyclohexyl)-propionic acids		[253]
1-Dodecene	Ni/silica gel	2-Methyldodecanoic acid + tridecanoic acid	28	[14]
1-Octadecene	$Ni(CO)_4$	2-Methyloctadecanoic acid	67	[14]
Cyclohexene	$Ni(CO)_4$	Cyclohexanecarboxylic acid	78	[14, 245, 259, 263]
Cyclohexene	$Co_2(CO)_8$	Cyclohexanecarboxylic acid	89	[261]
Cyclooctene	$Ni(CO)_4$	Cyclooctanecarboxylic acid	31	[264]
1,5-Cycloocta-diene	$Pd(P(C_6H_5)_3)_2Cl_2/HCl$	4-Cyclooctene-1-carboxylic acid	51	[253]
1,5-Cyclooctadiene	$Pd(P(C_6H_5)_3)_2Cl_2/HCl$	1,5-Cyclooctanedicarboxylic acid		[253]
Bicyclo-(2.2.1)-heptene		Bicyclo-(2.2.1)-heptane-2-carboxylic acid	80	[265]
cis, trans, trans-1,5,9-Cyclododeca-triene	$Pd(P(C_6H_5)_3)_2Cl_2$	Cyclododecadiene-carboxylic acid		[253]

Table 3.14 (continued)

Olefin	Catalyst	Products	Yield (%)	Ref.
Bicyclo (2.2.1)-heptadiene	$Pd(P(C_6H_5)_3)_2Cl_2/HCl$	Bicyclo-(2.2.1)-heptene-5-carboxylic acid	80	[265]
1,5,9-Cyclododeca-triene	$PdI_2/HI/I_2$	5,9-Cyclododecane-dicarboxylic acid + 9-cyclododecene-1,5-dicarboxylic acid		[254]
Cyclooctadiene		Cyclooctanecarboxylic acid		[266]
1,5-Hexadiene	$Ni(CO)_4$	2-Methyl-5-hexenoic acid	20	[141, 259]
Cyclododeca-triene	$Co/CoSO_4(5:1)$	Cyclododecanecarboxylic acid + cyclododecadiene-carboxylic acid	16	[267]
1,3-Butadiene	$[Pd(P(C_6H_5)_3]_2Cl_2/HCl$	2-Butene-I-carboxylic acid	70	[253]
1,3-Butadiene	$Pd/HCl/O_2$	2-Butene-I-carboxylic acid		[254]
1,3-Butadiene	$Ni(CO)_4$	2-(Carboxycyclohexyl)-propionic acid		[259]
1,3-Butadiene	$Pd[P(C_6H_5)_3]/HCl$	2-Butene-I-carboxylic acid		[253]
Undecylenic acid	$Ni(CO)_4$	1,12-Dodecanedicarboxylic acid	54	[14, 245]
Undecylenic acid	$Co_2(CO)_8$	Dodecanedicarboxylic acid	76	[261]
3-Butene-I-one-	$Ni(CO)_4$	Levulinic acid + 2 methyl-acetoacetic acid	20	[14]
Dihydrofuran	$Ni(CO)_4$	Tetrahydrofurancarboxylic acid		[14]
Styrene	$Pd[P(C_6H_5)_3]/HCl$	α-Phenylpropionic acid		[253]

Nickel catalysts

Reppe [14] employed $Ni(CO)_4$ stoichiometrically. The course of the reaction was very similar to that outlined in Sect. 3.6.1 for the alkynes. However, the hydrocarbonylation of alkenes [Eq. (60)] normally yields saturated products which include a large number of isomers. While the stoichiometric reaction ensues at 150 °C/50 bar with Ni, the corresponding catalytic reaction requires conditions of 250 °C/200 bar. On using Ni catalysts to hydrocarboxylate α-olefins, the α-methyl branched acid predominates (60–70%) in the product [25, 108].

In general, Ni catalyzes the double bond isomerization to a much lesser extent than Co.

$$CH_3CH=CH_2 + CO + H_2O \xrightarrow[HOAc]{Ni(CO)_4} \begin{array}{l} CH_3-CH_2-CH_2-COOH \\ \\ CH_3-\underset{\underset{CH_3}{|}}{CH}-COOH \end{array} \qquad (60)$$

In the presence of PPh_3, the above hydrocarboxylation [Eq. (60)] can be conducted using milder conditions and the isobutyric acid selectivity drops from ~70% to ca. 50% [109]. It was also established [110] that on hydrocarboxylating ethylene with carbon monoxide and water in the presence of Ni/I_2 at 195 °C/70 bar, the optimal I:Ni ratio was 1:2, the selectivity then being 99% which was considerably higher than with other Ni systems [110].

The addition of acid facilitates the $Ni(CO)_4$ catalyzed hydrocarboxylation of olefins to such an extent that the conversion can ensue at ambient pressures and temperatures [26, 111, 112]. Mild conditions are also possible when the $Ni(CO)_4$ stoichiometric hydrocarboxylation is conducted in the presence of UV light, enabling reaction to occur at 50–60 °C and atmospheric pressure [80], approaching the conditions required for the stoichiometric conversion with acetylene. Tables of various precursors and products have been published [1, 2, 9, 14].

Higher olefins (C_4–C_{18}) can be converted into acids using $Ni(CO)_4$ as catalyst at 240 °C/250 bar, producing the acids in 90% yield [23]. When olefins with internal double bonds are employed as precursors, the Ni catalyzed reaction yields exclusively branched acid products [25, 108]. As pure Ni is unsuitable as catalyst, promoters such as halides are essential to obtain suitable reaction rates [25, 108]. The catalytic activity increases in the series I > Br > Cl > F, the addition of $pyridine/PR_3/H_2/Cu$ also having a favorable effect [3]. Instead of using inorganic Ni salts, nickel can be introduced into the reaction as the carboxylate of the resultant acid.

Cobalt Catalysts

Cobalt compounds can participate in hydrocarboxylation reactions both stoichiometrically and in catalytic amounts [14]. As is known from hydroformylation reactions, cobalt catalysts can cause a displacement of the double bond. In the acidic product from the Co catalyzed hydrocarboxylation [Eq. (61)], the linear acid tends to predominate over the branched acid,

$$R-CH=CH_2 + CO + H_2O \xrightarrow[\substack{150-200\,°C \\ 150-250\,bar}]{Co_2(CO)_8} \begin{array}{l} R-CH_2-CH_2-COOH \\ \\ R-\underset{\underset{CH_3}{|}}{CH}-COOH \end{array} \qquad (61)$$

the total yield being usually 80–90%. Even with internal olefins, the linear acid is generally the main product. Furthermore, when an organic base such as pyridine is added to the $Co_2(CO)_8$ catalyzed hydrocarboxylation of α-olefins, the share of the linear acid in the reaction product can be increased to 80% [28–9, 284]. With an internal olefin as substrate, the linear acid's share is also enhanced on adding pyridine, Alkenes with an internal double bond are hydrocarboxylated at a slower rate than the α-olefins. While the addition of H_2 has a favorable effect on the rate, there is the danger that a competitive hydroformylation reaction occurs resulting in increased by-product formation. On carbonylating 1-butene with water and carbon monoxide in the presence of $Co_2(CO)_8/pyridine$, a 93.6% yield of pentanoic acid isomers was obtained of which ~77% were linear [Eq. (62)] [3].

$$H_3C-CH_2-CH=CH_2 + CO + H_2O \xrightarrow{\text{Co}_2\text{(CO)}_8\text{/pyridine}} H_3C-CH_2-CH_2-CH_2-CO_2H$$

(62)

This contrasts with the same reaction at 210 °C/250 bar, in the absence of pyridine, which produced n-pentanoic acid in only 56% yield [113].

Recently, there has been growing interest in the hydrocarboxylation of dienes with Co catalysts [Eq. (63)], the unsaturated acid being produced in high yield.

$$H_2C=CH-CH=CH_2 + CO + H_2O \xrightarrow[250\,\text{bar}/160-200\,°C]{\text{Co}_2\text{(CO)}_8\text{/pyridine}} \underset{95\%}{H_3C-CH=CH-CH_2-CO_2H}$$

(63)

The reactivity of the dienes decreases in the following sequence [5]:

1,3-butadiene, isoprene > 1,5-hexadiene > pentadiene > 2,4-hexadiene > 2,3-dimethyl-1,3-butadiene

Adipic acid is obtained in a reasonable yield from the double hydrocarboxylation of butadiene with carbon monoxide and water in the presence of cobalt hydridocarbonyl and pyridine at 210 °C and 250 bar [115, 116].

Palladium and Platinum Catalysts

Tsuji has published comprehensive reviews dealing with the utilization of palladium catalysts in the carbonylation of olefins [7, 42]. As relatively low temperatures are required with Pd catalysts, styrene and butadiene can be readily carbonylated without undergoing polymerization. Moreover, there is also little isomerization of the olefinic double bond.

$PdCl_2(PPh_3)_2/HCl$ possesses high catalytic activity when used in the carbonylation of substituted olefins [7, 42]. Pd/HCl is active at 100 °C and the relatively low pressure of 50 bar [6]. Moreover, alkenyl halides [Eq. (64)] can be hydrocarboxylated whilst retaining the double bond [41].

$$R-CH=CH-CH_2Cl + CO \xrightarrow{\text{Pd cat.}} RCH=CH_2-CH_2-COCl$$

$$\xrightarrow{\text{H}_2\text{O}} R-CH=CH-CH_2-CO_2H$$

(64)

For the effect of ligands on the selectivity of Pd catalysts, see Sect. 3.2.

Although Pt does not catalyze the hydrocarboxylation of non-terminal olefins, α-olefins can be converted into the corresponding acids under mild conditions [Eq. (65)] in good yields [4, 41].

$$H_3C(CH_2)_9-CH=CH_2 + CO + H_2O \xrightarrow[100\,°C]{\text{SnCl}_2/\text{H}_2\text{PtCl}_4} \underset{68\%}{H_3C(CH_2)_{11}-COOH}$$

(65)

Various Catalysts

Propanol can be yielded directly from ethylene, carbon monoxide and water with iron hydridocarbonyl as catalyst [1, 14]. Similarly, butanol was prepared in 90% yield from propene, carbon monoxide and water using a $Fe(CO)_5$ at 15 bar/100 °C [118].

Simple and complex Rh or Ir salts with PPh_3 ligands are suitable catalysts for hydrocarboxylating olefins at 50–300 °C and slightly elevated pressures [117, 271 –278]. A 40% yield of adipic acid was reported when a $RhCl_3/I_2$ catalyst was used to carbonylate 1,3-butadiene at 50–200 °C/75–100 bar [4].

An unusual conversion [Eq. (66)] with a two-component system, consisting of a saturated and an unsaturated hydrocarbon substrate, was observed with $Cu(CO)_n^+$ catalysts under mild conditions [62].

$$\text{(cyclohexane)} \quad + \quad H_3C(CH_2)_3\text{–}CH\text{=}CH_2 \quad + \quad CO \quad + \quad H_2O \xrightarrow{Cu(CO)_3^+} \quad \text{(cyclohexane)} \quad + \quad tert\text{–}C_7 \text{ acids} \quad (66)$$

This reaction only occurs with saturated hydrocarbons possessing a tert. carbon atom [62]. In hydrocarboxylation reactions, $Cu(CO)_n^+$ catalysis generally gives rise to a predominantly branched acid product. The yields are between 75–80% [61] (cf Sect. 3.3.6).

3.6.2.1.1 Oxidative Carbonylation of Alkenes

The oxidative hydrocarboxylation of olefins which occurs in the presence of a Pd(II)–Cu(II) catalyst system [13, 119] has already been discussed in Sect. 3.2 and 3.4. Acetic acid is frequently employed as solvent and, in the presence of a dehydrating agent such as Ac_2O, the ratio of β-acetoxypropionic acid to unsaturated acid can be decreased to 1 : 100 [120]. Furthermore, Ac_2O apparently also suppresses CO_2 formation [13].

The total acid yield is often as high as 80–85% with the $PdCl_2$–$CuCl_2$–LiCl-NaOAc catalyst system at ~140 °C/80 bar [120], the function of the Cu(II) salt being to oxidize catalytically Pd(O) to Pd(II) with molecular oxygen [13]. Other co-catalysts such as $FeCl_3$, hydroquinone, $AlCl_3$, $CaCl_2$, $MgCl_2$ have been used in place of $CuCl_2$. The Pd recovery from the final residue is facilitated when soluble or sparingly soluble Al, Ca, Mg salts are employed as co-catalysts [13].

The reaction is of commercial interest, being a potential route to acrylic acid directly from ethylene (cf Sect. 3.7) instead of acetylene which is a more expensive feedstock base.

3.6.2.2 Alkenes and Functional Derivatives in the Presence of Alcohols

This type of carbonylation, which is closely related to the hydrocarboxylation discussed in Sect. 3.6.2.1, can be represented as follows:

$$2\,R\text{–}CH\text{=}CH_2 + 2\,CO + 2\,R'OH \xrightarrow{cat.} R\text{–}CH_2\text{–}CH_2\text{–}COOR' + R\text{–}\underset{\underset{CH_3}{|}}{CH}\text{–}COOR' \quad (67)$$

R' = Alkyl

In general, the hydroesterification takes place at a slower rate than the hydrocarboxylation [14]. As it is closely allied to the hydrocarboxylation, the usual transition metal catalysts are employed. As before, catalyst systems which facilitate carbonylation under mild conditions are attractive as there is a marked reduction in by-product formation, e.g., polymerizations, ether formation etc. [7].

$Ni(CO)_4$ can be employed either stoichiometrically or in catalytic amounts at 180–200 °C and 100–200 bar, the ester yield being often as high as 90% [1]. During the last decade there has been little work done on the nickel catalyzed hydroesterification of simple olefins. Typical results are presented in Table 3.15.

Normally, cobalt catalysts (Table 3.2) are usually active at lower temperatures (100–260 °C) than their nickel counterparts [4]. With α-olefins and $Co_2(CO)_8$ as catalyst, the ester is obtained in high yield, the linear acid generally accounting for almost 75% of the product. With internal olefins, cobalt catalysts rapidly isomerize the double bond to the α-position [cf. Eq. (14)] which results in the linear ester being the main product from the carbonylation [1, 31, 124]. In addition, the formation of the linear ester is further encouraged (up to 80% selectivity) in the presence of pyridine [6] (cf Sect. 3.6.2.1). With Co as well as Ni the products are generally saturated esters [1].

Dienes have also been hydroesterified with Co catalysts leading to displacement of the remaining double bond [131].

The cobalt salt of the acid product has also frequently been employed as catalyst, producing the ester in up to 80% yield [132]. This property can be utilized to remove carboxylic acid impurities in the ester product. After the mixture is treated with a Co salt, cobalt carboxylate is fed to the reaction where it functions as hydroesterification catalyst [133].

During the last decade, Pd catalyzed hydroesterifications have been given a great deal of attention on account of their high selectivity at temperatures generally < 100 °C, resulting in low by-product formation [7, 283].

Interest in alternative olefin-based routes to methyl methacrylate (MMA) production has not only given impetus to oxidative carbonylation (cf Sect. 3.6.2.1.1) but also to a bis(triphenylarsine)$PdCl_2$/$AlCl_3$/triphenylarsine-catalyzed process [Eq. (68)] [285].

$$CH_3CH=CH_2 + CO + CH_3OH \xrightarrow{cat.} CH_3-\overset{\overset{\displaystyle CH_3}{|}}{C}H-CO_2CH_3 \tag{68}$$

$$CH_3-\overset{\overset{\displaystyle CH_3}{|}}{C}H-CO_2CH_3 + S \xrightarrow[cat.]{-H_2} CH_2=\overset{\overset{\displaystyle }{|}}{\underset{\underset{\displaystyle CH_3}{|}}{C}}-CO_2CH_3 + H_2S$$

DuPont's patent [285] claims a high yield of the branched ester in contrast to the usual Pd catalyzed carbonylations – the n : iso ratio being 1 : 11 in the above case. At 89% conversion, the selectivity reached 91.7%. The subsequent hydrogenation step produced methyl methacrylate in 89% selectivity [285].

By varying the ligand in the Pd complex, catalysts [Eq. (69)] can be prepared which exhibit high selectivity [134, 283]. Knifton [283] studied dispersions of ligand-

Table 3.15. Hydroesterification of alkenes with Ni catalysts

Precursors	Temp. [°C]	Pressure	Catalyst	Product	Yield	Ref.
$H_2C=CH_2 + CH_3(CH_2)_2CH_2OH$	240	200 bar	$Ni(CO)_4$	$H_3CCH_2C(=O)OBu$	72 %	[121]
$H_2C=CH–CH_2Cl + CH_3OH$	25	atmos.	$NiCl_2$/thiourea	$H_2C=CH–CH_2–COOCH_3$	63%	[122]
$H_2C=C(CH_3)–CH_2Cl + CH_3OH$	25	atmos.	$NiCl_2$/thiourea	$H_2C=C(CH_3)–CH_2–CO_2CH_3$	67%	[122]
$H_2C=C=CH_2 + CH_3OH$	100–250	5–100 bar	$Ni(CO)_4$	$H_2C=C(CH_3)–CO_2CH_3$	23.5%	[123]

Table 3.16. Hydroesterification of alkenes with Co catalysts

Precursors	Temp. [°C]	Pressure	Catalyst	Product	Yield	Ref.
$H_2C=CH_2 + CH_3OH$	300	250 bar	CoI_2	$H_3CCH_2CO_2CH_3$	26%	[129]
$H_2C=CH–CH_3 + CH_3OH$	135	250 bar	$Co_2(CO)_8$	$H_3C(CH_2)_2CO_2CH_3 +$ $H_3CCH(CH_3)CO_2CH_3$	85%	[128]
$H_2C=CHCH_2CH_3 + CH_3OH$	240–250	690–750 bar	Co propionate	$H_3C(CH_2)_3CO_2CH_3$	24.7%	[130]
$H_3C(CH_2)_5CH=CH_2 + CH_3OH$	170	200 bar	$Co_2(CO)_8$/N-methyl-pyrrolidine	$H_3C(CH_2)_7CO_2CH_3$	61.8%	[125]
				$H_3C(CH_2)_5CH(CH_3)CO_2CH_3$	20.1%	
				$H_3C(CH_2)_4CH(C_2H_5)CO_2CH_3$	9.5%	
				$H_3C(CH_2)_3CH(C_3H_7)CO_2CH_3$	8.6%	
$H_3C(CH_2)_7–CH=CH_2 + CH_3OH$	130	150 bar	$Co_2(CO)_8$/γ-picoline	$H_3C(CH_2)_9CO_2CH_3$	83%	[127]
$H_3C(CH_2)_9CH=CH_2 + CH_3OH$	150	140 bar	$HCo(CO)_4$	$n\text{-}C_{12}—CO_2CH_3$	56%	[127]
				yield improved on adding MeOH portionwise		

stabilized palladium (II) chlorides in quaternary Group Vb salts of trichlorostannate (II), e.g. $[PdCl_2(PPh_3)_2-10(Et_4)]N[SnCl_3]$ for the synthesis of linear carboxylic acid esters from α-alkenes via regioselective carbonylation.

$$C_5H_{11}-CH=CH_2 + CO + CH_3OH \xrightarrow[SnCl_2/70\,°C/141\ bar]{PdCl_2(PPh_3)_2} H_3C(CH_2)_6-CO_2CH_3 \qquad (69)$$

In the reaction [Eq. (69)], the linear ester is obtained in 93% yield [134].

In contrast to the other catalysts, olefins can frequently be hydroesterified with Pd catalysts at lower temperatures than the corresponding hydrocarboxylation reactions [135,136]. Tsuji, in particular, has been very active in this field (see his review of the hydro-esterification of dienes [131]).

Using various Pd catalysts at 25 °C/100 bar, 1,3-butadiene was hydroesterified yielding the following products [Eq. (70) [131].

$$\begin{array}{ll}
\begin{array}{l}
H_3C-HC=CH-CH_2-CO_2Et \\
\text{Ethyl 3-pentenoate}
\end{array} & \text{I} \\[2em]
\begin{array}{l}
H_3C-CH_2-\underset{\underset{CH_3}{|}}{CH}-CO_2Et \\
\text{Ethyl 2-methylbutyrate}
\end{array} & \text{II} \\[2em]
\begin{array}{l}
H_3C-CH_2-CH_2-CH_2-CO_2Et \\
\text{Ethyl n-valerate}
\end{array} & \text{III}
\end{array} \qquad (70)$$

H$_3$C

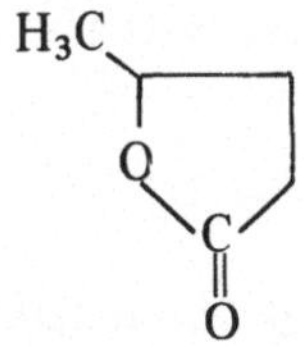

γ-Methylbutyrolactone   IV

Diesters   V

The effects of Pd catalysts on the product distribution can be appreciated from Table 3.17 [131]:

Table 3.17. Effect of various Pd catalysts on product distribution of butadiene carbonylation [131]

Substrates	Catalyst	Product Distribution					Total yield
		I	II	III	IV	V	
1,3-Butadiene, ethanol	Pd/Cl$_2$	92.7	3.8	1.5	0	2.0	23.7%
1,3-Butadiene, ethanol	PdCl$_2$	96.4	2.2	–	–	1.4	32.0%
1,3-Butadiene, ethanol	PdBr$_2$	92.2	1.7	2.7	–	3.4	29.5%
1,3-Butadiene, ethanol	PdI$_2$	36.8	15.8	15.6	12.4	19.8	72.5%

The bromide and chloride have thus similar catalytic activities (Table 3.17) whilst the relatively large amount of saturated esters (III + IV) with PdI_2 indicates the presence of $HPdI_2$. Similar reactions were conducted with isoprene, yielding ethyl 4-methyl-3-pentenoate (corresponds to I above) as the main product in 80% yield [7, 131]. For the effect of Pd-complexes on the selectivity of the styrene carbonylation [43] see Sect. 3.3.

Oxidative hydroesterifications [Eqs. (71, 72)] have also been successfully carried out, as for example in the hydrocarboxylation with a $PdCl_2$–$FeCl_3$, $PdCl_2$–$CuCl_2$ [137] or $PdCl_2$/benzoquinone catalyst (cf Sect. 3.6.2.1.1) [138].

$$H_2C=CH_2 + 2\,CO + 1/2\,O_2 + 2\,ROH \xrightarrow{\text{cat.}} \begin{array}{c} CH_2-CO_2R \\ | \\ CH_2-CO_2R \end{array} + H_2O \tag{71}$$

$$H_2C=CH_2 + EtOH + CO \xrightarrow[100\,^\circ C/70\,bar]{PdCl_2/benzoquinone} H_2C=CH-\overset{O}{\overset{\|}{C}}\underset{OEt}{\diagdown} + \begin{array}{c} CH_2-CO_2Et \\ | \\ CH_2-CO_2Et \end{array} + H_2O \tag{72}$$

Pt catalysts are also active under very mild conditions [139] and produce the linear ester [Eq. (73)] in very high yield [41, 283].

$$H_3C-(CH_2)_8-CH=CH_2 + CO + CH_3OH \xrightarrow[90\,^\circ C/200\,bar]{H_2PtCl_2\ \ SnCl_2} H_3C-(CH_2)_{10}-CO_2CH_3 \tag{73}$$
$$85\%$$

With α-olefins, linear ester selectivities as high as 98% have been reported using R_2PtCl_2 –$SnCl_2$ [139]. However, as mentioned in the previous Section, the Pt catalysts are unsuitable for non-terminal olefins [41, 139].

3.6.2.3 Alkenes and Functional Derivatives in the Presence of Nucleophiles other than Water or Alcohols

These reactions are closely allied to the hydrocarboxylations (Sect. 3.6.2.1) and hydroesterifications (Sect. 3.6.2.2), the corresponding catalysts and reaction conditions being similar (cf Table 3.14) [1].

The reactions [Eq. (74)] can be represented as follows:

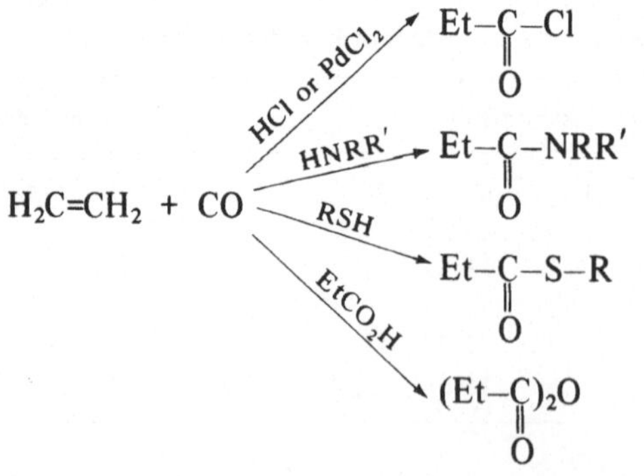

$$H_2C=CH_2 + CO \underset{\substack{\text{HCl or }PdCl_2 \\ HNRR' \\ RSH \\ EtCO_2H}}{\overset{}{\diagup \diagup \diagdown \diagdown}} \begin{cases} Et-\overset{O}{\overset{\|}{C}}-Cl \\ Et-\overset{O}{\overset{\|}{C}}-NRR' \\ Et-\overset{O}{\overset{\|}{C}}-S-R \\ (Et-\overset{O}{\overset{\|}{C}})_2O \end{cases} \tag{74}$$

Propionic anhydride formation during hydrocarboxylation in the presence of propionic acid was first reported by Reppe [14]. Nonanoic anhydride was recently obtained on carbonylating 1-octene in the presence of HOAc/Ac$_2$O. In this case, a PdCl$_2$–PPh$_3$ catalyst system was employed at 150 °C/21 bar and the ratio of linear: branched anhydride was 4:1. The yield of linear anhydride could be increased by maintaining a > 10% acid anhydride concentration in the reaction mixture [140]. In addition, Ni [1, 14], Co$_2$(CO)$_8$ · (PBu$_3$)$_2$ [141] and Rh$_2$O$_3$/EtI/H$_2$ [142] have also been employed in acid anhydride syntheses with good results.

Amide formation via carbonylation with Co [143, 144], Ni [14], Rh [145] and Pd [146] (oxidative carbonylation) has been carried out with various substrates (Table 3.14). Amines were also produced using a modified Rh catalyst in the presence of hydrogen [147] (cf Table 3.18).

The acid chloride of the corresponding olefin can result via carbonylation in the presence of hydrogen chloride [Eq. (75)] [150].

Table 3.18. Synthesis of amides via carbonylation of alkenes in the presence of amines

Substrates	Temp./Press.	Catalyst	Product(s)	Ref.
CH$_3$CH=CH$_2$ + PhNH$_2$	180°C/150 bar	Co$_2$(CO)$_8$	Et–C–NHPh $\parallel$ O CH$_3$CH—C–NHPh $\mid$ CH$_3$ $\parallel$ O	[143]
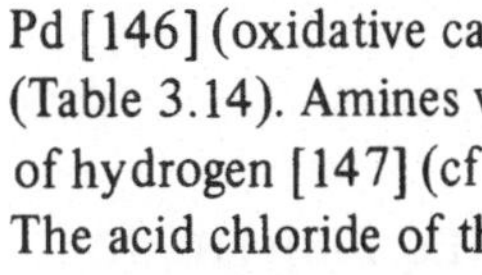 Ph, H, C=C, H, Br + pyrrolidine (N–H)	60°C	Ni(CO)$_4$	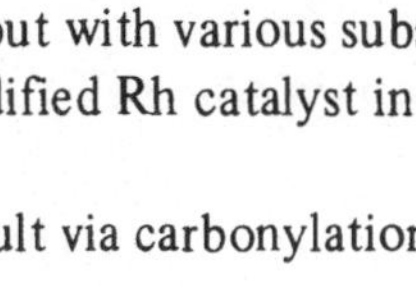 Ph, H, C=C, H, CON(pyrrolidine) 82%	[148]
CH$_3$–(CH$_2$)$_3$–CH=CH$_2$ + BuNH$_2$	200°C/150 bar	Co$_2$(CO)$_8$/ vinylpy-ridine	CH$_3$–(CH$_2$)$_5$–C=O, NHBu	[144]
H$_2$C=CH–CH=CH$_2$ + NH$_3$	125°C	Pd(NO$_3$)$_2$/ PPh$_3$	3,8-Nonadienonic acid amide	[149]
H$_2$C=CH$_2$ + PrNH$_2$ + O$_2$	–	PdCl$_2$ –CuCl$_2$	H$_2$C=C–C $\parallel$ O, NHPr	[146]
Bu–CH=CH$_2$ + H$_2$ + Me$_2$NH	–	Rh(CO)$_2$Cl/ p-toluidine	C$_7$H$_{15}$–N(CH$_3$)(CH$_3$)	[147]
H$_2$C=CH$_2$ + NH$_3$ + H$_2$O	–	Fe(CO)$_5$	PrNH$_2$ Pr$_2$NH Pr$_3$N	[14]

Pr = propyl, Bu = butyl

$$H_3C-(CH_2)_4-CH=CH_2 + CO + HCl \xrightarrow[90°C/210\ bar]{PdCl_2(PPh_3)_2} H_3C-(CH_2)_6-COCl \qquad (75)$$

Instead of HCl, $COCl_2$ can be employed at slightly higher temperatures [Eq. (76)], leading to chlorination of the β-carbon atom of the resultant acid chloride [151].

$$H_2C=CH_2 + COCl_2 + CO \xrightarrow[130-60\,°C/100-800\ bar]{PdCl_2(PPh_3)_2} ClCH_2-CH_2COCl \qquad (76)$$

$PdCl_2$ can also be used stoichiometrically as chlorinating agent [Eq. (77)] [119, 152].

$$H_2C=CH_2 + PdCl_2 + CO \xrightarrow[40-100\ bar]{20-100\,°C} ClCH_2CH_2COCl \qquad (77)$$

Interesting results were found when using CCl_4 as chlorine source [Eq. (78)] [10, 70].

$$RCH=CH_2 + CO + CCl_4 \xrightarrow[\substack{170-200\ bar \\ 50-140\,°C}]{Co_2(CO)_8} \begin{array}{l} R-\underset{\underset{COCl}{|}}{CH}-CH_2-CCl_3 \\[2mm] R-\underset{\underset{Cl}{|}}{CH}-CH_2-CH_3 \end{array} \qquad (78)$$

It was reported that increased CO pressure and lower temperature favor acid chloride formation [Eq. (78)] [10]. Using a $PdCl_2$–PPh_3–$SnCl_2$ catalyst system, a thiol ester was obtained in 78% yield [Eq. (79)] from ethyl mercaptan and 1-heptene [153].

$$H_3C-(CH_2)_4-CH=CH_2 + CO + EtSH \xrightarrow{Pd\ cat.} H_3C-(CH_2)_6\underset{\underset{O}{\|}}{C}-S-Et \qquad (79)$$

Nickel carbonyl also catalyzes the thiol ester formation, but the yields are generally only 15–30% [1, 14].

3.6.3 Carbonylation of Alcohols

$$ROH + CO + (H_2O) \xrightarrow{cat.} RCOOH \qquad (80)$$

$$2\ ROH + CO \xrightarrow{cat.} RCOOR + H_2O$$

There has been considerable interest in this aspect of carbonylation in view of the acetic acid manufacture from methanol [279–281]. The actual conditions depend on substrate, catalyst metal (Sect. 3.3), ligand, promoters and solvent (cf Sect. 3.5 and Table 3.16). The mechanism of the Rh catalyzed reaction has been treated in Sect. 3.2. Primary and secondary alcohols are suitable substrates, the tertiary alcohols being mainly dehydrated to the corresponding olefins [154, 155]. In addition, dicarboxylic acids can be obtained in good yields from diols [14] (cf Table 3.22).

When the ester is the desired product, the conversion is conducted in the absence of water [156]. A water : alcohol ratio of 1–2 : 1 encourages acid formation (with Ni catalysts) [156].

The usual catalysts employed are Co, Ni, Rh and Pd, Fe possessing low activity. Some conditions for the various systems are presented in Table 3.19.

Table 3.19. Reaction conditions with various catalysts

Catalyst metal	Pressure	Temperature
Co	70–700 bar	80–280 °C
Ni	40–300 bar	150–280 °C
Pd, Pt, Ir, Os, Ru	10–15 bar	200 °C
Rh	1–15 bar	150–80 °C

Some results of the alcohol carbonylation are shown in Table 3.20.

Table 3.20. Carbonylation of alcohols with various catalysts

Substrate	Temp. [°C]	Pressure [bar]	Catalyst	Product	Yield	Ref.
MeOH	250	680	$Co(OAc)_2/CoI_2$	Acetic acid	90%	[157]
MeOH	150	35	$Ni(OAc)_2/$ $SnPh_4/CH_3I$	Acetic acid + ester	61% (of product)	[268]
MeOH	200	10	$RhCl_3/C$	Acetic acid	88%	[175]
MeOH	190	52.5	$RhCl_2/(Ph_2P)_2$ Et	Acetic acid	95%	[286]
MeOH	175	1–15	Rh_2O_3/HI	Acetic acid	99%	[176]
Ethanol	250	120	$NiCl_2$–$CuCl$– clay	$EtCO_2H$ $EtCO_2Et$	76% 15.5%	[177]
2-Propanol	300	300	$Ni/Ni(NH_3)_6Cl_2$ NaI	Isobutyric acid	68%	[178]
2-Butanol	300	43	$Ni(CO)_4$, $NiCl_2/HCl$	2-Methylbutanoic acid	70%	[154]
2-Octanol	300	60	$Ni(CO)_4$, $NiCl_2$ /HCl	2-Methylocta-noic acid	76%	[154]
Benzyl alcohol	175	71.3	$RhCl_3$–$3H_2O/HI$	2-Phenylacetic acid	83%	[179]

3.6.3.1 Cobalt Catalysts

$$CH_3OH + CO \xrightarrow{\ Co_2(CO)_8\ } CH_3COOH \tag{81}$$

In the Co catalyzed carbonylation the main by-products are [Eq. (82)]:

$$CH_3COOH + CH_3OH \rightleftharpoons CH_3COOCH_3 + H_2O$$
$$2\ CH_3OH \rightleftharpoons CH_3-O-CH_3 + H_2O$$
$$CH_3OH + CO \rightleftharpoons HCOOCH_3$$
$$CO + H_2O \rightleftharpoons CO_2 + H_2$$

(82)

The ether and ester formation (as well as the water gas shift reaction) can be suppressed on adding H_2O. However, any resulting ester or ether is usually recycled to the reactor in commercial processes.

$HCo(CO)_4$, the actual catalytic species, is generated via the reaction:

$$2\ CoI_2 + 2\ H_2O + 10\ CO \rightarrow Co_2(CO)_8 + 4\ HI + 2\ CO_2$$
$$Co_2(CO)_8 + H_2O + CO \rightarrow 2\ HCo(CO)_4 + CO_2$$

(83)

Another important initial step is the formation of the alkyl iodide

$$ROH + HI \rightarrow RI + H_2O$$

(84)

A scheme [Eq. (85)] has been proposed for the $HCo(CO)_4$ catalyzed carbonylation [157].

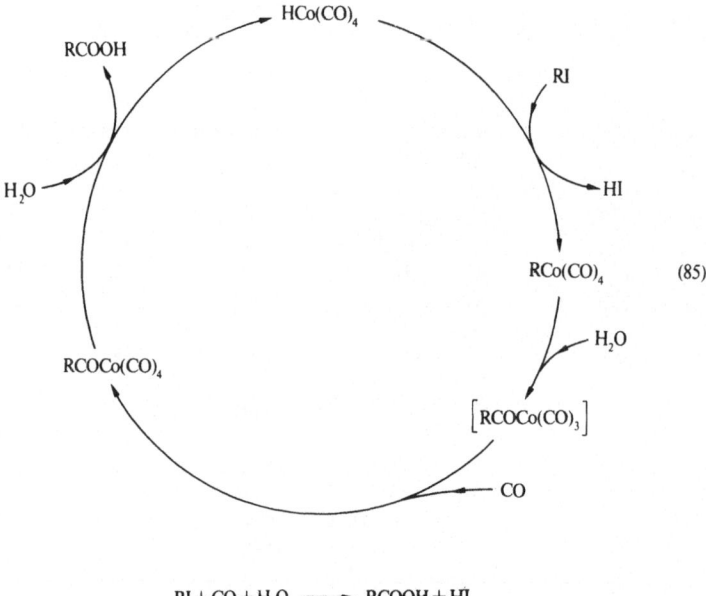

(85)

$$RI + CO + H_2O \longrightarrow RCOOH + HI$$

It was found that the catalytic activity of the cobalt catalyst was increased on introducing H_2 [158] or halides [1]. Studies on cobalt catalysis involving I_2 and additives such as Pd, Pt, Ir, Ru or Cu compounds have shown that conversions can occur at 80–200 °C and 70–300 bar [36] instead of 280 °C/680 bar. This work on improving the classical Co process is a consequence of the high cost of Rh coupled with its limited availability. The selectivity of the traditional MeOH carbonylation with Co is > 90%.

3.6.3.2 Rhodium Catalysts

Catalyst systems based on Rh represent the most effective way of manufacturing acetic acid under mild conditions, the selectivity being as high as 99%. This application of Rh catalysis has been reviewed on numerous occasions [8, 22, 73, 155, 164, 165]. For the mechanism of the Rh catalyzed carbonylation see Sects. 3.2 and 3.3.3.

Due to the high cost of rhodium and its efficiency as a catalyst, there have recently been a number of reports concerning heterogenized Rh systems (anchoring) [162, 269]. The trend is to eliminate the separation and corrosion problems encountered in homogeneous catalysis with Rh.

The rates of the homogeneous and heterogeneous systems (in MeOH carbonylation) were recently compared at 250 °C/atmospheric pressure (cf Table 3.21) [163].

Table 3.21. Comparison of performance of homogeneous and heterogeneous Rh catalysts

Catalyst	Rh content (wt.%)	Rate (g MeOAc/g Rh/hr)
$RhCl_3$	homogeneous	3000
$RhCl_3$-NaX	0.25	50
$RhCl(CO) (PPh_3)_2$-alumina	1.39	40
$Rh(NO_3)_3$/C	3	25

Thus, the rate of reaction of the heterogenized catalyst is not noticeably dependent on the Rh compound or the carrier material. The results can be explained by the supposition that only ~1% of Rh in the heterogenized catalyst is active [163].

The most suitable medium for the rhodium catalyzed carbonylation of methanol is acetic acid [20], the polarity of less polar solvents being adjusted accordingly. The best way of adding rhodium to the reaction system (cf Sect. 3.5) is as rhodium nitrate, which, in terms of conversion, gave results at least twice as good as any other Rh salt tested [20, 69]. The methanol conversion increased with rhodium concentration, reaching a maximum at 2% Rh [69].

3.6.3.3 Nickel Catalysts

A variety of nickel compounds [Eq. (85)], as well as the metal in the presence of iodine, have been used as catalysts. $Ni(CO)_4$ is generated from NiI_2 according to reaction (85a) [155]:

$$NiI_2 + H_2O + 5\,CO \longrightarrow Ni(CO)_4 + 2\,HI + CO_2 \qquad (85a)$$

Heck proposed a mechanism [Eq. (86)] for the Ni catalyzed carbonylation [16].

$$\begin{aligned}
ROH + HI &\rightleftharpoons RI + H_2O \\
Ni(CO)_4 + RI &\rightleftharpoons RNi(CO)_2I + 2\,CO \\
RNi(CO)_2I + CO &\rightleftharpoons RCONi(CO)_2I \\
RCONi(CO)_2I + CO &\rightleftharpoons RCOI + Ni(CO)_4 \\
RCOI + ROH &\rightleftharpoons RCOOR + HI
\end{aligned} \qquad (86)$$

Ni generally requires fairly high temperatures and pressures [156].

Halcon Int. Inc. (New York) reported that methanol can be carbonylated under mild conditions in the presence of nickel diacetate tetrahydrate/tetraphenyltin and methyl iodide [268]. When the molar ratio of iodide to methanol is at least $1:10$, a pressure as low as 35 bar can be applied at 150 °C. This pathway may well be very significant in the future as no costly noble metals are required.

Ni has often been employed in the carbonylation of diols (Table 3.22).

Table 3.22. Carbonylation of diols

Substrate	Temp. [°C]	Pressure	Catalyst	Product	Yield	Ref.
1,2-Ethanediol	250–60	200 bar	$Ni(CO)_4, I_2$	Succinic acid	15%	[14]
1,4-Butanediol	260	200 bar	$Ni(CO)_4, I_2$	Adipic acid	69%	[14]
1,4-Butanediol	200	70 bar	$RhCl_3/HI$	Adipic acid	38%	[180]
1,5-Pentanediol	250	200 bar	$Ni(CO)_4, I_2$	Pimelic acid	94%	[14]
1,6-Hexanediol	260	200 bar	$Ni(CO)_4, I_2$	Suberic acid	90%	[14]

3.6.3.4 Palladium Catalysts

Palladium-based catalysts can be used for the carbonylation of allyl alcohols, leading to a retention of the double bond in the products [46, 152, 166]. The main work carried out with Pd catalysts concerns their application in oxidative carbonylations [Eq. (87)]. The usual reaction conditions are 60–140 °C/80–100 bar [167–170].

$$2\,ROH + 2\,CO + 1/2\,O_2 \rightarrow (COOR)_2 + H_2O \qquad (87)$$

Various catalyst systems have been employed:

$PdCl_2–FeNO_3$	[167]	$PdCl_2–KNO_3$	[169]
$PdCl_2–La(NO)_3$	[168]	$PdCl_2–CuCl_2–LiCl_2/NH_3$	[170]

With the last catalyst system [170], dimethyl oxalate was obtained in 90% yield (100 bar/60 °C).

3.6.3.5 Various Catalysts

Iridium has successfully been used in place of Rh, producing comparable results [12, 171].

With $Cu(CO)_3^+$, yields as high as 80% have been obtained [172–73]. However, the H_2SO_4 concentration must be $\geqslant$ 80%, otherwise no reaction occurs [172–4] (room temperature and atmospheric pressure [174]). Trace amounts of linear acids are formed, from primary alcohols due to rearrangement [Eq. (88)] of the carbenium ion [173] (cf Sect. 3.3.6 and Chap. 5).

$$ROH \xrightarrow{H_2SO_4} ROH_2^+ \rightarrow R^+ + H_2O \rightarrow R'-\underset{\underset{R''}{|}}{\overset{\overset{R'''}{|}}{C}}{}^+ \xrightarrow{Cu(CO)_3^+} R'-\underset{\underset{R''}{|}}{\overset{\overset{R'''}{|}}{C}}-CO_2H \tag{88}$$

3.6.4 Carbonylation of Amines

The reactions can be summarized as follows [51]:

$$RNH_2 + CO \tag{89}$$

$$H-\underset{\underset{O}{\|}}{C}-NHR \qquad R-NH-\underset{\underset{O}{\|}}{C}-NH-R$$

Primary and secondary amines are suitable as substrates. Although aliphatic and aromatic amines usually give rise to N-formyl and urea derivatives respectively, on varying reaction conditions the urea can be obtained from the aliphatic amine in good yield [181].

The usual catalyst metals are Ni [182], Co [183], Pd [184], Rh [181], Mn [185], Fe [14], Hg [186], Cu [187], Ru [188], Se [189] and Ag [287].

The mechanism of the $Co_2(CO)_8$ catalyzed reaction has been studied by Orchin et al. [190].

In the presence of $Mn_2(CO)_{10}$, primary aliphatic amines yield 1,3-dialkylureas as the major product [185]. When butylamine is carbonylated with $[Rh(CO)_2Cl]_2$ as catalyst the following product distribution resulted [191]:

$$BuNH_2 + CO \xrightarrow{Rh(CO)_2Cl_2} H-\underset{\underset{O}{\|}}{C}-NHBu + BuNH-\underset{\underset{O}{\|}}{C}-NHBu \tag{90}$$
$$\qquad\qquad\qquad\qquad\qquad\qquad\quad 35\% \qquad\qquad\quad 65\%$$

In the presence of PMe_3 (ratio PMe_3 : catalyst = 6 : 1) at 160 °C/60 bar, N-butyl-formamide (Eq. 90) is exclusively obtained in ~100% yield [191].

When tertiary amines are carbonylated [Eq. (91), (92)] there is again a distinction between aliphatic and aromatic substrates [183]:

$$Bu_3N + CO \xrightarrow{Ni(CO)_4} H-\underset{\underset{O}{\|}}{C}-NBu_2 \tag{91}$$

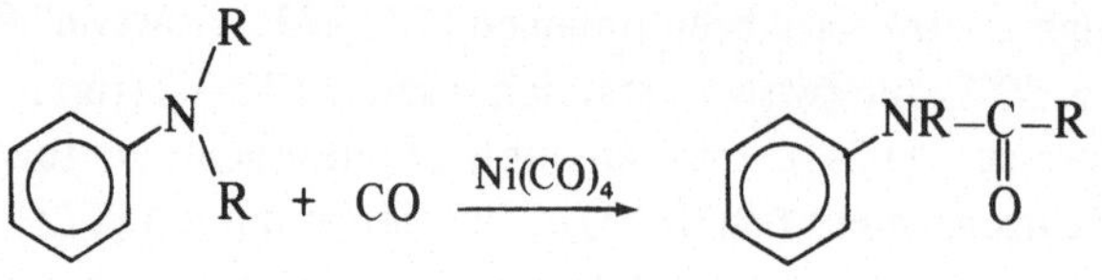

$$(92)$$

Some recent examples of carbonylations of amines with various catalysts are shown in Table 3.23.

Table 3.23. Carbonylation of amines

Substrate	Temp. (°C)	Pressure	Catalysts	Product	Yield	Ref.
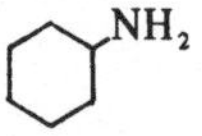 (cyclohexyl)NH$_2$	180–220	130 bar	Mn$_2$(CO)$_{10}$	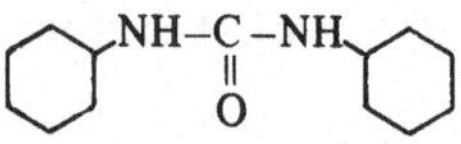 (cyclohexyl)NH–C(=O)–NH(cyclohexyl)	58%	[185]
BuNH$_2$	–	atmos.	Se	BuNHCNHBu (C=O)	100%	[192]
BuNH$_2$	150	70 bar	RhH(CO)$_3$– PPh$_3$	BuNHCNHBu (C=O)	20%	[181]
				H–C(=O)–NH–Bu	40%	
NHMe$_2$	110–160	60 bar	Cu(CN)$_2$	H–C(=O)–NMe$_2$	34%	[187]
NHEt$_2$	55–220	atmos.	Se	Et$_2$NCNEt$_2$ (C=O)	83%	[189]
 (piperidine)	280–320	100 bar	CuNaA-zeolite	Formyl-piperidine	95–98%	[288]
Me$_3$N	225	84–140 bar	Co$_2$(CO)$_8$	H–C(=O)–NMe$_2$	100%	[183]

3.6.5 Carbonylation of Ethers and Esters

3.6.5.1 Carbonylation of Carboxylic Acid Esters

Esters can be carbonylated with CO, generating acids [Eq. (93)]. A variety of catalysts are employed – mainly Ni, Co and Rh. When the carbonylation is conducted in the absence of water or another proton donor, then the product is usually the anhydride [Eq. (93)].

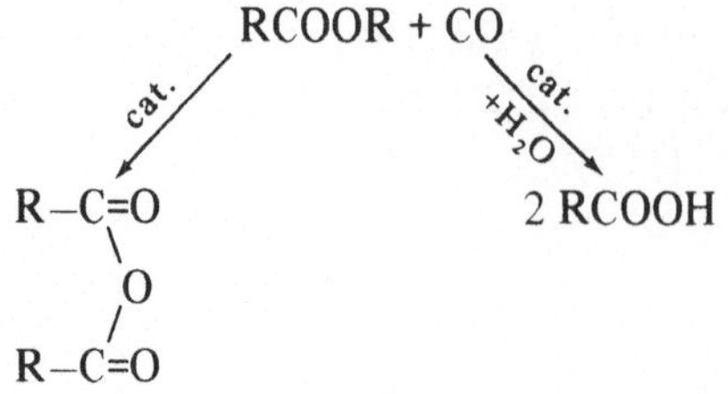

$$(93)$$

However, when the ester formula is $RCOOR'$ ($R \neq R'$) then a mixed anhydride results, or, in the presence of water, two different acids are produced [Eq. (94)] [156].

$$H_3C \overset{\displaystyle O}{\underset{\displaystyle OCH_2CH_3}{\overset{\|}{\diagup}\diagdown}} + H_2O + CO \xrightarrow[\text{Raney-Ni}]{\text{NiI}_2/\text{NaI}} H_3CCOOH + H_3CCH_2COOH \qquad (94)$$

Some results are shown in Table 3.24.

Table 3.24. Carbonylation of esters

Substrate	Temp.[° C]	Pressure (bar)	Catalyst	Product	Yield	Ref.
$H_3CCOOEt$	320–340	270–300	$Ni(CO)_4$	H_3CCOOH H_3CCH_2COOH	57% 39%	[156]
$H_3CCOOCH_3$	180–190	650–700	$NiBr_4$-$(Ph_3PBu)_2$	Ac_2O	43%	[193]
$EtCO_2Et$	180–190	650–700	$CoBr_2/Et_4NBr$	$(EtCO)_2O$	40%	[193, 194]
$H_3CCOOCH_3$	175	24.5	$RhCl_3/MeI$ $Cr(CO)_6$	Ac_2O	54%	[195]
$H_3CCOOCH_3$	230–240	70	$RhCl(CO)$ $(PPh_3)_2/MeI$	Ac_2O	50%	[196]
$HCOOCH_3$	220	300	HgI_2/N-methyl-pyrrolidone	H_3CCOOH	92%	[197]

3.6.5.2 Carbonylation of Ethers

The carbonylation of ethers gives rise either to esters or, if sufficient water is present, to carboxylic acids [Eq. (95)].

$$ROR \underset{\text{cat.}}{\overset{2\ CO + H_2O}{\diagup}} \quad \begin{array}{l} 2\ RCOOH \\[1em] RCOOR \end{array} \qquad \underset{\text{cat.}}{\overset{+\ CO}{\diagdown}} \qquad (95)$$

The usual catalysts are based on Co, Ni, Rh, Fe and Pd [155–56]. It was established that cyclic ethers tend to react more readily than the corresponding acyclic ones [14].

Table 3.25 presents some typical examples.

The epoxides are highly reactive and give rise to β-OH acids which can be dehydrated to the unsaturated acids [Eq. (96)] [1, 155].

$$H_2C \underset{\displaystyle O}{\diagdown\diagup} CH_2 + CO + H_2O \xrightarrow{\text{cat.}} \underset{\displaystyle OH}{CH_2-CH_2-COOH} \xrightarrow{-H_2O} H_2C{=}CH{-}CO_2H \qquad (96)$$

Table 3.25. Carbonylation of ethers

Substrate	Temp. [°C]	Pressure [bar]	Catalyst	Product	Yield	Ref.
Et_2O	270–300	240–350	$NiCl_2$	EtCOOEt	32%	[156]
				EtCOOH	68%	
Et_2O	320–340	310–400	$NiI_2/Ni(OAc)_2$	EtCOOH	81%	[156]
Et_2O	230	240	FeI_2/SiO_2	EtCOOEt	28.8%	[16]
				EtCOOH	10.1%	
Et_2O	170	150	$RhCl(PPh_3)_3$	EtCOOEt	16.3%	[199]
			MeI	AcOEt	15.8%	
Tetrahydro-furan	330–350	330–400	Ni_2S/Iso-propyl iodide	Adipic acid	64%	[156]

Cobalt catalysts have been frequently used for these conversions, yields generally lying between 11–40% [200, 201].

3.6.6 Carbonylation of Halides

The carbonylation of saturated, unsaturated, and aromatic halides using transition metal carbonyls has been discussed in several reviews [1, 5, 12, 202, 203] (for the mechanism of the Pd-catalyzed carbonylation of aromatic halides, see Sect. 3.2).

Various products can be obtained [Eq. (97)].

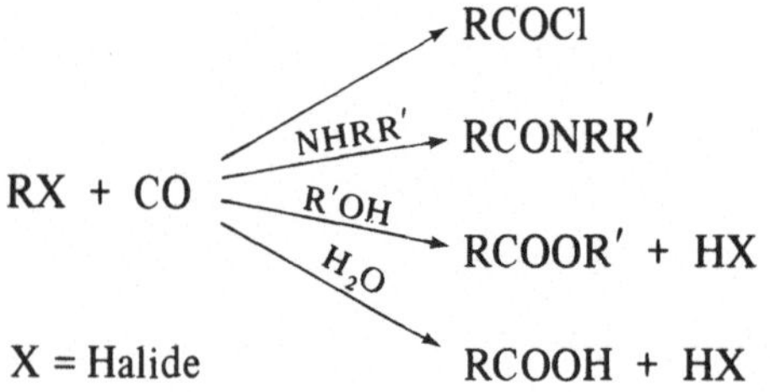

$$
\begin{array}{l}
\text{RX + CO} \\
\\
\text{X = Halide}
\end{array}
\quad
\begin{array}{l}
\xrightarrow{} \text{RCOCl} \\
\xrightarrow{\text{NHRR}'} \text{RCONRR}' \\
\xrightarrow{\text{R}'\text{OH}} \text{RCOOR}' + \text{HX} \\
\xrightarrow{\text{H}_2\text{O}} \text{RCOOH} + \text{HX}
\end{array}
\qquad (97)
$$

The usual catalysts are based on Ni, Co, Fe [208], Rh and Pd [203]. Not only Rh enables the conversion to be conducted under mild conditions but also cobalt in the form of $NaCo(CO)_4$ (cf Table 3.26) along with equimolar amounts of a tertiary amine or alcoholate [16, 207].

Table 3.26. Carbonylation of alkyl halides

Substrate	Temp. [°C]	Pressure	Catalyst	Product	Yield	Ref.
CH_3I/CH_3OH	0	atmos.	$NaCo(CO)_4$	$H_3CCOOCH_3$	80%	[207]
$n-C_5H_{11}I/CH_3OH$	50	atmos.	$NaCo(CO)_4$	$n-C_5H_{11}COOCH_3$	33%	[207]
$n-C_8H_{17}Cl/EtOH$	ambient	–	$Na_2Fe(CO)_4$	$n-C_8H_{17}COOEt$	89%	[208]
$n-C_5H_{11}Br/NHEt_2$	ambient	–	$Na_2Fe(CO)_4$	$n-C_5H_{11}CONEt_2$	80%	[208]

Other examples using more drastic reaction conditions with Ni and Co can be found in various reviews [5, 12, 203]. Good yields at atmospheric pressure and temperature < 100 °C have been reported with Pd catalysts [204].

A mechanism [Eq. (98)] was proposed for the Pd/(PPh$_3$) complex catalyzed alkoxycarbonylation in the presence of a tertiary amine (1 bar/100 °C) [17]:

$$RX + CO + R'OH + R_3''N \rightarrow RCOOR' + R_3''NHX$$

R = aryl, vinyl   X = Br, I

Mechanism: (98)

$$RX + Pd(CO)(PPh_3)_2 \rightarrow RPd(X)(CO)PPh_3 \rightleftharpoons RCOPd(X)(PPh_3)_2$$

$$\downarrow R'OH$$

$$Pd(CO)(PPh_3)_2 + HX \overset{CO}{\rightleftharpoons} RCoOR' + HPd(X)(PPh_3)_2$$

$$HX + R_3''N \rightarrow R_3''NHX$$

The mechanism of the cobalt [205] and nickel [14, 16, 206] catalyzed carbonylation of halides has also been studied [203]. There has been considerable activity in the aryl halide field. For typical results, see Table 3.27.

Ni catalysts help to convert allyl halides — in the presence of CO and an alcohol — into the ester of vinylacetic acid [Eq. (99)], the double bond being thereby retained [203, 217, 220].

$$H_2C=CH-CH_2Cl + CO + CH_3OH \xrightarrow{Ni\,cat.} H_2C=CH-CH_2-CO_2CH_3 + HCl \quad (99)$$

Table 3.27. Carbonylation of aryl halides

Substrate	Temp. (°C)	Pressure	Catalyst	Product(s)	Yield	Ref.
Chlorobenzene	305	330 bar	Ni–SiO$_2$	Ethyl benzoate	74.7%	[209]
				Benzoic acid	25.1%	
Methyl p-chloro-benzoate	50	5 bar	Co$_2$(CO)$_8$/NaOMe	Dimethyl terephthalate	92%	[210]
p-Chlorotoluene	50	0.5 bar	Co$_2$(CO)$_8$/NaOMe	Methyl p-toluate	82%	[211]
m-Dichloromethyl-toluene	55	5–10 bar	Co$_2$(CO)$_8$	Diethyl benzene-1,3-diacetate	65%	[212]
Bromobenzene	95–98	atmos.	Pd(PPh$_3$)$_4$/Bu$_4$NI	Benzoic acid	86%	[204]
Iodobenzene	60	14 bar	PdCl$_2$(PPh$_3$)$_2$/Et$_3$N	Methyl benzoate	85%	[213]
Chlorobenzene	–	–	PdCl$_2$/AlCl$_3$	Benzoyl fluoride	77.5%	[214]
Benzyl chloride	150	150 bar	Rh(CO)Cl(PPh$_3$)$_2$	Phenylacetyl chloride	33%	[215]
Benzyl chloride	55	1.7 bar	K[Fe(CO)$_3$NO]/NaOMe	Methyl phenyl-acetate	85%	[216]

Using $NiCl_2$, thiourea and Fe powder, the reaction can take place at room temperature and atmospheric pressure producing the ester in yields between 63–67% [122]. This is a useful route to β, γ-unsaturated acids. The reaction can also be conducted with insertion of alkenes [Eq. (100)] [218]:

$$H_2C{=}CH{-}CH_2Cl + H_2C{=}CH_2 + CO + H_2O \xrightarrow[\substack{30-50°C \\ 10-40\,bar}]{Ni\,cat.} H_2C{=}CH(CH_2)_3CO_2H + HCl \tag{100}$$

In this reaction, 5-hexenoic acid is obtained in 40–60% yield [218]. Other unsaturated systems (dienes or alkynes) can be substituted for ethylene in Eq. (100) [88, 119, 219], Pd catalysts being frequently employed at ambient conditions [119].

3.6.7 Carbonylation of Aldehydes

There is not a great deal of material available concerning this type of carbonylation [1]. The most suitable catalysts are the halides of Ni, Co or Fe, their activity decreasing in the same sequence [221].

$$H_2CO + H_2O + CO \xrightarrow[200°C/610\,bar]{NiI_2} HOCH_2COOH \tag{101}$$

In this reaction, glycolic acid is produced in 42% yield, and is accompanied by small amounts of formic acid and methanol [221].

3.6.8 Carbonylation of Aromatic Nitro Compounds

During the last decade, there has been very considerable industrial activity in this field due to the strong commercial interest in the carbonylation product – aromatic isocyanates – which are used in the manufacture of polyurethanes [15] (cf Sect. 3.7.4).

Attention has been focused on nitrobenzene and closely related compounds. The general equation for the reaction when neither water, hydrogen nor alcohol is present is as follows:

$$ArNO_2 + 3\,CO \rightarrow ArNCO + 2\,CO_2 \tag{102}$$

The catalysts are mainly based on Pd, Rh, Co or Fe, the pressures vary between 120–560 bar and the temperatures from 150 °C to 250 °C. Table 3.28 lists some recent examples. Most interest has been devoted to Pd catalyzed carbonylations. Aromatic nitro compounds can also be directly converted into 1,3-dialkylureas (cf. Sect. 6.4) using $Rh(CO)_{16}$ as catalyst [222].

Table 3.28. Carbonylation of aromatic nitro compounds

Substrate	Temp. [°C]	Pressure	Catalyst	Product	Yield	Ref.
Nitrobenzene	190	500 bar	Rh/C/FeCl$_3$	Phenyl isocyanate	35%	[223]
Nitrobenzene	190	95–130 bar	RhCl$_3$/PdCl$_2$/ PhCN	Phenyl isocyanate	58%	[221]
Nitrobenzene	190	200 bar	Co$_2$(CO)$_8$	Phenyl isocyanate	25%	[225, 226]
Nitrobenzene	190	130 bar	PdCl$_2$/V$_2$O$_5$	Phenyl isocyanate	56%	[227]
Nitrobenzene	210	120 bar	PdCl$_2$–C$_5$H$_5$N	Phenyl isocyanate	78%	[229]
2,4-Dinitro-toluene	190	227 bar	PdCl$_2$(CO)$_2$/ isoquinoline	2,4-Diisocyanato-toluene	49%	[228]
2,4-Dinitro-toluene	210	120 bar	PdCl$_2$–C$_5$H$_5$N	2,4-Diisocyanato-toluene	80.9%	[229]
2,4-Dinitro-toluene	high	high	PdCl$_2$-quinoline	2,4-Diisocyanato-toluene	80.6%	[230]
2,4-Dinitro-toluene + ethanol	150	380 bar	Pd/Al$_2$O$_3$	Diethyl toluene-2,4-dicarbamate	89%	[231]

3.7 Industrial Applications of Carbonylation Reactions

There has been considerable activity in the commercialization of the metal carbonyl catalyzed transformations. The traditional high pressure processes face competition from the lower pressure – usually Rh catalyzed – variants [15, 21, 232]. The use of acetylene as feedstock for the manufacture of acrylic acid and its esters is slowly being displaced by propene oxidation processes, in part because of the high cost of acetylene. With the same aim in view, work is also being conducted on olefin-based carbonylations (cf. Sects. 3.6.2.1.1 and 3.6.2.2).

The main areas of industrial interest are:

- Acrylic acid/ester production from acetylene (Table 3.29)
- Acetic acid production from methanol (Table 3.30)
- Butanol production from propene (Table 3.31)
- Tolylene diisocyanate manufacture from dinitrotoluene

Their relative economic impact is shown in Fig. 3.7.

While the overriding position of the Rh catalyzed methanol carbonylation can readily be appreciated, this has not prevented research into other routes employing less expensive catalysts.

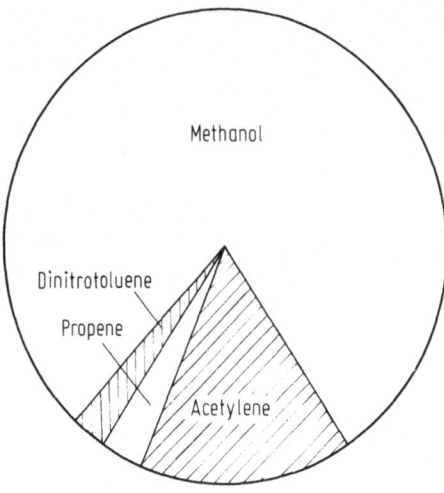

Fig. 3.7. Relative capacities of present and projected Reppe carbonylation processes – according to feedstock

3.7.1 Production of Acrylic Acid and its Esters

The available capacities of plants for the production of acrylic acid/acrylates are shown in Table 3.29.

Table 3.29. Carbonylation of acetylene

Process	Product	Company	Capacity t/a	Ref.
BASF High Pressure (cat.)	Acrylic acid	BASF (W. Germany)	132,000	[15]
BASF High Pressure	Acrylic acid	Dow-Badische (USA)	30,000	
Rohm & Haas	Acrylic acid	Rohm & Haas (USA)	140,000	[233]
(modified Reppe)	ester		(4 plants)	
Modified Reppe	Acrylic acid	Toagosei Chem. Ind. Ltd.	36,000	[4]

The catalytic process operates at 40–55 bar and between 180–205 °C using a $NiBr_2$-copper halide catalyst in THF. As the catalyst is employed in negligible amounts it is not recovered, the reaction product being worked up via distillation. The acrylic acid selectivities are 90% (based on C_2H_2) and 85% (based on CO). The dangers inherent in working with C_2H_2 under pressure were overcome by dividing the reaction chamber into smaller units thus enabling a rapid removal of any heat of decomposition from C_2H_2. The BASF process [1, 4, 11, 234] which has frequently been described is reproduced in Fig. 3.8 [234].

A new process developed by Union Oil [36, 235] involves the oxidative carbonylation of propene to acrylic acid using a $PdCl_2$–$CuCl_2$ catalyst at 70 bar/95 °C. The acrylic and β-acetoxypropionic acid selectivites are 75–80%. While there has been no commercial application so far, a process diagram was published recently [235].

There will undoubtedly be increased effort to find a non-acetylene based route to to acrylic acid.

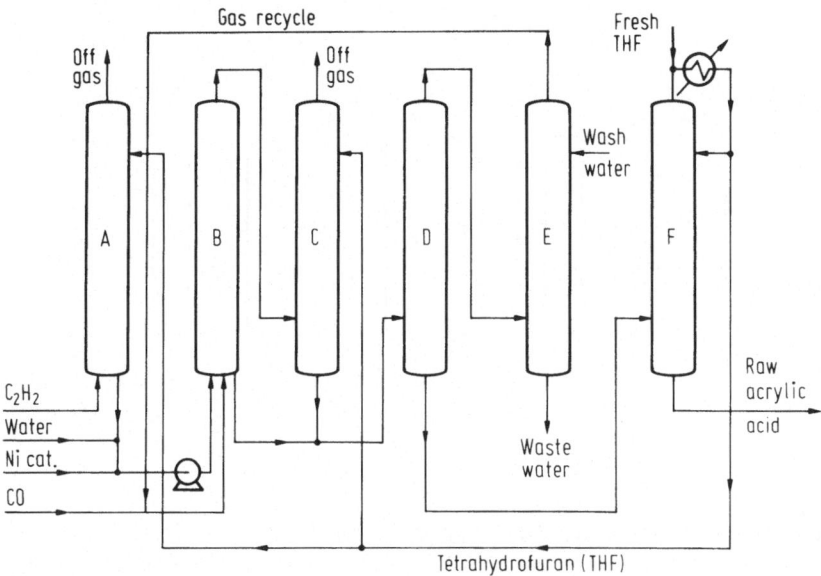

Fig. 3.8. BASF Process for the carbonylation of acetylene [234]
(A = Saturator, B = Reactor, C = Saturator, D = Degasser, E = Washing column, F = Distillation column).

3.7.2 Production of Acetic Acid

This expanding area of interest has been studied by BASF since the early twenties. The commercialization was hampered by the corrosion resulting from catalyst and product at the applied high temperatures and pressures ($250\,^\circ$C/680 bar with CoI_2) This problem was solved in the late 1950's when a Mo–Ni alloy (Hastelloy B and C). was developed for lining the reactor [1]. Acetic acid selectivities of 90% (based on CH_3OH) and 70% (based on CO) are now obtained. The organic by-products of the reaction (methyl acetate, dimethyl ether, formic acid) are generally recycled. After a 5 column distillation, acetic acid is obtained with a purity of 99.8%. A process diagram of the BASF process [11, 236] is shown in Fig. 3.9.

On account of the severe process conditions required by the cobalt catalyzed carbonylation, Monsanto carried out research on catalyst systems enabling milder conditions to be applied (cf Sects. 3.3.3 and 3.6.3). An acetic acid manufacturing process with a Rh/I_2 catalyst was evolved which produced selectivites of 99% based on CH_3OH and 90% based on CO [19]. The process could be conducted at lower temperatures (150–200 $^\circ$C) and pressures of a few bar. However, as rhodium is a very rare metal and roughly a thousand times more expensive than cobalt [107], particular care must be taken to ensure complete catalyst recovery. It has been suggested that the Monsanto catalyst recovery probably involves several steps due to the high volatility of the iodide promoter [21].

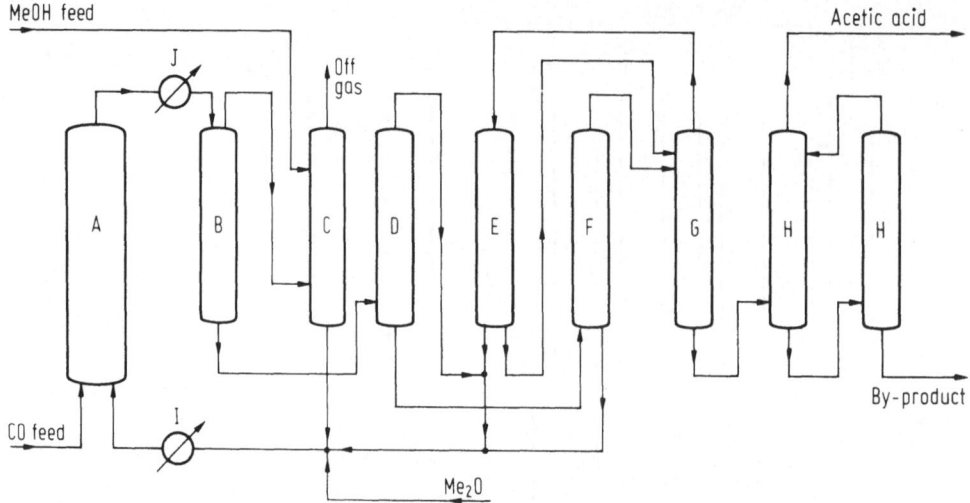

Fig. 3.9. BASF process for carbonylation of methanol [236, 279].
(A = Reactor, B = Separator, C = Wash column, D = Degasser, E = Separator, F = Catalyst separator, G = Dehydrater, H = Distillation units, I = Preheater, J = Cooler)

Moreover, as the Rh-complex/HI system is also highly corrosive, the reactor as well as recycle loops will have to be constructed from expensive alloys [21].

A flow diagram of the Monsanto process is shown in Fig. 3.10 [237, 239].

A further feature of this route is that even in the presence of large amounts of H_2, there is apparently no major by-product formation, in contrast to the Co catalyzed process [74].

Table 3.30 includes several projected plants which are to utilize the Monsanto process. The interest in the Rh catalyzed route has been very rapidly gaining momentum. Comparison of the traditional gas-phase BASF process with the liquid phase Monsanto process was recently subject of a review [239]. There have also been moves by BASF and Shell to improve the Co catalyzed process [36]. When additives such as Pd, Pt, Ir, Ru, or Cu with Co/I_2 are used, reaction occurs at 80–200 °C and 70–300 bar [36] (cf Sect. 3.6.3).

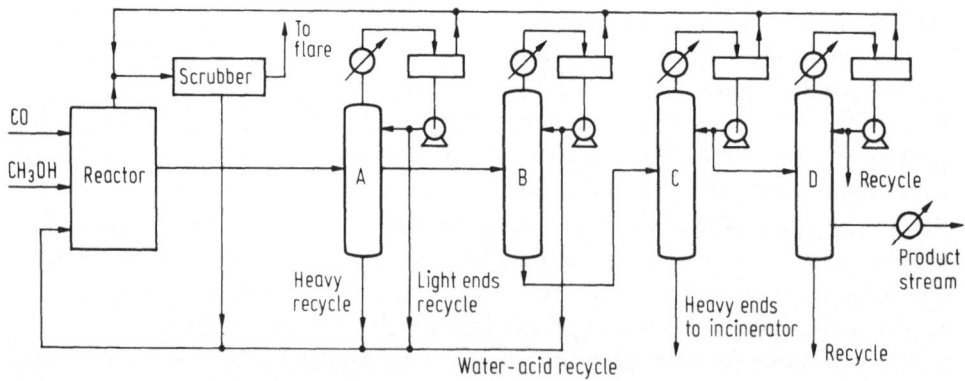

Fig. 3.10. Monsanto acetic acid process [237]
(A = Light ends column, B = Drying column. C = Product column, D = Finishing column)

Table 3.30. Carbonylation of methanol

Process	Product	Company	Capacity t/a	Ref.
BASF high pressure	Acetic acid	BASF (W. Germany)	35,000	[107]
BASF high pressure	Acetic acid	Borden Co. (USA)	52,000	[25]
Monsanto low pressure	Acetic acid	Monsanto (USA)	136,000	[232]
Monsanto low pressure	Acetic acid	Celanese	272,000[a]	[18, 21]
Monsanto low pressure	Acetic acid	MSK (Yugoslavia)	100,000[a]	[280]
Monsanto low pressure	Acetic acid	Rhône-Poulenc	225,000[a]	[281]
Monsanto low pressure	Acetic acid	USSR	150,000[a]	[281]
Monsanto low pressure	Acetic acid	BP	150,000[a] expansion	[281]
Monsanto low pressure	Acetic acid	Daicel (Japan)	150,000[a]	[279]
Monsanto low pressure	Acetic acid	NSC/MCI[b] (Japan)	200,000[a]	[279]

[a] under construction or planned
[b] NSC = Nippon Synthetic Chemical Industry Co. Osaka
  MCI = Mitsubishi Chemical Industries Ltd., Tokyo

3.7.3 Production of Butanol and Propionic Acid

Using a Ni propionate catalyst, ethylene can be hydrocarboxylated forming propionic acid in 95% yield (based on C_2H_4). The conditions of this liquid phase reaction are 200–240 bar and 270–320 °C, crude propionic acid also being used as solvent. The main by-products are ethane, higher carboxylic acids and CO_2

$$H_2C=CH_2 + CO + H_2O \xrightarrow{Ni(C))_4} CH_3-CH_2-COOH \tag{103}$$

Butanol can be manufactured at 100 °C/15 bar from propylene using $Fe(CO)_5$ as catalyst in the presence of N-n-propylpyrrolidine. This is the basis of the BASF process which produces an 85% yield of n-butanol and 15% isobutanol [107, 240–41]. The known capacities of the BASF processes based on propylene are presented in Table 3.31.

Table 3.31. Carbonylation of alkenes

Process	Product	Manufacturer	Capacity t/a	Ref.
BASF	Propionic acid	BASF (W. Germany)	30,000	[15]
BASF	Butanol	Japan Butanol Co.	30,000	[107]

Due to costly catalyst recycles and the required threefold CO quantity, this process is not very attractive when compared to the Oxo process for butanol production [107].

The potential significance of the oxidative carbonylation of alkenes leading to acrylates (cf Sect. 3.6.2.1.1) and the Dupont route from isobutene (cf Sect. 3.6.2.2) should not be underestimated.

3.7.4 Production of Tolylene Diisocyanates

Although there are considerable data in patents, details of Mitsui Toatsu's process for the manufacture of tolylene diisocyanate which is planned to go into production in 1982–83 have not been disclosed [279]. The projected capacity is 50,000 t/a [15, 279].

3.8 Concluding Remarks

The dramatic expansion in Monsanto's Rh catalyzed acetic process helps underline the present significance of the Reppe reactions – more than 40 years after the introduction of the term 'carbonylation'. Undoubtedly the next few years will bring an increase in the oil price and, coupled with the uncertain political situation in several of the main oil producing countries this will give an even greater boost to research – probably government sponsored – in the (coal-based) carbon monoxide field. Although the hydroformylation or Oxo process has until recently dwarfed the other carbonylation reactions (in terms of capacity) this is now no longer the case, as can be appreciated from Table 3.30.

Future commercial developments might well involve olefin-based acrylate production as well as new routes to acetic acid with less expensive catalysts or even the direct manufacture of acetic anhydride and/or VAM [279].

3.9 References

1. Falbe, J.: Carbon Monoxide in Organic Synthesis, Chapter 2, Springer Verlag, , New York 1970; Falbe, J.: Synthesen mit Kohlenmonoxid, Springer Verlag, Berlin-Heidelberg-New York (1967)
2. Kemball, C.: Catalysis, Vol. 1, Chapter 10, Chemical Society, London 1977
3. Falbe, J.: Methodicum Chimicum, Vol. 5, p. 530–534, 622–663, 711–713, Thieme Verlag, Stuttgart 1975
4. Ullmanns Encyklopädie der techn. Chemie, Vol. 7, 81-83, 91, 1974; Vol. 9, 156-170, 1975, 4th Edition, Verlag Chemie, Weinheim
5. Eidus, Ya. T., Lapidus, A. L. et al.: Russ. Chem. Rev., *42*, 199, 1973
6. Cassar, L. et al.: Synthesis, 509, 1973
7. Tsuji, J.: Adv. Org. Chem., *6*, 109, 150-198, (Ed. Taylor, E. and Winberg, C.), Wiley, New York 1969
8. Taqui Khan, M.M., Martell, A.E.: Homogeneous Catalysis by Metal Complexes, Vol. 1, Chapter 4, Academic Press, New York and London 1974
9. Wender, I., Pino, P.: Organic Synthesis via Metal Carbonyls, Vol. 2, 297-516, Wiley, New York 1977
10. Ryang, M.: Organomet. Chem. Rev. A, 5, 67-93, 1970
11. Winnacker, K.L., Küchler, L.: Chemische Technologie, Vol. 4, 18-19, 90-92, 99, 3rd Edition, Carl Hanser Verlag, Munich, 1972

12. Schrauzer, G.N.: Transition Metals in Homogeneous Catalysis, Chapter 5, Marcel Dekker Inc., New York 1971
13. Fenton, D.M., Olivier, K.L., Biale, G.: Amer. Chem. Soc., Petrol. Div. Preprint, *14*, (4), C77-C84, Sept. 1969
14. Reppe, W.: Liebigs Ann. Chem., *582*, 1-161, 1953
15. Weissermel, K., Arpe, H.J.: Industrial Organic Chemistry, Verlag Chemie, Weinheim, 1978
16. Heck, R.F.: J. Amer. Chem. Soc., *85*, 2013, 1963
17. Schoenberg, A., Bartoletti, I., Heck, R.F.: J. Org. Chem., *39*, 3318, 1974
18. Forster, D.: J. Amer. Chem. Soc., *98*, 846-848, 1976
19. Roth, J.F., Craddock, J.H. et al.: Chem. Technol., 600, 1971
20. Forster, D.: Inorg. Chem., *8*, 2556, 1969
21. Webber, K.M., Gates, B.C., Drenth, W.: J. Catalysis, *47*, 269-271, 1977
22. Hjortkjaer, J., Jensen, V.W.: Ind. Eng. Chem.; Prod. Res. Dev., *15*, 46-49, 1976
23. Germain, J.E.: Catalytic Conversions of Hydrocarbons, 278-287, Academic Press, New York (1969)
24. Tolman, C.A.: Chem. Rev., *77*, 313-348, 1977
25. Vogt, V.: Dissertation, Aachen, 1969
26. Chiusoli, G.P., Merzoni, S.: Chim. e Ind., *51*, 612, 1969
27. Kunichika, S. et al.: Bull. Chem. Soc. Japan, *44*, 3405, 1971
28. Seide, W.: Dissertation, Aachen, 1969
29. Imyanitov, N.S., Rudkovskii, D.M.: J. Appl. Chem. USSR, *41*, 157, 1968
30. Matsuda, A.: Kagaku Kogaku Kogyo, *25*, 1180-1184, 1974
31. Matsuda, A., Uchida, H.: Bull. Chem. Soc. Japan, *38*, 710, 1965
32. Matsuda, A.: Bull. Chem. Soc. Japan, *46*, 524, 1973
33. Lion Fat and Oil Co., JP-138704 (03.12.74)
34. Hershman, A., Forster, D. (Monsanto Co.): DE Offen. 2,439,951 (27.02.75)
35. Kiyonaga, Y., Yoshimura, Y. (Lion Fat and Oil Co.): JP Kokai, 76-52, 389, (01.11.74)
36. BASF, DE Offen. 2,303,271 (1973), Shell, DE Offen. 2,400,534 (1974)
37. Michalska, Z.M., Webster, D.E.: Chem. Technol., 117, 1975
38. Sato, T., Kono, T. (Ajinomoto Co.): JP Kokai, 74-53, 585 (28.10.72)
39. Hüttel, R.: Brennstoffchemie, *50*, 281, 331, 1969; Synthesis, 225, 1970
40. Fannin, L.W. et al. (Monsanto Co.): DE Offen. 2,358,410 (24.11.71); CA *81*, 91059, 1974
41. Kehoe, L.J., Schell, R.A.: J. Org. Chem., *35*, 2846, 1970
42. Tsuji, J.: Acc. Chem. Res., *2*, 144-52, 1969
43. Sugi, Y., Bando, K., Shin, S.: Chem. and Ind., 397, 1975
44. Schoenberg, A. et al.: J. Org. Chem., *39*, 3318-3326, 1974
45. Schoenberg, A., Heck, R.F.: J. Org. Chem., *39*, 3327, 1974
46. Bittler, K. et al.: Angew. Chemie., *80*, 352, 1968
47. Nozaki, K. (Shell Int. Research): DE Offen., 2,410,246 (05.03.73)
48. Hara, M., Ohno, K. et al. (Toray Ind. Inc.): JP 74-14,635 (23.10.70)
49. Tsuji, J.: Acc. Chem. Res., *6*, 8-15, 1973
50. US Sec. of Agric., US 3,928,231 (06.10.72)
51. Bird, C.W.: Chem. Rev., *62*, 283, 1962
52. Fenton, D.M. (Union Oil Co. of Calif.): US 3,661,949 (11.06.69)
53. Schrauzer, G.N.: Chem. Ber., *94*, 1891, 1961
54. Clark, J.C., Cookson, R.C.: J. Chem. Soc., 686, 1962
55. Suzuki, H. et al.: Chem. Lett., 641, 1975
56. Houben-Weyl, Vol. 5/1b, 1034-1040, Thieme Verlag, Stuttgart 1972
57. Usami, M. et al. (Toa Nenryo Kogyo K. K.): JP 71-06,494 (14.04.66)
58. Thompson, D.T.: Platinum Metals Rev., *19*, 88, 1975
59. Souma, Y., Sano, H.: Bull. Chem. Soc. Japan, *47*, 1717, 1974
60. Yoneda, N., Fukuhara, T. et al.: Chem. Lett., 607, 1974
61. Souma, Y. et al.: Bull. Chem. Soc. Japan, *49*, 3291, 1976
62. Souma, Y., Sano, H.: J. Org. Chem., *38*, 3633, 1973

63. Heck, R.F.: J. Amer. Chem. Soc., *94*, 2712, 1974
64. Braca, G.: Thesis, Pisa University, Feb. 1961
65. Natta, G., Albanesi, G.: Chimca e Ind., *48*, 1157-1161, 1966
66. Sbrana, G., Braca, G., Piacenti, F. et al.: Chim. e Ind., *54*, 117, 1972
67. Eastman Kodak Co., US 2,739,109 (1948)
68. Monsanto Co., DE Offen. 1,767,151 (1968), 1,941,449 (1969), 1,966,695 (1969)
69. Schultz, R.G., Montgomery, P.D.: Amer. Chem. Soc., Div. Petrol. Chem. Prepr., *17*, B13-B18, 1972
70. Susuki, T., Tsuji, J.: J. Org. Chem., 35, 2982, 1970
71. Bird, C.W., Cookston, R.C., Hudec, J.: Chem. and Ind., 20, 1960
72. Foa, M., Cassar, L.: Gazz. Chim. Ital., 102, 85, 1972
73. Hjortjaer, J., Jensen, O.R.: Ind. Eng. Chem., Prod. Res. Dev., *16*, 281-285, 1977
74. Chem. Engng. News, 19, 30.08.71
75. Yoshisato, E., Tsutsumi, S.: J. Org. Chem., *33*, 869, 1968
76. Bird, C.W., Briggs, E.M.: J. Chem. Soc. C, 1265, 1967
77. Suzuki, S., Uno, K. et al.: J. Chem. Soc. Japan, Ind. Chem. Sect., *55*, 718, 1952
78. Yakubovic, A., Volkova, E.: Dokl. Akad. Nauk SSSR, *84*, 1183, 1952
79. Jones, E.R.H., Shen, T.Y., Whiting, M.C.: J. Chem. Soc., 230-236, 1950
80. Tetteroo, J.M.J.: Dissertation, Aachen, 1965
81. See ref. 9, p. 477, 495
82. Ehrreich, J.E. et al.: Ind. Eng. Chem., Proc. Des. Dev., *4*, 77, 1965
83. Clark, H.C., Mittal, R.K.: Can. J. Chem., *51*, 1511, 1973
84. Jones, E.R.H. et al.: J. Chem. Soc., 763, 1951
85. Reppe, A., Magin, A.: US 2,604,490 (1951)
86. Wender, I., Pino, P.: Organic Synthesis via Metal Carbonyls, Vol. 1, 343-371, Wiley Interscience, New York 1968
87. Natta, G., Pino, P.: US 2,851,486 (1955)
88. Chiusoli, G.P., Cassar, L.: Angew. Chem. Int. Edit., *6*, 124, 1967; Angew. Chem. *79*, 177-86 (1967)
89. Sokol'skii, D.V., Khasanova, R.N.: SU 335,934 (29.07.68); CA *84*, 121177, 1976
90. Sokol'skii, D.V., Levchenko, L.V., Pridanova, Yu.A.: SU 447,028 (28.06.72); CA *84*, 122549, 1976
91. Kushnikov, Yu.A. et al.: Dokl. Vses. Konf. Khim. Atsetilena. 4th., *2*, 265-71, 1972; CA *79*, 115093, 1973
92. Sokol'skii, D.V. et al.: Katalitich. reaktsii u zhid k. faze. (Ch.3), 635-7, 1974; CA *83*, 113569, 1975
93. Thompson, D.T., Jackson, R.: DE Offen. 2,114,544 (25.03.70); CA *76*, 13931, 1972
94. Reppe, W., Stadler, R.: DE Offen. 942,809, 1953
95. Cassar, L., Foa, M.: Inorg. Chem. Nucl. Letters, *6*, 291, 1970
96. Kunichika, S. et al.: Bull. Chem. Soc. Japan, *41*, 390, 1968
97. Happel, J. et al.: Ind. Eng. Chem., Prod. Res. Dev., *14*, 44, 1975
98. Natta, G., Pino, P.: Chim. e Ind. (Milan), *34*, 449, 1952
99. Kaliya, O.L. et al.: Doklady Akad. Nauk. SSSR, *199*, 1321, 1971
100. Lines, B.C., Long, R.: Amer. Chem. Soc., Div. Petrol. Chem. Prepr., *14*, B159-B169, 1969; CA *73*, 130595, 1970
101. Heck, R.F.: J. Amer. Chem. Soc., *94*, 2712, 1972
102. Tsuji, J., Nogi, T. (Toyo Rayon Co. Ltd.): JP 68-09,044 (01.10.64)
103. Ajinomoto Co. Inc., FR 1,486,666, 1966
104. Albanesi, G.: Chem. e Ind. (Milan), *46*, 1169, 1964
105. Tsuji, J., Morikawa, M., Iwamoto, N.: J. Amer. Chem. Soc., *86*, 2095, 1964
106. Blomquist, A.T., Maitlis, P.M.: J. Amer. Chem. Soc., *84*, 2329, 1962
107. Falbe, J.: J. Organomet. Chem., *94*, 213-227, 1975
108. Levering, D.R., Glasebrook, A.R.: J. Org. Chem., *23*, 1836, 1958
109. Chiusoli, G.P., Cometti, G.: J. Chem. Soc., Chem. Comm., 1015, 1972; Chiusoli, G.P.: Acc. Chem. Res., *6*, 422, 1973

110. Hershman, A., Forster, D.: US 3,944,604; Chemtech, 140, 1977

111. Fell, B., Tetteroo, J.M.: Angew. Chem., *77*, 813, 1965

112. Bird, C.W. et al.: J. Chem. Soc., 410, 1963

113. Kubaev, B.E. et al.: SU 186,427 (30.03.65);

114. Kubaev, B.E., Imyanitow, N.S., Rudkovskii, D.M.: Prikl. Khim., *42*, 1149, 1969

115. Imyanitow, N.S. et al.: Zh. Prikl. Khim., *38*, 2558, 1965

116. Imyanitow, N.S., Rudkovskii, D.M.: Zh. Organ. Khim., *2*, 231, 1966

117. Monsanto, DE Offen. 1,941,501 (1961); DE Offen. 2,310,808 (1972); DE Offen. 2.324,765 (1972)

118. Hydrocarbon Process., 138, Nov. 1971

119. Medema, D. et al.: Inorg. Chim. Acta, *3*, 255, 1969

120. Fenton, D.M. et al.: Erdöl u. Kohle, *24*, 350, 1971

121. Kröper, H. et al. (BASF): DE Offen. 920,244 (1954); Z., 5896, 1955

122. Chiusoli, G.P. et al.: J. Chem. Soc. C, 2889, 1968

123. Kunichika, S., Sakakibara, Y. (Chiyoda Chem. Eng. and Construction Co. Ltd.): US 3,466,324 (09.10.69); GB 1,110,565 (14.07.66)

124. Pino, P., Ercoli, R.: Chim. e Ind. (Milan), *36*, 536, 1954

125. Harrison, A.C., Pugh, L.K. (ICI Ltd.): DE Offen. 2,108,422 (23.02.70); 2,023,690 (14.05.69)

126. Fanning, J.R. (Ethyl Corp.): US 3,976,670 (09.12.74); 3,935,228 (23.10.73)

127. Isa, H. et al. (Lion Fat and Oil Co. Ltd.): JP Kokai, 75-62,888 (06.10.73)

128. Natta, G. et al. (Montecatini): US 2.805.245, 1957

129. Bhaltacharyya, S.K., Nag, S.N.: Brennstoff-Chem., *43*, 114, 1962

130. DuPont, GB 651,853 (1951)

131. Hosaka, S., Tsuji, J.: Tetrahedron, *27*, 3821, 1971

132. Isa, H. et al. (Lion Fat and Oil Co. Ltd.): DE Offen. 2,263,907 (28.12.71)

133. Lion Fat and Oil Co. Ltd., JP 1054-510 (01.11.74)

134. Knifton, J.F. (Texaco Develop. Corp.): DE Offen. 2,303,118 (02.02.72)

135. BASF, DE Offen. 1,229,089 (1964); 1,249,867, (1964); GB 1,123,367 (1965)

136. BASF, DE Offen. 1,259,867 (1964)

137. Fenton, D.M., Steinwand, P.J.: J. Org. Chem. *37*, 2034, 1972

138. Fenton, D.M., Biale, G. (Union Oil Co. of Calif.): US 3,755,421 (27.06.68)

139. Knifton, J.F.: J. Org. Chem, *41*, 793, 2885, 1976

140. Fenton, D.M. (Union Oil Co. of Calif.): US 3,641,071 (11.06.69)

141. Forster, D., Morris, D.E. (Monsanto Co.): DE Offen. 2,263,442 (27.12.71)

142. Forster, D., Hershman, A., Paulik, F.E. (Monsanto Co.): DE Offen. 2,364,446 (26.12.72)

143. Pino, P. et al.: Chim. e Ind. (Milan), *50*, 106, 1968

144. Isa, H., Inagaki, T., Kadoya, I. (Lion Fat and Oil Co. Ltd.): JP Kokai 75-129,509 (30.03.74)

145. Iwashita, Y., Sakurabab, M.: J. Org. Chem, *36*, 3927, 1971

146. Biale, G. (Union Oil Co. of Calif.): US 3,523,971 (26.05.66)

147. Berthoux, J., Chevallier, Y., Martinand, J.P. (Rhône-Progil): FR 2,211,002 (20.12.72)

148. Corey, E.J., Hegedus, L.S.: J. Amer. Chem. Soc., *91*, 1233, 1969

149. Tsuji, J., Hara, M. (Toray Ind. Inc.): JP 73-11,036 (15.12.70)

150. Knifton, J.F. (Texaco Inc.): US 3,880,898 (03.08.73)

151. Prichard, W.W. (DuPont): US 3,681,449 (23.07.70)

152. Tsuji, J., Morikawa, M., Kiji, J.: Tetrahedron Lett. 1061, 1963

153. Knifton, J.F. (Texaco Inc.), US 3,933,884 (10.10.73)

154. Adkins, H., Rosenthal, R.W.: J. Amer. Chem. Soc., *72*, 4550, 1950

155. See ref. 9, p. 1-42

156. Eastman Kodak Co., US 2,739,169 (22.04.48)

157. Kutepow, N. v., Himmele, W., Hohenschutz, H.: Chem.-Ing.-Techn., *37*, 383, 1965; Hohenschutz, H., Kutepow, N. v., Himmele, W.: Hydrocarbon Process., *45*, 141, 1966

158. BASF, DE Offen, 902,495 (1951); 933,148 (1953)

159. Wender, I., Greenfield, H., Orchin, M.: J. Amer. Chem. Soc., *73*, 2656, 1951

160. Berty, J., Markó, L., Kalló, D.: Chem. Technol., *8*, 260, 1956

161. Mizoroki, T., Nakayama, M.: Bull. Chem. Soc. Japan, *37*, 236, 1964

162. Jarrel, M.S., Gates, B.C.: J. Catalysis, *40*, 255-67, 1975
163. Scurrel, M.S.: Platinum Metal Rev., *21*, 92-96, 1977
164. Brodzki, D. et al.: Bull. Soc. Chim. Fr., 61, 1976
165. Kujimoto, K., Tanemura, S., Kunugi, T.: Nippon Kaguku Kaishi, *2*, 167, 1977
166. Kutepow, N. v., Bittler, N., Neubauer, D. (BASF): DE Offen. 1,221,224 (1963)
167. Yamazaki, T. et al.: DE Offen. 2,514,685 (05.04.74)
168. Röhm GmbH, DE Offen. 2,238,837 (14.02.74)
169. Yamazaki, T. et al. (Ube Industries Ltd.): JP Kokai, 76-29.428 (03.09.74)
170. Montedison Spa., DE Offen. 2,601,139 (17.01.75)
171. Ichikawa, M., Uda, A., Tamarn, K.: JP Kokai 73-41,987 (14.10.71)
172. Souma, Y., Sano, H.: Oskaka Kogyo Gijutsu Shikensho Kiho, *25*, 163, 1974
173. Souma, Y., Sano, H.: Bull. Chem. Soc. Japan, *46*, 3237, 1973
174. Souma, Y., Sano, H.: Nikkakyo Geppo, *26*, 220, 1973
175. Uda, A., Tamarn, K. (Sagami Chem. Res. Center): JP Kokai, 73-18.217 (22.06.71)
176. Paulik, F.E., Roth, J.F.: J. Chem. Soc., Chem. Comm., 1578, 1968
177. Makhmudov, T.M. et al.: Dokl. Akad. Nauk. Uzb. SSR, *26*, 36-7, 1969
178. Hagemeyer, H.J.: US 2,739,169
179. Paulik, F.E. et al. (Monsanto): DE Offen. 1,941,449 (15.08.68)
180. Monsanto Co., GB 1,278,354 (15.08.69)
181. Tsuji, J., Iwamoto, N. (Toyo Rayon Co. Ltd.): JP 69-10,250 (16.09.66);
    Lassau, C., Chauvin, Y., Lefebvro, G.: DE Offen. 1,902,560 (22.01.68)
182. Kraus, T.C. (Olin Corp.): US 3,870,758
183. Nozaki, K. (Shell Oil Co.): US 3,407,231 (25.04.66)
184. Tsuji, J., Iwamoto, N.: J. Chem. Soc., Chem. Comm., 380, 1966
185. Fausto, C. (American Cyanamid Co.): US 3,316,297 (20.12.65)
186. Fenton, D.M. (Union Oil Co. of Calif.): US 3,277,061 (01.07.63)
187. Kobayaski, S., Hirota, K., Saegusa, T.: JP 69-26,447 (21.06.66)
188. Byerley, J.J. et al.: J. Chem. Soc., Chem. Commun., 1482, 1971
189. Tsutsumi, S., Sonoda, N. (Asahi Chem. Ind. Co. Ltd.): JP Kokai 72-42,721 (10.04.71)
190. Sternberg, H.W. et al.: J. Amer. Chem. Soc., *75*, 3148, 1953
191. Durand, D., Lassan, C.: Tetrahedron Lett., *28*, 2329-30, 1969
192. Sonoda, N. et al.: J. Amer. Chem. Soc., *93*, 6344, 1971
193. Bhattacharyya, S.K., Palit, S.K.: J. Appl. Chem., *12*, 174, 1961; Reppe, W. et al.:
    US 2,729,651 (1956)
194. Reppe, W., Friederich, H.: US 2,730,546 (1956)
195. Roth, J.F. et al.: Chem. Technol., *23*, 600, 1971
196. Kagawa, T., Kono, T., Hamaoka, T. (Ajinomoto Co. Inc.): JP Kokai, 75-30,820 (08.07.73)
197. Isogai, N. (Mitsubishi Gas Chemical Co. Inc.): JP 73-35,053 (30.12.70)
198. Aliev, Y.Y., Romanova, I.B.: Neftekhim. Akad. Nauk. Turk. SSR, 204, 1963
199. Matsui, Y., Konishi, S., Taniguchi, H. (Maruzen Oil Co. Ltd.): JP Kokai 76-91,205 (07.02.75)
200. Takegami, Y. et al.: Bull. Chem. Soc. Japan, *37*, 672, 1962
201. Eisenmann, J.L. et al.: J. Org. Chem. *26*, 2102, 1961
202. El-Chahawi, M., Prange, E.: Chem.-Ztg., *102*, 1-9, 1978
203. See ref. 9, p. 517–543
204. Montedison SpA, BE 803,119 (11.06.74)
205. Heck, R.F.: J. Amer. Chem. Soc., *90*, 5518, 1968
206. Mikitake, M., Mizoroki, T.: Bull. Chem. Soc. Japan, *43*, 569-72, 1970
207. Heck, R.F., Breslow, D.S.: J. Amer. Chem. Soc., *85*, 2779, 1963
208. Collmann, J.P. et al.: J. Amer. Chem. Soc., *95*, 249, 1973
209. Murty, K. et al.: J. Appl. Chem. Biotechnol., *26*, 135, 1976
210. Dynamit Nobel AG, BE 833,872 (30.09.74)
211. Redecker, K. et al. (Dynamit Nobel AG): DE Offen. 2,446,657 (30.09.74)
212. El-Chahawi, M. et al. (Dynamit Nobel AG): DE Offen. 2,410,782 (07.03.74)
213. Stille, J.K., Wong, P.K.: J. Org. Chem., *40*, 532-34, 1975
214. Pritchard, W.W. (DuPont): US 3,632,643 (26.06.67)

215. Tsuji, J., Ono, K. (Toyo Rayon Co. Ltd.), JP 69-17,128 (27.06.66)
216. BASF, DE Offen. 2,521,610
217. Chiusoli, G.P.: Chim. e Ind. (Milan), *1*, 503, 1969
218. Chiusoli, G.P., Cometti, G.: J. Chem. Soc., Chem. Comm., 1016, 1972
219. Chiusoli, G.P.: Angew. Chem., *72*, 74, 1960
220. Cassar, L., Foà, M.: Chim. e Ind. (Milan), *51*, 673, 1969
221. Bhattacharyya, S.K., Vir, D.: Adv. Catal., *9*, 625, 1957
222. Iqbal, A.F.M.: Helv. Chim. Acta, *55*, 2637, 1972
223. Hardy, W.B., Bennett, R.B.: Tetrahedron Lett., *11*, 961-2, 1967
224. Ottman, G.F. et al. (Olin Mathieson Corp.): FR 1,556,876 (06.02.68)
225. Yamahara, T., Usui, M. (Sumitomo Chem. Co. Ltd.): JP Kokai 74-135,943 (15.05.73)
226. Yamahara, T. et al. (Sumitomo Chem. Co. Ltd.): DE Offen. 2,334,532 (06.07.72)
227. Wilhelm, J. et al. (Olin Mathieson Corp.): FR 1,564,753 (28.02.67); CA *72*, 66602, 1970
228. Yamahara, T. et al. (Sumitomo Chem. Co. Ltd.): JP Kokai 73-64,048 (08.12.71)
229. Mizoguchi, Y. et al. (Asahi Chem. Ind. Co. Ltd.): JP Kokai 74-92,041 (03.09.74)
230. Yamahara, T., Deguchi, T. (Sumitomo Chem. Co. Ltd.): JP Kokai 75-49,253 (03.09.73)
231. Balling, P.J., Wiesner, G.L. (American Cyanamid Co.): FR Demande 2,008,365 (13.05.68)
232. Roth, J.F.: Platinum Met. Rev., *19*, 12, 1975
233. Imhausen License Study, Imhausen International, W. Germany, 1978
234. Hydrocarbon Process., 120, Nov. 1971
235. Olivier, K.L. et al.: Hydrocarbon Process., 95-8, Nov., 1972
236. Hydrocarbon Process., 92, Nov., 1973
237. Grove, H.D.: Hydrocarbon Process., 76, Nov., 1972
238. Union Oil, US 3,381,030, 1968
239. Cociasu, C.A.: Rev. Chim. (Bucharest), *24*, 875, 1974
240. Hydrocarbon Process., 138, Nov. 1971
241. Kindler, H., Eisfeld, K.: Ind. Chim. Belge, *32*, 650-3
242. Ziegler, K.: Naturforschung und Medizin in Deutschland (1939-46), Vol. 36, Präparative organische Chemie, Part 1, Dieterich'sche Verlagsbuchhandlung, Wiesbaden 1948; Hecht, O., Kröper, H., ibid, Part 1, Sect. 1, p 115-154; Kröper, H.: in Houben-Weyl, Vol. IV/2, 385-415, Georg Thieme Verlag, Stuttgart (1955), Eidus, Ya.T., Puzitskii, K.V.: Russ. Chem. Rev. *33*, 438 (1964)
243. Reppe, W.: Neue Entwicklungen auf dem Gebiet der Chemie des Acetylens und Kohlen-oxyds, 96-126, Springer-Verlag, Berlin-Göttingen-Heidelberg (1949)
244. Jones, E.R.H. et al.: J. Chem. Soc. 763 (1951)
245. DuPont, G., Pignaniol, P., Vialle, J.: Bull. Soc. Chim. France, 529 (1948)
246. Bergmann, E.D., Zimkin, E.: J. Chem. Soc. 3455 (1950)
247. Jones, E.R.H., Whitham, G.H., Whiting, M.C.: J. Chem. Soc., 4628 (1957)
248. Mueller, G.P., MacArthur, F.C.: J. Amer. Chem. Soc. *76*, 4621 (1954)
249. Jones, E.R.H., Shen, T.Y., Whiting, M.C.: J. Chem. Soc., 48 (1951)
250. Jones, E.R.H. et al.: J. Chem. Soc., 1865 (1954)
251. Schwartzmann, L.H., Rosenthal, R.W.: Abstr. ACS Meeting April 1959, p 57
252. Ashworth, P.J. et al.: J. Chem. Soc. 4633 (1957)
253. Kutepow, N.v. et al. (BASF): DE Offen. 1,229,089 (1966)
254. Kutepow, N.v. et al.: BE 679,611 (1966)
255. Gresham, W.R., Brooks, R.E. (DuPont): US 2,448,368
256. Gresham, W.R., Brooks, R.E. (DuPont): GB 631,001 (1950), US 2,549,453 (1951)
257. Hagemeyer, H.J.: US 2,739,169 (1952)
258. Larson, A.T.: US 2,448,375 (1945)
259. Reppe, W., Kröper, H.: DE Offen. 863,194 (1943)
260. Bhattacharyya, S.K., Sourirajan, J.: J. Sci. Ind. Res. (India) *11B*, 123 (1952)
261. Ercoli, R.: Chim. e Ind. (Milan) *37*, 1029 (1955)
262. Rosenthal, R.W. (Texas Co.): US 2,652,413 (1953)
263. Ercoli, R. et al.: Chim. e Ind. (Milan) *42*, 587 (1960)
264. Reppe, W. et al.: Liebigs Ann. Chem., *560*, 1 (1948)

265. Bird, C.W. et al.: J. Chem. and Ind. 20 (1960)
266. Detzer, H. et al. (BASF): BE 613,730 (1961)
267. Rull, T.: Mémoires présentes à la Société Chimique No 440, 2080 (1964)
268. Haglieri, A.N., Rizkalla, N. (Halcon Int.): DE Offen. 2,749,954 (8.11.76), 2,749,955 (08.11.76)
269. Christensen, B., Scurrell, M.S.: J. C. S., Faraday Trans. *73*, 2036 (1977)
270. Kurmeier, H.A.: Kontakte *1*, 3 (1978) (E. Merck, Darmstadt)
271. Imyanitov, N.S., Rudkovskii, D.M.: Kinet. Katal. *8*, (6), 1240 (1967)
272. Craddock, J.H. et al. (Monsanto Co.): DE Offen. 1,941,501 (19.02.70)
273. Fenton, D.M. (Union Oil Co. of California): US 3,637,833 (25.01.72)
274. Forster, G., Bethell, J.R. (BP Chemicals Ltd.): DE 2,101,909 (17.02.72)
275. Kutepow, N. v. (BASF): DE Offen. 2,037,782 (03.02.72)
276. Craddock, J.H. et al. (Monsanto Co.): US 3,816,488 (11.06.74)
277. Craddock, J.H. et al. (Monsanto Co.): US 3,989,747 (02.11.76)
278. Morris, D.E., Johnson, G.V. (Monsanto Co.): Mechanistic Studies of the Rhodium Iodide Catalyzed Carboxylation of Ethylene to Propionic Acid, Symposium on Rhodium in Homogeneous Catalysis, Veszprém, Hungary, Sept. 11-3 (1978)
279. Kohn, P.M.: Chem. Engng. *86*, Jan. 29, p. 49 (1979)
280. Chem. Market Reporter, p. 4, Feb. 26 (1979)
281. Europa Chemie *6*, 104 (1979)
282. Adv. Chem. Technol. *1*, (1), 1, 18 (1979); Chemical Technology Associaties, Wyckoff, N-J. USA
283. Knifton, J.F.: J. Amer. Oil Chem. Soc. *55*, 496 (1978)
284. Rybakov, V.A. et al.: Zh. Prikl. Khim. *49*, 1835-38 (1976)
285. DuPont, DE Offen. 2,739,096 (02.03.78)
286. Air Products and Chem. Inc., BE 862,823 (13.01.77)
287. Tsuda, T., Isegawa, Y. and Saegusa, T.: J. Org. Chem. *37*, 2670 (1972)
288. Nefedov, B.K. et al.: Izv. Akad. Nauk SSSR, Ser. Khim., *9*, 2119 (1974)
289. Haupt, H.J. et al.: Untersuchungen über Metallcarbonyle und CO Reaktionsmechanismen. Forschungsbericht des Landes Nordrhein-Westfalen No. 2721, Opladen 1978

4. Hydrogenation of the Carbon Monoxide

C.D. Frohning

4.1 Methanol Syntheses

4.1.1 General Remarks

1913 saw the first synthesis of methanol via the hydrogenation of carbon monoxide under pressure. The reaction, which was conducted with various catalysts at high pressures, yielded a mixture of alcohols and hydrocarbons [1]. A few years later it was established that iron, in the presence of carbon monoxide, led to the formation of volatile pentacarbonyl iron which was responsible for the rapid deactivation of the catalysts via carbon deposition. In the absence of iron [2], methanol could be more selectively obtained on using zinc oxides activated with chromic acid. This opened the way to industrial scale methanol production which was initiated in 1923 in Germany [3] and in 1927 in the USA [4].

In contrast to other carbon monoxide hydrogenations, the methane and methanol syntheses (cf Sect. 4.3) are highly selective. Around 1923, the previously mentioned processes along with others were further developed at varying intensity although certain similarities are unmistakable.

On doping the ZnO-Cr_2O_3 catalysts with alkali, the selectivity of the reaction was drastically reduced resulting in the formation of an alcohol mixture ($\leqslant C_6$), containing ~40% methanol. This reaction, known as the 'isobutyl oil' synthesis achieved some significance during the Second World War for the production of isobutanol and methanol [5, 6]. The detrimental properties of iron and alkali in the methanol synthesis were used to advantage in the catalysts for the Fischer-Tropsch synthesis (cf Sect. 4.4). These features were to be found in all efficient Fischer-Tropsch processes. There were many similarities in the above syntheses – gas phase reactions (with the exception of special processes), strongly exothermic conversions in which temperature control played a decisive role.

The initially used ZnO–Cr_2O_3 catalysts exhibited low specific activity, demanding high pressures (250–750 bar) and high temperatures (350–400 °C) for a sufficiently high conversion. Improved purification of the gas feed enabled the application of more specifically active catalysts based on copper, the danger of sulfur poisoning being largely eliminated. Consequently, pressure and temperature could be reduced to 50–100 bar and 250–300 °C respectively. Modern production units are exclusively based on this low or medium pressure process [7].

Concurrent with the constantly expanding methanol production, there were changes in the applications and significance of methanol. Initially, methanol was mainly limited to more or less classical spheres – as a solvent and in syntheses. The high selectivity and the available process know-how resulted in attention being drawn to other fossil sources (besides oil) to produce synthesis gas and methanol as intermediates for raw material and energy supply [8]. The cracking of methanol to form synthesis gas (syn gas) does not involve any basic technical problems while its application as a motor fuel or heating oil component brings up a series of unanswered questions – particularly with regard to the economic feasibility of the application. Although a 'methanol-based technology' discussed as a consequence of the oil crisis in 1973 is not to be expected in the near future, it cannot be excluded from considerations [9]. Methanol plants with reactor outputs around 2000 tons/day could be built using available know-how. Thus, there are no problems related to production on a scale we are accustomed to with oil-based hydrocarbons.

4.1.2 Reaction Mechanism

Methanol is manufactured via the hydrogenation of carbon monoxide or carbon dioxide.

$$CO + 2\,H_2 \rightleftharpoons CH_3OH \qquad \Delta H_{298} = -21.7 \text{ kcal } (-90.8 \text{ kJ}) \qquad (1)$$

$$CO_2 + 3\,H_2 \rightleftharpoons CH_3OH + H_2O \quad \Delta H_{298} = -11.8 \text{ kcal } (-49.8 \text{ kJ}) \qquad (2)$$

Various side reactions can occur depending on reaction conditions and catalysts:

$$CO + 3\,H_2 \rightleftharpoons CH_4 \quad + H_2O \quad \Delta H_{298} = -49.3 \text{ kcal } (-206.4 \text{ kJ}) \qquad (3)$$

$$CO_2 + 4\,H_2 \rightleftharpoons CH_4 \quad + 2\,H_2O \quad \Delta H_{298} = -39.4 \text{ kcal } (-164.9 \text{ kJ}) \qquad (4)$$

$$2\,CO + 2\,H_2 \rightleftharpoons CH_4 \quad + CO_2 \quad \Delta H_{298} = -59.1 \text{ kcal } (-247 \text{ kJ}) \qquad (5)$$

$$CO + H_2O \rightleftharpoons CO_2 \quad + H_2 \quad \Delta H_{298} = -9.9 \text{ kcal } (-41.5 \text{ kJ}) \qquad (6)$$

$$2\,CO + 4\,H_2 \rightleftharpoons (CH_3)_2O + H_2O \qquad (7)$$

or $\quad 2\,CH_3OH \rightleftharpoons (CH_3)_2O + H_2O \qquad (8)$

As the reactions (1) and (2) are exothermic and methanol formation is accompanied by a decrease in mole number, the highest methanol yields are obtained at low temperature and high pressure.

The position of thermodynamic equilibrium determines the maximum possible methanol yields. Earlier calculations were frequently made without sufficiently considering the actual behaviour of the reaction partners at operating pressure – resulting in considerable deviations. On employing fugacities or activities, which take the actual behaviour into account, the equilibrium conditions are rendered by Eq. (9) [10].

$$\frac{\gamma_{CH_3OH}}{\gamma_{CO} \cdot \gamma_{H_2}^2} = Kp \cdot p^2 \cdot \frac{f_{CO} \cdot f_{H_2}^2}{f_{CH_3OH}} \tag{9}$$

In Eq. (9), γ represents the mole fraction, p the total pressure and f the activity co-efficient. Assuming that the activity coefficients are independent of the content of the reaction mixture, the following Eq. results:

$$K'_\gamma = Kp \cdot \frac{p^2}{\pi_f} \tag{10}$$

where

$$\pi_f = \frac{f_{CH_3OH}}{f_{CO} \cdot f_{H_2}^2} \qquad \text{or} \qquad K'_\gamma = \frac{\gamma_{CH_3OH}}{\gamma_{CO} \cdot \gamma_{H_2}^2}$$

The functions π_f and K'_γ are shown in Figs. 4.1 and 4.2 [11]. Table 4.1 gives the calculated values for π_f, K'_γ and the corresponding maximum conversions at various pressures. The calculations are based on synthesis gas with stoichiometry as in Eq. (1). The temperature dependence of the methanol formation equilibrium is shown in Table 4.2.

Due to the decreasing value of π_f (Fig. 4.1) with rising pressure, the latter has a positive effect on the methanol concentration at equilibrium. It is advantageous to conduct the synthesis at low temperature in order to limit thermodynamically feasible (and even favoured) side reactions. The lower operational temperature limit is determined by the reaction velocity necessary for economic conversion. The former is largely dependent on catalyst activity.

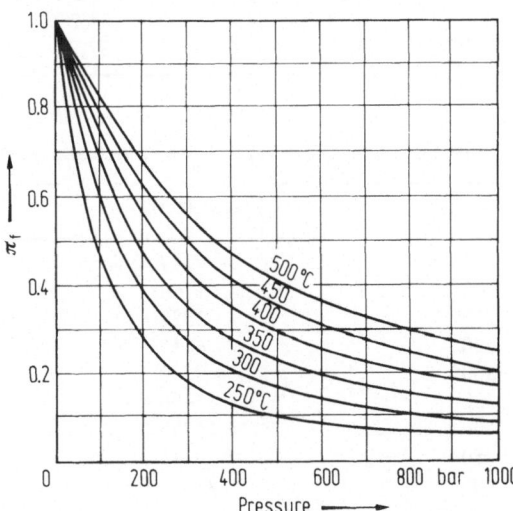

Fig. 4.1. Values of π_f [cf Eq. (10)] for the reaction $CO + 2\,H_2 \rightleftharpoons CH_3OH$ [11]

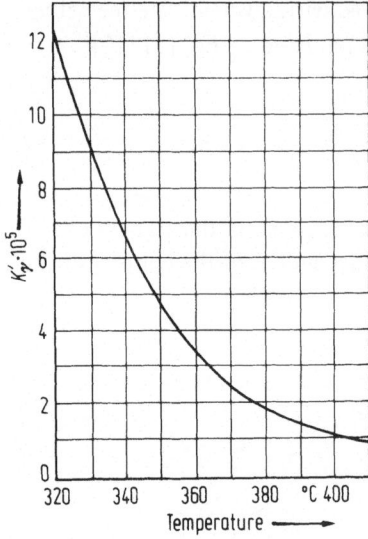

Fig. 4.2. Equilibrium constants K'_γ [cf Eq. (10)] for the reaction $CO + 2\,H_2 \rightleftharpoons CH_3OH$ [11]

Methanol can be synthesized from carbon dioxide either directly [Eq. (2)] or via a series of reactions including the reverse shift conversion [Eq. (6)]. The dependence of enthalpy (ΔH) and equilibrium constant (K_p) on temperature is shown in Table 4.3 for the hydrogenation of carbon monoxide or carbon dioxide (as ideal gas).

Table 4.1. Effect of pressure on methanol formation at 300 °C

Pressure bar	π_f	K'_γ	PCO	PH_2	PCH_3OH	Maximum Conversion %
10	0,96	0,0242	3,32	6,65	0,036	0
25	0,90	0,191	8,15	16,29	0,56	1,7
50	0,80	0,725	15,30	30,60	4,1	8,0
100	0,61	3,80	25,2	50,50	34,3	24,2
200	0,38	24,40	34,2	68,4	97,4	48,7
300	0,27	77,4	37,7	75,4	186,9	62,3

Table 4.2. Temperature dependence of methanol formation Equilibrium [11]

°C	$\Delta G°$ (cal)	K_p
0	− 7147	527,450
100	− 1766	10,84
200	+ 3822	$1,695 \times 10^{-2}$
250	+ 6617	$1,692 \times 10^{-3}$
300	+ 9530	$2,316 \times 10^{-4}$
350	+ 12400	$4,458 \times 10^{-5}$
400	+ 15279	$1,091 \times 10^{-5}$
450	+ 18148	$3,265 \times 10^{-6}$
500	+ 21023	$1,134 \times 10^{-6}$

Table 4.3. ΔH and log Kp for methanol formation

Temperature °C	25	200	300	400	500	1000	
$-\Delta H$ kcal/mol	21,54	23,10	23,74	24,24	24,60	25,09	[Eq. (1)]
$-\log K_p$	-4,352	1,678	3,565	4,924	5,950	8,719	
$-\Delta H$ kcal/mol	11,71	13,52	14,39	15,12	15,74	17,43	[Eq. (2)]
$-\log K_p$	0,662	4,051	5,174	6,009	6,657	8,500	

In the methanol synthesis, hydrogen and carbon monoxide are adsorbed or chemisorbed to varying degrees on the catalysts. The temperature dependence of the mixed adsorption increases above approximately 150 °C. This applies not only to the pure components (ZnO or Cr_2O_3) but also to the mixed oxide catalysts ($2\ ZnO \cdot 1\ Cr_2O_3$). However, with both ZnO and the mixed oxide catalyst, maximum adsorption occurs at 200 °C [11]. The ratio of the adsorbed quantities of H_2 and CO is approximately 1:1, being only slightly dependent on their partial pressures.

Much less work has been done on the mechanism of the surface reaction compared to, for example, the methane or Fischer-Tropsch synthesis. It is assumed that an oxygen-containing primary complex initially results from the reaction between a chemisorbed CO molecule (attached at the carbon atom) and a similarly chemisorbed hydrogen − possibly dissociated. The primary complex subsequently reacts with more hydrogen forming methanol [12, 13]. The same reaction mechanism apparently applies on using either copper- or zinc-containing catalysts.

The important, as well as rate determining, step is thought to be the reaction between the chemisorbed CO and H_2 molecules on the catalyst surface. Compared to the surface reaction, the transport of reactants, diffusion processes and desorption of products ensue rapidly and are thus not rate determining.

There were marked deviations in the energy of activation with various catalysts. 14 kcal/mol (approx. 59 kJ/mol) were calculated for a $Cu-Cr_2O_3$ catalyst (35 and 65 wt% resp.) in the range 290−325 °C, and 30 kcal/mol (approx. 126 kJ/mol) for a $ZnO-Cr_2O_3$ catalyst (89 and 11 wt.% resp.) in the range 325−380 °C [14]. These values readily explain the gradation in activity.

4.1.3 Reaction Conditions

The most important process variables in the methanol synthesis are pressure, reaction temperature, catalyst charge and synthesis gas content. They influence not only the design of the synthesis but also the economics of the process.

Catalyst activity determines the choice of pressure and reaction temperature. The catalyst ($ZnO-Cr_2O_3$), which was employed in the older high pressure process, had a low specific activity, necessitating temperatures of 350−400 °C for an adequate reaction velocity. Pressures of 250−350 bar (cf Sect. 4.1.2) are required to suitably adjust the methanol formation equilibrium in this temperature range. In special cases, pressures of 750 or 1000 bar were even applied. The choice of pressure is also determined by the synthesis gas content. A high total pressure, for example, is essential to increase the partial pressure with a low carbon monoxide concentration. While raising the pressure above 400 bar effects an improvement in the methanol yield, by-product formation also tends to increase [methane Eq. (3)–(5); dimethyl ether Eq. (7), (8)] and the costs for the syn gas compresssion also rise sharply [15].

From a reaction velocity standpoint, 230–300 °C are adequate for the low pressure process with copper-based catalysts [16]. As copper catalysts are more sensitive to thermal overload than the relatively robust $ZnO-Cr_2O_3$ catalyst, the useful temperature range is limited with adiabatic operation. The lowest feed temperature is determined by catalyst activity, while the thermal stability of the catalyst determines the highest outlet temperature. These aspects are, of course, insignificant for an isothermal process operation, the reaction temperature being adjusted to give a maximum methanol yield. The first low pressure processes were operated at 50 bar. Thereafter it was found to be advantageous, from the process engineering point of view, to raise the pressure to approximately 100 bar (medium pressure process). A further rise in pressure leads to increased thermal strain on the catalyst due to the greater conversion rate [17].

While increasing catalyst load, the specific methanol yield reaches a maximum, then drops off. This applies equally to the high and low pressure syntheses. When the residence time of the gas on the catalyst is reduced, the side reactions as well as the methanol formation decrease to a different extent, resulting in a fall in the methanol concentration in the product. Thus, for a given pressure, the temperature and catalyst load must be carefully modulated (cf Fig. 4.3). As only partial conversion of the carbon oxides can be achieved in a single pass, the residual gas is recycled after condensing out the resulting methanol. This helps facilitate improved heat recovery and maximum utilisation of the syn gas. The change in residence time and concentrations in the recycling operation must be taken in account.

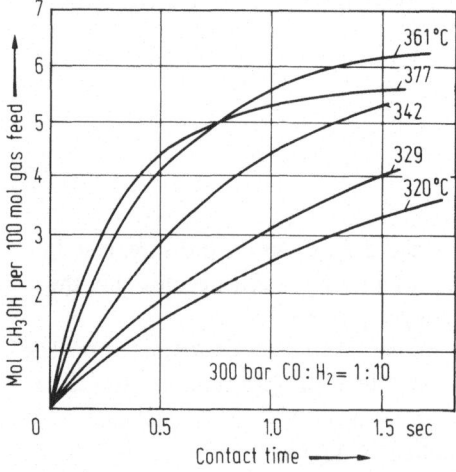

Fig. 4.3. Effect of temperature and contact time on methanol formation [13]

The relationship for the stoichiometric content of synthesis gas can be derived from Eq. (1)–(2).

$$\frac{H_2}{2\,CO + 3\,CO_2} = 1 \tag{11}$$

A H_2 excess, demanded by process engineering and kinetic factors, is always used, effecting an improvement in methanol quality and reaction velocity. Increasing H_2 excess suppresses the side reactions (methane and ether formation) which would otherwise cause a decrease in selectivity. Moreover, the reaction on the catalyst surface of the relatively strongly chemisorbed carbon monoxide is accelerated, leading to a decrease in residence time thus facilitating a higher catalyst load [18] (Fig. 4.4):

Carbon dioxide can, similar to a given quantity of inert gas, be left in the gas feed if it is within certain limits. In some cases it is even desirable. Temperature peaks are lowered due to carbon dioxide's relatively high specific heat and lower enthalpy of hydrogenation compared to carbon monoxide. However, it also inhibits the conversion of carbon monoxide as a result of competitive reactions on the catalyst surface (reverse shift conversion) to which it is strongly chemisorbed. In general, therefore, the carbon monoxide conversion is noticeably influenced by the CO_2:CO ratio (Fig. 4.5), necessitating a limitation of the CO_2 quantity by adjusting the amounts of fresh and recycle gas. In the gas feeds for the high pressure synthesis, approximately 2 vol% CO_2 are permissible, while with the low or medium pressure synthesis the corresponding figure is usually around 5 vol.%. The formation of dimethyl ether is suppressed by the presence of carbon dioxide.

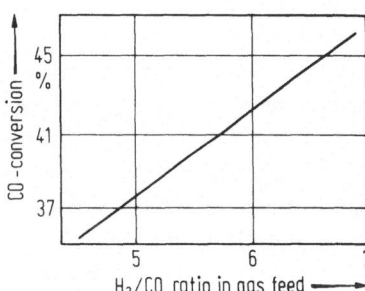

Fig. 4.4. Effect of H_2:CO ratio on CO conversion [7]

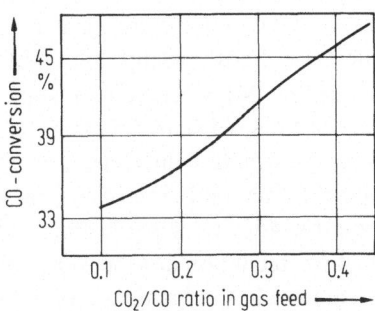

Fig. 4.5. Effect of CO_2:CO ratio on CO conversion [7]

The carbon monoxide content of the fresh gas is in the region of 5–25 vol.%. High CO contents not only adversely affect temperature control but also impair methanol quality (side reactions) and encourage the (undesired) formation of penta-carbonyliron. Consequently, special precautions must be taken to ensure utilisation of the basic advantages of a CO-rich gas. While low carbon monoxide content helps to avoid the above difficulties, the efficiency of the process is impaired (lower specific methanol production). Gas production, gas treatment (conversion, CO_2 separation) and synthesis (fresh gas : recycle gas ratio) must be carefully balanced against one another. The fresh gas composition can – depending on the operation – fluctuate within the following limits:

H_2	65–85 vol.%
CO	8–35 vol.%
CO_2	0.5–5.5 vol.%
CH_4	0.2–1.5 vol.%
$N_2 + Ar$	1.5–3.5 vol.%
O_2	traces

4.1.4 Catalysts

Only a few of the multitude of proposed catalysts for the methanol synthesis have achieved any significance. The following properties are necessary for an industrially viable catalyst:
good activity,
mechanical and thermal stability and, if possible,
marked resistance towards poisoning.

Multi-component catalysts have proved to be especially useful, particularly those with zinc oxide or copper oxide as the main component. Chromium oxide and other additives are generally also present – functioning as structural or electronic promoters.

Catalysts for the high pressure synthesis usually contain 60–65 wt% ZnO, 25–30 wt.% Cr_2O_3, 1–2 wt.% graphite and 8–10% H_2O [19]. They are often manufactured via precipitation of basic zinc chromates or via mechanical mixing of ZnO and chromic acid with subsequent drying and shaping to tablets or hollow cyclindrical pellets [20–22]. The mechanical stability of the shaped catalysts is very high. There is a connection between the specific catalyst surface and the catalytic activity. The activity increases almost linearly up to a BET surface of approximately 100 m^2/g (in reduced state) [23, 24], whereas with larger surfaces there is no essential increase.

In the solid state ZnO and Cr_2O_3 can react to form spinels, which possess not only a crystal structure different from pure ZnO but also exhibit low catalytic activity. As spinel formation [25] ensues at an increased rate above the Tammann temperature (approx. 450 °C), super-heating of the catalyst must be carefully avoided during manufacture, reduction or processing. Oxidized promoters (Al-, Th-, Zr-, Ta-, V-oxides) can be present in the catalysts either intra-crystalline (part of the ZnO crystal lattice) or extra-crystalline (Fe-, Mg-, Ca-, Cr-oxides). They affect the recrystallisation properties. The formation of dimethyl ether is encouraged by the dehydrative effect of Al_2O_3, while alkali metal oxides favour the formation of higher alcohols [7].

During the reduction, which can be conducted either externally or in the actual reactor, Cr^{VI} is largely converted into Cr^{III}. Undesired spinel formation ($ZnCr_2O_4$) must be avoided by careful temperature control [26]. During the reaction, the reduced catalyst undergoes an activation phase in which structural alteration can take place [27]. Initially, the specific catalyst efficiency increases, then slowly drops off after several months operation due to aging (poisoning). On account of their suscept-ibility towards sulfur poisons and thermal overload, the catalysts for the low pressure synthesis were virtually neglected at the outset. However, the significance of these catalysts increased as soon as purer synthesis gas (< 0.5 ppm S) became available. Modern processes exclusively employ copper catalysts which contain 30–80 wt.% CuO, 10–50 wt.% ZnO as well as variable amounts of the oxides of the elements Cr, Mn, V or Al. The favoured manufacture ensues via co-precipitation of the basic oxides whilst carefully controlling reaction conditions. After drying and calcination, the catalyst is shaped into tablets or hollow cylindrical pellets.

The thermal instability of the copper catalysts demands even greater precautions than with the $ZnO–Cr_2O_3$ catalysts. The Tammann temperature is around 190 °C for metallic copper, and above approximately 280 °C recrystallisation occurs relatively quickly [16]. The exothermic reduction is conducted as mildly as possible over a period of several days. During the synthesis, super-heating of the catalyst is avoided as far as possible using process engineering techniques [28]. The tendency of copper to form mixed crystals is less marked than with zinc oxide, as copper is present as the metal. The highly stable oxides present fulfil structural as well as electronic functions thereby influencing the catalyst activities and, more significantly, extending their lifetimes.

4.1.5 Processes

The operation of the high and low (medium) pressure processes are basically similar [11]. A typical feature is the high flow rate or low residence time of the gas on the catalyst, as otherwise side reactions (formation of dimethyl ether, methane or ammo-nia) and superheating can occur (cf Sect. 4.1.3). Partial conversion of the carbon monoxide (approx. 10–30% of the CO feed) is accepted and gas recycles are employed, the condensable products being separated before recycling. A portion of the reaction gas is removed from the process in order to limit the inert gas content [7] (cf. Fig. 4.6).

Fresh gas for the synthesis is highly purified beforehand, $ZnO–Cr_2O_3$ catalysts making lesser demands than Cu catalysts. The former can cope with sulfur contents between 20–30 ppm, whereas with the latter the maximum sulfur content is 1 ppm. The synthesis gas composition can be varied within fairly broad limits, the main demands being a sufficiently high H_2 content and an exact $CO:CO_2$ ratio (cf 4.1.3). The purified, pretreated synthesis gas is then brought up to reaction pressure.

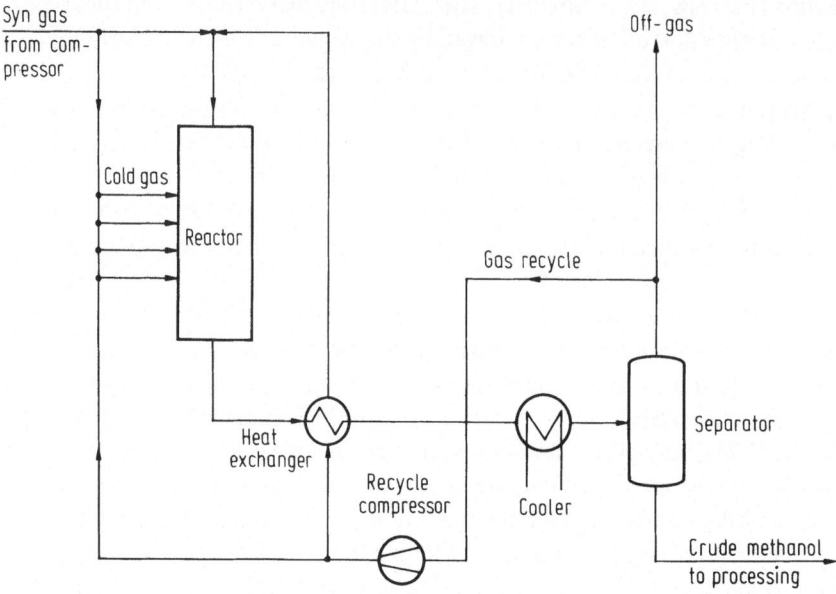

Fig. 4.6. Basic scheme for methanol synthesis [7]

Conditions	High pressure process	Low pressure process
p [bar]	250−350	100
T [°C]	330−400	220−270
		Cu/Zn/Cr
Catalyst	ZnO/Cr$_2$O$_3$	(Al; Mn. Cr; V; Ag)
Capacity of industrial plants [tons/annum]	∼670.000	∼1.700.00

The reactors are similar to those developed for ammonia synthesis and are either the packed bed or tubular bundle type. In the packed bed reactor, the catalyst is divided horizontally into several stages between which cold gas is introduced to facilitate cooling [29]. There is a zig-zag temperature profile along the whole length of the reactor with minima and maxima at the beginning and end of the individual stages, while the tubular bundle reactors exhibit a less marked temperature profile. There is a considerable number of variants which are characterized by a particular type of gas flow and heat removal. The choice of construction material is extremely important. Iron can form volatile carbonyls − depending on CO partial pressure and temperature − which then decompose in the reaction chamber to finely divided iron which is an excellent methanation catalyst. The resulting temperature increase can lead to either destruction of the catalyst or even the reactor. Copper-lined aggregates are therefore used at the high CO partial pressures necessary for improving reactor performance.

Besides heat removal, the flow resistance is an important aspect to be considered when determining the geometry of reactor and catalyst. The energy input for the recycling operation at high flow rates represents an economic factor which cannot be neglected. This energy requirement can be partly covered by the heat of reaction (steam turbines), the energy released in the synthesis being also partly used for methanol purification (distillations).

4.1.6 Economic Potential and Possible Developments in Methanol Synthesis

Fifty years of industrial application have brought the catalytic hydrogenation of carbon oxides to methanol to an advanced level of development. Even at the outset, high pressure synthesis presented a satisfactory route to the production of large quantities of methanol on account of the conversion rate, product quality and process operation. The transition to a low or medium pressure synthesis was based on the optimal balance of gas production, synthesis and product treatment — particularly with regard to investment costs and energy consumption (operating costs). This transition was first feasible after improved gas purification processes had been developed. Both processes exhibit very high selectivity — almost comparable to that of the methane synthesis (cf Sect. 4.2). The development of the methanol synthesis — taking catalyst and process engineering factors into consideration — can be regarded as being virtually complete. Correspondingly, there is little information about new processes (e.g., liquid phase synthesis) [30] or new catalysts (e.g., homogeneously catalysed reaction between carbon monoxide and water) [31]. The reaction of carbon dioxide and hydrogen to methanol — while first appearing attractive — is coupled to the availability of cheap hydrogen whose consumption, compared to carbon monoxide, is increased.

Methanol could well assume an important role (methanol-based technologies) [9, 33, 34] during the transition from oil to other energy and feedstock bases. The various processes considered for methanol conversion exhibit lower selectivity than the methanol syntheses. This is the case for the cracking of methanol to synthesis gas (CO_2 formation) or, especially when methanol is converted into hydrocarbons. In the latter example the dimethyl ether, formed as intermediate, is converted by dehydration into alkanes, alkenes and aromatics [35]. These secondary processes and the corresponding catalysts require not only extensive and involved development work but also scaling-up over several stages [36]. The question arises whether on combining the methanol synthesis and secondary reactions the resulting selectivity can also be attained in one stage or improved if the object is to manufacture chemical feedstocks (e.g., short chained olefins) or motor fuels (gasoline or aromatics). This aim represents a departure from the previous main requirement in the methanol synthesis — to suppress carbon chain formation as much as possible. The 'isobutyl oil' synthesis, already operated on a large scale, is a modified methanol synthesis of the above type whose industrial feasibility has been proved. A mixture of oxygen-containing products whose main components are methanol (40–60%), propanol (2–5%) and isobutanol (15–20%) [37] result on using alkalized catalysts for the methanol synthesis. The product content can be adjusted by varying the reaction parameters. The work up of the crude reaction mixture to hydrocarbons, possibly to alkenes, is perhaps less difficult and can be carried out in better yields than the conversion of pure methanol. However, the advantage of the total process could well be the application of substantially unmodified established technology in the synthesis stage.

4.2 Glycol Syntheses

4.2.1 General Remarks

Ethylene glycol possesses considerable importance as a component of polyesters and antifreeze agents. Currently, the synthesis is almost exclusively based on ethylene which, in the presence of Ag catalysts, can be converted into ethylene oxide via addition of oxygen. Thereafter, glycol is formed via hydrolysis of ethylene oxide — either in the presence of sulfuric acid or with water alone at raised pressure and temperature. The glycol syntheses have constantly grown in importance since their break through in the mid-fifties with the availability of oil-based ethylene and ethylene oxide. The world production was around 3.7 million tons in 1977.

In the USA (1940) a multi-stage glycol synthesis was used in which formaldehyde and synthesis gas were converted into ethylene glycol [38–41]. If formaldehyde is regarded as a secondary product of methanol — obtained from carbon monoxide and hydrogen (cf Sect. 4.1) — then the process can be divided into the following steps [42]:

$$2\,H_2 + CO \rightleftharpoons CH_3OH \tag{12}$$

$$CH_3OH + 1/2\,O_2 \rightleftharpoons HCHO + H_2O \tag{13}$$

$$HCHO + CO + H_2O \rightleftharpoons HOCH_2COOH \tag{14}$$

$$HOCH_2COOH + CH_3OH \rightleftharpoons HOCH_2COOCH_3 + H_2O \tag{15}$$

$$HOCH_2COOCH_3 + 2\,H_2 \rightleftharpoons HOCH_2CH_2OH + CH_3OH \tag{16}$$

$$4\,H_2 + 2\,CO + 1/2\,O_2 \rightleftharpoons HOCH_2CH_2OH + H_2O \tag{17}$$

In 1968, the plant was shut down due to competition from a more economic ethylene-based process.

The increasing availability of cheap synthesis gas and the rising price of ethylene and other oil-based hydrocarbons were the motives for the development of processes for the direct conversion of synthesis gas to ethylene glycol. Economic calculations [43] based on current costs make the direct synthesis appear attractive even with less than 100% selectivity. Syntheses will now be discussed in which ethylene glycol can be prepared in a single stage by catalytic hydrogenation of carbon monoxide.

4.2.2 Glycols via Hydrogenation of Carbon Monoxide with Cobalt Catalysts

As the synthesis of ethylene glycol, based on formaldehyde (or methanol), is not only technically involved but also gives rise to unsatisfactory yields of ethylene glycol, it was attempted to produce glycols via a single stage catalytic hydrogenation of carbon monoxide. This was achieved for the first time using cobalt-containing catalysts.

$$2\,CO + 3\,H_2 \quad \rightleftharpoons \quad HOCH_2CH_2OH \tag{18}$$

$$CO + 2\,H_2 \quad \rightleftharpoons \quad CH_3OH \tag{19}$$

$$2\,CH_3OH + CO \quad \rightleftharpoons \quad CH_3COOCH_3 + H_2O \tag{20}$$

$$2\,CH_3OH \quad \rightleftharpoons \quad CH_3OCH_3 + H_2O \tag{21}$$

$$HOCH_2CH_2OH + CH_3OH \quad \rightleftharpoons \quad HOCH_2CH_2OCH_3 + H_2O \tag{22}$$

$$HOCH_2CH_2OH + 2\,CH_3OH \quad \rightleftharpoons \quad CH_3OCH_2CH_2OCH_3 + 2\,H_2O \tag{23}$$

$$HOCH_2CH_2OH + CH_3COOCH_3 \quad \rightleftharpoons \quad HOCH_2CH_2OOCCH_3 + CH_3OH \tag{24}$$

On synthesizing methanol at pressures up to 1000 bar or more, monofunctional compounds (acetic acid, methyl acetate, dimethyl ether, etc.) were formed in increasing amounts, while di- or polyfunctional compounds were not present. The situation changes when higher pressures (2000–5000 bar) and a liquid reaction medium are employed. The conversion and selectivity of glycol formation can be increased using solvents which are either inert or react with the newly formed glycols.

Soluble cobalt salts, e.g. cobalt acetate, are suitable catalysts. A solution of 3.5 wt% anhydrous cobalt acetate in acetic acid catalyzes the conversion of synthesis gas ($H_2:CO$ ratio = 2:1 to 1:1) to a product consisting mainly of ethylene glycol diacetate and glycerol triacetate. In this reaction, which took place at 225–250 °C and 3000 bar, methyl-, ethyl-, n-propyl- and isopropyl acetate were obtained as by-products [39]. A complete break down of the resulting products cannot be given here. On employing an aqueous cobalt acetate solution as catalyst, free ethylene glycol (22% of glycol fraction) results instead of the acetate, along with a mixture of the mono- and diformates of ethylene glycol. At reaction pressures around 3000 bar, glycols and monofunctional compounds are obtained in roughly a 1:1 ratio while at around 1500 bar the ratio is approximately 3:1 in favour of the monofunctional compounds. Besides being influenced by pressure, the selectivity of the carbon monoxide hydrogenation is also affected by the solvent. In the series toluene-benzene-cyclohexane-water, the formation of monofunctional compounds decreases relative to glycol or polyol formation. However, the space-time yield also declines concurrently [44].

The mechanism of the cobalt-catalyzed glycol formation is still unsolved. It was not possible to prove the existence of formaldehyde – as primary product of the carbon monoxide hydrogenation – in the reaction mixture. However, the cobalt-catalyzed reaction of formaldehyde with carbon monoxide to hydroxyacetic acid, an intermediate in the multistage glycol process, is known [45]. The active form of the catalyst is formed under reaction conditions and cobalt salts, soluble in water or organic solvents, are equally effective. Supported cobalt or ruthenium-based catalysts also find application. However, they are less effective than soluble organic or inorganic cobalt salts. In the presence of water, the main portion of cobalt is homogeneously dissolved in the aqueous phase. When the latter is distilled in the presence of carbon monoxide, the resulting volatile cobalt carbonyls can be recycled to the reactor.

4.2.3 Glycols via Hydrogenation of Carbon Monoxide with Rhodium Catalysts

The second phase in the development of the glycol synthesis involved the use of rhodium as a catalyst instead of cobalt, facilitating an essential improvement in reaction control and selectivity. Typical characteristics remain

Conductance of the reaction in a solvent (even one stemming from the reaction)
Application of pressures between 50 and 3000 bar,
Temperatures in the region of 150 to 300 °C.
The reaction

$$n\,CO + (n+1)\,H_2 \rightarrow H(CHOH)_n H \tag{25}$$

can be catalyzed by a rhodium carbonyl complex:

$$H_x Rh\,(CO)_y\,L_z \qquad\qquad x = 0-1; y = 1-3; z = 1-3$$
$$x + y + z = 3-5 \tag{26}$$

L symbolizes an organic ligand [46]. The active form of the catalyst can either be formed externally or in situ. In the latter case, inorganic or organic rhodium compounds (oxides, fatty acid salts, carbonyls, etc.) are reacted in a solvent with carbon monoxide in the presence of a ligand, resulting in soluble complexes of the above composition. For example, finely divided dodecacarbonyltetrarhodium $Rh_4(CO)_{12}$ is dissolved or suspended in benzene and, in the presence of a ligand, converted into a soluble complex at 2–15 bar CO pressure and 30–100 °C. Synthesis gas can be used instead of carbon monoxide for complex formation if the CO partial pressure is adjusted to the previously mentioned level. Ligands with at least two Lewis base centres are used as stabilizing ligands. These ligands are capable of forming coordinative bonds and chelate structures with 4–6 atoms (including rhodium). Oxygen or nitrogen atoms are centres with Lewis base character which could be present in the form of ether, ester, carboxyl, imino or amino functions. Pyrocatechol,2-hydroxypyridine, picolinic acid, dioxan, tetrahydrofuran, 2,2'-dipyridyl or 1,10–phenanthroline are all possible ligands [46].

The hydrogenation of carbon monoxide, in the presence of Rh complexes, is carried out at 210–250 °C and a synthesis gas pressure (H$_2$:CO ratio approx. 1) of

3000–3500 bar. The reaction proceeds very slowly below 100 bar and above 3000 bar there is almost no increase in rate. The Rh concentration can vary between approximately 10^{-3} and 10^{-1} wt.% Rh (relative to the liquid reaction mixture). A variety of solvents can be employed: aromatics, ethers (preferably cyclic), esters, alcohols or even water. With progressing reaction (batch operation), reaction products can assume the role of a solvent. The selectivity of the carbon monoxide conversion to polyols is given by the following expression for S.

$$S = \frac{\text{parts by weight of resulting polyols}}{\begin{array}{l}\text{parts by weight of resulting oxygen-} \\ \text{containing compounds}\end{array}} \cdot 100 \, (\%)$$

The best values achieved for S were in the region of 60–70%, ethylene glycol, propylene glycol and glycerol being formed in the weight ratio $6:1:1$. The main by-products are methanol and methyl acetate, along with a small quantity of higher alcohols [47–59].

It was assumed that the structure of the active rhodium compound (a mononuclear complex) could be represented by the formula in Eq. (26) [46]. However, studies relating to the reaction mechanism and particularly concerning the effect of solvent on the course and rate of the reaction indicated ionic rhodium cluster compounds – the actual catalysts – must be present (Fig. 4.7) under reaction conditions [47]. The formation of these rhodium clusters, which ensues independently of composition of the initial rhodium compound, can be represented as follows [51]:

$$2 \, [Rh_6(CO)_{15}H]^- \underset{H_2}{\rightleftharpoons} [Rh_{12}(CO)_{30}]^{2-} \overset{CO}{\rightleftharpoons} [Rh_{12}(CO)_{34}]^{2-} \qquad (27)$$
$$\quad \text{I} \qquad\qquad\qquad \text{II} \qquad\qquad \text{III}$$

IR absorption techniques can be used to prove the existence of ionic cluster structures with Rh-Rh bonds under reaction conditions [53–55]. If a dissolved Rh compound (e.g., $Rh(CO)_2$-acetyl acetonate in tetrahydrofuran) is exposed at 150–250 °C to increasing syn gas pressures (and thus increasing CO partial pressures) then above ~30 bar CO partial pressure the $Rh_6(CO)_{16}$ cluster is formed. With increasing CO partial pressure, the $[Rh_{12}(CO)_{30}]^{2-}$ cluster (absorption ~1830 cm^{-1}) is formed from the inactive precursor I. In the presence of syn gas the cluster (II), which is in equilibrium with I and III (approx. 1870 and 1780 cm^{-1} respectively) has the ability to catalyze the carbon monoxide hydrogenation [55, 56]. The Lewis bases are important as counter ions to the cluster structures which they tend to stabilize via chelate formation, e.g. trialkanolamine borates, $B(OR)_3N$ [51]. These Lewis bases are not actually important for the catalysis – a fact which is underlined by the numerous proposed and actually suitable substances.

A particularly stable form of the catalyst is obtained in the presence of bis(triorganophosphine)iminium cations $(R_3P)_2N^+$ which can be synthesized via chlorination of the phosphine followed by reaction with hydroxylamine [53]:

$$2 \, R_3P + 2 \, Cl_2 \rightarrow 2 \, R_3PCl_2 \qquad\qquad \text{R = Alkyl, Aryl, Aralkyl} \qquad (28)$$

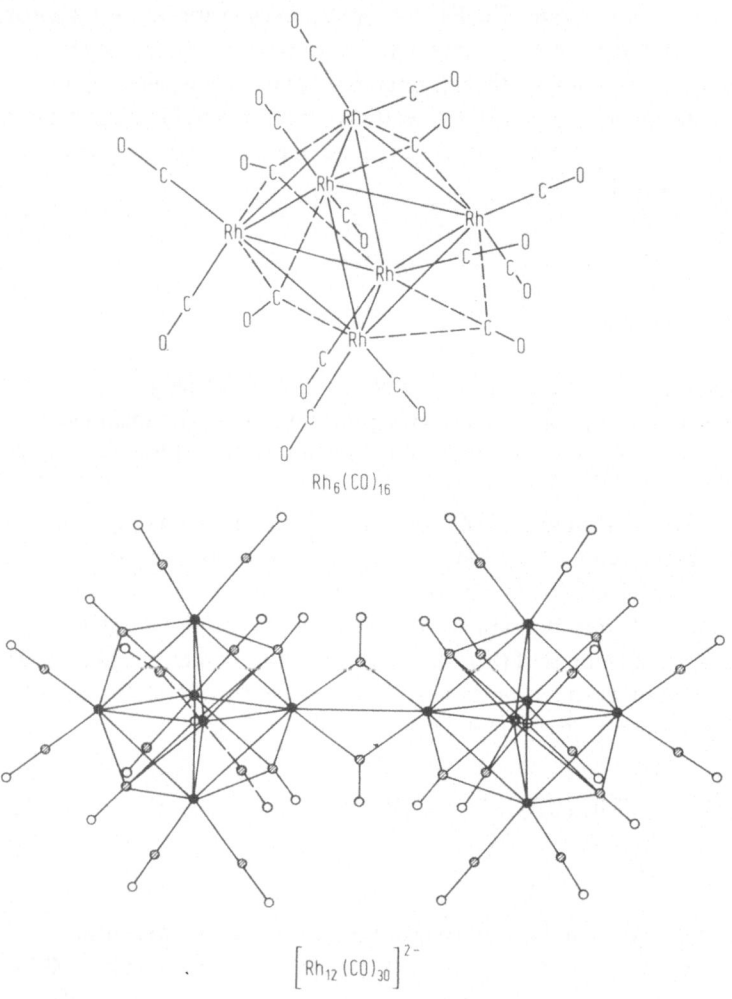

Fig. 4.7. Structure of rhodium carbonyl clusters [50]

$$2 R_3PCl_2 + R_3P + NH_2OH \cdot HCl \rightarrow (R_3P)_2N\,Cl + R_3PO + 4\,HCl \qquad (29)$$

In the presence of syn gas, the complex results from reaction between bis(triorgano-phosphine)imine chlorides and the Na salt of the rhodium cluster.

A number of modifications was proposed for the catalyst system based on rhodium carbonyl clusters. The addition of defined amounts of alkali metal cations (Na, K, Cs) increases the yield of polyols and enables the reaction temperature to be reduced from 375 to 275 °C (at a synthesis gas pressure below 1000 bar) [54, 56]. One disadvantage is the increased formation of monohydric alcohols, such as methanol, ethanol and propanol, which is encouraged by alkali metal additives. Even milder conditions (220 °C, 560 bar) are possible when quaternary ammonium or phosphor compounds are added to the rhodium carbonyl clusters.

The important role of the Rh cluster in the ethylene glycol synthesis is underlined by the fact that other metals — incapable of forming multi-nuclear carbonyl complexes under reaction conditions—do not catalyze the carbon monoxide hydrogenation. Ru, Ir, Pt, Pd, Pb, Sn, Cu, Cr and Pb are only slightly or even inactive, while cobalt, with lower activity and selectivity compared to rhodium, is able to catalyze the polyol synthesis.

However, a number of transition metals exhibit good catalytic activity when combined with rhodium. On reacting mixed multi-nuclear carbonyls of the general formula $[M_a^1 M_{4-a}^2 (CO)_{12}]$ $(M^1, M^2 = Co, Ir$ or $Rh)$ with aqueous metal salt solutions, the following complexes are obtained:

$$M^3[M_b^1 M_{12-b}^2 (CO)_{30}] \qquad\qquad b = 1-11 \qquad\qquad\qquad\qquad (30)$$
$$M^3 = \text{alkali, alkaline earth,}$$
$$\text{transition metal}$$

One example is $Ca[Ir_3Rh_9(CO)_{30}]$ which catalyzes the carbon monoxide hydrogenation under relatively mild conditions [57]. Rh carbonyl complexes of the following structure are also suitable:

$$M_2[Rh_{12}(CO)_{30}] \text{ or } M_2[Rh_{12}(CO)_{30}]_3 \qquad\qquad\qquad\qquad\qquad (31)$$

M represents a monovalent (Rh, Cu, Ag, Au, Ir, In, Tl) or a trivalent (Al, Ga, Ir, Sc, Y, Re) cation [58]. Compared to the 'pure' Rh carbonyl complexes, the more complicated structures of the multi-nuclear complexes possess higher activity. However, their selectivity with regard to ethylene glycol formation remains almost unchanged.

The choice of solvent for the ethylene glycol synthesis is determined not only by its ability to solvate the catalyst but also by its facile separation from the product [60]. As water results during the formation of monofunctional by-products (alcohols, esters), water-miscible solvents are advantageous in order to maintain a homogeneous reaction medium. However, they are not absolutely necessary. The trend towards using polar solvents is very apparent. Whereas initially toluene, benzene, cyclohexane or alcohols [46] were employed, recently tetrahydrofuran, tetraglyme (tetraethylene glycol dimethyl ether) sulfolanes (tetramethylene sulfone and derivatives) or lactones (especially γ-butyrolactone) have been used [47–52]. These solvents increase the stability of the catalyst components, losses being reduced via heterogenisation and repeated reuse of the catalyst.

The reaction mixture is generally separated by distillation. The low boiling components (alcohols, esters) can be recycled into the reaction where they serve as solvents or diluents or are partly converted to polyols. The recovery of the catalyst components can ensue via decomposition, concentration of the distillation residue or via extraction — if necessary in the presence of complexing agents [50]. If the catalyst is recycled, then its separation or concentration must ensue under mild conditions or in the presence of (stabilizing) complexing agents—to avoid decomposition of the carbonyl complexes [51].

4.2.4 Economic Potential and Possible Developments in the Glycol Synthesis

In 1972, a publication appeared describing the application of rhodium carbonyl complexes in the manufacture of ethylene glycol via the hydrogenation of carbon monoxide. Today the reaction conditions can be presented as follows:

Pressure	400–2000 bar
Temperature	190–240 °C
Solvent	tetraglyme, sulfolane
Rh concentration	0.05–0.3 wt.%
Specific conversion	50–500 g/h (product per l reaction solution and g Rh)
Selectivity S	$\leqslant$ 60–65% polyols

Pressures between 400–600 bar are the lower limit for an adequate reaction velocity, pressures between 1500–2000 bar are preferred. The selectivity of ethylene glycol formation decreases with increasing temperature, and monohydric alcohols mainly result – in particular, methanol. If low Rh concentrations are employed, then the most suitable solvent is a mixture of sulfolane and tetraglyme. The specific conversion was derived from batch tests. Depending on type and concentration of the catalyst and additives, the specific conversion can vary over a relatively wide area. The selectivity of ethylene glycol formation can be increased on raising catalyst concentrations, via the polarity of the solvent and on adding nitrogen or phosphor-containing complexing agents at otherwise constant conditions so that mixtures with approximately 60% ethylene glycol and 25% methanol can be obtained.

The selectivity and specific conversion of the reaction can be improved on increasing catalyst concentrations. Thus the demand for catalysts with improved specific activity can be conjectured. As no element has been found with comparable activity to rhodium – possibly in combination with other cluster-forming metals, it would be worthwhile to establish, whether the active multi-nuclear Rh clusters can be modified to yield compounds with greater hydrogenation activity capable of more rapidly converting the complex-bound carbon monoxide [61, 62]. The first steps in this direction are the combination of Rh with more electronegative metals and doping with Lewis bases [57, 58]. Tests involving the rate of ligand exchange and the solvent effect could make a contribution to the understanding of the solvent effect on the rate as well as on the selectivity of the synthesis [51]. A concentration gradient, analogous to heterogeneously catalyzed reactions involving cluster molecule and 'free' solvent, should be considered.

An essential problem associated with the industrial operation of the ethylene glycol synthesis is the high reaction pressure. There must be a minimum CO partial pressure to maintain stable complexes and an adequate concentration gradient to ensure that the reaction velocity is kept within a useful order of magnitude. The availability of stable carbonyl complexes (with high formation tendency) is a prerequisite for a lowering of the partial pressure at constant specific conversion.

4.3 Methane Syntheses

4.3.1 General Remarks

Methane formation via the hydrogenation of carbon monoxide belongs to the well-known heterogeneously catalyzed gas phase reactions. Up to 1960, this technique was used to free process gases (e.g. hydrogen or ammonia syn gas) from small amounts (0.1—2 vol.%) of carbon oxides. The latter were hydrogenated to methane, which could be regarded as inert as far as the usual applications were concerned. Town gas was also detoxicated via a purification process in which carbon monoxide was methanated. This treatment also resulted in an increase in calorific value.

The growing development and utilisation of natural gas reserves for the private and industrial sectors soon led to the recognition that supplies were limited. Particularly in the USA, natural gas quickly established itself as an important fuel and raw material with few pollution problems. This was also a consequence of the available infrastructure (pipeline system), which was readily utilized by industry. Natural gas is widely used not only as an energy source (up to 30% of primary energy consumption) but also as a feedstock for process gas.

In order to cover gaps in supply, an intermediate solution was found via a series of processes for the catalytic steam cracking (reforming) of liquid hydrocarbons to synthesis gas (containing methane) [63, 66—71]. The carbon monoxide and hydrogen present in the synthesis gas were then converted into methane in a second process step. Consequently, a gas with > 95% methane was made available — SNG (substitute natural gas). The best feedstock for reforming purposes is a mixture of light hydrocarbons (naphtha).

The alternative — both in the US and Europe — for a long term secure supply is the production of methane from coal-based synthesis gas. Therefore, parallel to the further development of the new generation of coal gasification processes, the hydrogenation of carbon monoxide (and carbon dioxide) is of industrial and commercial interest [64, 65].

The catalytic hydrogenation of carbon monoxide to methane using cobalt- or nickel-containing catalysts at 200—350 °C is a special case of the Fischer-Tropsch hydrocarbon synthesis [72]. The development of both processes displays recognizable similarities [73, 74]. In the methane synthesis, the cost of synthesis gas relative to the value of the final product is > 85% — an improvement on the Fischer-Tropsch synthesis (65—70%). Instead of the previous aim of complete hydrogenation of carbon monoxide at the most thermodynamically favourable and industrially suitable low temperature level, the current object is to divide the total reaction into steps, the first ensuing at the highest feasible temperature (350—500 °C) to utilize the reaction enthalpy at the highest possible level. After the main reaction, residual conversion takes place at low temperatures. An extreme case would be to employ synthesis gas as an energy source and the methane synthesis as a heat generator.

4.3.2 Reaction Mechanism

Stoichiometry and Thermodynamics

The fundamental equation for the reaction of carbon monoxide with hydrogen is:

$$CO + 3\,H_2 \leftrightharpoons CH_4 + H_2O \qquad \Delta H298 = \underline{-206.4\ kJ}\ (-49.3\ kcal) \qquad (32)$$

$$\Delta H_{298} = -49{,}3\ kcal\ (-206{,}4\ kJ)$$

The water formed reacts with carbon monoxide yielding carbon dioxide and hydrogen (shift conversion):

$$CO + H_2O \leftrightharpoons CO_2 + H_2 \qquad \Delta H298 = \underline{-41.5\ kJ}\ (-9.9\ kcal) \qquad (33)$$

$$\Delta H_{298} = -9{,}9\ kcal\ (-41{,}5\ kJ)$$

When carbon monoxide is completely converted to hydrogen and carbon dioxide, they react forming carbon monoxide and water once again. The stoichiometric equation for the carbon dioxide hydrogenation is:

$$CO_2 + 4\,H_2 \leftrightharpoons CH_4 + 2\,H_2O \quad \Delta H298 = \underline{-164.9\ kJ}\ (-39.4\ kcal) \qquad (34)$$

$$\Delta H_{298} = -39{,}4\ kcal\ (-164{,}9\ kJ)$$

The equilibria in reactions (32) to (34) are displaced to the left with rising temperature, while an increase in pressure causes displacement to the right. The side reactions are decomposition of carbon monoxide to carbon dioxide and elementary carbon [Boudouard reaction Eq. (35)] and hydrogenation of the deposited carbon to methane [Eq. (36)]:

$$2\,CO \leftrightharpoons CO_2 + C \qquad\qquad \Delta H298 = \underline{-171.7\ kJ}\ (-41\ kcal) \qquad (35)$$

$$\Delta H_{298} = -41\ kcal\ (-171{,}7\ kJ)$$

$$C + 2\,H_2 \leftrightharpoons CH_4 \qquad\qquad \Delta H298 = \underline{-73.7\ kJ}\ (-17.6\ kcal) \qquad (36)$$

$$\Delta H_{298} = -17{,}6\ kcal\ (-73{,}7\ kJ)$$

Equation (36) attains equilibrium slowly at the usual temperatures for the methane synthesis — similar to the (endothermic) steam gasification of carbon. Thus, when carbon deposition ensues it is almost irreversible and can block the catalyst. Attempts are made to counteract carbon deposition via addition of steam, via an adequate excess of hydrogen and via temperature control [80].

The thermodynamic equilibrium composition of the gas mixture leaving the reaction chamber depends on the composition of the gas feed, on pressure and on temperature. The mixture results from parallel and consecutive reactions. The calculation of equilibrium compositions, which are of considerable significance for reactor design and for controlling catalyst efficiency, is best carried out using a programme developed for the computing of simultaneous equilibria [75—79]. The minimisation of the free enthalpy of the total system has proved to be the most reliable calculating method, the (kinetically inhibited) deposition of free carbon either taken into account or neglected (cf.Sect. 4.4.2) [81]. The effect of the carbon dioxide content on the conversion of synthesis gas (with varying $H_2:CO$ ratio) at equilibrium is shown in Fig. 4.8, while the effect of temperature and pressure is presented in Fig. 4.9. The methane content at equilibrium increases with rising temperature, increasing carbon dioxide content in the fresh gas and with decreasing pressure, while the carbon monoxide and dioxide contents increase correspondingly.

Mechanism

The mechanism for the hydrogenation of carbon monoxide to methane has been discussed [22–24]. Chemisorption studies on carbon monoxide, hydrogen and mixtures of both gases on metal surfaces led to the following results [83]:

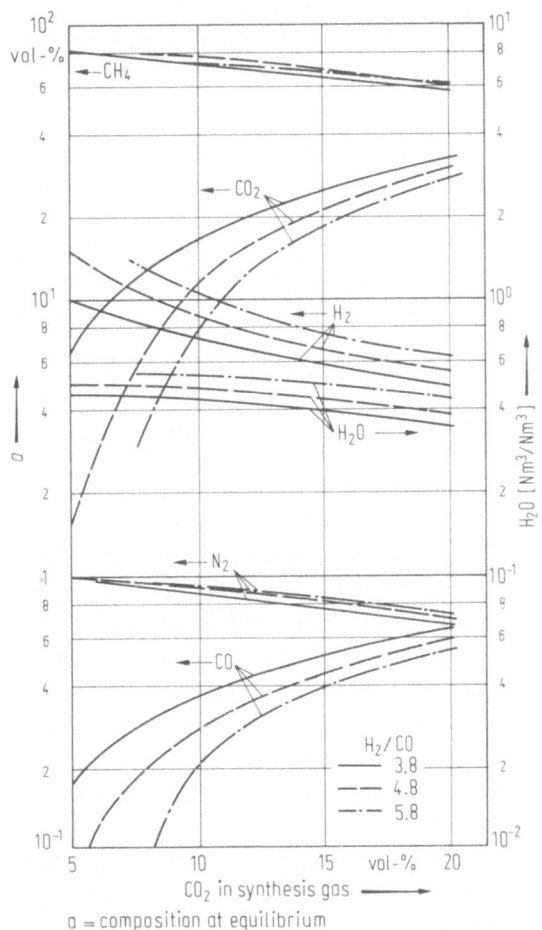

Fig. 4.8. Effect of carbon dioxide on the conversion of synthesis gas (varying $H_2:CO$ ratio)

$\square$ = composition at equilibrium

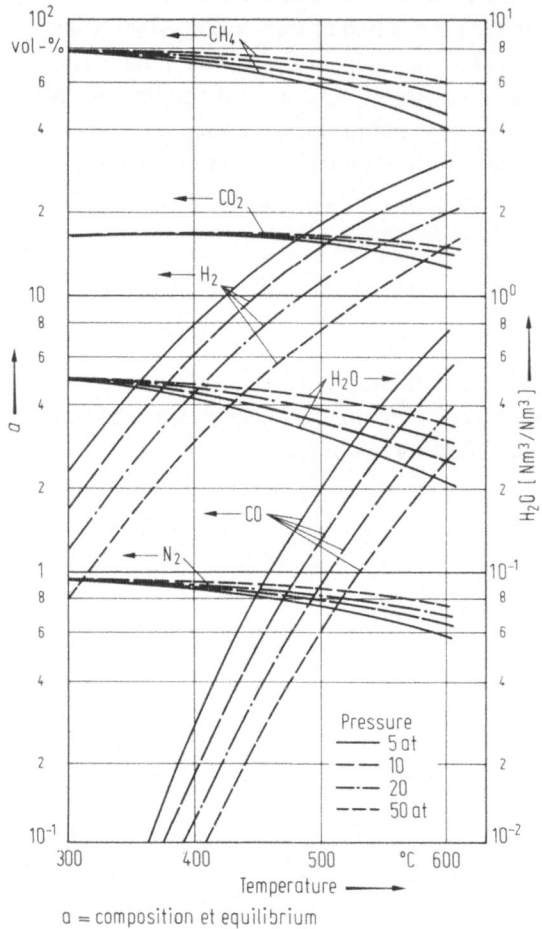

Fig. 4.9. Effect of temperature and pressure on the conversion of synthesis gas

a = composition et equilibrium

The exothermic chemisorption of carbon monoxide (approx. 120–210 kJ/mol – approx. 30–50 kcal/mol) allows a correlation to be made between the methanation activity of the metals of the 8th group of the Periodic System and their heat of chemisorption with CO (Fig. 4.10). The heat of chemisorption depends on the extent of coverage of the metal surface and decreases with increasing coverage.

For a 0.6 monolayer coverage the heat of CO chemisorption with Ru is 121 kJ/mol (approx. 30 kcal/mol), whereas at a higher coverage density only 90 kJ/mol (approx. 20 kcal/mol) was measured (LEED-, AES- and desorption measurements [94]). The corresponding heats of chemisorption of CO for polycrystalline nickel surfaces were 126 kJ/mol (approx. 30 kcal/mol) and 105 kJ/mol (approx. 25 kcal/mol), these values were determined by desorption and conductivity measurements [95]. Individual crystal surfaces can exhibit varying heats of CO chemisorption. A range between 142 and 167 kJ/mol (approx. 34–40 kcal/mol) was found for Pd crystal surfaces [96].

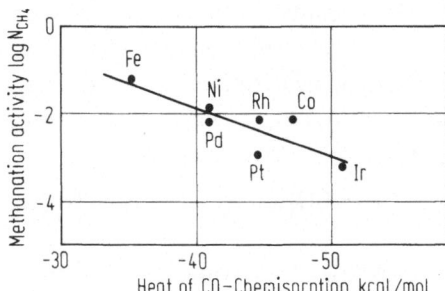

Fig. 4.10. Relationship between methanation activity and the heat of CO chemisorption [82]

Carbon monoxide appears to be partly dissociatively bound to various metals. IR studies indicated that two CO species were present with Ni, Pd, Pt [97] and Ru [98] which could be interpreted as being bound in either linear or bridged form [87–92]. The relative frequency of both species is a function of the CO partial pressure, the temperature, the type and degree of distribution of the metal. For example, with nickel a mainly undissociated (associated) CO bond is assumed below approximately 400 °C, whereas above 430 °C the bond is thought to be mainly dissociated. The transition from one species to the other is continuous. The assumption of a dissociated bonding state leads to the postulate involving carbide-like intermediate states [84, 85] in the methanation and in the Fischer-Tropsch synthesis and to analogies for the N_2 dissociation in the NH_3 synthesis [93]. The heat of chemisorption for hydrogen – assuming dissociated bonds on the metal surfaces – is approximately 70–90 kJ/mol (17–21 kcal/mol), markedly lower than the heat of chemisorption of carbon monoxide [99].

The interaction of chemisorbed carbon monoxide-hydrogen mixtures on the nickel surface - as determined by volumetric ad- and desorption measurements – differs in some respects from the results obtained with the iron catalysts in the Fischer-Tropsch synthesis [100]. In the mixed adsorption, the increase in heats of chemisorption and amount of gas suggest a similar structure to the proposed primary complex in the Fischer-Tropsch reaction. The species expected with iron catalysts could not be identified unequivocally [99, 101]. The desorption of chemisorbed gas mixtures supplies H_2 and CO in varying molar quantities in the absence of hydrocarbons and oxygen-containing compounds. The probability of the existence of oxygen-containing surface complexes is supported by IR measurements and desorption induced by electron impact at raised temperatures. The probability, based on these measurements, of the reaction between surface carbonyls and hydrogen increases in the series

Pd > Rh > Pt > Ru > Ir > Os,

which – with the exception of Ru – roughly corresponds to their methanation activity (cf Fig. 4.10). With simultaneous adsorption [83], the interaction of carbon monoxide and hydrogen on metal surfaces was confirmed by the increase in volume and rate.

The assumption that metal-oxygen and metal-carbon bonds can appear next to one another on metals suitable for methanation is supported by theoretical bonding studies. According to these calculations, stable oxygen-containing surface complexes are more likely with iron and cobalt than with nickel for which, in many respects, a (CH_2) grouping corresponds to experimental findings [102]. Based on the foregoing, the mechanism of the methane synthesis can be represented by the following scheme [82]:

$$H_{2(g)} + CO_{(g)} \qquad\qquad \rightleftharpoons H_2CO_{(ad)} \tag{37}$$

$$H_2CO_{(ad)} + y/2\, H_{2(ad)} \qquad \rightleftharpoons CH_{y\,(ad)} + H_2O_{(g)} \tag{38}$$

$$CH_{y\,(ad)} + H_{2(ad)} \qquad\qquad \rightleftharpoons CH_{4(g)} \tag{39}$$

Reaction (38) is the rate determining step. The available kinetic data do not permit an unequivocal decision to be made as to which of the proposed reaction sequences best represents the course of reaction.

Generally, the carbon dioxide [105, 107] hydrogenation is not included in mechanistic considerations as it is assumed that carbon monoxide, which is initially reformed via a reverse shift conversion [Eq. (33)], is subsequently hydrogenated [104]. Data obtained from Ni–Ce [108], Ni–Cr$_2$O$_3$ [106] and Ru catalysts [103] do not contradict this assumption.

Kinetics

Kinetic methanation data from literature sources are partly contradictory. This may be due to a basic difficulty associated with the exothermic reaction, the aging of the catalysts (poisoning, thermal aging) and the relatively high activation energy. Often formal kinetic principles for the rate of reaction are sufficient for reactor design; they can, however, considerably differ, from corresponding microkinetic data.

The following expression was used for measurements in a micro reactor (1 bar total pressure) involving comparative tests of various metals on a Al$_2$O$_3$ support:

$$N_{CH_4} = A \cdot e^{-\frac{E_A}{RT}} \cdot p_{H_2}^x \cdot p_{CO}^y \tag{40}$$

N_{CH_4} = rate of methane formation per active centre per second. The exponents x and y as well as the activation energy were determined (Table 4.4). The results below agree with other data in which the apparent order of reaction, with nickel catalysts, is positive for hydrogen (0.7–0.9) and slightly negative for carbon monoxide (−0.3 to −0.6).

The data, which were partly determined under ideal conditions (low total pressure, temperature range approximately 230–300 °C, differential reactor), frequently do not take into account the practical requirements of the methane synthesis (pressures up to 40 bar, temperatures > 300 °C, integrated reactors). Water and carbon dioxide influence the rate of reaction as shown in Eqs. (41)–(42) [109–114].

$$r = \frac{C_1\left(p_{CO_2} \cdot p_{H_2}^2 - \frac{1}{k} \cdot \frac{p_{CH_4} \cdot p_{H_2O}^2}{p_{H_2}^2}\right)}{(p_{H_2}^{0.5} + C_2 p_{CO_2} + C_3)^5} \tag{41}$$

Table 4.4. Kinetics of methane synthesis with metal-Al_2O_3 catalysts [Eq. (40)] [92]

Catalyst	E_A kcal/mol	x	y	A (molecules/centre $\cdot$ s)
5% Ru	24.2	1.6	−0.6	$5.7 \cdot 10^8$
15% Fe	21.3	1.14	−0.05	$2.2 \cdot 10^7$
5% Ni	25.0	0.77	−0.31	$2.3 \cdot 10^8$
2% Co	27.0	1.22	−0.48	$9.0 \cdot 10^8$
1% Rh	24.0	1.04	−0.20	$5.2 \cdot 10^7$
2% Pd	19.7	1.03	0.03	$1.2 \cdot 10^6$
17.5% Pt	16.7	0.83	0.04	$1.6 \cdot 10^4$
2% Ir	16.9	0.96	0.1	$1.4 \cdot 10^4$

$$r = \frac{k_1 \cdot p_{H_2}^{0.65} \cdot p_{CO}^{-0.25}}{1 + k_2 \cdot p_{H_2O}} \qquad (42)$$

On account of the large number of constants, equations such as (41) are suitable for approximating functions using a computer programme (Hougen-Watson). However, these constants have not more than a formal meaning. As expected, positive orders of reaction are obtained for carbon monoxide at very low carbon monoxide concentrations (purification of ammonia synthesis gas) [115].

$$r_{CO} = K \cdot P_{CO}^{0.7} \qquad (42)$$

The constant K includes the excess of hydrogen.

4.3.3 Reaction Conditions

There are always a number of secondary conditions which must be heeded in the industrial production of methane. While they limit variation of reaction conditions to a great extent, they are economically and technically necessary. The following requirements stem from gas generation and purification (or processing) which are relatively more involved than the actual methane synthesis:

higher degree of utilisation of the synthesis gas feed with regard to conversion and heat of reaction -

avoidance of energy-consuming process steps such as compression or intercooling

lowest possible investment and running costs, i.e. low volume of catalyst coupled with long service life

These partly opposing requisites demand compromises when considering reaction conditions.

The following aspects are important for the choice of reaction temperature:

above 200 °C, the active catalyst components (mainly metals from the 8th group of the Periodic System) facilitate sufficiently high rates of reaction for the carbon monoxide hydrogenation the lowest reaction temperature for metals which form volatile carbonyls is determined by the CO partial pressure

when the temperature is raised at a fixed pressure, the methane content, based on the thermodynamic equilibrium, and thus the gas quality, drops, thereby limiting the reaction or it must be divided into several steps with decreasing temperature gradient

the velocity of the Boudouard reaction, which is inhibited at low temperatures, increases irregularly above ~450 °C and can lead to rapid deactivation of the catalyst

the thermal efficiency of the process increases with temperature

the thermal stability of the metal particles and the catalyst supports declines rapidly above approximately 500 °C, causing the catalytic activity to drop due to sintering and crystallite growth.

With nickel catalysts, the usual reaction temperatures lie between approximately 280 and 500 °C, the lifetime of the catalysts being more than a year and in favourable cases even five or more years.

The lowest operating temperature is mainly determined by the CO partial pressure which causes catalyst corrosion via formation of volatile carbonyls. For nickel the relationship between temperature and partial CO pressure is presented in Table 4.5. Thus, for nickel or cobalt catalysts, reaction temperatures above 225 °C are employed.

The increasing pressure arising from the equilibrium conversion to methane via the carbon monoxide hydrogenation is limited by the generation and purification or processing (conversion, CO_2 wash) of the gas feed. An energy intensive intermediate compression of the gas feed should be avoided if at all possible. The conventional pressures between 20–25 bar will increase to 50–70 bar with the development of the new generation of gasification processes.

The catalyst load is determined by the heat removal system used in the process. With the adiabatic procedure, the gas temperature increases by around 54 °C per vol.% of methanated carbon monoxide. If adequate heat removal is available, then loads up to 1 000 volume gas/volume catalyst/hour are possible.

Due to its specific heat, the addition of steam to the gas feed reduces the temperature increase and counteracts carbon deposition. However, carbon monoxide conversion and displacement of equilibrium (in favour of methane formation) occur to a

Table 4.5. Dependence of nickel carbonyl equilibrium concentration on temperature [116]

Temperature °C	Mol fraction Ni(CO)$_4$ in equilibrium CO partial pressure (bar)			
	1	5	10	20
52	0.92	0.98	0.995	1.0
77	0.78	0.93	0.97	0.99
102	0.48	0.82	0.89	0.94
114	0.12	0.63	0.77	0.85

greater extent. If CO-rich gases, e.g., from coal gasification, are to be directly methanated, then steam must be introduced.

As methanation and conversion ensue at such high velocities, the mass- and heat-transfer data which could influence process design are of minor importance. Greater than 95% thermodynamic equilibrium is usually attained even when high linear gas velocities are employed.

4.3.4 Catalysts

All metals of the 8th group of the Periodic System catalyze the hydrogenation of carbon monoxide (to methane) to a varying extent.

If the CO conversion rate is standardized to conversions per active centre, then ruthenium — the metal with highest specific activity — has the value 325 and the slightly active iridium only 2 (converted CO molecules per active centre per second $\cdot$ 10^3 at 275 °C; cf Sect. 4.3.2). Nickel exhibits roughly twice the specific activity of cobalt. As the metal concentration, the support [118] and the type of manufacture can frequently displace the specific activity by an order of magnitude, these factors must be considered when conducting comparative activity tests.

The development of methanation catalysts, which ran roughly parallel to the development of Fischer-Tropsch catalysts, led to nickel being favoured as catalyst metal. For the methane synthesis and the retro-reaction — the cracking of hydrocarbons with steam to synthesis gas — other metals were only used as promoters. Despite their activity, ruthenium catalysts have not been widely used. Insufficient information is available regarding the suitability of molybdenum [177] or tungsten [119].

The properties of pure nickel catalysts can be basically altered on adding metallic or oxidized promoters. This encompasses their catalytic properties (activity, resistance to poisons) as well as their general applicability (thermal and mechanical stability, lifetime). Promoters are divided into two main types — electronic (energetic) and structural.

It is assumed that energetically effective promoters influence the catalytic properties of the metal (chemisorption, reaction) connected with electronic transitions, while structurally effective promoters assist in the development and maintainance (sintering, recrystallisation) of a stable surface. In addition to the promoters present in relatively minor quantities, the catalyst support can also affect the catalytic properties [118, 120].

The ready reduction of catalysts containing nickel oxide can be noticeably influenced by salts or oxides of other metals. For example, the nickel oxide reduction can be markedly facilitated if Pt, Pd or Cu [117] are present in certain concentrations. The presence of alkaline earth metal oxides [108] makes the catalyst more difficult to reduce. The reason for the difficult reducibility was found to lie with the oxides (Al_2O_3, Cr_2O_3, MgO, SiO_2) which are co-precipitated with nickel and form spinel compounds.

When manufacturing catalysts for the methane synthesis, the object is to produce large specific surfaces which should be sufficiently stable under the partly drastic reaction conditions. In order to achieve long life and uniform activity, the active metal content should not be too low. Promoters and supports must be chosen in a suitable ratio to the metal content. The following manufacturing processes — in order of importance — have found application:

mechanical mixing of metal salts (e.g. $Ni(NO_3)_2$) with activators (e.g. MgO, Cr_2O_3) and supports (e.g. Al_2O_3, SiO_2) followed by chemical combination via heat treatment (calcination)

impregnation of supports with metal and activator solutions and subsequent heat treatment

precipitation of metal and promoters (together or separate) in presence of supports

co-precipitation of metal, promoters and supports

The last processes ensure an extensive homogeneous distribution of all components and produce particularly effective catalysts [121, 125]. Previously, kieselguhr was favoured as support. Due to its limited chemical and thermal stability, it has been replaced by aluminum oxide or aluminum silicates. Spinels (cordierite or spodumen structure) are used for catalysts particularly subject to thermal stress. Besides several multi-element catalysts, the current trend is to catalysts containing approximately 25–30 wt.% nickel, 3–6 wt.% of a basic oxide and a thermally stable aluminosilicate.

Raney nickel catalysts, which did not possess any significance as methanation catalysts, have recently aroused interest in connection with new processes (cf Sect. 4.3.5). The possibility of hydrogenating carbon monoxide with metal clusters (e.g. $Ir_4(CO)_{12}$, $Os_3(CO)_{12}$, $Cp_2Ti(CO)_2$) [122–124] will only be referred to here. Ruthenium and nickel catalysts are suitable for the hydrogenation of carbon dioxide [85].

4.3.5 Processes

Processes for the hydrogenation of carbon monoxide to methane can be characterized by the type of heat removal and thus the shape and arrangement of the catalyst.

a) processes with fixed-bed catalysts with partial or complete heat removal via cooling elements and/or gas recycle

b) processes with fixed-bed catalysts, operated adiabatically

c) fluidized-bed processes with or without internal cooling and mobile catalyst

d) processes with suspended catalyst, the heat of reaction being removed via the suspension agent

The first industrial methanation plants were installed with water-cooled tubular bundle reactors. On account of the limited heat removal through the reactor wall, the efficiency of the reactor was improved on using a gas recycle with intermediate cooling. This process is limited by the temperature difference between catalyst and reactor wall which depends to a certain extent on the diameter of the tubes and the amount of recycle gas. The effect of greater heat transfer with decreasing diameter of the catalyst layer was recently the basis of a concept simultaneously developed in the USA and the USSR.

A nickel-aluminum alloy (Raney nickel) is introduced onto the wall of a large surface area heat exchanger by means of a flame spray or sintering [130]. Thereafter, the aluminum is removed by alkaline extraction. Then the nickel adhering to the wall can be treated with promoters, thereby facilitating a high specific gas load without superheating. This process, also known as the TWR (Tube Wall Reactor), has not been employed industrially [133].

The development of highly stable catalysts meant that the involved construction features for internal cooling were unnecessary and adiabatic operation was possible.

The forerunners of this development were reactors in which the catalyst bed was divided into sections, cold fresh gas being introduced after each stage. The gas temperature increases by 54 °C for each volume percent methane formed, thus with adiabatic operation only around 3–5 vol.% CO could be converted into CH_4 per reactor, as the currently available fixed-bed catalysts are active above approximately 250 °C and can be utilized up to approximately 500 °C. Consequently, multi-stage reactors connected in series are used [126–128, 132]. Two effects result on introducing steam into the first reactors of a cascade of this type. Some of the heat of reaction is taken up by the steam which converts part of the CO feed parallel to the methanation. The conversion can be controlled by quantity of steam and temperature (from first to last reactor drop from approximately 500 °C to approximately 250 °C [129, 131]). This facilitates the application of untreated CO-rich gases from coal gasifications. The processes operating according to variant a) require an initial conversion to adjust the H_2:CO ratio to around 3:1. The final gas is, after removal of water and carbon dioxide, brought to a low temperature level (cf Fig. 4.11) [134, 135].

The arrangement of the catalyst in a fluidized-bed and the inherent advantages of good mass- and heat-transfer contrast with the difficulties involved in process control. Thus, the methane synthesis is still to be commercially proved. Pilot plants are being operated in the USA and in West Germany [136, 137] (cf Fig. 4.13).

Operating the strongly exothermic methanation reaction in the presence of a liquid phase containing a suspended catalyst has often been considered as it offers fundamental advantages [138]. The plant concept is very simple in principle (cf Fig. 4.12). A nickel catalyst in tablet form (size approximately 2.5 · 4.5 mm) is covered with light oil and fed along with synthesis gas to a reactor. The catalyst bed then undergoes expansion. However, below a critical gas velocity (approximately 10–30 cm/s) – which depends on the density of the gas, catalyst and organic liquid – there is a distinct division between the catalyst and the supernatant light oil. For cooling purposes, the latter is continuously

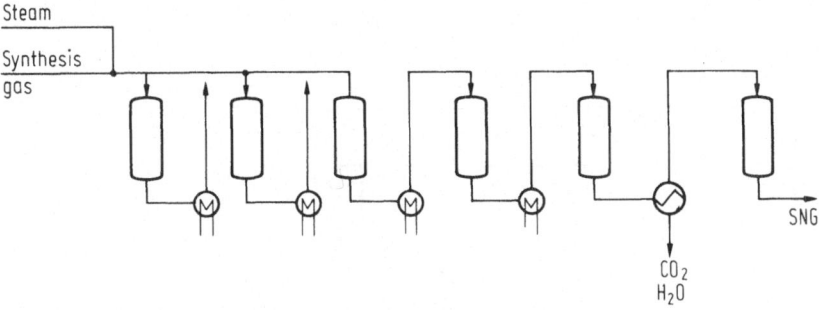

Fig. 4.11. Adiabatic reactors in series with internal conversion

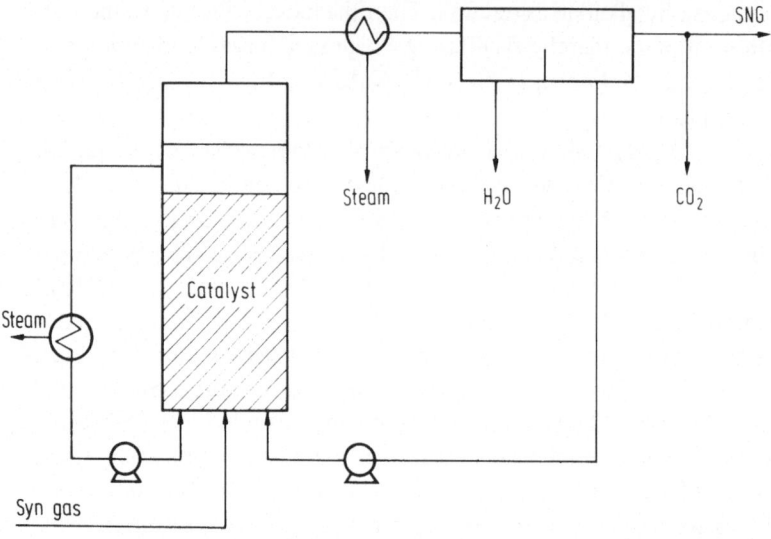

Fig. 4.12. Liquid phase methanation

pumped to an external heat exchanger. Water and the evaporated organic constituents, which are to be recycled, are removed from the product gas. After a CO_2 wash, the resulting gas, which is of virtually SNG quality, can, if necessary, be subjected to methanation once again.

During the development of this process it was found that CO_2 formation from the shift conversion can be largely suppressed if, at a pressure above 15 bar (e.g. 40–60 bar)

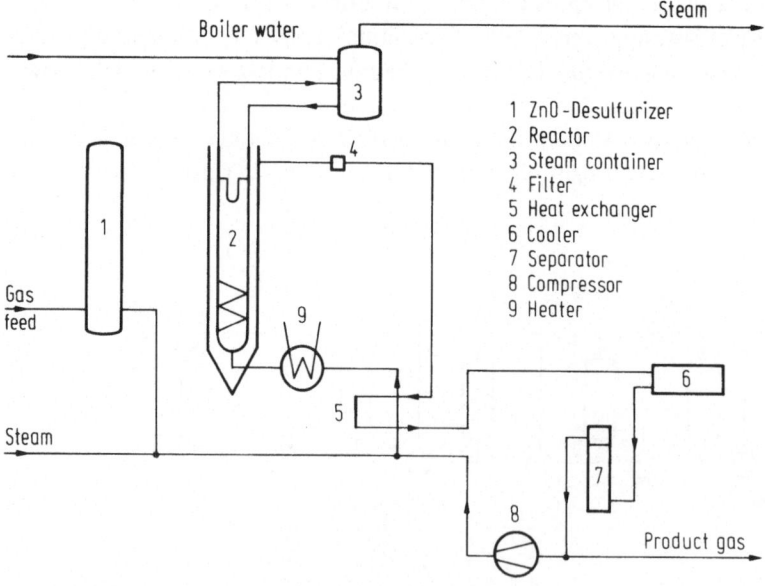

Fig. 4.13. Methanation via fluidized-bed process

the H_2:CO ratio is greater than 3 (e.g. Sect. 3.1) and 5–10 vol.% CO_2 are present in the gas feed. CO_2 formation could then be limited to approximately 3% of the CO conversion. It should also be possible to directly employ CO-rich gases, e.g. from coal gasification without previous conversion, enabling the process to be both highly selective and flexible [140].

A modification of the liquid phase process uses water instead of light oil as reaction medium [139]. A characteristic of this operation is the application of catalysts exhibiting high activity at low temperatures as the reaction pressure should not be too high. The favoured region is 18–25 bar, corresponding to temperatures between approximately 180 and 205 °C. Water and synthesis gas are fed together as trickle phase over a sub-divided fixed-bed catalyst. An uncertain factor is the lifetime of the catalyst in the presence of water at raised temperatures and pressures.

4.3.6 Economic Potential and Possible Developments in the Methane Synthesis

Due to its relative simplicity along with its periodic importance in the past, methanation catalysis has been studied frequently. There is a wide range of information available as far as choice of catalyst and process is concerned. The type of operation is somewhat restricted due to the unavoidable coupling of the methane synthesis to gas production. This applies equally to SNG (Substitute Natural Gas) manufacture as well as to the fine purification of, for example, ammonia synthesis gas. The methane synthesis must be adapted to the more involved synthesis gas production and not vice versa. The ideal case is, therefore, the highest possible flexibility of the manufacturing process enabling the facile treatment of synthesis gas, stemming from various gasification processes. The best possible utilisation of the heat released will gain particular importance in the future. This last aspect is the basis of current trends concerning the development of a new energy transport system involving circulating methane or synthesis gas whilst coupling nuclear process heat. The further development of high temperature nuclear reactors and the problems associated with energy transfer from reactor to methane cracking are rate determining. Consequently, even the operation of demonstration plants is hardly to be expected before the mid-eighties. Methane synthesis will enjoy rapidly increasing importance with the advance in development and testing of coal gasification plants of the new generation together with the predictable shortage and price increase of oil and oil-based products. Due to the special situation prevalent in the USA the lead will probably be taken there.

Nickel has dominated as the main catalyst component from the outset of the methane synthesis to the present although a variety of other metals exhibit better (e.g. Ru) or comparable (e.g. Co) suitability. The reason probably lies with the amount of information available about nickel catalysts, which enables all problems to be solved satisfactorily by modifying these catalysts via activators, supports or manufacturing processes. The high price of ruthenium, which limits its wider use, and the higher sensitivity of cobalt with regard to extreme conditions make both unattractive alternatives. Furthermore, while $MoSi_2$ is active at 500–600 °C, the conversion is only around 20% [141]. Progress could be expected when nickel catalysts can be replaced by those more or totally resistant to poisons. While the addition of other metals (W, Mo) [142] or oxides (MgO [144], Cr_2O_3, ZrO_2 [145]) can markedly improve the resistance of nickel towards sulfur poisoning, the raw synthesis gas must still be treated to remove all but traces of sulfur compounds. One of the tasks of catalyst development will be to improve the activity of new or known sulfidized catalysts to such an extent that they can replace nickel catalysts, enabling improved economic efficiency of the whole process via omission of purification steps. The adiabatic operation, which is being considered for newer methane processes (at high temperatures), is compatible with this aim.

The conventional methanation process with fixed-bed catalysts can be regarded as being safe and operational. Processes with mobile catalysts facilitate better heat transfer and thus a higher specific load as well as a relatively simple, and continuous, exchange of deactivated or poisoned catalyst. Fluidized-bed processes should have good prospects when sufficiently abrasion-resistant catalysts are available.

Fixed-bed processes with circulating organic heat exchangers or with water and synthesis gas in co-current as a trickle phase are probably the closest approach to isothermal operation. This ensures that the thermal catalyst load will be very low. Modifying the methane synthesis from a pure gas phase reaction to a gas-liquid phase reaction [147] displaces the rate determining step from the catalyst surface to interface between the phases, causing mass-transfer phenomena to become the dominant limiting factor. Bubble column reactors with a suspended catalyst have only been tested to a reactor volume of around 10 m^3, which is insufficient for industrial applications. We will have to wait and see whether the present development trends retain their basic advantages on an industrial scale. An alternative to the gas phase process could result from these developments. The same applies to reactions with fused salts [146].

In principle, suitable catalysts are available for all processes which are of interest. The various combinations of basic metal and activator(s) as well as the manufacturing variants can be tailored to meet particular requirements, so that solutions to all process problems should be found. So far, no methane production unit has been operated with the necessary output of at least 1 million Nm^3 methane/day required according to current conceptions.

4.4 The Fischer-Tropsch Hydrocarbon Synthesis

4.4.1 General Remarks

In 1922, F. Fischer and H. Tropsch [148] discovered that the heterogeneously catalyzed reaction between carbon monoxide and hydrogen yields mixtures of mainly linear alkanes and alkenes. The process, kown initially as the "Synthol process", was later designated the Fischer-Tropsch synthesis. It presented an alternative to the production of liquid hydrocarbons via coal hydrogenation and rapidly aroused interest in the chemical industry ('Oil' or 'Gasoline Synthesis'). From 1927 on, intensive work was carried out to solve the process engineering problems arising from the scaling up of the synthesis [149]. In 1934, the multi-stage atmospheric pressure process was further developed by Ruhrchemie AG. This work enabled start up of the first four plants for the production of 200,000 tons/annum hydrocarbons (motor fuels) [150] as early as 1936. In 1937, the atmospheric pressure process was replaced by a medium pressure variant which was modified in 1939 using a synthesis gas recycle. These improvements were utilized in the design of other plants resulting in the German production being around 600,000 tons/annum in 1944 [153].

Soon after the Second World War, the increasing availability of cheap oil-based hydrocarbons led to the shut down of production in the few German Fischer-Tropsch plants which were then operating [151]. Despite basic improvements in the process and catalysts, the production of motor fuels via carbon monoxide hydrogenation is still economically unattractive for Western countries. One exception is South Africa [152], where extremely low-cost coal is available. The production there is to be expanded from approximately 200,000 to around 2 million tons/annum hydrocarbons [156]. Using conventional and industrially tested processes, the Fischer-Tropsch synthesis can supply a particularly wide spectrum of hydrocarbons — extending from methane to hydrocarbon waxes. The by-products, usually only a few percent to the total yield, are oxygen-containing compounds (alcohols, aldehydes, ketones, acids, esters). The latter can be obtained as the main products if higher pressures, lower temperatures and oxidized catalysts are employed, the conversion then being known as the "Oxyl" synthesis [154]. The resulting hydrocarbons are usually non-branched, although sometimes methyl branching is present, and contain only traces of alicyclic or aromatic compounds. Depending on catalyst and reaction conditions, varying amounts of alkenes (40–85%) — mainly α-alkenes [154, 156] — result, whose content varies with C number. The products are virtually sulfur-free.

The ever-present aim during the development of the Fischer-Tropsch synthesis was to increase the selectivity of this type of carbon monoxide hydrogenation. In contrast to the methane and methanol syntheses, where by-products are formed in very small amounts, it has not been possible to bring the selectivity of the Fischer-Tropsch synthesis to a remotely comparable level and to limit formation of hydrocarbons or oxygen-containing compounds to a narrow carbon number range.

In the industrial Fischer-Tropsch synthesis, limited reaction control was achieved by choosing a particular catalyst and appropriate reaction conditions to preferentially produce either long-chained (wax-like) or short-chained (motor fuel applications) products via the gas phase fixed-bed [157] or entrained-bed [158] processes. In later stages, the product was separated and processed according to end use.

The reason for the low selectivity of the Fischer-Tropsch reaction is that the catalysts kinetically control the carbon monoxide hydrogenation and the reaction does not reach thermodynamic equilibrium as in the methane synthesis. Hydrocarbon formation ensues via a complicated, and only partially understood, network of primary and secondary reactions (cf Sect. 4.4.2.3) which occur to a varying extent depending on catalyst and reaction conditions (temperature, partial pressures or concentration of feedstocks, intermediates, products, mass- and heat-transfer, etc.). Recently resumed research and development work have shown, however, that composition and topography of the catalyst as well as reaction conditions — after careful balancing — can lead to a marked increase in selectivity, e.g. with regard, to formation of short-chained alkene-rich hydrocarbon mixtures (cf Sect. 4.4.6).

The Fischer-Tropsch synthesis was discovered using iron catalysts which permitted carbon monoxide to react with hydrogen at atmospheric pressure. The development then made a detour as cobalt catalysts were developed for the atmospheric operation and utilized in the first production units. The use of iron catalysts, which only possess sufficient stability and life at raised pressure, was of interest once the plants were converted to medium pressure operation. The suitability of the iron catalysts was closely studied leading to their exclusive use in industrial plants. The "Arge high load synthesis" (gas phase, fixed-bed) employs a precipitated iron catalyst and the "Kellogg-Sasol synthesis" (gas phase, entrained-bed) uses an iron catalyst (manufactured via fusion). The fixed-bed synthesis, which operates between 220–250 °C, yields diesel oil and wax-like hydrocarbons, while the entrained-bed synthesis at 320–350 °C produces mainly hydrocarbons in the gasoline range [159–161]. Both processes operate at 22–26 bar.

On account of the rising cost and scarcity of oil-based hydrocarbons, the Fischer-Tropsch process is becoming economically more attractive [162, 165, 166]. The prerequisite is the availability of cheap synthesis gas from fossil sources. Currently a series of improved synthesis gas processes are being developed or are at the testing stage [163, 164]. At the present time, the production of motor fuels via the Fischer-Tropsch synthesis is uneconomic so that, initially, attention will be concentrated on the manufacture of higher grade feedstocks for the chemical industry [165]. To achieve this aim, it will be necessary to considerably increase the selectivity of conventional processes and to remove low value products, thus alleviating the synthesis from certain economic and technical constraints. Recent studies have indicated that the development of improved catalysts will lead to new concepts for the reaction, facilitating the more selective production of short chained olefins – mainly ethylene and propylene – via a single stage carbon monoxide hydrogenation.

4.4.2 Reaction Mechanism

4.4.2.1 Stoichiometry

The hydrogenation of carbon monoxide to hydrocarbons proceeds according to a series of complicated parallel and consecutive reactions whose relative velocity depends on catalyst type and reaction conditions. The stoichiometry of the hydrocarbon formation can be derived from two basic reactions:

$$CO + 2\,H_2 \;\rightarrow\; (-CH_2-) + H_2O \qquad \Delta H_R\,(227\,°C) = -39.5\;\text{kcal}\,(-165\;\text{kJ}) \qquad (43)$$

$$CO + H_2O \;\rightarrow\; H_2 + CO_2 \qquad \Delta H_R\,(227\,°C) = -9.5\;\text{kcal}\,(-39.8\;\text{kJ}) \qquad (44)$$

The actual Fischer-Tropsch reaction [Eq. (43)] is best catalyzed using cobalt while, with iron catalysts, the shift conversion [Eq. (44)] increases in importance as a secondary reaction. Thus, for iron catalysts the net equation is as follows [Eq. (45)]:

$$2\,CO + H_2 \;\rightarrow\; (-CH_2-) + CO_2 \qquad \Delta H_R\,(227\,°C) = -48.9\;\text{kcal}\,(-204.7\;\text{kJ}) \qquad (45)$$

The maximum yield is 208.5 g alkene C_nH_{2n} per Nm^3 synthesis gas for complete conversion. On combining Eq. (43) – (44), two further net equations for the hydrocarbon synthesis are obtained:

$$3\,CO + H_2 \rightarrow (-CH_2-) + 2\,CO_2 \quad \Delta H_R\,(227\,^\circ C) = -58.4\ \text{kcal}\ (-244.5\ \text{kJ}) \quad (46)$$

$$CO_2 + 3\,H_2 \rightarrow (-CH_2-) + 2\,H_2O \quad \Delta H_R\,(227\,^\circ C) = -29.9\ \text{kcal}\ (-125.2\ \text{kJ}) \quad (47)$$

The reactions (43) to (47) may be accompanied by the following side reactions:

$$CO + 3\,H_2 \rightarrow CH_4 + H_2O \quad \Delta H_R\,(227\,^\circ C) = -51.3\ \text{kcal}\ (-214.8\ \text{kJ}) \qquad (48)$$

$$2\,CO + 2\,H_2 \rightarrow CH_4 + CO_2 \quad \Delta H_R\,(227\,^\circ C) = -60.8\ \text{kcal}\ (-254.1\ \text{kJ}) \qquad (49)$$

$$2\,CO \rightarrow C + CO_2 \quad \Delta H_R\,(227\,^\circ C) = -32.0\ \text{kcal}\ (-134\ \text{kJ}) \qquad (50)$$

Oxygen-containing compounds (aldehydes, ketones, alcohols, acids and esters) result in addition to alkanes and alkenes (see below). The ratio in which CO and H_2 are consumed can vary from around $1:2$ to approximately $2:1$, depending on which of the above equations best represents the overall reaction. In order to obtain optimal hydrocarbon yields, the feed and consumption ratio of carbon monoxide to hydrogen must be the same. Each deviation between feed and consumption ratio will lead to reduced yields [153].

The carbon monoxide hydrogenation to methane and the decomposition of carbon monoxide to elementary carbon and carbon dioxide are undesired side reactions. Methane results mainly with cobalt or nickel catalysts [Eqs. (48)–(49)] at higher temperatures ($> 300\,^\circ C$, entrained-bed synthesis) and from carbon dioxide with iron catalysts [Eq. (45)]. The carbon resulting from the Boudouard reaction [Eq. (50)] may block the active catalyst centres. However, the thermodynamically probable carbon separation is frequently limited kinetically; consequently, it only occurs to a very minor extent under normal conditions.

4.4.2.2 Thermodynamics

The thermodynamic probability of formation of individual products in a system of coupled reactions can be established by calculating the simultaneous equilibria [167–170]. The thermodynamics of the Fischer-Tropsch synthesis are based on the assumption that selected individual reactions ensue independently of one another. The temperature dependence of the free enthalpy of formation for several hydrocarbons and hypothetical intermediates are shown in Fig. 4.14.

According to the above, methane formation, via carbon monoxide hydrogenation, is thermodynamically favoured in the range 50 to 350 $^\circ C$. While the tendency for hydrocarbon formation above C_1 increases with rising temperature, it is of minor

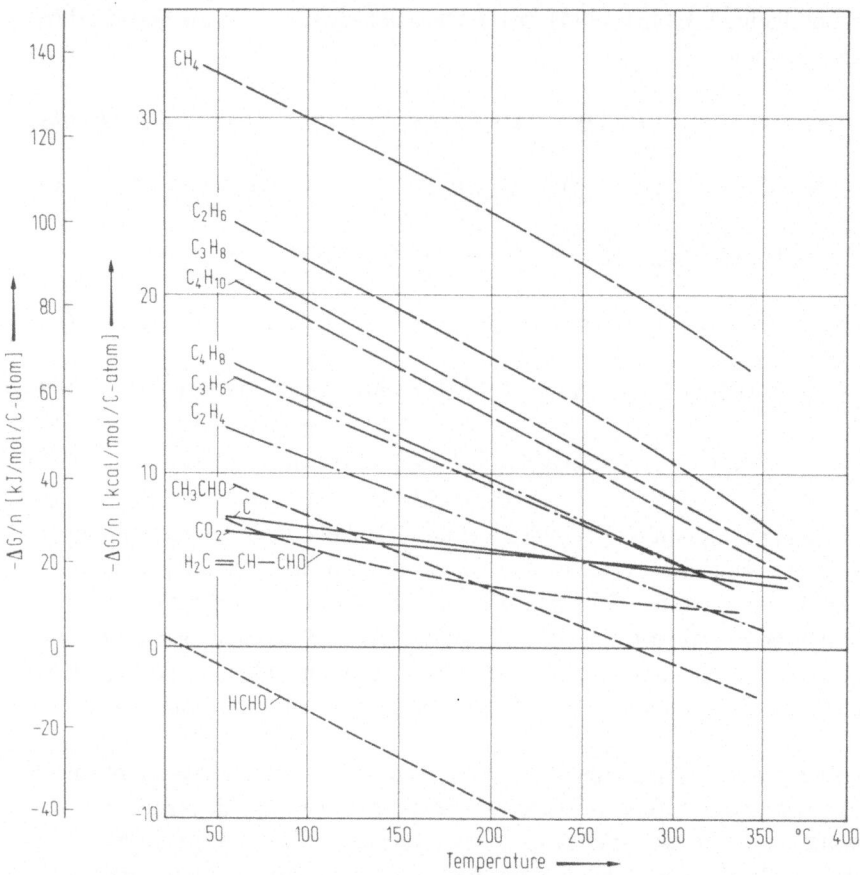

Fig. 4.14. Temperature dependence of the free enthalpies in Eqs. (43)–(50) [153]

importance in the range under consideration. With increasing temperature, the reaction enthalpies of the side reactions (Boudouard reaction, shift conversion), leading to free carbon and carbon dioxide, decrease slightly relative to hydrocarbon formation and thereby become more favoured.

The estimated composition of the Fischer-Tropsch products, based on thermodynamically probable equilibrium contents, differs considerably from the actual product composition. The selectivity of the various Fischer-Tropsch syntheses is thus determined by a sequence of kinetically controlled reaction steps which can be influenced by reaction conditions and catalyst properties.

4.4.2.3 Mechanism

The initial step in the carbon monoxide hydrogenation is the simultaneous chemisorption of carbon monoxide and hydrogen on the catalyst surface [171]. Carbon monoxide is attached to the metal via carbon, the carbon-oxygen bond being thereby weakened, facilitating the formation of a primary complex via reaction with hydrogen. This primary complex, which is regarded as the precursor in the constructive reaction leading to the formation of hydrocarbon chains, has been frequently studied [172]. With iron catalysts, the following results suggest the appearance of an enolic primary

complex of the structure $M = C\begin{smallmatrix} \nearrow H \\ \searrow OH \end{smallmatrix}$ [153]:

> IR spectra of chemisorbed carbon monoxide-hydrogen mixtures [175]
>
> sorption measurements: when chemisorbed carbon monoxide and hydrogen are desorbed they are released in the ratio $1:1$, even when gas mixtures of different contents were absorbed [171]
>
> calorific measurements: the heat of chemisorption of a carbon monoxide-hydrogen mixture is higher than the corresponding values for the individual gases. Consequently, the quantity of chemisorbed gas is greater than the sum of the quantities of the individual gases. The difference between the sum of the heats of absorption of the individual gases compared to the heat of absorption for the mixture corresponds to the calculated enthalpy of formation of the enolic primary complex (6 kcal/mol) [176–178]
>
> mass spectrometry: formaldehyde could be proved to be the primary product of the synthesis [179]

Several of these findings are supported by results based on model studies (Extended-Hückel theory) according to which the C–H–O complexes with cobalt or iron catalysts were estimated to be stable, whereas methylene complexes are more likely with nickel catalysts [174]. The calculated and measured heats of chemisorption are in relatively good agreement [173]. Two mechanisms were postulated and discussed for the individual steps on the catalyst surface (primary complex, chain initiation, chain growth, chain termination) after Fischer suggested that the intermediate carbides (from CO decomposition) were hydrogenated. According to Anderson [172], the reaction begins with the cleavage of water and the formation of a C–C bond between two enolic primary complexes, arising from the chemisorption of hydrogen and carbon monoxide (see below). A carbon atom is simultaneously released from the catalyst surface by hydrogenation (chain initiation). Water is cleaved from the resulting C_2 complex, which is then detached from the catalyst surface via hydrogenation (chain growth). Growth is continued at the hydroxyl group at the end of the chain. The desorption of the non-hydrogenated intermediate leads to aldehyde formation [179] and in consecutive reactions to the formation of alcohols, carboxylic acids or esters (cf Fig. 4.15).

Hydrocarbons may be formed either via dehydration of alcohols or via cleavage of adsorbed complexes. Chain initiation can start with alcohols, if they are attached to the catalyst in their enolic form [180, 181], as well as with alkenes which can react with the primary complexes. Alkene polymerisation is another possibility for chain growth, while the cracking of intermediates will reduce the chain length of the products.

Pichler [182] proposed a mechanism, based on cobalt and iron hydrocarbonyls. Chain growth ensues via insertion of carbon monoxide between the catalyst and the terminal hydrogenated carbon atom. Initially an aldehydic surface complex results

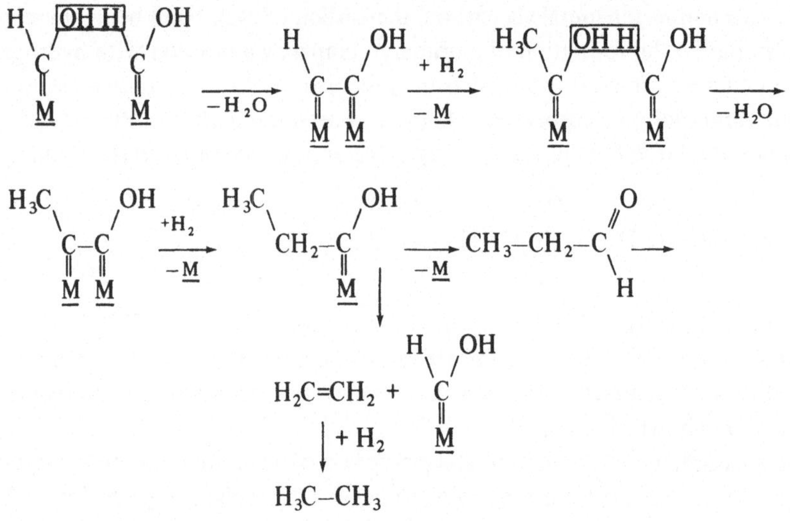

Fig. 4.15. Mechanism of the Fischer-Tropsch synthesis [172]

which, after hydrogenation and dehydration, is converted into a methyl group via a methylene complex (cf Fig. 4.16). The methyl group is directly attached to a metal atom of the catalyst (chain initiation). The methyl group can either be converted into methane (via hydrogenation) or into a new aldehydic complex (via CO insertion in the metal-carbon bond) which leads to chain growth [184]. Chain extension continues via the interplay of hydrogenation, dehydration and CO insertion until desorption ensues leading to alkanes, alkenes or oxygen-containing products [183].

At present it is not possible to unequivocally favour one of the above mechanisms. The Anderson mechanism is supported by experimental data (see above) confirming the existence of the primary complex, comparison of calculated and actual product distribution as well as studies relating to the presence of alcohols, aldehydes or alkenes in the Fischer-Tropsch products [185, 186]. The main arguments confronting the Pichler mechanism are the instability of metal hydrocarbonyls under the prevailing Fischer-Tropsch conditions and the intricacy of the insertion reaction. Recent proposals attempt to circumvent the difficulties inherent in both postulates by assuming that the primary product is either an acyl complex or a M−CH group [181]. Hydrocarbon chains could then be formed via hydrogenation and dehydration or via a carbene-like insertion reaction.

4.4.2.4 Kinetics

Due to the complexity of the Fischer-Tropsch reaction coupled with the large number of variables, it has not been possible to formulate a general equation for the macrokinetics. However, reaction velocity equations have been advanced for individual catalysts and processes under defined conditions [187, 188]. When the H_2:CO ratio in fresh gas is between 3 and 1, the Fischer-Tropsch reaction (iron-catalyzed) is first order rela-

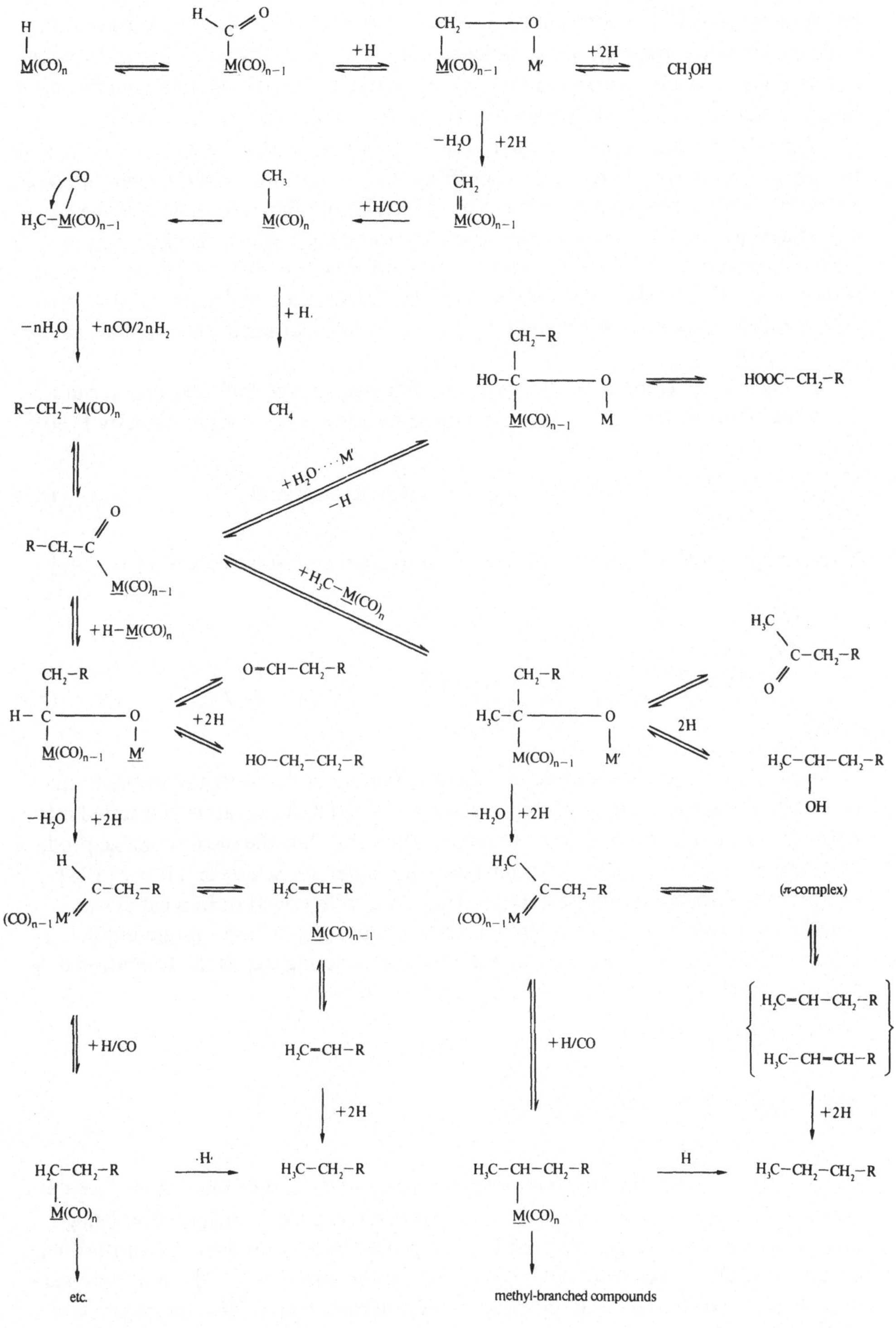

$$\underline{M}(CO)_n \; \underset{H}{\overset{}{|}} \;\rightleftharpoons\; \underline{M}(CO)_{n-1}\;\overset{H}{\underset{}{C}}\!\!=\!\!O \;\;\overset{+H}{\rightleftharpoons}\;\; \underline{M}(CO)_{n-1}\!-\!CH_2\!-\!O\!-\!M' \;\;\overset{+2H}{\rightleftharpoons}\;\; CH_3OH$$

$$-H_2O \;\Big|\; +2H$$

$$H_3C\!-\!\underline{M}(CO)_{n-1}\overset{CO}{\nwarrow}\;\;\longleftarrow\;\; \underset{\underline{M}(CO)_n}{\overset{CH_3}{|}}\;\;\overset{+H/CO}{\longleftarrow}\;\; \underset{\underline{M}(CO)_{n-1}}{\overset{CH_2}{\|}}$$

$$-nH_2O \;\Big|\; +nCO/2nH_2 \qquad\qquad \Big|\; +H\cdot$$

$$R\!-\!CH_2\!-\!M(CO)_n \qquad\qquad CH_4$$

$$\underset{M}{\overset{CH_2-R}{|}} \;\; HO\!-\!\underset{\underline{M}(CO)_{n-1}}{C}\!-\!O \;\;\rightleftharpoons\;\; HOOC\!-\!CH_2\!-\!R$$

$$R\!-\!CH_2\!-\!\underset{\underline{M}(CO)_{n-1}}{C}\!\!=\!\!O \qquad \overset{+H_2O\cdots M'}{\underset{-H}{\rightleftharpoons}} \qquad \overset{+H_3C-\underline{M}(CO)_n}{\longrightarrow}$$

$$\Big|\; +H\!-\!\underline{M}(CO)_n$$

$$\underset{\underline{M}(CO)_{n-1}\;\underline{M'}}{\overset{CH_2-R}{\overset{|}{H\!-\!C\!-\!O}}} \quad\overset{+2H}{\rightleftharpoons}\quad \begin{array}{l} O\!=\!CH\!-\!CH_2\!-\!R \\[4pt] HO\!-\!CH_2\!-\!CH_2\!-\!R \end{array} \qquad \underset{M(CO)_{n-1}\;M'}{\overset{CH_2-R}{\overset{|}{H_3C\!-\!C\!-\!O}}} \quad\overset{2H}{\rightleftharpoons}\quad \begin{array}{l} \overset{H_3C}{\underset{O}{\diagdown}}C\!-\!CH_2\!-\!R \\[8pt] H_3C\!-\!CH\!-\!CH_2\!-\!R \\ \qquad\quad\;| \\ \qquad\quad OH \end{array}$$

$$-H_2O \;\Big|\; +2H \qquad\qquad\qquad\qquad -H_2O \;\Big|\; +2H$$

$$\underset{(CO)_{n-1}M'}{\overset{H}{\diagdown}}C\!-\!CH_2\!-\!R \;\;\rightleftharpoons\;\; \underset{\underline{M}(CO)_{n-1}}{\overset{H_2C=CH-R}{|}} \qquad \underset{(CO)_{n-1}M}{\overset{H_3C}{\diagdown}}C\!-\!CH_2\!-\!R \;\;\rightleftharpoons\;\; (\pi\text{-complex})$$

$$\Big\updownarrow \qquad\qquad\qquad\qquad\qquad\qquad\qquad \Big\updownarrow$$

$$\Big|\; +H/CO \qquad H_2C=CH-R \qquad\qquad \Big|\; +H/CO \qquad\qquad \left\{\begin{array}{l} H_2C=CH-CH_2-R \\[4pt] H_3C-CH=CH-R \end{array}\right\}$$

$$\Big|\; +2H \qquad\qquad\qquad\qquad\qquad\qquad\qquad\qquad \Big|\; +2H$$

$$\underset{\underline{M}(CO)_n}{\overset{H_2C-CH_2-R}{|}} \;\overset{\cdot H\cdot}{\longrightarrow}\; H_3C\!-\!CH_2\!-\!R \qquad \underset{\underline{M}(CO)_n}{\overset{H_3C-CH-CH_2-R}{|}} \;\overset{H}{\longrightarrow}\; H_3C\!-\!CH_2\!-\!CH_2\!-\!R$$

$$\Big\downarrow \qquad\qquad\qquad\qquad\qquad\qquad \Big\downarrow$$

etc.   methyl-branched compounds

Fig. 4.16. Mechanism of the Fischer-Tropsch synthesis [153]

tive to the H_2 partial pressure and almost zero order relative to the CO partial pressure. Besides such rate equations, which basically only take the partial pressures of hydrogen and carbon monoxide into account, it was attempted to include the rate retarding effect of steam and carbon dioxide in expressions for the reaction velocity (see below).

The rate determining step of the Fischer-Tropsch reaction is unlikely to be the mass-transfer of reactants and products to and from the catalyst surface. Of greater influence is apparently the high activation energy (20—25 kcal/mol). But independent of the process, the catalyst particle is encapsulated by a hydrocarbon layer which also fills the pores. With increasing pore depth, the concentration gradient formed between reactants and products is kept in equilibrium by diffusion, thereby coupling reaction and mass-transfer. It is fairly probable that the rate determining step is the latter effect at the internal catalyst surface [189].

In the Fischer-Tropsch reaction with cobalt catalysts, the shift conversion can be neglected, allowing the following expression to be derived for reaction velocity [190]:

$$r = \frac{k \cdot P^2_{H_2}}{P_{CO}} \qquad\qquad E_A = \text{approximately 21 kcal/mol} \qquad\qquad (51)$$

A semi-empirical expression was proposed for precipitated iron catalysts [191, 192]

$$r = K_1 \frac{P^m_{H_2}}{P_{CO}} \; \frac{1}{1 + K_2 \left(\dfrac{\bar{P}_{CO_2} + P_{H_2O}}{P_{CO} + P_{H_2}} \right)^n} \qquad \begin{array}{l} m = 1-2 \\ n = 4-7 \end{array} \qquad (52)$$

Equation (52) allows for the effect of the $CO:H_2$ ratio in the fresh gas. In the formation of hydrocarbon chains (via constructive reactions) a carbon atom is inserted into a growing chain. Formally, this is a polymerisation and thus the most probable product distribution can be estimated using the laws formulated by Schulz and Flory [193]. The product distribution found in larger scale units with cobalt or iron catalysts roughly corresponds to the calculated, thereby supporting the assumption implicit in the above rate equations; namely, that the rate determining step is the formation of a primary complex from carbon monoxide and hydrogen.

4.4.3 Reaction Conditions

The production of motor fuels has remained the supreme aim of the Fischer-Tropsch synthesis. Other products formed alongside were either used as such or after being processed. They were largely regarded as being undesired by-products. As motor fuel composition allows a certain leeway, demands on the selectivity of the synthesis were nothing like as high as in other processes involving carbon monoxide hydrogenation. If the Fischer-Tropsch synthesis is to assume the role of supplier of chemical or petro-chemical feedstocks, then its low selectivity compared to conventional processes will be a central problem whose solution will be decisive for economic operation. The following summary of Fischer-Tropsch reaction conditions should therefore also consider potential reaction control which is of importance for adjusting selectivity.

Table 4.6 shows the general effect of reaction parameters on the formation of alkanes or alkenes of varying chain length. In addition to the influence of the catalyst product distributions are determined by thermodynamic and kinetic factors which can have a varying effect on the operative range of the catalyst.

The main secondary constraints are temperature and pressure, applicable for a given catalyst, and the parameters associated with process control (synthesis gas content, gas velocity, residence time, heat- and mass-transfer, concentration gradient, etc.). As the catalysts used in the Fischer-Tropsch synthesis are tailored for particular processes, it is difficult to generalize about the complex relationships applying in each case.

Tables 4.7 and 4.8 summarize operating conditions and typical hydrocarbon compositions from several industrial Fischer-Tropsch processes. The optimal and usual temperatures for cobalt-catalyzed processes lie (depending on age and activity) between 170 and 215 °C, the corresponding values with iron catalysts being between 220 and 350 °C. As in all kinetic equations, the reaction rate and thus the space-time yield increase with rising temperature and pressure. The individual reactions leading to hydrocarbon formation have varying activation energies. Higher temperatures can cause an increase in rate of reactions with higher activation energies, thus increasing the selectivity. The average molecular weight of the products decreases with increasing temperature, in agreement with the previously discussed ideas about a sequence of hydrocarbon formation reactions, as the rise in temperature increases the probability of desorption of intermediates, thus limiting growth reactions.

The effect of increasing temperature has several effects on the different processes. In the gas phase fixed-bed process, the increase from 213 to 247 °C causes the wax

Table 4.6. Effect of reaction parameters on product composition in the Fischer-Tropsch synthesis [194]

Process	ARGE		Kellogg or Synthol	
Temperature °C	220–240		320–330	
Pressure bar	26		22	
H_2/CO ratio	1,7:1		3:1	
Product distribution	wt% total	wt% olefins	wt% total	wt% olefins
C_1	7.8	–	13.1	–
C_2	3.2	23	10.2	43
C_3	6.1	64	16.2	79
C_4	4.9	51	13.2	76
C_5–C_{11}	24.8	50	33.4	70
C_{12}–C_{20}	14.7	40	5.1	60
C_{20}	36.2	~ 15	–	–
Alcohols, Ketones	2.3	–	7.8	–
Acids	–	–	1.0	–

Table 4.7. Typical hydrocarbon mixtures from several industrial Fischer-Tropsch processes [157]

Process	Cobalt atmospheric pressure synthesis[b]	Cobalt medium pressure synthesis[a]	Iron medium pressure synthesis[b]	ARGE high load synthesis[c]
Average product composition				
Alkanes/alkenes				
C_1/C_2 hydrocarbons	–	–	5/-	7/-
C_3/C_4 hydrocarbons	8/6	6/4	2/6	5/5
Fractions 30–165 °C	29.5/17.5	19.5/6.6	4.1/8.7	8.5/8.5
165–230 °C	14/3	21.5/2.5	2.4/3.8	5/3.5
230–320 °C	10/1	11/2	8/7	7.6/4.4
320–460 °C	8/-	17/1	16/-	23/-
> 460 °C	3/-	10/-	37/-	18/0-
Oxygen-containing compounds	–	–	–	4
Load (Nm^3/m^3 catalyst · hr)	70–100	100–110	100–110	500–700
Conversion (CO +H$_2$) %	90–95	90–95	85	73
Yield (gC$_{2+}$ per Nm^3 CO + H$_2$ feed)	150–160	150–160	170	140

[a] three stage operation   [b] two stage operation   [c] single stage operation

Table 4.8. Selectivity and reaction conditions of entrained-bed synthesis [158]

Product	% of Converted Carbon
Methane	10
Ethylene	4 } 10
Ethane	6
Propene	12 } 14
Propane	2
Butene	8 } 9
Butane	1
C_5–C_{12}	39
C_{13}–C_{18}	5
C_{19}–C_{21}	1 } 11
C_{22}–C_{30}	3
C_{31}Sn	2
n.a.c[a]	6
Carboxylic acids	1
Temperature °C	300–340
Pressure (bar)	20–23
H$_2$/CO, fresh gas	2.4–2.8
H$_2$(CO, total syn gas	5–6
(CO + H$_2$) conversion (%)	77–85
Recycle ratio	2–2.4
Operational period (d)	approx. 40

[a] n.a.c. = neutral organic oxygen-containing compounds in aqueous product phase

content (mp > 320 °C) to drop from 47% to 17% of the total product. With growing catalyst life, the necessary temperature increase also affects the selectivity of the entrained-bed process (Table 4.9). However, catalyst alterations can also contribute to this effect.

Table 4.9. Effect of operational period on product content (selectivity and olefin content in relation to catalyst life) [158]

Product	Start of operational period		End of operational period		Average	
	% C	Olefin content %	% C	Olefin content %	% C	Olefin content %
Methane	7		13		10	
C_2	7	57	12	25	10	40
C_3	11	90	16	80	14	85
C_4	8	87	11	81	9	85
Light oil	46		39		43	
Heavy oil	14		2		7	
n.a.c.[a]	6		6		6	
Carboxylic acids	1		1		1	

[a] n.a.c. = neutral organic oxygen-containing compounds in aqueous product phase

The rate of the side reactions also increases with rising temperature. This increase is often so drastic that the process operation becomes restricted due to, for example, the cracking of carbon monoxide to carbon and carbon dioxide [Eq. (50)]. Carbon deposition and methane formation increase with growing partial pressures of carbon monoxide and hydrogen respectively. The tendency of the catalyst metals to cause cleavage limits the permitted partial pressure of carbon monoxide, the hydrogen partial pressure being limited by the tendency to form methane. However, high partial pressures of hydrogen are basically desired due to their positive effect on the reaction rate (cf Sect. 4.4.2) and space-time yield. The alkene content decreases on increasing pressure with cobalt catalysts, while with iron catalysts there is only a minor effect on the alkene content of the products. In addition, the formation of oxygen-containing compounds increases with growing pressure (border-line cases – methanol, oxyl synthesis). The carbon monoxide-hydrogen consumption ratio is generally not essentially affected by pressure as the shift conversion ensues without change in mol number.

A $H_2:CO$ ratio of 2:1 is recommended with cobalt catalysts, as the synthesis, in accordance with Eq. (43), takes place with formation of water. A lowering of the $H_2:CO$ ratio is therefore inadvisable; in addition, cobalt catalysts (in contrast to iron catalysts) are only capable of catalyzing the shift conversion to a limited extent. Various synthesis gas contents are common with iron catalysts. The gas phase fixed-bed process operates at a $H_2:CO$ ratio of around 1.7 which approaches the recommended composition. However, H- or CO-rich gases can also be processed ($H_2:CO$ from approximately 0.9 to > 2.5). In contrast, due to the high reaction temperature, the entrained-bed process is operated with H-rich gas (H_2:CO approximately 5–6) in order to suppress carbon deposition and to compensate for the low specific activity of the catalyst. The rate determining step is the reaction between the strongly chemisorbed carbon monoxide and the weakly chemisorbed hydrogen. Consequently, the total reaction rate is largely determined by the partial pressure of the hydrogen. The methane and short chain alkane content increase with growing partial pressure of hydrogen. With iron catalysts, the product distribution can be varied to a very limited extent via the synthesis gas content. Hydrogen deficiency in the primary reaction [Eq. (43)] can be compensated by the subsequent shift conversion [Eq. (44)].

The increase in specific catalyst load or decrease in residence time of synthesis gas on the catalyst hinders the secondary reactions of the primary products – whose concentrations on the catalyst surface are in accordance with adsorption and desorption equilibria – and thus the synthesis of higher molecular hydrocarbons. Reactions involving cleavage of oxygen, hydrogenation of alkenes, polymerisation and cracking become less favoured. Consequently, briefer average residence times result in an increase in the alcohol and alkene contents and in a reduction of the average molecular weight of the products. A structured catalyst surface which encourages mass-transfer intensifies the effect of the residence time while the specific conversion efficiency of the catalyst generally recedes.

The effect of water and carbon dioxide in synthesis gas is limited to the rate of the total reaction. In general, it has no influence on the selectivity. When the gas feed passes through the catalyst bed, the rate lowering effects of increasing concentrations of water and carbon dioxide, together with the simultaneous changes in carbon monoxide and hydrogen concentrations cause an alteration in selectivity. A similar effect can ensue via backmixing in processes with mobile catalysts (entrained-bed, fluidized-bed, suspension). Synthesis gas content, residence time, residence time distribution and recycle operation are of special importance when considering control of optimal selectivity. For example, the fixed-bed synthesis can be operated in a reactor in which the composition of the synthesis gas can be regulated on feeding fresh gas after each stage or in a recycle operation where the recycled mixture is freed from components which react to long chain hydrocarbons in subsequent conversion. The average chain lengths of the products can be limited in both ways.

All catalysts and processes which facilitate the rapid attainment of adsorption and desorption equilibria are potentially suitable for the synthesis of short-chained hydrocarbons. Catalysts with mainly external surfaces are preferable as fine structured catalysts with large internal surfaces (pore system) yield longer chained products under comparable conditions [195]. High gas flow rates and brief residence times also encourage the formation of short-chained products. Thus catalyst, process and desired product spectrum must be balanced against one another to attain optimal results.

4.4.4 Catalysts

As approximately 4,000 publications [157] and patents relating to Fischer-Tropsch catalysts existed already 1954, only a broad outline of their development can be presented here.

The first step in the reaction is the chemisorption of carbon monoxide and hydrogen. Transition metals with $3d$ and $4f$ levels available for bonding are particularly suitable for the chemisorption step. Of all the metals active in the Fischer-Tropsch synthesis (iron, cobalt, nickel and ruthenium [196]), only iron is currently of importance as it has advantages over cobalt with regard to conversion rate, selectivity and flexibility [202]. With nickel, the carbon monoxide hydrogenation yields mainly methane, while with ruthenium (at raised pressure), high molecular weight alkanes (polymethylene) can be synthesized. At low pressure with ruthenium on a special TiO_2 support, short-chained hydrocarbons can be obtained; however, the $CO-H_2$ conversion is maximum 10% [197]. Other platinum metals (Ir, Pd, Pt, Rh, Os) exhibit low specific activity [200, 201].

The suitability of catalysts for the Fischer-Tropsch synthesis seems to be dependent on the presence of metallic or metal-like structures (e.g. carbides, nitrides) as well as on at least minor hydrogenative activity and the ability to form metal carbonyl complexes [199, 203, 204]. Metals such as vanadium, manganese, titanium, molybdenum or chromium have not found any application as base metals as only their oxides are stable under reaction conditions. However, the above elements, along with a number of others, have achieved industrial significance as promoters for cobalt [198] or iron catalysts [205]. Work on suitable catalysts is hampered by lack of knowledge about the actual catalytically active species.

While with nickel and cobalt catalysts, the activity is thought to stem from the metallic components, with iron catalysts the active catalytic phases could not always be unequivocally identified. Iron exhibits a marked tendency to form mixed phases consisting of iron, iron oxide and carbon whose presence could be ascertained under reaction conditions as well as in the fresh catalyst after reduction with hydrogen and subsequent treatment with carbon monoxide. Iron catalysts undergo an alteration in phase composition with lifetime. The former is characterized by a decrease in the carbide and an increase in the oxide content. The free carbon content increases simultaneously [206-210].

The possibility of forming intercalation structures with metalloids (e.g. borides, boron carbides, sulfocarbides, carbonitrides) which is presented by the electronic structure of iron and the combination of iron with base metals, possessing unfilled d orbitals, has been insufficiently studied. Thus, there is considerable scope to influence the selectivity of the Fischer-Tropsch synthesis by modifying iron catalysts.

As with other catalysts, promoters for the Fischer-Tropsch catalysts can be roughly divided into two groups according to their mode of action:

Structural promoters [210-213] affect the formation and stabilize the resulting catalyst structure, in particular its surface.

Electronic promoters can intensify or weaken the interplay between catalyst and reactants.

Structural and electronic effects on the activity and selectivity of a catalyst cannot always be strictly separated. Under certain circumstances they may result from a single promoter (activator).

The specific lattice deformations caused by (difficult-to-reduce) oxides such as SiO_2, Al_2O_3, MgO and ThO_2 are stabilized under reaction conditions. The addition of these oxides during catalyst manufacture — by precipitation or sintering — increases not only the dispersity of the base metal, but also the total surface area and the pore volume. In addition, the union of crystallites, whilst under thermal strain, is largely suppressed. Furthermore, the distribution of the pore radii can be controlled within certain limits [155].

The use of alkali metal salts as electronic promoters for the Fischer-Tropsch synthesis has been known since the outset. Alkali metal cations function as electron donors with metallic iron, facilitating chemisorption of carbon monoxide via the $3d$ electrons of iron. This results in a strengthening of the Fe—C and a weakening of the C—O bond [217]. The following effects can be attained via the doping of iron catalysts with alkali metal ions:

the amount of chemisorbed carbon monoxide and the heat of chemisorption increase, the converse applies to hydrogen [218]

the electron affinity of iron is lowered [219]

the rate of the carbon monoxide consuming reactions increases (synthesis, shift-conversion, carbide formation, carbon deposition) [212]

the average molecular weight of the products increases [155]

The decreased amount of chemisorbed hydrogen present, along with the increased amount of chemisorbed carbon monoxide, results in a drop in methane formation and an increase in the yield of alkenes and oxygen-containing products. While the addition of alkali metal ions assists selectivity control, copper only effects a lowering of the reduction temperature of the catalyst and does not noticeably affect its selectivity [206, 211].

Cobalt catalysts are preferentially manufactured via precipitation. The catalyst used in the German plants had the following content:

100 pbw[1] Co, 8 pbw MgO, 5 pbw ThO$_2$ and 200 pbw Kieselguhr.

The currently interesting catalysts, which are based solely on iron, can be manufactured via precipitation, sintering or melting of oxide mixtures. Precipitated catalysts exhibit high specific surfaces, large pore volumes and high specific activities. On the other hand, sintered catalysts, and to an even greater extent, fused catalysts, possess a low specific surface area, small or even no pore volume and low specific activity [157]. Precipitated catalysts are therefore predestined for low temperature operation (e.g. 220—280 °C), while the low specific activity simultaneously demands and facilitates the application of fused catalysts at high temperatures (e.g. 280—350 °C) where the reaction with precipitated catalysts goes out of control.

4.4.5 Processes

During the 50 years of development of the Fischer-Tropsch synthesis there have been numerous process technologies, the basic difference being the heat removal (approximately 600 kcal/Nm^3 converted synthesis gas). The processes can be classified as follows:

---

1   pbw = parts by weight

fixed-bed reactors with a stationary charge to which synthesis gas is fed
entrained-bed or fluidized-bed reactors with catalysts agitated by synthesis gas
liquid phase reactors with either fixed or mobile catalyst beds
The characteristic process data are presented in Table 4.10. Industrially speaking,
only the gas phase fixed-bed process (Fig. 4.17) and the entrained-bed process (Fig.
4.18) have achieved significance. The ARGE process, developed by Ruhrchemie-

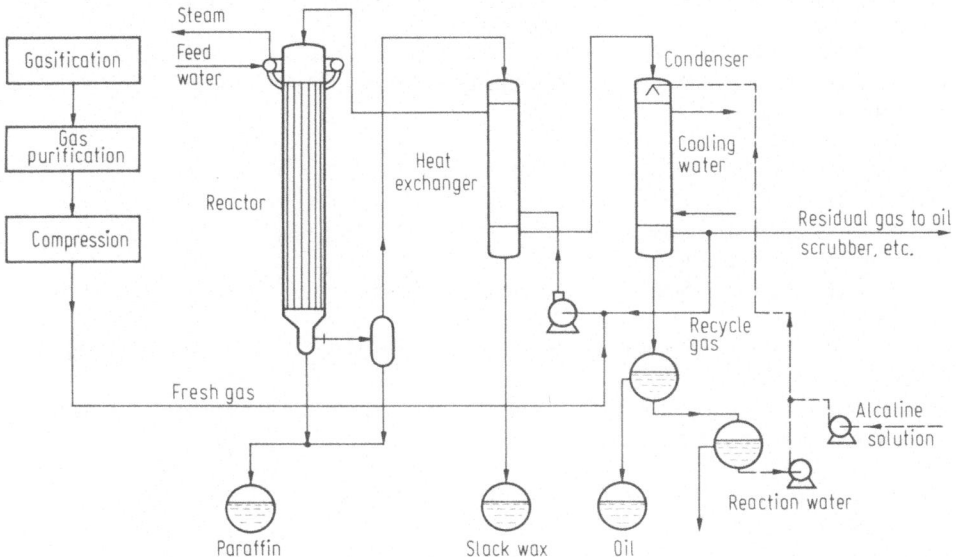

Fig. 4.17. Process scheme for the gas phase fixed-bed process

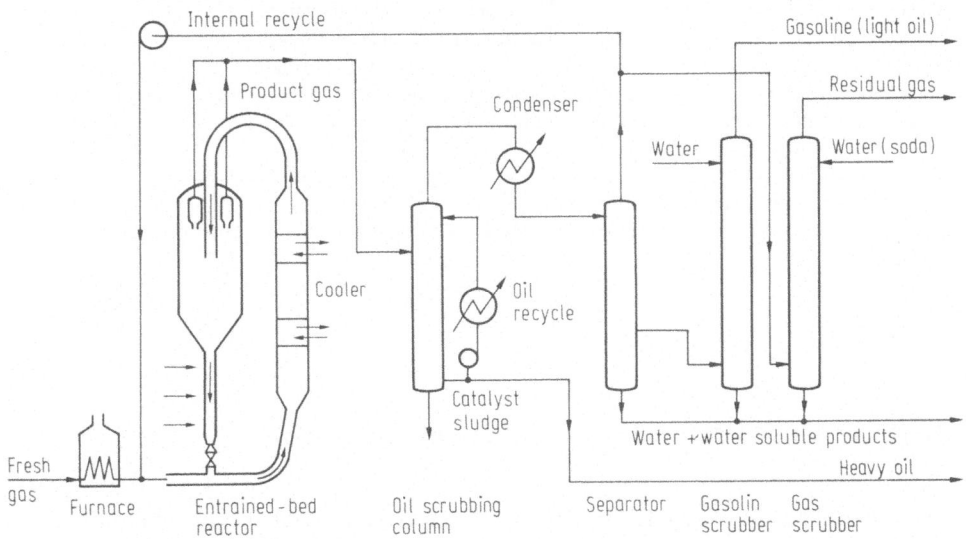

Fig. 4.18. Process scheme for entrained-bed synthesis [158]

Lurgi, and the entrained-bed processes, developed and licenzed by M.W. Kellogg, are operated by Sasol in South Africa. Iron catalysts are utilized in both processes.

Table 4.10. Characteristic data of different Fischer-Tropsch processes [72]

Characteristic data	Reactor type	Gas/solid			Gas/solid/liquid
		Fixed bed	Entrained fluid bed	Fluidized bed	Bubble reactor
Heat transfer velocity or heat removal through transferring surfaces		Slow	Medium up to high	High	High
Actual heat conductivity within the system		Poor	Good	Good	Good
Max. reactor diameter as limited by heat removal		Approx. to 8 cm[b]		No limitation	
Pressure drop at high gas velocity		Small	Medium	High	Medium up to high
Residence time distribution of the gaseous phase		Narrow	**Narrow**	**Broad**	Narrow up to medium
Axial mixing of the gas		Small	Small	Large	Small up to medium
Axial mixing of the solid catalyst		None	Small	Large	Large
Catalyst concentration as volume portion of solid $(1 - \epsilon)$[a]		0.55−0.7	0.01−0.1	0.3−0.6	Up to max. 0.6
Particle size range of the solids, mm		1−5	0.01−0.5	0.03−1	0.01−1
Mechanical stress of the solid by shock or friction		None	Great	Great	Small
Catalyst losses		Non	2−4% per day due to abrasion	Not recoverable discharge due to abrasion	Small
Regenerability or exchangeability of the catalyst during synthesis		Interruption of synthesis necessary	Without interruption of synthesis by continuous purge and feed		

[a] ϵ = Relative, solid free particle interspace
[b] A small increase seems to be possible if the heat transfer can be increased by higher gas velocities

4.4.6 Economic Potential and Possible Developments

The use of the Fischer-Tropsch synthesis to produce coal-based heating oil or motor fuels, as an alternative to mineral oil, can be only economic in countries with access to extremely cheap coal. As far as is known, the sole commercial plant is situated in

South Africa at Sasolburg were a second unit is currently being constructed. The new plant, which is to have a planned capacity of 2.2 million tons/annum primary products, is expected to go on stream in 1981. At the moment, planning-stage talks are being held about the possibility of erecting a motor fuel production plant in Australia.

Using the Fischer-Tropsch synthesis to manufacture substitutes for oil-based chemical feedstocks looks more economically promising than the production of fuel oil or motor fuels. This aspect could justify the application of more expensive coals.

Two ways of applying the FischerTropsch synthesis to produce chemical feedstocks have been mainly studied:
The manufacture of a naphtha substitute (which could be processed to olefins in conventional cracking plant) and
the direct synthesis to ethylene, propylene and to a certain extent, butylenes via the single stage reaction between carbon monoxide and hydrogen.

The production of C_5–C_{11} hydrocarbons as a naphtha substitute does not introduce any basic problems and would be feasible by slightly displacing the product spectrum from the entrained-bed process. However, the (actually desired) olefin content of the primary product (Table 4.9) must be hydrogenated — demanding a separate step — before being used as feedstock in cracking plants, thereby making the economics of the whole process unfavourable. However, numerous research and development projects have recently been started which already indicate ways in which the problem could be solved [220–223]. The direct synthesis of short-chained alkenes — in particular ethylene — via the hydrogenation of carbon monoxide demands the development of new highly selective catalysts and their adaption to special process conditions [224]. Data in older literature about the highly selective formation of ethylene via metallic or oxidized catalysts could not be confirmed [225–229, 233]. Recent work has concentrated on matching the metal with activators and structure [222, 230–232] in order to restrict the product spectrum to short-chained hydrocarbons with high alkene content [234, 235]. Neither nickel nor cobalt can be noticeably improved in this way, either by the type of catalyst manufacture or on adding promoters. Despite possessing high specific activites, mainly saturated products result with these metals. On the other hand, with iron catalysts, containing varying amounts of elements of the IVth to the VIIth transition group, it was possible to limit the product spectrum and to obtain short-chained alkenes in relatively high yield [222]. Some results are presented in Table 4.11 [220].

The addition of titanium oxide, vanadium oxide, molybdenum oxide, tungsten oxide or manganese oxide to iron effects a lowering of the alkane yield and, together with the high reaction temperature, leads to an almost complete suppression of hydrocarbon formation above C_5 [236]. The activity loss (compared to pure iron catalysts) stemming from the above activators can be only partially compensated by manufacturing techniques. Consequently, temperatures above 300 °C are generally necessary for satisfactory conversions. The lifetime of this catalyst, in its most active and selective state, is insufficient as relatively rapid deactivation ensues via carbon deposition. Selectivity improvements via the addition of cobalt or manganese to iron catalysts have also been reported [222].

In experiments aimed at increasing catalyst selectivity, a series of questions remain unanswered with regard to the main effects. The type and amount of activator probably

357

Table 4.11. Reaction conditions, yields and product distribution with the new selective Fischer-Tropsch catalysts I to V [220]

Product yield (g/Nm³ Synthesis gas feed)	I	II	III	IV	V
CH_4	22.1	20.2	21.1	36	28
C_2H_4	11.5	16.2	21.2	3.1	24
C_3H_6	14.3	25.3	31.7	10.8	27
C_4H_8	11.5	16.2	16.9	15.2	17
C_5H_{10}				8.4	8
$\Sigma C_2-C_4(C_5)$-Alkenes	37.3	57.7	69.8	37.5	76
$\Sigma C_2-C_4(C_5)$-Alkenes	8.1	9.3	8.5	58	26
Process Reaction conditions	Gas phase fixed-bed			Liquid phase	
Temperature	360	340	280	290	350
Pressure	10	10	10	12	20
CO:H₂ ratio	1	1	1	1.5	1
Yield (g/Nm³) (without CH_4 incl. C_5+)	178.4	132	103	95.5	127
Conversion (%)	87.6	63.1	49.5	91	83

affect the sorption properties of carbon monoxide and hydrogen. This aspect has been fairly well studied. On the other hand, corresponding measurements about the, at least relative, ad- and desorption equilibria of reactive intermediates have been almost ignored. Besides the basic catalyst composition, the type of manufacture is apparently also significant. For example, with catalysts obtained from the thermal decomposition of complex salts on supports with large surface area, relativity high yields of lower alkanes were obtained [234, 235], while attempts to limit growth using molecular sieves as supports (shape selective catalysts) – analogous to steam cracking – were unsuccessful [237]. According to current ideas, lower alkenes are thought to be intermediates from synthetic reactions on the catalyst surface which must be suppressed as far as possible to control selectivity [239]. The topography of the catalyst surface and the related mass-transfer possibilities play an important role [237]. Their effect has not been clarified to any extent.

The external reaction conditions which favour mass-transfer – high temperature, brief residence time of high flow rate, low pressure – also cause an increase in selectivity [238].

The main objective for future developments is to increase selectivity via a narrowing of the resulting product spectrum. Concepts about the 'actual' reaction mechanism and primary and secondary steps as well as selectivity control are by no means uniform.

4.5 Polymethylene Synthesis

4.5.1 General Remarks

During the initial studies about the suitability of the platinum metals as catalysts for the (atmospheric pressure) hydrogenation of carbon monoxide [241], ruthenium was found to be the most active; however, methane was virtually the only reaction product. The excellent methanation properties of ruthenium, which are superior to nickel in terms of activity, were a deciding factor in its application in the carbon monoxide hydrogenation. The development of the polymethylene synthesis was based on the superior properties of ruthenium compared to the other noble metals under pressure [242]. With Pt, Pd and Ir only a small amount of hydrocarbons higher than methane, was formed, while with Os, liquid and solid hydrocarbons result. Rh yields low molecular oxygen-containing products, while at sufficiently high pressure with Ru, the formation of higher paraffins is favoured at the expense of methane formation. Consequently, they can become the main product from the carbon monoxide hydrogenation. Ru is the only noble metal suitable for the formation of higher hydrocarbons with melting points up to 131 °C and average molecular weights around 23,000. Their physical properties closely resemble those of polyethylene manufactured according to the Ziegler process. The term 'polymethylene' has been coined for the high molecular weight hydrocarbons synthesized directly via carbon monoxide hydrogenation [243].

Although the polymethylene synthesis was discovered in 1938 [240] — much earlier than the polyethylene synthesis — it has made no industrial impact. There are probably two main reasons — the initially used catalysts possessed only minor activity, permitting only very low space-time yields. In addition, pressures up to 2,000 bar were necessary, in order to produce noticeable amounts of polymethylene along with other products [244].

The development of more active catalysts facilitated a lowering of the reaction temperature whilst maintaining an adequate synthesis gas conversion. This enabled a basic improvement in the selectivity of polymethylene formation. Later work, which led to improvements in the synthesis, ran roughly concurrent with the development of the polyethylene synthesis and its industrial realisation. Thereafter, the polymethylene synthesis remained far behind the high and low pressure polyethylene processes in terms of development, work on the former being continued by only a few research groups [245–248]. Industrial application has recently been reconsidered and initial research work has been started.

4.5.2 Reaction Mechanism

Under pressures and temperatures favourable for the polymethylene synthesis, carbon monoxide hydrogenation takes place largely according to the following Eq.

$$n\,CO + 2\,n\,H_2 \rightleftharpoons n\,CH_2 + n\,H_2O \qquad (53)$$

The reaction enthalpy tends to a limiting value with increasing chain length of the resulting hydrocarbons – amounting to around 35 kcal ($\sim$146 kJ) per mol of resulting CH_2 group. The side reactions (cf. Sect. 4.4.2), which occur during the Fischer-Tropsch hydrocarbon synthesis, are of minor importance with ruthenium catalysts. The shift conversion which ensues with iron catalysts

$$CO + H_2O \rightleftharpoons CO_2 + H_2 \qquad\qquad \Delta H_{298} = -9.8 \text{ kcal/mol} \qquad\qquad (54)$$
$$(-41.0 \text{ kJ/mol})$$

or the deposition of elementary carbon via decomposition of carbon monoxide (Boudouard reaction) consumes only a small portion of the carbon monoxide feed as a consequence of the relatively low reaction temperature ($< 200\,^{\circ}C$). At temperatures above $200\,^{\circ}C$, carbon monoxide can be used instead of synthesis gas if the reaction is conducted in aqueous suspension.

$$3\,n\,CO + n\,H_2O \rightleftharpoons n\,CH_2 + 2\,n\,CO_2 \qquad\qquad (55)$$

The shift conversion [Eq. (54)] occurs before the actual synthesis.

Proposals for the mechanism of the polymethylene synthesis with ruthenium catalysts exhibit marked analogies to the mechanisms (cf. Sect. 4.4.2) discussed for the Fischer-Tropsch synthesis. So far, no unequivocal decision can be made in favour of one of the suggested interpretations which allows ready explanation of all experimental findings. Enolic complexes or carbonyl structures $M\,(CO)_x$ (x = unspecified number of CO ligands) can both be considered for the primary step in chain growth which is preceded by the competitive chemisorption of hydrogen atoms and carbon monoxide molecules.

M = metal atom of catalyst

The tendency of ruthenium carbonyls to form during the carbon monoxide hydrogenation with Ru catalysts at high CO partial pressure and low temperature (e.g. 300 bar, $120\,^{\circ}C$) is without doubt greater than that of iron carbonyls during the Fischer-Tropsch synthesis (e.g. 10 bar, $300\,^{\circ}C$) [250]. As ruthenium carbonyls do not catalyze the polymethylene synthesis [249], their formation must be suppressed to ensure trouble-free operation without catalyst losses. Ruthenium carbonyls and ruthenium carbonyl-hydrogen compounds have been found in the products of the polymethylene synthesis. These compounds become discoloured on longer exposure to air [243, 249]. The colourless components were assumed to be $Ru(CO)_5$ or $H_2Ru(CO)_4$, the coloured ones being apparently $Ru_3(CO)_{12}$ (orange) or $H_4Ru_3(CO)_{11}$ (yellow).

Ruthenium catalysts, which were completely reduced to the metal under mild conditions, exhibited almost no tendency to form carbonyls under reaction conditions. It was assumed that, while surface carbonyl complexes result with such catalysts, they did not lead to removal of metal atoms and thus to corrosion of the catalyst [243, 249].

According to this theory the reaction mechanism can be intepreted in the following manner (Fig. 4.19) in contrast to the above mechanism involving enolic primary complexes.

The group R, which initially results or stems from the reaction mixture, is readsorbed then extended by one carbon atom via insertion of a CO molecule into the bond at a surface metal atom. The intermediate can be converted into an aldehyde via partial- or into an alcohol via total hydrogenation then desorbed. This explains the relatively high content of oxygen-containing compounds in the low molecular products from the polymethylene synthesis. The desorption of the olefin intermediate (resulting from partial hydrogenation or dehydration of the intermediate complex) necessary for the growth steps in the Fischer-Tropsch synthesis is apparently almost insignificant as the products are virtually olefin-free. The high hydrogenative activity of ruthenium catalysts and low reaction temperatures, which do not encourage desorption of the olefinic intermediates, may be the reasons.

The composition of the low molecular products from the polymethylene synthesis alters in a characteristic manner with reaction temperature. Although polymethylene is

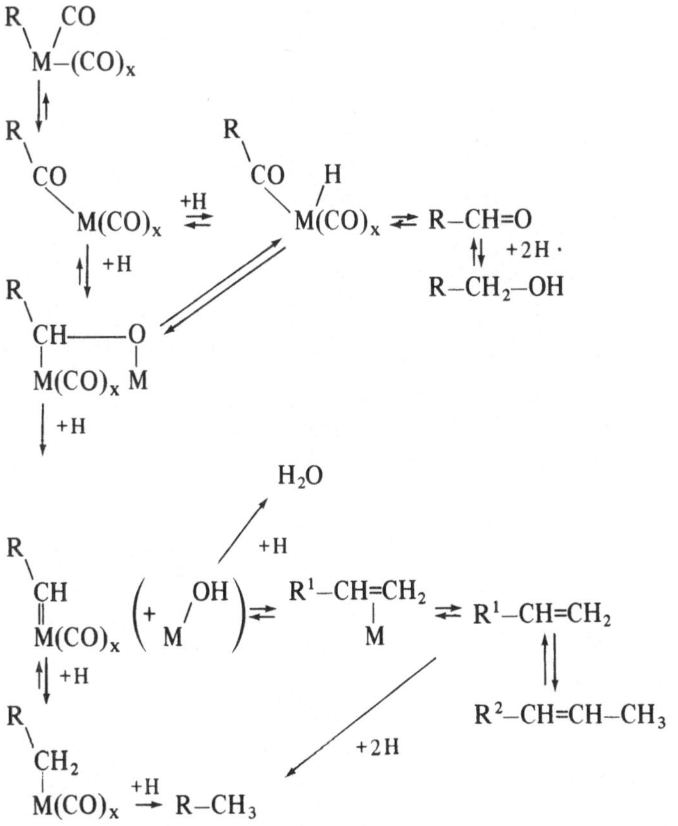

Fig. 4.19. Mechanism of the polymethylene synthesis [244, 251]

M = metal atom from catalyst surface

X = unspecified number of CO ligands

the main product obtained at low reaction temperatures, the content of substances with functional groups (e.g. aldehydes, alcohols, carboxylic acids) in the low molecular products is very high at 100 °C (Fig. 4.20). At 120 °C, the paraffin content increases noticeably at the expense of compounds with functional groups (Fig. 4.21), then at 140 °C the latter are almost completely suppressed in favour of the paraffins (Fig. 4.22).

The long-chained hydrocarbons, formed at higher temperatures, are also free of functional groups.

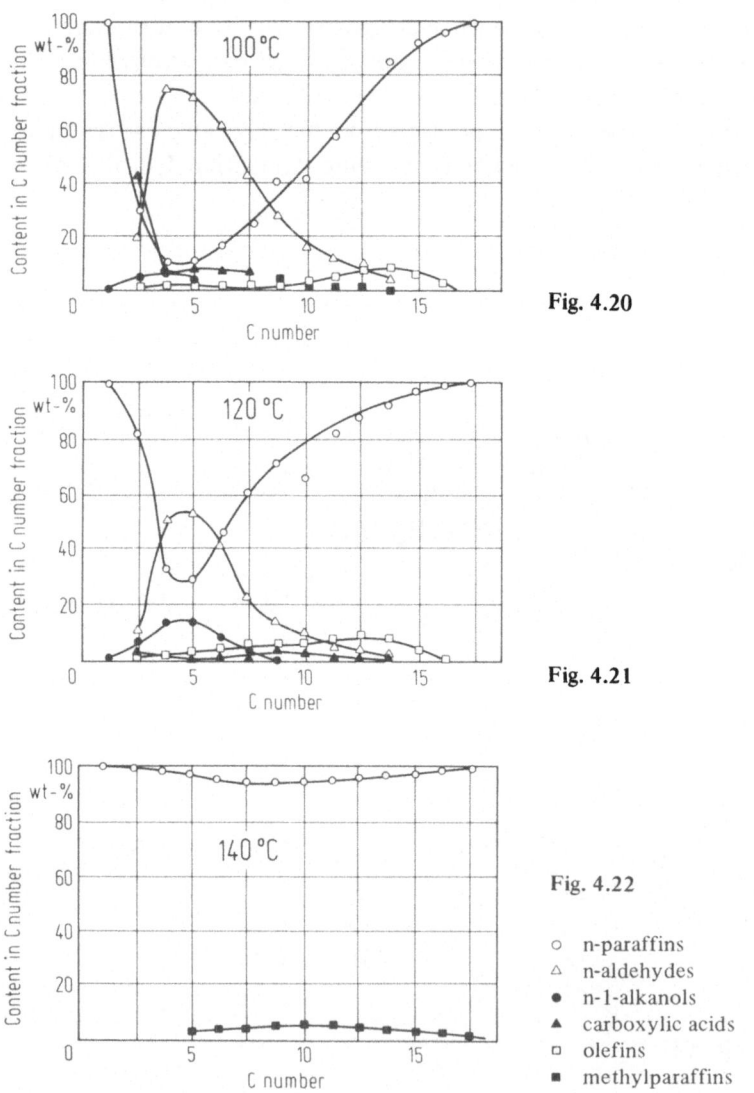

Fig. 4.20

Fig. 4.21

Fig. 4.22

   ○ n-paraffins
   △ n-aldehydes
   ● n-1-alkanols
   ▲ carboxylic acids
   □ olefins
   ■ methylparaffins

Fig. 4.20–4.22. Content of various classes of compounds (according to C number) on altering reaction temperature [244]

It is assumed that the growing probability (with an increasing chain length) of readsorption of intermediates with reactive groups, causes further growth steps which are terminated by the irreversible hydrogenation step. Thus, saturated high molecular weight hydrocarbons dominate in the product [244].

The product distribution and its dependence on reaction conditions can also be explained using carbonyl structures other than the ones mentioned. For example, it could ensue via hydrogenation of chemisorbed carbon monoxide to CH_2 groups, which could be converted into longer chained products via polymerisation, thereby circumventing the mechanistically complicated insertion step. A transition between the two different mechanisms is worth considering on account of the narrow temperature range (approx. 40 °C) in which the relative content of the functional group compounds (in the individual C number fractions) are drastically lowered (Fig. 4.20–4.22). The reason being that in the Fischer-Tropsch synthesis, which is often used for mechanistic comparison, an equivalent alteration in the product spectrum is unknown. Thus, for both syntheses, the structure of the primary complex and its secondary reactions must be established.

4.5.3 Catalysts

Ruthenium distinguishes itself from other catalyst metals in that it catalyzes the hydrogenation of carbon monoxide at temperatures as low as 100 °C. In addition, when high pressures are applied, high molecular weight hydrocarbons result. No other catalyst metal exhibited comparable properties under like conditions. The catalysts initially tested during the development of the polymethylene synthesis only possessed sufficient activity when the temperature was slightly below 200 °C. It was soon recognized that the selectivity of the reaction to long-chained paraffins could be increased not only on using higher pressures but also by lowering the reaction temperature. The object then became the development of more active ruthenium catalysts [254]. Consequently, the lowest temperature for the synthesis became a measure of catalyst performance. Now, conventional catalysts are active below 100 °C [253].

The best catalytic properties are exhibited by highly dispersed metallic ruthenium. In catalyst manufacture, metallic ruthenium is melted in a fused KOH/KNO_3 mixture, where it is converted into soluble K_2RuO_4 which is reductively precipitated to RuO_2 in aqueous methanolic solution [252, 255, 256]. The Ru(IV) oxide hydrate is active above 140 °C and exhibits good properties over a long period. In addition, the activity of Ru oxide hydrate is increased after being exposed to γ-radiation [253] not, however, when the metal (from the reduction of the oxide hydrate) is exposed [249].

Studies relating the ease of reduction of Ru(IV) oxide hydrates with synthesis gas (under pressure) led to the conclusion that below 140 °C incomplete reduction ensues and Ru carbonyls result. Carbonyl formation decreases with the degree of reduction of ruthenium and the catalytic activity increases concurrently [257]. Thus a careful pre-reduction of the catalyst is advantageous. Totally reduced catalysts tend not to form carbonyls, even at the high CO partial pressures at which carbonyls are stable.

Very active catalysts (temperature ~90 °C) were obtained on distilling RuO_4 followed by reduction in aqueous solution to Ru(IV) oxide hydrate. Thereafter, the oxide hydrate was thoroughly heated in vacuum before being reduced with hydrogen to the metallic catalyst. Very active catalysts were also obtained on reducing Ru(III) chloride hydrate with hydrogen; however, there was a tendency to form colloidal suspensions. Very active supported catalysts were also manufactured by decomposing Ru carbonyls in the presence of oxidized supports (SiO_2, Al_2O_3) or active carbon [254].

The effects of carrier materials and promoters on the catalyst performance, as is known from the Fischer-Tropsch synthesis and hydrogenations, are not found with ruthenium catalysts in the polymethylene synthesis. Electronic or structural factors which might facilitate the influencing of selectivity or activity have not been established. The performance of the catalyst in the polymethylene synthesis is determined by a high degree of reduction and a high dispersity of ruthenium (surface area = $10-20$ m^2/g).

Catalyst developments led to considerable improvements in their performance. Initially, conversions of approximately 2–3 g CO/g Ru · h were obtained at 180 °C/ 100 bar, which rose to around 15 g CO/g Ru · h at 120 °C/1,000 bar. The catalyst life-time is impaired by halogen or sulfur compounds, by incomplete reduction and by surface blockage via high molecular compounds. In one case, the synthesis was operated for over a year and a half using the same catalyst charge.

4.5.4 Reaction Conditions

Methane is the main product on employing Ru catalysts at pressures up to ~30 bar. Low temperature and high pressure are essential in order to kinetically and thermo-dynamically suppress chain termination reactions.

While the reaction velocity decreases with falling temperature, the selectivity of formation of high molecular products increases. Reaction temperatures around 90 °C are possible with very active catalysts. The specific carbon monoxide conversion and the selectivity of formation of high molecular products increase with rising pressure. The usual pressures employed are 1,000–2,000 bar. While there is only a very slight rise in reaction velocity above approximately 2,000 bar, there is an increase in the average molecular weight of polymethylene.

Carbon monoxide can be hydrogenated either in the presence or absence of solvents. If temperatures above the melting points of the resulting long-chain paraffins are used, then they melt covering the catalyst surface and retarding the chemisorption of the synthesis gas components. The presence of a solvent ($C_{10}-C_{20}$ hydrocarbons), subject to good mixing, is favoured as it improves the diffusion controlled mass-transfer in the three phase reaction system. The catalyst is either in a fixed-bed cooled by circulating solvent, the high molecular components being extracted from the catalyst surface or is suspended in a stirring vessel from which part of the solvent-product mixture is continuously extracted. The resulting separated catalyst is recycled to the reaction chamber with fresh solvent.

The synthesis gas conversion depends on pressure, temperature, mixing and catalyst activity. The specific conversion is around 0.05–0.1 g CO/g Ru · h at 1,000 bar/

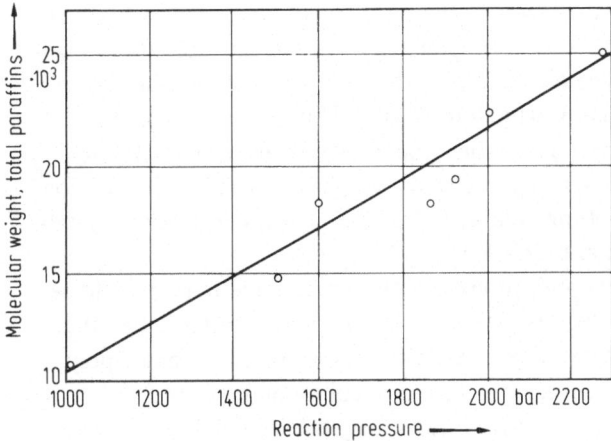

Fig. 4.23. Average molecular weight of resulting paraffins (polymethylene) in relation to pressure [244]

120 °C (agitated autoclave). The resulting product consists of approximately 14% CH_4, 25% short-chained and approximately 60% long-chained products – relative to the converted carbon. On raising the temperature to 140 °C, the specific CO conversion is doubled and the content of the long-chained products drops to around 30% (relative to converted carbon) in favour of the formation of methane and other lower molecular products. Specific conversion rates (relative to polymethylene) in the range 0.19/g Ru · h are feasible.

4.5.5 Economic Potential and Possible Developments in the Polymethylene Synthesis

The polymethylene synthesis has only been studied on laboratory scale by a few research groups. Some typical features are shown below [244]:

Reaction temperature	100–120 °C
Synthesis gas pressure	1000–2000 bar
Catalyst	metallic Ru
CO conversion	0.05–0.2 g CO/g Ru · h
Selectivity (% of converted C)	
Methane	5–15
Liquid saturated hydrocarbons	5–12
Alcohols/aldehydes	6–9
Solid saturated hydrocarbons	60–75
CO_2	1–2

Up to 40% of the solid saturated hydrocarbons ($\sim$25% of converted carbon) have melting points between 129–134 °C – corresponding to an average molecular weight between 5000–23000. Approximately 8–10% of the converted carbon is in the fraction with an average molecular weight above 20000. This polymethylene has similar physical data and possesses comparable properties to Ziegler's high density polyethylene which also exhibits a similar molecular weight distribution. In addition, the densities ($\sim$0.975 g/cm^3), melting points ($\sim$132–138 °C) and degrees of crystallinity closely resemble one another [258–261].

If one assumes, that as far as applications are concerned, polyethylene could be substituted by polymethylene, which is currently not completely certain, then the carbon monoxide hydrogenation could present an alternative to the oil-based polyethylene manufacture. The basic advantage, the use of almost all types of fossil carbon to produce carbon monoxide or synthesis gas, and thus greater flexibility in choice of feedstocks, is confronted by the insurmountable disadvantages inherent in the polymethylene synthesis – high pressure, low selectivity of formation of value products and low space-time yield. Compared to the established polyethylene processes, the polymethylene synthesis would appear, at present, to have little prospect of achieving an economic breakthrough. However, future results could well introduce new factors which might make the synthesis economically attractive.

So far, no metal other than ruthenium has proved to be sufficiently active and selective for the carbon monoxide hydrogenation. Experiments aimed at modifying ruthenium catalysts with promoters or supports have been disappointing [262].

4.6 References

1. DRP 293,787 (1913) BASF Inv. Mittasch, A., Schneider, C.
2. Hirst, L. L. in Lowry, H. H.: Chemistry of Coal Utilisation, John Wiley & Sons, London 1945
3. Schmidt, J.: Das Kohlenoxid, Akadem. Verlagsges. Geest & Portig KG, Leipzig 1950
4. Kasten, M. L., Dudley, J. F., Troeltzsch, J.: Ind. Eng. Chem. *40*, 2230 (1948)
5. Winnacker-Weingärtner: Chem. Technologie, Organ. Technologie I, 459, Carl-Hanser-Verlag, München 1952
6. Winnacker-Küchler: Chem. Technologie, Vol. 3, Organ. Technologie I, 437, Carl-Hanser-Verlag, München 1959
7. Marschner, F., Möller, F. W., Schulze-Bentrop, R. in Falbe, J. (Ed.): Chemierohstoffe aus Kohle, 300, Georg-Thieme-Verlag, Stuttgart 1977
8. Schwarzmann, M.: Chem.-Ing.-Techn. *47*, 56 (1975)
9. Mills, G. A., Harney, B. M.: Chemtech. *4*, 26 (1974)
10. Ewell, R. H.: Ind. Eng. Chem. *32*, 147 (1940)
11. Palm, A. in Winnacker-Küchler: Chem. Technologie, Vol. 3, Organ. Technologie, 359, Carl-Hanser-Verlag, München 1971
12. Natta, G., Pino, P., Mazzanti, G., Pasquon, L.: Chim. e Ind. *35*, 705 (1953)
13. Natta, G., Mazzanti, G., Pasquon, L.: Chim. e Ind. *37*, 1015 (1955)
14. Natta, G. in Emmett, P. H. (Ed.): Catalysis, Bd. 3, 349, Reinhold Publ. Corp., New York 1955
15. Bolton, D. H.: Chem.-Ing.-Techn. *41*, 129 (1969)
16. Liebgott, H. E., Herbert, W., Baron, G.: Erdöl-Kohle-Erdgas-Petrochemie *25*, 75 (1972)

17. Rogerson, P. L.: Chem. Engng. *80*, 112 (1973)
18. Shah, M. J., Stillmann, R. E.: Ind. Eng. Chem. *62*, 59 (1970)
19. Natta, G.: Giorn. Chimici *12*, 1 (1930)
20. Storch, H. H.: J. physic. Chem. *32*, 1743 (1928)
21. Münzing, E.: Chem. Techn. *16*, 98 (1964)
22. Münzing, E.: Chem. Techn. *17*, 460 (1965)
23. Gosh, J. C., Sastri, M. V., Kamath, G. S.: J. Chim. physique *49*, 502 (1952)
24. Brunauer, S., Emmett, P. H., Teller, E.: J. Amer. Chem. Soc. *60*, 309 (1938)
25. Yamaguchi, S.: Z. physik. Chem. 7, 115 (1956)
26. Spindler, H.: Chem. Techn. *18*, 463 (1966)
27. Uchida, H., Oba, M., Araki, M.: Bull. Chem. Soc. Japan *38*, 1993 (1965)
28. Kotowski, W.: Chem. Techn. *15*, 204 (1963)
29. Faltin, H.: Die Technik *3*, 462 (1948)
30. US 4.031.123 Chem. Systems Inc. (Inv. Espino, R. L., Pletzke, T. S.)
31. Mauldin, C. H.: C. A. *87*, 101880 p (1977)
32. Ogino, Y., Tani, M.: J. Chem. Soc. Japan 1883 (1975)
33. Soedjanto, P., Schaffert, F. W., Mason, N. C. M.: GWF-Gas, Erdgas *116*, 279 (1975)
34. Winter, C., Kohl, A.: Chem. Eng. *80*, 233 (1973)
35. Chang, C. D., Silvestri, A. J.: J. Catalysis *47*, 249 (1977)
36. Meisel, S. L., Mc Cullough, J. P., Lechthaler, C. H., Weisz, P. B.: Chemtech *6*, 86 (1976)
37. Rottig, W. in Falbe, J. (Ed.): Chemierohstoffe aus Kohle, 323, Georg-Thieme-Verlag, Stuttgart 1977
38. GB 655,237 (1948) E. I. Du Pont, Inv. Gresham, W. F.
39. US 2,636,046 (1948) E. I. Du Pont, Inv. Gresham, W. F.
40. US 2,534,018 (1949) E. I. Du Pont, Inv. Howk, B. W., Hagar, G. F.
41. US 2,570,792 (1949) E. I. Du Pont, Inv. Gresham, W. F.
42. Rottig, W. in Falbe, J. (Ed.): Chemierohstoffe aus Kohle, 328, Georg-Thieme-Verlag, Stuttgart 1977
43. Spitz, P. H.: Chemtech. *7*, 295 (1977)
44. US 2,451,33 (1945) E. I. Du Pont, Inv. Gresham, W. F., Brooks, R. E.
45. Cornils, B. in Falbe, J. (Ed.): Chemierohstoffe aus Kohle, 331, Georg-Thieme-Verlag, Stuttgart 1977
46. US 3,833,634 (1972), Union Carbide, Inv. Pruett, R. L., Walker, W. E.
47. US 3,878,214 (1973), Union Carbide, Inv. Walker, W. E., Brown, E. S., Pruett, R. L.
48. US 3,878,290 (1973), Union Carbide, Inv. Walker, W. E., Brown, E. S., Pruett, R. L.
49. US 3,878,292 (1973), Union Carbide, Inv. Walker, W. E., Brown, E. S., Pruett, R. L.
50. US 3,940,432 (1974), Union Carbide, Inv. Walker, W. E., Cropley, J. B.
51. US 3,944,588 (1975), Union Carbide, Inv. Kaplan, L.
52. DE-OS 2.743.630 (1977) Union Carbide, Inv. Kaplan, L.
53. US 3,948,965 (1974), Union Carbide, Inv. Cawse, J. N.
54. US 3,952,039 (1974), Union Carbide, Inv. Walker, W. E., Bryant, D. R., Brown, E. S.
55. US 3,957,857 (1974), Union Carbide, Inv. Pruett, R. L., Walker, W. E.
56. Chine, P., Martinengo, S.: Inorg. Chim. Acta *3*, 299 (1969)
57. US 3,974,259 (1974), Union Carbide, Inv. Brown, E. S.
58. US 3,989,799 (1974), Union Carbide, Inv. Brown, E. S.
59. US 4,013,700 (1974), Union Carbide, Inv. Cawse, J. N.
60. US 4,001,289 (1974), Union Carbide, Inv. Dougherty, S. J., Wolls, R. C.
61. Muetterties, E. L.: Science *196*, 839 (1977)
62. Homogene Kohlenmonoxid-Hydrierung mit Übergangsmetall-Katalysatoren; Berger, M.: Dissertation RWTH Aachen 1977
63. DE-AS 1,545,463 (1966) Metallgesellschaft AG Inv. Baron, G. et al.
64. Keen, D., Parry, M. R.: Erdöl, Kohle, Erdgas, Petrochem. *28*, 137 (1975)
65. Liesen, K.: GWF-Gas, Erdgas *115*, 369 (1974)
66. Richardson, J. T.: Hydrocarbon Proc. *52*, 91 (1973)
67. Schulz, G., Gründler, K. H., Hiller, H.: Chem.-Ing.-Techn. *45*, 704 (1973)

68. Crossland, S.: Hydrocarbon Proc. *51*, 89 (1972)
69. Hart, F.E., Baker, N.C., Williams, I.: Hydrocarbon Proc. *51*, 94 (1972)
70. Jockel, H., Triebskorn, B.E.: Hydrocarbon Proc. *52*, 93 (1973)
71. Schulz, G., Hiller, H.: Erdöl, Kohle, Erdgas, Petrochem. *26*, 11 (1973)
72. Fischer, F., Tropsch, H.: Ber. *59*, 830 (1926)
73. Kölbel, H. in Winnacker, K. u. Küchler, L.: Chemische Technologie; Vol. 3, Organ. Technologie, 439, Carl-Hanser-Verlag, München 1959
74. Roelen, O. et al. in Ullmanns Encyklopädie d. techn. Chemie, 3. edit., S. 684, Urban & Schwarzenberg, München-Berlin 1957
75. Edminster, W.C.: Hydrocarbon Proc. *52*, 109 (1973)
76. Ludwig, F.: Erdöl, Kohle, Erdgas, Petrochem. *25*, 711 (1972)
77. Pattas, E., Neumann, K.-K.: Chemie-Techn. *2*, 215 (1973)
78. Neumann, K.-K.: Chemiker-Ztg. *97*, 492 (1973)
79. Stein, W.A.: Chemiker-Ztg. *97*, 85 (1973)
80. Frohning, C.D., Hammer, H. in Falbe, J. (Ed.): Chemierohstoffe aus Kohle, 174 f., Georg-Thieme-Verlag, Stuttgart 1977
81. Rostrup-Nielsen, J.R.: J. Catalysis *27*, 343 (1972)
82. Vannice, M.A.: J. Catalysis *37*, 462 (1975)
83. Vannice, M.A.: Catal. Rev. - Sci. Eng. *14*, 153 (1976)
84. Wentreek, P.R., Wood, B.J., Wise, H.: J. Catalysis *43*, 363 (1976)
85. Araki, M., Ponec, V.: J. Catalysis *44*, 439 (1976)
86. Sexton, B.A., Somorjai, G.A.: J. Catalysis *46*, 167 (1977)
87. Eischens, R.P., Pliskin, W.A., Frances, S.A.: J. Phys. Chem. *60*, 194, (1956)
88. Blyholder, G.: J. Phys. Chem. *68*, 2772 (1964)
89. Ozin, G.A.: Acc. Chem. Res. *10*, 21 (1977)
90. Hulse, J.E., Moskovits, M.: Surf. Sci. *57*, 125 (1976)
91. Darydow, A.A., Bell, A.T.: J. Catalysis *49*, 332 (1977)
92. Darydow, A.A., Bell, A.T.: J. Catalysis *49*, 345 (1977)
93. Jones, A., McNicol, B.D.: J. Catalysis *47*, 384 (1977)
94. Kraemer, K., Menzel, D.: Ber. Bunsenges. Phys. Chem. *78*, 591 (1974)
95. Wedler, G., Papp, H.: Z. Phys. Chem. (N. F.) *82*, 195 (1972)
96. Conrad, H. et al.: Surf. Sci. *43*, 462 (1974)
97. Eischens, R.P.: Adv. Catalysis *10*, 1 (1958)
98. Dalla Betta, R.A.: J. Phys. Chem. *79*, 2519 (1975)
99. Wedler, G., Papp, H., Schroll, G.: J. Catalysis *38*, 153 (1975)
100. McKee, D.W.: J. Catalysis *8*, 240 (1967)
101. Blyholder, G., Neff, L.D.: J. Catalysis *2*, 138 (1963)
102. Kölbel, H., Tillmetz, K.D.: J. Catalysis *34*, 307 (1974)
103. Lunde, P.J., Kester, F.L.: Ind. Eng. Chem., Process Des. Develop. *13*, 27 (1974)
104. Allen, D.W., Yen, W.H.: Chem. Eng. Progr. *69*, 75 (1973)
105. Müller, J., Pour, V., Regner, A.: J. Catalysis *11*, 326 (1968)
106. Hille, J.: Chem. Techn. *21*, 357 (1969)
107. Bareicki, J. et al.: Chem. Techn. *29*, 497 (1977)
108. Luengo, C.A. et al.: J. Catalysis *47*, 1 (1977)
109. Binder, G.G., White, R.R.: Chem. Engng. Progr. *44*, 553 (1950)
110. Ryborz, H.: Diplomarbeit Techn. Universität Berlin 1962
111. Schoubye, P.: J. Catalysis *14*, 238 (1969)
112. Saletore, D.A., Thomson, W.J.: Ind. Eng. Chem., Process Des. Dev. *16*, 70 (1977)
113. Ollis, D.F., Vannice, M.A.: J. Catalysis *38*, 514 (1975)
114. Palmer, R.L., Vroom, D.A.: J. Catalysis *50*, 244 (1977)
115. Randhava, S.S., Camara, E.H., Amirali, R.: Ind. Engng. Prod. Res. Develp. *8*, 482 (1969)
116. Allen, D.W., Yen, W.H.: Chem. Eng. Progr. *69*, 75 (1973)
117. DE-OS 2,606,755 (1976) Harshaw Inv. Alcorn, W.R., Cullo, L.A.
118. DE-OS 2,531,411 (1975) Rhein. Braunkohlenwerke AG Inv. Förster, F. et al.
119. Kelley, R.D., Madey, T.E., Yates, J.T.: J. Catalysis *50*, 301 (1977)

120. DE-OS 2,510,164 (1975) Rhein. Braunkohlenwerke AG Inv. Fremery, M., Kühn, R., Förster, F. 1975
121. Kaempfer, K.: Erdöl, Kohle, Erdgas, Petrochem. *28*, 388 (1975)
122. Thomas, H.G., Beier, B.F., Muetterties, E.L.: J. Amer. Chem. Soc. *98*, 1296 (1976)
123. Muetterties, E.L.: Science *196*, 839 (1977)
124. Hoffmann, J.C. et al.: J. Amer. Chem. Soc. *99*, 5829 (1977)
125. US 4,022,810 (1976) Gulf Research and Developm. Co., Inv. Kobylinski, T.P. Swift, H.E.
126. DE-OS 2,518,872 Lummers Co. (Inv. Talbert, S., Weiss, A.J.) 1975
127. DE-OS 2,201,278 Metallgesellschaft AG (Inv. Möller, F.W., Müller, W.-D., Heidl, H.) 1972
128. DE-OS 2,200,004 Metallgesellschaft AG (Inv. Liebgott, H.) 1972
129. U-P 3,511,624 Gas Council (Inv. Humphries, K.J., Yarwood, T.A.) 1967
130. DE-OS 2,552,645 u. DE-OS 2,552,646 (1975) Instytut Nawozow Sztucznyck Inv. Golebiowski, A. et al.
131. CA 1,003,216 (1973) Bechtel Internat. Corp. Inv. Galstaun, L.S.
132. US 4,005,996 (1975) El Paso Natural Gas Co. Inv. Hausberger, A.L., Hämmons, G.A.
133. Hammer, H.: Erdöl, Kohle, Erdgas, Petrochem. *30*, 132 (1977)
134. DE-OS 2,462,153 (1974) Metallgesellschaft AG Inv. Müller, W.-D., Möller, F.W., Jockel, H.
135. DE-OS 2,705,673 (1977) Davy Powergas Ltd. Inv. Hardwick, W.E.
136. Lommerzheim, E.: GWF Gas/Erdgas *118*, 417 (1977)
137. Anderlohr, A., Hedden, K.: GWF Gas/Erdgas *118*, 422 (1977)
138. Langensiepen, H.-W., Hammer, H.: Chem.-Ing.-Techn. *46*, 1051 (1974)
139. DE-OS 2,432,885 (1974) Davy Powergas Ltd. Inv. Harris, N., Fowler, R.
140. Blum, D.B., Frank, M.E., Scherwin, M.B.: Synthetic Pipeline Gas Symposium, 28.–30.10.1974, Chicago
141. US 3,996,256 (1975) Shell Oil Co. Inv. Slaugh, L.H.
142. DE-OS 2,624,396 (1976) H. Topsoe A/S Inv. N.N.
143. DE-OS 2,619,325 (1976) Ford-Werke AG Inv. Mordicai, S.
144. DE-OS 2,461,482 (1974) BASF AG Inv. Broecker, F.J., Schwarzmann, M., Kaempfer, K.
145. DE-OS 2,528,148 (1975) E.I. Du Pont de Nemours Inv. Stiles, A.B.
146. DE-OS 2,603,892 (1976) Shell Intern. Res. Maatschappij BV Inv. Kiovsky, T.E., Wald, M.M.
147. DE-OS 2,506,199 (1975) Air Products and Chemicals, Inv. Upson, L.L.
148. Fischer, F., Tropsch, H.: Brennstoff-Chem. *4*, 276 (1923)
149. Roelen, O. in Ullmanns Encyklopädie der technischen Chemie, Edition 3, Vol. 9, S.684, Urban & Schwarzenberg, München/Berlin 1957
150. Kölbel, H. in Winnacker, K. und Küchler, L.: Chemische Technologie, Vol. 3, S. 439, Carl-Hanser-Verlag, München 1959
151. Kölbel, H.: Chem.-Ing.-Techn. *29*, 505 (1957)
152. Rousseau, P.E., Merve, J.W. van der, Louw, J.D.: Brennstoff-Chem. *44*, 36 (1963)
153. Kölbel, H., Ralek, M. in Falbe, J. (Ed.): Chemierohstoffe aus Kohle, Georg-Thieme-Verlag, Stuttgart 1977
154. Palm, A. in Winnacker-Küchler: Chemische Technologie, Vol. 3, Org. Technologie I, S.389, Carl-Hanser-Verlag, München 1971
155. Schulz, H. und Cronjé, H. in Ullmanns Encyklopädie der technischen Chemie, Vol. 14, S. 329, Verlag Chemie, Weinheim 1977
156. Hoogendoorn, J.: Gas, Wärme Intern. *25*, 283 (1976)
157. Frohning, C.D., Rottig, W., Schnur, F. in Falbe, J. (Ed.): Chemierohstoffe aus Kohle, Georg-Thieme-Verlag, Stuttgart 1977
158. Schulz, H. in Falbe, J. (Ed.): Chemierohstoffe aus Kohle, Georg-Thieme-Verlag, Stuttgart 1977 1977
159. Hydrocarbon Proc. *54*, 119 (1975)
160. Asinger, F.: Die Petrolchemische Industrie, Part I, p.95–107, Akademie-Verlag Berlin 1971
161. O. Roelen in Ullmanns Encyklopädie der technischen Chemie, Vol. 9, p.684, Verlag Urban & Schwarzenberg, München-Berlin 1957
162. O'Hara, J.B., Cumare, F.E., Rippes, S.M.: Coal Proc. Technol. 83 (1975)
163. Eickhoff, H.-G., Kugeler, K.: Erdöl, Kohle, Erdgas, Petrochem. *28*, 375 (1975)

164. Baron, G. et al. in Falbe, J. (Ed.): Chemierohstoffe aus Kohle, Georg-Thieme-Verlag Stuttgart 1977
165. Schulze, J.: Chem. Ind. *30*, 74 (1978)
166. Stand and Possible Developments in the Fischer-Tropsch Synthesis – Study compiled by Ruhrchemie AG, commissioned by the W. German Minister for Research and Technology 1976
167. Tillmetz, K.D.: Chem.-Ing.-Techn. *48*, 1065 (1976)
168. Christoffel, E., Surjo, I., Baerns, M.: Chemiker-Ztg. *102*, 19 (1978)
169. Anderson, R.B., Lee, C.-B., Machiels, J.C.: Can. J. Chem. Eng. *54*, 590 (1976)
170. Stein, W.A.: Chemiker-Ztg. *98*, 446 (1974)
171. Kölbel, H., Roberg, H.: Ber. Bunsenges. phys. Chem. *81*, 634 (1977)
172. Storch, H., Golumbic, H., Anderson, R.B.: The Fischer-Tropsch and Related Syntheses, J. Wiley, New York 1951
173. Kölbel, H., Tillmetz, K.D.: Ber. Bunsenges. phys. Chem. *76*, 1156 (1972)
174. Kölbel, H., Tillmetz, K.D.: J. Catalysis *34*, 307 (1974)
175. Blyholder, G., Neff, L.D.: Z. phys. Chem. *66*, 1664 (1962)
176. Kölbel, H., Roberg, H.: Ber. Bunsenges. phys. Chem. *75*, 1100 (1971)
177. Kölbel, H., Patschke, G., Hammer, H.: Brennstoff-Chem. *47*, 4 (1966)
178. Kölbel, H., Patschke, G., Hammer, H.: Z. physik. Chem. N. F. *48*, 3 (1966)
179. Kölbel, H., Hanus, D.: Chem.-Ing.-Techn. *46*, 1042 (1974)
180. Deluzarche, D. et al.: Tetrahedron Letters Nr. 9, 797 (1977)
181. Joyner, R.W.: J. Catalysis *50*, 176 (1977)
182. Pichler, H., Buffleb, H.: Brennstoff-Chem. *21*, 273 (1940)
183. Schulz, H.: Erdöl, Kohle, Erdgas, Petrochem. *30*, 123 (1977)
184. Pichler, H., Schulz, H.: Chem.-Ing.-Techn. *42*, 1162 (1970)
185. Schulz, H., Zein El Deen, A.: Fuel Proc. Techn. *1*, 31 (1977)
186. Schulz, H., Rao, B.R., Elstner, M.: Erdöl, Kohle, Erdgas, Petrochem. *23*, 651 (1970)
187. Karn, F.S., Schultz, J.F., Anderson, R.B.: Ind. Eng. Chem., Prod. Res. Dev. *4*, 275 (1965)
188. Dry, M.E., Shingles, T., Boshoff, L.J.: J. Catalysis *25*, 99 (1972)
189. Dautzenberg, F.M. et al.: J. Catalysis *50*, 8 (1977)
190. Brötz, W.: Z. Elektrochem. *5*, 301 (1949)
191. Tramm, H.: Chem.-Ing.Techn. *24*, 237 (1952)
192. Brötz, W., Rottig, W.: Z. Elektrochem. *56*, 896 (1952)
193. Henrici-Olivé, G., Olivé, S.: Angew. Chem. *88*, 144 (1976)
194. Anderson, R.B., Hofer, L.E., Storch, H.H.: Chem.-Ing.-Techn. *30*, 560 (1958)
195. Brötz, W., Spengler, H.: Brennstoff-Chem. *31*, 97 (1950)
196. Goodman, D.W. et al.: J. Catalysis *50*, 279 (1977)
197. US 4,042,614 and US 4,042,615 (1976) Exxon Inv. Vannice, M.A., Garten, R.L.
198. US 4,039,302 (1976) Battelle Inv. Khera, S.S.
199. DE-OS 2,644,185 (1976) Shell Inv. Masters, C., Doorn, J.A. van
200. US 3,941,819 (1974) Exxon Inv. Vannice, M.A., Garten, R.L.
201. Vannice, M.A.: J. Catalysis *37*, 449 (1975)
202. Shah, Y.T., Perrotta, A.J.: Ind. Eng. Chem., Prod. Res. Dev. *15*, 123 (1976)
203. Eischens, R.P., Pliskin, W.A., Frances, S.A.: J. phys. Chem. *60*, 194 (1956)
204. Blyholder, G.: J. phys. Chem. *68*, 2772 (1964)
205. Shah, Y.T., Perrotta, A.J.: Ind. Eng. Chem., Prod. Res. Dev. *15*, 123 (1976)
206. Kölbel, H., Langheim, R.: Erdöl, Kohle, Erdgas, Petrochem. *2*, 544 (1949)
207. Pichler, H., Merkel, H.: Brennstoff-Chem. *31*, 33 (1950)
208. Cohn, E.M., Hofer, L.J.: J. Amer. Chem. Soc. *72*, 4662 (1950)
209. Anderson, R.B., Hofer, L.J., Cohn, E.M., Seligman, B.: J. Amer. Chem. Soc. *73*, 944 (1951)
210. Dry, M.E.: Brennstoff-Chem. *50*, 193 (1969)
211. Anderson, R.B. in Emmett, H.P.: Catalysis, Vol. 4, Reinhold Publ. Corp., New York 1956
212. Dry, M.E., Shingles, T., Botha, C.S. van: J. Catalysis *17*, 341, 347 (1970)
213. Dry, M.E., Plessis, J.A.K. du, Leuteritz, G.M.: J. Catalysis *6*, 194 (1966)
214. Rähse, W., Schneidt, D.: Ber. Bunsenges. phys. Chem. *77*, 127 (1973)

215. Dry, M.E., Ferreira, L.C.: J. Catalysis 7, 352 (1967)
216. Dry, M.E. et al.: J. Catalysis 15, 190 (1969)
217. Kölbel, H., Schneidt, D.: Erdöl, Kohle, Erdgas, Petrochem. 30, 139 (1977)
218. Kölbel, H., Haubold, H.: Z. Elektrochem. 65, 421 (1961)
219. Kölbel, H., Müller, W.K.H.: Ber. Bunsenges. phys. Chem. 67, 212 (1963)
220. Büssemeier, B., Frohning, C.D., Cornils, B.: Hydrocarbon Proc. 55, 105 (1976)
221. Frohning, C.D., Cornils, B.: Hydrocarbon Proc. 53, 143 (1974)
222. DE-AS 2,507,647 (1976) Inv. Kölbel, H., Tillmetz, K.D.
223. Cornils, B., Büssemeier, B., Frohning, C.D.: Erdöl, Kohle, Erdgas, Petrochem. 30, 137 (1977)
224. Kitzelmann, D., Vielstich, W., Dittrich, T.: Chem.-Ing.-Techn. 49, 463 (1977)
225. GB 833,976 (1953) Inv. Peters, K.
226. JP 319,160 (1959) Kurasluke Rayon Co., Inv. Tsutsumi, S.
227. Orlow, E.: J. russ. physik. Ges. 40, 1588 (1908)
228. DBP 896,338 (1949) BASF AG Inv. Noonenmacher, H.
229. AT 171,701 (1950) Inv. Asboth, K.
230. DE 2,536,488 (1976) Ruhrchemie AG Inv. Büssemeier, B. et al.
231. DE-OS 2,518,982 Ruhrchemie AG Inv. Rottig, W. 1975
232. DE-AS 2,518,964 Ruhrchemie AG (Inv. Büssemeier, B. et al.) 1975
233. DE-AS 2,149,161 Sagami Chemical Res. (Inv. Ichikawa, M. et al.) 1971
234. DE-OS 2,546,587 Hoechst AG (Inv. Vogt, W., Glaser, H., Koch, J.) 1975
235. DE-AS 2,653,986 Hoechst AG (Inv. Vogt, W., Glaser, H., Koch, J.) 1976
236. US 2,490,488 Phillips Petroleum (Inv. Stewart, S.G.) 1947
237. Abdullahad, I., Ralek, M.: Erdöl, Kohle, Erdgas, Petrochem. 25, 187 (1972)
238. Dry, M.E.: Ind. Eng. Chem., Prod. Res. Dev. 15, 282 (1976)
239. Schulz, H.: Erdöl, Kohle, Erdgas, Petrochem. 29, 570 (1976)
240. Fischer, F., Tropsch, H., Dilthey, P.: Brennstoff-Chem. 6, 265 (1925)
241. Pichler, H.: Brennstoff-Chem. 19, 226 (1938)
242. Fischer, F., Bahr, T., Meusel, H.: Brennstoff-Chem. 16, 466 (1935)
243. Pichler, H. et al.: Makromol. Chem. 70, 12 (1964)
244. Schulz, H. in Falbe, J. (Ed.): Chemierohstoffe aus Kohle, Georg-Thieme-Verlag, Stuttgart 1977
245. Guyer, A. et al.: Helv. Chim. Acta 38, 798 (1955)
246. Guyer, A., Jutz, J., Guyer, P.: Helv. Chim. Acta 38, 971 (1955)
247. Guyer, A., Guyer, P., Thomas, D.: Helv. Chim. Acta 42, 481 (1959)
248. Kölbel, H., Müller, W.H.E., Hammer, H.: Makromol. Chem. 70, 1 (1964)
249. Pichler, H. et al.: Brennstoff-Chem. 48, 266 (1967)
250. Pichler, H.: Brennstoff-Chem. 33, 289 (1952)
251. Pichler, H., Schulz, H.: Chem.-Ing.-Techn. 42, 1162 (1970)
252. Gutbier, A., Trenkner, C.: Z. Anorg. Chem. 45, 166 (1905)
253. Pichler, H., Firnhaber, B.: Brennstoff-Chem. 44, 33 (1963)
254. Pichler, H., Burgert, W.: Brennstoff-Chem. 49, 1 (1968)
255. Pichler, H., Buffleb, H.: Brennstoff-Chem. 21, 283 (1940)
256. Kölbel, H., Battacharyya, K.K.: Liebigs Ann. Chem. 618, 67 (1958)
257. Pichler, H., Bellstedt, F.: Erdöl, Kohle, Erdgas, Petrochem. 26, 560 (1973)
258. Pichler, H.: Chem. Techn. 18, 392 (1966)
259. Keller, A.: Kolloid-Z. 165, 18 (1959)
260. Bellstedt, F.: Diss. Univ. Karlsruhe 1971
261. Pichler, H., Kater, E.: Brennstoff-Chem. 50, 373 (1969)
262. Karn, F.S., Schultz, I.F., Anderson, R.B.: Ind. Eng. Chem., Prod. Res. Dev. 4, 265 (1965)

5. Koch Reactions

H. Bahrmann

5.1 Introduction

The synthesis of carboxyl acids from olefins, carbon monoxide and water in the presence of metal carbonyls has already been discussed (cf Chap. 3).
The same reactants also form carboxylic acids with an acidic catalyst. For comprehensive reviews, see [40, 127, 211, 212, 214, 215].

The various milestones in the development of the Koch synthesis can be represented as follows:

1. High pressure synthesis (DuPont process 1933)
2. Medium pressure synthesis (Koch-Haaf synthesis 1955)
3. Normal pressure synthesis (Koch-Haaf synthesis in presence of metals of Group-IB 1973).

In the DuPont process [1–10], the reaction components are introduced together. However, as the reaction requires drastic conditions (500–1000 bar/100–350 °C) this hampered industrial realization.

The decisive breakthrough was made by Koch, Gilfert and Huisken [11, 12] (1955) who used a two-stage operation (medium pressure synthesis). In the first stage, the olefin reacts with the acid catalyst and carbon monoxide in the absence of water, then in the second step the complex formed by the olefin, carbon monoxide and the acid catalyst is hydrolyzed.

$$
\underset{\underset{CH_3}{|}}{H_2C{=}C{-}CH_3} \quad \xrightarrow[\text{2.}\,H_2O]{\text{1.}\,H_2SO_4/CO} \quad \underset{\underset{CH_3}{|}}{\overset{\overset{CH_3}{|}}{H_3C{-}C{-}COOH}} \tag{1}
$$

The reaction occurs at temperatures between −20 to +80 °C and pressures up to 100 bar. Generally H_2SO_4, H_3PO_4, HF or Lewis acids such as BF_3 are employed as catalysts. Under these conditions, nearly all olefins and a great number of dienes, unsaturated esters, unsaturated and saturated alcohols and diols, reactive alicyclic compounds, halogenated compounds, certain amines, esters and aldehydes react to form the corresponding carboxylic acids. The Koch process is currently being operated in two commercial plants by Esso and Shell (cf Sect. 5.7, p.406).

In 1965, Haaf developed a pressure-free variant with formic acid as CO source [13, 14]. However, due to the high cost of concentrated formic acid, this method has

372

been restricted to the laboratory. Recent work [15, 16, 33] conducted by a Japanese group could be of great significance i.e., the normal pressure synthesis in the presence of catalytic amounts of metals of Group IB of the Periodic Table.

Basically, the relatively mild reaction condition make industrial realization appear favorable. However, as in all new commercial variants, besides many minor considerations, a satisfactory solution to the catalyst separation and recycling problems must be found.

Olah's work [17] in 1963 concerning the possibility of directly observing the alkyl cations in the NMR spectrum via stabilization in strong Lewis acids such as SbF$_5$ has given rise to a new branch of chemical structure research. Results of these studies could well have an impact on the Koch synthesis.

5.2 Reaction Mechanism

Koch's basic reaction mechanism [8] for the carboxylic acid synthesis has been generally accepted.

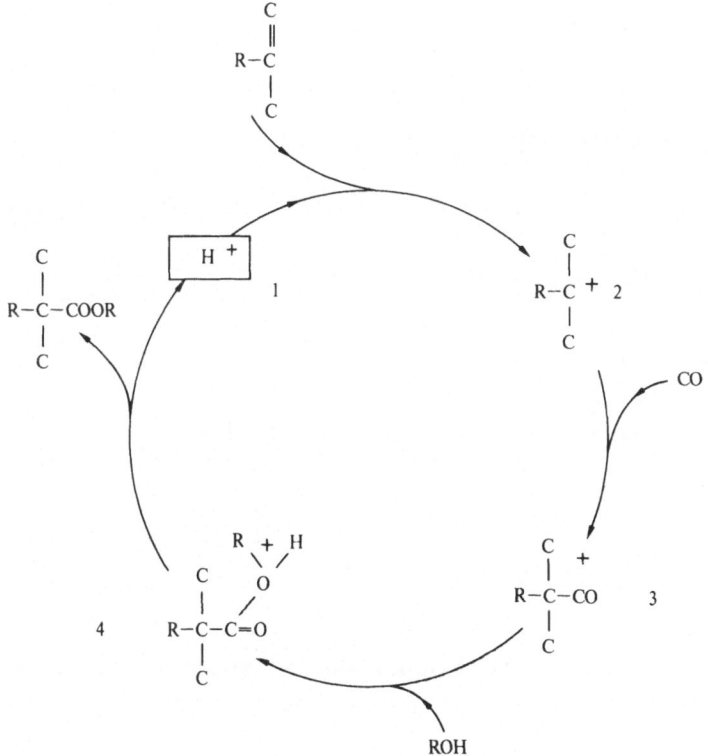

Fig. 5.1. Basic reaction mechanism of the Koch synthesis

After initial formation of a carbenium ion **2** from, for example, an olefinic starting material with a proton from the acidic catalyst **1**, subsequent addition of CO to **2** gives rise to the acylium cation **3**. When the latter reacts with alcohol or water, the carboxylic ester or acid is formed, releasing the active catalyst (proton **1**).

The comparatively large variation of possible precursors and products is due to the multitudinous ways in which the carbenium ion can be formed or in which it can react. The conversion is often made even more complex on account of a series of side reactions, which occur parallel to the main reaction. For this reason, the most important carbenium ion reactions [49] (in the Koch synthesis) are discussed in the next Sect.

5.2.1 Formation of Intermediate Carbenium Ions 2 from Various Precursors

Olefins (addition of a proton) [18]

$$R-CH=CH_2 \xrightarrow{H^+} R-\overset{+}{C}H-CH_3 \tag{2}$$

Alcohols (addition of a proton and dehydration) [18]

$$ROH \xrightarrow[2.-H_2O]{1.+H^+} R^+ \tag{3}$$

Halides (addition of a proton, dehydrohalogenation) [13]

$$R-X \xrightarrow[-HX]{H^+} R^+ \tag{4}$$

Halides (displacement of halide) [160]

$$R-X \xrightarrow{+SbCl_5-liqSO_2} [R^+][SbCl_5X]^- \tag{5}$$

Aliphatic branched hydrocarbons (carbon-hydride shift to proton [19]

$$R-H \xrightarrow[-H_2]{H^+} R^+ \tag{6}$$

Alicyclic compounds

a) Disproportionation to aliphatic branched hydrocarbons and carbenium ions [13, 19, 65]

$$2 \text{ cyclo}-C_6H_{12} \xrightarrow[-iC_6H_{14}]{HF/SbF_5} [\text{cyclo-}C_6H_{11}]^+ [SbF_6]^- \tag{7}$$

b) Addition of proton and ring cleavage [13, 64]

$$\begin{array}{c} CH_2 \\ / \quad \backslash \\ CH_2 - CH_2 \end{array} \xrightarrow{H^+} CH_3-\overset{+}{C}H-CH_3 \qquad (8)$$

Alicyclic hydrocarbons and branched aliphatic hydrocarbons (carbon-hydride shift leading to formation of another carbenium ion) [16, 20–28, 84]

$$R-H \underset{-R'H}{\xrightarrow{R'^+}} R^+ \qquad (9)$$

Aromatic compounds (Gattermann-Koch synthesis) – carbenium ion complex from electrophilic aromatic substitution [29, 30, 31]

$$Ar \xrightarrow{H^+BF_4^-} [Ar]^+\,[BF_4]^- \qquad (10)$$

Aldehydes (addition of proton) [32, 62]

$$\begin{array}{l} R \\ \quad \backslash \\ \quad C{=}O \\ \quad / \\ H \end{array} \xrightarrow{H^+} H-\overset{+}{\underset{R}{\underset{|}{C}}}-OH \qquad (11)$$

Alkylenehalides (cleavage of the halogen-carbon bond and formation of resonance-stabilized allyl cation)

$$R-CH_2{=}CH-CH_2-X \underset{-X^-}{\longrightarrow} [R-CH{=}CH-CH_2^+ \leftrightarrow + CH_2{=}CH-\overset{+}{C}H-R] \qquad (12)$$

α, β-unsaturated ketones (via diprotonation) [52]

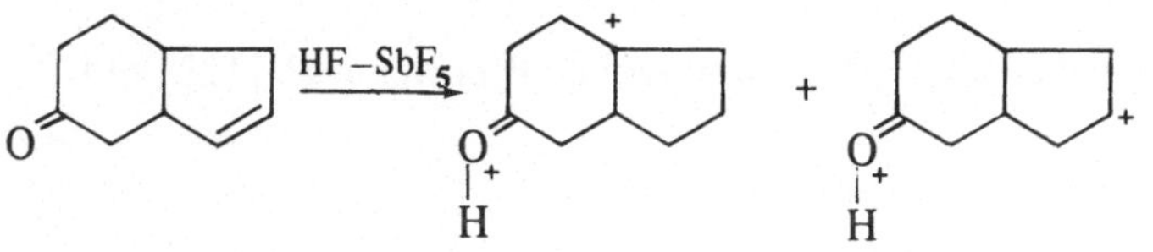

(13)

Alkynes (→ vinyl cations) [47]

$$H_3C-C{\equiv}C-CH_3 \xrightarrow{FHSO_3SbF_5^-} \begin{array}{l} H_3C \\ \quad \backslash \\ \quad C{=}\overset{+}{C}-CH_3 \\ \quad / \\ H \end{array} \qquad (14)$$

Ethers [85]

$$(CH_3)_3C-O-CH_3 \xrightarrow{H^+} (CH_3)_3-C-\overset{\overset{H}{|}}{\underset{+}{O}}-CH_3 \rightarrow (CH_3)_3C^+ + CH_3OH \qquad (15)$$

5.2.2 Reactions of Intermediate Carbenium Ions 2

Isomerization/rearrangement of the C skeleton

Isomerization is an important side reaction of the intermediate carbenium ions (cf 2 in Fig. 5.1) under the conditions of the Koch synthesis. It leads to the formation of stable tertiary carbenium ions [37]. For example, on carbonylating 1-hexene [34] in the presence of $BF_3 \cdot H_2O$, the isomeric carbenium ions listed in Fig. 5.2 must be formed – as can be appreciated from the resulting carboxylic acids.

Information about the relative ratio of hydride shift to alkyl group migration can be gleaned from the weight ratio of the secondary and tertiary carboxylic acids. The ratio is strongly dependent on the reaction conditions and catalyst concentration [57].

The isomerization of cyclic carbenium ions almost invariably leads to the formation of the stable 5- and 6-membered ring carbenium ions.

Rearrangements involving adamantyl [24, 35, 56, 59, 60, 193, 194] and diadamantanyl [195] cations are discussed in the literature. In addition, the formation of tertiary cyclohexyl cations [36] from cyclohexyl carbenium ions has also been reported.

Two mechanisms were proposed for the isomerization:
a 1,2-hydride or alkyl shift (Wagner-Meerwein rearrangement) [36, 127]

R = H (hydride shift)
  = Alkyl (migration of alkyl group)

   a 1,3-shift via protonated cyclopropanes proposed by Brouwer et al. [128] and Kell et al. [39]

Products from an intermediate of this type may give rise to methyl branching at the carbon adjacent to the carboxylated carbon [127].

Using mass spectroscopy data, Kell and McQuillin [39] found a general relationship between the products from the alkene and substituted cyclopropane precursors. This is the basis of the supposition that the isomerization (particularly of higher n-olefins) ensues via a protonated cyclopropane which 'rolls' along the alkane chain (cf Fig. 5.3).

The isomerization of alkyl substituted cyclohexanes (according to the same mechanism) has been dealt with in the Ref. [55].

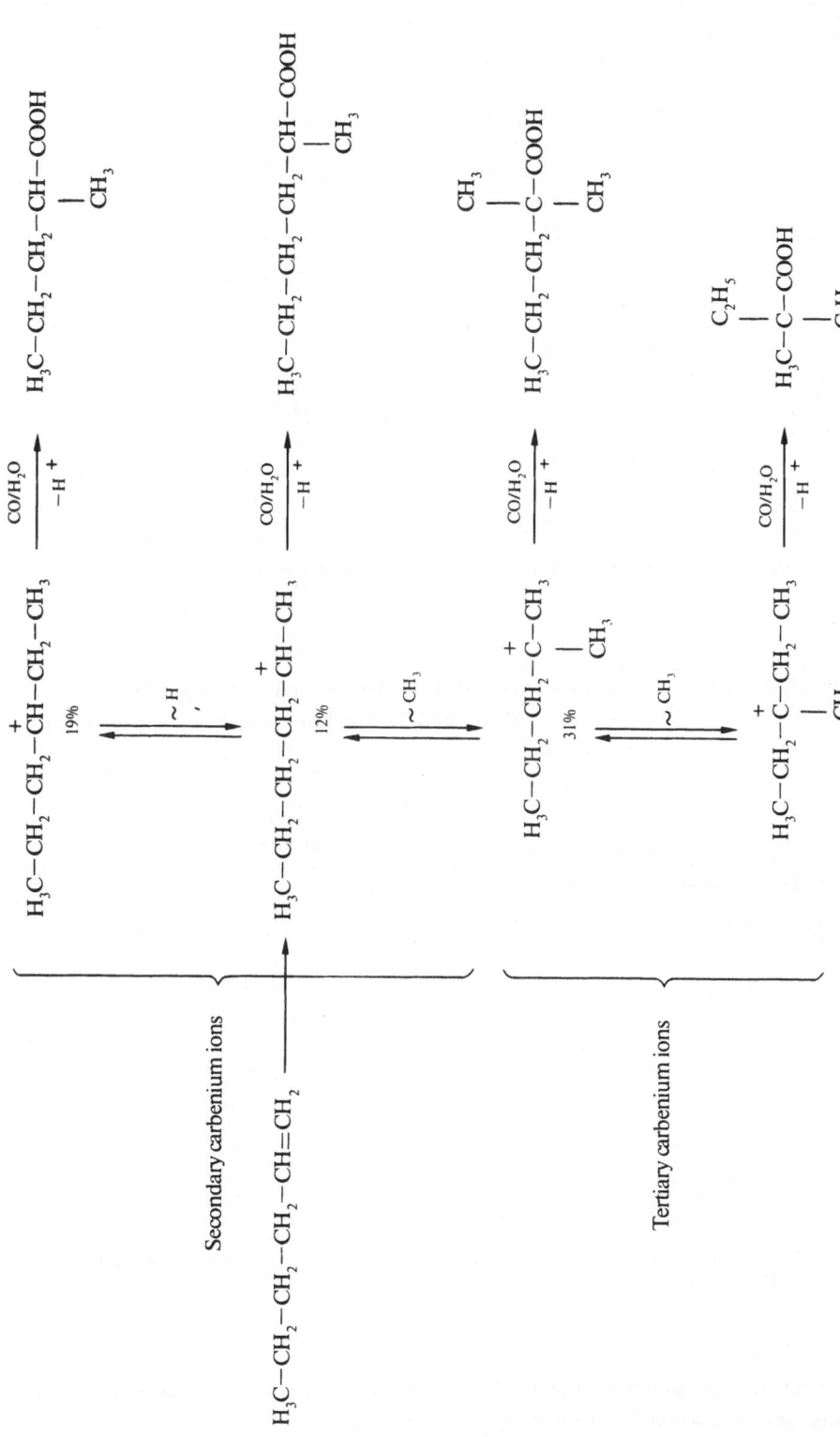

Fig. 5.2. Isomerization of carbenium ions [34]

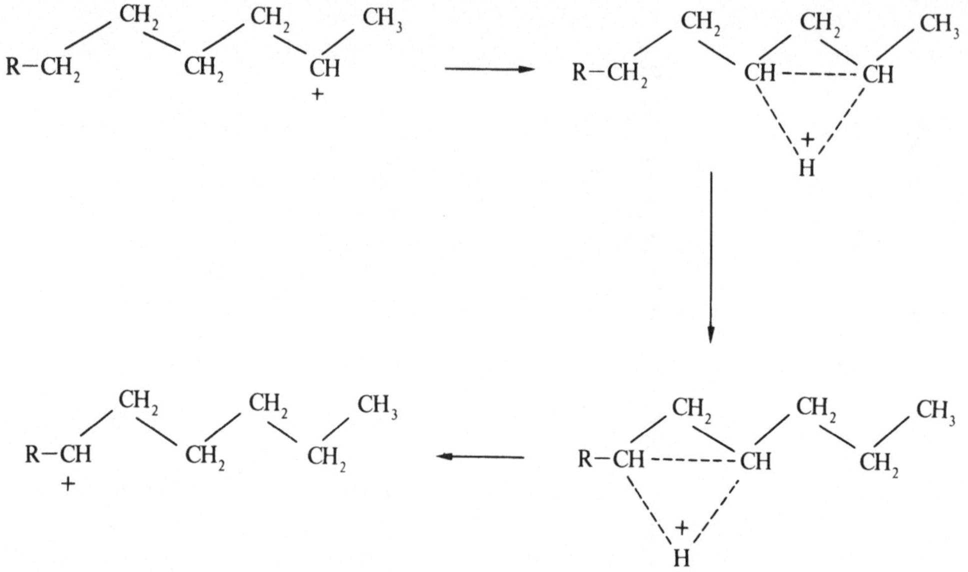

Fig. 5.3. Isomerization of carbenium ions by 1,3-shifts via protonated cyclopropane

The interpretation of the 1,2-hydride shift involves a transition state with an energetically unfavorable carbenium ion possessing a terminal positive charge (cf **4** in Fig. 5.4).

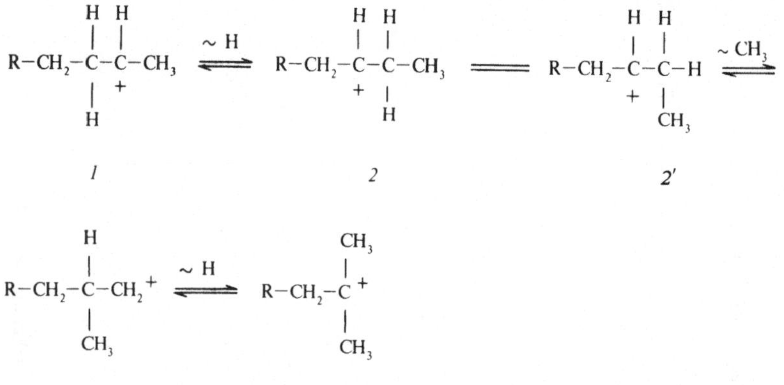

Fig. 5.4. Isomerization of secondary carbenium ions to tertiary carbenium ions via the 1,2-alkyl or hydride shift mechanism

As a whole, the second interpretation is more satisfactory as it attempts to explain the transition from secondary to tertiary carbenium ions.

378

Olefin Formation

via loss of a proton:

Even with nonolefinic components, olefin formation leads to additional by-products via side reactions (e.g., oligomerization) [36].

via depolymerization [38]:

On carbonylating diisobutylene and isobutylene with H_3PO_4/BF_3-H_2O and isopentane [25], identical products result at 125 °C/120 bar CO pressure [38]. The resulting product spectrum can be explained by means of the following polymerization–depolymerization equilibrium

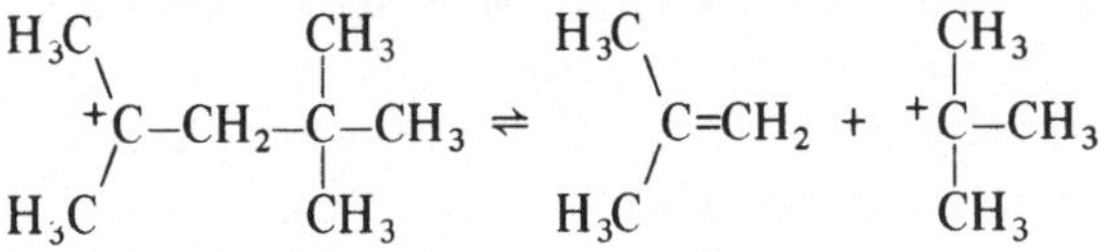

$$\text{(16)}$$

Disproportionation

As can be appreciated from product formation on carboxylating 2-methyl-1-butene, the depolymerization of the intermediate C_{10} cation does not take place with reformation of the initial C_5 units but disproportionation ensues producing larger and smaller fragments relative to the starting materials [40, 63, 167].

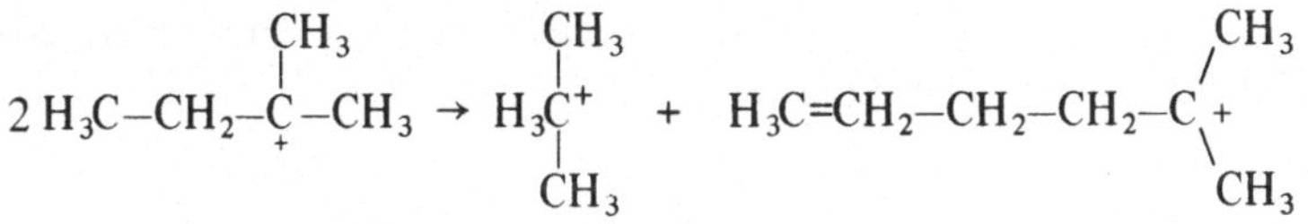

$$\text{(17)}$$

$$C_{10}^+ \rightarrow C_4^+ \text{ (or } C_4\text{-alkene)} + C_6\text{-alkene (or } C_6^+)$$

Besides trimethylacetic acid, C_5- and C_8-carboxylic acids are obtained as by-products on carboxylating isobutylene and its oligomers in the presence of BF_3/H_2O at 80–140 °C/50 bar CO pressure [43]. Similar results were found with the normal pressure operation in the presence of elements of Group IB of the Periodic Table [138].

The formation of the C_7 moiety is thought to proceed as follows.

$$C_8^+ \rightarrow 2\,C_4^+ \text{ (or } C_4\text{-alkene)} \qquad \text{(18)}$$

$$C_8^+ + C_4\text{-alkene} \rightarrow C_{12}^+ \qquad \text{(19)}$$

$$C_{12}^+ \rightarrow C_5^+ \text{ (or } C_5\text{-alkene)} + C_7\text{-alkene (or } C_7^+) \qquad \text{(20)}$$

The carbonylation of 3-methyl-3-pentanol yields considerable amounts of 2,2-dimethylbutane- and propionic acid as by-product, indicating that the disproportionation does not solely take place "stoichiometrically" [33].

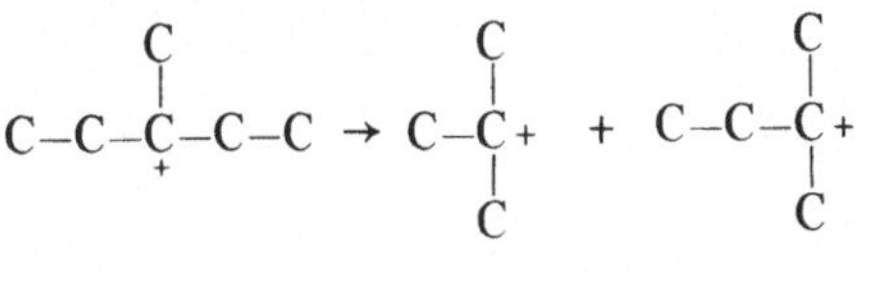

$$\text{(21)}$$

Oligomerization (Via of an Alkene)

$$H_3C-\underset{\underset{\displaystyle CH_3}{|}}{\overset{\overset{\displaystyle CH_3}{|}}{C}}=CH_2 \;\; + \;\; +\underset{\underset{\displaystyle CH_3}{|}}{\overset{\overset{\displaystyle CH_3}{|}}{C}}-CH_3 \;\; \rightleftharpoons \;\; H_3C-\underset{\underset{\displaystyle CH_3}{|}}{\overset{\overset{\displaystyle CH_3}{|}}{\overset{+}{C}}}-CH_2-\underset{\underset{\displaystyle CH_3}{|}}{\overset{\overset{\displaystyle CH_3}{|}}{C}}-CH_3 \tag{22}$$

The proton- or carbenium ion catalyzed oligomerization of initial [33, 38, 40, 41, 50, 51] and intermediate olefins [36, 58, 154] is a very important side reaction, which in the case of isobutylene, with BF_3/H_2O at 80–140 °C/50 bar, leads to the formation of the C_9 or C_{13} acid via diiso- or triisobutylene [43].

On the other hand, no C_9 acid was produced with the same olefinic feedstock in the presence of H_3PO_4 at 125 °C/120 bar CO pressure [38].

The carbonylation of cyclopentanol at 125 °C/100 bar CO pressure yields a cis/trans mixture of (bicyclic) decalincarboxylic acid [58].

The planned dimerization of cyclohexene and lower aliphatic olefins is claimed in a patent filed by Phillips Petroleum Co. [163].

Cracking of the Carbon Skeleton (Cleavage of Tertiary Alkyl Groups)

On reacting 2,8-dimethyl-2,8-nonanediol and 2,9-dimethyl-2,9-decanediol according to the formic acid method, besides the dicarboxylic acids, 1-methylcyclohexanecarboxylic acid was mainly formed. The reaction sequence [42] has been interpretated as follows: initially cyclization occurs leading to a dialkyl branched cyclohexyl ring which, after isomerization, splits off a tertiary butyl group.

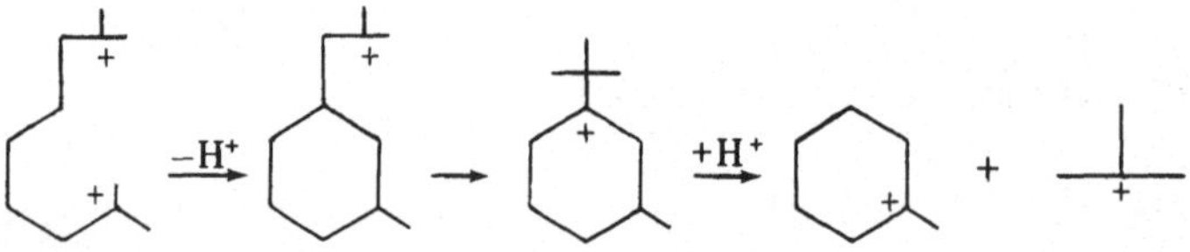

The formation of stable cations of five- or six-membered rings is typical for reactions involving cyclic compounds (cf Sect. 5.2.2).

The cleavage of other alkyl groups has also been treated in the Ref. [55].

5.2.3 Formation of Intermediate Acyl Cations

Complex acyl cations are directly formed during the electrophilic carbonylation of aromatic compounds and various amines in the presence of halogens [45] or during the reaction between alkyl halides and CO in $SbCl_5$–liq. SO_2 [160]. The acyl cations can then be decomposed in the usual manner to acids or esters by reacting them with nucleophiles such as water or alcohol. The conversion takes place with cleavage of HX

from halo-carbenium ions [45]

$$Br_2 + SbCl_5 \rightarrow Br^+ (SbCl_5Br)^- \xrightarrow[\text{liq.}SO_3]{CO} (BrCO)^+ (SbCl_5Br)^- \tag{23}$$

$$(BrCO)^+ (SbCl_5Br)^- \xrightarrow{\text{halogen exchange}} (ClCO)^+ (SbCl_4Br_2)^- \qquad (24)$$

$$(ClCO)^+ + C_6H_6 \rightarrow \left[\begin{array}{c} \text{H} \\ \text{CO} \\ \text{Cl} \end{array} \right]^+ \qquad (25)$$

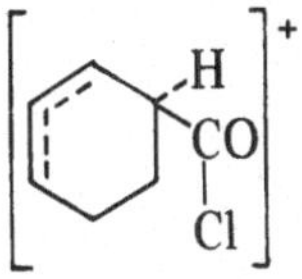

Evidence for the presence of halo-oxo carbenium ions – $(CLCO)^+$ – was provided, for example, by the reaction in which ethyl chloroformate ($ClCOOC_2H_5$) was formed in the presence of ethyl alcohol.

from alkyl halides (direct addition of CO) [160]

$$RX + CO \xrightarrow{\text{SbCl}_5\text{-liq.SO}_2} (RCO)^+ (SbCl_5X)^- \qquad (26)$$

The rapid formation of the acyl cation produced transiently during the reaction prevents isomerization and other side reactions thereby leading to a selective carboxylation.

vinyl acyl cations from vinyl cations [47, 49]

$$\qquad (27)$$

from nonclassical carbenium ions (norbonyl cation) [48].

The CO attack is kinetically controlled and takes place stereospecifically at one side.

5.2.4 Reactions of the Intermediate Acyl Cations (3 in Fig. 5.1)

Decarboxylation/recarboxylation

$$R_3CCO^+ \rightleftharpoons R_3C^+ + CO$$

In analogy to Pratt's acid chlorides findings [46], the position of the equilibrium must be favorable where planar sp^2 carbenium ions can be formed. In this instance not more than two of the bonds of the carbon atom should be rigidly fixed.

If steric hindrance is present, then ketone formation may ensue. In certain circumstances, the reaction can take a very stereochemical course, CO addition occuring at the less hindered side of the carbenium ion [48, 52].

Besides high stereo-selectivity, a high regio-selectivity was recently observed in the reaction involving α,β-unsaturated ketones in HF/SbF$_5$ [52] (cf Fig. 5.5). As a consequence of the rapid attainment of equilibrium between 1 and 2 [53] and the very slow decarbonylation of 4 [54], the overall process leads to products stemming from 4.

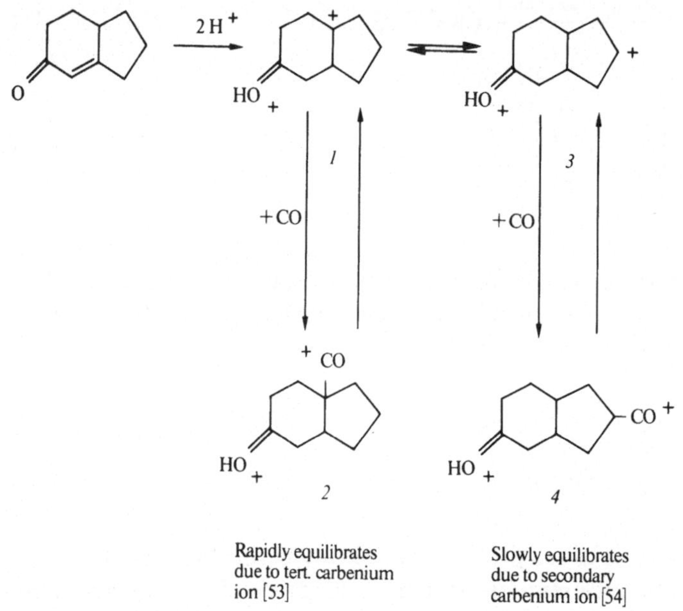

Fig. 5.5. Carbonylation of an α, β-unsaturated ketone with high regio- and stereo-selectivity [52]

Ketone formation [21, 46]

$$H_3C-\underset{\underset{R'}{|}}{\overset{\overset{R}{|}}{C}}-\overset{\overset{O}{\|}}{C_+} + H-\underset{\underset{R'''}{|}}{\overset{\overset{R''}{|}}{C}}-CH_3 \rightarrow H_3C-\underset{\underset{R'}{|}}{\overset{\overset{R}{|}}{C}}-\overset{\overset{O}{\|}}{C}-\underset{\underset{R'''}{|}}{\overset{\overset{R''}{|}}{C}}-CH_3 + H^+ \qquad (28)$$

Besides iso-branched hydrocarbons, other nucleophilic compounds such as benzene, cyclohexane, and methylcyclohexane can react with acylium ions if they are present as a complex e.g., $[C_6H_{11}CO^+ \ SbF_6^-]$ [19].

5.2.5 Secondary Reaction of Carboxylic Acids

Internal ester formation [51]

$$\underset{\underset{COOH}{|}}{C-C-C-C-C} + C=C-C_3 \rightarrow \underset{\underset{COOC_5}{|}}{C-C-C-C-C}$$

The formation of internal esters is apparently favored with lower olefins (ethylene and propylene) and the catalyst $[H_3O] \ [BF_4]$ [50].

Ketone formation [19]

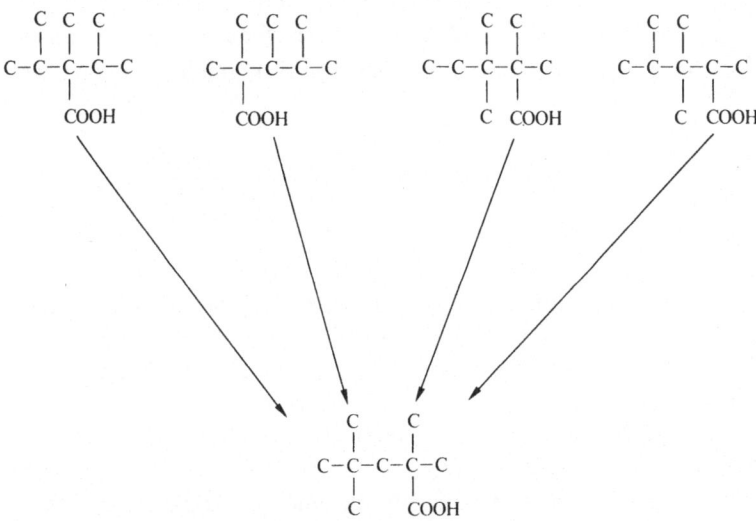

HR can represent various nucleophilic components for example − iso-C_6H_{13} or −C_6H_5.

5.2.6 Retro-Koch Reaction

Several recent reports [25, 36, 44] suggest that the Koch synthesis is reversible and apparently ensues via formation of the most thermodynamically stable carbenium ion or carboxylic acid. Yoneda [44] found on making a structural analysis of the C_9 acids from diisobutylene and BF_3−2 H_2O, that although certain acids are initially formed, they decompose during the reaction in favor of a particular C_9 acid (Fig. 5.6).

Additional data obtained via H/D exchange studies with carboxylic acids in D_2SO_4−HCOOH can be found in the Ref. [36].

On employing 1-isopropylcyclohexanecarboxylic acids as well as 2-cyclohexyl-2-methylpropanoic acid in the Koch synthesis with 96% H_2SO_4, it was established that a certain ratio of the acids sets in depending on the reaction conditions (see for scheme 5.1).

Fig. 5.6. Rearrangement of tertiary C_9-acids during longer reaction periods

Paatz und Weisgerber [19] made similar observations. They found that on allowing a mixture of 2 mole methylcyclopentane and 1 mole SbF_5 in HF to stand for 24 h, cyclohexyl isohexyl ketone was formed in 75% yield. In this case, the acylium hexa-fluoroantimonate (from the retro-cleavage of the initially resulting cyclohexanoic acid) reacts with the primary resulting isohexane (cf Sects. 5.2.1 and 5.2.4 for mechanism).

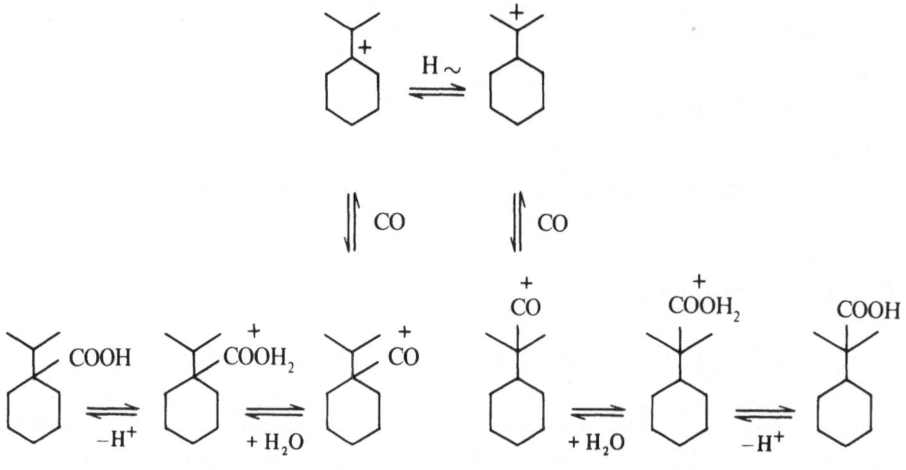

Scheme 5.1. Decarbonylation equilibria with 1-isopropylcyclohexanecarboxylic acid and 2-cyclo-hexylpropanoic acid in the Koch synthesis [55]

5.2.7 Heterogeneous Variants of the Koch Synthesis

There is little information available about the application of heterogeneous catalysts. However, the carbonylation of methanol with CO yielding acetic acid with a H_3PO_4-on-charcoal catalyst is discussed in the Ref. [61]. In addition, the reaction of 1-octene in the presence of a solid H_3PO_4–BF_3 catalyst on a Kieselguhr or graphite support has also been reported [75].

5.3 Catalysts

As can be seen from Fig. 5.1, all substances which can supply protons or assist in the formation and stabilization of carbenium ions are suitable catalysts for the Koch reaction. Falbe [18] and Möller [98] made earlier summaries of suitable catalyst systems. These included the classical Broensted acids such as H_2SO_4, HF, H_3PO_4 as well as in combination with Lewis acids. The combinations with BF_3 are more significant than those with $SbCl_5$, $AlCl_3$ and the other Broensted acids. This is probably due to the apparently more facile catalyst recycle in the industrial application of these systems.

5.3.1 Based on BF_3

The system BF_3/H_2O and methanol usually requires CO pressures around 100 bar and temperatures of 100 °C. Generally, the conditions are somewhat milder with H_2SO_4 and H_3PO_4 (25–100 bar CO pressure and 20–100 °C). According to recent Russian reports [70, 76, 86], the addition of acetic acid/chloroacetic acid apparently gives rise to a more active, as well as a more selective catalyst system. Very mild conditions (25 bar CO/< 0 °C) are possible with aromatic compounds and HF [29, 90]. Good yields of the methyl ester are obtained in the presence of BF_3–CH_3OH [99, 68]. There are contradictory reports about the effect of the catalyst on the isomeric distribution, while earlier work by Möller [92] indicated that BF_3-containing catalysts caused less isomerization of secondary alcohols compared to H_2SO_4 and in addition more secondary alcohols were formed compared to tertiary. Eidus et al. [68] found that BF_3-containing catalysts were the most selective and active with regard to the formation of acids and esters with a quaternary C atom. Mutsubara [71] established that with $BF_3 \cdot H_2O$ and isobutylene the selectivity to pivalic acid was higher than with H_2SO_4.

Depending on the weight ratio, the BF_3/H_2O system can either be homogeneous (ratio 1 : 1) or heterogeneous (ratio 1 : 2). The heterogeneous system also exhibited higher activity [78]. With the tertiary system BF_3/HF/H_2O and BF_3/HF/CH_3OH, hydroxonium and methoxonium tetrafluoroborate (H_3O or CH_3OH_2–BF_4), respectively, are formed. They exhibit high catalytic activity even for difficult – to – carbonylate substrates such as ethylene and propylene. In addition, they apparently also facilitate mild continuous reaction control [50, 77].

5.3.2 Based on H_2SO_4

With H_2SO_4 and aliphatic olefins or alcohols, the reaction conditions are noticeably milder compared to the H_3PO_4-catalyzed conversions [57].

The catalyst concentration generally lies around > 90% as at this point the activity generally recedes and the hydration of the olefinic starting materials (to the alcohols) increases markedly [126]. With long-chained unsaturated fatty acids, the conversion can occur at 1 bar CO pressure [108] even in the absence of formic acid. However, the appearance of strong foaming is disadvantageous [126] (larger reactor volume necessary). Therefore, apart from the formic acid method, pressures between 10 and 85 bar CO pressure are generally applied at temperatures between −10 and +75 °C. Even with a slight excess of formic acid, sulfate formation may be avoided when another acid such as phosphoric acid or chloroacetic acid is present [129].

According to a patent filed by the Amer. Cyanamid Co. [118], sulfate esters are claimed as desired intermediates as apparently in this way a complete catalyst recycle can be realized. This is particularly applicable when – in contrast to the usual method – the water of reaction is present from the start, necessitating product separation via azeotropic distillation. However, this attractive feature is compensated by the drastic reaction conditions i.e., 50–100 °C and 210–560 bar CO pressure. In contrast, according to a Hungarian patent [109], diisobutylene can be reacted at 25–30 °C/80 bar CO pressure in recycled acid consisting of the resulting acid, water, hexanes and 60–80% H_2SO_4. New reports have appeared in the literature covering the formic acid [27, 35, 103, 112, 124, 130] and the H_2SO_4 methods [41, 57, 71, 100–102, 106–108, 110, 111, 114–117, 119–123, 126, 148, 164, 165, 221].

5.3.3 Based on H₃PO₄

The H_3PO_4/BF_3 catalyst system [95–97] permits excellent separation of the reaction products and was thus favored for industrial application (cf Shell process [97, 217]). On the other hand, the application of pure H_3PO_4 compared to H_2SO_4 would necessitate more severe reaction conditions (75–200 bar instead of 70–80 bar and 125–150 °C instead of 10–50 °C). Thus, these features coupled with the comparatively low yields have prevented commercial application of the H_3PO_4 process. The higher temperature level encourages ester formation, particularly with an alcoholic feedstock [154]. Product distribution and yield can be influenced by the H_3PO_4 concentration [38] which generally lies between 60–100%. At higher concentration, the acid content of the mixture can be increased [154] and the oligomerization repressed [58]. Aliphatic and alicyclic olefins can be directly converted into their corresponding methyl esters [155, 157] with $H_3PO_4/HCOOH$ and CH_3OH.

5.3.4 Based on HF

The HF catalyst is employed in pure form as well as in aqueous solution, the applied conditions generally lying between 25–80 bar and 20–60 °C. With the aqueous system, HF can be diluted up to approx. 50% [149, 153]. Very good yields i.e., > 95% (relative to the olefin) [51] were obtained at higher catalyst concentrations e.g., HF : olefin : H₂O = 10 : 1 : 1 [153]. The low boiling point of hydrogen fluoride ($\sim$20 °C) led to proposals for catalyst separation via a pressure distillation [149, 153, 202].

As $BF_3 \cdot 2 H_2O$ enables catalyst precipitation via simple phase separation, this procedure was chosen by Enjay Chem. Corp. for commercial application [216] (cf Sect. 5.7).

The tendency to form stable HF-adducts complicates the industrial application as these complexes must be decomposed during catalyst recycle. However, they also permit – via isolating, separation and cleavage of the adduct – realization of a loss-free catalyst recycle [90, 148]. The corrosion problems must be mentioned at this juncture as they have probably prevented the construction of larger plants due to the high cost of special construction materials. Besides Hastelloy C steel [85] and tantalum [140], stainless steel [153] was also recommended for this purpose.

5.3.5 Based on SbF₅/SbCl₅

Generally, liquid SO_2 is employed at very low temperatures (to −70 °C) [45, 158, 159, 160, 162] in combination with $FHSO_3$ [47] or in strong mineral acids with SbF_4Cl [120].

The apparently better complexing ability of the intermediate hexahaloantimonate compared to Lewis acids such as BF_3 led, according to recent work [160, 162], to the irreversible formation of 1 : 1 donor-acceptor complexes with alkyl halides. These complexes stabilize the intermediate acyl complex, thereby facilitating a very selective reaction together with a high yield.

5.3.6 Various

The applications of fluorosulfonic and fluorophosphoric acids have been treated in the Ref. [120]. Apparently, trifluoromethanesulfonic acid/alcohol causes the olefin substrate to be converted into the corresponding dimeric acid at 85 bar (20–30 °C). The reaction ensues via a planned intermediate dimerization of the olefinic precursor [163].

The application of $AlCl_3$ has also been discussed in a report [30].

5.3.7 Normal Pressure Synthesis in the Presence of Elements of Group IB

Besides the formic acid method, the Koch synthesis can also be conducted at normal pressure in the presence of elements of Group IB of the Periodic System (cf p. 225). In this case, the heavy metals form di- [131], tri- [15, 33] and tetracarbonyl complex cations [131, 140, 168] in a strongly acidic reaction medium. These cation complexes serve as solvent for CO.

$$Me^+ \xrightarrow{\quad n\ CO \quad} Me(CO)_n^+ \rightleftharpoons MeCO^+ + n-1\ CO$$

Me = metal
n = 2 (Ag) or 3–4 (Cu)

The reaction conditions are very mild (e.g., 0–35 °C) and largely correspond to those of the formic acid method. As a consequence of the partial CO deficiency (cf effect of reaction conditions – Sect. 5.4), the initially resulting carbenium ions are not immediately trapped. Thus, they are able to undergo the various rearrangements to a considerable extent (cf Sect. 5.2). The product therefore mainly consists of tertiary carboxylic acids. While the yields are relatively high (60–98%), the selectivity with certain compounds is disappointing compared to the pressure variants.

While Souma [131] found no differences in the catalytic behavior of $Cu(CO)_4^+$ and $Ag(CO)_2^+$ in BF_3/H_2O solution, according to Yoneda higher carboxylic acid yields can be obtained with $Cu(CO)_4^+$. These findings are in accord with the greater solubility of CO in the copper complex. The concentration of the heavy metal catalyst lies around 0.2 mole/l feedstock [33, 16]. Cu carbonyls are more stable than Ag carbonyls [169].

While the Cu- and Ag-carbonyl complexes can be homogeneously dissolved in BF_3–H_2O, $FHSO_3$, CF_3SO_3H and HF, with H_2SO_4 they form a (heterogeneous) suspension [169]. This explains the considerably higher yields with diisobutylene and CuI catalysts in BF_3/H_2O and BF_3/H_3PO_4 compared to 96% H_2SO_4 [138]. Apparently, sulfuric acid is less suitable than other acids for the normal pressure Koch synthesis in the presence of Cu or Ag.

When the oxygen content of CO lies around 5%, this is said to be advantageous for the normal pressure synthesis [136, 145]. High yields (> 95%) can also be achieved on stabilizing the Cu carbonyls with metallic copper or after terminating the reaction after 50% CO consumption [136]. Proposed catalyst recycles are base on:

the addition of alkali metal ions to the catalyst solution (phase separation via decanting feasible) [142, 138]

solvent extraction [133, 135]

For test for presence of metal carbonyl complexes cf Ref. [169].

5.4 Effect of Temperature and Pressure

Temperature and pressure markedly affect the reactants, intermediates (carbenium, acylium and carboxylium ions) and final products. In many cases, an inversion of product formation takes place.

The position and rate of attainment of equilibrium of the following important reactions are determined by the temperature:

dehydration/hydration and esterification/saponification (with alcoholic starting materials)

isomerization of the carbenium ions [34, 69]

oligomerization/depolymerization of carbenium ions [43, 147]

carboxylation/decarboxylation (relates to equilibria of acyl and carboxyl ions).

In addition, physical effects such as the solubility of CO in the reaction mixture also affect the equilibria. With regard to the relationship between temperature and pressure, many effects which are apparently due to temperature can also be interpreted as being a consequence of a change in CO pressure.

In general, higher yields and a more uniform product are achieved at high CO pressures. This is due to the trapping of the carbenium ion (via transformation into acyl complexes) which represses the attainment of the isomerization and oligomerization equilibria thereby preventing the formation of a series of by-products [11, 51, 80].

While the carboxylation-decarboxylation equilibrium is almost unaffected by CO pressure, the temperature plays a significant role (cf Sect. 5.2.6 – Retro-Koch reaction).

Examples

The effect of pressure can be appreciated from Table 5.1.

At pressures higher than 20 bar, C_6 acids result exclusively from 2-pentene whereas at lower pressures higher carboxylic acids are obtained. At 1 bar pressure, the latter acids account for 90% of the product.

The fundamental importance of the CO pressure or CO concentration in the reaction mixture is also impressively underlined by Haaf's results based on the "formic acid method [14]".

Table 5.1. Influence of CO pressure on 2-pentene reaction (catalyst 96% H_2SO_4)

CO pressure (atm)	Yield of carboxylic acids (mole; %)	Composition of carboxylic acids according to the number of C-atoms					Composition of C_6-acids (mole; %)	
		C_6	C_{11}	C_{10}	C_{21}	C_{26}	tert.	sec.
1	31.5	11.1	25.6	19.8	18.7	24.5	100	–
5	59.0	24.3	24.7	18.2	32.8	–	100	–
10	77.5	95.0	5.0	–	–	–	79.7	20.3
20	78.0	100	–	–	–	–	64.5	35.5
30	88.0	100	–	–	–	–	60.9	40.0

Table 5.2. Influence of stirring velocity on yield of carboxylic acids in the Koch synthesis using formic acid as CO source

Starting material	% Yield (first figure obtained with moderate stirring 20 cpm; figures in brackets with vigorous stirring)
2-Pentanol	30 (79) 2,2-Dimethylbutanoic acid
	26 (1) 2-Ethylbutanoic acid
	26 (1) 2-Methylpentanoic acid
2-Heptanol	17 (29) 2,2-Dimethylhexanoic acid
	28 (48) 2-Ethyl-2-methylpentanoic acid
	9 (3) 2-Propylpentanoic acid
	23 (1) 2-Ethylhexanoic acid
	8 (1) 2-Methylheptanoic acid
Cyclohexanol	14 (61) 1-Methylcyclopentanecarboxylic acid
	75 (8) Cyclohexanecarboxylic acid

Haaf stated that less effective stirring of the reaction mixture can result in an supersaturation of carbon monoxide in the range of 10^2. This increase is sufficient to convert large amounts of the initial secondary carbenium ions to secondary carboxylic acids before rearrangement to tertiary carbenium ions can occur (cf Table 5.2).

The drastic displacement of the ratio of secondary to tertiary carboxylic acids (e.g., with cyclohexanol) at the same temperature is particularly interesting.

It can therefore be assumed that the previous relationship found between temperature and ratio of tertiary secondary carboxylic acids [34, 170, 171] (at higher temperatures – more tertiary acids) is probably largely due to a lowering of the CO concentration in the reaction mixture instead of increased isomerization to the more stable tertiary carbenium ion with rising temperature (cf Table 5.3).

Gushchin's studies on the conversion of 1-hexene in the presence of $BF_3 \cdot H_2O$ [34] must also be examined in light of the above. Gushchin found the tertiary acid content in the product increased on raising the temperature (20 → 100 °C) and on decreasing the CO pressure (85 → 27 bar).

Table 5.3. Temperature dependence of the ratio of tertiary to secondary carboxylic acids; CO-pressure 100 bar

Olefin	H_2SO_4 (96%)				$BF_3 \cdot CH_3OH$			
	tert. acid (%)		sec. acid (%)		tert. acid (%)		sec. acid (%)	
	−5 °C	+15 °C	−5 °C	+15 °C	−5 °C	+15 °C	−5 °C	+15 °C
1-Hexene	43	58	57	42	22	35	78	65
1-Heptene	54	68	46	32	26	42	74	58
1-Octene	50	64	50	36	30	43	70	57
1-Nonene					34	44	66	56
1-Decene					33	50	67	50

Table 5.4. Influence of temperature on reaction of cyclohexene (cyclohexene: $HCOOH : H_3PO_4$: CH_3OH = 1 : 5 : 8 : 8) [157]

Reaction conditions		Reaction products; Yield in %			
Temp. °C	Pressure bar	⬡-COOCH₃	⬠ COOCH₃ CH₃	Higher esters	Σ
30	1	35	60	5	95
50	1	41	47	12	88

The significance of the CO concentration in the reaction mixture is also supported by recent findings by Ordyan [157] cf Table 5.4.

The marked effect of temperature can also be observed in the esterification-saponification equilibrium with alcohols. Usually, more esters result at lower temperatures, while at higher temperatures virtually only acids are formed (cf Table 5.5). However, the question remains unanswered as to whether the primary resulting esters are rapidly saponified at higher temperatures or whether oxonium complexes with ROH are unstable.

5.5 Solvents and Diluents

The catalyst solution which is used in excess generally also serves as solvent. Solvents are employed in particular in the commercial reactivation of recycled catalysts, the addition ensues either during the reaction [138], whereby yield and selectivity are simultaneously increased [27, 41, 71, 100, 101, 109, 119, 122, 124, 125, 132, 133, 147, 162, 163, 172, 173] or later to extract the resulting acids [135].

According to Komatsu [122], with H_2SO_4 the solvent effects are related to the acidity of the catalyst (H_2SO_4), because when pivalic acid or $Cl_2C=CHCl$ was used as

Table 5.5. Effect of temperature and pressure on the reaction of isobutanol and isobutene [80] (catalyst: $BF_3 \cdot 2\,H_2O$)

I. Starting material	Reaction conditions Temperature °C	Pressure bar	Composition of product (yield in %) C–C–COOH (C,C branches)	C–C–COOR (C,C branches)	C–C–C–C–COOH (C,C branches)	C–C–C–C–COOR (C,C branches)	C_{13}-acids	Σ %
Isobutanol	110	20	57	25	7	5	–	94
	110	100	91	Sp	5	Sp	–	96
	60	100	4	71	3	4	–	82
Isobutene	30	40	38	–	21	–	14	73
	30	100	71	–	13	–	11	95
	50	100	47	–	15	–	7	69

391

solvent, the acidity was clearly related to the pivalic acid yield. For discussions of the mechanism of these solvent effects cf Ref. [122].

In contrast to the above, Gulf disclosed in a patent that with higher α-olefins (> C_{16}) higher yields of carboxylic acids were obtained in the *absence* of a solvent [165].

The following compounds were proposed as solvents: saturated hydrocarbons such as n-hexane [71, 109, 119, 133, 147], cyclohexane [27, 132, 134, 138], methylcyclohexane [133], benzene [133, 135], chlorobenzene [134, 138, 173], chloroform [100, 133, 173], trichloroethylene [101, 122], tetrachloroethylene [41, 124, 125], methylene chloride [162], trifluoro- and trichloroethane [163], carbon tetrachloride [125], fluorobenzene [90] or mixtures thereof.

5.6 Carbonylation of Particular Compounds

The general rules discussed in the reaction mechanism Sect. also apply to the various starting materials. Particular features stemming from a special structure of the reactant will be briefly outlined. The literature referred to in the tables must be consulted, if further information is required.

5.6.1 Olefins and Dienes

A mixture of secondary and tertiary carboxylic acids results from n-olefins. With "internal" (non-terminal) olefins only tertiary carboxylic acids are produced. In the last case, the position of the double bond has no effect on the isomeric distribution. With alicyclic olefins, the course of reaction is determined by the stability of the intermediate carbenium ion (dependent on ring size: tendency to form stable 5- and 6-membered rings) in addition to the usual reaction parameters.

A number of rules have been derived for the carbonylation of cycloolefins according to the "formic acid method" [13, 14] (deviations from the rules relating to isomer distribution resulting from supersaturation of the reaction mixture have been discussed in Sect. 5.4)

1. Tertiary acids are formed nearly exclusively, except from those cycloolefins that are unable to form these acids because of ring strain.
2. Branched cyclohexane ring systems are formed if possible. The ratio of secondary to tertiary acids obtained from cyclic olefins of different ring size can be seen in Fig. 5.7.

Cyclopentene reacts exclusively to give cyclopentanecarboxylic acid. Formation of methyl-substituted four-membered rings has not been observed. Also, cyclohexene reacts to give nearly 90% cyclohexanecarboxylic acid, whereas cycloolefins containing 8, 9 and 10 C-atoms in the ring generally do not form secondary acids. The performance of the methyl-substituted cycloalkenes can be appreciated from Fig. 5.8.

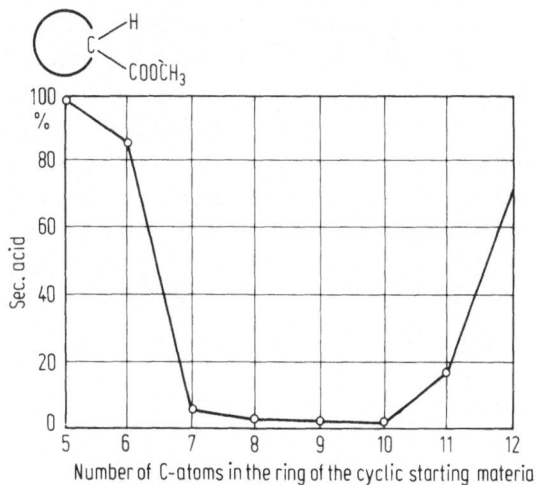

Fig. 5.7. Methyl esters from cycloalkenes

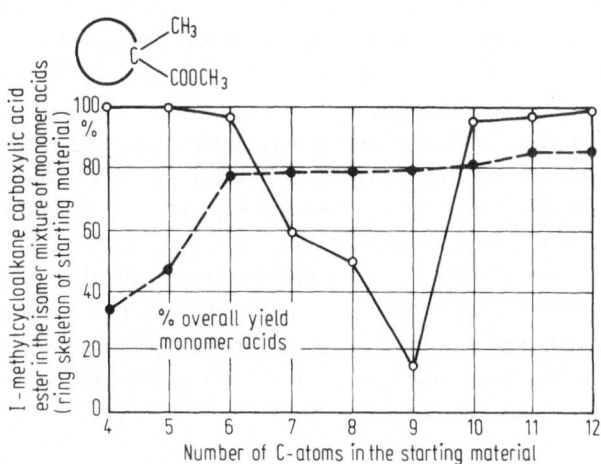

Fig. 5.8. Methyl esters from 1-methylcycloalkenes (catalyst $BF_3 \cdot CH_3OH$, 150 bar CO, 15 °C)

Branching at the ring skeleton favors formation of carboxylic acids with retention of the original ring size. Cycloolefins containing 4, 5, 6, 10, 11 and 12 C-atoms in the ring react exclusively with retention of the ring skeleton. Of the C_7 to C_9 cycloolefins (capable of isomerization), the nine-membered ring has the greatest stability. In the case of four- and five-membered rings, dimerization of the starting material is the preferred reaction.

Diolefins with widely separated substituted double bonds react to give dicarboxylic acids in low yields [42]. Thus, from 2,11-dimethyl-1,11-dodecadiene, 2,11-dimethyl-dodecanedicarboxylic acid is obtained in approximately 30% yield [183]. Under the conditions of the Koch synthesis, conjugated dienes polymerize in the presence of the strong acid catalyst [42].

Highly unsaturated cyclic compounds likewise do not react to give di- or tri-carboxylic acids, but tertiary monocarboxylic acids are obtained with CO/H_2O via a transannular reaction with hydride transfer and formation of bi- or tricyclic systems at the bridgehead. Thus, 1,5-cyclodecadiene reacts to give a mixture of cis- and trans-decalincarboxylic acids [184] and 1,5,9-cyclododecatriene reacts to give a mixture of isomeric tertiary perhydroacenaphthenecarboxylic acids. Further data can be found in the literature i.e., Refs. [15, 25, 38, 41, 43, 44, 52, 57, 58, 68–71, 73–76, 79–81, 84, 86, 89, 91, 101, 104, 106, 109, 114, 115, 117–120, 122–125, 130, 131, 133, 134, 138, 140, 142, 143, 145, 146, 148, 149, 152, 153, 155–157, 163, 172, 174, 196] as well as in Table 5.6.

5.6.2 Alcohols and Diols

In most cases, higher yields are obtained starting from alcohols, since carbenium ions are formed more slowly and therefore dimerization is suppressed. In general, conversion and yield increase with rising C number. Thus, methanol [185, 186] is converted into acetic acid under severe conditions (cf Table 5.7). Propanol can be reacted under pressure but no reaction occurs under the conditions of the normal pressure carbonylation with formic acid [180]. While there is no structural rearrangement to tertiary carbenium ions with n-butanol and H_2SO_4, it ensues to 50% in the presence of Cu or Ag carbonyls [16, 33]. The normal pressure carbonylation with Ag/Cu^+ carbonyls yields exclusively tertiary carboxylic acids with all alcohols $\geq C_5$ [33].

Tertiary alcohols are converted very easily into carbenium ions, Therefore, they are suitable for carbonylation with formic acid. The reaction of alcohols of the neopentyl-glycol type is of interest because fo their structure, which is not comparable to that of olefins. Also, these alcohols form no primary carboxylic acids only tertiary acids result via skeletal rearrangement [14].

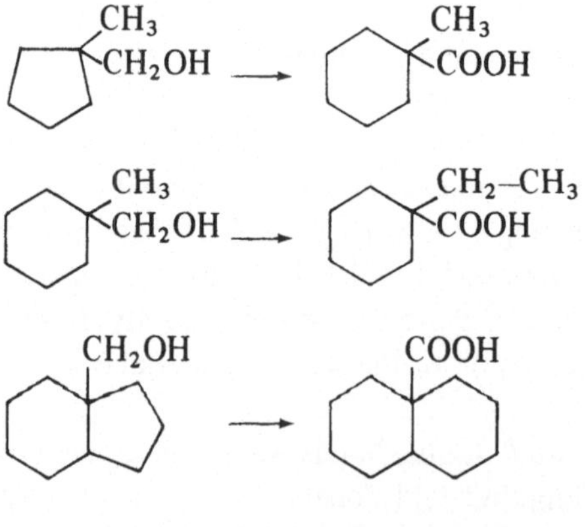

Ditertiary diols are converted into dicarboxylic acids if the reactive centers are widely separated [42]. Monocarboxylic acids are obtained exclusively if the branchings are four CH_2-groups distant (see Table 5.7). A distance of 5 to 8 C-atoms favors forma-

Table 5.6. Hydrocarboxylation of olefins and dienes via the Koch route

Starting material	Catalyst	Reaction products	Yield (%)	Ref.
Ethylene	HF + BF$_3$	Propionic acid	30	[175]
	(H$_3$O)(BF$_4$)	Ethyl α-methyl-α-ethyl-butyrate	~30	[100], 77c]
Propene	BF$_3$ · 2 ClCH$_2$COOH · H$_2$O	Isobutyric acid	100	[86]
Propene	97% H$_2$SO$_4$	Isobutyric acid	90	[118, 176]
1-Butene 2-Butene	97% H$_2$SO$_4$	2-Methylbutyric acid	90–95	[176]
Isobutene	BF$_3$ · 2 H$_2$O	Pivalic acid (72.5%) C$_9$-acid (13%) higher molecular weight acids (14.5%)		[177]
	82–88% H$_2$SO$_4$	Pivalic acid		[102, 173]
Isobutene	HF	Pivalic acid	90–98	[152]
2-Methyl-1-butene	BF$_3$ · 2 H$_2$O	Ethyldimethylacetic acid	75	[177]
2-Methyl-1-butene	BF$_3$ · 2,5 H$_2$O	2,2-Dimethylbutyric	93	[79]
2,3-Dimethyl-1-butene	H$_2$SO$_4$	2,2,3-Trimethylbutyric acid	70	[176]
2-Ethyl-1-butene		2-Methyl-2-ethyl-butyric acid		[176]
1-Pentene	HF/H$_2$O	2,2-Dimethylbutyric acid 2-Ethylbutyric acid 2-Methylvaleric acid	36 37 27	[51]
1-Pentene	BF$_3$ · 2,5 H$_2$O	2,2-Dimethylbutyric acid	92	[82]
2-Pentene	97% H$_2$SO$_4$	C$_5$/C$_6$-acids		[178]
2-Methyl-1-pentene	H$_2$SO$_4$/BF$_3$/H$_3$PO$_4$	2,2-Dimethylvaleric acid C$_{13}$-acids	80	[176]
3-Methyl-2-pentene		2-Methyl-2-ethyl-butyric acid		[176]
2-Ethyl-1-pentene	BF$_3$ · 2 H$_2$O	C$_6$-acids	62	[177]
1-Hexene	H$_3$PO$_4$	C$_7$-acids (87%) C$_{10}$-acids (13%)		[179]
1-Hexene	BF$_3$ · H$_2$O	C$_7$-tert. carboxylic acids (52) C$_7$-sec. carboxylic acids (31)	83	[34]
1-Hexene	> 90% H$_2$SO$_4$, Cu(I)	C$_7$-acids	96	[135]
1-Hexene	H$_2$SO$_4$ 58%-Ag(I)	C$_7$-acids 2,2-Dimethyl-heptanoic acid (61)	96	[16] [16]
Diisobutylene	82–88% H$_2$SO$_4$	Isononanoic acid		[102]
Diisobutylene	BF$_3$ · H$_2$O/Cu(I)	C$_5$-acids (42) C$_6$-C$_8$-acids (12) C$_9$-acids (45) C$_{>10}$-acids (1)	95	[136]
Propylene tetramer	(H$_3$O)(BF$_4$)	C$_{13}$-acid	96	[50]
Cyclopentene	H$_2$SO$_4$/HCOOH	Cyclopentanecarboxylic acid cis-Decalincarboxylic acid (9:1)	6	[180]
Cyclohexene	H$_2$SO$_4$/HCOOH	1-Methylcyclopentane-carboxylic acid	75	[180]
Cyclohexene	H$_2$SO$_4$ Ag(I)	1-Methylcyclopentane carboxylic acid	84	[16]

Table 5.6 (continued)

Starting material	Catalyst	Reaction products	Yield (%)	Ref.
1-Octene	H_2SO_4 58% Ag(I)	2-Methyl-2-ethyl-hexanoic acid (24)	97	[16]
		2-Methyl-2-propyl-pentanoic acid (12)		
Cyclooctene	H_2SO_4/HCOOH	1-Ethylcyclohexane carboxylic acid, higher molecular weight acids (36:64)	45	[181]
2-Ethylhexene	$BF_3 \cdot H_3PO_4$	2-Methyl-2-ethylhexane carboxylic acid	83	[72, 77b]
2,7-Dimethyl-4-octene	H_2SO_4 98,5%/HCOOH	2,2,7,7-Tetramethyl-suberic acid	32	[132, 210]
2,7-Dimethyl-octadiene	H_2SO_4 98%	2,2,7,7-Tetramethyl-suberic acid		[132, 210]
1-Hexadecane	H_2SO_4 97%	C_{17}-acids	74	[165]
2,5-Dimethyl-1,5-hexadiene	H_2SO_4	2,2,5,5-Tetramethyl-adipic acid	2–3	[177]
		Mono carboxylic acids	20–30	
Dihydrodicyclo-pentadiene	H_2SO_4	Exo-Tricyclo-(5.2.1.0)-decene-2-carboxylic acid	74	[182]
2,11-Dimethyl-dodecadiene		2,11-Dimethyldodecane-2,11-dicarboxylic acid	30	[177, 98]
C_{17}-Diene-mixts.	strong acids	C_{19} dicarboxylic acids		[197]
Cyclohexene MeOH	$F_3C SO_3H$	1-Methyl-bicyclohexyl carboxylat	73	[163]

tion of dicarboxylic acids. An optimum is observed with a distance in the range of 6 or 7 C-atoms [183].

Main products in the carbonylation of primary diols are ω-unsaturated monocarboxylic acids, ω-hydroxy acids and lactones [42, 187, 204]. However, Schauerte [183] succeeded in carbonylating 1,10-decanediol to 2,2,7,7-tetramethylsuberic acid together with other products.

Lactones are the main products in the reaction of carbon monoxide and diols with structures that favor lactone formation. Thus, 2,5-dimethyl-2,5-hexanediol reacts to give the corresponding C_9-lactone [187].

Koch syntheses are accompanied by isomerization so that in most cases the reaction products are more numerous than the starting material. On the other hand, the tendency to undergo isomerization in the case of alicyclic alcohols favor homogenization of the reaction products. Thus, a mixture of hydrogenated cresols (1,2-, 1,3- and 1,4-cresol) as well as 1-methyl-1-cyclohexanol forms only one main product — 1-methylcyclohexanecarboxylic acid. Similar results are obtained with hydrogenated xylenols.

On carbonylating cyclopentanol in the presence of H_3PO_4, a cis/trans mixture of decalincarboxylic acids results [58].

Additional information can be found in the Refs. [35, 38, 57, 58, 68, 112, 131, 139, 151, 154, 164] and Table 5.7.

Table 5.7 Hydrocarboxylation of alcohols and diols via Koch synthesis route

Starting material	Catalyst	Temp. ($^{\circ}$C)	Yield of carboxylic acids (%)	Distribution of reaction products (%)	Ref.
Methanol	BF$_3$	160–200 (765–1120 bar)	92	Acetic acid	[185, 186]
	H$_3$PO$_4$				[61]
n-Propanol	H$_2$SO$_4$			Isobutyric acid	[180]
Isopropanol	H$_2$SO$_4$			Isobutyric acid	[180]
Isopropanol	H$_2$SO$_4$/CH$_3$OH	45–48	89	77% Me$_2$CHCOOMe 12% Me$_2$CHCOOH	[111]
1-Butanol[a]	H$_2$SO$_4$	30	36	85% 2-Methylbutyric acid 15% C$_9$-acids	[180]
1-Butanol	H$_2$SO$_4$/Cu$^+$	30 (1 atm)	22 23	Pivalic acid 2-Methylbutyric acid	[33]
1-Butanol	H$_2$SO$_4$/Ag$^+$	30	29 17	Pivalic acid 2-Methylbutyric acid	[16]
2-Butanol[a]	H$_2$SO$_4$	20	43	100% 2-Methylbutyric acid	[180]
2-Butanol	H$_2$SO$_4$/Cu$^+$	30 (1 bar)	28 19	Pivalic acid 2-Methylbutyric acid	[33]
tert-Butanol[a]	H$_2$SO$_4$	10–15	78	95% Pivalic acid	[180, 77c]
tert-Butanol	HF	25–150		Pivalic acid	[151]
tert-Butanol	H$_2$SO$_4$/Cu$^+$		52	Pivalic acid	[139]
tert-Butanol	BF$_3$·H$_2$O/Ag$^+$	35	74 10	C$_5$-acids C$_9$-acids	[146]
Isobutanol	H$_2$SO$_4$	60	89	Pivalic acid	[116]
Isobutanol	H$_2$SO$_4$/Cu$^+$	30 (1 bar)	56	Pivalic acid	[33]
Isobutanol	BF$_3$·H$_2$O	100	>89	Isobutyl pivaloate	[88]
Isobutanol	BF$_3$·2 H$_2$O	110	91	Pivalic acid	[80]
Isobutanol	BF$_3$·2 H$_2$O	60	71	Isobutyl pivaloate	[80]
1-Pentanol[a]	H$_2$SO$_4$	10	76	77% 2,2-Dimethylbutyric acid 23% higher acids mainly C$_{11}$-acids	[180]
1-Pentanol	BF$_3$·H$_2$O	100	73	2,2-Dimethyl-butyric acid	[82]

Table 5.7 (continued)

Starting material	Catalyst	Temp. (°C)	Yield of carb-oxylic acids (%)	Distribution of reaction products (vol-%)	Ref.
1-Pentanol	H_2SO_4	20–40 (38–47 bar)	85	58 wt-% 2,2-Dimethylbutyric acid-methyl ester 27 wt-% 2-ethylbutyric acid methyl ester	[191] [192]
2-Pentanol[a]	H_2SO_4	5	81	79% 2,2-Dimethylbutyric acid 21% higher acids mainly C_{11}-acids	[180]
3-Methyl-1-butanol	$BF_3 \cdot H_2O$	100	82 10 5	2,2-Dimethylbutyric acid C_5-Ester of 2,2-Dimethylbutyric acid sec. acids	[78]
2-Methyl-2-butanol[a]	H_2SO_4	5–10	73	10% Pivalic acid 42% 2,2-Dimethylbutyric acid 12% C_7-acids 36% higher acids mainly C_{11}-acids	[180]
2,2-Dimethyl-1-propanol[a]	H_2SO_4	22–26	83	100% 2,2-Dimethylbutyric acid	[180]
2,2-Dimethyl-1-butanol[a]	H_2SO_4	10–25	80	63% 2-Methyl-2-ethylbutyric acid 37% 2,2-dimethylvaleric acid	[180]
1-Hexanol	H_2SO_4/Cu^+	30 (1 bar)	60 25 4 2	2,2-Dimethylpentanoic acid 2-Methyl-2-Ethylbutyric acid 2,2-Dimethylbutyric acid Pivalic acid	[33]
1-Hexanol	H_2SO_4/Ag^+	30	79 19	2,2-Dimethylpentanoic acid 2-Methyl-2-ethylbutyric acid	[16, 145]
Cyclohexanol	H_2SO_4/Cu^+	30 (1 bar)	80	1-Methylcyclopenanoic acid	[33]
Cyclohexanol	H_2SO_4/Ag^+	30	81	1-Methylcyclopentanoic acid	[16]
2,2-Dimethyl-pentanol[a]	H_2SO_4	28–26	82	65% 2-Methyl-2-ethylvaleric acid 35% 2,2-dimethyl-caproic acid	[180]
1,1-Diethyl-propanol[a]	H_2SO_4	5–10	64	8% C_6–C_7-acids 45% C_8-acids 9% C_9–C_{10}-acids 38% C_{15}-acids	[180]
1-Octanol	H_2SO_4/Cu^+	30	95	75% 2,2-Dimethyl-heptanoic acid 25% 2-Methyl-2-ethylhexanoic acid	[141]
1-Decanol	H_2SO_4/Cu^+	30 (1 bar)	75 15 5	2,2-Dimethylnonanoic acid 2-Methyl-2-ethyloctanoic acid 2-Methyl-2-propyl-heptanoic acid	[33]
2,3,3-Trimethyl-2-butanol[a]	H_2SO_4	10	88	100% 2,2,3,3-Tetramethylbutyric acid	[180]
1-Adamantyl-methanol		25 – 15		Homoadamantane-1-carboxylic acid Homoadamantane-3-carboxylic acid	[194]

Table 5.7 (continued)

Starting material	Catalyst	Temp. (°C)	Yield of carboxylic acids (%)	Distribution of reaction products (vol-%)	Ref.
$\alpha,\alpha,\alpha',\alpha'$-Tetramethylbutanediol-1,4	BF$_3$/acid	150		$\alpha,\alpha,\alpha',\alpha'$-Tetramethyl-adipinic acid	[83]
Hexanediol	H$_2$SO$_4$			13% C$_8$-Dicarboxylic acid 33% 2,2,4-trimethylbutyrolactone	[187]
2,5-Dimethyl-2,6-hexanediol	H$_2$SO$_4$	20		44% C$_{10}$-Diacid 23.5% C$_9$-lactone	[187]
1-Hydroxy methyl-1-methyl-cyclopentane[a]	H$_2$SO$_4$	19–23	95	1-Methylcyclohexanecarboxylic acid	[188, 180]
1-Hydroxy methyl-1-methyl-cyclohexane	H$_2$SO$_4$	20–25	86	1-Ethylcyclohexane carboxylic acid	[188, 180]
5-Hydrindanol	H$_2$SO$_4$	10–15	56	100% cis Hydrindane-8-carboxylic acid	[189]
β-Decalin-5-ol	H$_2$SO$_4$		95	Decalin-9-carboxylic acid	[188, 190]
cis-8-Hydroxy-methyl hydrindane[a]	H$_2$SO$_4$	15–20	68	Decalin-9-carboxylic acid	[189]
1,10-Decanediol[a]	H$_2$SO$_4$	25	7	2,2,7,7-Tetramethylsuberic acid	[183]
2,9-Dimethyl-decanediol-2,9[a]	H$_2$SO$_4$	0–5	87	63% 2,2,9,9-Tetramethylsebacic acid	[183]
Mixtures of cis/trans 4-tert-Butyl-1-methyl-cyclohexanol	SbCl$_5$/SO$_2$ SO$_2$(liq.)	–70	92	Cis-4-tert-Butyl-1-methyl-cyclohexanoic acid	[162]

[a] With formic acid

5.6.3 Paraffins

In accordance with the general discussions in Sect. 5.2, paraffinic hydrocarbons are accessible to the Koch synthesis on forming carbenium ions via ring cleavage (with cyclic compounds), via disproportionation and via carbon-hydride shift. Examples of ring cleavage with alicyclic compounds are summarized in Table 5.8.

Table 5.8. Hydrocarboxylation of reactive alicyclic compounds

Starting material	Catalyst	Temp. (°C)	Reaction product	Yield %	Ref.
Cyclopropane	H_2SO_4	20	Isobutyric acid	80	[64]
Methylcyclopropane	H_2SO_4	30	2-Methylbutyric acid	78	[64] [13]
1,1-Dimethylcyclopropane	H_2SO_4	50	2,2-Dimethylbutyric acid	60	[64]
n-Propylcyclopropane	H_2SO_4	20	2,2-Dimethylvaleric acid and 2-methyl-2-ethylbutyric acid	82	[13]
Cyclobutane	H_2SO_4	50	2-Methylbutyric acid	30	[64]
Methylcyclobutane	H_2SO_4	70	2-Methylvaleric acid	17	[64]
Methylcyclopentane	$FSO_3H/$ SbF_5/Cu	0	1-Methylcyclopentanecarboxylic acid	45	[137]
			Cyclohexanecarboxylic acid	40	
Methylcyclopentane	HF/SbF_5	0–20	Cyclohexanecarboxylic acid	85	[19]
			1-Methylcyclopentanoic acid	5 (crude)	

The reactivity of small strained rings decrease in the series cyclopropane > alkyl-substituted cyclopropane > cyclobutane > alkyl-substituted cyclobutane [64]. Cyclopentane does not react under the action of sulfuric acid but cyclopentane as well as cyclohexane react in the presence of HF/SbF_5 [65], which, compared to H_2SO_4, is a more effective agent for yielding carbenium ions from paraffins.

Examples of the Koch synthesis involving iso-paraffins and carbon-hydride shifts are summarized in Table 5.9.

Low molecular iso-paraffins in sulfuric acid are especially suited as hydride ion donors. Under the conditions of the Koch synthesis, tertiary carboxylic acids are formed [20, 21, 22–24].

In general, higher yields can be achieved with the normal pressure carbonylation when conducted in the presence of elements of Group 1b (cf Ref. [26]).

Starting from adamantane and tert-butanol (as hydride ion acceptor), adamantane-carboxylic acid is formed in 80% yield. An 85% yield of adamantanedicarboxylic acid is reported in a DuPont patent [67].

The tendency to undergo the hydride transfer reaction (and therefore the yield of carboxylic acids) falls on passing from lower to higher homologs of iso-paraffins. Of course the yield largely depends on the structure of the hydride ion acceptor. The following may be used as hydride ion acceptors: olefins, alcohols, or alkyl chlorides.

5.6.4 Unsaturated Carboxylic Acids

Generally speaking, unsaturated carboxylic acids are very suitable starting materials for the Koch snythesis [8, 65, 147, 180, 183]. The industrially interesting higher carboxylic acids, e.g., based on oleic acid, can be converted at normal pressure in the presence of 97% H_2SO_4 (cf Refs. [108, 199, 200, 201] and Table 5.10).

Table 5.9. Hydrocarboxylation of isoparaffins by hydride transfer

Starting material	Catalyst	Olefins or alcohols (hydride ion acceptor)	Temp. (°C)	Ref.	Yield of carboxylic acids (%) from	
					Isoparaffin	Olefin or alcohol
Isopentane	H_2SO_4	Cyclohexene	15–25[a]	[24]	37% 2,2-Dimethylbutyric acid	52% Cyclic C_7-acids (90% methylcyclopentane-carboxylic acid)
2-Methylpentane	H_2SO_4	Cyclohexene	15–25	[24]	21% C_7-acids (no cyclic acids)	59% C_7-acids (cyclic)
2,3-Dimethylbutane	H_2SO_4	Methylcyclohexene	15–25	[24]	22% 2,2,3-Trimethylbutyric acid	33% 1-Methylcyclohexane-1-carboxylic acid
2,3-Dimethylbutane	H_2SO_4	1-Heptene	15–25	[24]	31% 2,2,3-Trimethylbutyric acid	46% C_8-acids
Methylcyclopentane	H_2SO_4	Propene	15–25	[24]	28% 1-Methylcyclopentane-1-carboxylic acid	1% Isobutyric acid
n-Hexane	H_2SO_4	Cyclohexene	15–25	[24]		79% Cyclic C_7-acids (no aliph. acids)
Methylcyclopentane	H_2SO_4	Isopropanol	15–25	[24]	18% 1-Methylcyclopentane-1-carboxylic acid	1% Isobutyric acid
Cyclohexane	H_2SO_4	Tert-butanol	15–25	[24]	7% C_6- and C_7-acids, no aliph. C_7-acids	54% Pivalic acid
Cyclohexane	H_2SO_4	2-Butanol	15–25	[24]	16% 1-Methylcyclopentane-1-carboxylic acid 3% cyclohexanecarboxylic acid	13% 2-Methylbutyric acid
trans Decalin	H_2SO_4	2-Butanol	15–25	[24]	8% Decalin-9-carboxylic acid (mainly cis)	12% 2-Methylbutyric acid
2-Methylpentane	H_2SO_4/Cu^+	Octene-1	35 (1 bar)	[28]	53% 2,2-Dimethyl-pentanoic acid 17% 2-Methyl-2-ethylbutanoic acid	10–15% C_9-acids
Methylcyclohexane	H_2SO_4/Ag^+	1-Hexene	30 (1 bar)	[16]	45% Methylcyclohexanecarboxylic acid	33% tert-C_7-acids
Methylcyclohexane	H_2SO_4/Ag^+	1-Hexanol	30 (1 bar)	[16]	42% Methylcyclohexane-carboxylic acid	31% tert-C_7-acids
Methylcyclohexane	H_2SO_4/Ag^+	1-Octene	30 (1 bar)	[16]	72% Methylcyclohexa-carboxylic acid	23% tert-C_9-acids

Table 5.9 (continued)

Starting material	Catalyst	Olefins or alcohols (hydride ion acceptor)	Temp. (°C)	Ref.	Yield of carboxylic acids (%) from	
					Isoparaffin	Olefin or alcohol
1,4-Dimethyl-cyclohexane	H_2SO_4/Ag^+	1-Octene	30 (1 bar)	[16]	60% 1,4-Dimethylcyclo-hexanecarboxylic acid (cis: trans = 1:1)	30% tert-C_9-acids
Methylcyclopentane	H_2SO_4/Ag^+	1-Octanol	30 (1 bar)	[16]	60% Methylcyclopentane carboxylic acid	27% tert-C_9-acids
n-Octane	H_2SO_4/Ag^+	1-Hexanol	30 (1 bar)	[16]	0% tert-C_9-acids	75% tert-C_7-acids
4-Homoisotwistane-(tricyclo-[5,3,1,0^{3,8}] undeane)	$H_2SO_4/$ HCOOH	tert-Butanol	10–15	[27, 198]	63% 4-Homoisotwistane-3-carboxylic acid	no ref.
Isopentane	$BF_3 \cdot H_2O$	Isobutane		[84]	42% 2,2-Dimethylbutyric acid	C_5-acids
Isopentane	$BF_3 \cdot H_2O$	Diisobutylene		[25]	2,2-Dimethylbutyric acid	Pivalic acid
trans Decalin	H_2SO_4	tert-Butanol	15–25^a	[24]	5% Decalin-9-carboxylic acid	47% Pivalic acid 19% C_6, C_7-C_9-acids
Methylcyclopentane	H_2SO_4	tert-Butanol	15–25	[24]	46% 1-Methylcyclopentane-1-carboxylic acid	7% Pivalic acid
Methylcyclohecane	H_2SO_4	tert-Butanol	15–25	[24]	72% 1-Methylcyclohexane-1-carboxylic acid	16% Pivalic acid
Methylcyclohexane	H_2SO_4	2-Pentanol	15–25	[24]	36% 1-Methylcyclohexane-1-carboxylic acid	30% 2,2-Dimethylbutyric acid
Methylcyclohexane	H_2SO_4	1,3-Butadiene	15–25	[24]	15% 1-Methylcyclohexane-1-carboxylic acid	
C_{13}-acid from propene tetramer	H_2SO_4	tert-Butanol	15–25	[24]	6% C_{14}-Dicarboxylic acids	30% Pivalic acid 43% C_{13}-acid
1,4-Dimethylcyclo-hexanecarboxylic acid	H_2SO_4	tert-Butanol	15–25	[24]		71% Pivalic acid
2-Methylbutane	H_2SO_4	tert-Butanol	20	[24]	57% 2,2-Dimethylbutyric acid methyl ester 1-methylcyclohexane-carboxylic acid (ratio ester to acid 1.4:1)	11% Pivalic acid methyl pivalate

^{a} Without pressure, formic acid method, in the experiments with 2-methylbutane and methylcyclohexane methanol was added in the 2nd reaction step

Table 5.10. Carboxylation of various unsaturated carboxylic acids

Starting material	Temp. °C	Catalyst	Reaction products (distribution)		Yield	Ref.
Acrylic acid	40	H_2SO_4/SO_3	Succinic acid		85	[105, 65, 113]
Oleic acid	10–20	H_2SO_4	Mixture of carboxystearic acids			[108, 126]
Oleic, tall oil fatty and partially hydrogenated tall oil fatty acid	20–25	H_2SO_4	25% secondary C_{19}-dicarboxylic acids			[121]
			75% tertiary C_{19}-dicarboxylic acids		83	
Undecylenic acid	3–16	H_2SO_4	*49% secondary diacids*			[183]
			α-Methylundecane diacid	9%		
			α-Ethylsebacic acid	25%		
			α-n-Propylazelaic acid	15%		
			41% tert. diacids			
			α,α-Dimethylsebacic acid	20%		
			α-Methyl-α-ethylazelaic acid	10%		
			α-Methyl-α-n-propylsuberic acid	8%		
			α-Methyl-α-n-butylpimelic acid	3%		
Undecylenic acid	3–16	$H_2SO_4/$ HCOOH	*100% tert. diacids*			[183]
			α,α-Dimethylsebacic acid	47%		
			α-Methyl-α-ethylazelaic acid	24%		
			α-Methyl-α-n-propylsuberic acid	23%		
			α-Methyl-α-n-butylpimelic acid	6%		

Table 5.11. Carbonylation of halogenated compounds [99]

Starting material	Reaction product	Yield (%)	Ref.
Methallyl chloride	Chloropivalic acid	70	[99]
Methallyl bromide	Bromopivalic acid	90	[99]
β-Citronellyl chloride	2,2,6-Trimethyl-8-chloro-1-octanoic acid	90	[99]
β-Citronellyl bromide	2,2,6-Trimethyl-8-bromo-1-octanoic acid		[99]
8-Bromo-1-octene	2,2-Dimethyl-7-bromo-heptanoic acid 2-Methal-2-ethyl-6-bromo-hexanoic acid		[99]
β-Chloromethyl-2-propanol	Chloropivalic acid	68	[87]
2,3,5,6-Tetra-methyl-1,4-di-chloromethyl-benzene	2,3,5,6-Tetra-methyl-1,4-benzene-diacetic acid		[100]

The HF-catalyzed carbonylations of oleic, eruic and isoundecenoic acids have also been discussed in a report [202].

5.6.5 Halogenated Compounds

Halogenated olefins [99] or alcohols [87] can be converted into α,α-dialkyl-β-halocarboxylic acids. If severe reactions conditions [158, 180] are applied (or tertiary alkyl halides [203] are employed) then carboxylic acids can be yielded from saturated instead of unsaturated halogenated aliphatic or aromatic compounds.

Halogenated adamantyl derivatives [103] as well as alkyl chlorosulfites or diakyl sulfites [159] also undergo the Koch reaction.

Various modern variants of the Koch reaction (cf Sects. 2–3 – Reaction mechanism) involve – depending on precursor – intermediate halogen additions and cleavages. The electrophilic carbonylation of aromatic compounds and amines is possible with the system Br_2–$SbCl_5$–CO–liq.–SO_2 [45]

$$X_2 + CO + HR' + HR'' \xrightarrow[\text{liq.SO}_2]{\text{SbCl}_5} 2\,HX + R'COR''$$

$R' = C_6H_5-, C_6H_5NH-$
$R'' = HO-, RO-, C_6H_5NH-$

Table 5.12. Electrophilic carbonylation of aromatic hydrocarbons and various amines in Br_2–$SbCl_5$–CO–liq. SO_2-system: Ethyl ester formation [45]

Starting material	Reaction products	Yield
Benzene	Ethyl benzoate	31
	Benzene (recovered)	58
Toluene	o-Toluic acid ethyl ester	8
	p-Toluic acid ethyl ester	13
	Monohalogenated toluenes	33
	Dihalogenated toluenes	35
Cyclohexylamine	C_6H_{11} NH COO C_2H_5	34
	Cyclohexylamine (recovered)	60
n-Hexylamine	n-C_6H_{13}NH COO C_2H_5	37
	n-Hexylamine (recovered)	62
Diethylamine	(C_2H_5) N–COOC_2H_5	77

The acylation of various nucleophiles can take place with the pivaloyl chloride – $SbCl_5$ complex derived from the reaction of tert-butyl chloride and CO in the $SbCl_5$–liq.SO_2 system [160].

$$R'X + CO + HR'' \xrightarrow[\text{liq.SO}_2]{\text{SbCl}_5} R'COR'' + HX$$

$R' = \text{tert.}-C_4H_9$
$R'' = HO-, RO-, \text{cyclohexenyl}-. \text{anilyl}-$

Table 5.13. Acylation of various nucleophiles with tert-butyl chloride in $SbCl_5-liq.SO_2$ system [160]

Nucelophile	Reaction product	Yield
Water	Pivalic acid	42
Ethyl alcohol	Ethyl pivaloate	49
Cyclohexene	$(CH_3)_3CCO-$	69
1-Methylcyclohexene	$(CH_3)_3CCO-$ with CH_3	40
Aniline	$(CH_3)_3 CCONH C_6H_5$	22

Table 5.14. Carbonylation of aldehydes

Starting material	Catalyst	Reaction products	Temp. °C	Yield mol %	Ref.
Acetaldehyde	HF	α-Hydroxypropionic acid	20	80	[150]
Paraformaldehyde	H_2SO_4/SO_3	glycolic acid	70	~100	[32]
aq. Formaldehyde	H_2SO_4/Ag^+	glycolic acid (polymer)	30	85	[144]
Formaldehyde/ AcOH	mineral acids	α-Acetoxyacetic acid	150	12–52	[62]
Formaldehyde/ CH_3OH	HF	43% methyl methoxyacetate 42% methoxyacetic acid 14% methyl hydroxyacetate 1% hydroxyacetic acid	10	–	[166]
CHO	H_2SO_4	CH_3 COOH COOH	40	57	[110]

5.6.6 Other Starting Materials

Besides esters, N-tert-alkylacyl amines and aldehydes (cf Table 5.14), compounds such as tertiary hexyl formate, tertiary hexyl acetate [180] and isobutyl formate [210] have been employed in the Koch reaction. A selection of recent work in this sector — comprising the carbonylation of aromatic compounds, enamines [130], esters, epoxides, and substituted amides — is summarized in Table 5.15.

Table 5.15. Carbonylation of various structures

Starting material	Catalyst	Reaction products	Temp. °C	Yield %	Ref.
Aromatics					
Biphenyl	AlCl$_3$	4-Phenylbenzaldehyde	45	~ 75	[30]
Toluene	HF · BF$_3$	p-Tolylaldehyde	155	90–95	[29, 90]
Perhydroindane-4-en-5-one	HF/SbF$_5$	2-β-Methoxycarbonyl-cis-(8βH, 9βH)-perhydroindan-5-one	0	60	[52]
Methyl tert-butyl ether	HF	Methyl pivaloate	60	95	[85]
Isobutylene oxide	H$_2$SO$_4$	α-Hydroxyisobutyric acid		~ 40	[107]
Methylcyclopentane	HF/SbF$_5$	Cyclohexyl isohexyl ketone	0	75	[19]
Methylcyclopentane/benzene	HF/SbF$_5$	Cyclohexyl phenyl ketone	0	63	[19]
Methylcyclohexane/methylcyclopentane	HF/SbF$_5$	Cyclohexyl methyl-cyclohexyl ketone	0	82	[19]
N-(hydroxymethyl)-benzamide	H$_2$SO$_4$/CH$_3$COOH	Hipparic acid	10		[161]
N-(hydroxymethyl)-phthalimide	H$_2$SO$_4$/CH$_3$COOH	N-phthaloylglycine	10	88	[161]

5.7 Industrial Applications and Economic Aspects

The most important commercial application of the Koch synthesis is the estimated 60,000 t/a DuPont plant for the production of glycolic acid from CO and formaldehyde. Thereafter follows the relatively less significant olefin conversions which are operated by Shell [97, 217] (Pernes, Holland) and Enjay Chem. Corp. [216] (Baton Rouge, USA). Schering and a Japanese company have licensed the process. Both Shell and Enjay use the process to produce – in the main – long chained Koch acids, each company possessing a capacity of around 5,000 t/a. Shell uses C$_8$–C$_{11}$ fractions as feedstock to manufacture C$_7$–C$_{11}$ carboxylic acids (Versatic acids) whereas Enjay employs propylene oligomers, which can be more readily carbonylated, to produce the "Neo-acids". The usual industrial reaction conditions are ca. 40–70 °C and 70–100 bar carbon monoxide pressure with H$_3$PO$_4$/BF$_3$/H$_2$O in the ratio 1:1:1 (Shell) or with BF$_3$ · 2 H$_2$O (Enjay) as catalyst. Fig. 5.9 presents a simplified flow sheet of the Shell Versatic acid plant.

The plant operates continuously. The crude acids pass a purification step before distillation.

The simplified flow diagram presented in Fig. 5.10 (Enjay) illustrates the basic process steps: – complex reaction, degasing, hydrolysis, and purification fractionation.

Schering's new process [50], based on oxonium tetrafluorocarbonate, has so far remained at the pilot stage.

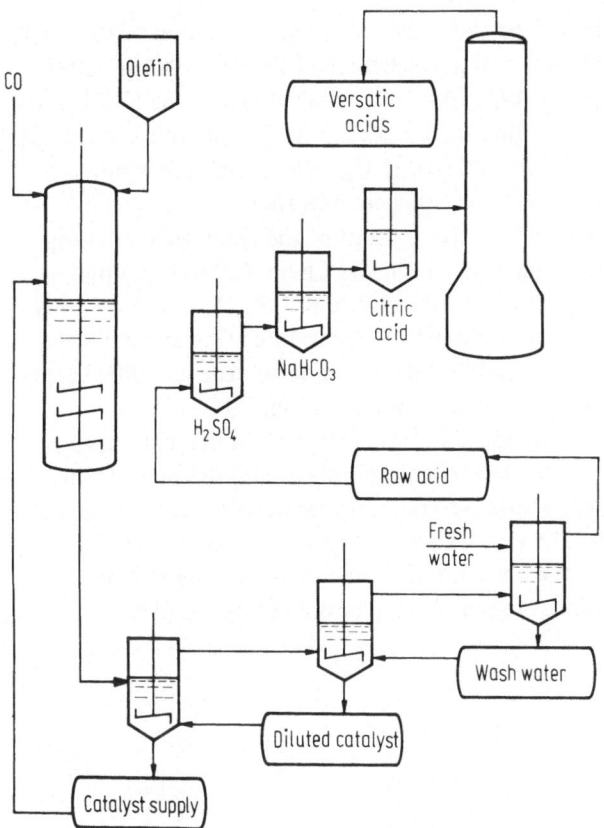

Fig. 5.9. Flow sheet of Shell Versatic acid unit

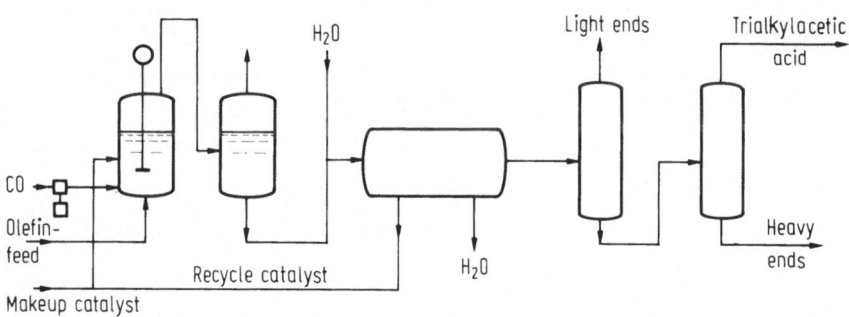

Fig. 5.10. Flowsheet of Enjay process for production of Neo-acids

Although initially great hopes were placed in the production and marketing of pivalic acid, this basic Koch acid has remained commercially insignificant (annual estimated production 1,000 – 2,000 t/a). Pivalic acid is produced by Shell and Enjay via a "blocked out operation" in large plants for Versatic- or Neo-acids.

However, a series of new publications about the industrial application of the Koch synthesis indicates i.e., reaction control, reactor technology [29, 85, 140, 196, 209], catalyst separation [90, 117, 113, 133, 142, 148, 149], catalyst processing [220] and product work-up [101, 124, 206–208] that there is a great deal of interest particularly in pivalic acid derivatives. The situation could well change when Shell's new process based on MTBE (methyl tert-butyl ether) is industrially operated.

Another reaction with future potential – on account of the abundance of cheap toluene – is the carbonylation of toluene (Mitsubishi Gas Chem. Co) to p-toluoylaldehyde (PTAL) [29, 31, 90] and its subsequent oxidation [218, 219] to terephthalic acid (for polyester production). The new conversion – a modified Gattermann-Koch reaction – is catalyzed by HF/BF_3. The yield is 96% based on toluene and 98% based on CO. A flow diagram of the PTAL process is presented in Fig. 5.11.
The acids resulting from the Koch synthesis [211–215] are used in the manufacture of esters, which are characterized by their excellent hydrolytic and thermal stability, and for the production of resins and paints. Several heavy metal salts are employed to a considerable extent as siccatives. The vinyl esters – obtained on reacting the Koch-acids with acetylene or ethylene in the presence of Pd compounds – are also of interest as they are important intermediates for emulsion paints and internal plasticizers for polymers.

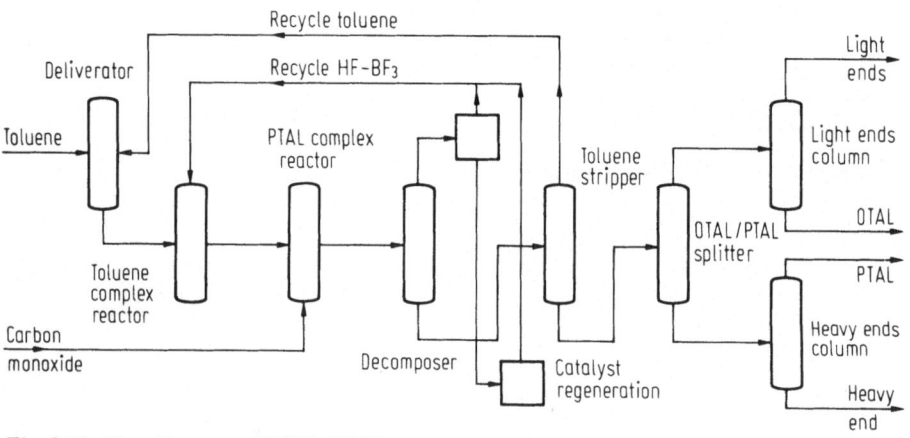

Fig. 5.11. Flow diagram of MGC's PTAL process

5.8 References

1. U. S. 1924766 (1932)
2. U. S. 1924767 (1932)
3. U. S. 1957939 (1934)
4. Larson, A. T., Vail, W. E. (Du Pont): U. S. 1924765 (1932)
5. Vail, W. E. (Du Pont): U. S. 1924764 (1932)
6. Carpenter, G. B. (Du Pont): U. S. 1924763 (1932)
7. Woodhouse, J. C. (Du Pont): U. S. 1924762 (1932)
8. Koch, H.: Fette, Seifen, Anstrichmittel 59, 493 (1957)

9. Ford, T.A.: U. S. 2491131 (1947)

10. Roland, I.R., Wilson II, I.D.C., Hanford, W.E.: J. Am. Chem. Soc. *72*, 2122 (1956)

11. Koch, H., Gilfert, W.: Part of the work of Koch, H., in: Brennstoff-Chem., *36*, 321 (1955)

12. Huisken, W.: Diplomarbeit, Universität Bonn (1952)

13. Haaf, W.: Brennstoff-Chem. 45 (7), 209 (1964)

14. Haaf, W.: Chem. Ber. *99*, 1149 (166)

15. Souma, Y., Sano, H., Yoda, J.I.: J. Org. Chem. *38* (11) 2016-2020 (1973)

16. Souma, Y., Sano, H.: Bull. Chem. Soc. Jap. *47* (7) 1717-1719, (1974)

17. Olah, G.A. et al.: J. Amer. Chem. Soc. *86*, 1360 (1964)

18. Falbe, J.: Carbon Monoxide in Organic Synthesis. p 131 ff. New York: Springer Verlag 1970

19. Paatz, R., Weisgerber, G.: Chem. Ber. *100*, 984 (1967)

20. Friedman, B.S., Cotton, S.M.: J. Org. Chem. *27*, 481 (1962)

21. Friedman, B.S. (Sinclair Refining): U.S. 2874186 (1959)

22. Schneider, A. (Sun Oil): U.S. 2864858 (1958)

23. U. S. 2864859 (1958)

24. Haaf, W., Koch, H.: Liebigs Ann. Chem. *638*, 122 (1960)

25. Matsubara, M. et al.: J. Chem. Soc. Ind. Chem. Sect. (Kogyo Kagaku Zasshi) *72*, 12, 2581-2586 (1969)

26. Souma, Y., Sano, H.: J. Org. Chem. *38* (20), 3633-3635 (1973)

27. (Kao Soap KK): JP 76.011.752 (1976)

28. (Agency of Ind. Sci.): JP 9.024.911 (1974), Appl. JP 064 213, 26.06.72

29. Fujiyama, Susuma, Kasahara Togomi (Mitsubishi Gas Chem. Inc.): DE-OS 2.422.197, 28.11.1974; JP. Appl. 73.51.832, 10.05.1973

30. Lachmann, B., Rosenkranz, H.J. (Bayer AG): DE-OS 2 413 892 (02.10.1973 Appl. P 24 13 892.9 (22.03.1974)

31. Fujiyama, Susuma, Kasahara, Toyomi: Hydrocarb. Proc. *57*, 147 (1978)

32. Kuraishi, Michio (Mitsubishi Gas Chem. Co. Inc.): JP Kokai 74 55,617 (30.05.1974) Appl. 72 98 641 (03.10.1972)

33. Souma, Yoshie, Sano, Hiroshi: Bull. Chem. Soc. Jap. *46*, 3273 (1973)

34. Gushchin, P.P., Lebeder, E.V., Pivovarova, Y.E.: Neftekhimiya *12* (3) 383 (1972), Chem. Inf. 41 (1972)

35. Alford et al.: Perkin Transactions (21) 2707 (1972)

36. Peters, J.A., Rog, J., Van Bekkum, H.: Recl. Trav. Chim. Pays-Bas 93 (9-10) 248, (1974)

37. Hine, J.: Reaktivität und Mechanismus in der org. Chimie, p. 303, Stuttgart: Georg Thieme Verlag 1960

38. Eidus, K. et al.: Zh. Org. Khim *4* (1), 36 (1968)

39. Kell, D.R., McQuillin, F.J.: J. Chem. Soc., Perkin Trans 1 (16), 2096 (1972)

40. Mayar, Y.: L'Ind. Chim. *53*, 214 (1966)

41. Yeomans et al. (BP Chemicals Ltd.): DE-OS 2 062 365 (18.12.1970)

42. Schauerte, K.H., Koch, H.: Brennstoff-Chem. *49* (9) 264 (1968)

43. Yoneda, W. et al.: Bull. Jap. Pat. Inst. *14* 2 178 (1972); Chem. Inf. 21 (1973)

44. Yoneda et al.: Nippon Kagaku Kaishi (8) 1475 (1972)

45. Yoshimura, M., Namba, T., Tokura, H.: Tetrahedron Lett. 2287 (*1973*)

46. Pratt, D.G., Rothstein, E.: J. Chem. Soc. (C) 2548 (*1968*)

47. Hogeveen, H., Roobeck, C.F.: Tetrahedron Lett. 3343 *1971*

48. Hogeveen, H., Roobeck, C.F.: Tetrahedron Lett. 4941 *1969*

49. For reviews see Hanack, M.: Accounts Chem. Res. *3*, 209 (1970); Richey, H.G., Richey, J.M. In: Carbonium Ions (eds.). (Olah, G.A., Schleyer, P.v.R.) Vol. 2. New York: Interscience Publishers 1970

50. Pawlenko, S.: Chem.-Ing.-Techn. *40* (1/2) 52 (1968)

51. Narrell, J.R.: J. Org. Chem. (12) 1971 (1972)

52. Coustard, J.M., Jacquesy, J.C.: J. Chem. Res. S (11), 281 (1977)

53. Hogeveen, H., Baardman, F., Roobeck, C.F.: Rec. Trav. Chim. *89*, 227 (1970)

54. Hogeveen, H., Roobeck, C.F.: Rec. Trav. Chim. *89*, 1121 (1970)
55. Peters, J.A. Bekkum, H. van: Recl. Trav. Chim. Pay-Bas *90*, (1) 65-80 (1971)
56. Peters, J.A., Bekkum, H. van: Recl. Trav. Chim. Pays-Bas *92* (3) 379-92 (1973)
57. Eidus, J.T. et al.: Neftekhimiya 8 (3) 313 (1968)
58. Eidus, J.T. et al.: Z. Org. Chim. 4 (7) 1214 (1968)
59. Stepanov, F.N., Guts, S.R.: Izv. Akad. Nauk SSSR, Ser. Khim (2) 439 (1970)
60. Godleski, St.A. et al.: Chem. Ber. 107 (4) 1257 (1974)
61. Krause, A.: Rocz. Chem. *43* (1) 223 (1969)
62. Kurkov, V.P., Laporte, S.J., Toland, W.G. (Chevron Research Co.): US 38 01 627 (02.04.1974)
63. Müller, K.E.: Diss. T. H. Aachen (1954)
64. Falbe, J., Pantz, R., Korte, F.: Chem. Ber. *97*, 3088 (1964)
65. Weintraub, L., Vitoha, J.F., Limon, R.: Chem. Ind. (London) 185 (1965)
66. Puzitskii, K.V., Eidus, Ya.T., Pyabora, K.G.: Zh. Obshch. Khim *33*, 3278 (1963)
67. Lamola, A.A. (Du Pont): Ger. Pat. 12 19 400 (1963)
68. Eidus, Ya.T., Puzitskii, K.V., Yung-Ping Yong.: Inv. Akad. Nauk SSSR, Ser. Khim (2) 428 (1972)
69. Lebedev, E.V. et al.: Neftepererab. Neftekhim (8) 7 (1972)
70. Pirozhkov, S.D. et al.: Izv. Akad. Nauk SSSR, Ser. Khim (7) 1534 (1976)
71. Matsubara, M.: Hokkaido-Ritsu Kogyo Shikenjo Hokoku (1968) (Pub. 1970) 83-91
72. Matsubara, M.: Hokkaido-Ritsu Kogyo Shikenjo Hokoku (1968) (Pub. 1970) 102-10
73. Matsubara, M.: Hokkaido-Ritsu Kogyo Shikenjo Hokoku (1968) (Pub. 1970) 111-24
74. Matsubara, M. et al.: Kogyo, Kagaku Zasshi 73 (10) 2147-51 (1970)
75. Matsubara, M.: Hokkaido-Ritsu Kogyo Shikenjo Hokoku 1966-1967 (Pub. 1969) 43-8
76. Pirozhkv, S.D. et al.: USSR 504.751, 28.02.1976, (Appl. 2,002,783 11.03.74)
77. a) Pawlenko (Schering AG).: US 33 49 107 (18.06.62 and 16.11.65 issued 24.10.67);
    b) Pawlenko (Schering AG).: DE-OS 1 211 621 (23.01.1961) (03.03.1966 – Ger. Ausleg.);
    c) Pawlenko (Schering AG).: DE-OS 1 212 061 (12.08.61) DE-AS 10.03.66
78. Bin, Y.Y., Puzitskii, K.V., Eidus, Y.T.: Izv. Akad. Nauk SSSR, Ser. Khim (8) 1779 (1971)
79. Eidus, Y.T., Puzitskij, K.V., Yong, Y.P.: Izv. Akad. Nauk SSSR, Ser. Khim. (7) 1673 (1970)
80. Bin et al.: Izv. Akad. Nauk SSSR, Ser. Khim. (2) 424 (1970); Erdöl u. Kohle *23* (12) 832 (1970)
81. Gushchin, P.P. et al.: Nefteperenab (Kiev) 12, 98 (1975)
82. Eidus et al.: Jeftekhimiya *12* (5) 754 (1972); Chem. Inf. *4*, 3 (1973)
83. Tanomura, M., Kau, S. (Kuraray Co. Ltd.).: JP 73 40 334 (30.11.1973) Appl. 69,55,914 (15.07.1969)
84. Matsubara et al.: Kogyo Kagaku Zasshi *72* (9) 1999-2004 (1969)
85. Shell AG, NL. Appl. 69 16 889 (13.05.1970); Eur. Chem. News *18* (452), 32 (1970)
86. Pirozhkov, S.D. et al.: C.A. *88*, 151986 u (1978)
87. Komatsu, Y. et al. (Maruzen Oil Co. Ltd.): JP 71 08 964 (06.03.1971) Appl. 21.02.1968
88. Yong, Y.P., Puzitskij, K.V., Eidus, Y.T.: Izv. Akad. Nauk SSSR, Ser. Khim. (2) 424 (1970)
89. Paulis et al.: Shell Intern. Res. Maatsch. N. V., GB 11 82 519 (15.11.1968) (issued 25.02.1970)
90. (Mitsubishi Gas Chem. Ind.): DE 2,559,164 (08.07.1976); JP 148 840 (27.12.74)
91. Matsubara, M.: Hokkaido-Ritsu Kogyo Shikenjo Hokoku (1968) (Pub. 1970) 92-101
92. Möller, K.E.: Angew. Chem. *73*, 767 (1961)
93. Falbe, J. et al.: Chem. Ber. *97*, 863 (1964)
94. Rohloffs, G., Pawlenko, St. (Schering AG): US 3 099 687 (1963)
95. Koch, H., Huisken, W. (Studienges. Kohle): DE 972 291 cf. GB 798 065 (1958)
96. Koch, H., Huisken, W. (Studiengesellschaft Kohle m. b. H.): DE 973 077 cf. GB 798 065 (1958)
97. Regimbeau, P. & Boisse, L.A. de (Shell AG): FR 1 252 675 (1960)
98. Möller, K.E.: Brennstoff-Chem. *45*, 129 (1964)
99. Möller, K.E.: Brennstoff-Chem. *47*, 10 (1966)
100. Friedman, B.S. (Atlantic Richfield Co.): US 3 708 530 (02.01.1973) (Appl. 05.01.1970)

101. Kumatsu, Y. et al. (Maruzen Oil Co. Ltd.): JP 73 16 897 (25.05.1973) (Appl. 12.05.1969)
102. Anderson, Frantze (Gulf Res. u. Dev. Co.): US 3 167 585 (26.01.1965) (Appl. 25.08.1960)
103. Stepanov, F.N. et al.: Zh. org. Khim 10 (2) 234, *1974*
104. Ordjan et al.: Izv. Akad. Nauk Anm.: SSR Chim Nauk *18*, 578 (1965)
105. Vitcha, J.F., Weintraub, L. (Air Reduction C. Inc.): US 3,341,578 – 12.09.1967
     (Appl. 30.10.1963)
106. Friedman, B.S., Nason, St. B. (Atlantic Richfield Co.): US 3,781,342 (25.12.1973)
     (Appl. 14.03.1973)
107. Komatsu, Y. et al. (Maruzen Oil Co.): JP 71 41 525 (08.12.1971) (Appl. 30.08.1069)
108. Roe, E.T. et al. (U. S. Secretary of Agriculture); US 3.170.939 – 23.02.1965
     (Appl. 20.06.1960)
109. Laky, J. et al.: HU 4679 28.07.1972 (Appl. 09.04.1970)
110. Himmele, W. (BASF): DE 1,227,010 – 20.10.1966 (Appl. 07.02.1964)
111. Schultz, H.S. (GAF-Corp.): US 3,449 408 – 10.06.1969 (Appl. 02.11.1966)
112. Seeliger, W., Witte, H.: Liebigs Ann. Chem. *755*, 163 (1972)
113. Sugita, N., Yasutomi, T., Takezaki, Y.: Bull. Jap. Petrol. Inst. *12*, 66 (1970); C.A. *73*, 65971u
     (1970)
114. Matsubara, M. et al.: Nenryo Kyokaishi *48* (505) 286 (1969)
115. Hartle, R.J. (Gulf Res. Dev. Co.): US 3 964 368 – 04.02.1975 (Appl. 05.06.1973)
116. Falbe, J. et al. (Ruhrchemie AG): FR Demande 2 017 568 (22.05.1970);
     DE App. 07.09.1968
117. Trocsanyi, T., Laky, J.: Ropa Uhlie 17 (6) 312 (1975)
118. American Cyanamid Co., FR 2 135 729 – 19.01.1973 (Appl. 26.01.1971)
119. Onopohenko, Anatoli, Schulz, J.G.D. (Gulf Res. Dev. Co.): US 3 870 734 – 11.03.1975
     (Appl. 05.06.1973)
120. (Shell N. V.), NL 6717436 – 24.06.1969 (Appl. 20.12.1967)
121. Lawson, N.E., Ching, T.T., Slezak, F.B.: J. Am. Oil Chem. Soc. *54* (6) 215 (1977)
122. Komatsu, Y., Tamura, T., Fujii, K.: Maruzen Sekiyo Gihi *21*, 51 (1976)
123. Kurhajec (Shell Oil Co.): Appl. 347622 – 31.07.62 cited in Sittig, Org. Chem. Proc.
     Encyclopedia 1969, p. 532,
124. BP Chem. Ltd., DE 20 61 549 (24.06.1971)
125. BP Chem. Ltd., GB 12 19 109 – 13.01.1971 (Appl. 30.12.1967)
126. Matsubara, M., Susaki, M., Ohtsuka, H.: Kogyo Kagaku Zasshi *71*, 1179 (1968)
127. Pryde, E.H., Frankel, E.N., Cowan, J.C.: J. Am. Oil Chem. Soc. *49* (8) 451 (1972)
128. Brouwer, D.M., Oelderik: Rec. Trav. Chim. Pays-Bas *87*, 721 (1968)
129. Gurriloff, A., Dusant, F. (N. V. Bougies de la Cour et de Roubaix Oedenkoven): BE 650,876
     (1964); NL-Appl. 64 09 056 (1965)
130. Mambudiry, M.E.N., Krishna Rao, G.S.: Tetrahedron Lett. *46*, 4707 (1972)
131. Souma, Y., Sano, H.: Bull. Chem. Soc. Jap. *49* (11) 3296 (1976)
132. BP Chem. (UK) Ltd., FR 2014433 – 17.04.1970, Appl. GB 031222 (06.01.1969)
133. Souma, Yoshie, Sano, H. (Agency of Ind. Sci. and Technol.): JP Kokai 7156409 –
     18.05.1976 (App. 14.11.1974)
134. Mitsubishi Gas Chem. Ind., JP 0062927 (29.05.1975), Appl. JP 113483 (09.10.1973)
135. Sano, Hiroshi, Sama, Y. (Agency of Ind. Sci. and Technol.): JP 7403,511 – 26.01.1974
     (Appl. 1067477 – 31.07.1970)
136. Matsushima, Y. et al. (Toa Nenryo Kogyo K. K.): JP Kokai 75134992 (25.10.1976)
     Appl. 7441753 (16.04.1974)
137. Souma, Y., Sano, H.: Osaka Kogyo Gigutsu Shikensko Kiho *28* (2) 144 (1977)
138. Yoneda, N. et al.: Bull. Chem. Soc, Jap. 51 (8) 2347 (1978)
139. Souma, Y., Sano, H.: Nikkakyo Geppo 26 (5) 220 (1973)
140. (Toa Nenryo Kogyo KK), NL 7310-874, JP-Appl. 27.02.73 (022691), 078401 (11.02.1974)
141. Agency of Ind. Sci. Technol., JP 7320530 (21.6.73) Appl. JP 061267 (1.8.1969)
142. Agency of Ind. Sci. Technol., JP 1041321 (7.4.76), Appl. JP 114913 (5.10.74)
143. Agency of Ind. Sci. Technol., JP 1041320 (7.4.76) Appl. JP 114912 (5.10.74)
144. Montedison Spa, DE 2334073 (12.02.76), Appl. IT 025846 (1.8.74)

145. Sano, H., Soma, Y. (Agency of Ind. Sci. and Technol.): JP 7335055 (25.10.1973) Appl. 7010218 (4.2.1970)
146. Matsushima, Y. et al. (Toa Nenryo Kogyo K. K.): JP Kokai 75 123 613 (29.09.1975) Appl. 74 28 624 (14.03.1974)
147. Koch, H., Schauerte, K.: Brennstoff-Chem. *46*, 392 a (1965)
148. Komatsu, Y. et al.: Bull. Jap. Pet. Inst. 16 (2) 124 – 31 (1974)
149. Armour Ind. Chem. Co., US 3 661 951 (09.05.72) Appl. US 876 203 (01.12.69) and 74 76 13 (15.07.68)
150. Shigeto, Sizuki (Chevron Res. Co.): US 3 948 986 (06.04.76) Appl. 542 138 (20.01.75)
151. (Shell AG) NL 69 17 064 (13.11.1969), App. 15.11.1968; Eur. Chem. News *18* (452), 32 (1970)
152. Shell AG, NL 69 07 233 (12.05.69), Appl. GB 31.05.68 and 24.06.68
153. Miller jr, E.J., Mais, A. (Armour Ind. Chem. Co.): FR 1 477 330 (14.04.67), Appl. US 26.04.65 (450 836)
154. Ejdus, J.T. et al.: Z. org. Chim. *4*, 580 – 84 (1968)
155. Ordyan, M.B. et al.: Izv. Akad. Nauk SSSR. Ser Khim 1116 *1971* (5); I. org. Chim. *36*, 256 (1971)
156. Eidus et al.: Z. org. Chim. *4* (3) 376 (1968)
157. Ordyan, M.B. et al.: Izr. Akad. Nauk SSR, Ser Khim, *3*, 555 (1972)
158. Jojima et al.: Bull. Chem. Soc. Jap. *44* (7) 2001 (1971)
159. Nojima et al.: Chem. Lett. (11) 1137 (1972)
160. Nojima et al.: Chem. Lett. (11) 2233 (1972)
161. Giordano, C.: Gazz. Chim. Ital. *102*, 2, 167 (1972)
162. Yoshimura, M., Nojima, M., Tokura, N.: Bull. Chem. Soc. Jap. *46* (7) 2164 (1973)
163. Phillips Petroleum Co., US 3 965 132 (22.06.76) Appl. US 4 759 77 (03.06.74)
164. Gaspard, S. et al.: Acad. Sci Sér C 281 22, 925 (1975)
165. Onopchenko, A., Schulz, J.G.D. (Gulf Res. Dev. Co.): Appl. CE 28 11 867 (18.03.78) US Appl. 24.03.1977
166. Chevron Research Co., US 3 948 977 (06.04.76) US Appl. 53 25 63 (13.12.74)
167. Olah, G.A., Halpern, Y.: J. Org. Chem. *36* (16) 2354 (1971)
168. Matsushima, Y. et al.: Chem. Lett. 433 (1973)
169. Souma, Y., Iyoda, J., Sano, H.: Inorg. Chem. *15* (4), 968 (1976)
170. Möller, K.E.: Angew. Chem. *75*, 1098 (1963)
171. Koch, H. et al.: Addition to DE 942 987 under St. 85 34 IV/b/120 (1954)
172. Maruzen Oil Co. Ltd., JP 13 23 413 (13.07.73), JP-Appl. 08 54 03 (25.10.1969)
173. Takeshi Chiba Tamura Youji Komatsu; Masafumi Hatsutori; Fuji Ichihana and Ichikawa Kinya Maruzen Oil Company Ltd.): Appl. DE 19 21 223 (25.04.68), Prior. 25.04.1969
174. Matsubara, M. et al.: J. Chem. Soc. Jap. (Ind. Chem. Soc.), *73*, 2147 (1970)
175. Koch, H., Möller, K.E. (Studiengesellschaft Kohle): DE 942 987 (1956)
176. Koch, H.: Riv. Combust. *10*, 77 (1956)
177. Koch, H., Möller, K.E.: DE 1 095 802 (1958)
178. Koch, H. et al.: DE 972 315 (1959)
179. Koch, H.: US 3 061 621 (1962)
180. Koch, H., Haaf, W.: Liebigs Ann. Chem. *618*, 251 (1958)
181. Falbe, J. (Ruhrchemie AG): BE 718 856 (1969)
182. Koch, H., Haaf, W.: Liebigs Ann. Chem. *638*, 111 (1960)
183. Schauerte, K.H.: Dissert. T. H. Aachen 1962
184. Möller, K.E.: Angew. Chem. *75*, 1122 (1963)
185. Takezaki, Y. et al.: Bull. Jap. Petrol. Inst. *2*, 94 (1960)
186. Takezaki, Y.: Kogyo Kagaku Zasshi *60*, 1038 (1957)
187. Benedictis, A.D., Furman, K.E. (Shell Oil Co.): US 2 913 489 (1961)
188. Koch, H., Haaf, W.: Angew. Chem. *70*, 311 (1958)
189. Koch, H., Haaf, W.: Chem. Ber. *94*, 1252 (1961)
190. Christol, H., Solladie, G.: Bull. Soc. Chim. Fr. 3193, *1966*
191. Eidus, Y.T., Kaal, T.A.: J. Gen. Chem. USSR *34*, 3447 (1964)

192. Eidus, Y.T., Kaal, T.A.: J. Gen. Chem. USSR *35*, 119 (1965)
193. Vais, J., Burkhard, J., Landa, St.: Z. Chem. *9*, (7) 268 (1969)
194. Langhals, H., Ruechardt, Ch.: Chem. Ber. *107* 1245 (1974)
195. Gund, T.M., Nomura, M., Schleyer, P.V.R.: J. Org. Chem. *39* (20), 2987 (1974)
196. Wesselingh, J.A. (Shell Oil Co.): US 3 691 230 (12.09.72) Appl. 69 10 591 (10.07.69)
197. Barr, P.A., Foglia, T.A., Schmeltz, I.: J. Am. Oil. Chem. Soc. *52*, (10) 407 (1975)
198. Takaishi, N. et al.: J. Chem. Soc., Chem. Commun. (10) 371-2 (1975)
199. Roe, E.T., Swern, D.: J. Amer. Oil Chemists Soc. *37*, 661 (1960)
200. Roe, E.T., Swern, D. (Secretary of Agriculture): US 3 169 140 (1965)
201. Roe, E.T., Swern, D. (Secretary of Agriculture): US 3 270 035 (1966)
202. Miller Jr., E.J., Mais, A., Say, D. (Armour Co.): US 3 481 977 (1969) FR 1 477 301 (1967)
203. Palit, S.K., Purohit, G.B., Murty, K.R.: Indian J. Technol. *6* (11) 323 (1968);
     Chem. Titles (4) 119 (1969)
204. Souma, Y., Iyoda, J., Sano, H.: Osaka Kogyo Gijutsu Shikensho Kiho 1977 *28* (2) 131 (1977)
205. (Technologiechemie GmbH Verfahrenstechnik), DE 1 593 575 (11.10.73)
206. (BP Chem. Ltd.), BE 76 64 29 (28.10.71)
207. Komatsu, Y. et al. (Maruzen Petrochem. Co. Ltd.): JP 73 00 807 (11.01.1973)
     Appl. 69 91 31 (07.02.69)
208. (Maruzen Oil Co. Ltd.) JP 71 35 724 (20.10.71); JP Appl. 06 35 53 (04.09.68)
209. Falbe, J. et al. (Ruhrchemie AG): FR-Appl. 2 017 568 (22.05.70); DE-OS 1 793 369
     (07.09.68)
210. Cedric, W., Holmes, N. (Imperial Chemical Industries Ltd.): GB 906 109 (19.09.62)
211. Falbe, J. (Ruhrchemie AG, Ger.): Ullmanns Encykl. Techn. Chem. 3rd ed. Ergänzungsband
     P 131-132. München, Berlin, Vienna: Urban & Schwarzenberg 1970
212. Othmer, Kirk: Encyclopedia of chemical Technology. 2rd ed., Vol. 8, p 854, Interscience
     Publishers, 1965
213. Frankel, E.N., Thomas, F.L.: J. Amer. Oil Chem. Soc. *50*, 39 (1973)
214. Bednarcyk, N.E., Erickson, W.L.: Fatty acids, Synthesis and Applications London:
     Noyes Data Corp. 1973
215. Asinger, F.: Die petrochemische Industrie, Vol. 2, 1180-92, Akademie Verlag Leipzig 1971
216. Ellis, W.J., Ronning, C.: Hydrocarbon Process. Petrol Refiner *44*, 139 (1965)
217. Dam, J. van, Waale, M.J.: Chim. Ind. *90*, 511 (1963)
218. Mitsubishi Gas Chem. Ind. BE 837 945 (14.05.76), JP Appl. 01 21 28 (28.01.75)
219. Mitsubishi Gas Chem. Ind. BE 838 583 (28.05.76), JP Appl. 01 93 12 (14.02.75)
220. Smith, G.M., Mantius, E.: Chem. Eng. Progr. *74*, (9) 78 (1978)
221. Friedman, B.S., Nason, St.B. (Atlantic Richfield Co.): US 3 781 342 (25.12.1973);
     Appl. 340 990 (14.03.73)

6. Ring Closure Reactions with Carbon Monoxide

A. Mullen

6.1 Introduction

Ring closures with CO present a route for the synthesis of heterocyclic and other cyclic compounds from relatively simple precursors [1–4].

2-Mercaptoethanol has recently been subjected to a carbonylation under mild conditions producing ethylenethiocarbonate [Eq. (1)] in 60% yield [5]

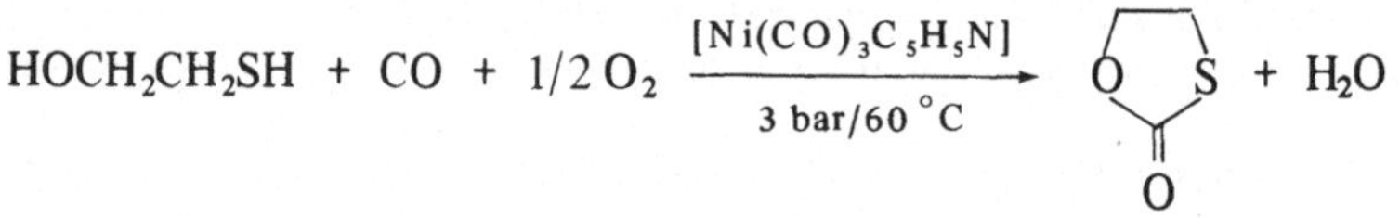

$$\text{HOCH}_2\text{CH}_2\text{SH} + \text{CO} + 1/2\,\text{O}_2 \xrightarrow[\text{3 bar/60 °C}]{[\text{Ni(CO)}_3\text{C}_5\text{H}_5\text{N}]} \quad \begin{array}{c} \\ O \quad S \\ \parallel \\ O \end{array} + \text{H}_2\text{O} \tag{1}$$

Evidence was found to suggest that the reaction takes place in at least two steps (cf Sect. 6.4.10.3) [5].

As the metal carbonyl-catalyzed conversions with acetylene, carbon monoxide and hydrogen or water leading to hydroquinone, cyclopentadienone

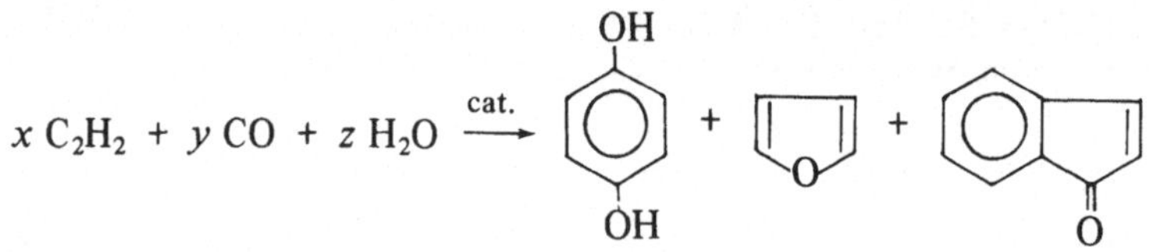

$$x\,\text{C}_2\text{H}_2 + y\,\text{CO} + z\,\text{H}_2\text{O} \xrightarrow{\text{cat.}} \begin{array}{c} \text{OH} \\ \bigcirc \\ \text{OH} \end{array} + \begin{array}{c} \\ \boxed{}_O \end{array} + \begin{array}{c} \\ \bigcirc \\ O \end{array} \tag{2}$$

hydrinone etc. are well documented [2, 4, 6, 7] [Eq. (2)] and there have been no commercial developments in this field, this aspect will not be treated in this review.

6.2 Reaction Mechanism

The pathway of the ring closure reaction is undoubtedly analogous to the mechanism of the hydroformylation. While three mechanisms are presented here, others are to be found in Sect. 6.4, together with the synthetic routes.

It has been proposed that the actual catalytic species is HM(CO)_x, the hydrocarbonyl, since with the exception of nickel, the catalytically active metals for these

414

reactions all have the ability to generate hydrocarbonyls (Co, Rh, Fe) [1]. For example, the inactivity of $Co_2(CO)_8$ in catalyzing the reaction between imides and CO at low temperatures contrasts with the facile ring closure catalyzed by $HCo(CO)_4$ under the same conditions [8].

The mechanism of Eq. (3) was recently outlined for the conversion of an unsaturated amide into a succinimide [1, 9, 22].

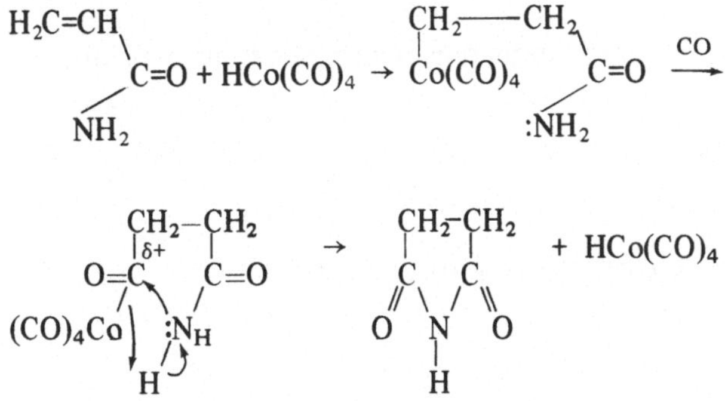

$$
\begin{array}{c}
H_2C=CH \\
\diagdown \\
C=O + HCo(CO)_4 \\
/ \\
NH_2
\end{array}
\rightarrow
\begin{array}{c}
CH_2\text{——}CH_2 \\
| \quad \diagdown \\
Co(CO)_4 \quad C=O \\
: NH_2
\end{array}
\xrightarrow{\ CO\ }
$$

$$
\begin{array}{c}
CH_2\text{—}CH_2 \\
|_{\delta+} \quad | \\
O=C \quad C=O \\
/ \quad \diagdown \\
(CO)_4Co \quad :N_H \\
| \\
H
\end{array}
\rightarrow
\begin{array}{c}
CH_2\text{—}CH_2 \\
| \quad | \\
C \quad C \\
\diagup\diagdown\diagup\diagdown \\
O \quad N \quad O \\
| \\
H
\end{array}
+ HCo(CO)_4
\tag{3}
$$

Cobalt hydridocarbonyl, stemming from $Co_2(CO)_8$, reacts with the C–C double bond which then undergoes CO insertion in the resulting metal-carbon bond. By means of a concerted reaction, $HCo(CO)_4$ is eliminated with formation of succinimide.

Tsuji suggested the formation of an intermediate complex [Eq. (4)] in the palladium-catalyzed carbonylation of azobenzene [10].

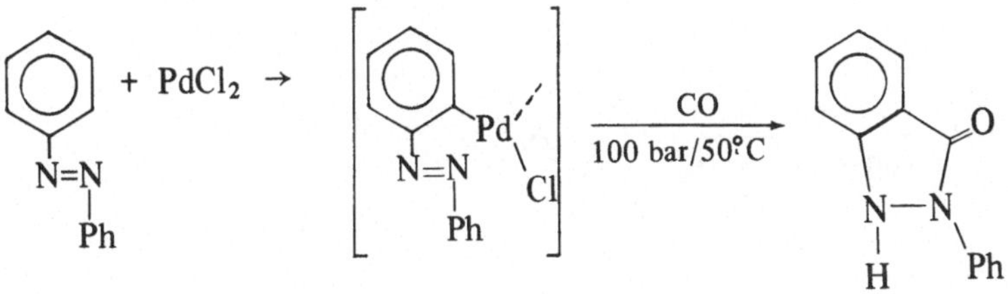

$$
\tag{4}
$$

The stable σ-complex reported previously gave a high yield of 2-phenyl-3-indazolone [10]. With substituted azobenzenes, ring closure was facilitated by the presence of groups which had a +I or +M effect, i.e., electron releasing. When nitro groups were present no cyclization ensued.

The following mechanism [Eq. (5)] was proposed for lactone formation from iodobenzene, styrene and nickel carbonyl as stoichiometric reagent [11].

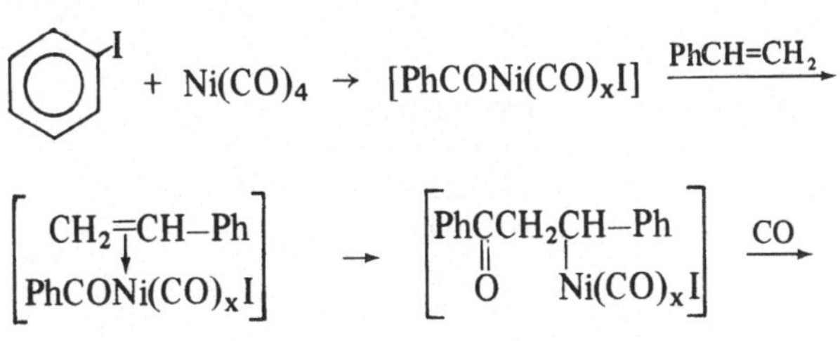

$$
+ Ni(CO)_4 \rightarrow [PhCONi(CO)_xI] \xrightarrow{PhCH=CH_2}
$$

$$
\begin{bmatrix} CH_2\text{=}CH\text{–}Ph \\ | \\ PhCONi(CO)_xI \end{bmatrix}
\rightarrow
\begin{bmatrix} Ph\overset{\parallel}{C}CH_2CH\text{–}Ph \\ O \qquad Ni(CO)_xI \end{bmatrix}
\xrightarrow{\ CO\ }
\tag{5}
$$

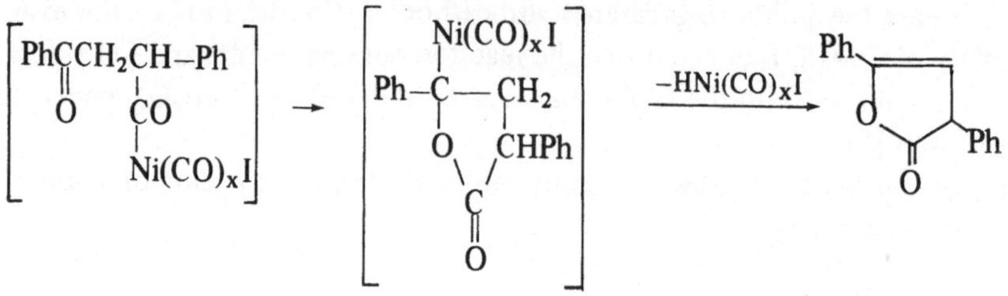

This synthesis is the first example of the formation of a lactone from an olefin.

6.3 Catalysts, Reaction Conditions and Solvents

6.3.1 Catalysts

The usual catalysts are cobalt carbonyls [1] or cobalt salts [12], palladium compounds [10, 13], iron carbonyls, rhodium carbonyls [14] and nickel in some cases [15, 16]. Nickel, for example, catalyzes the ring closure reaction with 2-mercaptoethanol [Eq. (1)] [5].

There have also been a number of interesting conversions with cyclic ethers which give rise to lactones in 35–60% yield on using $Co(OAc)_2$ as catalyst [4]. Under similar conditions, $Ni(CO)_4$ gave 9% yield of lactone [Eq. (6)] [7].

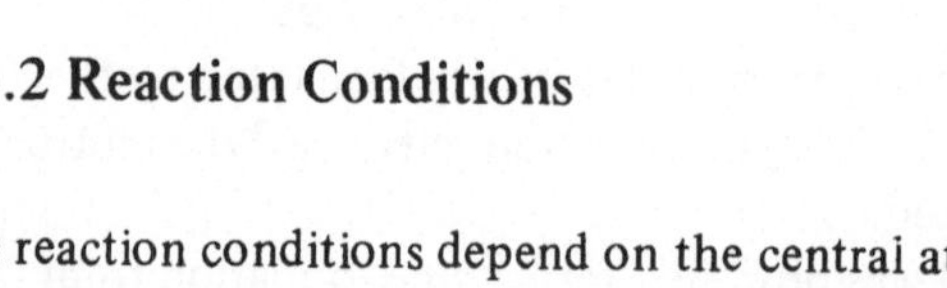

$$\text{(6)}$$

6.3.2 Reaction Conditions

The reaction conditions depend on the central atom and the ligands of the catalysts, solvents as well as the substrate. Some typical conditions are:

Catalyst metal	Temperature (°C)	Pressure (bar)
Rh	100–150	130–250
Co	100–250	100–300
Pd	50–100	3–100

6.3.3 Solvents

Aliphatic and aromatic hydrocarbons are suitable solvents, however, when the reactants are very polar, cyclic ethers and ketones can be used. In some cases, a conversion only occurs when pyridine is used as solvent.

In acetone, allyl chloride, acetylene and CO give lactone derivatives [17], whereas in alcohol the hexadienoate ester is formed [Eq. (7)] [8].

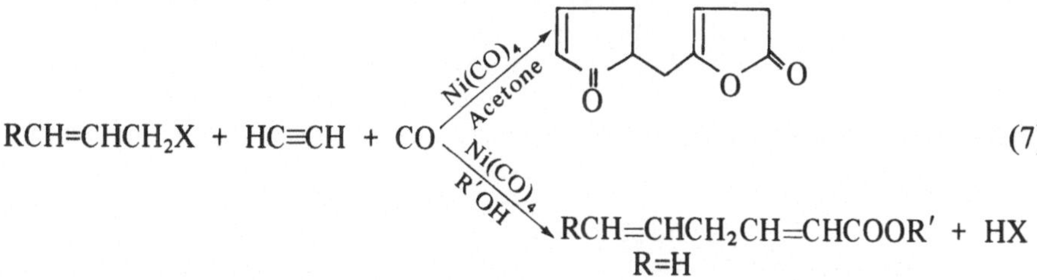

$$RCH=CHCH_2X + HC{\equiv}CH + CO \begin{array}{c} \xrightarrow{\begin{subarray}{c} Ni(CO)_4 \\ Acetone \end{subarray}} \\ \\ \xrightarrow[\begin{subarray}{c} Ni(CO)_4 \\ R'OH \end{subarray}]{} \end{array} \qquad (7)$$

$$RCH=CHCH_2CH=CHCOOR' + HX$$
$$R=H$$

γ-Lactone formation via the Rh-catalyzed carbonylation of butadiene [19, 20] was hindered when an alcohol was used as solvent.

While the cyclization of benzaldehyde anil with carbon monoxide to 2-phenylphthalimidine [21] occurs smoothly in non-polar solvents (yield = 80%), no phthalimidine formation was reported with solvents such as THF or ether [21].

6.4 Ring Closure Reactions with CO and Various Substrates

6.4.1 Formation of Imides

Using a $Co_2(CO)_8$ catalyst at 100–300 bar/150–300 °C, α,β-unsaturated amides can be carbonylated leading to the formation of a five-membered heterocyclic ring, an imide, in good yield [Eq. (8)] [1, 9, 22]

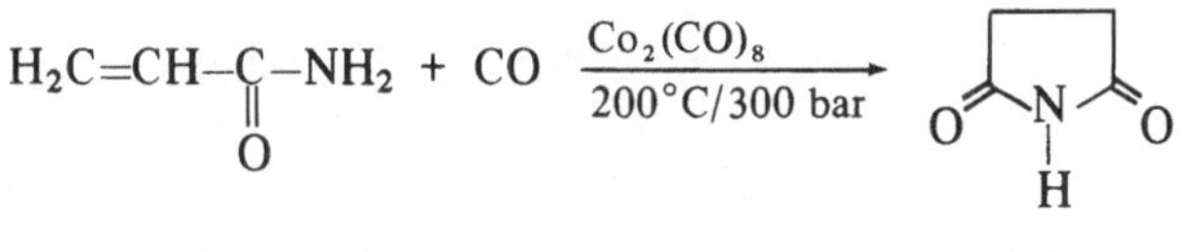

$$H_2C=CH-\underset{\underset{O}{\|}}{C}-NH_2 + CO \xrightarrow[200°C/300\ bar]{Co_2(CO)_8} \qquad (8)$$

β,γ-Unsaturated amides produce glutarimides in accordance with Eq. (9).

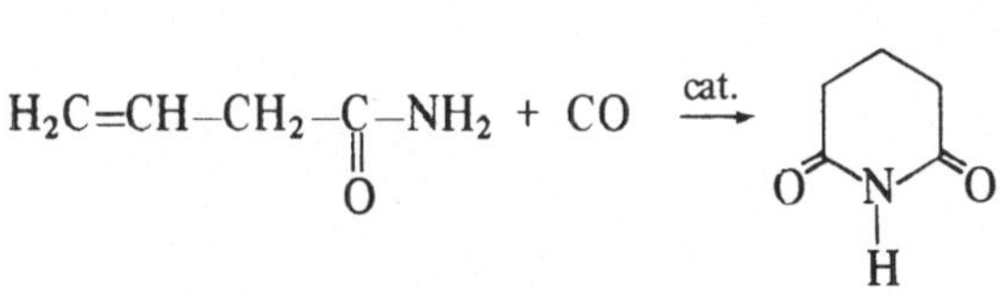

$$H_2C=CH-CH_2-\underset{\underset{O}{\|}}{C}-NH_2 + CO \xrightarrow{cat.} \qquad (9)$$

When a β-substituted acrylamide such as crotonamide is used as substrate, a mixture of products results [1] [Eq. (10)].

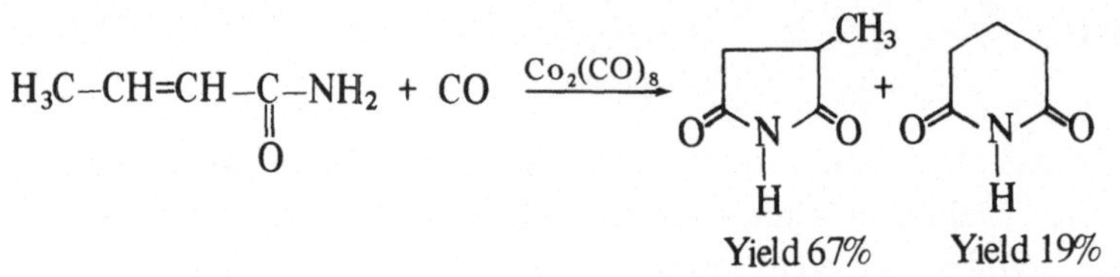

$$H_3C-CH=CH-\underset{\underset{O}{\|}}{C}-NH_2 + CO \xrightarrow{Co_2(CO)_8}$$

Yield 67%   Yield 19%   (10)

The cobalt carbonyl's ability to isomerize C—C double bonds is obviously responsible for the formation of the six-membered ring [1].

As expected, N-substituted imides can be obtained on using N-substituted unsaturated amides as precursors (Table 6.1).

Table 6.1. Imide formation via carbonylation of unsaturated amides with $Co_2(CO)_8$ at $160-280\,^\circ C/$ $100-300$ bar

Precursor	Product(s)	Yield (%)	Ref.
Acrylamide	Succinimide	82	[22]
N-Butylacrylamide	N-Butylsuccinimide	72	[22]
N-Hexylacrylamide	N-Hexylsuccinimide	77	[22]
N-Benzylacrylamide	N-Benzylsuccinimide	92	[22]
Methacrylamide	α-Methylsuccinimide	68	[22]
Crotonamide	α-Methylsuccinimide	67	[1]
	Glutarimide	19	[1]

Five- or six-membered bicyclic imides result from unsaturated alicyclic amides. The product depends on the pattern of unsaturation of the substrate [1] [Eq. (11)–(12)]

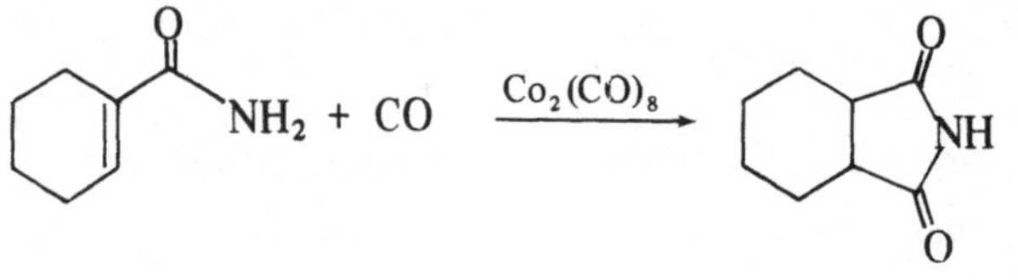

$$\text{(structure)} \quad NH_2 + CO \xrightarrow{Co_2(CO)_8} \quad \text{(structure)} \quad (11)$$

$$\text{(structure)} + CO \xrightarrow{Co_2(CO)_8} \quad \text{(structure)} \quad (12)$$

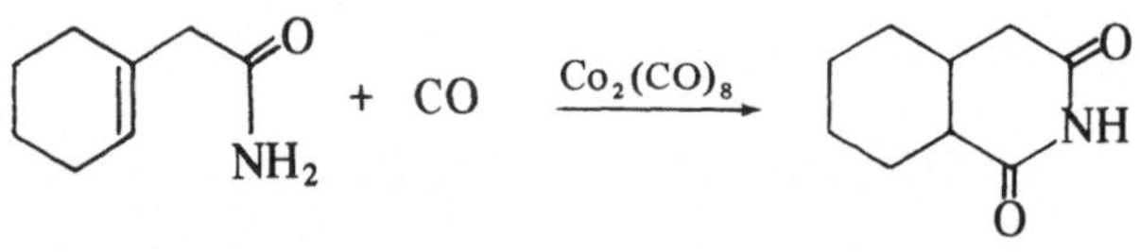

As well as benzamide and other aromatic amides, unsaturated N,N-dialkylamides were also found to be unsuitable for conversion into the imides.

6.4.2 Formation of Lactams

Lactams can be prepared by the (transition) metal carbonyl catalyzed carbonylation of unsaturated amines, of cyclic amines as well as when ammonia and alkene derivatives are used as substrates.

Unsaturated amines were readily carbonylated and cyclized by means of a cobalt carbonyl catalyst at 125–250 °C/60–300 bar [1, 27, 88] [Eq. (13)]. Cobalt is the most suitable catalytic metal, Rh and Ru both possessing lower activities [23, 24],

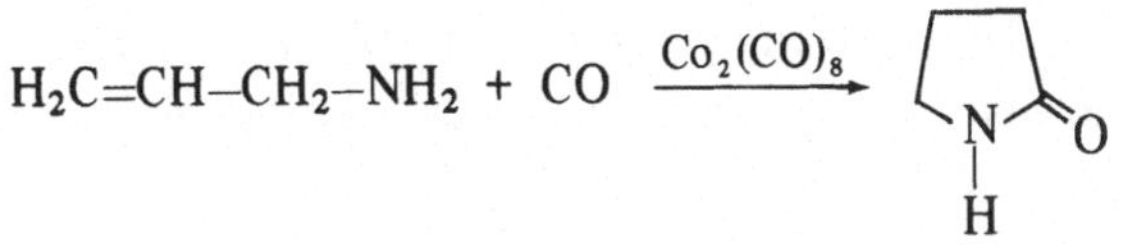

$$H_2C=CH-CH_2-NH_2 + CO \xrightarrow{Co_2(CO)_8} \qquad\qquad (13)$$

When homologues of allylamine are employed, a mixture of 5- and 6-membered lactams results [Eq. (14)] [1, 88].

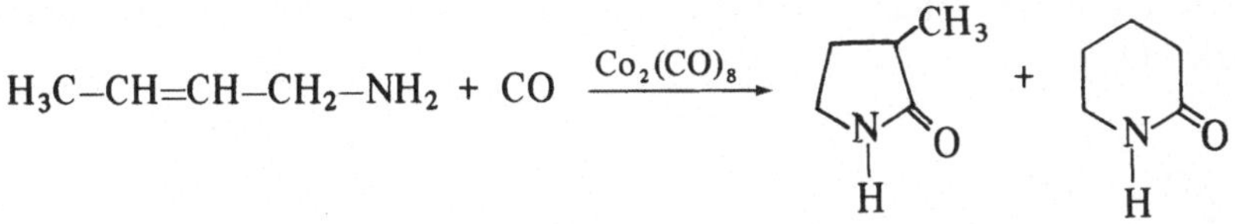

$$H_3C-CH=CH-CH_2-NH_2 + CO \xrightarrow{Co_2(CO)_8} \qquad + \qquad\qquad (14)$$

As well as the expected product, 3-methyl-2-pyrrolidone, 2-piperidone probably arises as a consequence of the cobalt-catalyzed isomerization of the double bond in the substrate.

N-Alkylallylamines lead to the formation of N-alkylpyrrolidones in good yield [1] (Table 6.2).

Table 6.2. Pyrrolidone formation via carbonylation of unsaturated amines in the presence of $Co_2(CO)_8$

Precursor	Product	Yield (%)	Ref.
N-Methylallylamine	N-Methylpyrrolidone	78	[88]
N-Ethylallylamine	N-Ethylpyrrolidone	61	[88]
N-Isobutylallylamine	N-Isobutylpyrrolidone	61	[88]
N-Octylallylamine	N-Octylpyrrolidone	98	[88]
N-Dodecylallylamine	N-Dodecylpyrrolidone	47	[88]
N-Phenylallylamine	N-Phenylpyrrolidone	26	[26]

The carbonylation is not restricted to aliphatic amines, as can be appreciated from the following carbonylation of o-aminostyrene [27, 28] [Eq. (15)]:

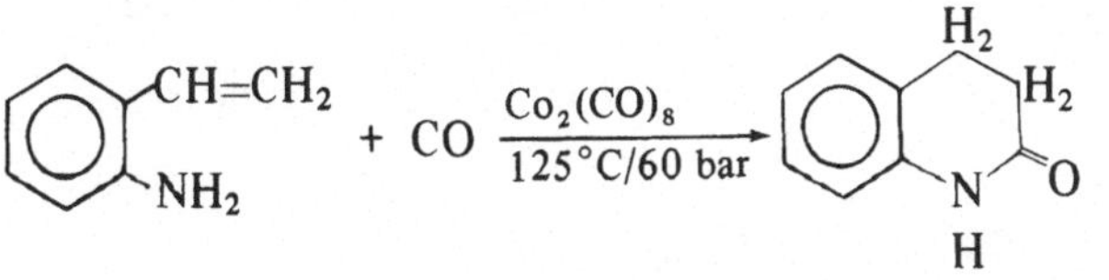

$$\xrightarrow[125°C/60\ bar]{Co_2(CO)_8} \qquad\qquad (15)$$

In addition, cyclopropylamine was readily carbonylated using $Rh_6(CO)_{16}$ at 100–140 °C/130–150 bar, producing N-cyclopropylpyrrolidone [29] in 92% yield [Eq. (16)]:

419

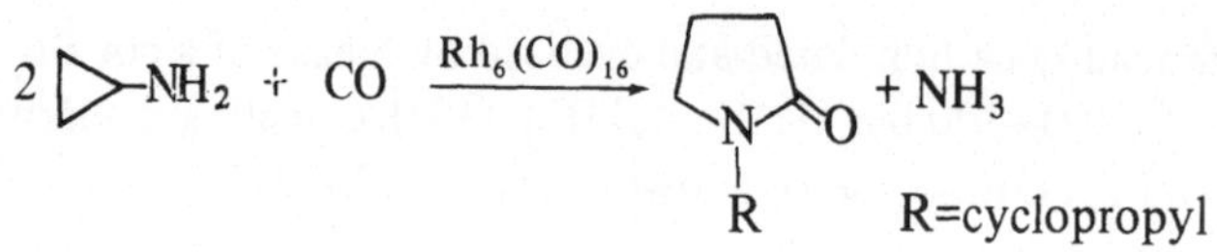

$$2 \; \triangleright\!-NH_2 \; + \; CO \; \xrightarrow{Rh_6(CO)_{16}} \; \underset{\underset{R}{|}}{\text{[pyrrolidone ring]}} \!=\! O \; + \; NH_3 \qquad (16)$$

$$R = \text{cyclopropyl}$$

Traces of the N-n-propyl and N-allyl derivative were also found [29]. 2-Pyrrolidone can also be synthesized on reacting allyl chloride, ammonia and CO in the presence of a cobalt chloride/PPh$_3$ catalyst system at 250 °C/150 bar [28] [Eq. (17)].

$$H_2C=CH-CH_2Cl + NH_3 + CO \rightarrow 2\text{-pyrrolidone} + HCl \qquad (17)$$

When propene is added to the above precursors, the cobalt carbonyl/quaternary ammonium base catalyst system leads to the formation of an eight-membered lactam ring [Eq. (18)] [31].

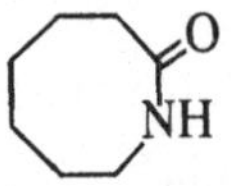

$$\underset{\text{NH}}{\bigcirc}\!\!=\!\!O \qquad (18)$$

Substituted 2-pyrrolidones were also generated on carbonylating lower primary amines in the presence of conjugated dienes. The reaction is catalyzed by Co or Rh carbonyls along with alkali metal hydroxides [30].

6.4.3 Formation of Lactones

6.4.3.1 Lactones via Carbonylation of Unsaturated Alcohols, Esters or Acids

In the presence of CO and Co$_2$(CO)$_8$, or RhCl$_3$ as catalyst, 2,3- and 3,4-unsaturated alcohols yield the corresponding γ- or δ-lactones [1, 89].

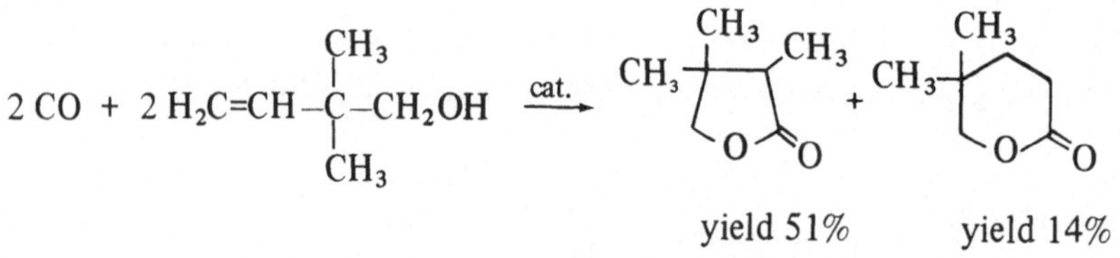

$$2\,CO \; + \; 2\,H_2C=CH-\underset{\underset{CH_3}{|}}{\overset{\overset{CH_3}{|}}{C}}-CH_2OH \; \xrightarrow{cat.} \qquad (19)$$

yield 51%   yield 14%

The usual conditions are 125–250 °C/70–300 bar [8, 32].

The competing reaction is the isomerization of the unsaturated alcohol to the aldehyde, e.g., with allyl alcohol only ~2% γ-butyrolactone is obtained, 50% of the starting material being isomerized to propanal. However, when acetonitrile is used as solvent in the presence of pyridine, then γ-butyrolactone results in 60% yield [32].

Unsaturated alcohols which have an alkyl substituent at the carbon atom adjacent to the C—OH group give better yields of lactone, since isomerization does not ensue.

The relative proportion of γ- to δ-lactone generally varies, the 6-membered ring being favored by alkyl substitution in the 3 position. However, γ-isocaprolactone was obtained in good yield from 3-methyl-3-buten-1-ol using RhCl$_3$/CH$_3$I as catalyst [33] [Eq. (20)].

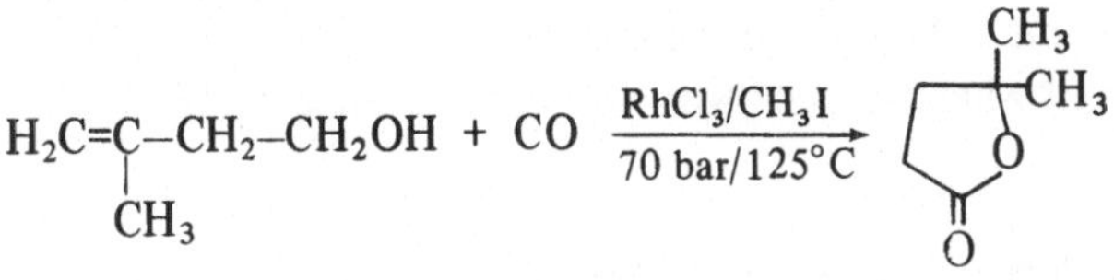

$$H_2C=C-CH_2-CH_2OH + CO \xrightarrow[70\ bar/125°C]{RhCl_3/CH_3I}$$
$$\underset{CH_3}{}$$

(20)

yield 60%

Although unsaturated secondary alcohols can only be carbonylated to the corresponding lactones in low yield (due to isomerization to the corresponding ketone) their tertiary isomers give better results [1, 89].

While lactones are generally accessible via condensations at much milder conditions than are required for carbonylations, various mono and bicyclic unsaturated lactones are most conveniently prepared via these single-step reactions.

Recently, thiourea/PdCl$_2$ mixtures were used to catalyze the formation of α-methylene-γ-lactones from acetylenic alcohols [Eq. (21)] [34].

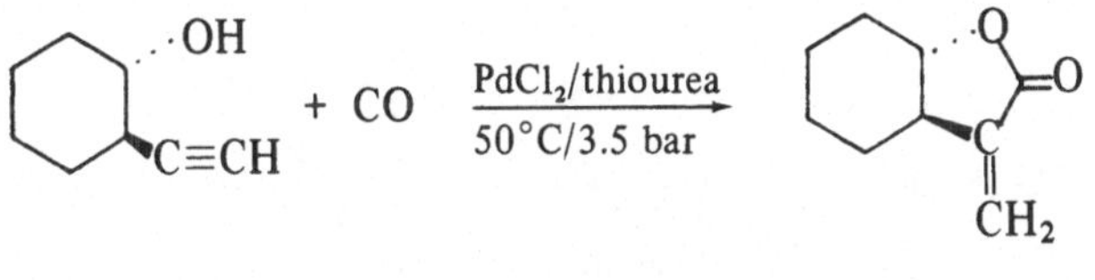

$$+ CO \xrightarrow[50°C/3.5\ bar]{PdCl_2/thiourea}$$

(21)

The α-methylene-γ-butyrolactone was obtained in 94% yield. Compounds of this type, exhibiting trans fusion, are biologically interesting and may have pharmaceutical applications [34].

Lactones can also be prepared by carbonylation of unsaturated esters [35—37], acids [37, 38], aldehydes [39] as well as on using o-cresol [40] as substrate [Eq. (22)]:

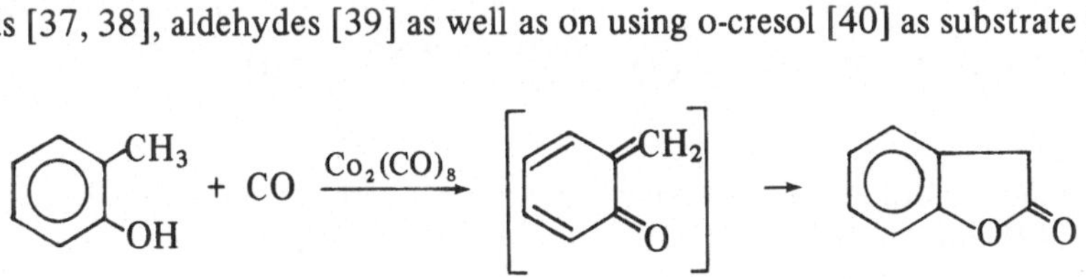

$$+ CO \xrightarrow{Co_2(CO)_8}$$

(22)

6.4.3.2 Lactones via Carbonylation of Alkenes or Alkynes

The carbonylation of acetylenes yields lactones [Eq. (23)] with various catalyst systems based on, for example, Pd [10], Co [41] or Fe [42].

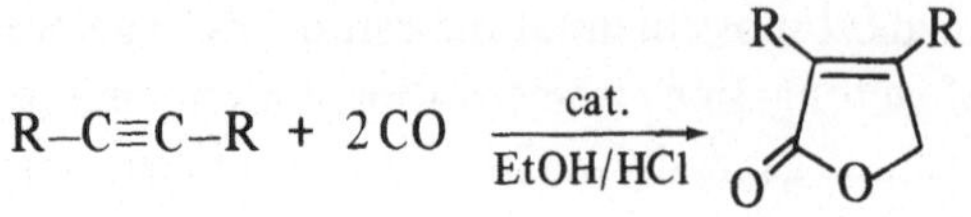

$$R-C\equiv C-R \ + \ 2\,CO \xrightarrow[\text{EtOH/HCl}]{\text{cat.}} \quad \text{(structure)} \tag{23}$$

With diphenylacetylene as substrate, the $PdCl_2$ catalyzed reaction [10], which took place at 100 °C/100 bar, was found to be solvent dependent (Table 6.3). The HCl concentration has a direct effect on the amount of lactone formed.

No carbonylation occurred when benzene was employed as solvent [43, 44].

Table 6.3. Carbonylation of diphenylacetylene in various solvents

Precursor	Temp. (°C)	Pressure (bar)	Catalyst	Solvent	Product	Yield (%)
Diphenylacetylene	100	100	$PdCl_2$	HCl/EtOH	Diphenylcrotonolactone	66
Diphenylacetylene	100	100	$PdCl_2$	EtOH	Diphenylcrotonolactone	20
Diphenylacetylene	100	100	$PdCl_2$	HCl/MeOH	Diphenylcrotonolactone	60

Palladium catalysts have frequently been used to catalyze the carbonylation of alkenes and dienes, leading to lactone formation. These conversions generally take place under severe conditions (50–1000 bar/80–200 °C) [45–47].

α-Methyl-γ-butyrolactone was obtained on carbonylating 1,3-butadiene with CO in the presence of $RhCl_3$ [49].

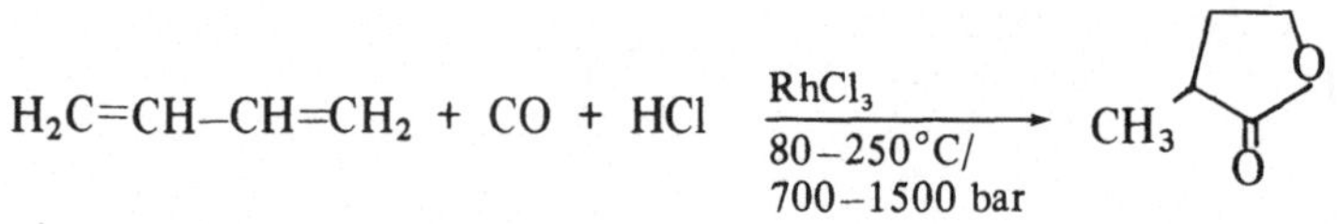

$$H_2C=CH-CH=CH_2 \ + \ CO \ + \ HCl \xrightarrow[\substack{80-250°C/ \\ 700-1500 \ bar}]{RhCl_3} \quad CH_3\text{(structure)} \tag{24}$$

1,5-Hexadiene gives a 95% yield of α-ethyl-γ-valerolactone at 20 °C/13 bar using a Cu catalyst [50].

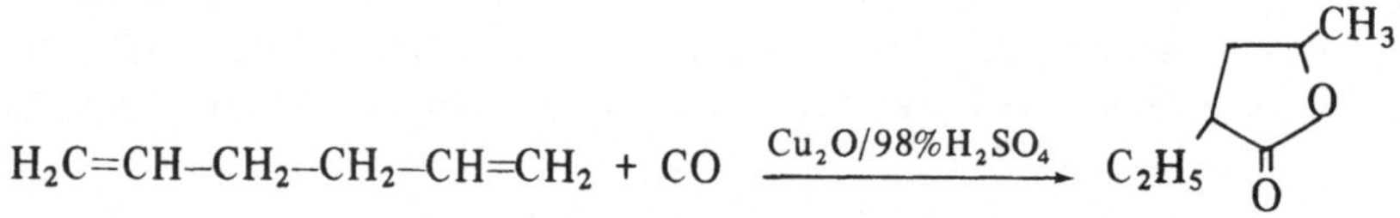

$$H_2C=CH-CH_2-CH_2-CH=CH_2 \ + \ CO \xrightarrow{Cu_2O/98\%H_2SO_4} C_2H_5\text{(structure)}CH_3 \tag{25}$$

The relatively mild conditions of this conversion contrast with the more drastic conditions with Rh [49] and Pd [48].

6.4.3.3 Lactones via Carbonylation of Cyclic Ethers or Epoxides

Cyclic ethers such as trimethylene oxide (oxetane) have been converted in good yields (35–60%) into the corresponding lactones [Eq. (26)] via metal carbonyl (Co, Ni) catalyzed carbonylations [7, 12].

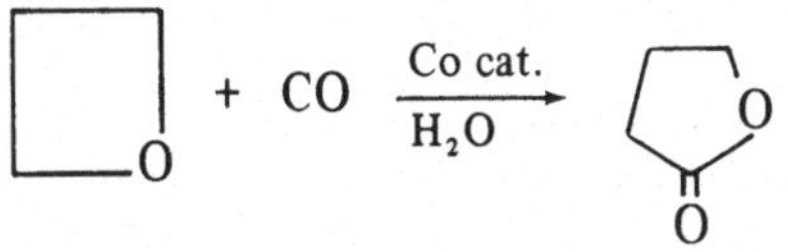

$$\square + CO \xrightarrow[\text{H}_2\text{O}]{\text{Co cat.}} \text{(lactone)} \qquad (26)$$

The Co-catalyzed reaction generally takes place between 200–250 bar/200–250 °C whereas with $Ni(CO)_4$, carbonylation occurs at 60 bar/250 °C [12].

The following reaction steps were proposed for the cobalt carbonyl catalyzed carbonylation [51] [Eq. (27)]:

$$\square + HCo(CO)_4 + CO \rightarrow HO–CH_2–CH_2–CH_2–COCo(CO)_4 \qquad (27)$$

An elegant δ-lactone synthesis was reported by Aumann and Ring [52], who employed isoprene oxide as precursor [Eq. (28)]. A different type of unsaturation arose depending on the metal carbonyl catalyst employed [52].

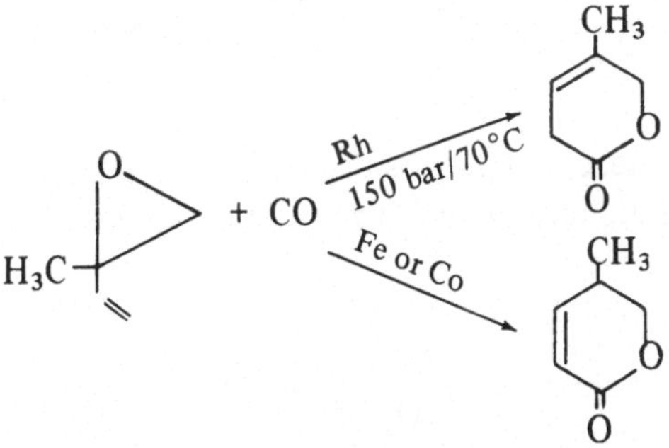

$$(28)$$

$$(29)$$

The presence of an alkyl group at the epoxy function (as above) was found to favor lactone formation, otherwise the competitive reactions (non-cyclic ketone formation, polymerization) occur to a greater extent [52].

6.4.3.4 Lactones via Carbonylation of Alkyl, Allyl or Acyl Halides

$$RC\equiv CR + R'CH_2X + 2\,CO \xrightarrow{\text{cat.}} \text{(lactone)} + [HX] \qquad (30)$$

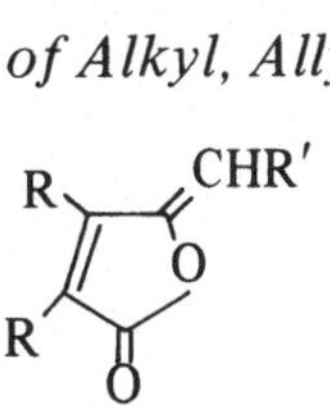

The reaction [Eq. (30)] which facilitates the synthesis of unsaturated lactones under mild conditions employs either cobalt or nickel carbonyl as catalyst [53]. Disubstituted acetylenes give good yields. The lactone yield depends fairly strongly on the acetylenic substrate. The cobalt carbonyl complex initially formed does not readily undergo a base-catalyzed elimination unless R' is an activating group.

Lactones can also be synthesized from allyl halides and acetylene [Eq. (31)] [4, 96]:

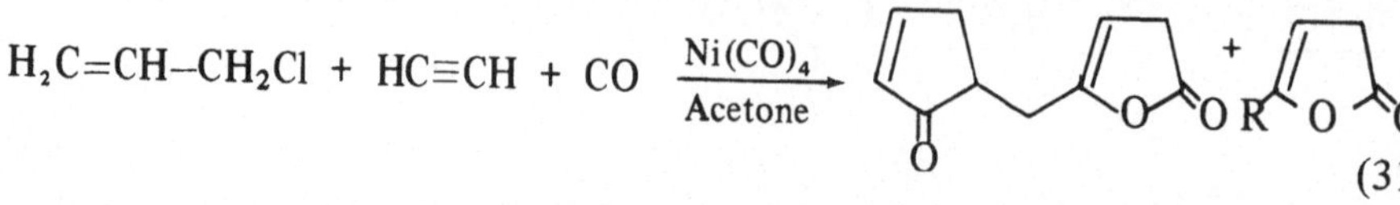

$$H_2C=CH-CH_2Cl + HC\equiv CH + CO \xrightarrow[\text{Acetone}]{Ni(CO)_4}$$

(31)

When an alcohol is used as solvent in the above reaction [Eq. (31)], a linear unsaturated ester is obtained [4]. Carbonylations with allyl chloride were studied by Chiusoli [56]. As discussed in Sect. 6.2, olefins and phenyl halides can be carbonylated with stoichiometric amounts of Ni(CO)$_4$ at 50–60 °C to produce lactones in yields between 19–30% [Eq. (32)] [11].

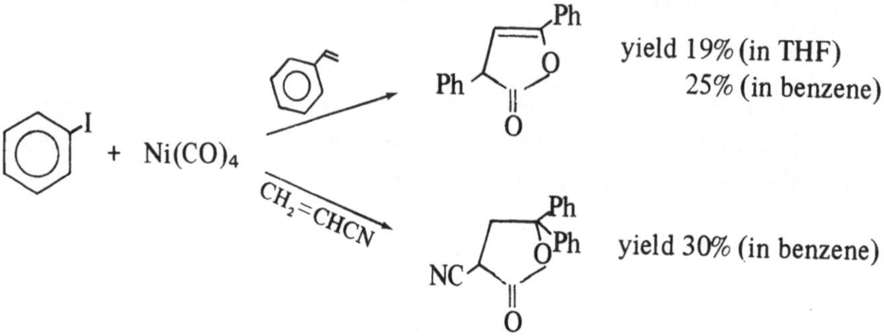

yield 19% (in THF)
25% (in benzene)

(32)

yield 30% (in benzene)

Butadiene as well as cyclohexene did not react [11].

In the presence of Ni(CO)$_4$, acyl halides react with CO and alkynes to yield lactones [Eq. (33)] [57].

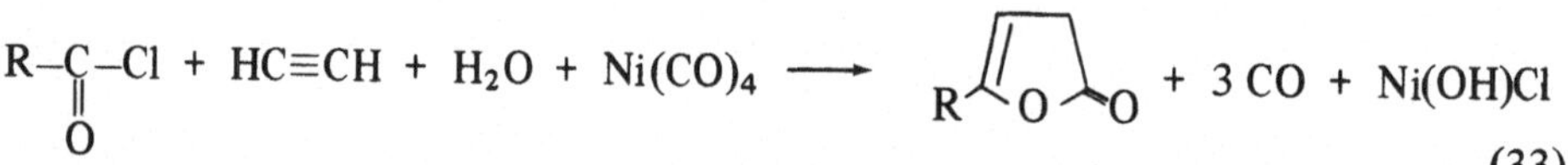

$$R-\underset{O}{\underset{\|}{C}}-Cl + HC\equiv CH + H_2O + Ni(CO)_4 \longrightarrow \quad + 3\,CO + Ni(OH)Cl$$

(33)

The lactone formation is promoted by ketonic solvents [57]. Unsaturated γ- as well as ε-lactones [Eq. (34)] [58] were obtained with esters or ethers as solvent and low CO concentrations at 40–50 °C.

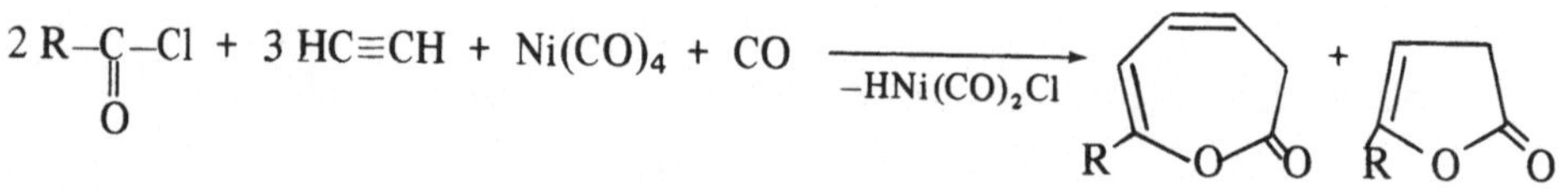

$$2\,R-\underset{O}{\underset{\|}{C}}-Cl + 3\,HC\equiv CH + Ni(CO)_4 + CO \xrightarrow[-HNi(CO)_2Cl]{} \quad +$$

(34)

6.4.4 Formation of Phthalimidines from Schiff Bases or Aromatic Nitriles

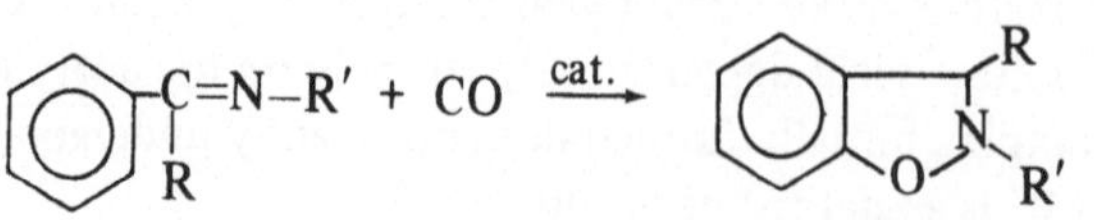

$$\underset{R}{\overset{C=N-R'}{\bigcirc}} + CO \xrightarrow{cat.}$$

(35)

In the presence of metal carbonyl catalysts, Schiff bases of aromatic and aliphatic amines produce a wide variety of phthalimidines in good yield (65–85%) [1, 3, 59, 60, 90, 91]. $Co_2(CO)_8$ has frequently been employed as catalyst as $Fe(CO)_5$ is less active, $Ni(CO)_4$ being apparently inactive [1].

The cobalt-catalyzed carbonylation, which generally takes place at 200–230 °C/ 100–200 bar, is inhibited by polar solvents such as ethanol, tetrahydrofuran and water [61, 62]. Consequently, benzene or toluene has often been employed as reaction medium [59, 60].

Aromatic substrates with substituents in the para position of the aniline derivative (OH, Cl, OCH_3) or in the para position of the aldehyde ring [OH, $N(CH_3)_2$] also gave good yields of phthalimidines [64, 95]. As anticipated, strongly deactivating groups such as NO_2 led to lower yields. Extensive tables of various substrates and products with Co and Pt catalysts have been compiled by Falbe [1] and Bird [2].

Recently, palladium acetate has been used by Heck [63] to catalyze phthalimidine formation under very mild conditions [Eq. (36)]:

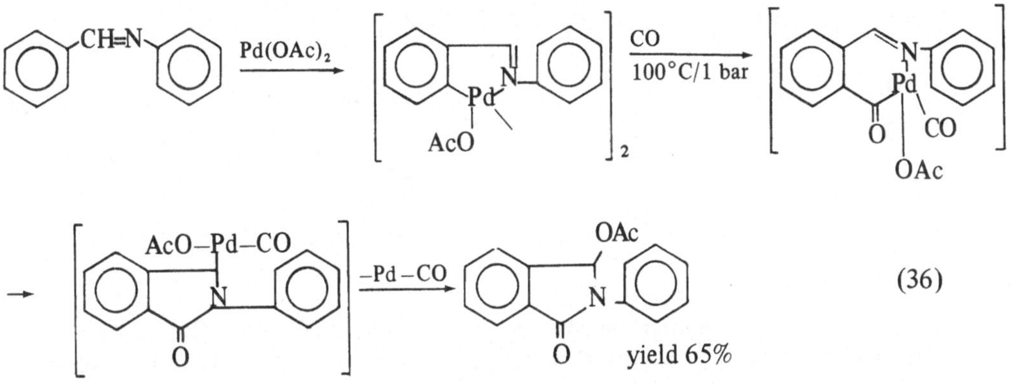

(36)

Aromatic nitriles are also suitable substrates for phthalimidine synthesis [Eq. (37)]. The reaction is thought to proceed via a Schiff base intermediate [1]. This is underlined by the observation that better yields are obtained on employing synthesis gas instead of CO [65]. The presence of pyridine was also found to be beneficial [3].

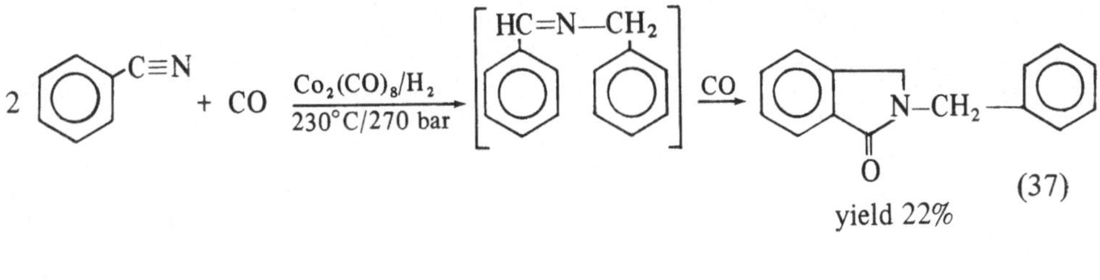

(37)

yield 22%

6.4.5 Formation of Phthalimidines from Aromatic Ketoximes, Phenylhydrazones, Semicarbazones or Azines

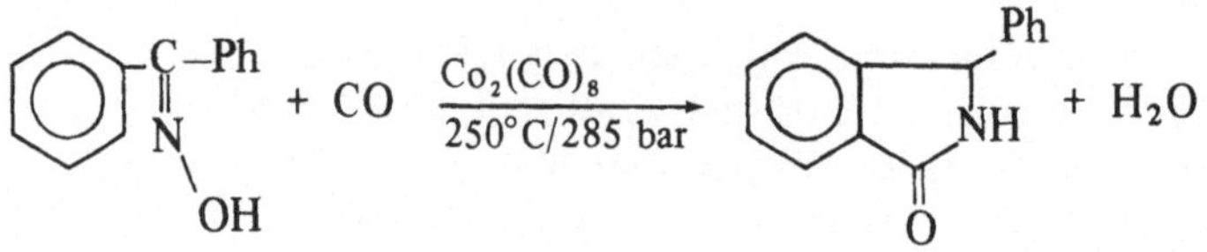

$$(38)$$

Phthalimidines [Eq. (38)] are yielded when aromatic ketoximes react with carbon monoxide in the presence of hydrogen (~1.5%). The amount of hydrogen can have a marked influence on the course of the reaction. The reaction is believed to pass through the N-hydroxyphthalimidine step which then undergoes hydrogenation [1, 66]. Better selectivities were attained with diarylketoximes than with alkyl-aryl-ketoximes, the yield in the former case often lying between 80–86% [1].

The N-hydroxyphthalimidines themselves have defied synthesis [66], probably on account of the ready hydrogenation of the N–O bond.

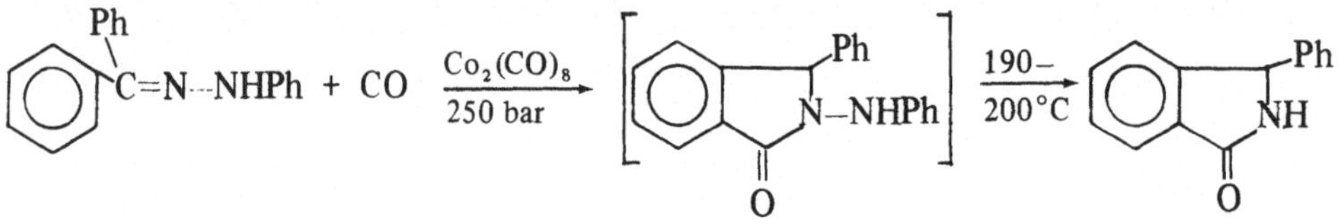

$$(39)$$

Benzophenone phenylhydrazone gives rise to 3-phenylphthalimidine [Eq. (39)] at 190–200 °C, while at 230–240 °C 3-phenyl-N-(N'-phenylcarbamoyl)-phthalimidine results via CO insertion into the N–N bond.

4-Methoxybenzophenone phenylhydrazone yielded two isomeric reaction products [Eq. (40)]. The ring closure does not occur exclusively with the unsubstituted ring as with the Schiff bases but takes place to an equal extent with both rings [1]. $Co_2(CO)_8$ is the most frequently reported catalyst for this carbonylation.

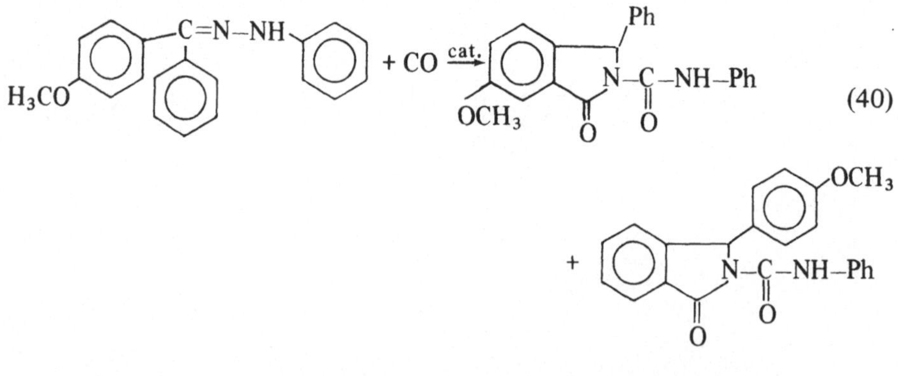

$$(40)$$

Recently, Heck [63] described the application of a palladium acetate complex which facilitated the synthesis of phthalimidine derivatives in 61% yield [Eq. (41)]:

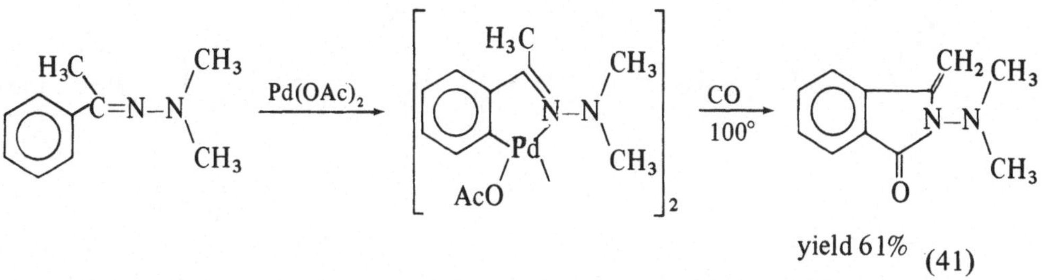

yield 61% (41)

Tables of various substrates and products of the $Co_2(CO)_8$-catalyzed carbonylation of hydrazones have been compiled by Falbe [1].

The semicarbazones of aromatic ketones react with carbon monoxide (containing hydrogen) to form phthalimidines [Eq. (42)]. The carbonylation is generally conducted in the presence of $Co_2(CO)_8$.

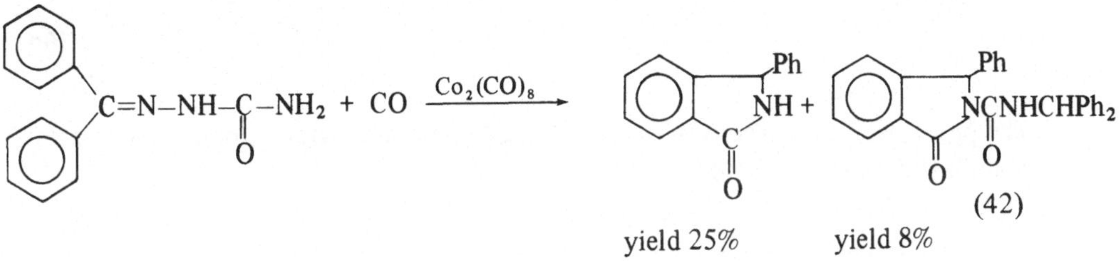

yield 25%   yield 8%

(42)

As in previous cases, the probable intermediate, N-ureidophthalimidine [Eq. (43)], has not been isolated possibly on account of hydrogenation [1].

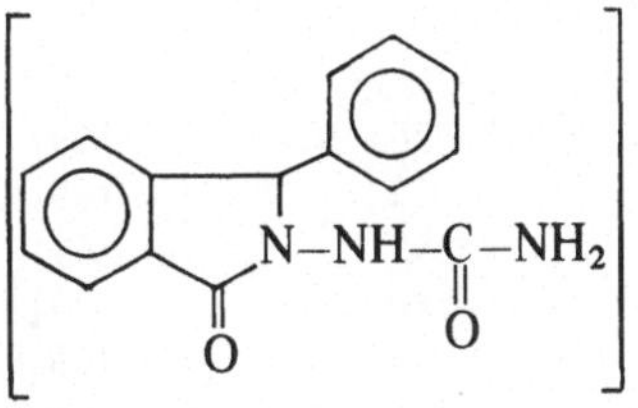

(43)

Heck [63] reported the carbonylation of the benzaldazine-Pd(OAc)$_2$ complex at 100 °C [Eq. (44)] which resulted in the formation of 3-acetoxy-2-benzalimidophthalimidine in 48% yield.

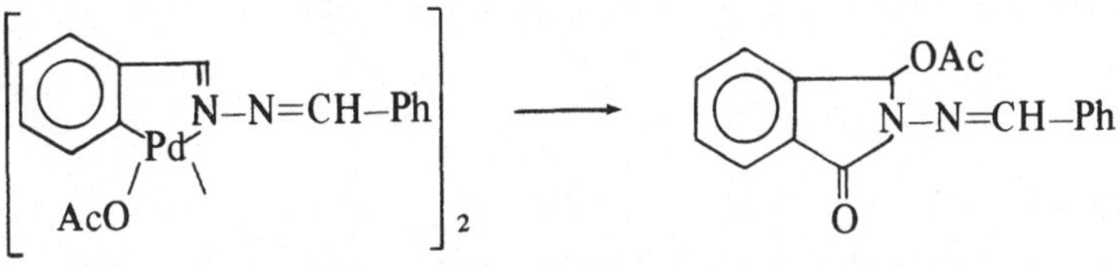

(44)

When heated above their melting points, benzophenone semicarbazones decompose into benzophenone azines [67], which react with CO to produce phthalimidines in high yield [Eq. (45)] [68].

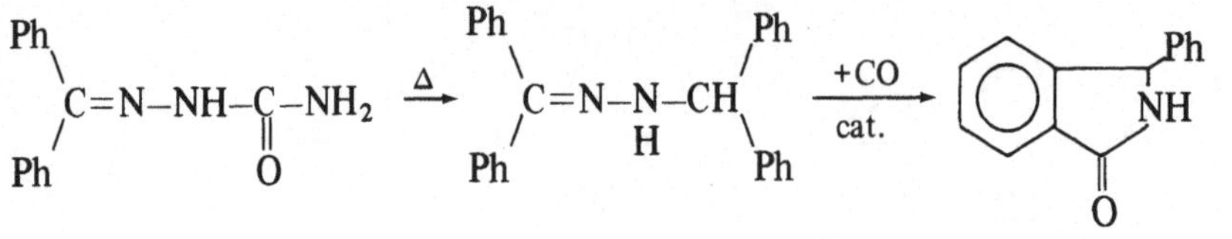

$$\underset{Ph}{\overset{Ph}{\diagdown}}C=N-NH-\underset{O}{\overset{\|}{C}}-NH_2 \xrightarrow{\Delta} \underset{Ph}{\overset{Ph}{\diagdown}}C=N-\underset{H}{\overset{|}{N}}-CH\underset{Ph}{\overset{Ph}{\diagup}} \xrightarrow[\text{cat.}]{+CO} \text{[structure]} \quad (45)$$

yield 70%

Under usual carbonylation conditions, o-phthalaldehyde [69] reacts with CO and aromatic nitro compounds in the presence of Rh carbonyls to yield N-substituted phthalimidines [Eq. (46)]. Pyridine is employed as solvent, whilst in benzene no reaction ensues [69].

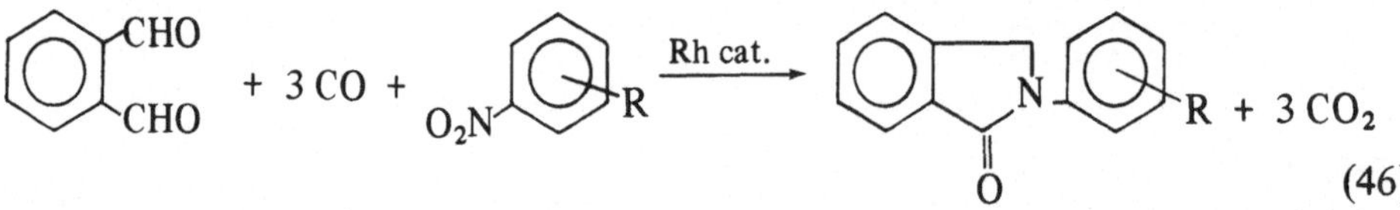

$$\text{[structure]}\overset{CHO}{\underset{CHO}{}} + 3\,CO + \underset{O_2N}{}\text{[structure]}R \xrightarrow{\text{Rh cat.}} \text{[structure]}N\text{[structure]}R + 3\,CO_2 \quad (46)$$

The products of the carbonylation of phthalaldehyde and various aromatic nitro compounds [Eq. (46)] are listed in Table 6.4.

Table 6.4. Products of carbonylation of phthalaldehyde and various nitro compounds [Eq. (46)]

Nitro derivative	Product	Yield (%)	Ref.
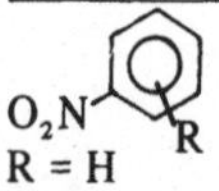 O_2N—[structure]R			
R = H	N-Phenylphthalimidine	58	[69]
R = p-OCH$_3$	N-(p-Methoxyphenyl)phthalimidine	42	[69]
R = m-Cl	N-(m-Chlorophenyl)phthalimidine	55	[69]
R = m-CH$_3$	N-(m-Tolyl)phthalimidine	40	[69]
R = p-CH$_3$	N-(p-Tolyl)phthalimidine	46	[69]
R = p-C$_6$H$_5$	N-(p-Diphenyl)phthalimidine	60	[69]

N-substituted phthalimidines can also be prepared on carbonylating tertiary benzyl amines in the presence of Pd(OAc)$_2$ at 30–200 °C/1 bar [63, 70].

6.4.6 Formation of Indazolones and Quinazolines from Azobenzenes

Azobenzenes are suitable precursors for both indazolones and quinazolines [Eq. (47)] [1]:

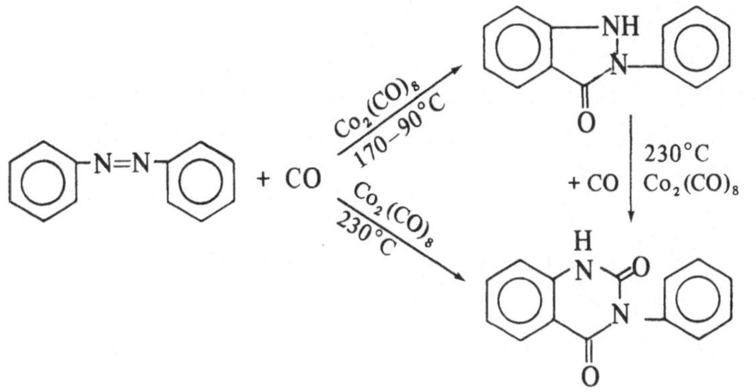

$$\text{(47)}$$

2-Phenylindazolone was obtained in 49% yield from azobenzene [Eq. (47)] [60, 62] (190 °C/150 bar). A non-polar medium (aromatic hydrocarbon) was used as solvent. On raising the temperature to 220–230 °C, 3-phenyl-2,4-dioxo-1,2,3,4-tetrahydroquinazoline was produced via insertion of carbon monoxide [Eq. (47)]. It was possible to convert the indazolone into the quinazoline on heating to 230 °C, making it probable that the indazolone is an intermediate in the formation of quinazoline. The best catalyst metals were Co, Fe, and Pd, Ni apparently being less suitable.

With substituted azobenzenes, ring closure occurred at the ring containing an electron-releasing substituent while electron deficient groups prevented cyclization [60, 62].

Tables of substrates and products from the $Co_2(CO)_8$ catalyzed formation of indazolones and quinazolines via carbonylation have been compiled by Falbe [1].

Recently, interest has been centered on Pd-catalyzed reactions. Tsuji prepared 2-aryl-3-indazolones in good yield on carbonylating azobenzene–$PdCl_2$ complexes in H_2O or ethanol at 50 °C/100 bar [Eq. (4)] [10, 43]. Heck obtained 2-phenylindazolone in 77% yield on reacting an azobenzene-$PdCl_2$ complex in xylene at 100 °C/1 bar with CO [Eq. (48)] [54].

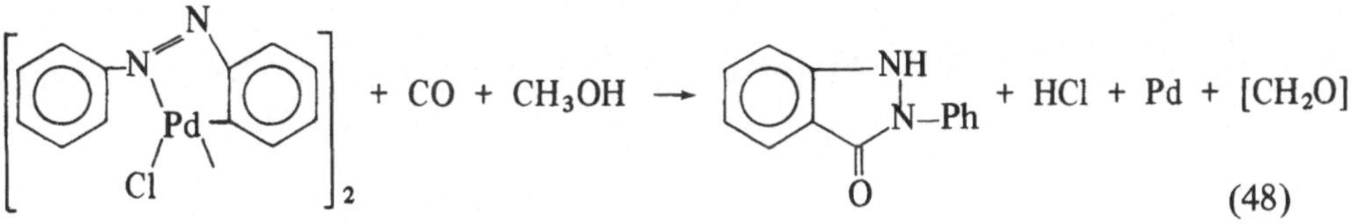

$$\text{(48)}$$

6.4.7 Formation of Indones

The carbonylation of substituted cumulenes [Eq. (49)] in the presence of a $Co_2(CO)_8$ catalyst results in indone formation [1].

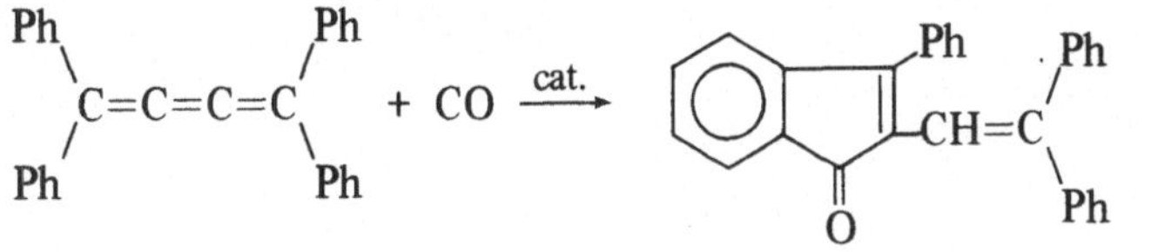

$$\text{(49)}$$

2-(β,β-diphenylvinyl)-3-phenylindone was obtained in 70% yield on carbonylating tetraphenylbutatriene (230–250 °C/150 bar) with $Co_2(CO)_8$ as catalyst and benzene as solvent [Eq. (49)] [71]. In the presence of water, the reaction led to the formation of 2-(β,β-diphenylethyl)-3-phenylindanone, apparently on account of hydrogenation of the two olefinic bonds [Eq. (50)] [1]:

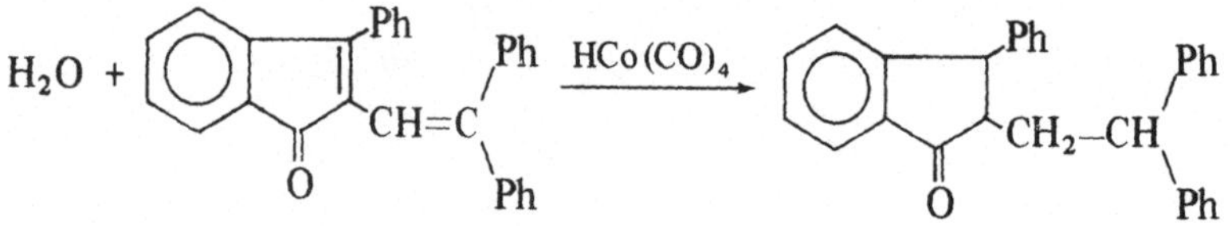

$$(50)$$

Using the same reaction conditions but in the absence of water, the carbonylation of tetraphenylallene gave three products [Eq. (51)] [1, 92]:

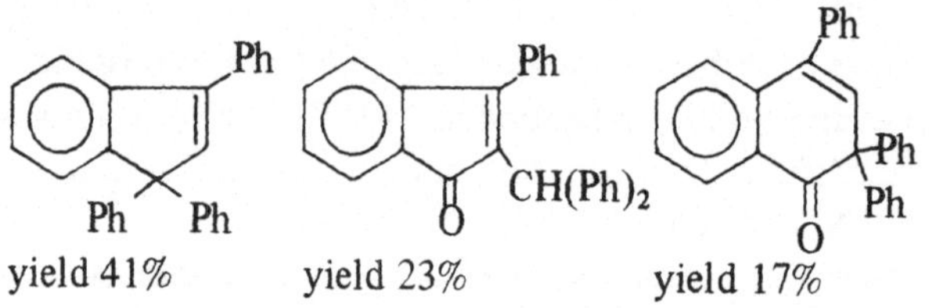

$$(51)$$

yield 41%   yield 23%   yield 17%

Indanones can be prepared under mild conditions via an $AlCl_3$-catalyzed carbonylation [Eq. (52)–(53)] involving benzene and saturated or unsaturated alkyl halides [72].

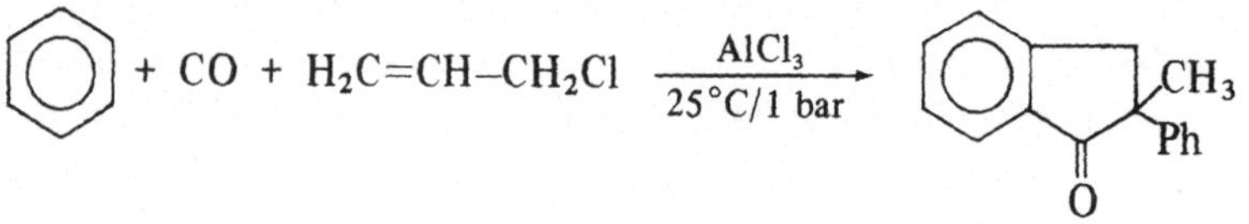

$$(52)$$

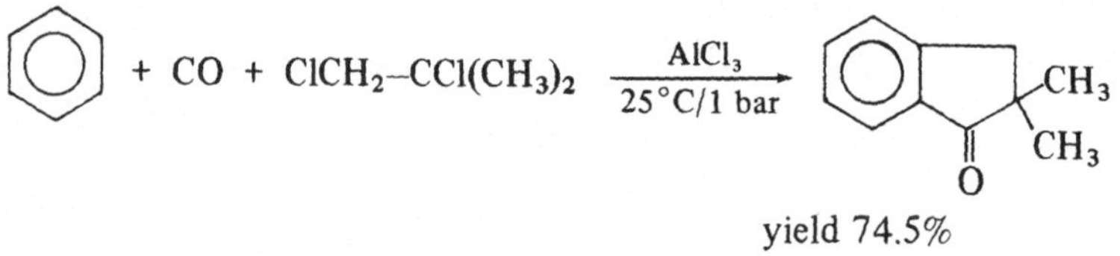

$$(53)$$

yield 74.5%

In a similar reaction, 1,3,5-trineopentylbenzene was converted (under mild conditions) into 2,2,3-trimethyl-5,7-dineopentyl-1-indanone [Eq. (54)] [73].

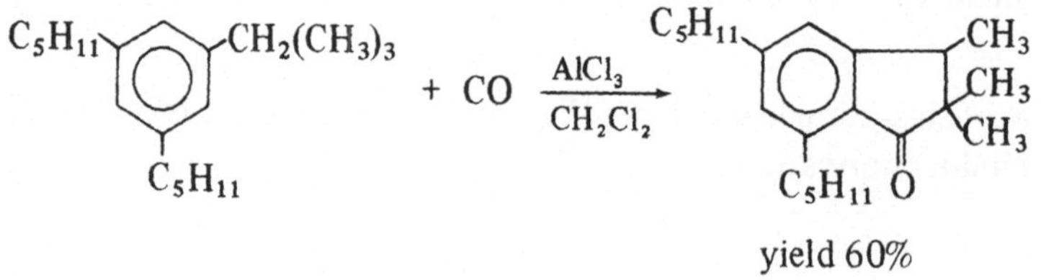

$$(54)$$

yield 60%

6.4.8 Formation of Cyclic Ketones from Dienes

In the presence of a cobalt or palladium catalyst, non-conjugated dienes are carbonylated to saturated and unsaturated cyclic ketones [Eq. (55)] [1, 93]

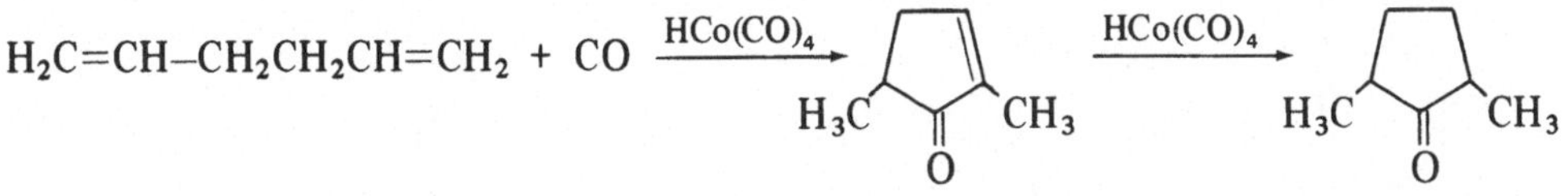
H₂C=CH–CH₂CH₂CH=CH₂ + CO
HCo(CO)₄
H₃C
CH₃
O
HCo(CO)₄
H₃C
CH₃
O

yield 35% yield 6%

(55)

The yield is largely dependent on the number of carbon atoms between the unsaturated C–C bonds. Ring closure is favored when the double bonds are separated by two carbon or other atoms.

The mechanism of the $Co_2(CO)_8$ catalyzed ring closure probably proceeds via the following steps [1, 93]:

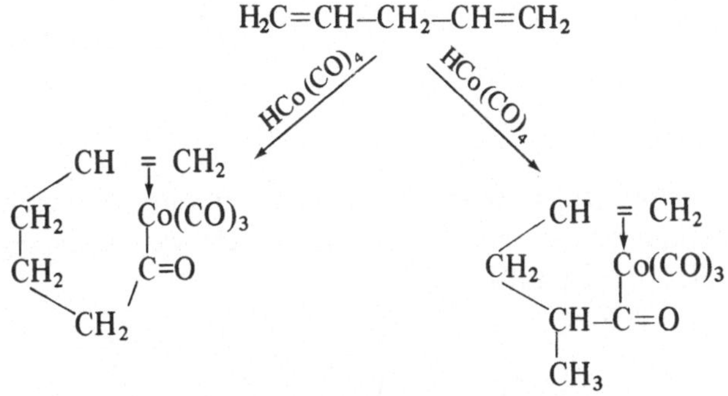
H₂C=CH–CH₂–CH=CH₂
HCo(CO)₄
HCo(CO)₄
CH = CH₂
CH₂
Co(CO)₃
CH₂
C=O
CH₂
CH = CH₂
CH₂
Co(CO)₃
CH–C=O
CH₃

(56)

On elimination of $HCo(CO)_3$, ring closure occurs yielding the cyclic ketones.

Only 2,5-dimethylcyclopent-2-enone is produced (in low yield ~6%) from the palladium-catalyzed carbonylation of 1,5-hexadiene [Eq. (57)]. The conditions are much more drastic compared to the cobalt-catalyzed conversion.

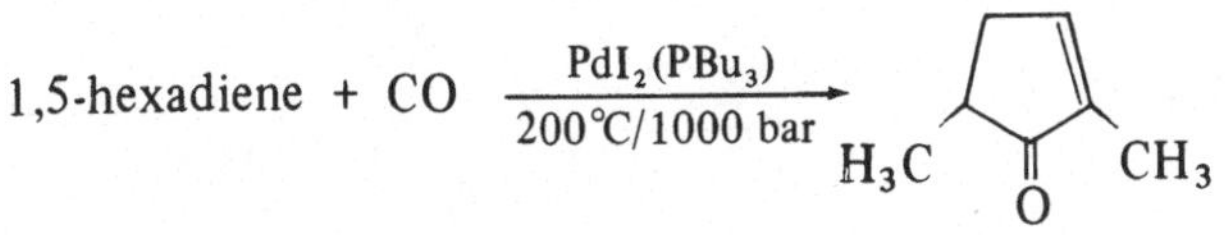
1,5-hexadiene + CO
PdI₂(PBu₃)
200°C/1000 bar
H₃C
CH₃
O

(57)

Using the same reaction conditions, the Pd catalyst was employed in the carbonylation of cycloocta-1,5-diene [Eq. (58)] producing bicyclo[3.3.1]nona-2-en-9-one in 40–45% yield [75].

(58)

The following mechanism is proposed for the carbonylation [Eq. (59)] [1, 10, 75].

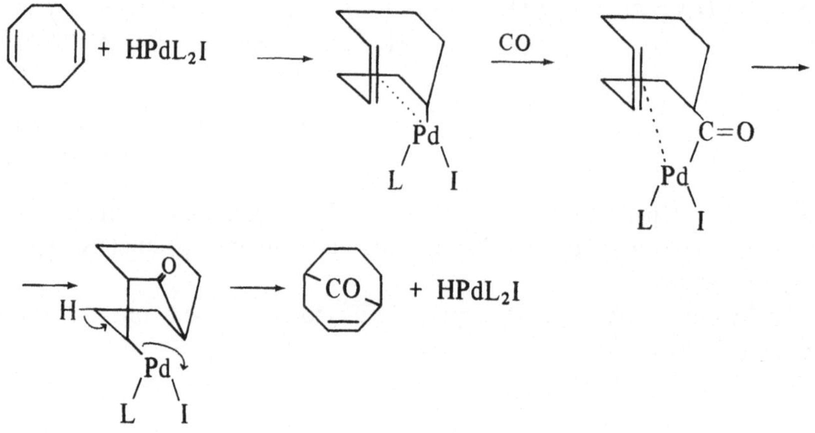

(59)

L = Ligand

In accordance with the introductory remarks about the most suitable arrangement of double bonds in the substrate, best results were obtained when there was a C_2 linkage between the unsaturated bonds [10] [Eq. (60)–(62)]:

$$H_2C=CH-CH_2-CH=CH_2 + CO \xrightarrow[\text{ROH}]{\text{Pd cat.}}$$

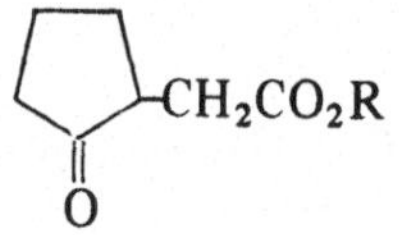

(60)

yield 5–10%

$$H_2C=CH-CH_2-CH_2-CH=CH_2 + CO \xrightarrow[\text{ROH}]{\text{Pd cat.}}$$

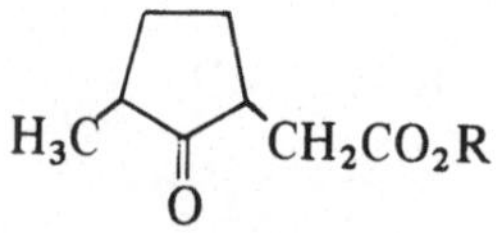

(61)

yield 40–50%

$$H_2C=CH-CH_2-CH_2-CH_2-CH=CH_2 + CO \xrightarrow{\text{Pd cat.}} \text{only traces of ketonic product}$$

(62)

$Ni(CO)_4$ reacts in stoichiometric amounts with 1,5-hexadiene at 50–70 °C in acetone [Eq. (63)] [76]. No ring closure occured with 1,7-octadiene as substrate [76].

432

$$H_2C=CH-CH_2-CH_2-CH=CH_2 \xrightarrow[50-70°C]{Ni(CO)_4}$$ (63)

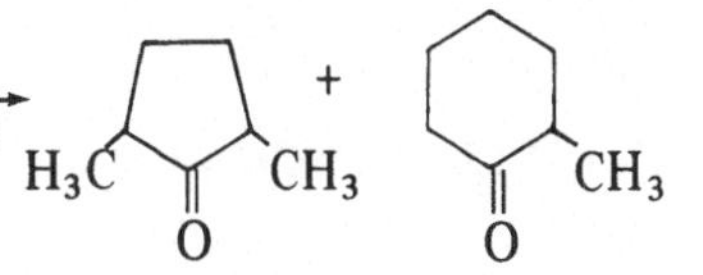

yield 44%   yield 24%

Chiusoli [77] found that cyclopentanone derivatives could be obtained in satis-factory yields on carbonylating a two-component system in the presence of a $Ni(CO)_4$ catalyst system [Eq. (64)] [77].

$$H_3C-CH=CH-CH_2-Cl + H_2C=CH-CH_2-CH_2-CH=CH_2 + CO + H_2O$$

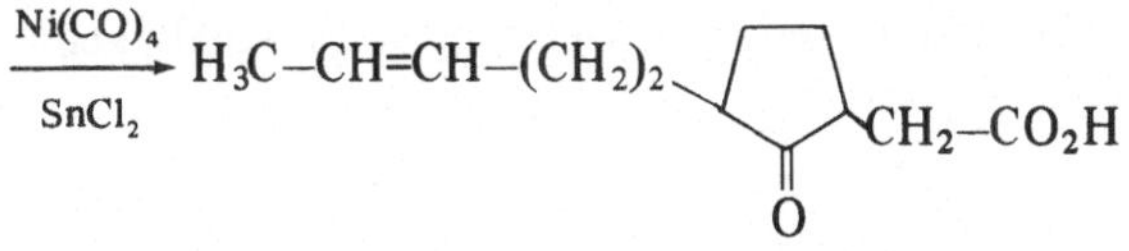

$$\xrightarrow[SnCl_2]{Ni(CO)_4} H_3C-CH=CH-(CH_2)_2 \qquad CH_2-CO_2H$$ (64)

yield 33%

With allyl chloride instead of 1-chloro-2-butene the yield dropped to 16%.

A low yield of 3-cyclopentenone resulted from the $Fe(CO)_3/AlCl_3$-catalyzed ring closure reaction [Eq. (65)] with butadiene (under mild conditions) [78].

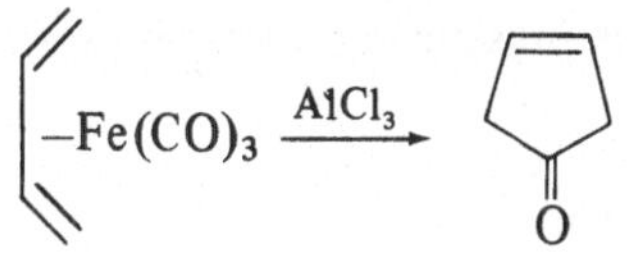

$$-Fe(CO)_3 \xrightarrow{AlCl_3}$$ (65)

6.4.9 Formation of Phenols from Allyl Halides and Alkynes

These $Ni(CO)_4$-catalyzed carbonylations are thought to pass through the following steps [1, 9] [Eq. (66)]:

$$H_2C=CH-CH_2Cl + HC≡CH + CO \xrightarrow{Ni(CO)_4} H_2C=CH-CH_2-CH=CHCONiCl$$ (66)

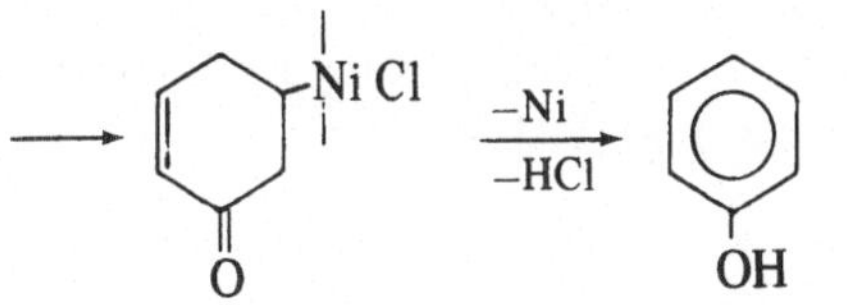

$$\longrightarrow \xrightarrow[-HCl]{-Ni}$$

The yield of phenol can be slightly improved on using aprotic solvents [1, 9]. m-Cresol [Eq. (67)] can be prepared in 77% yield via the carbonylation of methallyl chloride and acetylene (in acetone) using a multi-component catalyst [79–80].

433

$$
\underset{\substack{|\\ H_2C=C-CH_2Cl}}{\overset{CH_3}{}} + HC\equiv CH + CO \xrightarrow[20\,°C]{Ni(CO)_4/NaI/Fe/thiourea} m\text{-Cresol} \qquad (67)
$$

A 34% yield of 2-hydroxy-4-methylbiphenyl resulted on employing phenylacetylene and methallyl chloride as substrates [Eq. (68)] [81].

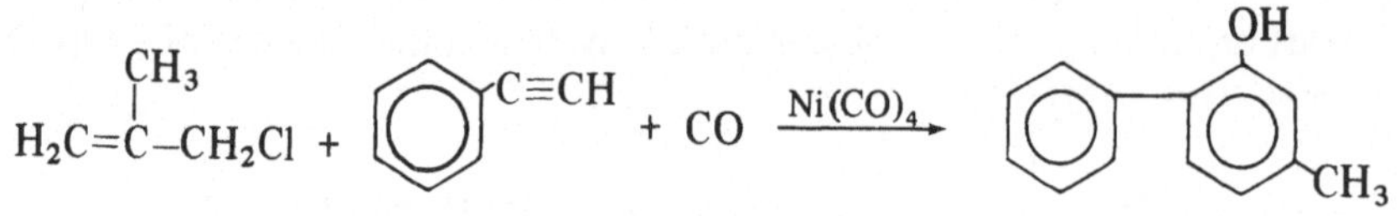

$$
\underset{\substack{|\\ H_2C=C-CH_2Cl}}{\overset{CH_3}{}} + \underset{}{\text{⬡}-C\equiv CH} + CO \xrightarrow{Ni(CO)_4} \text{⬡—⬡} \qquad (68)
$$

o-Cresol is gained on carbonylating methylacetylene together with allyl chloride, while 2-butyl-5-methylphenol can be synthesized from 1-hexyne and methallyl chloride [1, 4].

6.4.10 Other Carbonylation Reactions Leading to Heterocyclic Compounds

6.4.10.1 Formation of Oxygen-Containing Ring Systems

Dihydrofurans [Eq. (69)] can be prepared in a 60–70% yield by carbonylation of homologues of allyl alcohol at relatively mild conditions in the presence of a Rh catalyst [82].

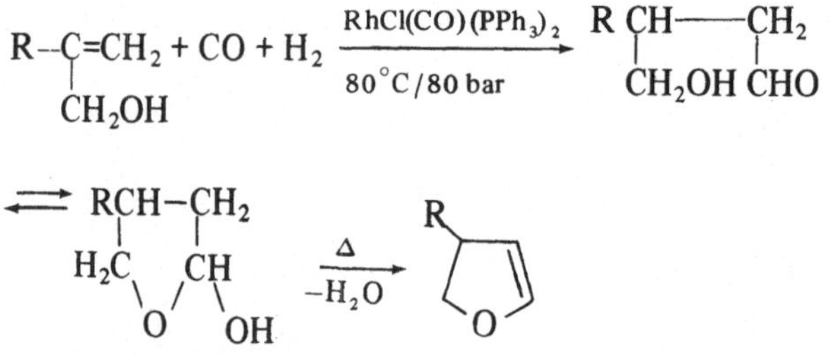

$$
\underset{\substack{|\\ CH_2OH}}{\overset{}{R-C=CH_2}} + CO + H_2 \xrightarrow[80\,°C\,/\,80\,bar]{RhCl(CO)(PPh_3)_2} \underset{\substack{|\\ CH_2OH\ CHO}}{\overset{}{R\ CH-\!\!-\!\!-CH_2}} \qquad (69)
$$

$$
\rightleftharpoons \underset{\substack{|\ \ \ \ |\\ H_2C\ \ \ CH\\ \diagdown\!O\!\diagup\ OH}}{\overset{}{RCH-CH_2}} \xrightarrow[-H_2O]{\Delta} \overset{R}{\text{◁O}}
$$

This route was proposed by Pino in 1951, but it proved to be unsuitable with Co catalysts due to the high degree of isomerization of the alkyl substituted allyl alcohols which led to the formation of several by-products [83]. The rhodium-catalyzed carbonylation produced the desired γ-OH aldehyde in 80–90% yield, making this pathway attractive for the preparation of optically active monoalkyl substituted 2,3-dihydrofurans (asymmetric carbon atom directly attached to the furan ring) [83].

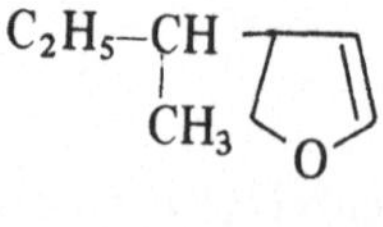

$$
\underset{\substack{|\\ CH_3}}{\overset{}{C_2H_5-CH}}\text{—◻O}
$$

The catalyst was also used to prepare dialkyl substituted furan and pyran derivatives [Eq. (70)] [84]:

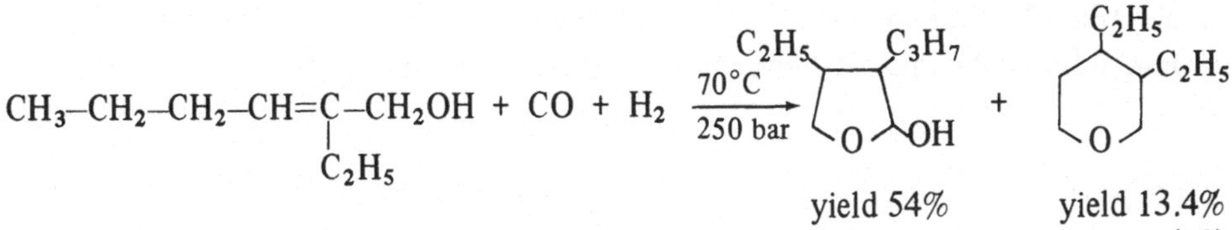
CH3–CH2–CH2–CH=C–CH2OH + CO + H2
C2H5
70°C
250 bar
C2H5
C3H7
O
OH
+
C2H5
C2H5
O
yield 54%
yield 13.4%

(70)

2-Hydroxytetrahydrofuran can be synthesized by carbonylating allyl alcohol in the presence of $RhHCO(PPh_3)_3/PPh_3$ at 93 °C/14 bar [85].

Co catalysts have, however, been used with aryl-substituted allyl alcohols [Eq. (71)] producing low yields (~23%) of 3-substituted tetrahydrofurans along with several by-products [86].

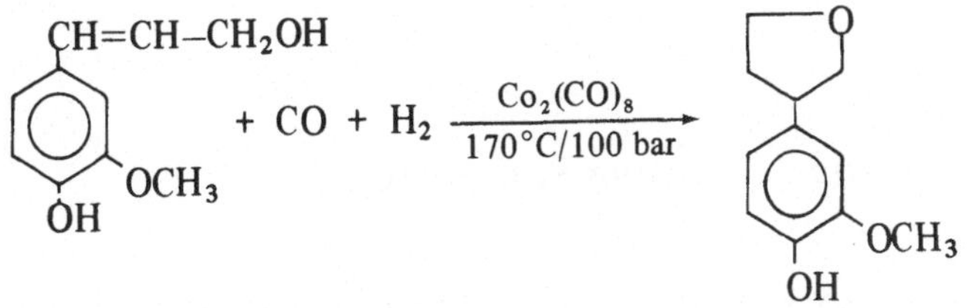
CH=CH–CH2OH
OCH3
OH
+ CO + H2
Co2(CO)8
170°C/100 bar
O
OCH3
OH

(71)

6.4.10.2 *Formation of Nitrogen-Containing Ring Systems*

α-Olefins can be carbonylated in the presence of NH_3 to produce the corresponding 2,4,5-trialkylimidazoles [Eq. (72)] in yields between 50–60% [14].

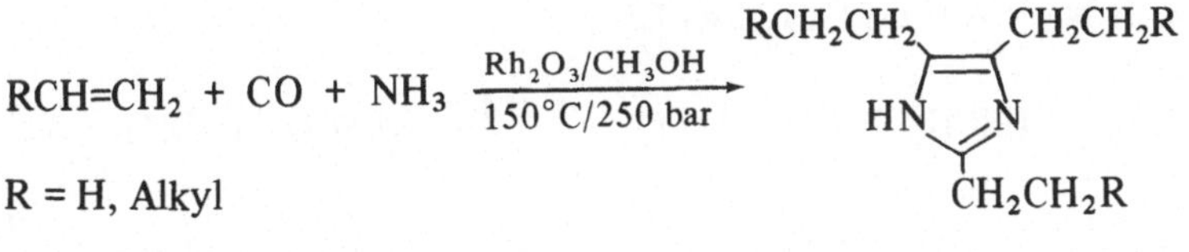
RCH=CH2 + CO + NH3
Rh2O3/CH3OH
150°C/250 bar
R = H, Alkyl
RCH2CH2
CH2CH2R
HN
N
CH2CH2R

(72)

This synthesis enables imidazole derivatives to be produced in a single step. With ethylene, propene and 1-butene the yields of the corresponding imidazoles were 59%, 59%, and 40% resp. [14].

Substituted dihydropyridines [Eq. (73)] can be prepared via the carbonylation of alkenes in the presence of NH_3 and a modified Rh catalyst. The conversion is conducted under more severe conditions than the previous imidazole synthesis [87].

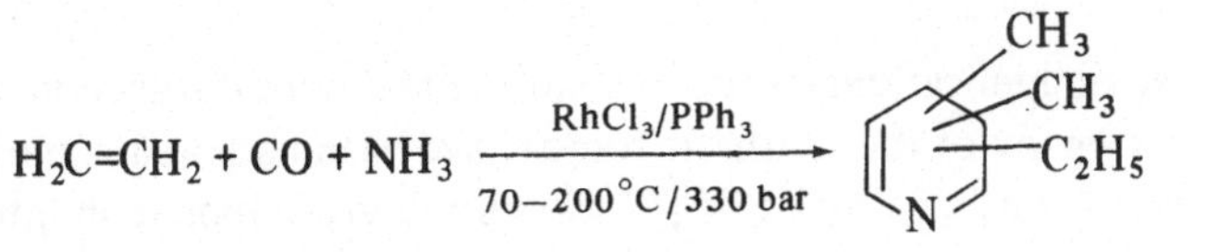
H2C=CH2 + CO + NH3
RhCl3/PPh3
70–200°C/330 bar
CH3
CH3
C2H5
N

(73)

Urazoles [Eq. (74)] are obtained by carbonylating methylhydrazine in the presence of selenium in DMF, the urazole derivative being the main product [25].

$$CH_3NHNH_2 + Se + CO \rightarrow [CH_3NHNH_3]^+ \ [CH_3NHNHCOSe]^- \xrightarrow[+CH_3NHNH_2]{CO+O_2}$$

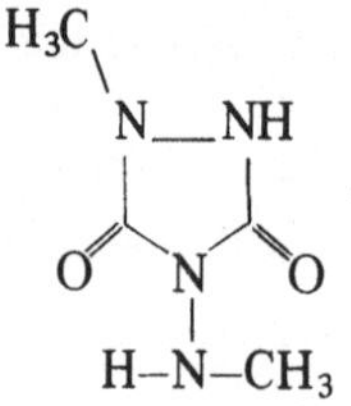

$$+ (CH_3NHNH)_2CO \tag{74}$$

6.4.10.3 Formation of Sulfur-Containing Ring Systems

2-Mercaptoethanol [Eq. (75)] was subjected to carbonylation at 60 °C/3 bar in the presence of a Ni(CO)$_4$ catalyst, leading to the formation of cyclic O,S-ethylenethiocarbonate in 60% yield [5].

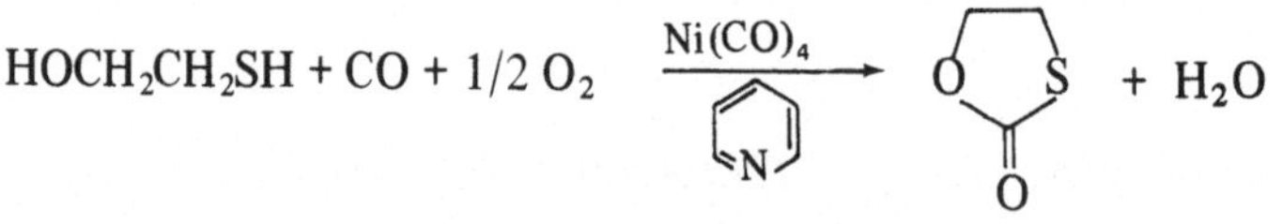

$$HOCH_2CH_2SH + CO + 1/2 \ O_2 \xrightarrow{Ni(CO)_4} \qquad + H_2O \tag{75}$$

The reaction was conducted in pyridine. Based on IR spectra and other evidence, the following reaction route [Eq. (76)] was proposed [5].

$$Ni(CO)_4 + C_5H_5N \rightarrow Ni(CO)_3C_5H_5N + CO$$

$$Ni(CO)_3C_5H_5N + 2 \ HSCH_2CH_2OH + 1/2 \ O_2 \xrightarrow{-C_5H_5N} Ni(SCH_2CH_2OH)_2 + 3 \ CO + H_2O$$

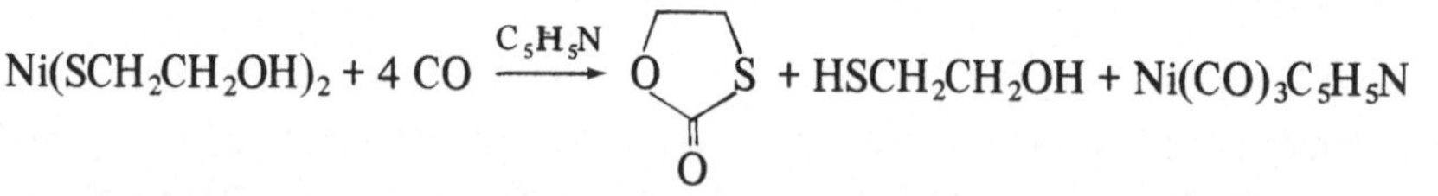

$$Ni(SCH_2CH_2OH)_2 + 4 \ CO \xrightarrow{C_5H_5N} \qquad + HSCH_2CH_2OH + Ni(CO)_3C_5H_5N \tag{76}$$

6.5 Commercial Applications

There have been no industrial applications of the ring closure reactions outlined in the previous sections. However, many of the synthetic routes have been patented — indicating commercial interest. As outlined by Falbe [1], N-methylpyrrolidone is an interesting product with a wide variety of applications including extractive distillation of alkenes, alkynes and aromatics. It can readily be synthesized via the carbonylation of allyl chloride and methylamine. Succinimides, which can be prepared via the carbonylation of acrylamide, find application in organic syntheses as well as being plant growth stimulators. American Cyanamid [95] disclosed the preparation of anthraquinone [Eq. (77)] in 90% yield from benzophenone and CO.

$$+ \; CO \quad \xrightarrow[220-250°C/22-26\,bar]{CuCl_2} \qquad\qquad (77)$$

Instead of benzophenone, benzene can be employed as precursor [95]. The advantages of this procedure are reported to be the high yields, inexpensive starting materials coupled with the relatively simple work-up of the product. The current world capacity of anthraquinone, which amounts to approx. 30,000 t/a, is undergoing growth, due to demand from the dye sector. However, this carbonylation route will have to compete with other processes using precursors such as phthalic anhydride and benzene, naphthoquinone and butadiene (Bayer) or styrene (BASF).

6.6 References

1. Falbe, J.: Carbon Monoxide in Organic Synthesis, Chapter 4, Springer-Verlag, Berlin, Heidelberg, New York (1970)
2. Bird, C.W.: Chem. Rev., 62, 283-302 (1962)
3. Schrauzer, G.N.: Transition Metals in Homogeneous Catalysis, Chapter 5, Marcel Dekker Inc., New York (1971)
4. Wender, I., Pino, P.: Organic Syntheses via Metal Carbonyls, Vol. 2, Wiley, New York (1977)
5. Koch, P., Perrotti, E.: J. Organomet. Chem. 81, 111 (1974)
6. Germain, J.E.: Catalytic Conversions of Hydrocarbons, 278, Academic Press, New York (1969)
7. Reppe, W.: Liebigs Ann. Chem. 582, 1 (1953)
8. Falbe, J.: Angew. Chem. 78, 532 (1966); Intern. Edit. 5, 435 (1966)
9. Falbe, J., Korte, F.: Angew. Chem. 71, 291 (1962); Int. Edit. 1, 266 (1966)
10. Tsuji, J.: Adv. Org. Chem., 6, 150–198; Ed. Taylor, E., Winberg, C., Wiley, New York (1969)
11. Yoshisato, E., Ryang, M., Tsutsumi, S.: J. Org. Chem., 34, 1500 (1969)
12. Nienburg, H., Elsehnig, G.: DE Offen. 1,066,572 (1959); CA 55, 10323 h (1961)
13. Noge, T., Tsuji, J.: Tetrahedron, 4099 (1969)
14. Iwashita, T., Sakuraba, M.: J. Org. Chem., 36, 3927 (1971)
15. Ryang, M.: Organomet. Chem. Rev. A, 5, 67-93 (1970)
16. Ryang, M., Toyoda, Y., Murai, S. et al.: J. Org. Chem., 38, 62 (1973)
17. Cassar, L., Chiusoli, G.P.: Tetrahedron Lett., 3295 (1965)
18. Chiusoli, G.P.: Angew. Chem., 74, 72 (1960)
19. US 3,065,242 (1962)
20. US 3,161,672 (1964)

21. Murahashi, S.: J. Amer. Chem. Soc., 77, 6403 (1955)
22. Falbe, J., Korte, F.: Chem. Ber., 95, 2680 (1962)
23. Atlantic Richfield Co., US 3,714,185 (25.09.70)
24. Alderson, T., Thomas, J.C. (DuPont), US 3,040,090 (14.04.59)
25. Kondo, K., Sonoda, N., Sakurai, H.: Synth. Commun., 5, 131-5 (1975)
26. Falbe, J., Korte, F.: Abstracts, IUPAC Congress, London (1962)
27. Soder, G.: DE Offen. 1,620,391 (13.07.66)
28. Prince, R.F. (Atlantic Richfield Co.): US 3,637,743 (28.08.68)
29. Iqbal, A.F.M.: Tetrahedron Lett., 3381 (1971)
30. Sentralinst. Indforskning, No. 73/2670 (27.06.73)
31. Soder, G.: DE Offen. 1,695,761 (18.08.67)
32. Matsuda, A.: Bull. Chem. Soc. Japan, 41, 1876 (1968); JP 17,900 (1969)
33. Stapp, P.R. (Phillips Petroleum Co.): US 3,952,020 (26.11.73); CA 85, 32450t (1976)
34. Norton, J.R., Shenton, K.E., Schwartz, J.: Tetrahedron Lett., 51-54 (1975)
35. Farbwerke Hoechst AG, NL 7,304,827 (12.04.72)
36. Fernholz, H., Freudenberger, D. (Farbwerke Hoechst AG): DE Offen. 2,303,997 (27.01.73)
37. Shell Int. Research N. V., BE 616,141 (08.10.62)
38. Toa Gosei Chem. Ind. Co., JP 66-21,603 (21.05.63)
39. Himmele, W. (BASF): DE Offen. 1,240,848 (02.11.63)
40. Holmquist, H.E.: J. Org. Chem., 34, 4164 (1969)
41. Albanesi, G., Tovaglieri, M. (Lonza AG): CH 376,900 (16.08.58)
42. Matsuda, T., Kondo, H., Nakamura, N.: Kogyo Kagaku Zasshi, 74, 1135 (1971)
43. Tsuji, J.: Acc. Chem. Research, 2, 144 (1969)
44. Tsuji, J., Nogi, T.: J. Amer. Chem. Soc., 88, 1289 (1966)
45. ICI, NL 6,511,995 (17.09.64)
46. BASF, NL 6,516,507 (19.12.64)
47. ICI, NL 6,613,339 (21.09.66)
48. Green, M., Hancock, R.I.: J. Chem. Soc. A, 109 (1968)
49. Alderson, T., Engelhardt, V.A. (DuPont): US 3,065,242 (23.02.60)
50. Japan Bureau Ind. Tech., JP 74-61,166 (12.10.72)
51. Heck, R.F.: J. Amer. Chem. Soc., 85, 1460 (1963)
52. Aumann, R., Ring, H.: Angew. Chem., 89, 47-48 (1977)
53. Heck, R.F.: Organic Syntheses via Metal Carbonyls. Editors: Wender, I., Pino, P., Vol. 1, p. 373, Wiley, New York (1977)
54. Heck, R.F.: J. Amer. Chem. Soc., 85, 3381 (1963)
55. Bannister, W.B., Green, M., Haszeldine, R.N.: J. Chem. Soc., Inorg., Phys., Theoret. (2), 194–196 (1966)
56. Chiusoli, G.P., Cassar, L.: Angew. Chem. Intern. Edit., 6, 124 (1967)
57. Cassar, L.: Corsi Semin. Chim., 67-68 (1968)
58. Foa, M., Cassar, L., Tacchi Venturi, M.: Tetrahedron Lett., 1357-61 (1968)
59. Toray Ind. Inc., JP 70-4,055 (13.12.66)
60. Horiie, S., Murahashi, S., Jo, T.: Bull, Chem. Soc. Japan 33, 81, 247-51 (1960)
61. Murahashi, S., Horiie, S.: Ann. Rep. Sci. Works, Fac. Sci. Oska Univ., 7, 89 (1959)
62. Murahashi, S., Horriie, S., Jo, T.: Nippon Kagaku Zasshi, 79, 72, 75 (1958)
63. Thompson, J.M., Heck, R.F.: J. Org. Chem., 40, 2667 (1975)
64. Murahashi, S., Horiie, S.: J. Amer. Chem. Soc., 78, 4816 (1956); ibid 77, 6403 (1955)
65. Bruin, P., Oosterhaf, H.A. et al.: VIIth FATIPEC Congress (1964), p. 49-60, Verlag Chemie, Weinheim (1964)
66. Rosenthal, A., Astbury, R.F., Hubscher, A.: J. Org. Chem., 23, 1037 (1958)
67. Borsche, W., Merkwitz, C.: Chem. Ber., 37, 3180 (1904)
68. Rosenthal, A., Millward, S.: Can. J. Chem., 41, 2504 (1963)
69. Iqbal, A.F.M.: Chemtech, 556-572 (1974)
70. Toyo Rayon, JP 69-16,648 (13.09.67)
71. Kim, Pu-Jun, Hagihara, N.: Mem. Inst. Sci. Ind. Res. Osaka Univ., 24, 133-39 (1967); CA 68, 2982z (1968)

72. Bruson, H.A., Plant, H.L.: J. Org. Chem., *32,* 3356-62 (1967)

73. Dahlberg, E., Martinson, P., Olsson, K., Acta Chem. Scand., Ser. B.28, 1143 (1974)

74. Klemchuk, P.R.: US 2,995,607 (1962)

75. Brewis, S., Hughes, P.R.: J. Chem. Soc., Chem. Commun., 6 (1966)

76. Fell, B., Seide, W., Asinger, F.: Tetrahedron Lett., 1003 (1968)

77. Chiusoli, G.P., Cometti, G., Bellotti, V.: Gazz. Chim. Ital., *103,* 4-5, 569 (1973)

78. Johnson, B.F.G., Lewis, J., Thompson, D.J.: Tetrahedron Lett. 3789 (1974)

79. Cassar, L., Foa, M., Chiusoli, G.P. (Montecatini Edison SpA): DE Offen. 2,118,111 (16.04.70)

80. Cassar, L., Foa, M., Chiusoli, G.P.: Organomet. Chem. Syn., *1,* 302 (1971)

81. Chiusoli, G.P., Cassar, L.: Angew. Chem., *79,* 181 (1967)

82. Botteghi, C., Consiglio, G., Ceccarelli, G., Stefani, A.: J. Org. Chem., *37,* 1835 (1972)

83. Pino, P.: Gazz. Chim. Ital., *81,* 625 (1951)

84. BASF, DE Offen. 2,410,156 (02.03.74)

85. Firestone Tire and Rubber Co., DE Offen. 2,649,900 (05.11.75)

86. Nahum, L.S.: J. Org. Chem., *33,* 3601 (1968)

87. Union Oil Co. of California, US 3,679,689 (06.07.70)

88. Falbe, J., Korte, F.: Chem. Ber. *98,* 1928 (1965)

89. Falbe, J., Schulze-Steinen, H.J., Korte, F.: Chem. Ber., *98,* 886 (1965)

90. Horiie, S., Murahashi, S.: Bull. Chem. Soc. Japan *33,* 247 (1960)

91. Prichard, W.W. (DuPont): US 2,841,591 (1958)

92. Kim, P., Hagihara, N.: Bull. Chem. Soc. Japan *38,* II 2022

93. Heck, R.F.: J. Amer. Chem. Soc. *85,* 3116 (1963)

94. Arzoumanidis, G.G., Rauch, F.C. (American Cyanamid Co.): US 3,932,474 (28.05.74); CA *84,* 121559 (1976)

95. Taqui Khan, M.M., Martell, A.E.: Homogeneous Catalysis by Metal Complexes, Chapter 4, Academic Press, New York (1974)

96. Eidus, Ya.T., Lapidus, A.L., Nefedov, B.K., Russ. Chem. Rev., *42* (3) 199 (1973)

Subject Index

methacrolein 154, 156
methacrylamide 418
methacrylates 118, 152, 270
methacrylic acid 154, 263
methallyl
– alcohol 136
– chloride 433, 434
methanal carbonylation 290
methanation catalysts 335
– in liquid phase 337
– process 340
methane 93, 336
– content 329
– syntheses 327
methanol 86, 128, 244, 259, 287, 298
– carbonylation 243
– –based technologies 319
– formation equilibrium 311
– syntheses 309
4-methoxybenzophenonephenylhydrazone
  426
2-(3-methoxy-4-hydroxyphenyl)-butanal 136
N-(p-methoxyphenyl)phthalimidine 428
β-methoxypropionaldehyde 123
methyl acetate 299
methylacetylene(s) 270, 434
methyl acrylate 86, 119, 120, 269
N-methylallyamine 419
methylamine(s) 292, 437
methyl benzoate 295
N-α-methyl benzylsalicyl aldimine 136
α-methyl branched acid 277
2-methylbutadiene 104
2-methyl-2,3-butadiene-1-carboxylic acid 267
2-methylbutanal 107, 150
3-methylbutanal 23, 99, 150
(S)-2-methylbutanal 136
(–)-(R)-2-methylbutanal 135
(+)-(S)-2-methylbutanal 135
2-methylbutanoic acid 287
2-methylbutanol 107, 117
3-methyl-1-butanol 98, 142
2-methyl-1-butene 98
2-methyl-2-butene 98
3-methyl-1-butene 98, 135
2-methyl-3-buten-2-ol 108
2-methylbut-3-en-2-ol 112
3-methyl-1-buten-3-ol 136
3-methylbut-2-en-1-ol 112
3-methyl-3-buten-1-ol 108, 421
3-methylbut-3-en-1-ol 112
3-methylbutyl formate 140
α-methylbutyraldehyde 110
α-methyl-γ-butyrolactone 422
γ-methylbutyrolactone 283

α-methylcinnamic acid 263
methyl p-chlorobenzoate 295
α-methylcrotonaldehyde 110
methyl crotonate 270
3-(4-methyl-3-cyclohexene-1-yl)-butyraldehyde
  102
methylcyclohexylanisylphosphines 56
4-methyl-2,3-dihydrofuran 136
5-methyl-2,3-dihydrofuran 136
2-methyldodecanoic acid 276
4-methylene-1-butene-2-carboxylic 264
α-methylene-γ-butyrolactone 264, 421
methylene complexes 345
α-methylene-γdimethyl-γ-butyrolactone 264
α-methylene-γ-lactones 421
α-methylene-γ-methyl-γbutyrolactone 264
α-methylene-δ-valerolactone 264
methyl ethyl ketone 86, 156, 260, 262
– – – (2-butanone) 153
– α-formalpropionate 120
– β-formylpropionate 119, 120
methylformylundecanoate 119
methyl fuels 240
(+)(S)-5-methyl-1-heptene 9
2-methylhexanal 62, 135
4-methylhexanal 135
2-methylhexanoic acid + heptanoic acid 276
3-methylhexanol 98
5-methylhexanol 98
(–)(S)-4-methyl-1-hexene 9
(+)-S-3-methyl-1-hexene-3-d 10
4-methyl-1-hexen-3-ol 136
2-methyl-5-hexenoic acid 277
methylhydrazine 436
4-methyl-2-hydroxytetrahydrofuran 136
5-methyl-2-hydroxytetrahydrofuran 136
methyl iodide 249, 290
methyl isobutyl ketone 89
methyl isobutyrate 154
methyl linolate 120
methyl methacrylate (MMA) 25, 31, 53, 118,
  120, 152, 176, 269, 270, 281
2-methyl-3-(3-methoxy-4-hydroxyphenyl)-
  propanal 136
methyl (1-naphthyl) acetaldehyde 98
2-methyloctadecanoic acid 276
3-methyl-2,6-octadiene 101
2-methyloctanal 135
(S)-2-methyloctanal 107
3-methyloctan-1-al-7-ol 113
2-methyloctanoic acid 276, 287
methyl oleate 9, 119, 120
– 2,4-pentadienoate 269
4-methyl-2,3-pentadiene-1-carboxylic acid
  267

J. Falbe

Carbon Monoxide in Organic Synthesis

Translator: Ch. R. Adams
1970. 21 figures. IX, 219 pages
ISBN 3-540-04814-6

"This is a good translation by Dr. C. R. Adams of Shell Devt Co., Emeryville, of the 1967 German edition, brought well up to date. ...The present text, however, stands apart from the half-dozen other reviews on carbonylation because of the emphasis on industrial practice, including catalyst composition and recovery; it also has a page or two in each chapter on application and economics. Much valuable information is condensed into 62 tables and into a number of diagrams correlating pressure, temperatures and yields.
...this is an excellent, comprehensive reference book for anyone interested in catalytic carbonylation and in industrial synthesis based on CO. It is beautifully produced and acceptable value for money."

Chemistry in Britain

J. Falbe

Synthesen mit Kohlenmonoxyd

1967. 20 Abbildungen. VIII, 212 Seiten
(Organische Chemie in Einzeldarstellungen, Band 10)
ISBN 3-540-03947-3

"...Der auf diesem Gebiet Tätige oder Interessierte muß dem Autor sehr dankbar dafür sein, daß dieser Stoff in so übersichtlicher Weise geordnet und dargestellt wurde...Das Buch, das sich auch durch übersichtlichen Druck und gute Ausstattung auszeichnet, erweist sich inzwischen in zahlreichen Industrie- und Hochschullaboratorien als unentbehrlicher Helfer."

Chemiker-Zeitung

Springer-Verlag
Berlin
Heidelberg
New York

Polymers Properties and Applications

Editorial Board: H.-J. Cantow,
H. J. Harwood, J. P. Kennedy,
A. Ledwith, J. Meißner,
S. Okamura, G. Olivé,
S. Olivé

Volume 3: A. Knop, W. Scheib

Chemistry and Application of Phenolic Resins

1979. 111 figures, 88 tables. XIII, 269 pages
ISBN 3-540-09051-7

The authors present the current theory of phenolic resin chemistry and the technical application of phenolic resins, based on day-to-day experience in research, production and marketing, and against the background of economic relevance. Where the first fully synthetic polymers (phenolic resins) stand today and what their future is are subjects of discussion. Looking back at their development, it is shown that after a wide variety of adaptions, they remain technically and economically irreplaceable products with potential for further market growth and a commensurate appreciation of their value. This book will be greatly appreciated by chemists, engineers, marketing professionals, and students.

Volume 2: H.-H. Kausch

Polymer Fracture

1978. 180 figures, 23 tables. X, 332 pages
ISBN 3-540-08786-9

"Kausch,... is well known for his work on polymer morphology and molecular mechanics as well as his research on the strength of materials. The avowed aim of this book is to connect the more conventional statistical and continuum mechanics interpretation of fracture phenomena to the newer spectroscopic studies of highly stressed polymeric chains and the kinetics of their rupture. Relating the literature on the observed modes of viscoelasticity and irreversible deformation from polymer morphology and solid-state physics, Kausch explains the behavior and rupture of polymeric materials in terms of molecular slip and breakage processes. This leads to interesting, methodical and well-thought-out interpretations of fracture toughness, crack propagation rates and fatigue of all major polymer systems... Thus, the book is an outstanding contribution to our understanding of the role of chain ruptures during mechanical failure... every student and practitioner of polymer science and engineering should find this book to be a valuable resource for his work."
Physics Today

Springer-Verlag
Berlin
Heidelberg
New York

Volume 1: B. Rånby, J. F. Rabek

ESR Spectroscopy in Polymer Research

1977. 356 figures, 29 tables. XIV, 410 pages
ISBN 3-540-08151-8

"...This book is a remarkable example for the successful combination of simplicity and clarity in its tutorial parts and of depth and width whenever and wherever is presents the state of the art...As ultimate and very gratifying reward for his investment the reader gets no less than 2519 references to the literature in excellent alphabetical order. Scientists who already work with ESR will be greatly assisted in their efforts by this book; those who do not yet use this method will have an easy time to learn and use it. All of them will be grateful to the authors for this exceptional addition to our scientific literature."
J. Polymer Science